Natural Pesticides and Allelochemicals

Natural pesticides and allelochemicals are used for protection against invasive plants, but when released into the environment they can have both positive and negative effects on plants. This book discusses a holistic and sustainable approach that balances effective pest management with minimizing environmental impacts, promoting biodiversity, and ensuring food safety and quality. It brings together proposals to help improve the quality of management and production of healthier foods utilizing compounds of natural origin. The authors provide a broad and diverse picture of the applications of terpenoids in plant safety and the possibilities for innovative biotechnological approaches for their extraction.

Features:

- Presents a comprehensive resource on recent advances in natural pesticides and new allelochemicals for crop protection.
- Discusses natural herbicides, sustainable agriculture, and bioeconomic processes.
- Explains the challenges of synthetic pesticides and their costs to human and environmental health.
- Covers different aspects of natural pesticides such as their sources, development, application, and toxicity.
- Helps professionals and scholars involved in chemical technology, biotechnology, and agriculture gain a thorough understanding of crop protection practices tailored to specific crop types, regional conditions, and pest pressure.

This is a great reference for researchers, academics, students, and professionals involved with or interested in agriculture and the environment, pest control, environmental chemistry, biology, food science, and forest engineering.

Natural Pesticides and Allelochemicals

Advances and Trends in Crop Protection

Edited by
Mozaniel Santana de Oliveira, Leo M.L. Nollet,
Ravendra Kumar, Eloisa Helena de Aguiar Andrade,
and Antônio Pedro da Silva Souza Filho

CRC Press is an imprint of the
Taylor & Francis Group, an **informa** business

Designed cover image: Shutterstock

First edition published 2025
by CRC Press
2385 NW Executive Center Drive, Suite 320, Boca Raton FL 33431

and by CRC Press
4 Park Square, Milton Park, Abingdon, Oxon, OX14 4RN

CRC Press is an imprint of Taylor & Francis Group, LLC

ISBN: 978-1-032-73272-5 (hbk)
ISBN: 978-1-032-73276-3 (pbk)
ISBN: 978-1-003-46342-9 (ebk)

DOI: 10.1201/9781003463429

Typeset in Times
by Apex CoVantage, LLC

It is with great pleasure that I dedicate this book to my esteemed parents, Maria de Oliveira and Manoel de Oliveira; my seven siblings; my cherished wife, Joyce Fontes; and my beloved daughter, Isabela de Oliveira. It is essential to acknowledge the pivotal role of Leo M.L. Nollet in the success of this book. His invaluable collaboration and guidance have been instrumental in the realization of this academic endeavor.

Dr. Mozaniel Santana de Oliveira

I appreciate Mozaniel's collaboration. It was a pleasure to resolve all the issues that arose and bring this book to a successful conclusion.

Leo M.L. Nollet

Contents

SECTION 2 Essential Oils and Allelochemicals

SECTION 3 Pesticides Based on Natural Products

SECTION 4 Natural Herbicides

SECTION 5 Seaweeds and Vermicompost

Preface

The growing demand for sustainable agricultural practices has highlighted the importance of natural products, especially allelochemicals and plant-derived pesticides, in modern crop protection strategies. In recent years, environmental concerns and the health risks associated with synthetic pesticides have pushed researchers and professionals to seek eco-friendly alternatives. This book, *Natural Pesticides and Allelochemicals: Advances and Trends in Crop Protection*, is a collaborative effort of leading scientists and experts, aiming to provide an in-depth analysis of the latest developments and applications of natural pesticides and allelochemicals.

Divided into 21 chapters, the book is organized into five main sections, each addressing key aspects of crop protection using natural products. Section 1: Pest Management—Sustainability Versus Unsustainability—Allelochemicals explores the role of allelochemicals in pest management, with a focus on sustainability. Topics include the toxicity of plant protection products, both chemical and biological, and their environmental and health impacts. The section examines the balance between the use of natural and synthetic products, highlighting their safe application in crop protection. Section 2: Essential Oils as Allelochemicals emphasizes the relevance of essential oils as important allelopathic tools for pest control. The chapters discuss the phytotoxic potential and antimicrobial properties of essential oils against plant pathogens, as well as their advantages as eco-friendly alternatives to synthetic pesticides in sustainable agricultural systems. Section 3: Pesticides Based on Natural Products focuses on pesticides derived from natural sources, such as phenolic compounds, terpenes, and azadirachtin (extracted from the neem plant). The chapters explore the potential of these natural products as effective alternatives to conventional pesticides, discussing their mechanisms of action and benefits in terms of reduced toxicity and environmental impact. They also cover microbial-based pesticides and their emerging role in integrated pest management strategies. Section 4: Natural Herbicides addresses the use of natural herbicides, investigating compounds that inhibit the growth of unwanted plants in agricultural systems. Topics include the herbicidal properties of various natural substances, including volatile and non-volatile terpenes, and their application in weed management. This section highlights the effectiveness of natural herbicides in reducing reliance on synthetic chemicals and their potential role in sustainable farming practices. Section 5: Seaweeds and Vermicompost deals with marine products and vermicompost as candidate pesticide agents.

This book provides an innovative and comprehensive perspective on the use of natural compounds in agriculture, significantly contributing to the development of more sustainable agricultural solutions. We hope this work will serve as a valuable resource for researchers, professionals, and students involved in agricultural sciences, plant biology, and environmental sustainability.

We are deeply grateful to all the authors who contributed their knowledge and dedication to making this work possible. Their innovative approaches will certainly enrich the field of crop protection and the advancement of sustainable agriculture.

Editors
Dr. Mozaniel Santana de Oliveira
Dr. Leo M.L. Nollet
Dr. Ravendra Kumar
Dr. Eloisa Helena de Aguiar Andrade
Dr. Antônio Pedro da Silva Souza Filho

About the Editors

Dr. Mozaniel Santana de Oliveira holds a bachelor of science degree in chemistry from the Federal University of Pará (UFPA), where he graduated in 2014. He holds a master of science (2016) and a doctor of philosophy (PhD) (2018) in food science and technology from the same institution. His doctoral dissertation and thesis centered on the extraction of essential oils from diverse plant matrices employing supercritical fluids and conventional techniques, with potential applications in agriculture and pharmacology. Dr. Oliveira has a notable academic output, with 217 published papers, including 100 articles in international journals. He has authored or edited 10 international books published by esteemed publishers. Additionally, he has contributed 48 chapters to international books and 62 abstracts to conferences. His work encompasses the extraction of secondary metabolites from medicinal and aromatic plants employing supercritical fluids, hydrodistillation, steam distillation, and solvent extraction techniques. Additionally, he has expertise in separation methods, including classical chromatography (open column) and thin-layer chromatography. Dr. Oliveira is a specialist in natural product chemistry, with a particular focus on essential oils and the identification of secondary metabolites employing modern analytical techniques. His research interests include food science and technology, prospecting for volatile bioactive molecules, biotechnology of natural products, allelopathy, and applications in agriculture seeking new allelochemicals. Additionally, he is engaged in research in pharmacology, specifically the search for new antibacterial, antiviral, antifungal, antiparasitic, anti-inflammatory, and phytotherapeutic agents. His work also encompasses medicinal applications such as antiproliferative, antioxidant, and anti-inflammatory agents, with a focus on in vitro, in vivo, and in silico studies. Dr. Oliveira currently serves as a permanent professor in the Postgraduate Program in Biological Sciences, with an emphasis on tropical botany, at the Federal Rural University of the Amazon (UFRA), in collaboration with the Museu Paraense Emílio Goeldi (MPEG). Additionally, he serves as a collaborating professor for the Postgraduate Programs in Food Science and Technology and Pharmaceutical Sciences at UFPA. He has considerable editorial experience, having served as an ambassador for Bentham Science Publishers and an academic mentor for Web of Science, in addition to acting as a reviewer of over 440 articles for 70 international journals, and serves on the editorial boards of numerous academic journals. To date, he has edited over 270 scientific articles and is currently a postdoctoral fellow in the Graduate Program in Biological Sciences, with an emphasis on tropical botany, at UFRA/MPEG.

Leo M. L. Nollet earned an MS (1973) and PhD (1978) in biology from the Katholieke Universiteit Leuven, Belgium. He is an editor and associate editor of numerous books. He edited for M. Dekker, New York—now CRC Press of Taylor & Francis Publishing Group—the first, second, and third editions of *Food Analysis by HPLC* and *Handbook of Food Analysis*. The last edition is a two-volume book. Dr. Nollet also edited the *Handbook of Water Analysis* (first, second, and third editions) and *Chromatographic Analysis of the Environment*, third and fourth editions (CRC Press). With F. Toldrá, he coedited two books published in 2006, 2007, and 2017: *Advanced Technologies for Meat Processing* (CRC Press) and *Advances in Food Diagnostics* (Blackwell Publishing—now Wiley). With M. Poschl, he coedited the book *Radionuclide Concentrations in Foods and the Environment*, also published in 2006 (CRC Press). Dr. Nollet has also coedited with Y. H. Hui and other colleagues on several books: *Handbook of Food Product Manufacturing* (Wiley, 2007), *Handbook of Food Science,*

Technology, and Engineering (CRC Press, 2005), *Food Biochemistry and Food Processing* (first and second editions; Blackwell Publishing—now Wiley—2006 and 2012), and the *Handbook of Fruits and Vegetable Flavors* (Wiley, 2010). In addition, he edited the *Handbook of Meat, Poultry, and Seafood Quality*, first and second editions (Blackwell Publishing—now Wiley—2007 and 2012). From 2008 to 2011, he published five volumes on animal product-related books with F. Toldrá: *Handbook of Muscle Foods Analysis*, *Handbook of Processed Meats and Poultry Analysis*, *Handbook of Seafood and Seafood Products Analysis*, *Handbook of Dairy Foods Analysis* (second edition in 2021), and *Handbook of Analysis of Edible Animal By-Products*. Also, in 2011, with F. Toldrá, he coedited two volumes for CRC Press: *Safety Analysis of Foods of Animal Origin* and *Sensory Analysis of Foods of Animal Origin*. In 2012, they published the *Handbook of Analysis of Active Compounds in Functional Foods*. In a coedition with Hamir Rathore, *Handbook of Pesticides: Methods of Pesticides Residues Analysis* was marketed in 2009, *Pesticides: Evaluation of Environmental Pollution* in 2012, *Biopesticides Handbook* in 2015, and *Green Pesticides Handbook: Essential Oils for Pest Control* in 2017. Other finished book projects include *Food Allergens: Analysis, Instrumentation, and Methods* (with A. van Hengel; CRC Press, 2011) and *Analysis of Endocrine Compounds in Food* (Wiley-Blackwell, 2011). Dr. Nollet's other projects include *Proteomics in Foods* with F. Toldrá (Springer, 2013) and *Transformation Products of Emerging Contaminants in the Environment: Analysis, Processes, Occurrence, Effects, and Risks* with D. Lambropoulou (Wiley, 2014). In the series Food Analysis & Properties, he edited (with C. Ruiz-Capillas) *Flow Injection Analysis of Food Additives* (CRC Press, 2015) and *Marine Microorganisms: Extraction and Analysis of Bioactive Compounds* (CRC Press, 2016). With A.S. Franca, he coedited *Spectroscopic Methods in Food Analysis* (CRC Press, 2017), and with Horacio Heinzen and Amadeo R. Fernandez-Alba he coedited *Multiresidue Methods for the Analysis of Pesticide Residues in Food* (CRC Press, 2017). Further volumes in the series Food Analysis & Properties are *Phenolic Compounds in Food: Characterization and Analysis* (with Janet Alejandra Gutierrez-Uribe, 2018); *Testing and Analysis of GMO-Containing Foods and Feed* (with Salah E. O. Mahgoub, 2018); *Fingerprinting Techniques in Food Authentication and Traceability* (with K. S. Siddiqi, 2018); *Hyperspectral Imaging Analysis and Applications for Food Quality* (with N.C. Basantia and Mohammed Kamruzzaman, 2018); *Ambient Mass Spectroscopy Techniques in Food and the Environment* (with Basil K. Munjanja, 2019); *Food Aroma Evolution: During Food Processing, Cooking, and Aging* (with M. Bordiga, 2019); *Mass Spectrometry Imaging in Food Analysis* (2020); *Proteomics in Food Authentication* (with S. Ötleş, 2020); *Analysis of Nanoplastics and Microplastics in Food* (with K.S. Siddiqi, 2020); *Chiral Organic Pollutants, Monitoring and Characterization in Food and the Environment* (with Edmond Sanganyado and Basil K. Munjanja, 2020); *Sequencing Technologies in Microbial Food Safety and Quality* (with Devarajan Thangardurai, Saher Islam, Jeyabalan Sangeetha, 2021); *Nanoemulsions in Food Technology: Development, Characterization, and Applications* (with Javed Ahmad, 2021); *Mass Spectrometry in Food Analysis* (with Robert Winkler, 2022); *Bioactive Peptides from Food: Sources, Analysis, and Functions* (with Semih Ötles, 2022); and *Nutriomics: Well-Being through Nutrition* (with Devarajan Thangadurai, Saher Islam, and Juliana Bunmi Adetunji, 2022). In 2023 he published *Analysis of Naturally Occurring Food Toxins of Plant Origin* with Javed Ahmad.

His most recently published books are *Biopesticides Handbook, Second Edition* (edited with Showkat Mir); *Handbook of Seafood and Seafood Products Analysis* (with F. Toldrá); *Flavoromics, an Integrated to Flavor and Sensory Assessment* (with M. Bordiga); *Analysis of Food Spices* (with J. Ahmad and J. Ahamad); *Global Regulations of Medicinal, Pharmaceutical, and Food Products* (with F. Ali); and *Bioactive Compounds from Food Benefits and Analysis* (with J. Ahmad).

Dr. Ravendra Kumar did his MSc (2006) and PhD (2010) in agricultural chemicals from G.B. Pant University of Agriculture & Technology, Pantnagar, U.S. Nagar, Uttarakhand, India, in the field of natural product chemistry. He qualified for the ICAR NET in 2010. He was a recipient of the UGC Fellowship, New Delhi (India) during his PhD. Dr. Kumar has nearly five years of industrial research experience in the extraction, separation, and purification of nutraceutical and pharmaceutical components at India Glycols Limited, Dehradun, India. Since November 2015, he has worked as an assistant professor in the Department of Chemistry, C.B.S. & H., G.B. Pant University of Agriculture & Technology, Pantnagar, U.S. Nagar, Uttarakhand, India. During this period, he published over 90 high-quality research and review articles in peer-reviewed journals of national and international repute. He has also published 22 book chapters for edited books and book series. He is the editor/co-editor of two books, *Biorationals and Biopesticides: Pest Management* (de Gruyter, 2024) and *Medicinal Plants: Chemical, Biochemical, and Pharmacological Approaches* (IntechOpen, 2024). Dr. Kumar guided ten MSc and two PhD students in the field of natural product chemistry. At present, he is guiding four MSc and two PhD students in the field of plant chemistry. He received the Young Scientist Award at the first Annual Conference of the Society for Himalayan Action Research and Development (SHARAD) on recent advances in chemicals and nanoscience at H.N.B. Garhwal University (Central University), Pauri Campus, Garhwal, 2018. He received the Excellence in Research Award for Outstanding Contribution in the field of natural product chemistry at the first International Agriculture Conference on Natural vs Organic Farming in Context to Bharatiya Agriculture at Gujrat Natural Farming and Science University, Anand, Gujrat, India, 2023. Dr. Kumar has membership in different chemical societies and journals and has delivered over ten lectures at national and international conferences in India and abroad. At present, he is handling two national projects in the field of natural product chemistry, which are funded by UCB and UCOST, Uttarakhand, India. Dr. Kumar is also a Co-PI in a central pollution control board (CPCB), New Delhi, an India-funded project. His current research interests lie in natural plant products with bioactivity toward insects, nematodes, and fish pathogenic bacteria, including essential oils, terpenes, flavonoids, alkaloids, polyphenols, cyanogenic glucosides, polyketides, and their nanoformulated products.

Dr. Eloisa Helena de Aguiar Andrade holds a degree in pharmacy (1980) and a specialization in biochemistry (1982) from the Federal University of Pará (UFPA). She has a master of science in chemistry of natural products (1992) and a doctor of philosophy in chemistry (2008), both from UFPA. She currently holds the position of Full Researcher II in Botany Coordination at the Museu Paraense Emílio Goeldi, where she is responsible for the Adolpho Ducke Multiuser Laboratory (LAD). Additionally, she is an associate professor at the Faculty of Chemistry of UFPA and a professor in the postgraduate programs in chemistry, tropical botany (UFRA/MPEG), and biodiversity and biotechnology at Rede Bionorte. She served as the coordinator of the Pará State Pole of the Postgraduate Program in Biodiversity and Biotechnology (PPGBIONORTE-PA) for the 2016–2020 and 2022–2024 academic periods. She has authored over 500 scientific contributions, including articles, conference papers, book chapters, and books. She has extensive experience in the field of chemistry, with a particular focus on natural product chemistry. Her research expertise encompasses various techniques, such as gas chromatography, gas chromatography coupled with mass spectrometry, and the analysis of volatile and fixed (derivatized) chemical constituents.

Dr. Antônio Pedro da Silva Souza Filho is a distinguished scholar in the field of agricultural sciences, holding a degree in agricultural engineering from the Federal Rural University of the Amazon (1977). He subsequently obtained a master of science in animal science from the Federal University of Lavras (1987) and a doctor of philosophy in animal science from the São Paulo State University Júlio de Mesquita Filho (1995). Dr. Souza Filho is currently retired from his role as a senior researcher at the Brazilian Agricultural Research Corporation (Embrapa Amazônia Oriental), where he has dedicated his career to investigating fundamental issues at the interface between chemistry and plant biology. His extensive experience in the field of chemical ecology is primarily centered on the isolation, identification, and characterization of the bioherbicide and biofungicide activity of chemical substances derived from plants and organisms such as endophytic fungi and phytopathogens. His research interests include allelopathy, an ecological phenomenon involving the release of chemicals by plants that influence the growth of other plants in their vicinity, as well as the development of agricultural biodefensives that offer sustainable alternatives to conventional chemical pesticides. Moreover, Dr. Souza Filho explores the enhancement of chemical substances' biological activity, aiming to augment the efficacy and safety of these compounds in the management of agricultural crops. Dr. Souza Filho's work not only advances scientific research but also promotes sustainable agricultural practices, which are crucial for environmental preservation and food security. His contributions have been acknowledged for his aptitude in translating research into practical applications, positively impacting agriculture in the Amazon region and beyond.

Contributors

Lorena C. Albernaz
Universidade de Brasília, Laboratório de Farmacognosia
Campus Universitário Darcy Ribeiro
Brasília, Brazil

Ammar Al-Farga
Department of Biochemistry
Faculty of Science
Alexandria University of Jeddah
Alexandria, Egypt

Fahd Mohammed Abd Al Galil
Department of Biology
Faculty of Applied Sciences
Umm Al-Qura University
Makkah, Saudi Arabia

Mariam S. Al-Ghamdi
Department of Biology
Faculty of Applied Sciences
Umm Al-Qura University
Makkah, Saudi Arabia

Ayah Talal Abdullah Al-Kilani
Nutritional Biochemistry
Faculty of Pharmacy
Petra University
Amman, Jordan

Abdulrahman A. Almadiy
Department of Biology
Faculty of Arts and Sciences
Najran University, Najran, Saudi Arabia

Ellem Suane Ferreira Alves
Universidade Federal do Pará
Programa de Pós-Graduação em Biodiversidade e Biotecnologia
Belém, Brazil

Saiqa Andleeb
Microbial Biotechnology and Vermitechnology Laboratory
Department of Zoology, University of Azad Jammu and Kashmir
Muzaffarabad, Pakistan

Sushila Arya
Department of Chemistry
College of Basic Science and Humanities
G.B. Pant University of Agriculture and Technology
Pantnagar, U.S. Nagar, Uttarakhand, India

Gloria Stefany Avendaño-Mora
Grupo de Investigación en Bioquímica y Microbiología (GIBIM)
Escuela de Química
Universidad Industrial de Santander
Parque tecnológico y de Investigaciones
Piedecuesta, Santander, Colombia

Avnee
Chaudhary Sarwan Kumar Himachal Pradesh Krishi Vishwavidyalaya
Palampur, Himachal Pradesh, India

Vitor Cunha Baia
Departamento de Química
Universidade Federal de Viçosa (UFV)
Viçosa-MG, Brazil

Dev Karan Bairwa
College of Agriculture
Kotputli (SKN Agriculture University, Jober)
Machhar Khani, Rajasthan, India

Pooja Bargali
Department of Chemistry
College of Basic Sciences and Humanities
G.B. Pant University of Agriculture and Technology
Pantnagar, Uttarakhand, India

Fernanda Wariss Figueiredo Bezerra
Programa de Pós-graduação em Ciência e Tecnologia de Alimentos (PPGCTA)
Universidade Federal do Pará (UFPA)
Belém, Pará, Brasil

Kangjam Bumpy
Pandit Deen Dayal Upadhyay Institute of Agricultural Sciences
Utlou, Bishnupur, Manipur

Fábio José Bonfim Cardoso
Laboratório de Química Multidisciplinar
Universidade Federal do Pará
Tucuruí, Para, Brasil

Márcia Moraes Cascaes
Universidade Federal do Pará. R. Augusto Corrêa
Belém, Brazil

Cristiane Isaac Cerceau
Departamento de Química
Universidade Federal de Viçosa (UFV)
Viçosa-MG, Brazil

Chandini
Bihar Agricultural University
Sabour, Bihar, India

Márlia Coelho-Ferreira
Campus de Pesquisa—Museu Paraense Emílio Goeldi
Belém, Brazil

Leonardo Souza da Costa
Laboratório Adolpho Ducke
Museu Paraense Emílio Goeldi
Belém, Brazil

Tatiani da Luz Silva Vasconcelos
Departamento de Química
Universidade Federal de Viçosa (UFV)
Viçosa-MG, Brazil

Marcilene Paiva da Silva
Laboratório Adolpho Ducke
Museu Paraense Emílio Goeldi
Belém, Brazil

Antônio Pedro da Silva Souza Filho
(Retired) Embrapa Amazônia Oriental
Belém—PA, Brazil

Ahmad Dar Showket
Faculty of Forestry (Entomology)
Division of Social and Basic Sciences
Sher-e-Kashmir University of Agricultural Sciences and Technology of Kashmir
Srinagar, India

Eliza de Jesus Barros dos Santos
Laboratório Adolpho Ducke
Museu Paraense Emílio Goeldi
Belém, Brazil

Eloisa Helena de Aguiar Andrade
Laboratório Adolpho Ducke
Museu Paraense Emílio Goeldi
Belém, Brazil

Celeste de Jesus Pereira Franco
Laboratório Adolpho Ducke
Museu Paraense Emílio Goeldi
Belém, Brazil

Mozaniel Santana de Oliveira
Laboratório Adolpho Ducke
Museu Paraense Emílio Goeldi
Belém, Brazil

Paula Maria Correa de Oliveira
Universidade Federal do Pará
Programa de Pós-Graduação em Biodiversidade e Biotecnologia
Belém, Brazil

Anderson de Santana Botelho
Laboratório Adolpho Ducke
Museu Paraense Emílio Goeldi
Belém, Brazil

Yendrembam K. Devi
Department of Entomology
Lovely Professional University
Punjab, India

Dahiphale Kalyan Devidas
Oilseeds Research Station
Latur, Maharashtra, India

Karyme do Socorro de Souza Vilhena
Universidade Federal do Pará. R. Augusto Corrêa
Belém, Brazil

Laboratório Adolpho Ducke
Museu Paraense Emílio Goeldi
Avenida Perimetral
Belém, Brazil

Jonny E Duque
Unidad de Investigación y Desarrollo en Alimentos (UNIDA)
Tecnológico Nacional de México/Instituto Tecnológico de Veracruz
Veracruz, Mexico

Laila Salmen Espindola
Universidade de Brasília, Laboratório de Farmacognosia
Campus Universitário Darcy Ribeiro
Brasília, Brazil

Oberdan Oliveira Ferreira
Universidade Federal do Pará
Belém, Brazil

Niraj Guleria
Mountain Agricultural Research and Extension Station
CSKHPKV, Salooni, Chamba, Himachal Pradesh, India

Wajid Hasan
Subject Matter Specialist Entomology
Krishi-Vigyan Kendra
Bihar Agricultural University
Jahanabad, Bihar, India

Kounser Javeed
Faculty of Horticulture FOA Wadura Sopore
Sher-e-Kashmir University of Agricultural Sciences and Technology of Kashmir
Srinagar, India

Shivang Joshi
Department of Chemistry
G.B. Pant University of Agriculture and Technology
India

Tanuja Kabdal
Department of Chemistry
College of Basic Sciences and Humanities
G.B. Pant University of Agriculture and Technology
Pantnagar, Uttarakhand, India

Himani Karakoti
Department of Chemistry
College of Basic Sciences and Humanities
G.B. Pant University of Agriculture and Technology
Pantnagar, Uttarakhand, India

Puneet Kaur
Department of Agronomy, Punjab Agricultural University
Ludhiana, India

Iram Khurshid
Ph.D. Scholar, Zoology
Kashmir University, Higher Secondary School
Jammu and Kashmir, India

Deepak Kumar
Research and Development Unit
Shri Ram Solvent Unit E
xtractions Pvt. Ltd
India

Randeep Kumar
Bihar Agricultural University
Sabour, Bhagalpur, Bihar, India

Ravendra Kumar
Department of Chemistry
College of Basic Science and Humanities
G.B. Pant University of Agriculture and Technology
Pantnagar, U.S. Nagar, Uttarakhand, India

Amrutha Lakshmi M
Division of Crop Improvement and Pest Management
ICAR-Central Arid Zone Research Institute
Jodhpur, India

Jonathan Lara-Sanchez
Unidad de Investigación y Desarrollo en Alimentos (UNIDA)
Tecnológico Nacional de México/Instituto Tecnológico de Veracruz
Veracruz, Mexico.

João Paulo Viana Leite
Professor at the Department of Biochemistry and Molecular Biology—UFV
Departamento de Bioquímica e Biologia Molecular
Viçosa—MG, Brazil

Iram Liaqat
Microbiology Laboratory, Department of Zoology
Government College University
Lahore, Pakistan

I Yimjenjang Longkumer
Department of Entomology
School of Agriculture
Lovely Professional University
Phagwara, Punjab, India

Sonu Kumar Mahawar
Department of Chemistry
College of Basic Sciences and Humanities
G.B. Pant University of Agriculture and Technology
Pantnagar, Uttarakhand, India

Guneshori Maisnam
Amity Institute of Organic Agriculture
Amity University
Noida, Uttar Pradesh, India

Ritu Mawar
Division of Crop Improvement and Pest Management
ICAR-Central Arid Zone Research Institute
Jodhpur, India

Stelia Mendez-Sanchez
Grupo de Investigación en Bioquímica y Microbiología (GIBIM)
Escuela de Química
Universidad Industrial de Santander
Piedecuesta, Santander, Colombia

Sakshi Negi
Department of Chemistry
College of Basic Science and Humanities
G.B. Pant University of Agriculture and Technology
Pantnagar, U.S. Nagar, Uttarakhand, India

Leo M.L. Nollet
University College Ghent
Gent, Belgium

Arley Rey Páez
PhD student in the Postgraduate Program in Applied Biochemistry—UFV Departamento de Bioquímica e Biologia Molecular
Campus Universitáriom
Viçosa—MG, Brazil

Om Prakash
Department of Chemistry
College of Basic Sciences and Humanities
G.B. Pant University of Agriculture and Technology
Pantnagar, Uttarakhand, India

SS Rana
Mountain Agricultural Research and Extension Station, CSKHPKV
Salooni, Chamba, Himachal Pradesh, India

Sumit Saini
Department of Entomology
CCSHAU Research Station
Kaul (Kaithal) Haryana

Elson Santiago Alvarenga
Departamento de Química
Universidade Federal de Viçosa (UFV)
Viçosa-MG, Brazil

Saranya R
ICAR-Central Arid Zone Research Institute
Jodhpur (Rajasthan), India

Hamdy A. Shaaban
National Research Center
Chemistry of Flavours and Aroma Department
El-Behoose St. Dokki Giza Egypt

Irsa Shafique
Microbial Biotechnology and Vermitechnology Laboratory
Department of Zoology, University of Azad Jammu and Kashmir
Muzaffarabad, Pakistan

Manisha Sharma
SKN College of Agriculture
SKNAU, Jobner Jaipur Rajasthan

Baljinder Singh
Chaudhary Sarwan Kumar Himachal Pradesh Krishi Vishwavidyalaya
Palampur, Himachal Pradesh, India

Hemant Kumar Singh
Krishi Vigyan Kendra
India

K. N. Singh
Department of Plant Molecular Biology and Genetic Engineering
N.D. University of Agriculture and Technology
India

M. K. Singh
Research and Development Unit
Shri Ram Solvent Unit Extractions Pvt. Ltd
India

Mohamed H. Soliman
National Research Center
Chemistry of Flavours and Aroma Department
El-Behoose St. Dokki Giza Egypt

João Paulo Barreto Sousa
Universidade de Brasília, Laboratório de Farmacognosia
Campus Universitário Darcy Ribeiro
Brasília, Brazil

Luis C Vesga
Grupo de Investigación en Bioquímica y Microbiología (GIBIM)
Escuela de Química
Universidad Industrial de Santander
Colombia

Alan K. Watson
Department of Plant Sciences
Macdonald Campus of McGill University
Canada

Syed Naseem Zaffar G
Faculty of Forestry
Head Division of Social and Basic Sciences
Sher-e-Kashmir University of Agricultural Sciences and Technology of Kashmir
Srinagar, India

Acknowledgments

We would like to thank all the contributors to this project, each chapter author and other contributors.

I would like to thank CRC Press for their support during the execution of the project. I would also thank the Emílio Goeldi Museum of Pará, especially to the Adolpho Ducke Laboratory—LAD.

Editor Dr. Mozaniel Santana de Oliveira thanks Programa de Desenvolvimento da Pós-Graduação Pós-Doutorado Estratégico, as well as CAPES for the scholarship (process number: (88887.852405/2023-00).

Introduction

Natural Pesticides and Allelochemicals: Advances and Trends in Crop Protection

Mozaniel Santana de Oliveira, Leo M.L. Nollet, Ravendra Kumar, Eloisa Helena de Aguiar Andrade, and Antônio Pedro da Silva Souza Filho

0.1 BACKGROUND

The evolution of humanity is inextricably linked to the consumption of food, which has been a fundamental aspect of human existence since the earliest periods of history. In the earliest stages of human evolution, our ancestors were primarily gatherers, and it was only later that protein-rich foods became a part of their diet. The discovery of the reproductive function of seeds prompted a shift in human behavior, leading to the abandonment of nomadic lifestyles and the establishment of permanent settlements on land. This marked the advent of agriculture, which is estimated to have originated approximately 12,000 years ago. The realization of the significance of this multiplication process—through which a single seed could yield 10, 20, 30, or more grains for sustenance—was pivotal in facilitating the transition from a nomadic to a sedentary way of life.

The formation of colonies, villages, agro-villages, and cities can be traced back to the advent of agricultural activity. However, as populations expanded, so too did the pressure to produce more food. This resulted in the necessity of not only incorporating new areas of cultivation but also adopting innovative techniques to ensure and increase food production. Consequently, issues pertaining to biotic agents detrimental to crops, including insects, fungi, bacteria, and weeds, have emerged. This has necessitated the implementation of measures by farmers to prevent a decline in production.

Over time, humans developed the capacity to observe, establish appropriate agricultural practices, and select plant species that were conducive to their needs. Additionally, they sought to incorporate methods and tools that would facilitate labor and ensure enhanced productivity. Consequently, significant innovations were developed that enhanced production efficiency, including the introduction of the plow, which is believed to have originated in 5000 BC. By the conclusion of the 19th century, the integration of mechanical processes into agricultural practices had become a firmly established phenomenon. As these facilities emerged, humans felt more comfortable increasing production and productivity, thereby meeting ever-increasing demands. The challenge of maintaining crops with minimal losses due to biotic agents has consistently been and will continue to be a fundamental aspect of agricultural practice. In the absence of this, the ability to guarantee that each human being receives 2,500 kcal per day would be severely compromised. Moreover, the production of crops that are highly susceptible to biotic agents, including tomatoes, peppers, potatoes, and others, would be unfeasible. These factors are of particular importance in tropical regions, where the prevalence of pests is enhanced by favorable environmental conditions.

Given that biotic agents represent the primary biological and economic constraints on agricultural development, particularly in tropical soils, where they constitute the predominant component

DOI: 10.1201/9781003463429-1

of crop maintenance costs, the control and management of these agents are of paramount importance for the success of enterprises and the return on investments.

Currently, the main method used to control agricultural pests is the use of chemical products, such as herbicides, fungicides and insecticides. At the beginning of this fight, control was done with ash, sulfur, copper, and arsenic, among others. Agricultural pesticides, strictly speaking, emerged at different times in the history of agriculture. For example, the first insecticide used efficiently was sulfur based, around 4,500 years ago, by the Sumerians. Compounds based on mercury and arsenic were used to control lice 3,200 years ago by the Chinese. The use of inorganic copper as a fungicide in the early 19th century for seed treatment was an important milestone in the control of fungal diseases. In 1882, a major event was the launch of the Bordeaux mixture in France to control downy mildew in grapevines. In the case of weeds, manual pulling was the most widely used method and is still widely used today. Organic herbicides emerged in 1932, with the launch of 3,5-dinitro-o-cresol (DNOC), but the greatest advances were achieved during World War II (1942–1944), notably 2,4-D, which marked the beginning of the Age of Chemical Agriculture.

The success of modern agriculture can be attributed to two fundamental factors: the utilization of growth stimulants and pesticides. This combination has ensured that the demands of both the world population and the population growth rate have been met. However, the use of these components, especially pesticides, is undergoing a process of exhaustion. This is due not only to the emergence of resistant biotic agents but also to the impact on natural resources, such as water, the depreciation of the quality of consumed products, and the compromise of wildlife and human health. While these issues have become part of the societal discourse, the solution to these demands is confronted with the lack of viable alternatives that can meet all these needs. The complete removal of existing products from the market, as proposed by some, would result in greater hardship than benefit for society. To envisage an agricultural system capable of meeting food demands without the requisite safeguards for the control of harmful biotic agents would be to court disaster. In the initial year following the cessation of pesticide use, production would decline by 40% to 50%. This decline would continue to increase in subsequent years. Crops that are highly dependent on pesticides, such as tomatoes, peppers, potatoes, strawberries, and others, would have to be abandoned, which would have significant implications for the human diet.

In light of the necessity to propose alternatives to the prevailing model of regulating biotic agents that are economically viable and in accordance with societal expectations, a number of studies have been conducted globally. These studies have focused on the identification of novel chemical compounds with the potential for direct utilization or incorporation into efficacious formulations for the control and management of these pests. A number of plant species, including neem, sunflower, timbó, and garlic vine, have demonstrated potential for use in this context. Moreover, other sources of chemical molecules, including fungi, bacteria, propolis, geopropolis, and algae, also demonstrate considerable potential. Significant progress has been made in recent years, yet the path forward remains lengthy. Nevertheless, a solution to the current predicament is already in development, and its eventual efficacy will become evident with time.

In light of the challenges posed by the use of synthetic agricultural pesticides, the search for more sustainable solutions has assumed greater prominence. In this context, allelochemicals, which are natural substances produced by plants, fungi, bacteria, and other organisms, emerge as a promising alternative. When released into the environment, these compounds have the potential to inhibit the growth of pests, diseases, and weeds without the adverse effects commonly associated with synthetic chemicals.

The future of sustainable agriculture appears to be inextricably linked to the use of allelochemicals. In addition to being biodegradable and less environmentally damaging, they can play a pivotal role in controlling harmful biotic agents without compromising the health of ecosystems and humans. Research is already underway to develop products based on allelochemicals extracted from plants such as neem, sunflower, timbó, and garlic vine, which have demonstrated efficacy in the management of agricultural pests and diseases.

In addition to their role in the preservation of natural resources, allelochemicals can also be seen as a response to the growing societal demands for the implementation of more sustainable and safer agricultural practices. In contrast to traditional chemical pesticides, which have been observed to induce resistance in pests and compromise biodiversity, allelochemicals offer a more balanced and adaptive approach to pest control. This could represent a novel model of agriculture, in which the utilization of natural resources becomes a facilitator, rather than an impediment, for the advancement of more resilient and enduring food production systems.

Nevertheless, for this transition to occur on a significant scale, it is imperative to expand research on the production, formulation, and application of these compounds. The prospective utility of allelochemicals is contingent upon their accessibility and economic viability, which would render them a viable alternative to the prevailing synthetic chemical model.

In light of these considerations, an agricultural scenario based on sustainable allelochemicals points to a revolution in pest control and crop management, promoting a more balanced, efficient, and environmentally harmonious approach to agriculture. As has been demonstrated by the advances made thus far in the prospecting of new natural molecules, the continuation of this research promises innovative solutions to the challenges of the 21st century.

I wish you an enjoyable perusal of the forthcoming chapters this book.

Section 1

Pest Management—Sustainability Versus Unsustainability—Allelochemicals

1 Synthetic Pesticides and Sustainable Food Products

Elson Santiago Alvarenga, Cristiane Isaac Cerceau, and Vitor Cunha Baia

1.1 INTRODUCTION

Agriculture has been an essential pillar of human civilization, providing sustenance and nutrition for an ever-expanding population. However, such an undertaking is not without significant challenges, one of the most pressing being the continued threat of agricultural pests, which can decimate entire harvests, compromising not only food security[1] but also the economic and social stability of communities dependent on agricultural production[2].

In this context, synthetic pesticides emerge as indispensable tools for protecting crops and, therefore, for maintaining global food security[3]. These compounds, developed through scientific and technological advances, offer unparalleled effectiveness in eradicating and controlling agricultural pests, allowing crop yields to be maximized and losses resulting from infestations to be minimized.

The growth of the world population constitutes an exacerbating factor of this problem, imposing an increasing demand for food. Estimates project that the global population will surpass the 9 billion mark by the middle of the 21st century[4, 5], placing additional pressure on already overburdened agricultural production systems. Efficient use of pesticides becomes not only a question of productivity but also of food security and economic stability.

However, the indiscriminate and inappropriate use of synthetic pesticides has raised increasing concerns due to their potential negative impacts on the environment and human health. The contamination of water resources, bioaccumulation, and the resistance developed by some pest species are just some of the side effects associated with this practice[6–8]. Given this scenario, there is an urgent need to adopt more selective and sustainable approaches to pest control, aiming to mitigate such undesirable consequences.

In recent decades, the area of agrochemical synthesis has benefited significantly from the advancement of computational technologies, particularly with regard to theoretical calculations and molecular docking[9–11]. These tools offer an accurate and cost-effective approach to designing and optimizing more selective and environmentally safe pesticides, allowing for the reduction of the use of harmful products without compromising pest control effectiveness.

Therefore, this chapter seeks to report some scientific studies on this topic, considering the intersection between the growing demands of modern agriculture and the imperatives of sustainability, in addition to highlighting the importance of synthetic pesticides in ensuring global food security while promoting the need for the development and adoption of more selective and sustainable approaches to pest control.

1.2 INSECTICIDES

Insecticides are compounds developed to kill or repel insects that threaten cultivated plants, thereby reducing economic losses and helping to ensure food security. Since their introduction, synthetic insecticides have been instrumental in increasing agricultural efficiency. However, their use has

DOI: 10.1201/9781003463429-3

raised environmental and public health concerns due to the persistence of some compounds and the resistance developed by certain insect populations[12, 13].

Insecticides encompass a wide range of compounds that can be categorized based on their chemical structures and mechanisms of action[14, 15]. Among these, amides, pyrethroids, lactones, and their derivatives stand out. Each of these classes has distinct characteristics that influence its effectiveness, selectivity, and environmental impact. Amides, for example, are known for their ability to inhibit enzymatic processes in insects, while pyrethroids, analogues of natural compounds, act predominantly on the nervous system[16]. Lactones, in turn, offer a versatile platform for the development of new insecticides due to their cyclic structure and potential for chemical modification. In this first section, we will explore some scientific work aimed at the synthesis of potential insecticides using these classes of compounds.

1.2.1 Amides

Amides hold significant importance, being one of the most common organic functional groups in nature, largely due to their presence in a wide variety of secondary metabolites. Beyond their occurrence in natural products, amides are found in over 25% of all existing drugs, underscoring their immense industrial and economic relevance. The synthesis of amides is particularly notable for its association with antibacterial, insecticidal, and herbicidal activities.

Nine amides[17] from potassium sorbate were synthesized in just three steps, as shown in Scheme 1. In a nutshell, potassium sorbate (1) was suspended in dichloromethane and acidified with hydrochloric acid to yield (2*E*,4*E*)-hexa-2,4-dienoic acid (2). Subsequently, oxalyl chloride was added to the solution of (2*E*,4*E*)-hexa-2,4-dienoic acid in anhydrous dichloromethane (DCM) at room temperature, resulting in acyl chloride (3) (a green oil). Finally, the acyl chloride was dissolved in anhydrous DCM, and the corresponding amine was added to the ice-cooled reaction mixture (0°C), leading to the formation of amides (5-12).

The amides were evaluated against *Ascia monuste* insect pest, a destructive pest of kale vegetables, and in favor of predator fire ant *Solenopsis saevissima* and pollinator *Tetragonisca angustula.*

The biological assay was carried out in three stages. In the first, amides with insecticidal activity against *A. monuste* were selected; in the second, the dose–mortality curves were estimated for the amides selected in the first stage, and, finally, the time–mortality curves were also estimated.

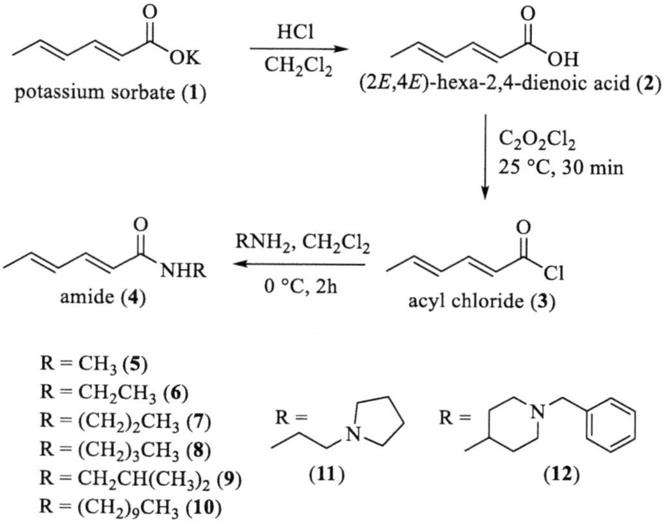

potassium sorbate (**1**) — HCl, CH_2Cl_2 → (2*E*,4*E*)-hexa-2,4-dienoic acid (**2**) — $C_2O_2Cl_2$, 25 °C, 30 min → acyl chloride (**3**) — RNH_2, CH_2Cl_2, 0 °C, 2h → amide (**4**)

R = CH_3 (**5**)
R = CH_2CH_3 (**6**)
R = $(CH_2)_2CH_3$ (**7**)
R = $(CH_2)_3CH_3$ (**8**)
R = $CH_2CH(CH_3)_2$ (**9**)
R = $(CH_2)_9CH_3$ (**10**)
R = (**11**)
R = (**12**)

SCHEME 1 Synthetic route for the preparation of the amides from potassium sorbate.

The amides (7) and (8) were the substances that caused the highest mortality of *A. monuste* (96 and 93%, respectively, at a dose of 30 mg of substance per g of insect). These amides presented a fast-acting control of *A. monuste* (less than 48 h), being indicated in situations of pest outbreaks that cause great economic losses if not controlled quickly.

In this study, the insecticidal activity of amides exhibited no correlation with their chemical structures, despite their similarities. Generally, compounds with lower molecular weight and reduced water solubility tend to display greater insecticidal potency due to enhanced penetration into the insect cuticle. However, in this instance, such a correlation was not observed due to the absence of a connection between the solubility and molecular weight of amides and their insecticidal activity.

Aguiar and collaborators[18] conducted the synthesis of 14 chiral amides and evaluated their insecticidal activity against *Rhyzopertha dominica* as well as their phytotoxicity to wheat seeds (*Triticum sativum*).

The synthetic route (Scheme 2) to obtain chiral amides: Commercially available D-mannitol (13) was acetylated using anhydrous acetone and zinc chloride to form compound (14) (87% yield). Oxidative cleavage of diacetal (14) by sodium periodate produced 2,3-*O*-isopropylidene-D-glyceraldehyde (15). A Wittig reaction of aldehyde (15) with (carbethoxymethylene)triphenylphosphorane yielded a mixture of esters (16-*Z*) and (17-*E*) in a 4:1 ratio with a 63% yield. Subsequently, the (16-*Z*) ester was hydrolyzed, followed by in situ formation of the anhydride (19) using methyl chloroformate in the presence of triethylamine. Finally, compound (19) underwent aminolysis (several different amines), leading to the formation of amides (20a–20n) (Table 1.1).

It is worth mentioning that the ylide (carbethoxymethylene)triphenylphosphorane, utilized in the conversion of compound (15) into compounds (16-*Z*) and (17-*E*), was previously synthesized, as depicted in Scheme 3.

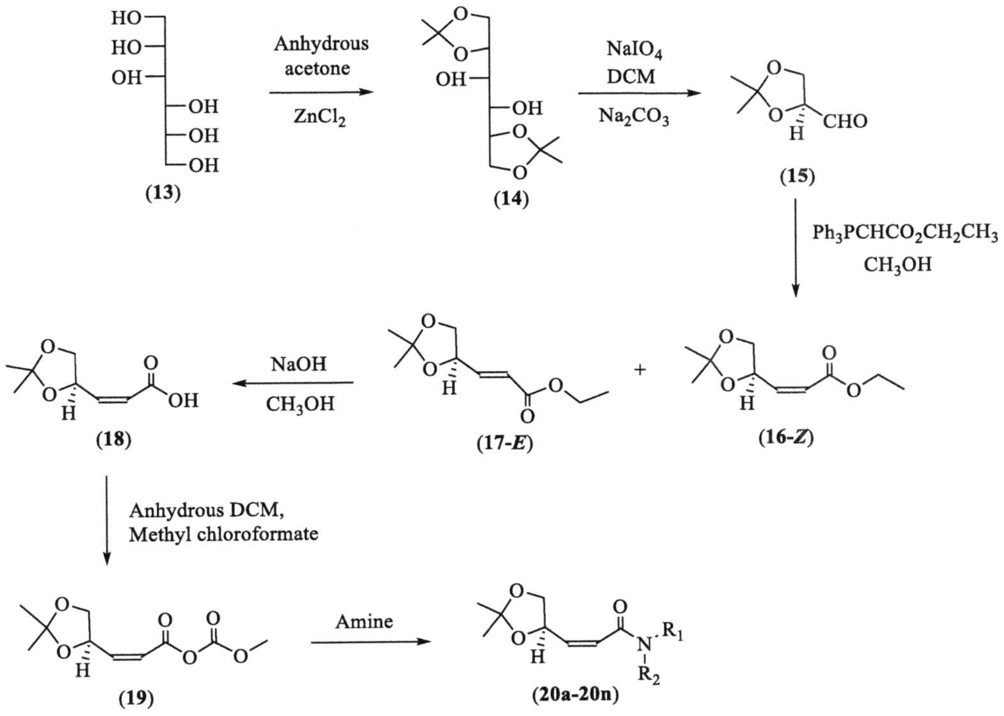

SCHEME 2 Synthetic route for the preparation of the amides from D-mannitol.

TABLE 1.1
Amines Used in the Reaction, Resulting in the Formation of 14 Chiral Amides (20a-20n)

Amines	Amides	
 H_2N	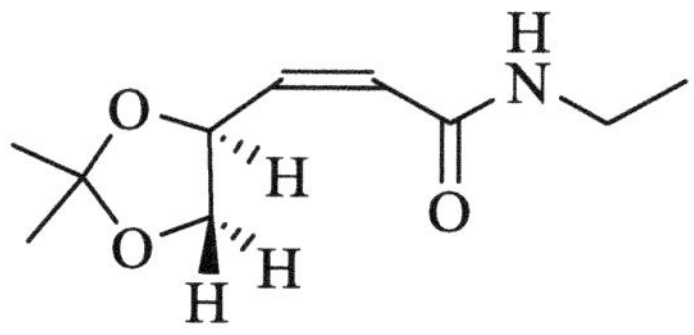	(20a)
 H_2N	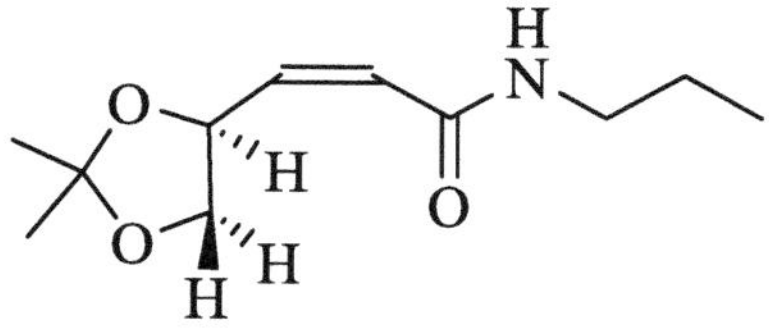	(20b)
 H_2N	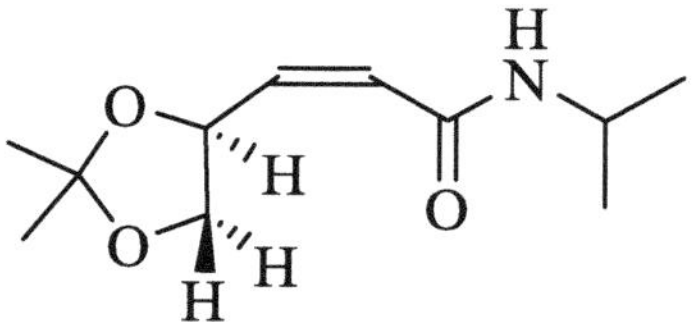	(20c)
 H_2N	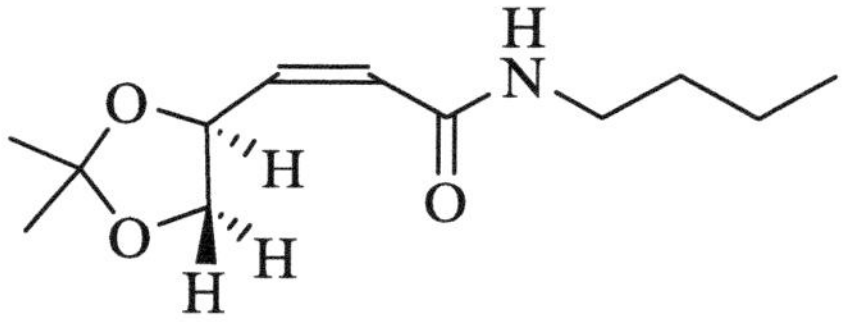	(20d)
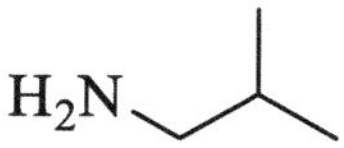 H_2N	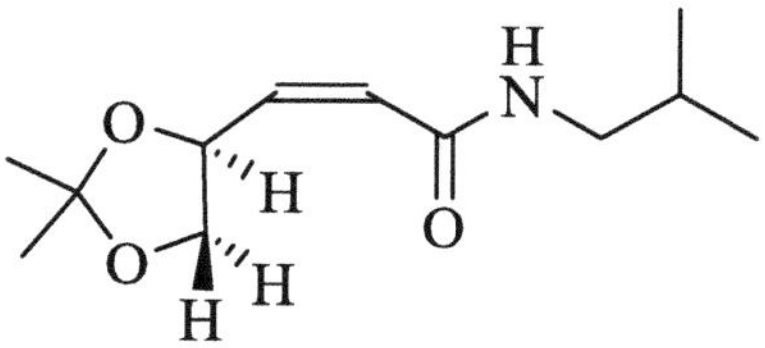	(20e)
 H_2N	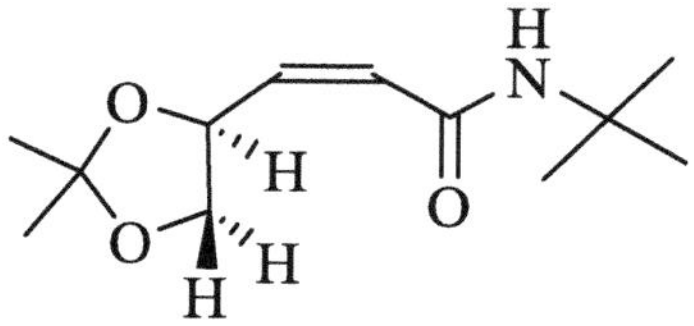	(20f)
 H_2N	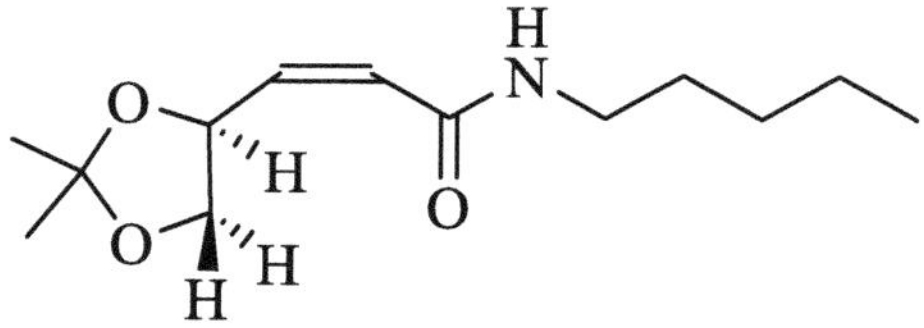	(20g)
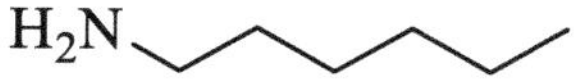 H_2N	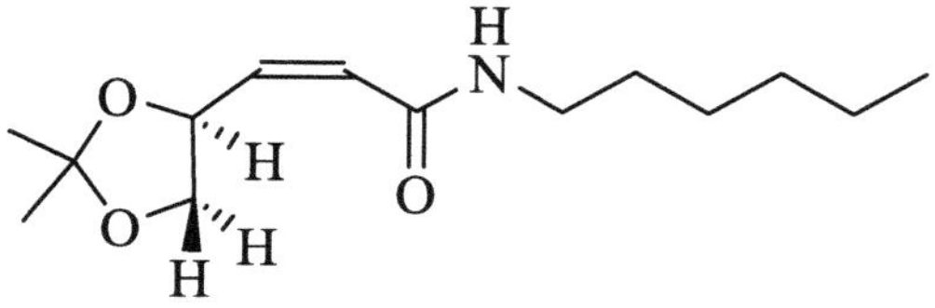	(20h)

(Continued)

TABLE 1.1 *(Continued)*
Amines Used in the Reaction, Resulting in the Formation of 14 Chiral Amides (20a-20n)

Amines	Amides	
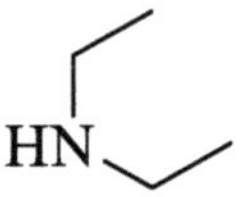 HN	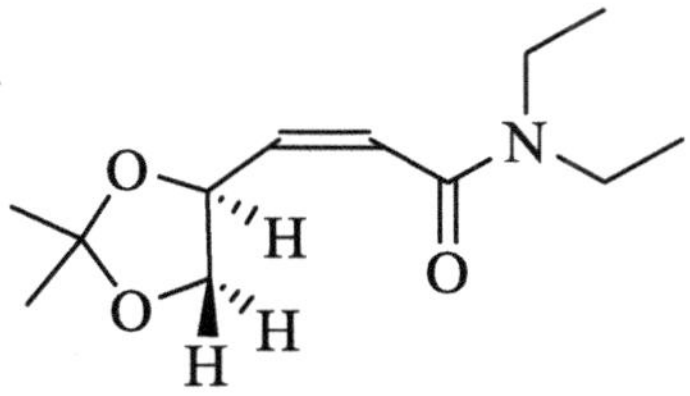 O, N, O, O, H, H, H	(20i)
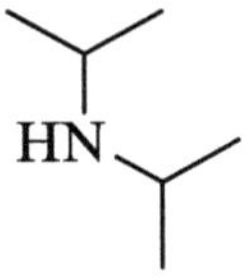 HN	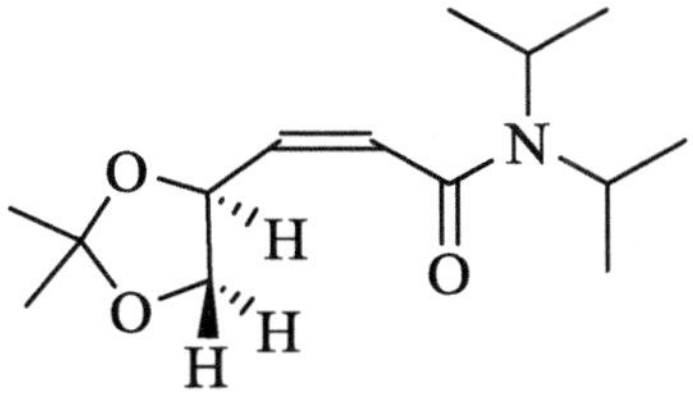 O, N, O, O, H, H, H	(20j)
 HN	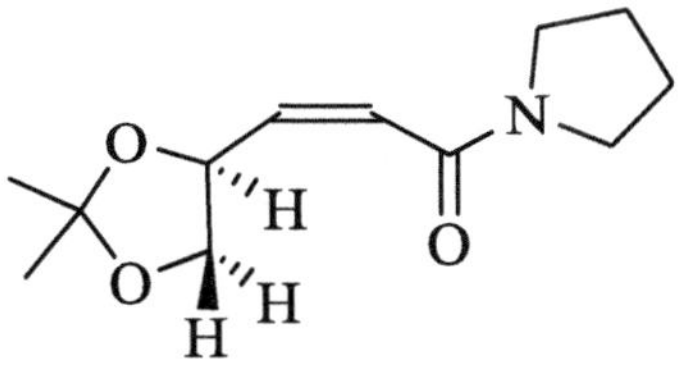 O, N, O, O, H, H, H	(20k)
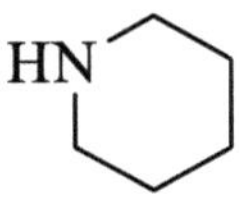 HN	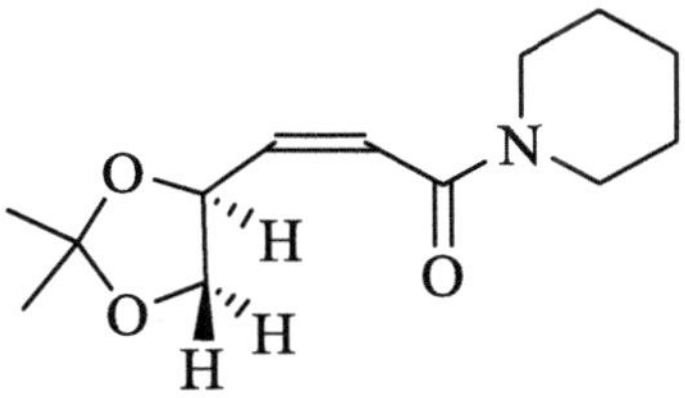 O, N, O, O, H, H, H	(20l)
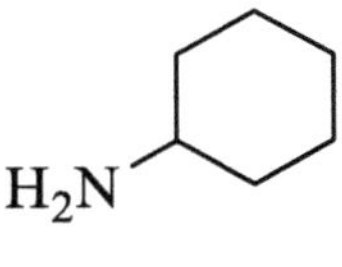 H_2N	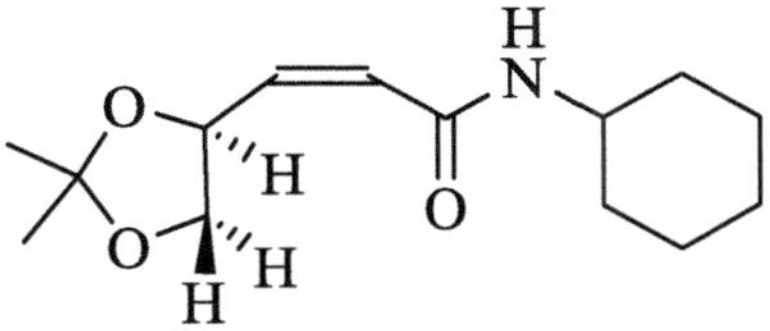 H, N, O, O, O, H, H, H	(20m)
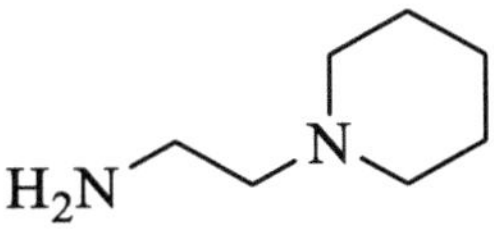 H_2N, N	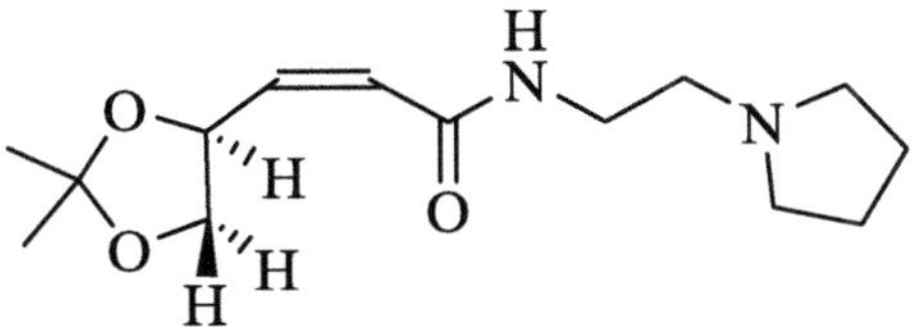 H, N, N, O, O, O, H, H, H	(20n)

Br, O, O — PPh_3 / Toluene → $\overset{\ominus}{Br}Ph_3\overset{\oplus}{P}$, O, O (Wittig salt) — NaOH / H_2O → Ph_3P, O, O (Ylide)

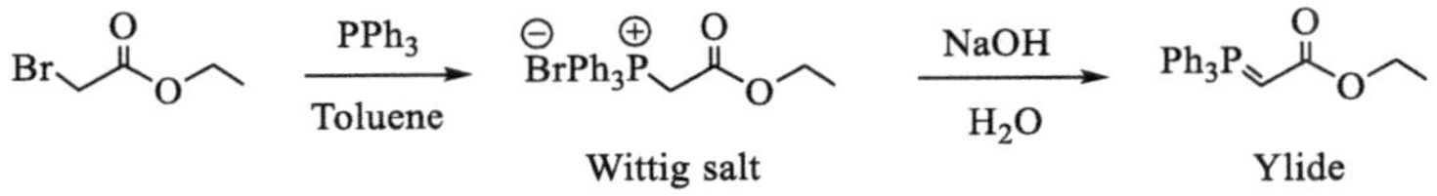

SCHEME 3 Synthesis of ylide (carbethoxymethylene)triphenylphosphorane.

(21) $\xrightarrow[CH_2Cl_2]{C_2O_2Cl_2}$ (22) $\xrightarrow{\text{amines (a-i)}}$ Dienamides (23a-23i)

SCHEME 4 Synthetic route for the preparation of the dienamides from sorbic acid.

The bioassays with *R. dominica* adults were conducted in three stages. First, screening was performed to select the most active amides. In the second stage, dose–mortality curves were estimated for the compounds selected in the initial screening (20i and 20j). In the third and final stage, the speed of action of these compounds against *R. dominica* was determined. The results were compared to those obtained with the commercial insecticide bifenthrin.

The phytotoxic effects of the most active amides on wheat seeds were evaluated using five concentrations (500, 400, 300, 200, and 100 μmol L^{-1}). The seeds were treated with 5 mL of 0.3% (v/v) aqueous dimethylsulfoxide (DMSO) for the negative control and with the pre-emergent commercial herbicide *S*-metolachlor for the positive control.

In the screening assay, compounds (20i) and (20j) caused 100% and 87% mortality of *R. dominica*, respectively. These values did not differ significantly from the mortality caused by bifenthrin (75%). Amide (20i) exhibited toxicity (LD_{50} = 27.98 μmol g^{-1}, Cl_{95} = 25.14–30.71) and speed of action (LT_{50} = 22 h, IC95 = 19.34–24.66) similar to amide 8j (LD_{50} = 29.37 μmol g^{-1}, Cl_{95} = 27.43–31.09, and TL_{50} = 19 h, Cl_{95} = 17.05–20.95) against the pest. Both amides inhibited wheat growth by less than 44%. Among the tested amides, only (20i) and (20j) were effective in controlling *R. dominica* and did not exhibit considerable phytotoxicity towards wheat seeds. Therefore, these amides show promise as insecticides for managing *R. dominica*.

The toxicity of new synthetic dienamides[19] was also evaluated against the insect pest *Diaphania hyalinata*, and the selectivity of these substances was evaluated for the predator *Solenopsis saevissima*.

In the synthetic route to obtain dienamides (23a–23i), sorbic acid (21) was converted into the more reactive acid chloride (22) by reaction with oxalyl chloride in DCM before reaction with the amine (Scheme 4).

The amines used in the reaction and the corresponding dienamides formed are presented in Table 1.2.

Four bioassays were performed. Initially, dienamides that resulted in high mortality in *D. hyalinata* were selected. In the second bioassay, dose–mortality curves were constructed for the selected dienamides. The third bioassay established survival curves for *D. hyalinata* and determined the time required to kill 50% of the population. The fourth bioassay assessed the selectivity of the compounds towards the predator *S. saevissima*. The most effective compound, (2*E*,4*E*)-*N*-butylhexa-2,4-dienamide (23d), killed 95% of the melonworm *D. hyalinata* and less than 10% of the natural enemy *S. saevissima*. This compound exhibited superior results compared to the commercial insecticide malathion. Dienamides (23c), (23d), and (23f) were shown to be selective, effectively targeting the pest while sparing the beneficial insect.

1.2.2 Pyrethroids

Pyrethroids constitute a widely used class of synthetic insecticides, modeled after pyrethrins, natural products found in flowers of the genus *Chrysanthemum*. They are known for their high effectiveness and relatively low toxicity to mammals, making them valuable tools for controlling agricultural pests. Pyrethroids act predominantly on the insect nervous system, altering the normal function of sodium channels in the membranes of nerve cells[16]. This disruption causes neuronal hyperexcitability, resulting in paralysis and the death of the insect, an effect known as "knockdown". This mechanism of action makes pyrethroids a compelling choice in many pest management programs.

TABLE 1.2
Amines Used in the Reaction with Acid Chloride, Leading to the Formation of Nine Dienamides (23a–23i)

Amine	Dienamides
NH_2CH_3 (a)	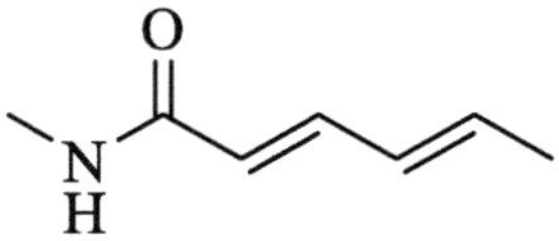 (23a)
$NH_2CH_2CH_3$ (b)	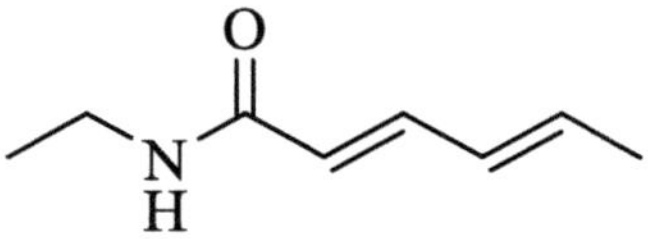 (23b)
$NH_2(CH_2)_2CH_3$ (c)	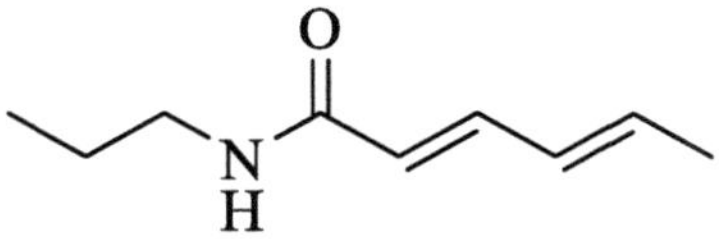 (23c)
$NH_2(CH_2)_3CH_3$ (d)	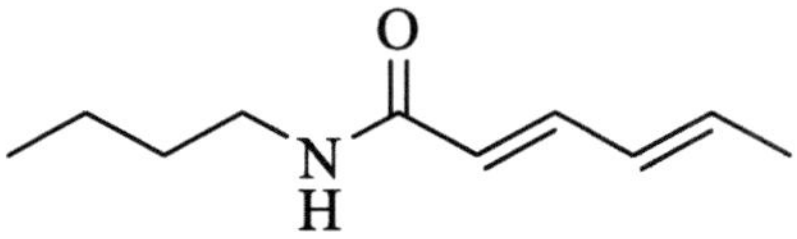 (23d)
$NH_2CH(CH_3)_2$ (e)	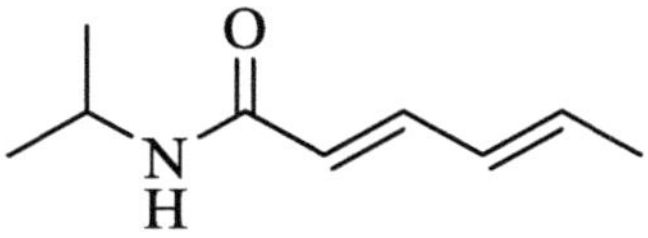 (23e)
$NH_2CH_2CH(CH_3)_2$ (f)	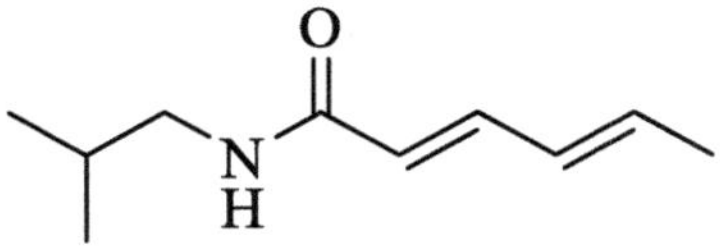 (23f)
$NH_2(CH_2)_9CH_3$ (g)	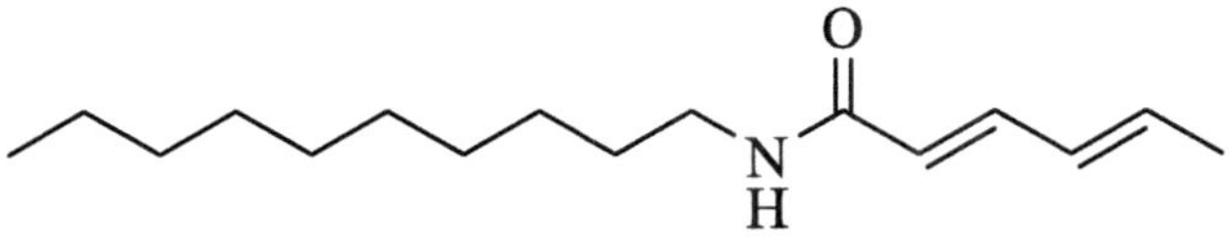 (23g)
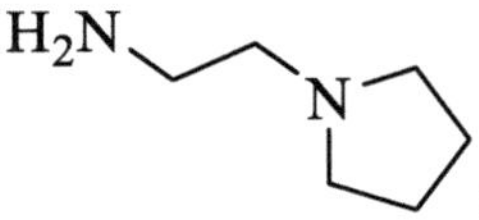 (h)	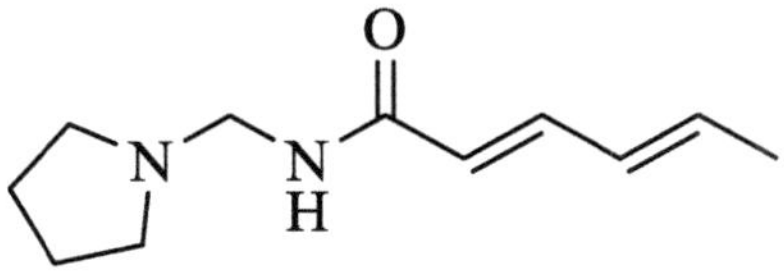 (23h)
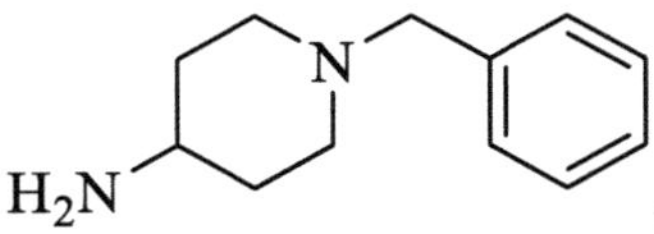 (i)	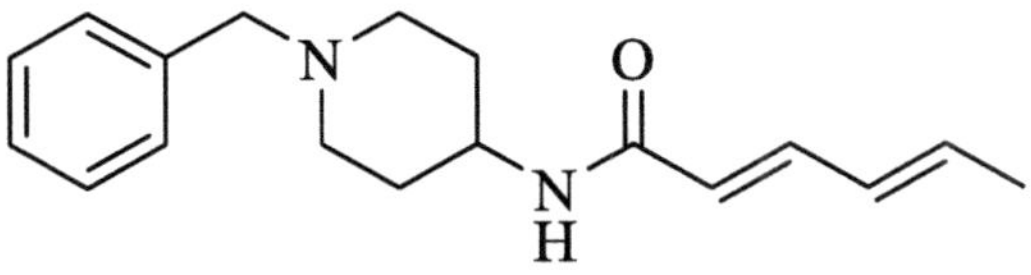 (23i)

Furthermore, pyrethroids are notable for their structural versatility, offering ample potential for modifications to increase their efficacy. This approach was adopted by Silvério and collaborators [20] in their scientific study entitled "Synthesis and Insecticidal Activity of New Pyrethroids". In this study, researchers synthesized ten new molecules belonging to the pyrethroid class and tested them for insecticidal activity against five insect species.

The synthesis of these molecules followed Scheme 5, which entailed the formation of an acetal from D-mannitol (24) in the presence of anhydrous acetone and zinc chloride as a catalyst, yielding diol (25) with an 87% yield. Subsequently, this compound underwent oxidative cleavage using sodium metaperiodate, resulting in the formation of 2,3-O-isopropylidene-D-glyceraldehyde (26). The heating reaction of this aldehyde with methoxycarbonylmethyldene(triphenyl)phosphorane yielded a mixture of alkenes (*Z*-27) and (*E*-28), with a yield of 65%. Enantioselective cyclopropanation of the

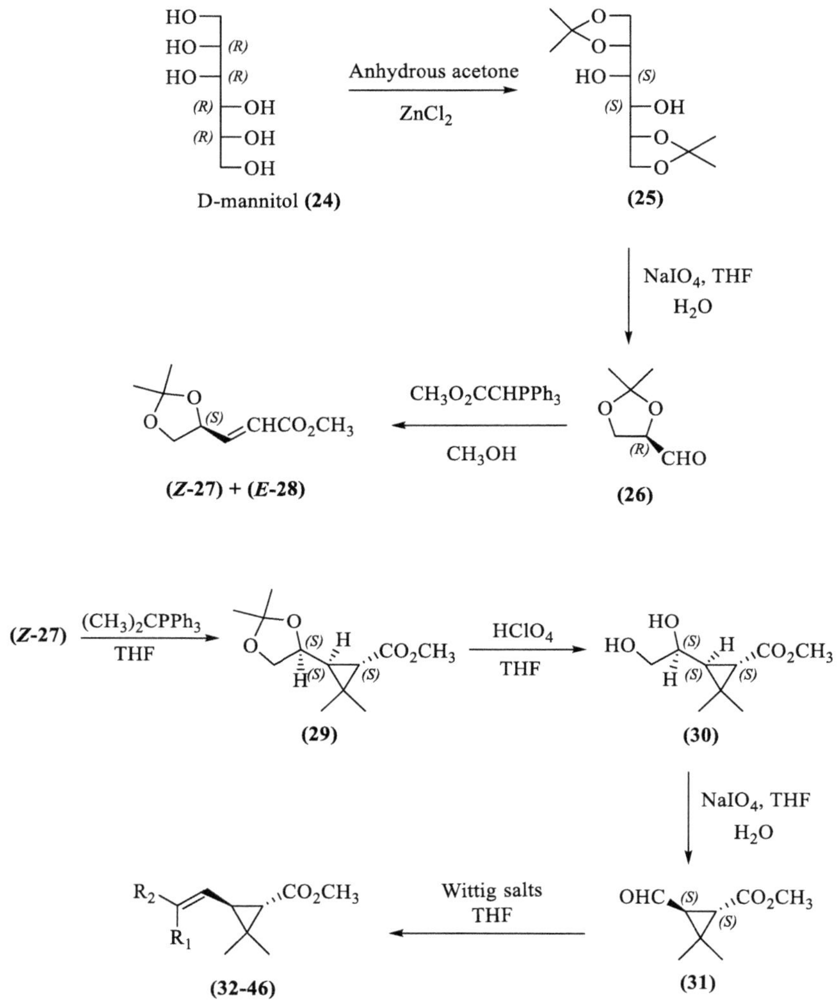

D-mannitol **(24)** — Anhydrous acetone, $ZnCl_2$ → **(25)**

(25) — $NaIO_4$, THF, H_2O → **(26)**

(26) — $CH_3O_2CCHPPh_3$, CH_3OH → **(*Z*-27) + (*E*-28)**

(*Z*-27) — $(CH_3)_2CPPh_3$, THF → **(29)** — $HClO_4$, THF → **(30)**

(30) — $NaIO_4$, THF, H_2O → **(31)** — Wittig salts, THF → **(32-46)**

(32) $R_1 = p$-NO_2Ph; $R_2 = H$
(33) $R_1 = H$; $R_2 = p$-NO_2Ph
(34) $R_1 = p$-BrPh; $R_2 = H$
(35) $R_1 = H$; $R_2 = p$-BrPh
(36) $R_1 = p$-ClPh; $R_2 = H$
(37) $R_1 = H$; $R_2 = p$-ClPh
(38) $R_1 = p$-FPh; $R_2 = H$
(39) $R_1 = H$; $R_2 = p$-FPh
(40) $R_1 = Ph$; $R_2 = H$
(41) $R_1 = H$; $R_2 = Ph$

(42) $R_1 = p$-$ArNO_2$; $R_2 = H$ (*cis*) and *trans*
(43) $R_1 = p$-ArBr; $R_2 = H$ (*cis*) and *trans*
(44) $R_1 = p$-ArCl; $R_2 = H$ (*cis*) and *trans*
(45) $R_1 = p$-ArF; $R_2 = H$ (*cis*) and *trans*
(46) $R_1 =$ -Ar; $R_2 = H$ (*cis*) and *trans*

SCHEME 5 Synthetic route for the preparation of the pyrethroids from D-mannitol.

(Z-27) alkene with isopropylidene(triphenyl)phosphorane produced compound (29), with an 81% yield. Hydrolysis of this compound with perchloric acid produced compound (30) and its oxidative cleavage with sodium metaperiodate resulted in the formation of compound (31), with a yield of 52%. Finally, a Wittig reaction was employed to produce the ten new alkenes (32-46), with yields varying between 60 and 95%. The compounds were characterized using FTIR, NMR, and mass spectrometry.

All compounds exhibited insecticidal activity against second instar larvae of *Ascia monuste orseis*, *Tuta absoluta*, and *Musca domestica*; second instar nymphs of *Periplaneta americana*, and adults of *Sitophilus zeamais*. Compound (36) showed better insecticidal activity, eliminating 90% of *A. monuste orseis* and 100% of *T. absoluta* and *P. americana*.

The compounds developed in this study were also subjected to toxicity analyses on three species of social insects: *Protonectarina sylveirae*, *Solenopsis saevissim*, and *Tetragonisca angustula*.[21] A primary concern in pest control is the use of selective pesticides, capable of targeting only intended organisms without harming non-target ones. Preserving insects beneficial to the ecosystem, such as the predators *P. sylveirae* and *S. saevissima*, along with the pollinator *T. angustula*, is essential for maintaining biodiversity. Therefore, seeking more selective insecticides is crucial. The results indicated low toxicity of these compounds for the tested species, suggesting that the studied pyrethroids represent a promising avenue for future research.

Another interesting study on pyrethroids was conducted by Moreno and collaborators.[22] The authors evaluated the toxicity of pyrethroids (42-46) (Scheme 5) with modifications to the acid portion against the insect pests *A. monuste* and *D. hyalinata*. They also assessed the selectivity of these pyrethroids for the predatory ant *S. saevissima* and the pollinator *T. angustula*.

For the bioassays, second instar larvae of *A. monuste* and *D. hyalinata*, adults of the predatory ant *S. saevissima*, and the pollinating stingless bee *T. angustula* were used. Four types of bioassays were performed. In the first bioassay, racemic mixtures with insecticidal activity against *A. monuste* and *D. hyalinata* were selected. In the second bioassay, dose–mortality curves were estimated for racemic mixtures applied to the pest insects (*A. monuste* and *D. hyalinata*). The third bioassay assessed the selectivity of the racemic mixtures for the predator *S. saevissima* and the pollinator *T. angustula*. In the fourth bioassay, isomer toxicities in *A. monuste* and *D. hyalinata* were compared.

All synthesized pyrethroids resulted in high (100%) and rapid mortality (stable LD_{50} after 12 hours) in *D. hyalinata* and *A. monuste*. Permethrin, a commercial insecticide, was more potent in controlling *A. monuste* and *D. hyalinata* than the five new pyrethroids, as it exhibited a lower LD_{50} value. In *A. monuste*, the *trans*-pyrethroid isomer (45) demonstrated toxicity comparable to permethrin. For *D. hyalinata*, the *trans*-pyrethroid (42) and *cis*-pyrethroid (43) isomers were as toxic as permethrin. When applied to pest insects (*A. monuste* and *D. hyalinata*), the new pyrethroids exhibited LD_{50} values greater than or equal to those obtained when applied to non-target insects (*S. saevissima* and *T. angustula*), indicating a lack of selectivity. Due to their low selectivity, these new pyrethroids should be applied based on principles of ecological selectivity to minimize impacts on non-target organisms such as *S. saevissima* and *T. angustula*.

Moreno and collaborators[23] also studied the effect of a racemic mixture of these five pyrethroids (42-46, Scheme 5) against *Tuta absoluta*, a neotropical oligophagous insect that targets solanaceous crops, particularly tomatoes. It is one of the primary pests affecting tomatoes in most South American countries and poses a significant threat to production in Europe, Africa, and the Middle East. The larvae of *T. absoluta* attack tomato plants in all stages of growth. The most common damage caused by this pest includes the formation of large galleries on the leaves, apical buds, and stems, as well as burrowing into both green and ripe fruits.

The bioassays were conducted in three stages. First, the mortality caused by a dose of E/Z mixtures of the new pyrethroids to the *T. absoluta* pest was evaluated. Second, dose–mortality curves of *E/Z* mixtures of pyrethroids for *T. absoluta* were estimated. Finally, the mortality caused by the *cis* and *trans* isomers of the new pyrethroids for *T. absoluta* was assessed.

The *E/Z* mixtures of the five pyrethroids caused high mortality levels, similar to the commercial insecticide permethrin (100%), indicating their potential as effective pesticides for use

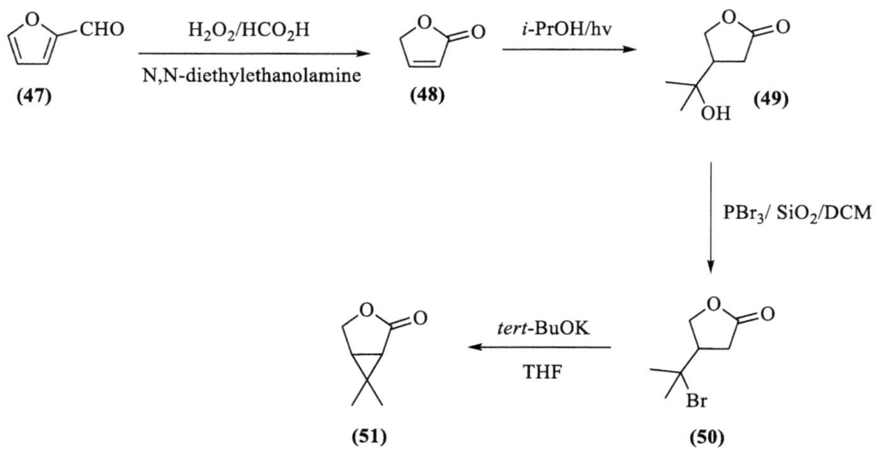

SCHEME 6 Synthesis of 6,6-dimethyl-3-oxabicyclo[3.1.0]hexan-2-one (51).

in agricultural crops. Additionally, all *E*/*Z* mixtures demonstrated a rapid action speed (less than 12 hours), ensuring swift control of *T. absoluta*, as their LD_{50} values remained unchanged even 12 hours post-application. Although the five *E*/*Z* pyrethroid mixtures were less potent than permethrin, which has a lower LD_{50} for *T. absoluta*, the *cis* isomer of pyrethroid (43) exhibited comparable mortality rates to permethrin, effectively causing plague-like mortality. Consequently, these new pyrethroids can be utilized as *E*/*Z* mixtures or as isomers with greater activity.

Another structural platform of interest in pesticide synthesis is the compound 6,6-dimethyl-3-oxabicyclo[3.1.0]hexan-2-one. The oxabicyclo portion of this compound can serve as a building block to obtain cis-chrysanthemic acid, from which many pyrethroids used in agriculture are derived. Oxabicyclolactone was the subject of studies by Alvarenga and collaborators[24], who synthesized and evaluated the bioactivity of this compound using furfural as a starting material.

Scheme 6 illustrates the synthetic steps of oxabicyclohexanone (51). Initially, furfural (47) was subjected to an oxidation reaction with performic acid in the presence of N,N-diethylethanolamine, resulting in the butenolide furan-2(5H)-one (48) in 40% yield. Then, the butenolide was irradiated in isopropyl alcohol with four low-pressure mercury lamps, forming compound (49) in 100% yield. This compound was treated with phosphorus tribromide (PBr_3) and silicon dioxide (SiO_2) in DCM to obtain compound (50) in 50% yield, which, in turn, was treated with potassium tert-butoxide in anhydrous THF, resulting in 6,6-dimethyl-3-oxabicyclo[3.1.0]hexan-2-one (51) in 40% yield.

Furthermore, the researchers synthesized five new pyrethroids following the synthetic procedure used by Silvério and collaborators[25]. The insecticidal activity of all compounds was evaluated through biological assays with *Acanthoscelides obtectus*, adults of *S. zeamais*, second instar larvae of *A. monuste orseis*, and second instar nymphs of *P. americana*, pests commonly found in Brazil. The results indicated that all compounds showed significant insecticidal activity.

1.2.3 Lactones and Derivates

Lactones are known for their broad spectrum of bioactivity, including insecticidal activity.[26] The versatility of lactones as a structural platform for new insecticides lies in their ability to interact with multiple biological targets in insects, providing an effective approach to pest control. The chemical modification of lactones allows for the optimization of essential characteristics such as selectivity, potency and stability, making them ideal candidates for the development of safer and more effective insecticides.

Teixeira et al.[27] synthesized ten novel lactones and evaluated their insecticidal activity against the pest insect *D. hyalinata*. Additionally, the impact of these compounds on the predator fire ant *S. saevissima* and the pollinator *T. angustula* was assessed.

The synthetic route for the lactones begins with the cycloaddition reaction of furan-2(5*H*)-one (52) and cyclopentadiene, forming the *endo*-(53) and *exo*-(54) adducts. This reaction proceeds with regio- and stereoselective control, but without asymmetric induction. In the next stage, the double bond of lactones (53) and (54) is oxidized with *meta*-chloroperbenzoic acid, yielding only epoxides (55) (98% yield) and (56) (83% yield) with stereospecific control. Halogenantion of unsaturated lactone (54) proceeded stereoselectively, yielding only the *trans*-1,2-dihalogenated products (59a, 59b, 60a, 60b), as expected (Scheme 7).

Biological tests were conducted using second instar larvae of *D. hyalinata* and adults of *S. saevissima* and *T. angustula*. Initially, a screening was performed to identify the most active lactones against *D. hyalinata*, with piperine serving as a positive control. Subsequently, the most active lactones were tested to determine the dose–mortality and time–mortality responses (using doses equivalent to the LD_{90} obtained for the three most active lactones) for *D. hyalinata*. Finally, the LD_{90} of the most active lactones for *D. hyalinata* was applied to assess the selectivity of these compounds towards non-target insects, *T. angustula* and *S. saevissima*.

The results showed that formulations (59a) + (59b), (60a) + (60b), and (60b) had high mortality rates, 96.3%, 91.3%, and 84.8%, respectively, against the pest *D. hyalinata*. In addition, it was observed that formulations such as (60a) + (60b) were selective, showing lower toxicity to the ant *S. saevissima* and the bee *T. angustula*.

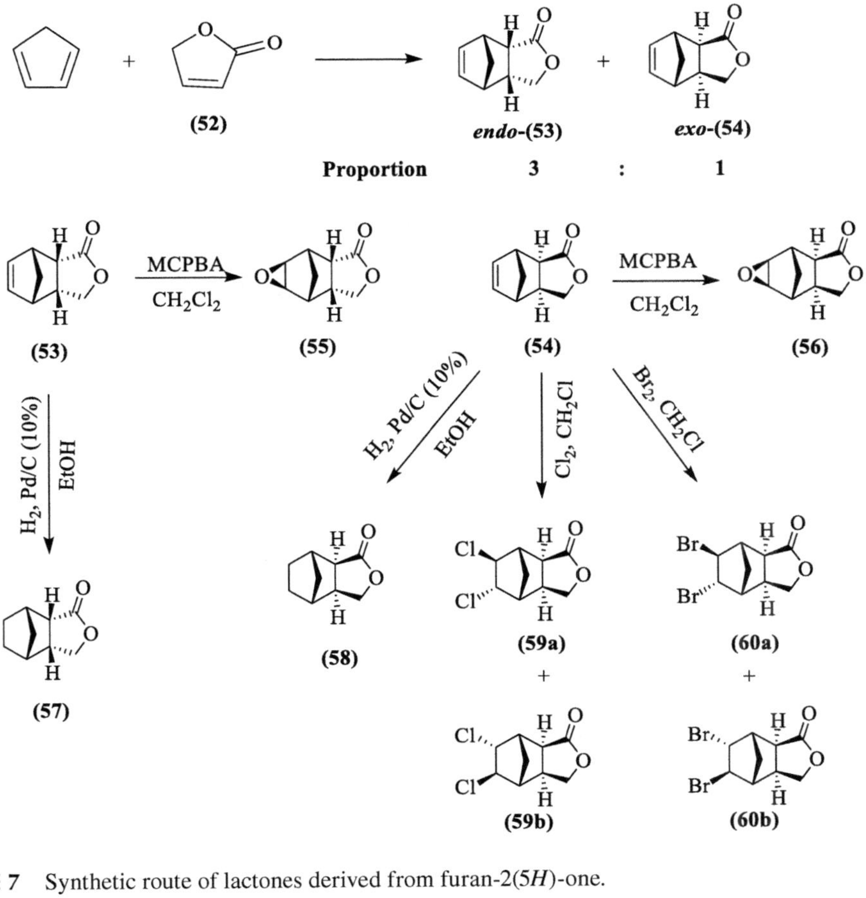

(52) *endo*-(53) *exo*-(54)

Proportion 3 : 1

(53) MCPBA, CH_2Cl_2 (55)

(54) MCPBA, CH_2Cl_2 (56)

(53) H_2, Pd/C (10%), EtOH (57)

(54) H_2, Pd/C (10%), EtOH (58)

(54) Cl_2, CH_2Cl (59a) + (59b)

(54) Br_2, CH_2Cl (60a) + (60b)

SCHEME 7 Synthetic route of lactones derived from furan-2(5*H*)-one.

The insecticidal activity of compounds (52), endo-(53), exo-(54), (55), (56), (57), and (58) was also evaluated against the spotted wing drosophila, *Drosophila suzukii*, and their selectivity in relation to *Trichopria anastrephae*.[28]

D. suzukii is a significant pest that reduces the productivity of fleshy fruits in the neotropical region. Strawberry plants are the primary hosts of *D. suzukii*. The females pierce the fruits to lay eggs, causing injuries that allow fungi and bacteria to enter, rendering the fruits unsuitable for consumption and processing.

Initially, the toxicity of lactone derivatives in adult *D. suzukii* was evaluated over a 24-hour period. For the lactone derivatives that resulted in more than 80% mortality in this initial test, further tests were conducted to determine concentration-mortality curves. The most toxic compounds for *D. suzukii* (52 and *endo*-53) were then used to evaluate toxicity in adults of the parasitoid *T. anastrephae*. Subsequently, molecular docking was employed to identify the potential target receptors of compounds (52) and *endo*-(53) in *D. suzukii* and *T. anastrephae*.

Compounds (52) and *endo*-(53) exhibited mortality rates greater than 75% at a concentration of 3 g L^{-1} in adult *D. suzukii* and showed no statistically significant differences in toxicity compound (52): LC_{50} = 1.04 (1.01–1.08) g L^{-1}; *endo*-(53): LC_{50} = 1.13 (1.07–1.18) gL^{-1}). Furthermore, compounds (52) and *endo*-(53) did not affect the ability of *T. anastrephae* to parasitize *D. suzukii* pupae when exposed to LC_{90} (compound 52: 1.46 g L^{-1}; *endo*-(53): 1.91 g L^{-1}). After 120 hours, the survival rate of all insects exposed to these compounds was greater than 80%, despite a lower overall survival capacity compared to individuals not exposed to the lactone derivatives.

Molecular docking revealed that compounds (52) and *endo*-(53) can effectively kill *D. suzukii* by targeting TRP channels and GABA receptors. This selective efficacy against *D. suzukii* can be attributed to the expression of a specific type of TRP channel (TRPM) in the fly, which facilitates more stable molecular interactions compared to the TRP channels (TRPC) expressed in the parasitoid.

Isobenzofuran-1(3H)-ones, commonly known as phthalides, are a class of organic compounds characterized by a fused bicyclic structure consisting of a benzene ring fused to a lactone. Phthalides are notable for their presence in a wide variety of natural products and for their importance in medicinal chemistry due to their diverse biological activities. These compounds exhibit a range of pharmacological properties, including anti-inflammatory, antimicrobial, and anticancer effects, making them valuable for drug development. In addition, phthalides exhibit phytotoxic, fungicidal, and insecticidal activities and can be used to construct potentially active molecules for use as agricultural pesticides.

Resende and team [29] evaluated the efficacy of 16 synthetic molecules, including 13 phthalides and 3 precursors, in controlling the melonworm (*D. hyalinata*), a key pest of cucurbit crops of economic importance in Brazil. The selectivity of the promising insecticides for the beneficial organisms *S. saevissima* and *T. angustula* was also evaluated, as well as the phytotoxicity of the insecticides on *Cucumis sativus*.

Compounds (61) and (62) (Figure 1.1) showed the greatest efficacy in controlling the insect pest *D. hyalinata*, with mortality rates of 91%, respectively. Selectivity tests with non-target organisms indicated that these compounds were selective in favor of the predator *S. saevissima* but were not selective for the pollinator *T. angustula*, causing a high mortality rate in this species. This result

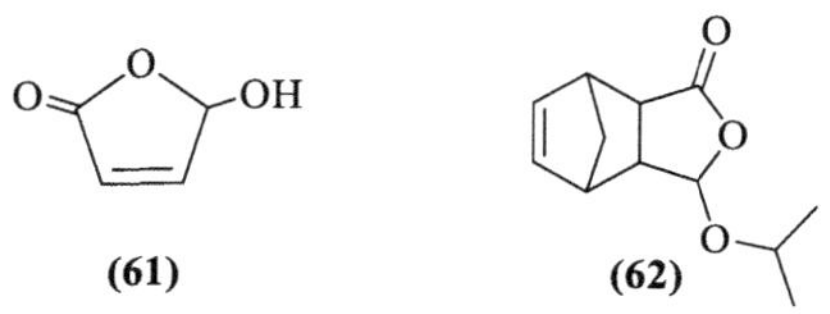

O O OH

(61)

O O O

(62)

FIGURE 1.1 Structures of the most active compounds.

suggests the need for careful application strategies to protect this beneficial insect. Phytotoxicity tests against the host plant *C. sativus* revealed low toxicity, indicating the potential for using these compounds without harming the host plants.

Eighteen other compounds, including γ-lactones (63-70) (Figure 1.2) and phthalide analogues (71–80) (Figure 1.3), were synthesized and tested against the stored grain pest *R. dominica*, an internal grain feeder that infests wheat and other cereals in silos, including corn, rice, and other substrates containing starch.[30]

In summary, the synthesis of these compounds proceeds as follows: 5-hydroxyfuran-2(5*H*)-one was synthesized through the photooxidation of furfural and subsequently used to produce the lactones (63) and (64). These lactones then underwent further reactions and chemical modifications to yield the saturated lactones (65)–(70).

The Diels-Alder reaction between cyclopentadiene and α,β-unsaturated γ-lactones (5-hydroxyfuran-2(5*H*)-one, and compounds (63) and (64) were selected as the pivotal step for constructing the bicyclic framework of phthalides. This reaction resulted in the formation of the adducts (71), (75), and (78). These adducts subsequently underwent hydrogenation to yield compounds (72), (76), and (79); epoxidation to produce acetate (74) and benzofuranone (80); and bromination to form compounds (73) and (77).

The biological assay was carried out in three stages, following the procedure previously described for *Ascia monuste*. In the screening bioassay, the compounds (77) and (79) exhibited mortality rates comparable to that of the commercial insecticide Bifenthrin (90%). The lethal time (LT_{50}) and dose (LD_{50}) required to eliminate 50% of the *R. dominica* population was determined for the most effective

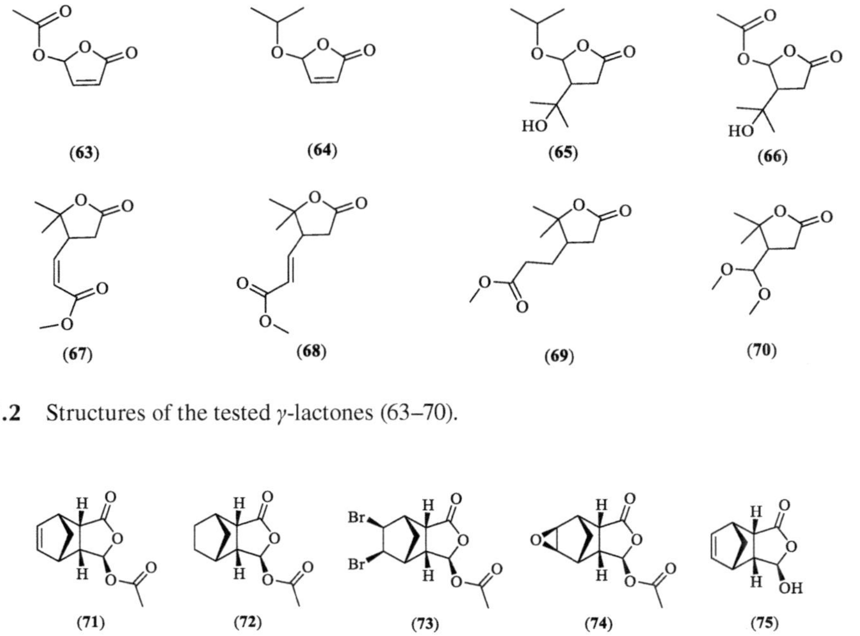

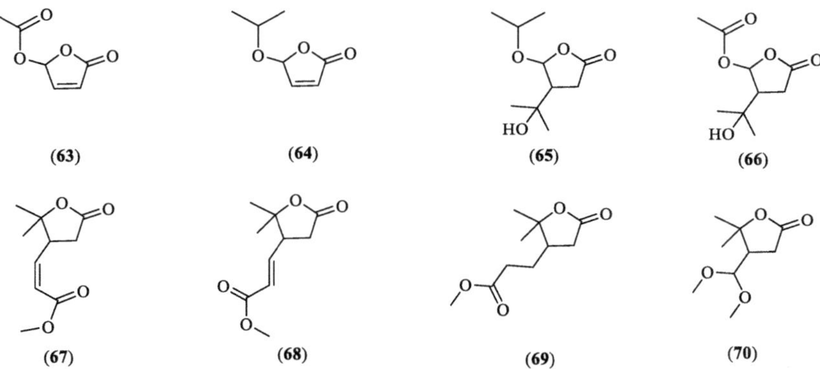

FIGURE 1.2 Structures of the tested γ-lactones (63–70).

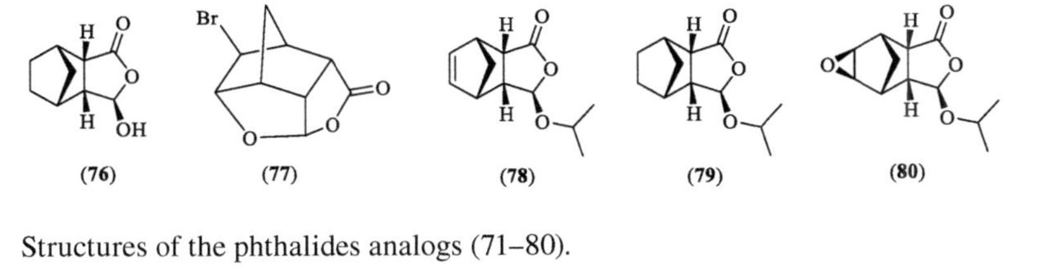

FIGURE 1.3 Structures of the phthalides analogs (71–80).

phthalides (77) and (79). Compound 77 (LD_{50} = 1.97 μg g^{-1}) is four times more toxic than Bifenthrin (LD_{50} = 9.11 μg g^{-1}). Both compounds exhibited an LT_{50} with a value equal to 24 hours. When administered at sublethal doses, both phthalides (particularly compound 77) reduced the emergence of the initial offspring of *R. dominica*. These findings underscore the potential of phthalides (77) and (79) as precursors for the development of insecticides for controlling *R. dominica*.[31]

1.3 HERBICIDES

Agricultural production is constantly challenged by numerous threats, of which weeds are one of the most pressing. Weeds compete with crops for nutrients, sunlight and water, significantly reducing crop yields and posing a constant threat to agricultural production[32, 33].

In this context, herbicides play a critical role in modern agriculture, enabling effective control of unwanted weeds. They are chemical agents designed to suppress weed growth, allowing crops to thrive and reach their maximum yield potential. Herbicides have been a valuable tool in agricultural management, as they help to increase crop yields and keep food prices affordable.

However, the indiscriminate and inappropriate use of herbicides can have serious negative impacts on the environment and human health. Non-selective herbicides can damage non-target plants and, in some cases, pose risks to fauna, including mammals that share the agricultural ecosystem. In addition, the persistence of herbicides in the soil and their leaching into water resources pose significant challenges for long-term sustainability.

Faced with these concerns, agrochemical research has increasingly focused on developing selective herbicides that are more effective and less harmful to mammals and the environment. In this section, we will explore some scientific publications on the synthesis of potential herbicides and the main biological activities found.

1.3.1 Cyclic Imides

Cyclic imides, composed of the structural unit -CO-N(R)-CO-, where R can be a hydrogen atom, an alkyl group, or an aryl group, represent a group of compounds with significant importance in various fields, including agrochemistry, pharmacology, and industry. Particularly noteworthy is their use as herbicides due to their high biological potential.

The synthesis of cyclic imides is achievable through various methods, one of which involves the utilization of dicarboxylic acids as initial reagents in the presence of ammonia, with heating at moderate temperatures, not exceeding 200°C. Another notable strategy is the Diels-Alder cycloaddition reaction, which was employed by Torrent and Alvarenga[34] in the synthesis of cyclic imides using maleic anhydride (81) and isoprene (82), followed by the nucleophilic addition of *m*-anisidine, as illustrated in Scheme 8. The *cis* geometry of the *cis*-carbamoyl-carboxylic acid (84) obtained in this reaction was influenced by the stereoselectivity of the Diels-Alder reaction. Subsequent treatment of this compound with sulfuric acid in methanol resulted in a mixture of isomers, which, in turn, underwent oxidation with methachloroperbenzoic acid (MCPBA), producing two epoxides derived from 5-methylhexahydroisoindol-1,3-dione. The epoxides were denoted as (87a) and (87b), with a yield of 19% and 66%, respectively. Their structures were confirmed using nuclear magnetic resonance spectroscopy, supported by theoretical calculations.

The herbicidal potential of these compounds was evaluated through biological tests using lettuce (*Lactuca sativa*), cucumber (*Cucumis sativus*), beggartick (*Bidens pilosa*), and sorghum seeds (*Sorghum bicolor*), with the commercial herbicide *S*-metolachlor as a positive control and a 0.3% (v/v) aqueous DMSO solution as a negative control. The results indicated that all the compounds affected the growth of both the aerial and root parts of the lettuce seeds, with greater inhibition observed in the roots. Notably, compound (87a) exhibited superior performance compared to compound (87b). In the case of cucumber seeds, all the compounds interfered with development, albeit

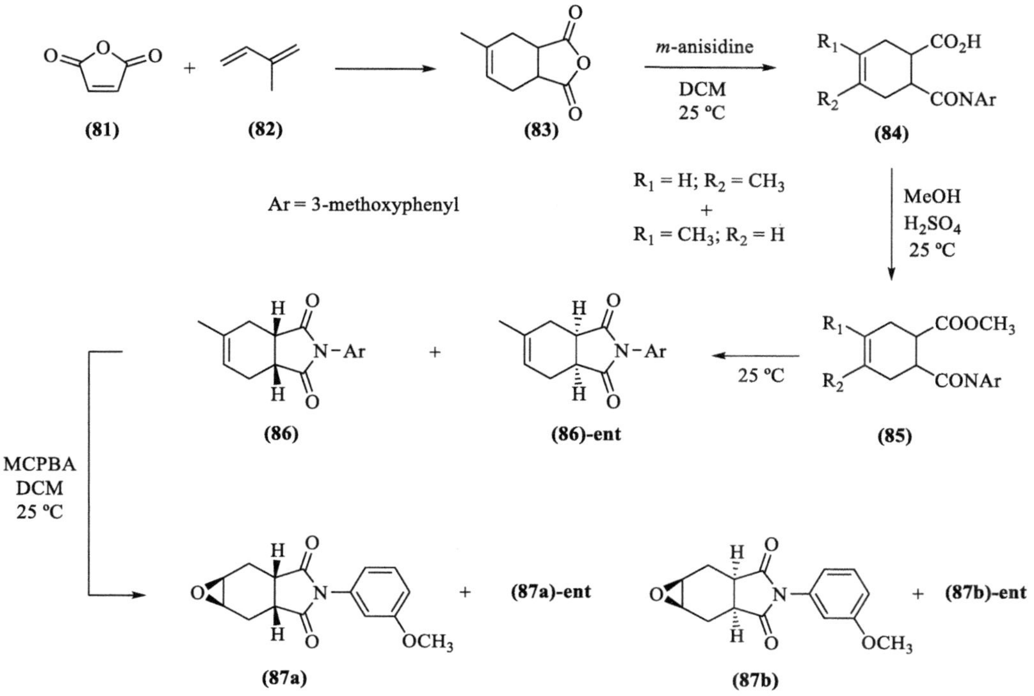

SCHEME 8 Synthesis of cyclic imides.

less effectively compared to the commercial herbicide. On the other hand, tests with sorghum seeds suggested more promising biological activity for compound (87a), especially at concentrations of 500 μM and 150 μM. Finally, experiments with the weed beggartick revealed better inhibition when compound 87b was used, particularly at a concentration of 500 μM. These results reaffirm the significant potential of cyclic imides as active ingredients for the research and development of new herbicides.

1.3.2 Lactones and Derivates

The presence of the lactone ring in many compounds of natural and synthetic origin is associated with a variety of biological activities, including herbicidal activity[35, 36]. Studies suggest that γ-lactones, especially the α,β-unsaturated ones, have greater efficacy in inhibiting weed growth compared to their saturated counterparts.

Resende and collaborators[37] synthesized 14 γ-lactones, 9 of which are new compounds derived from furfural (Scheme 9). The synthetic methodology involved several reactions, including photo-oxidation, photochemical addition, and acid-catalyzed rearrangement (Scheme 9). Initially, compound (88) was obtained through photo-oxidation catalyzed by rose bengal of furfural and was subsequently protected to generate compounds (89) and (90). These compounds were then subjected to photochemical addition with isopropyl alcohol, resulting in compounds (91) and (92) with yields of 91% and 99%, respectively. The photochemical addition of methanol to lactone (89) produced ester (93) with a significant yield of 58%. Under different acidic conditions, lactone (91) underwent rearrangement, giving rise to lactones (94), (95), and (96). Treatment of compound (94) with the phosphonate carbanion, prepared *in situ* from the reaction between potassium tert-butoxide and trimethyl phosphonoacetate, led to alkenes (97) and (98). Catalytic hydrogenation of alkenes (97)

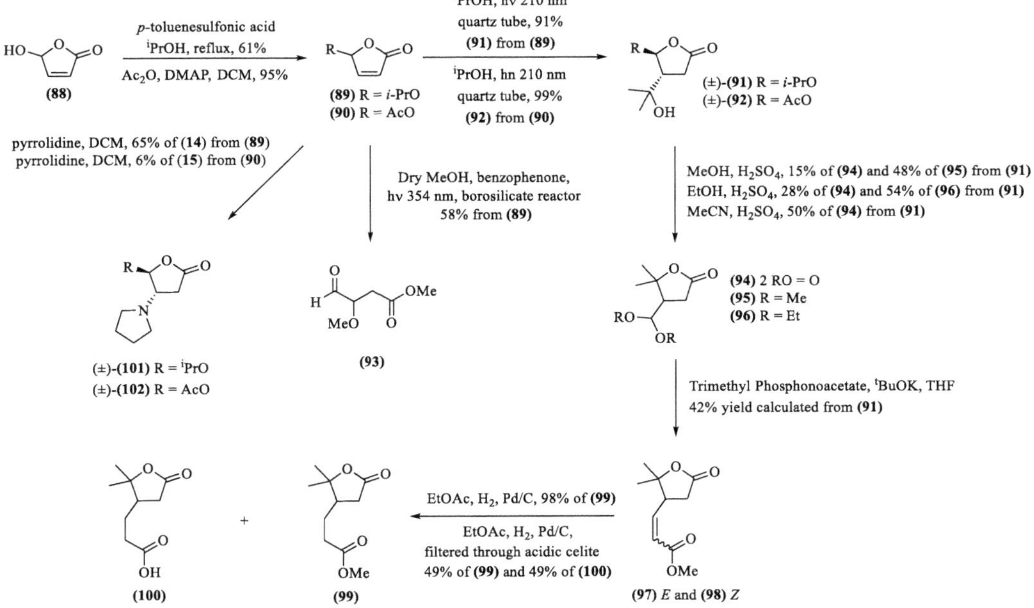

SCHEME 9 Synthesis of lactones.

and (98) resulted in compounds (99) and (100). Finally, compounds (101) and (102) were obtained through the conjugate addition of pyrrolidine to compounds (89) and (90), respectively.

Bioassay results demonstrated that most of the synthesized γ-lactones exhibited variable phytotoxic activity. In general, α,β-unsaturated lactones showed greater phytotoxic activity. Among the compounds tested, three stood out for their high inhibition of wheat coleoptile growth: one with 51% inhibition, another with 76%, and a third with 68%, all at a concentration of 1000 μmol L^{-1}. However, after dilution to 100 μmol L^{-1}, most compounds did not exhibit statistically significant inhibition relative to the control. These results suggest that although the presence of the γ-lactone ring is crucial for bioactivity, phytotoxic efficacy also depends on other structural factors, such as lipophilicity and the presence of specific functional groups.

Most of the herbicides currently available on the market are synthetic in nature and represent an effective approach to combating weeds. However, growing concerns about environmental issues, mammalian toxicity, and the development of weed resistance have led the scientific community to seek pesticide formulations that are more selective and less toxic to mammals and the environment.

One strategy in this direction is based on the synthesis of compounds inspired by natural products, which consists of modifications to chemical structures isolated from natural products in order to obtain formulations with biological activity. Isobenzofuranones are notable examples of natural products with biological activity described in the literature.

Teixeira and co-workers[38] published an interesting study on the phytotoxic activity of α,β-unsaturated lactones, tetrahydroisobenzofuran-1(3*H*)-ones, and hexahydroisobenzofuran-1(3*H*)-ones (Figure 1.4) on seeds of cucumber (*C. sativus*), sorghum (*S. bicolor*), and onion (*Allium cepa*).

The compounds of interest were derived from rac-(3a*R*,4*S*,7*R*,7a*S*)-3a,4,7,7a-tetrahydro-4,7-metanoisobenzofurano-1(3*H*)-one (104), and rac-(3a*S*,4*S*,7*R*,7a*R*)-3a,4,7,7a-tetrahydro-4,7-methanoisobenzofuran-1(3*H*)-one (105), which were synthesized through a cycloaddition reaction between cyclopentadiene and furan-2(5*H*)-one (103). Epoxidation reactions were also carried out, yielding products (106) and (107); hydrogenation (products 108 and 109); chlorination (110, 111a, and 111b); and bromination (112a and 112b), as shown in Scheme 10. Five

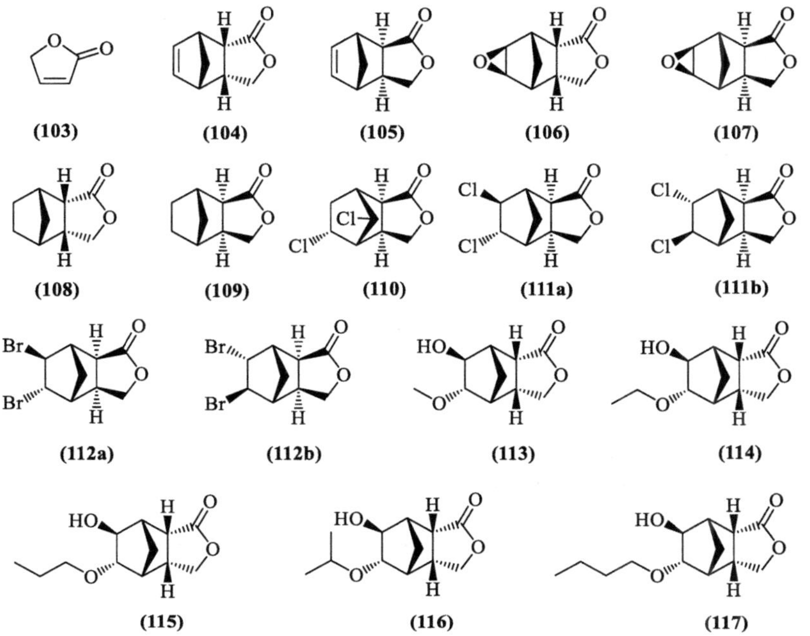

(103) (104) (105) (106) (107)

(108) (109) (110) (111a) (111b)

(112a) (112b) (113) (114)

(115) (116) (117)

FIGURE 1.4 Chemical structures of all compounds evaluated in the bioassay.

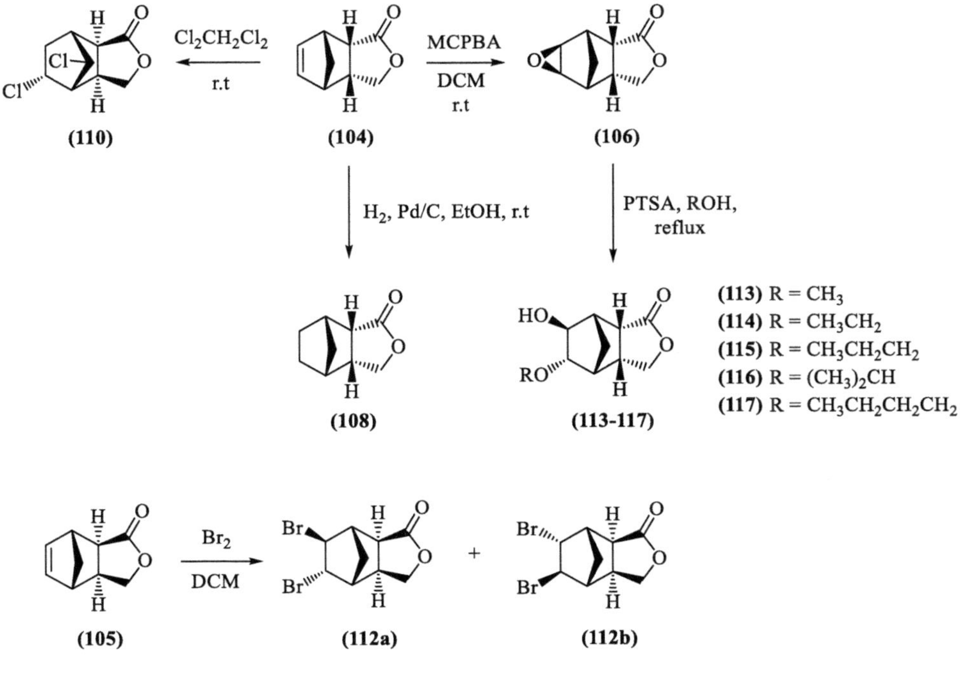

(110) ← $Cl_2CH_2Cl_2$, r.t — (104) — MCPBA, DCM, r.t → (106)

(104) → H_2, Pd/C, EtOH, r.t → (108)

(106) → PTSA, ROH, reflux → (113-117)

(113) R = CH_3
(114) R = CH_3CH_2
(115) R = $CH_3CH_2CH_2$
(116) R = $(CH_3)_2CH$
(117) R = $CH_3CH_2CH_2CH_2$

(105) → Br_2, DCM → (112a) + (112b)

SCHEME 10 Synthesis of isobenzofuranones.

compounds (113-117) were prepared by reacting (106) with methanol (product 113, 73% yield), ethanol (product 114, 74% yield), propan-1-ol (product 115, 51% yield), propan-2-ol (product 116, 46% yield), and butan-1-ol (product 117, 51% yield) in anhydrous DCM catalyzed by PTSA.

Biological assays conducted with cucumber and onion seeds demonstrated that compounds (112a) and (112b) exhibited inhibitory activity comparable to the herbicide (*S*)-metolachlor, employed as a control, within the concentration range of 50 to 1000 μM.

Compounds (103)–(110), (112a), (112b), and (113)–(117) were tested, in addition to mixtures (111a)/(111b) (2:1) and (112a)/(112b) (1:1). Compounds (112a) + (112b) and (112b) were the most active, inhibiting up to 79.4% of *S. bicolor* shoots, comparable to the 85% inhibition by (*S*)-metolachlor. In *C. sativus*, (112a) and the mixture (112a) + (112b) showed significant root inhibition. For *A. cepa*, (112a) + (112b) inhibited sprouts more effectively than (*S*)-metolachlor.

Almeida and collaborators[39] reported the synthesis of several analogs of nostoclides, lactones naturally found in the lichen *Peltigera canina* Willd, and their effect on the germination and growth of *Physalis ixocarpa* (dicotyledon) and *Lolium multiflorum* (monocotyledons).

The synthesis (Scheme 11) commenced with the treatment of 2-furyl-*N*,*N*,*N*,*N*-tetramethyl-phosphorodiamidate (118) with n-butyl lithium, followed by the addition of benzyl bromide, resulting in the formation of substituted furan (119). Upon treatment with formic acid, lactone (120) was produced with a total yield of 52%. Subsequently, this lactone was treated with TBDMSOTf in the presence of TEA and aldehyde. The elimination of the TBDMSO group from intermediate (121) was achieved by adding DBU to the crude reaction mixture, leading to the formation of alkenes (122a–122f) in variable yields. Intermediate TBDMS derivatives (121b, 121c, and 121g) were isolated during the reaction.

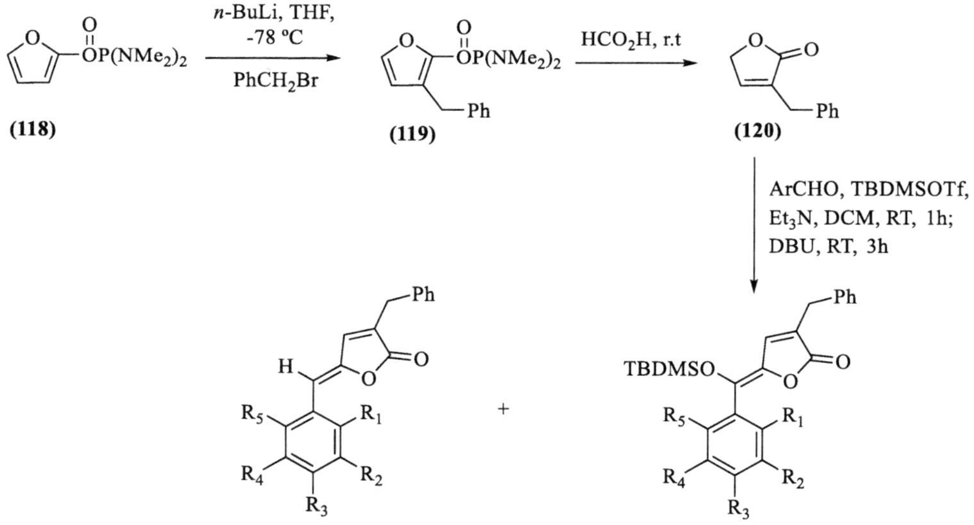

(118) $\xrightarrow{n\text{-BuLi, THF, -78 °C; PhCH}_2\text{Br}}$ (119) $\xrightarrow{\text{HCO}_2\text{H, r.t}}$ (120) $\xrightarrow{\text{ArCHO, TBDMSOTf, Et}_3\text{N, DCM, RT, 1h; DBU, RT, 3h}}$ (122) + (121)

(122a) $R_1 = R_2 = R_4 = R_5 = H, R_3 = NMe_2$ (71%)
(122b) $R_2 = R_4 = H, R_1 = R_3 = R_5 = OCH_3$ (23%)
(122c) $R_2 = R_3 = R_5 = H, R_1 = R_4 = OCH_3$ (26%)
(122d) $R_1 = R_2 = R_3 = R_4 = R_5 = H$ (42%)
(122e) $R_1 = R_4 = R_5 = H, R_2 = R_3 = OCH_3$ (10%)
(122f) $R_1 = R_3 = R_4 = R_5 = H, R_2 = Br$ (31%)
(122g) $R_2 = R_3 = R_4 = H, R_1 = R_5 = Cl$ (0%)

(121b) (51%)
(121c) (5%)
(121g) (6%)

SCHEME 11 Synthesis of lactones (5a-5g).

Compounds (122d) and (122e) were found to inhibit the growth of monocotyledonous weeds (*L. multiflorum*) while also inducing root growth. Lactone (122f) exhibited growth inhibition only in dicotyledonous weed species (*P. ixocarpa*). These findings demonstrate that synthetic lactones interfere in some manner with the plant growth process, affecting the development of roots and shoots, as well as the photosynthesis system.

Secondary metabolites constitute an important class of compounds naturally synthesized by plants, fungi, bacteria, and other organisms as part of their secondary biochemistry, often fulfilling roles in defense against predators, competition for resources, or symbiotic interactions. Many of these compounds exhibit diverse biological activities, including herbicidal properties.

Moraes and collaborators[40] synthesized new derivatives of the secondary metabolite platensimycin, employing α-santonine (123) as the raw material. To accomplish this, they initiated the process with a solution of α-santonine in anhydrous acetonitrile, subjecting it to degassing using a nitrogen flow. Subsequently, the solution was irradiated by six low-pressure mercury lamps. Following this, the solvent was removed under vacuum, and the resulting residue was purified by column chromatography, utilizing silica gel as the stationary phase and eluting with hexane/ethyl acetate. This procedure resulted in the production of lumisantonine (124) (Scheme 12), as a white solid, with a yield of 91%. In subsequent steps, lumisantonine underwent reaction with methylamine, ethylamine, propylamine, butylamine, and pyrrolidine in dichloromethane to yield various amides (125). These compounds underwent evaluation of their biological activity against *S. bicolor*, *A. cepa*, *C. sativus*, *Solanum lycopersicum*, and *Bidens pilosa*, demonstrating a promising inhibitory activity on seed germination.

Alvarenga and collaborators[41] also synthesized α-santonin derivatives and evaluated their phytotoxic activity on cucumber (*C. sativus*) and sorghum (*S. bicolor*) plants. The synthetic route employed was similar to previous studies, varying the solvent, temperature, radiation source, and type of reactor. α-santonin was irradiated with high- and low-pressure mercury lamps in different solvents (acetonitrile, acetic acid, and a water/acetic acid mixture). Various derivatives, including lumisantonine and different lactones, were obtained through photochemical reactions. Detailed synthesis conditions and yields are presented in Scheme 13.

Phytotoxicity tests on cucumber and sorghum seedlings were conducted with all compounds at a concentration of 4×10^{-4} M. Compounds (126), (127), (129), and (131) stimulated cucumber root growth, while the others caused inhibition, with compound (128) causing the greatest inhibition (48.1%). Compound (126) stimulated root growth by 48.2%, contrasting with the 27.9% inhibition by

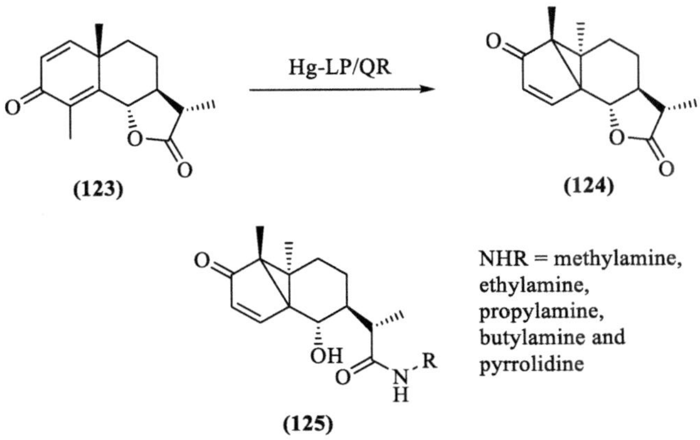

NHR = methylamine, ethylamine, propylamine, butylamine and pyrrolidine

SCHEME 12 Synthesis of lumisantonine (124), and platensimycin analogues (125). Hg-LP/QR: low-pressure mercury lamp/quartz reactor.

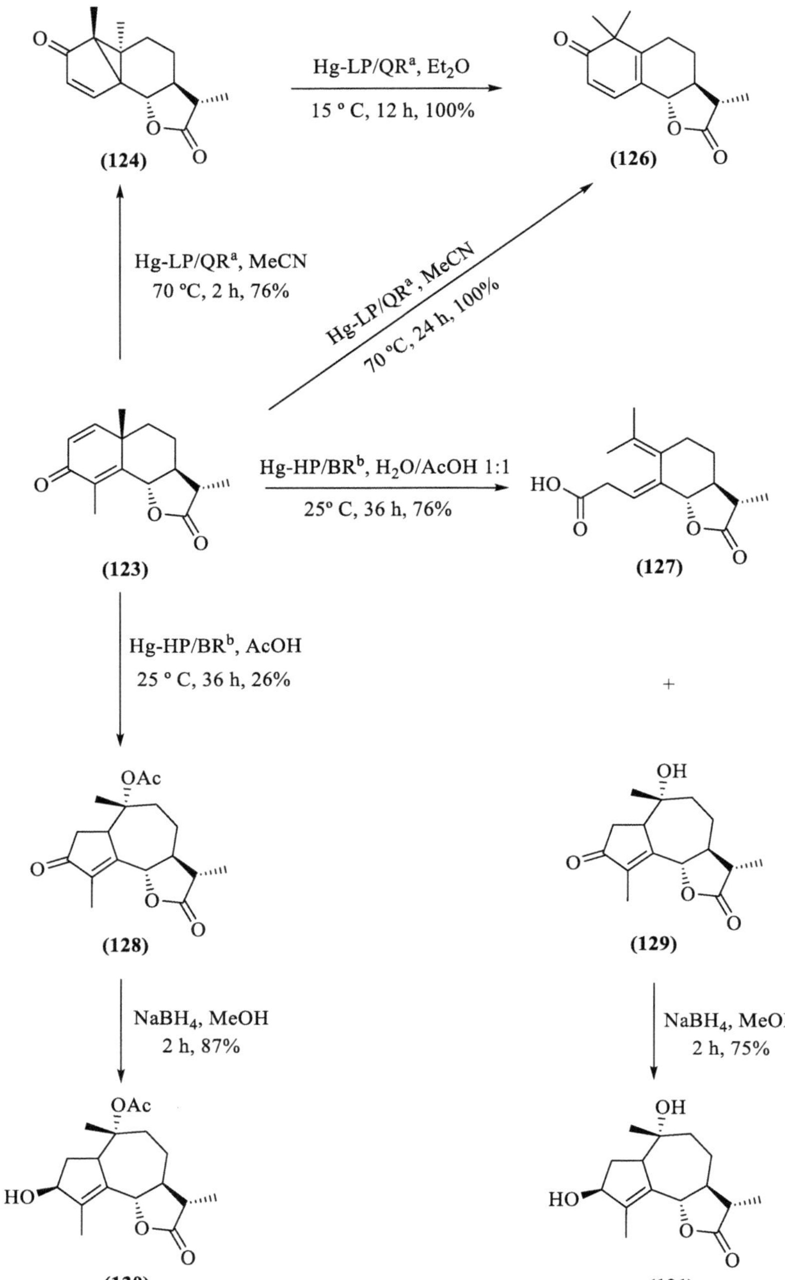

SCHEME 13 Synthesis of lactones (126-131). [a]Hg-LP/QR: low-pressure mercury lamp/quartz reactor. [b]Hg-HP/BR: high-pressure mercury lamp/borosilicate reactor.

α-santonin (123). In sorghum, all compounds inhibited root growth, with compounds (127), (128), (130), and α-santonin (123) showing the greatest inhibition (84.0%). Compounds (128) and (130) exhibited significant inhibitory activity. Diol (131) showed greater inhibitory activity (76%) than lactone (129) (60%) in sorghum, indicating a preliminary structural correlation between α-santonin derivatives and herbicidal activity.

1.3.3 Anilides

Losses in agricultural production due to weeds can vary significantly depending on the crop, region, and management practices adopted. Therefore, the selection of appropriate strategies for pest control in agricultural crops is fundamental. It is imperative that these strategies strike a balance between their effectiveness in combating pests and minimizing adverse environmental impacts. In this context, anilides have emerged as a class of chemical compounds of extreme importance due to their versatility and effectiveness in controlling weeds, providing an essential service to humanity.

Anilides are a class of amides with broad applications in the chemical and pharmaceutical industries. Their significance is underscored by the diverse biological activities they exhibit, including antibacterial[42], antifungal[43, 44], larvicidal[45], insecticidal[46], and herbicidal properties[43, 47, 48]. They are easily synthesized through one-pot reactions involving aromatic amines (anilines) with acyl chlorides or carboxylic acid anhydrides. Commercial herbicides such as *S*-metolachlor, alachlor, and butachlor belong to the anilide class and act as cell division inhibitors.

Sartori and collaborators[49] conducted remarkable work involving the synthesis of anilides through multicomponent reactions. Nine different anilides (133a–133i, Scheme 14) were synthesized by reacting sorbic acid with anhydrous dichloromethane and oxalyl chloride. The reaction mixture was stirred for 30 minutes and concentrated under vacuum. The resulting residue (a yellow oil) was then dissolved in anhydrous dichloromethane and cooled in an ice bath. The anilines were added to the round bottom flask, and the mixture was stirred for 2 hours. The solvent was removed under reduced pressure, and the residue was purified by silica gel chromatography.

Another nine anilides (Scheme 15) were synthesized using one-pot reactions. In this process, anilines were added to a solution of 97% (v/v) hexanoic anhydride in anhydrous dichloromethane. The reaction mixture was stirred under nitrogen atmosphere for 2 hours and then treated with saturated sodium bicarbonate solution. After separating the organic phase in a separating funnel, the aqueous phase was extracted with three portions of dichloromethane. The combined organic phases were then washed with HCl 1 M and dried over anhydrous sodium sulfate. Subsequently, the mixture was filtered, and the solvent was evaporated under reduced pressure. The residue obtained was purified by silica gel flash column chromatography, resulting in the production of anilides (135a–135i).

All tested anilides inhibited the growth of *A. cepa* shoots, with compounds (135d) and (135e) showing the most significant inhibition at 66% and 63%, respectively, reaching their highest efficacy at a concentration of 500 μM. These compounds also exhibited high inhibition of *A. cepa* root growth, surpassing the commercial herbicide containing (*S*)-metolachlor. In *L. sativa*, compounds (133e), (135d), (135e), and (135g) significantly inhibited both shoot and root growth, with (133e) exceeding the effectiveness of the commercial herbicide. Compounds such as (133i), (135a), (135c), and (135i) also showed significant root inhibition, while (133f) stimulated root growth at lower concentrations. In *C. sativus*, compounds (135d), (135e), and (135h) were highly effective in inhibiting

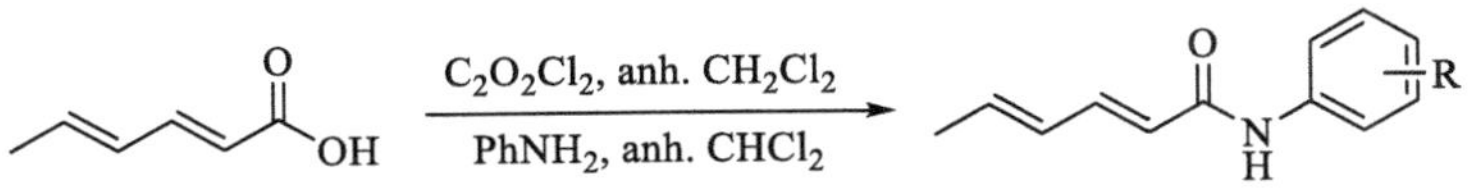

$C_2O_2Cl_2$, anh. CH_2Cl_2 / $PhNH_2$, anh. $CHCl_2$

(132) → **(133)**

(133a) (R = 4-CH_3)
(133b) (R = 4-OH)
(133c) (R = 4-OCH_3)
(133d) (R = 3-OCH_3)
(133e) (R = 3-NO_2)
(133f) (R = 2-F)
(133g) (R = 3-F)
(133h) (R = 4-F)
(133i) (R = 3,4-dichloro)

SCHEME 14 Synthesis of the anilides (133a–133i).

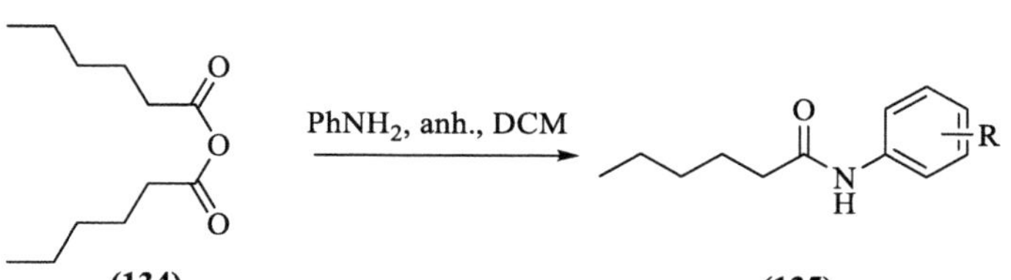

PhNH$_2$, anh., DCM

(134) (135)

(135a) (R = 4-CH_3)
(135b) (R = 4-OH)
(135c) (R = 4-OCH_3)
(135d) (R = 3-OCH_3)
(135e) (R = 3-NO_2)
(135f) (R = 2-F)
(135g) (R = 3-F)
(135h) (R = 4-F)
(135i) (R = 3,4-dichloro)

SCHEME 15 Synthesis of the anilides (135a–135i).

both shoots and roots. For *B. pilosa*, compounds (135e) and (133b) were the most effective, suggesting their potential as selective herbicides, particularly (133b), which inhibited *B. pilosa* without affecting other crops. These results indicate that anilides, especially those with substituents in the *meta* position of the aromatic ring, are promising candidates for the development of new herbicides.

In silico studies suggested histone deacetylase as the probable enzymatic target site in plants for these substances. The affinities of the most active anilides for the binding sites of this enzyme were equal to or greater than those calculated for its inhibitors.

1.3.4 Benzoxazole Derivatives

Many chemical substances possess versatile properties that can be applied in various contexts. For instance, an active ingredient developed to treat diseases in humans could be adapted to protect plants against pests. Aromatic heterocyclic compounds are renowned for their wide range of biological activities. Within this class, benzoxazole is a nucleus present in several natural products and synthetic medicines, which exhibit antiviral, antibacterial, antifungal, and anticancer activities. Due to their low cytotoxicity and ease of synthesis, benzoxazole derivatives have the potential to be valuable in the development of new agrochemicals. However, their application in agrochemical research has received little attention to date.

Sangi and collaborators[50] conducted the synthesis and phytotoxicity assessment of benzoxazole derivatives. The synthetic procedure involved double vinyl replacement reactions assisted by microwave irradiation. Compounds (138a–138d) (Scheme 16) were obtained by the reaction between 1,1-bis-methylsulfanyl-2-nitroethylene (136) and different disubstituted 2-aminophenols (137). Furthermore, 2-nitromethylene-oxazolidine (140) was obtained by the same synthetic method through the reaction between compound (136) and 2-aminoethanol (139).

The research group also carried out the reduction of the nitro group in the benzoxazole molecule (138a) in a hydrogen atmosphere, catalyzed by palladium, resulting in the compound benzoxazol-2-ylmethanamide (141) (Scheme 17).

Bioassays were conducted to evaluate the phytotoxicity of the compounds on species of *A. cepa*, *S. lycopersicum*, *C. sativus*, and *S. bicolor*. All compounds demonstrated biological activity in seed germination, with compound (138b) being more phytotoxic than (*S*)-metolachlor.

1.3.5 Dienamides and Epoxy Derivatives

The use of *in silico* approaches has become increasingly relevant in the planning and development of pesticides. These techniques involve utilizing computational methods and molecular modeling to predict the behavior and interaction of chemical compounds with biological targets, such as enzymes specific to agricultural pests.

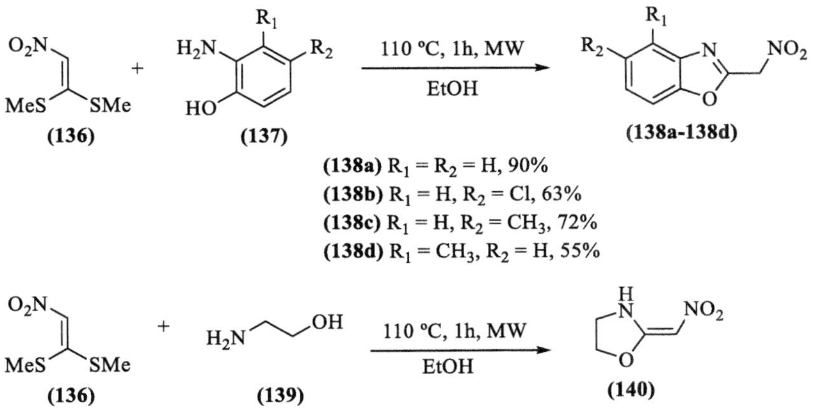

(138a) $R_1 = R_2 = H$, 90%
(138b) $R_1 = H$, $R_2 = Cl$, 63%
(138c) $R_1 = H$, $R_2 = CH_3$, 72%
(138d) $R_1 = CH_3$, $R_2 = H$, 55%

SCHEME 16 Synthesis of benzoxazole derivatives (138a–138d), and 2-nitromethylene-oxazolidine (140).

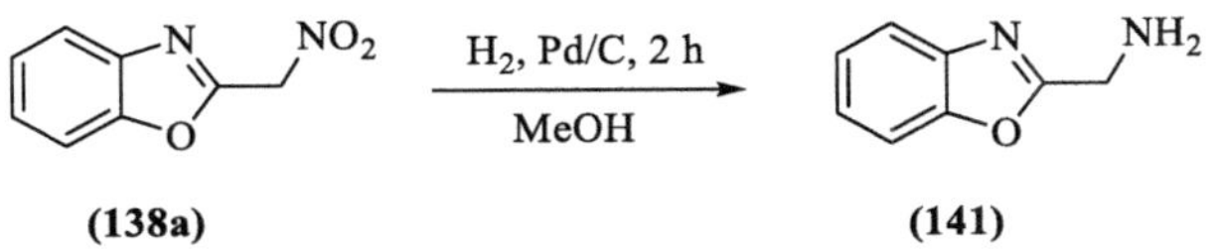

SCHEME 17 Synthesis of benzoxazol-2-ylmethanamine (141).

One of the advantages of these approaches is their ability to significantly reduce the time and costs associated with the pesticide development process. By conducting computer simulations, scientists can analyze a wide variety of candidate molecules in a short period, identifying those with the highest potential for biological activity against desired targets.

Furthermore, *in silico* approaches enable the rational optimization of chemical structures, facilitating the creation of more selective and less toxic compounds for non-target organisms and the environment. This contributes to the development of safer and more effective pesticides, thereby minimizing the adverse impacts associated with chemical use in agriculture.

In summary, the use of *in silico* approaches represents a powerful and promising tool in pesticide development, offering a faster, more efficient, and sustainable approach to addressing the challenges of crop protection against agricultural pests.

A scientific study employing this approach, conducted by Ramos and collaborators[51], underscored the impact of dienamides and epoxy derivatives on the germination of sorghum, onion, cucumber, lettuce, and beggartick seeds. Potential herbicides were synthesized from sorbic acid, following the synthetic procedure outlined in Scheme 18.

To obtain sorbic acid (142), 1 mmol of potassium sorbate and 1 $mol.L^{-1}$ hydrochloric acid were mixed under stirring at room temperature, followed by extraction with DCM. The dienamides were prepared via a one-pot reaction, where sorbic acid was converted into sorbic chloride through treatment with thionyl chloride. The sorbic chloride reacted with various anilines to yield dienamides (143a-143h). These dienamides were then subjected to regioselective epoxidation by reaction with MCPBA in DCM, yielding epoxides (144a–144f).

The phytotoxicity of each compound was evaluated against the five plant species, with the experiment conducted in four different concentrations and in triplicate, with 20 seeds in each analysis. All synthesized compounds exhibited activity in seed germination. Compounds (143c), (143d), (144b), (144e), and (144f) stood out, showing phytotoxic activity superior to the commercial herbicide (*S*)-metolachlor.

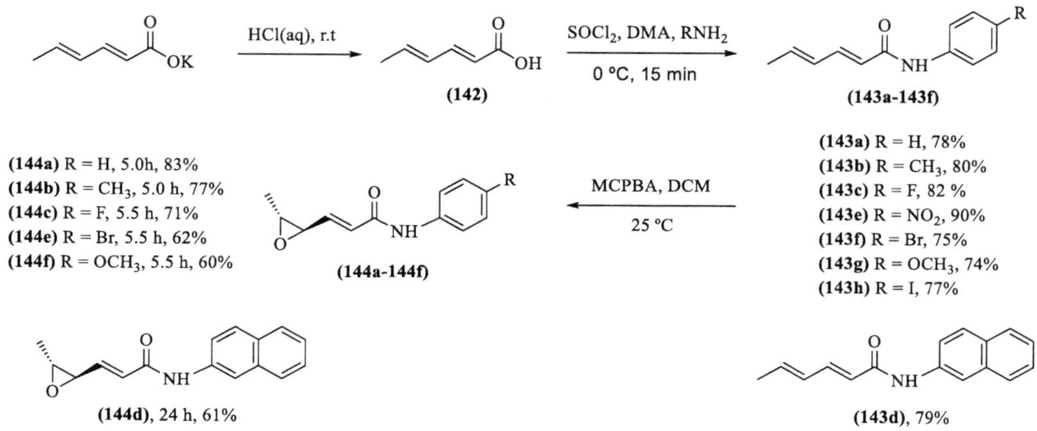

HCl(aq), r.t
(142)
$SOCl_2$, DMA, RNH_2
0 °C, 15 min
(143a-143f)

(143a) R = H, 78%
(143b) R = CH_3, 80%
(143c) R = F, 82 %
(143e) R = NO_2, 90%
(143f) R = Br, 75%
(143g) R = OCH_3, 74%
(143h) R = I, 77%

MCPBA, DCM
25 °C
(144a-144f)

(144a) R = H, 5.0h, 83%
(144b) R = CH_3, 5.0 h, 77%
(144c) R = F, 5.5 h, 71%
(144e) R = Br, 5.5 h, 62%
(144f) R = OCH_3, 5.5 h, 60%

(144d), 24 h, 61%
(143d), 79%

SCHEME 18 Synthesis of the dienamides (143a–143f) and epoxy derivatives (144a–144f).

To elucidate the biological targets of the tested compounds, a pharmacophoric investigation was conducted, indicating the proteins tyrosine phosphatase 1c84 and mitogen-activated kinase 6sot as potential receptors for the compounds. These proteins appear to be important in explaining the experimental results, as they are the only proteins for which similar sequences were found in plants, especially in crops of *Allium* spp., *Cucumis* spp., *Lactuca* spp., and *Sorghum* spp.

The results of molecular docking calculations suggested that substances (143c) and (143d) could interact with protein tyrosine phosphatase and mitogen-activated protein kinase, indicating the possibility of complex formation between the substances.

1.3.6 N-Phenylnorbornenesuccinimide Derivatives

Another interesting study on the molecular docking of potential herbicides was conducted by Gomes and collaborators[52]. Entitled "*N*-Phenylnorbornenesuccinimide Derivatives, Agricultural Defensive, and Enzymatic Target Selection", the work aimed to synthesize derivatives of *N*-phenylnorbornenesuccinimide and conduct both biological tests and *in silico* research.

The synthetic strategy, represented in Scheme 19, began with the reaction of endobicyclo[2.2.1]hept-5-en-3a,7a-dicarboxylic anhydride (145) with aromatic amines, resulting in *N*-phenylnorbornenesuccinic acids (146–155) with yields of 75% to 90%. Cyclization of compounds (146–155) in the presence of acetic anhydride and sodium acetate produced *N*-phenylnorbornenesuccinimides (156–164), with yields ranging from 65% to 89%. These imides were then subjected to epoxidation, generating *N*-phenylepoxynorbornanesuccinimides (165–173) with yields ranging from 60% to 90%. The structures of all the molecules synthesized are shown in Figure 1.5.

Phytotoxicity tests were conducted on *A. cepa*, *B. pilosa*, *C. sativus*, *S. bicolor*, and *S. lycopersicum*. All compounds demonstrated phytotoxic activity, inhibiting the growth of seedlings of the tested species. The commercial herbicide *S*-metolachlor was used as a positive control. Compound (167) was the most effective, inhibiting 100% of the growth of all species at all tested concentrations.

Compound (167) was used as a ligand in molecular docking calculations. The pharmacophoric research aimed to identify possible biological targets for the compound, suggesting that the protein cyclophilin D (6ra1), whose amino acid sequence is very similar to various enzymes found in plants, could form a complex with the tested ligand.

The function of cyclophilins is still being studied, but recent research indicates that they play roles in hormone signaling and plant adaptation to abiotic stress[53]. The molecular docking results

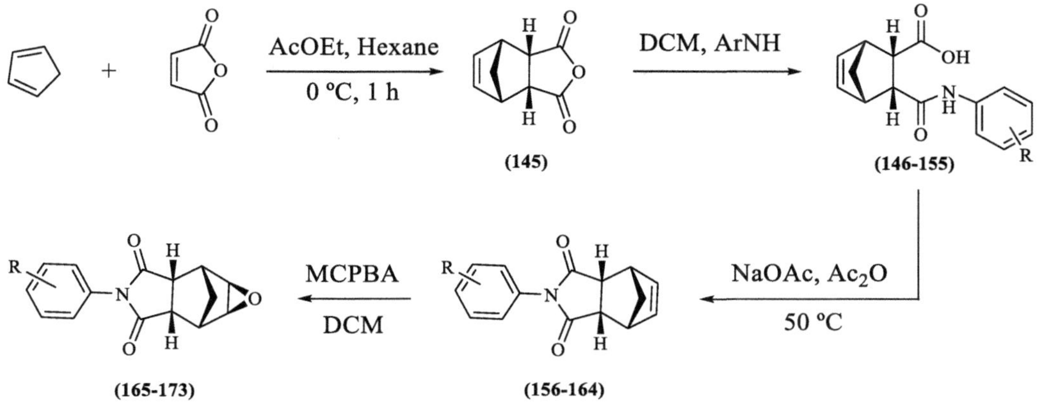

AcOEt, Hexane
0 °C, 1 h
(145)
DCM, ArNH
(146-155)
NaOAc, Ac_2O
50 °C
(156-164)
MCPBA
DCM
(165-173)

SCHEME 19 Synthesis of the *N*-phenylnorbornenesuccinimide derivatives.

suggest that N-phenylnorbornenesuccinimide derivatives could serve as inspiration in the search for new compounds with herbicidal activity.

1.3.7 Benzoquinones

Benzoquinones are organic compounds that belong to the class of quinones, which are cyclic derivatives of ketones. The chemical structure of benzoquinones consists of an unsaturated ring of six carbon atoms with two carbonyl groups in adjacent or opposite positions. They can be found in several natural sources, functioning as secondary metabolites produced by plants, fungi, bacteria, and some species of the animal kingdom.[54–56]

Sorgoleone (Figure 1.6) is an example of this class of compounds, found in the root exudates of sorghum (*Sorghum* spp.). It plays an important role in allelopathy, the process by which a plant releases chemicals that influence the germination and growth of other plants around it.[57–59]

Sorgoleone, due to its phytotoxicity, can serve as a structural platform for the development of synthetic herbicides. This was the proposal presented by Lima and collaborators, who synthesized new benzoquinones inspired by sorgoleone and evaluated their phytotoxicity in four plant species: *C. sativus*, *Brachiaria decumbens*, *Hyptis lophanta*, and *Euphorbia heterophylla*.[60] The synthetic procedure is described in the following and presented in Scheme 20.

Commercially available 3,5-dimethoxybenzyl alcohol (174) was used as the starting material for the synthesis of new sorgoleone analogues. It was converted to aldehyde (175) in 88% yield via Swern oxidation[61]. Subsequently, the aldehyde (175) was transformed into alcohol (176) in 88% yield using a Grignard reaction with 1-bromooctane. Attempts to prepare the hydrocarbon (179) directly from the alcohol (176) in a one-pot reaction failed.

Alcohol (176) was converted to hydrocarbon (179) via alkene (177), obtained in 74% yield with *p*-toluenesulfonic acid (PTSA). Hydrogenation of the alkene (177) with 10% palladium on carbon provided the compound (179) in 93% yield. This compound was oxidized with chromic anhydride in acetic acid, resulting in quinone (180) in 69% yield. Thiele acetoxylation of the quinone (180) produced triacetate (181) in 82% yield.

The attempt to convert triacetate (181) to quinone (183) resulted in a yield of only 8%, with the unexpected formation of the acetylated compound (182) at 37%. Due to the low yield, compound (182) was treated with DBU in THF under nitrogen, increasing the yield of quinone (183) to 64%.

Benzyl alcohol (174) was then used to prepare more polar quinones, analogues of sorgoleone (Scheme 21). This alcohol was converted to ethers (184)–(186) in good yields, which were

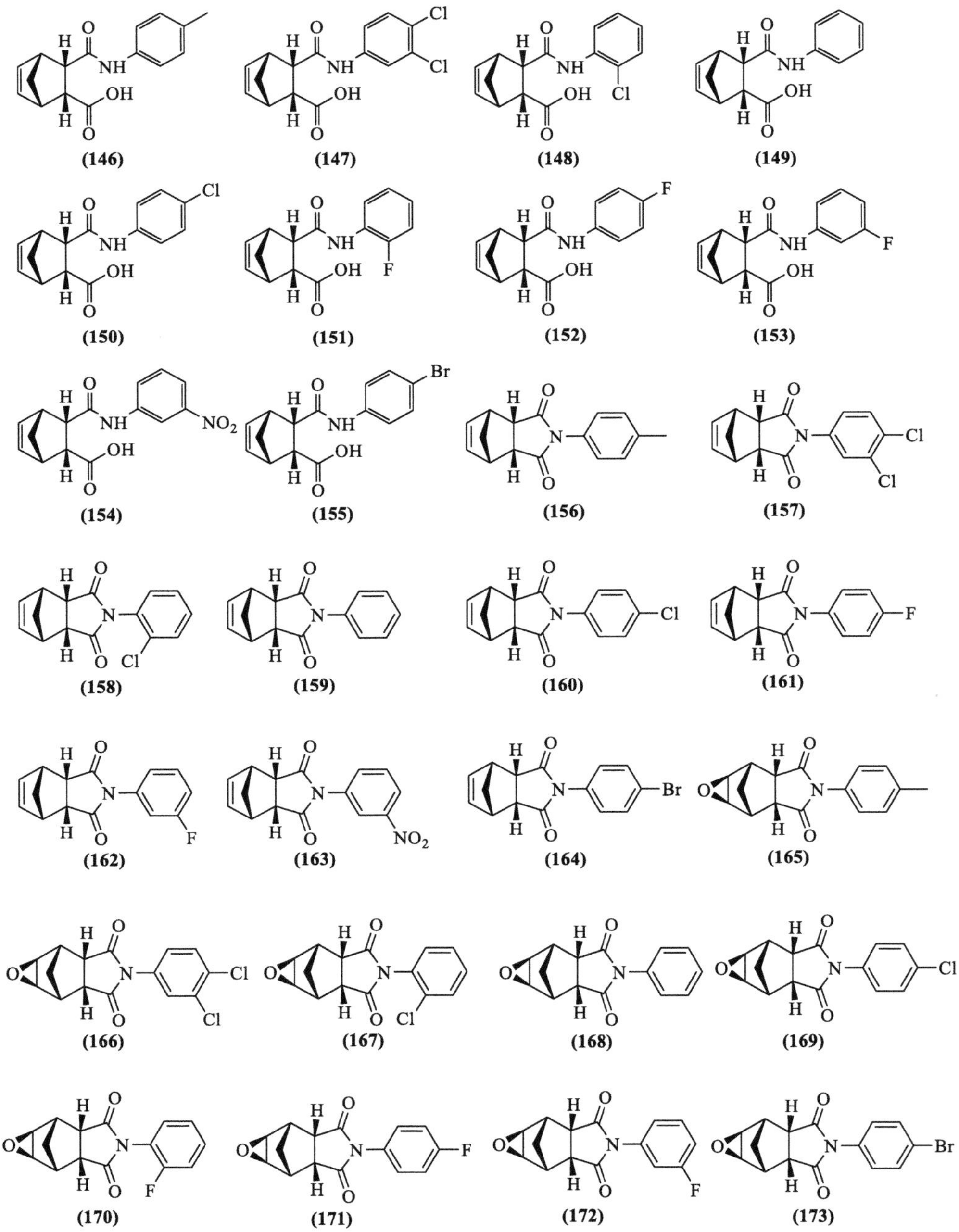

FIGURE 1.5 Structures of the *N*-phenylnorbornenesuccinimide derivatives.

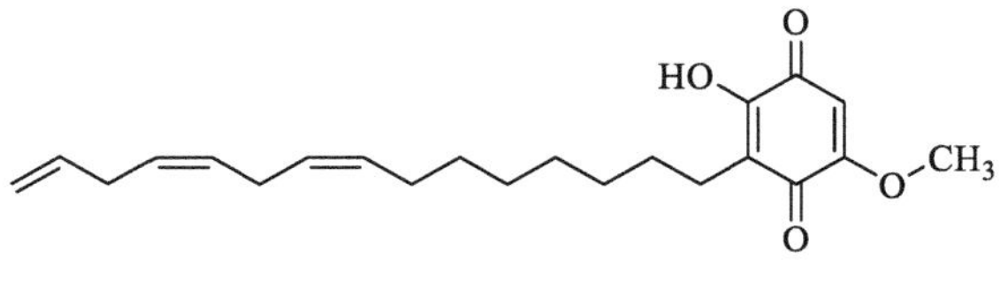

FIGURE 1.6 Structure of sorgoleone.

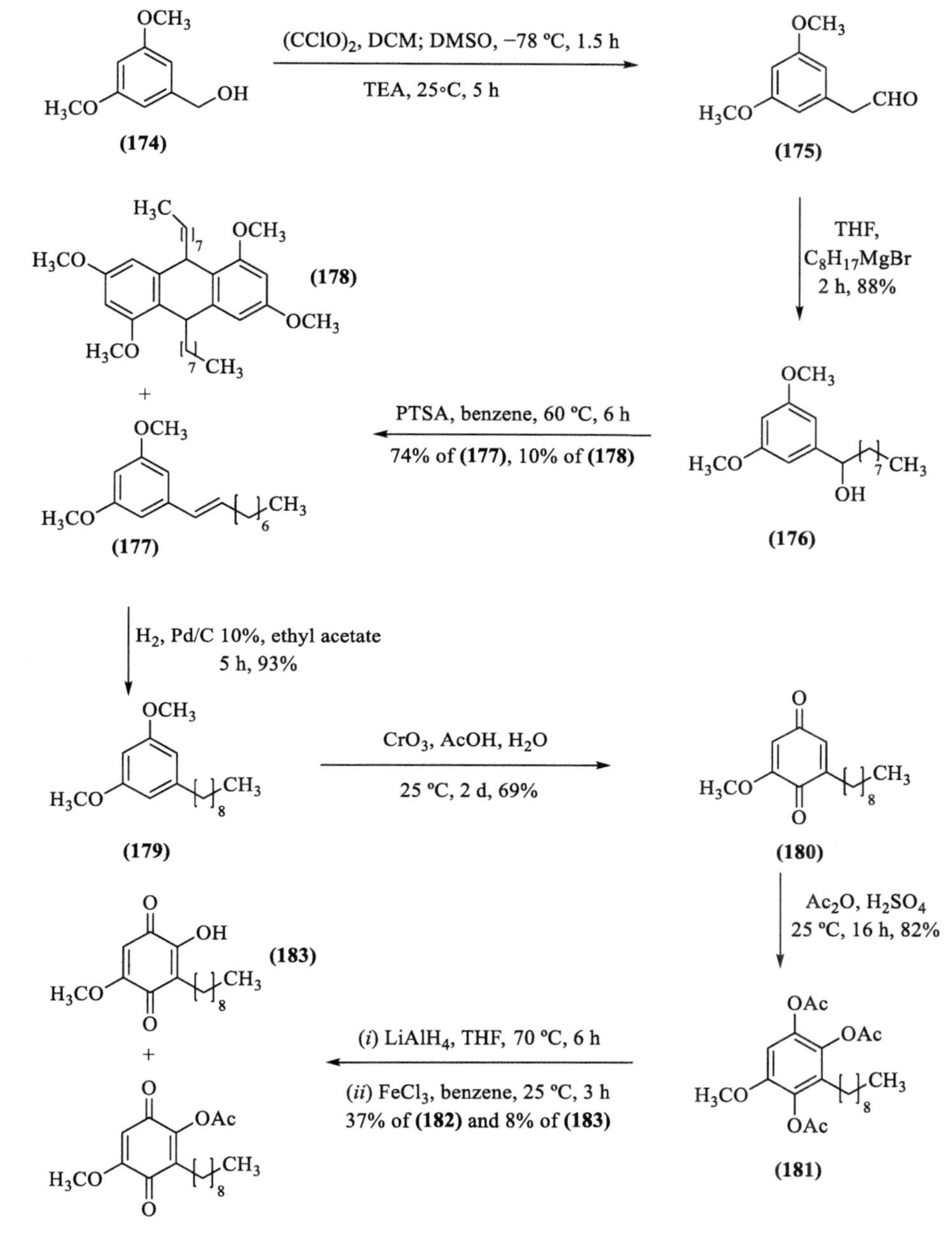

SCHEME 20 Synthesis of benzoquinones (180–183).

subsequently transformed into the corresponding quinones (187)–(189) using the same methods described for the synthesis of quinone (180).

The effects of sorgoleone, and compounds (180), (182), (183), (184)–(189) at a concentration of 5.5 μg.g^{-1} on the development of aerial parts and roots of *C. sativus*, *E. heterophylla*, *H. lophanta*, and *B. decumbens* were evaluated. For *C. sativus*, only compound (180) caused significant inhibition (8.52%) in the development of aerial parts, proving more active than sorgoleone. For roots,

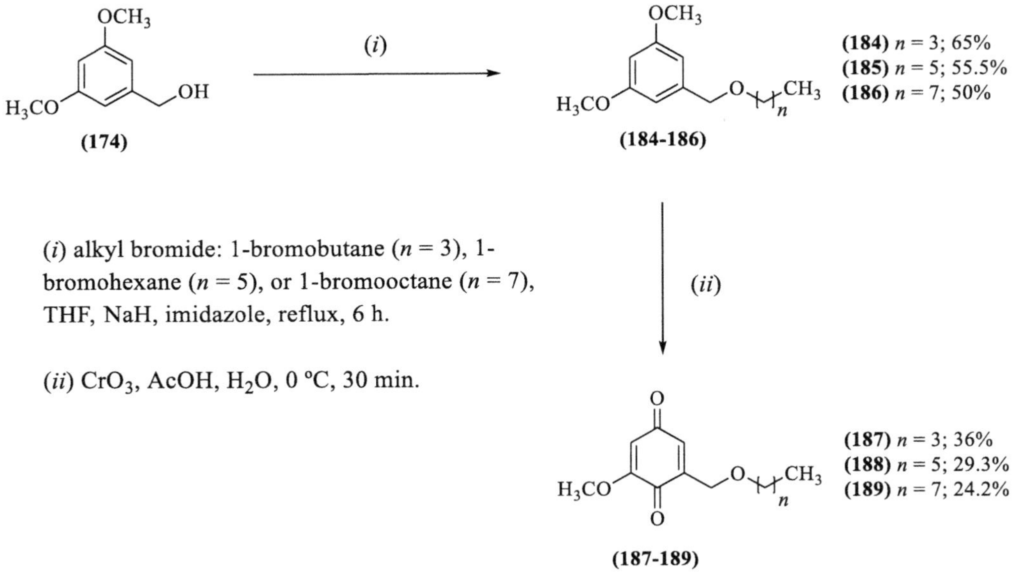

(174)

(184-186)

(184) $n = 3$; 65%
(185) $n = 5$; 55.5%
(186) $n = 7$; 50%

(*i*) alkyl bromide: 1-bromobutane ($n = 3$), 1-bromohexane ($n = 5$), or 1-bromooctane ($n = 7$), THF, NaH, imidazole, reflux, 6 h.

(*ii*) CrO_3, AcOH, H_2O, 0 °C, 30 min.

(187) $n = 3$; 36%
(188) $n = 5$; 29.3%
(189) $n = 7$; 24.2%

(187-189)

SCHEME 21 Synthesis of quinones (184)–(186) and (187)–(189).

sorgoleone, 182, and 185 caused significant inhibition, with acetate 182 being more active than sorgoleone. Compound (189) had a slight stimulant effect (4.14%).

In *E. heterophylla*, only compound (18) presented significant activity, inhibiting root growth by 11.86%. For *H. lophanta*, neither compound affected root development, but compounds (180) and (183) significantly (8.47%) inhibited aerial parts, although not significantly different from sorgoleone (1.69%). Compound (187) slightly stimulated the growth of aerial parts (8.47%). In *B. decumbens*, only compound (180) caused significant inhibition (18.92%) in root development.

Although no compound caused very high inhibition in the development of the tested plants, some, especially compound (180), were more active than natural sorgoleone. These results indicate that the triene unit in the sorgoleone side chain, which is difficult to synthesize, is not required for biological activity. In general, the introduction of the oxygen atom into the side chain in compounds (187)–(189) had little effect on activity.

1.4 CONCLUSION

This chapter presented a comprehensive analysis of the crucial role of synthetic pesticides, including insecticides and herbicides, in protecting agricultural crops and ensuring global food security. We discussed the efficacy of these products in pest control and their selectivity concerning predators and pollinators, as well as their positive impact on agricultural productivity. However, we also highlighted the challenges and concerns associated with the continuous and often indiscriminate use of these compounds.

The resistance developed by weeds and insects to existing pesticides has driven the search for new bioactive substances. In this context, we explored the insecticidal potential of amides, dienamides, imides, pyrethroids, lactones, and their derivatives, as well as the herbicidal potential of cyclic imides, lactones and their derivatives, anilides, benzoxazole derivatives, dienamides and epoxy derivatives, *N*-phenylnorbornenesuccinimide derivatives, and benzoquinones. We emphasized their advantages in terms of easy synthetic access and the possibilities for structural modifications to improve efficacy and reduce toxicity and environmental persistence.

The adoption of more selective and less harmful pesticides is crucial for mitigating the negative effects of bioaccumulation and environmental contamination. Computational technologies, such as theoretical calculations and molecular docking, play a significant role in the design and optimization of new compounds, offering more sustainable solutions for pest control.

REFERENCES

1. World Health Organization. UN Report: Global Hunger Numbers Rose to as Many as 828 million in 2021. https://www.who.int/news/item/06-07-2022-un-report-global-hunger-numbers-rose-to-as-many-as-828-million-in-2021 (accessed May 21, 2024).
2. Woodhill, J.; Kishore, A.; Njuki, J.; Jones, K.; Hasnain, S. Food Systems and Rural Wellbeing: Challenges and Opportunities. *Food Secur*, **2022**, *14* (5), 1099–1121. https://doi.org/10.1007/s12571-021-01217-0.
3. Popp, J.; Pető, K.; Nagy, J. Pesticide Productivity and Food Security. A Review. *Agron Sustain Dev*, **2013**, *33* (1), 243–255. https://doi.org/10.1007/s13593-012-0105-x.
4. Gu, D.; Andreev, K.; Dupre, M. E. Major Trends in Population Growth Around the World. *China CDC Wkly*, **2021**, *3* (28), 613. https://doi.org/10.46234/CCDCW2021.160.
5. United Nations. Population. https://www.un.org/en/global-issues/population#:%E2%88%BC:text=The%20world (accessed May 21, 2024).
6. Tison, L.; Beaumelle, L.; Monceau, K.; Thiéry, D. Transfer and Bioaccumulation of Pesticides in Terrestrial Arthropods and Food Webs: State of Knowledge and Perspectives for Research. *Chemosphere*, **2024**, *357*, 142036. https://doi.org/10.1016/j.chemosphere.2024.142036.
7. Shamim, A.; Neelam, K.; Kamaal, S.; Ali, A.; Ahmad, M. Removal of Pesticide Pollutants from Aqueous Waste Utilizing Nanomaterials via Photocatalytic Process: A Review. *Int J Environ Sci Technol*, **2024**, *21* (4), 4653–4684. https://doi.org/10.1007/s13762-023-05341-6.
8. Khursheed, A.; Rather, M. A.; Jain, V.; Wani, A. R.; Rasool, S.; Nazir, R.; Malik, N. A.; Majid, S. A. Plant Based Natural Products as Potential Ecofriendly and Safer Biopesticides: A Comprehensive Overview of Their Advantages over Conventional Pesticides, Limitations and Regulatory Aspects. *Microb Pathog*, **2022**, *173*, 105854. https://doi.org/10.1016/j.micpath.2022.105854.
9. Pinto, B. N. S.; Teixeira, M. G.; Alvarenga, E. S. Synthesis and Structural Elucidation of a Phthalide Analog Using NMR Analysis and DFT Calculations. *Magn Reson Chem*, **2020**, *58* (6), 559–565. https://doi.org/10.1002/mrc.4976.
10. Martins, L. M. O. S.; Santos, J. O.; Hoye, T.; Alvarenga, E. S. Synthesis of a Novel Naphthalenone Endoperoxide and Structural Elucidation by Nuclear Magnetic Resonance Spectroscopy and Theoretical Calculation. *Magn Reson Chem*, **2022**, *60* (1), 139–147. https://doi.org/10.1002/mrc.5195.
11. Hou, Y.; Bai, Y.; Lu, C.; Wang, Q.; Wang, Z.; Gao, J.; Xu, H. Applying Molecular Docking to Pesticides. *Pest Manag Sci*, **2023**, *79* (11), 4140–4152. https://doi.org/10.1002/ps.7700.
12. Plimmer, J. R.; Gammon, D. W. Insecticides: Overview and Introduction. In *Encyclopedia of Agrochemicals*; John Wiley & Sons, Ltd, **2003**. https://doi.org/10.1002/047126363X.agr367.
13. Buckle, A. Rodenticides and Insecticides. In *The Biocides Business*; John Wiley & Sons, Ltd, **2002**; pp. 267–286. https://doi.org/10.1002/352760197X.ch12.
14. Araújo, M. F.; Castanheira, E. M. S.; Sousa, S. F. The Buzz on Insecticides: A Review of Uses, Molecular Structures, Targets, Adverse Effects, and Alternatives. *Molecules*, **2023**, *28* (8). https://doi.org/10.3390/molecules28083641.
15. Matsumura, F.; Ghiasuddin, S. M. Evidence for Similarities between Cyclodiene Type Insecticides and Picrotoxinin in Their Action Mechanisms. *J Environ Sci Health B*, **1983**, *18* (1), 1–14. https://doi.org/10.1080/03601238309372355.
16. Xiaoman Liu Aocheng Cao, D. Y. C. O. Q. W.; Li, Y. Overview of Mechanisms and Uses of Biopesticides. *Int J Pest Manag*, **2021**, *67* (1), 65–72. https://doi.org/10.1080/09670874.2019.1664789.
17. Lopes, M. C.; Alvarenga, E. S.; Aguiar, A. R.; Dos Santos, I. B.; Silva, G. A.; De Paula Arcanjo, L.; Picanço, M. C. Insecticidal Activity of Dienamides on Cabbage Caterpillar and Beneficial Insects. *Quim Nova*, **2018**, *41* (4), 375–379. https://doi.org/10.21577/0100-4042.20170192.
18. Aguiar, A. R.; Alvarenga, E. S.; Silva, E. M. P.; Farias, E. S.; Picanço, M. C. Synthesis, Insecticidal Activity, and Phytotoxicity of Novel Chiral Amides. *Pest Manag Sci*, **2019**, *75* (6), 1689–1696. https://doi.org/10.1002/ps.5289.
19. Aguiar, A. R.; Alvarenga, E. S.; Lopes, M. C.; Dos Santos, I. B.; Galdino, T. V.; Picanço, M. C. Active Insecticides for Diaphania Hyalinata Selective for the Natural Enemy Solenopsis Saevissima. *J Environ Sci Health B*, **2016**, *51* (9), 579–588. https://doi.org/10.1080/03601234.2016.1181897.

20. Silvério, F. O.; de Alvarenga, E. S.; Moreno, S. C.; Picanço, M. C. Synthesis and Insecticidal Activity of New Pyrethroids. *Pest Manag Sci*, **2009**, *65* (8), 900–905. https://doi.org/10.1002/ps.1771.
21. Moreno, S. C.; Picanço, M. C.; Silvério, F. O.; De Alvarenga, E. S.; Carvalho, G. A. Toxicity of New Pyrethroids to the Social Insects Protonectarina Sylveirae, Solenopsis Saevissima and Tetragonisca Angustula. *Sociobiology*, **2009**, *54* (3), 893–905.
22. Moreno, S. C.; Silvério, F. O.; Lopes, M. C.; Ramos, R. S.; Alvarenga, E. S.; Picanço, M. C. Toxicity of New Pyrethroid in Pest Insects Asciamonuste and Diaphania Hyalinata, Predator Solenopsis Saevissima and Stingless Bee Tetragonisca Angustula. *J Environ Sci Health B*, **2017**, *52* (4), 237–243. https://doi.org/10.1080/03601234.2016.1270681.
23. Moreno, S. C.; Silvério, F. O.; Picanço, M. C.; Alvarenga, E. S.; Pereira, R. R.; Santana Júnior, P. A.; Silva, G. A. New Pyrethroids for Use Against Tuta Absoluta (Lepidoptera: Gelechiidae): Their Toxicity and Control Speed. *J Insect Sci*, **2017**, *17* (5), 6–11. https://doi.org/10.1093/jisesa/iex072.
24. Alvarenga, E. S. de; Carneiro, V. M. T.; Resende, G. C.; Picanço, M. C.; Farias, E. D. S.; Lopes, M. C. Synthesis and Insecticidal Activity of an Oxabicyclolactone and Novel Pyrethroids. *Molecules*, **2012**, *17* (12), 13989–14001. https://doi.org/10.3390/molecules171213989.
25. Silvério, F. O.; de Alvarenga, E. S.; Moreno, S. C.; Picanço, M. C. Synthesis and Insecticidal Activity of New Pyrethroids. *Pest Manag Sci*, **2009**, *65* (8), 900–905. https://doi.org/10.1002/ps.1771.
26. Guillermo Federico Padilla-Gonzalez, F. A. dos S.; Costa, F. B. Da. Sesquiterpene Lactones: More Than Protective Plant Compounds with High Toxicity. *CRC Crit Rev Plant Sci*, **2016**, *35* (1), 18–37. https://doi.org/10.1080/07352689.2016.1145956.
27. Teixeira, M. G.; Alvarenga, E. S.; Pimentel, M. F.; Picanço, M. C. Synthesis and Insecticidal Activity of Lactones Derived from Furan-2(5H)-One. *J Braz Chem Soc*, **2015**, *26* (11), 2279–2289. https://doi.org/10.5935/0103-5053.20150217.
28. Mantilla Afanador, J. G.; Araujo, S. H. C.; Teixeira, M. G.; Lopes, D. T.; Cerceau, C. I.; Andreazza, F.; Oliveira, D. C.; Bernardi, D.; Moura, W. S.; Aguiar, R. W. S.; et al. Novel Lactone-Based Insecticides and Drosophila Suzukii Management: Synthesis, Potential Action Mechanisms and Selectivity for Non-Target Parasitoids. *Insects*, **2023**, *14* (8). https://doi.org/10.3390/insects14080697.
29. Resende, G. C.; Alvarenga, E. S.; de Araújo, T. A.; Campos, J. N.; Pincanço, M. C. Toxicity to Diaphania Hyalinata, Selectivity to Non-Target Species and Phytotoxicity of Furanones and Phthalide Analogues. *Pest Manag Sci*, **2016**, *72* (9), 1772–1777. https://doi.org/10.1002/ps.4210.
30. Farias, E. S.; Araújo, T. A.; Resende, G. C.; Campos, J. N. D.; Pimentel, M. F.; Alvarenga, E. S.; Picanço, M. C. Toxicity and Sublethal Effects of Phthalides Analogs to Rhyzopertha Dominica. *Chem Biodivers*, **2019**, *16* (3), e1800557. https://doi.org/10.1002/cbdv.201800557.
31. Farias, E. S.; Araújo, T. A.; Resende, G. C.; Campos, J. N. D.; Pimentel, M. F.; Alvarenga, E. S.; Picanço, M. C. Toxicity and Sublethal Effects of Phthalides Analogs to Rhyzopertha Dominica. *Chem Biodivers*, **2019**, *16* (3). https://doi.org/10.1002/cbdv.201800557.
32. Das, T. K.; Behera, B.; Nath, C. P.; Ghosh, S.; Sen, S.; Raj, R.; Ghosh, S.; Sharma, A. R.; Yaduraju, N. T.; Nalia, A.; et al. Herbicides Use in Crop Production: An Analysis of Cost-Benefit, Non-Target Toxicities and Environmental Risks. *Crop Prot*, **2024**, *181*, 106691. https://doi.org/10.1016/j.cropro.2024.106691.
33. Tshewang, S.; Sindel, B. M.; Ghimiray, M.; Chauhan, B. S. Weed Management Challenges in Rice (Oryza Sativa L.) for Food Security in Bhutan: A Review. *Crop Protection*, **2016**, *90*, 117–124. https://doi.org/10.1016/j.cropro.2016.08.031.
34. Torrent, K. B. A.; Alvarenga, E. S. Synthesis and Identification of Epoxy Derivatives of 5-Methylhexahydroisoindole-1,3-Dione and Biological Evaluation. *Molecules*, **2021**, *26* (7). https://doi.org/10.3390/molecules26071923.
35. Nicotra, V. E.; Gil, R. R.; Oberti, J. C.; Burton, G. Withanolides with Phytotoxic Activity from Jaborosa Caulescent Var. Caulescens and J. Caulescens Var. Bipinnatifida. *J Nat Prod*, **2007**, *70* (5), 808–812. https://doi.org/10.1021/np070030g.
36. Queiroz, S. C. N.; Cantrell, C. L.; Duke, S. O.; Wedge, D. E.; Nandula, V. K.; Moraes, R. M.; Cerdeira, A. L. Bioassay-Directed Isolation and Identification of Phytotoxic and Fungitoxic Acetylenes from Conyza Canadensis. *J Agric Food Chem*, **2012**, *60* (23), 5893–5898. https://doi.org/10.1021/jf3010367.
37. Resende, G. C.; Alvarenga, E. S.; Galindo, J. C. G.; Macias, F. A. Synthesis and Phytotoxicity of 4,5 Functionalized Tetrahydrofuran-2-Ones. *J Braz Chem Soc*, **2012**, *23* (12), 2266–2270.
38. Teixeira, M. G.; Alvarenga, E. S.; Lopes, D. T.; Oliveira, D. F. Herbicidal Activity of Isobenzofuranones and in Silico Identification of Their Enzyme Target. *Pest Manag Sci*, **2019**, *75* (12), 3331–3339. https://doi.org/10.1002/ps.5456.

39. Barbosa, L. C. A.; Demuner, A. J.; de Alvarenga, E. S.; Oliveira, A.; King-Diaz, B.; Lotina-Hennsen, B. Phytogrowth- and Photosynthesis-Inhibiting Properties of Nostoclide Analogues. *Pest Manag Sci*, **2006**, *62* (3), 214–222. https://doi.org/10.1002/ps.1147.
40. Moraes, F. C.; Alvarenga, E. S.; Amorim, K. B.; Demuner, A. J.; Pereira-Flores, M. E. Novel Platensimycin Derivatives with Herbicidal Activity. *Pest Manag Sci*, **2016**, *72* (3), 580–584. https://doi.org/10.1002/ps.4028.
41. Alvarenga, S.; Barbosa, C. A.; Saliba, W. A.; Arantes, F. F. P.; Demuner, J. Síntese e Avaliação Da Atividade Fitotóxica de Derivados Da α-Santonina. *Quim Nova*, **2009**, *32* (2), 401–406.
42. Liu, K.-M.; Liu, K.-J. Lipase-Catalyzed Synthesis of Palmitanilide: Kinetic Model and Antimicrobial Activity Study. *Enzyme Microb Technol*, **2016**, *82*, 82–88. https://doi.org/10.1016/j.enzmictec.2015.08.017.
43. Zhang, A.; Yue, Y.; Yang, J.; Shi, J.; Tao, K.; Jin, H.; Hou, T. Design, Synthesis, and Antifungal Activities of Novel Aromatic Carboxamides Containing a Diphenylamine Scaffold. *J Agric Food Chem*, **2019**, *67* (17), 5008–5016. https://doi.org/10.1021/acs.jafc.9b00151.
44. Zhao, Y.; Zhang, A.; Wang, X.; Tao, K.; Jin, H.; Hou, T. Novel Pyrazole Carboxamide Containing a Diarylamine Scaffold Potentially Targeting Fungal Succinate Dehydrogenase: Antifungal Activity and Mechanism of Action. *J Agric Food Chem*, **2022**, *70* (42), 13464–13472. https://doi.org/10.1021/acs.jafc.2c00748.
45. Nakagawa, Y.; Akagi, T.; Iwamura, H.; Fujita, T. Quantitative Structure-Activity Studies of Benzoylphenylurea Larvicides: VI. Comparison of Substituent Effects among Activities against Different Insect Species. *Pestic Biochem Physiol*, **1989**, *33* (2), 144–157. https://doi.org/10.1016/0048-3575(89)90005-9.
46. Zhang, L.; Chen, R.; Li, X.; Xu, X.; Xu, Z.; Cheng, J.; Wang, Y.; Li, Y.; Shao, X.; Li, Z. Synthesis, Insecticidal Activities, and 3D-QASR of N-Pyridylpyrazole Amide Derivatives Containing a Phthalimide as Potential Ryanodine Receptor Activators. *J Agric Food Chem*, **2022**, *70* (39), 12651–12662. https://doi.org/10.1021/acs.jafc.2c03971.
47. Selby, T. P.; Satterfield, A. D.; Puri, A.; Stevenson, T. M.; Travis, D. A.; Campbell, M. J.; Taggi, A. E.; Hughes, K. A.; Bereznak, J. Bioisosteric Tactics in the Discovery of Tetflupyrolimet: A New Mode-of-Action Herbicide. *J Agric Food Chem*, Null. https://doi.org/10.1021/acs.jafc.3c01634.
48. Huffman, C. W.; Allen, S. E. Herbicidal Activity, Molecular Size vs. Herbicidal Activity of Anilides. *J Agric Food Chem*, **1960**, *8* (4), 298–302. https://doi.org/10.1021/jf60110a012.
49. Sartori, S. K.; Alvarenga, E. S.; Franco, C. A.; Ramos, D. S.; Oliveira, D. F. One-Pot Synthesis of Anilides, Herbicidal Activity and Molecular Docking Study. *Pest Manag Sci*, **2018**, *74* (7), 1637–1645. https://doi.org/10.1002/ps.4855.
50. Sangi, D. P.; Meira, Y. G.; Moreira, N. M.; Lopes, T. A.; Leite, M. P.; Pereira-Flores, M. E.; Alvarenga, E. S. Benzoxazoles as Novel Herbicidal Agents. *Pest Manag Sci*, **2019**, *75* (1), 262–269. https://doi.org/10.1002/ps.5111.
51. Ramos, D.; Alvarenga, E.; Silva, J.; Cerceau, C.; Sartori, S.; Carneiro, V.; Oliveira, D. Study of the Interference in Target Seed Germination Caused by Dienamides and Epoxy Derivatives, in the Search for New Herbicides. *J Braz Chem Soc*, **2023**. https://doi.org/10.21577/0103-5053.20230113.
52. Gomes, S. F.; Alvarenga, E. S.; Baia, V. C.; Oliveira, D. F. N-Phenylnorbornenesuccinimide Derivatives, Agricultural Defensive, and Enzymatic Target Selection. *Pest Manag Sci*. https://doi.org/10.1002/ps.8031.
53. Singh, H.; Kaur, K.; Singh, M.; Kaur, G.; Singh, P. Plant Cyclophilins: Multifaceted Proteins with Versatile Roles. *Front Plant Sci*, **2020**, *11*, 1–30. https://doi.org/10.3389/fpls.2020.585212.
54. Galberto, J.; Costa, M.; Deusdênia, O.; Pessoa, L.; José, F.; Monte, Q.; Menezes, E. A.; Leda, T.; Lemos, G. Benzoquinonas, Hidroquinonas e Sesquiterpenos de *Auxemma Glazioviana*. *Quim Nova*, **2005**, *28* (4), 591–595.
55. de Carvalho, N. K.; da Silva Mendes, J.; da Costa, J. G. Quinones: Biosynthesis, Characterization of 13C Spectroscopical Data and Pharmacological Activities. *Chem Biodivers*, **2023**, *20* (12), e202301365. https://doi.org/10.1002/cbdv.202301365.
56. Telma, L. G. L.; Monte, F. J. Q.; Santos, A. K. L.; Fonseca, A. M.; Santos, H. S.; Oliveira, M. F.; Costa, S. M. O.; Pessoa, O. D. L.; Braz-Filho, R.; Braz-Filho, R. Quinones from Plants of Northeastern Brazil: Structural Diversity, Chemical Transformations, NMR Data and Biological Activities. *Nat Prod Res*, **2007**, *21* (6), 529–550. https://doi.org/10.1080/14786410601130604.
57. Tibugari, H.; Chiduza, C. Sorghum Allelopathy under Field Conditions May Be Caused by a Combination of Allelochemicals. *Cogent Food Agric*, **2024**, *10* (1), 2324528. https://doi.org/10.1080/23311932.2024.2324528.

58. Mareya, C. R.; Tugizimana, F.; Steenkamp, P.; Piater, L.; Dubery, I. A. Lipopolysaccharides Trigger Synthesis of the Allelochemical Sorgoleone in Cell Cultures of *Sorghum Bicolor. Plant Signal Behav*, **2020**, *15* (10), 1796340. https://doi.org/10.1080/15592324.2020.1796340.
59. Tibugari, H.; Chiduza, C.; Mashingaidze, A. B.; Mabasa, S. High Sorgoleone Autotoxicity in Sorghum (*Sorghum Bicolor* (L.) Moench) Varieties That Produce High Sorgoleone Content. *South African Journal of Plant and Soil*, **2020**, *37* (2), 160–167. https://doi.org/10.1080/02571862.2020.1711539.
60. Lima, L. S.; Barbosa, L. C. de A.; de Alvarenga, E. S.; Demuner, A. J.; da Silva, A. A. Synthesis and Phytotoxicity Evaluation of Substituted *Para*-Benzoquinones. *Aust J Chem*, **2003**, *56* (6), 625–630.
61. Mancuso, A. J.; Huang, S.-L.; Swern, D. Oxidation of Long-Chain and Related Alcohols to Carbonyls by Dimethyl Sulfoxide "Activated" by Oxalyl Chloride. *J Org Chem*, **1978**, *43* (12), 2480–2482. https://doi.org/10.1021/jo00406a041.

2 Toxicity of Chemical and Biological Plant Protection Products to Agriculture

Hamdy A. Shaaban and Mohamed H. Soliman

2.1 INTRODUCTION

Environmental pollution includes pesticide chemicals, heavy metals, and fertilizers. Pesticides are chemical compounds that are used to kill pests that cause damage of crops, such as insects, rodents, fungi, and unwanted plants (weeds). Over 1000 different pesticides are used around the world. Pesticides are used in public health to kill vectors of disease, such as mosquitoes, and in agriculture to kill pests. Some pesticides produce acute toxic effects because of their corrosive or irritant properties. These can result in respiratory, skin or eye irritation or damage. The term pesticide covers a wide range of compounds including insecticides, fungicides, herbicides, rodenticides, molluscicides, nematicides, plant growth regulators, and others. Among these, organochlorine (OC) insecticides, used successfully in controlling a number of diseases, such as malaria and typhus, were banned or restricted after the 1960s in most of the technologically advanced countries. The introduction of other synthetic insecticides—organophosphate (OP) insecticides in the 1960s, carbamates in 1970s, and pyrethroids in 1980s and the introduction of herbicides and fungicides in the 1970s–1980s contributed greatly to pest control and agricultural output. Ideally a pesticide must be lethal to the targeted pests but not to non-target species, including humans. Unfortunately, this is not the case, so the controversy of use and abuse of pesticides has surfaced. The rampant use of these chemicals, under the adage, "if little is good, a lot more will be better" has played havoc with human and other life forms. Some can cause severe burns or permanent blindness. Chemicals with these irritant or corrosive properties stimulated enormously economic development and rapid growth in many fields, such as agriculture and industry, but the environment was becoming more polluted [1]. Environmental pollutants are toxic substances that enter the environment from both anthropogenic and natural sources. Certain environmental processes, such as synthetic industries, coal conversion, and waste burning, result in hazardous problems for a biotic element (water, air, and soil) and biotic communities (animals, plants and humans). Usually, environmental toxicants include pesticides and heavy metals and threaten the entire ecosystem, seriously damaging its function and structure. Naturally, heavy metals are metals with a high atomic weight and a density greater than 5 g/cm^3. Compared with their physical properties, the chemical characteristics of heavy metals are the most practical aspects. Environmental toxicity exceeding standard maximum residue limits (MRLs) has received heightened consideration from think tanks worldwide. Cadmium (Cd), lead (Pb), copper (Cu), and zinc (Zn) cause an alarming combination of environmental and health problems [2]. Heavy metals arise from many sources, such as industry, mining, and agriculture. In terms of the sources in the agricultural sector, these can be categorized into fertilization, pesticides, livestock manure, and wastewater. Recently, the risk of heavy metal pollution in the environment has been increasing rapidly and creating turmoil, especially in the agricultural sector, by accumulating in the soil and in plant uptake. The heavy metal contamination problem has become urgent and needs radical and practical solutions to

DOI: 10.1201/9781003463429-4

reduce the hazards as much as possible. Even though heavy metals are needed for several organs of both plants and humans, they become toxic when their concentration exceeds the prescribed level. Many studies have been done in this area, reporting that the primary sources of heavy metals are agriculture, mining, agrochemicals, and industry. A study done by Xiao et al. [3] reported that agriculture and industry significantly influence heavy metal pollution in agricultural soil and plants, especially soils near cement and electroplating factories. In other words, the soil surface is a fertile place for storing heavy metals and then transferring them to the plants by absorption along with water through the roots, followed by the vascular system.

Heavy metal accumulation can be described as an aggregation of elements in the ecosystem. Plant roots are the essential point of contact for heavy metal ions transmitted from the soil. They tend to stabilize and connect the pollutants in the soil, therefore reducing their bioavailability. The mechanisms of heavy metals transmission to plants include (i) phytoextraction (subprocess of phytoremediation in which plants eliminate hazardous components from contaminated soil), (ii) phyto-stabilization (immobilization and reduction of the mobility of heavy metals in soil), and (iii) rhizo-filtration (a form of phyto-remediation to use plant roots to absorb the toxic substances). These metals cause damage to plants and extend to harm human health through transference in the food chain [2]. Recently, due to the rapid evolution of technology, the ecosystem and humans have been exposed to many types of chemical toxicants, in particular pesticides (herbicides, insecticides, and fungicides) [1]. Scientists have defined pesticides as synthesized chemical compounds used in many areas, including in the agricultural sector, to control pests. Therefore, pesticides are considered efficient, economical, and effective weapons in integrated pest management systems (IPMs). The uncontrolled use of pesticides causes their bioaccumulation in food chains, which leads to high risk to mammals and other non-target organisms. In addition, the direct or indirect effect of pesticides on non-target organisms leads to an imbalance of the surrounding ecosystem. Moreover, the pesticide residues remain in the plant parts, soil, and air and even penetrate into water. Such residues are considered one of the most destructive threats the ecosystem faces; these can exist in the environment for a long time, with carcinogenic effects. The deleterious health problems caused by toxicants are increasing due to their penetration and accumulation through the food chain, and their persistence in the ecosystem. Such contaminants can cause acute and chronic diseases in the human body, such as lung cancer, renal dysfunction, osteoporosis, and cardiac failure. Tong et al. [4] mentioned the degree of human health threats posed by heavy metals in China's urban areas during the period 2003–2009. The results showed that human health risk reports for heavy metals suggested that absorption was the primary route of exposure that has harmful effects on human health. The accumulation of heavy metals in internal human tissues can affect the central nervous system and act as a pseudo-co-factor or promotor of some health problems, such as seizures (epilepsy), headache, and coma. Heavy metal contamination is considered a health threat to both adults and children. Pesticides are also hazardous to humans and other living organisms through contaminated food, water, or inhalation of contaminated air. Exposure (whatever the level) to pesticides is hazardous to the behavior and physiology of humans. Furthermore, pesticides are linked with a wide range of diseases, such as hypersensitivity, cancer, asthma, and hormonal disturbances. In addition, they can also lead to congenital disabilities, reduce birth weight, and even cause death.

Overall, in light of the aforementioned ecological risks, this review covers the sources of heavy metals, the classification of pesticides, and the types of both toxicants. It also discusses the properties of agricultural soils that change due to the contamination of heavy metals and pesticides. In addition, the harmful effects on different plant species ranging from infection to death are discussed. The novelty of this study is to provide an integrated synthesis of knowledge on the complete pathway of both heavy metals and pesticides starting from their various sources, accumulation in soil and plants, and then reaching human beings. In addition, the synergistic and antagonistic interactions between heavy metals and pesticides and their combined toxicity in soil, plants, and humans are reported.

1. **Pesticide toxicity.**

Any foreign substance entering the body causes damage to organisms or it is a substance foreign to the body that causes an imbalance in physiological processes or causing the death of the organism [1].

2.2 BENEFIT OF PESTICIDES

The primary benefits are the consequences of the pesticides' effects—the direct gains expected from their use. For example, the effect of killing caterpillars feeding on the crop brings the primary benefit of higher yields and better quality of cabbage. The three main effects result in 26 primary benefits ranging from protection of recreational turf to saved human lives. The secondary benefits are the less immediate or less obvious benefits that result from the primary benefits. They may be subtle, less intuitively obvious, or of longer term. It follows that for secondary benefits it is therefore more difficult to establish cause and effect, but nevertheless they can be powerful justifications for pesticide use. For example, the higher cabbage yield might bring additional revenue that could be put towards children's education or medical care, leading to a healthier, better-educated population. There are various secondary benefits identified, ranging from fitter people to conserved biodiversity.

1. Improving productivity
2. Pest control
3. Protection of crop losses/yield reduction
4. Vector disease control
5. Quality of food

2.3 EFFECT OF PESTICIDE TOXICITY ON AGRICULTURE SOIL AND PLANTS

The excessive and uncontrolled use of pesticides on different crop species leads to harmful effects on beneficial biota, including honey bees, predators, birds, plants, small mammals, and humans. In addition, these ramifications create an imbalance in the biodiversity of the entire ecological system [1]. Many systemic pesticides, derivatives, and metabolites are investigated to be moderately safe to beneficial biota, especially beneficial insects, after their direct contact with such toxicants when they feed on plant tissues. As systemic pesticides can contaminate the floral and extrafloral nectar parts when transmitted systemically through the plant's vascular system, it leads to high percentages of mortality to honeybees and nectar-feeding parasitoids.

Moreover, many pesticides, such as chlordane, dieldrin, hexachlorobenzenethiobencarb, and endrin, resist degradation (persistent organic pollutants) and remain in the environment for a long time. Furthermore, persistent pesticide residues can be bio-accumulated and reach up to a bio-concentration more than 70,000-fold compared with the original concentration. A study done by Sharma and Agrawal [5] investigated the residues of five neonicotinoids (thiamethoxam, imidacloprid, acetamiprid, clothianidin, and thiacloprid) that were detected in different honey samples collected from different countries. The concentrations of residues depended on the used amount of pesticides (excessive or moderate use), which reflect the accumulation and the toxic effects of such residues on pollinators (honeybees) and other beneficial organisms. The mechanistic pathway of pesticides is starting from the time of application followed by photodegradation, absorption by the plant parts (stem, leaves, or fruit), or sorption at the soil level. Once the pesticides reach the soil, they undergo several biodegradation processes: chemical decomposition (pH, humidity, and temperature) and biological degradation (microorganisms' enzymes). The pesticide residues and degradation by-products uptake through roots via xylem to entire plant parts, causing deleterious effects to soil and plant. These effects include overproduction of reactive oxygen species (ROS), oxidative stress, DNA damage, photosynthetic blockage, necrosis, chlorosis, and leaf twisting, and ultimately end with plant death (Figure 2.1).

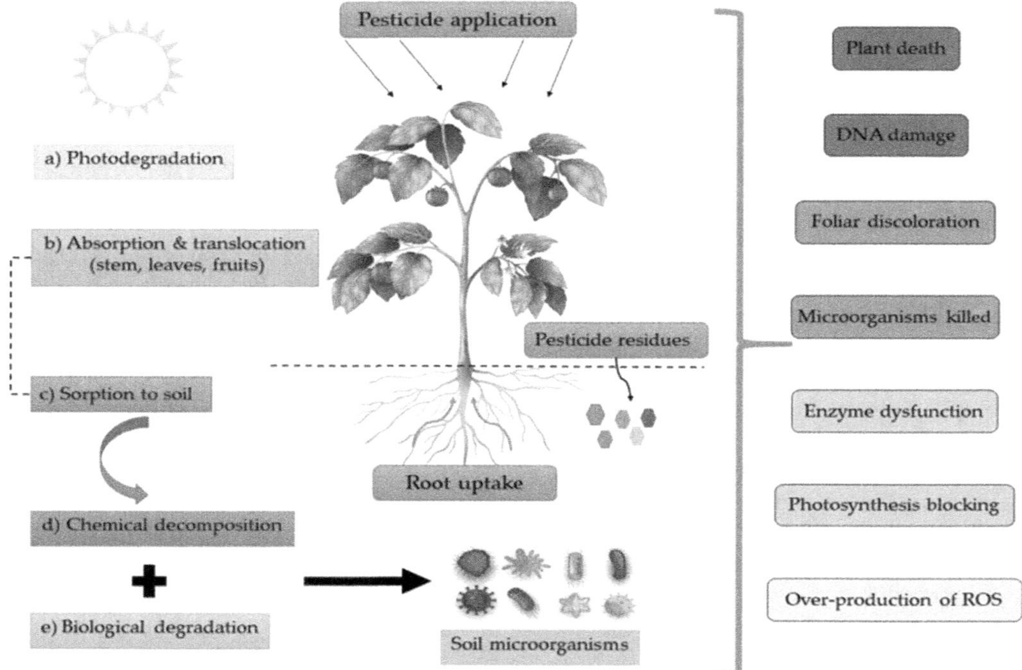

FIGURE 2.1 Mechanistic pathway of pesticide toxicity in soil and plants.

2.4 EFFECT OF PESTICIDE TOXICITY ON AGRICULTURAL SOIL

Many pesticides are being used extensively in the agricultural field to prevent pest damage and improve crop production without considering their harmful effects. As a result of that uncontrolled use, pesticide residues significantly congregate in the soil and increase the contamination, which is directly or indirectly harmful to fauna and flora [1]. On the soil level, pesticides can alter the physico-chemical and biological properties of the soil. They can also ultimately disturb microbial activity. Filimon et al. [6] studied the negative impacts of two insecticides (Cypermethrin and Thiomethoxam) on the soil to test a range of physical parameters and determine biochemical and microbial activities. The results showed that Thiomethoxam leads to a decrease in phosphatase activity by 6.5% compared with control. The number of nitrifying bacteria significantly decreased to 58.1%. Physico-chemical properties were positively correlated with phosphatase, urease, and dehydrogenase activities and negatively correlated with the aerobic nitrogen-fixing bacteria (*Azothobacter vinellandi*) and nitrifying-bacteria. Moreover, Cypermethrin leads to a decrease in the activity of the dehydrogenase enzyme by 32.8%. Likewise, the number of nitrifying bacteria decreased by 74%. Recently, Al-Ani et al. [7] determined the influence of two insecticides, Miraj (alpha-cypermethrin 10%) and Malathion (50% WP), on soil microorganisms (actinomycetes, fungi, and bacteria) and CO_2 production. The results revealed that CO_2 production decreased significantly for both insecticides. At week seven, the CO_2 production values were 32% and 36% for Miraj insecticide concentrations 100 and 200 ppm, respectively, while for Malathion application with 50, 100, and 200 ppm, the CO_2 production values decreased by 42, 45, and 52%, respectively. These results agree with Yousaf et al. [8], who presented the poisonous nature of insecticides to soil microorganisms due to the reduction in the production rate of CO_2. Goswami et al. [9] showed similar data that showed applying high concentrations of Cypermethrin insecticide leads to a severe impact on soil respiration and biomass. The effect of organo-phosphorus insecticides (dimethoate, diazinon, and Malathion) on soil's microbial communities during 24, 48, and 72 h was studied. The results reflected that microbial growth was significantly inhibited according to the concentration and the exposure time. The repeated application of chlorpyrifos, Malathion, lindane, and endosulfan insecticides reduced the nitrification and denitrification processes in the soil even when these insecticides were applied at the field-recommended doses. Many earlier studies focus on the hazardous effects of different insecticides on various plant species and agricultural soils. Carbamate pesticides inhibit the activity of the nitrogenase enzyme of *Azospirillum* sp. Furthermore, they suppress the growth of various types of soil fauna. Wang et al. [10] stated that imazetapir, carbendazim, and thiram pesticides reduced nitrogenase activity in some plant species, including *Rhizobium trifolii*, *R. leguminosarum*, and *Sinorhizobium melilot* in cultivated samples and under field conditions. Quinalphos decreases soil nitrification and ammonification processes.

The application of herbicides, especially glyphosate-based herbicides, causes various risks on microbial fauna depending on the application period. Some indirect risks to biodiversity occurred due to the alternation in the physiological and biosynthetic mechanisms of soil ecosystems. Some combinations of herbicides with heavy metals and inorganic fertilizers inhibit the functions of microbial soil communities. For example, Arif et al. [11] reported that Bromoxynil and Methomyl herbicides decrease the oxidation reaction of methane (CH_4) to CO_2. 2,4-D decreases the activity of nitrogenase, phosphatase enzymes, and hydrogen photoproduction of purple non-sulfur bacteria and harms the activities of *Rhizobium* sp. Moreover, glyphosate leads to a reduction in phosphatase enzyme activity and the growth activity of *Azotobacter.*

Fungicides are classified as the third most broadly used pesticide group after the insecticides and herbicides that are being used effectively nowadays for crop protection [1]. Furthermore, they cause harmful effects on non-target organisms such as soil microbial communities and influence soil biochemical processes like respiration. Baćmaga et al. [12] found that the urease enzyme was the most sensitive to the excessive exposure of azoxystrobin. Baćmaga et al. [12] reported that Falcon 460 EC fungicide significantly suppresses the activity of alkaline phosphatase, acid phosphatase, catalase, urease, and dehydrogenase enzymes in soil samples. In addition, overexposure to these active ingredients causes harmful effects on soil-dwelling microbial communities. Fungicides can neutralize soil enzyme activity due to the alleviation of some substances, such as compost and manure. Saha et al. [13] reported that tebuconazole has short-term hazardous effects on soil enzymatic activities (arylsulfatase, phosphatase, urease, and fluorescein). Baćmaga et al. [14] stated that the Chlorothalonil affected the soil microbial communities and the biochemical properties. Moreover, it led to the stimulation of heterotrophic and actinobacteria. The hazardous effects of Chlorothalonil were observed with applications higher than the recommended doses.

2.5 EFFECT OF PESTICIDE TOXICITY ON PLANTS

Plant transpiration facilitates the absorption of pesticides, which are soluble in soil, into all parts of the plant [1]. Pesticide translocation occurs through the root system, followed by the vascular system. The presence of their metabolites in the plant vascular system is determined by factors like their reactions with soil and plant, doses of applied pesticide, biochemical and physicochemical properties of pesticides, and the mechanism of pesticide entry. The deleterious effects of pesticides on plants can be detected as chlorosis, burns, leaf twisting, stunting, and necrosis. The exposure to organophosphorus insecticide chlorpyrifos suppressed the nitrogen metabolism and growth of *Vigna radiata* L. (mung bean). Parween et al. [15] investigated chlorpyrifos insecticide's metabolism and the response of the anti-oxidative enzymatic system of *Vigna radiata* L. during the different stages of growth after the application. The results showed that chlorpyrifos increased the rate of lipid peroxidation and proline content at a concentration of 1.5 mm during the post-flowering time. Moreover, ascorbate and glutathione levels significantly declined during all developmental stages. The activity of antioxidant enzymes increased with all concentrations during the pre-flowering stage. Sharma et al. [16] found that Imidacloprid insecticide caused a reduction in the levels of many phytochemical substances in *Brassica juncea*, the mustard plant.

Spraying herbicides around the vegetative parts of plants negatively affects the flowering and seed production of plants [1]. Such changes cause pigment discoloration and affect the antioxidant enzymes involved in the defense system, lipid peroxidation, and endogenous hormone levels of non-target plants. Kaya and Doganlar [17] reported that the application of imazapic herbicide induces some phytotoxic effects for tobacco plants, including carotenoids, jasmonic acid, antioxidant enzyme activity, and malondialdehyde contents. Recently, Fernandes et al. [18] investigated the effect of glyphosate-based herbicides (GBH) on a non-target plant (*Medicago sativa* L.). The results reflected an increase in lipid peroxidation that led to the suppression of root and shoot growth. Overall, GBH-contaminated soils can destroy the development of non-target plants.

The excessive use of fungicides exerts risky influences on plants (during different growth stages) that cannot be eliminated directly. The Falcon 460 EC fungicide showed adverse effects on root elongation and seed germination of some plant species, *Sorgo saccharum*, *Lepidium sativum*, and *Sinapsisalba*. In addition, the most inhibitory effects were observed on *S. alba*. Hydrogen peroxide levels were increased significantly in tomato plants when exposed to different concentrations of Chlorothalonil fungicide. Furthermore, Xia et al. [19] reported negative effects on stomatal conductance, photosynthetic rate, and cellular CO_2 of the photosynthetic system in cucumber plants. Ijaz et al. [20] examined the activity of five fungicides involved in triazole and strobilurin classes on two growth stages (course of pod development and green floral bud stages). The findings indicated that there was a significant influence on the leaf area. Carbendazim has the most negative bearing on seed germination and a reduction in the pea's net growth (*Pisum sativum* L.). Three tested fungicides (Carbendazim, kitazin, and hexaconazole) caused considerable consequences like cellular damage, cytotoxicity, and retardation in root morphology. Also, due to fungicide stress, photosynthetic pigment formation and grain production was prohibited. Moreover, some morphological disturbances and alternations in the stomatal behavior of pea leaves were seen.

2.6 HEAVY METALS

2.6.1 Sources of Heavy Metals

Scientists divide the sources of heavy metals into two major groups: natural and anthropogenic sources. Natural sources include sedimentary rocks, volcanic eruptions, soil formation, and rock weathering, while anthropogenic sources include industry, agriculture, mining, and domestic effluents [2]. However, pollution indicators are an effective mechanism for characterizing soil anthropogenic and gynogenic pollution. Nevertheless, it should be noted that source apportionment may be difficult in many cases, despite sophisticated research techniques. Alloway [21] discussed the different sources of heavy metals and their origin variation, which include sedimentation of aerosol particles, raindrops containing heavy metal, and agrochemicals. Although the study reported many types of metals, it mainly focused on Cd, Pb, Cu, and Zn, which are the same heavy metals in this study.

2.6.2 Natural Sources of Heavy Metals

Among the natural sources of heavy metals, igneous and sedimentary rocks are considered the most common. The concentration ranges (ppm) of heavy metals in the igneous and sedimentary rocks are listed in Table 2.1. It has been found that elements that exist in one rock type have varying proportions, and proportions of different elements vary from one rock type to another. Heavy metal concentration can be determined according to the type of rocks and the surrounding ecosystem conditions [2]. In addition, soil formation is also considered one of the main reasons for heavy metal accumulation besides river sediments.

TABLE 2.1
Heavy Metal Concentrations (ppm) in Igneous and Sedimentary Rocks

Metal	Basaltic Igneous	Granite Igneous	Shales and Clays	Black Shales	Sandstone
Cd	0.006–0.6	0.003–0.18	0.0–11	<0.3–8.4	–
Pb	30–160	4–30	18–120	20–200	–
Cu	48–240	5–140	18–180	34–1500	2–41
Zn	2–18	6–30	16–50	7–150	<1–31

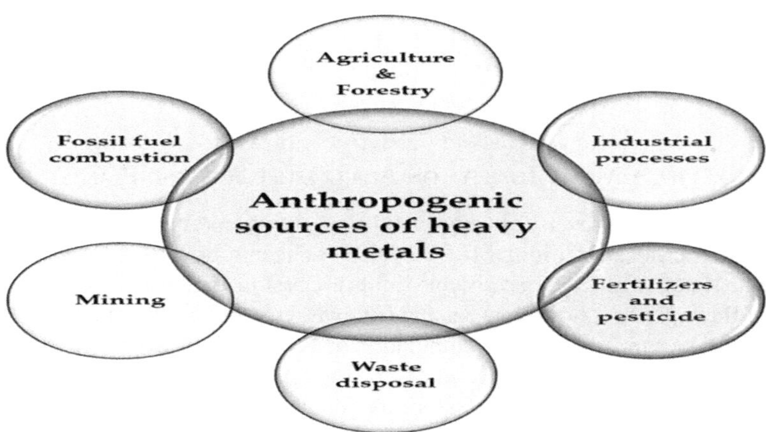

FIGURE 2.2 Anthropogenic sources of heavy metal pollution.

2.6.3 Anthropogenic Sources of Heavy Metals

Industries, agriculture, mining, and wastewater are deemed anthropogenic sources of heavy metals (Figure 2.2). These sources significantly lead to the elevation of heavy metal concentration and pollution in the ecosystem, such as smelting that results in releasing Cu, Zn, and As; insecticides that contribute to releasing As; burning of fossil fuels that produces Hg; and car exhaust that assists in releasing Pb [2]. In addition, daily human activities, such as farming, industrial processes, and manufacturing, impair the balance of the biosphere.

2.6.4 Agricultural Sources of Heavy Metals

Agroecosystems are usually affected by many types of pollutants, including agricultural pollutants, which are known as biotic and abiotic byproducts of farming practices. These pollutants usually cause contamination and degradation of the surrounding agroecosystem. Among the agricultural sources of heavy metals, fertilizers, pesticides, and sewage sludge are the most common. The toxic heavy metals vary in nature and the way of accumulation, whether in soil or plants.

2.6.4.1 Fertilizers as a Source of Heavy Metal Accumulation in Agricultural Soil and Plants

Fertilizers supply different necessary nutrients to enhance plant growth and productivity and increase the soil organic matter. Thus, fertilizers improve soil fertility. Fertilizers can be classified into organic (natural) and inorganic (synthetic) fertilizers. Organic or bio-fertilizers are produced after the anaerobic digestion (AD) process in the form of ammonium fertilizers (sulfate and nitrate). Inorganic fertilizers, also known as chemically manufactured/synthetic fertilizers, are a mix of inorganic substances and chemical materials. Fertilizers, including organic and inorganic elements, are responsible for producing heavy metals in the soil. Table 2.2 presents a comparison in heavy metal concentrations between worldwide and the European Union (EU), and also how heavy metals vary in different types of fertilizers. Phosphorus is widely used in fertilizer synthesis; simultaneously, it plays a significant role in heavy metal accumulation through its application to the soil. Water-insoluble phosphorus fertilizers have been shown to produce phosphate rocks, which play a major role in the immobilization of metals by precipitation as metal phosphates in the soil. Excessive use of fertilizers for a long time, resulting in heavy metal accumulation in agricultural soils, reduces soil fertility and consequently decreases plant growth and productivity. Cu, Zn, and Cd have a higher accumulation potential in agricultural soil due to the long-term use of fertilizers. Phosphate fertilizers, liming materials, and bio-fertilizers are the main types of inorganic fertilizers that contribute to the release of heavy metals in agricultural soil and are then taken up by plants. Therefore, they enter into the food chain and reach animals and humans.

2.6.5 Effect of Heavy Metal Toxicity on Agriculture Soil and Plants

As everyone knows, a high level of heavy metal concentrations influences both soil and plants. The WHO has set permissible limits/MRLs of their concentrations ($mg \cdot Kg^{-1}$) in soil and plants (Figure 2.3) [21]. Metals with high permissible limits are assumed to be safe. In soil, the permissible limits of Pb are the highest, followed by Zn and Cu, while the permissible limits of Cd are the lowest. These limits' values mean that the accumulation of Cd in the soil, even at a lower concentration, is more toxic than Cu, Zn, and Pb, while in plants, the limit of Cu is the highest, followed by Pb, Zn, and Cd. Unlike in soil, Cu has the safest limits, followed by Pb and Zn, while Cd accumulation in plants is the most serious.

2.6.5.1 Effect of Heavy Metal Toxicity on Agricultural Soil

Heavy metals are considered a part of the soil; however, they cause severe damage to the soil and plants when they are highly concentrated. Therefore, they are assumed to be toxicants [2]. Chrastný et al. [22] studied the distance between the pollution source and contaminated soils around a mining area near Olkusz town, Upper Silesia, South Poland. The studied soils were agricultural and forest soils. Their results showed that both soils suffered from the smelting processes, but at various levels. The heavy metal Cd, Pb, and Zn concentrations were found to be 200, 25, and 20 $mg \cdot Kg^{-1}$, respectively. A study done by Raţiu et al. [23] investigated heavy metal accumulation and their different concentrations around the Tisza River and its tributaries. Their findings stated that the concentration of Cd, Pb, Cu, and Zn in different studied areas ranged from (1.3–21), (38 — 3630),

TABLE 2.2
Comparison between Heavy Metal Concentrations ($mg \cdot Kg^{-1}$) in Different Types of Fertilizers and Livestock Manure around the World and in the EU

Heavy Metals	P Fertilizers		N Fertilizers		Lime Fertilizers		Manure Fertilizers	
	Worldwide	EU	Worldwide	EU	Worldwide	EU	Worldwide	EU
Cd	0.1–170	13	0.05–8.5	0.9	0.04–.01	0.2	0.3–0.8	–
Pb	1.0–300	26	1.0–15	2.0	2.0–125	5.6	2.0–60	–
Cu	7.0–225	13	2–1450	1.9	20–1250	8.2	6.6–350	–
Zn	50–1450	236	1.0–42	5.0	10–450	22	15–250	–

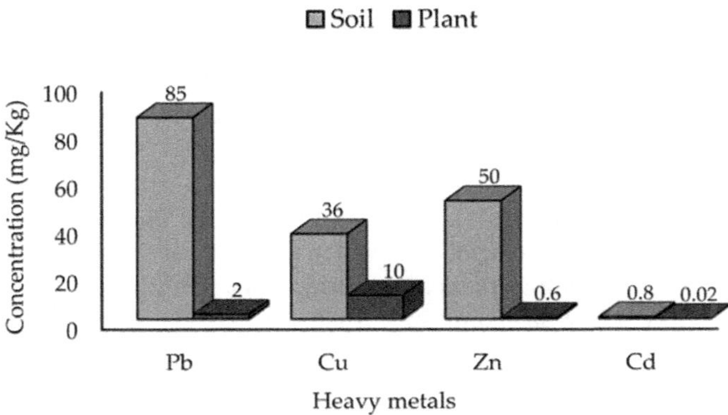

FIGURE 2.3 Permissible limits for concentrations of heavy metals in soil and plants (modified after WHO [21])

(54 − 4850), (200 − 770) mg·Kg^{-1}, respectively. These results revealed that the concentrations of the studied metals exceeded the permissible limits, indicating their toxicity. Low availability of macro-nutrients and soil acidity are among the main problems associated with accumulative heavy metal toxicity. These problems need to be overcome for the success of the phytostabilization process to remediate the contaminated soils. A comprehensive study was done by Sutkowska et al. [24], who assessed the content of heavy metals and their source apportionment in the Upper Silesia industrial region, southern Poland. They used different developed pollution indicators to determine the heavy metal content in various soil layers. Their results stated that the concentration of heavy metal was in the following order from highest to lowest: Pb > Cd > Zn in the shallow layers, while in the deep layers, it was Zn > Cd > Pb. The concentration of all metals exceeded the geochemical background levels, indicating the high toxicity of the studied area.

2.6.5.2 Effect of Cadmium Toxicity on Agricultural Soil

Among the heavy metals, Cd accumulation in the soil is a ubiquitous problem because of the advanced agriculture technology, economic revolution, and industry's rapid development. Usually, soil pH and the content of organic matter are the major factors affecting Cd accumulation. With the decrease of soil pH, Cd bioavailability increases, indicating a defect in soil properties. In 2015, Zeng et al. [25] studied the accumulation of heavy metals in three agricultural areas of Hunan province, China. The study investigated the content of heavy metals in each area. The results stated that Cd and Hg recorded the highest mean values among three tested areas (1.40 and 14.9 mg·Kg^{-1}, respectively), which exceeded the Chinese environmental quality standards for soil. In contrast, as a western country example, Chrastný et al. [26] studied the Cd isotope composition in Olkusz. The study was carried out for two meadows and three forest soil profiles influenced by various contamination sources. Their findings indicated that the upper soils in the forest soil profile revealed that Cd isotope compositions were the heaviest, while the lightest were in the deeper soil humus layer. A comprehensive study done by Sharma et al. [5] investigated the toxic effect of Cd on paddy soil properties. They used different kinetic and sigmoid dose-response models to determine ecological doses of Cd. Their results stated that Cd caused inhibitory effects on soil microbial activities, microbial growth, and microbial metabolic processes. Raiesi and Sadeghi [27] also studied the interactive consequences of Cd and salinity on soil microbes and enzymatic activities. Their findings reported the synergistic negative effect of Cd and salinity on soil properties. Moreover, their combined effect caused a reduction of microbial respiration and the content of microbial biomass in soil. Oumenskou et al. [28] measured the heavy metal contamination in agricultural soil using a GIS-based approach. Their results revealed that their concentration exceeded the limits set by

WHO and FAO. Cd is highly mobile in the soil, and consequently this resulted in high toxicity that affected the essential microorganisms, inhibited microbial activities, and absorbed the organic matter in the soil, as well as changing the physicochemical characteristics.

2.6.5.3 Effect of Lead Toxicity on Agricultural Soil

Pb has been listed as a hazardous heavy metal pollutant due to its high toxicity [2]. Long-term exposure to low concentrations of Pb leads to high toxic levels. The main source of Pb contamination in the soil is its geogenic contribution, which reduces soil microbial activities. There are several effects of Pb on soil, such as reducing soil nutrients, microbial diversity, and soil fertility. Furthermore, earthworms (*Eisenia fetida*) are usually affected by Pb toxicity, which may cause earthworm mortality. Reducing Pb bioavailability in the soil through phytoremediation or phytostabilization strategies is an important issue that should be a focus of concern. Many studies were done on the effect of Pb on agricultural soil in different geographical locations. Ijaz et al. [29] reported the impact of soil properties on absorption and retention of Pb. The results indicated that soil pH and cation exchange capacity were the important parameters influenced by Pb accumulation. In addition, Kumar et al. [30] stated that a negative correlation between Pb solubility and soil pH was found, indicating that the accumulation of Pb in the soil causes a defect in the plant absorption system from the soil. Pb also affects soil sorption capacity and humic acid content in the soil. Khan et al. [31] studied the single and joint effects of Pb and Cd on the soil microbial communities and some enzymatic activities. The findings showed that microbial communities were extremely affected by the contamination. In addition, the inhibition of enzyme activity has also been noted. The combined toxic effects of Pb and Cd were obvious on the number of bacteria and actinomycetes, which were notably reduced. Naturally, some essential soil factors control the mobility and bioavailability of Pb in the soil, such as soil pH, organic matter, ionic exchange capacity, and texture, which are affected by Pb accumulation in the soil.

2.6.5.4 Effect of Copper Toxicity on Agricultural Soil

Cu is an important micronutrient that is essential and necessary for plants. In addition, it is a significant element in the soil. Cu in supra-optimal levels may cause toxicity [2]. In agricultural soil, Cu availability is usually affected by several factors, such as soil pH, since its availability is usually higher in acidic soil than in alkaline and organic matter. The high accumulation rate of Cu in the soil is often due to the use of Cu-based fungicides or because of other agricultural activities. Naturally, the range of Cu concentration in agricultural soil is between 5 and 30 $mg{\cdot}Kg^{-1}$, but this level depends on the condition and the soil's location. As an example of a European country, Caetano et al. [32] studied the Cu toxicity in natural soil in Portugal; the ecotoxicological assessment reported a negative correlation between the Cu concentration and urease activity. Urease is one of the extracellular enzymes that break down the organic matter of soil. The results mentioned previously were consistent with the study done by Gülser et al. [33]. Numerous studies found that Cu toxicity could significantly inhibit soil microbial activities. Cu toxicity can also destroy cell membranes and cause protein denaturation in microbes. Wang et al. [34] studied the toxic effect of Cu on soil microorganisms and microbial biomasses. It was found that the microorganisms extremely affected by the toxicity were in the following order: bacteria > actinomycetes > fungi. A study done by Frenk et al. [35] demonstrated the negative effect of nanoscale Cu in the form of copper oxide (CuO) on the microbial groups of soil, such as Rhizobiales. Although the applied CuO concentration was only 1%, it caused a significant decrease in oxidation potential and changed the community formation.

2.6.5.5 Effect of Zinc Toxicity on Agricultural Soil

Zn is an important micronutrient, promoting plant growth hormones and proteins [2]. It has an active role in plants' metabolic and physiological processes since it is involved in sugar consumption. However, the danger of Zn toxicity is exhibited in its adverse effect on the soil microorganisms that contribute to enhancing soil fertility and structure [2]. Zn toxicity has a notable relationship with soil enzyme active sites, replacing certain cations that are crucial for cell performance. Moreover, Barman et al. [36] stated that Zn deficiency affects soil characteristics such as pH, the content of organic matter, and bicarbonate content and impedes the role of Mg and Fe in the soil.

2.6.6 Effect of Heavy Metal Toxicity on Plants

Naturally, plants require essential elements for growth. Although these trace elements are essential, exposure to heavy metals can severely damage plants. The effect of heavy metals on plants starts in the rhizosphere, where metalliferous minerals and substances interact with root exudates. Cabala and Teper [37] studied the characteristics of the rhizosphere soils polluted by Zn–Pb mining and thus their negative effect on plant roots. The carbonate formations transpiring on plant roots indicate vital oxidation and dissolution of minerals in the rhizospheres. These processes have been found to increase the metal ion concentrations in solutions of the rhizosphere. Cd toxicity causes a deficiency of minerals in plants [2]. A high concentration of Pb can cause different physiological and biochemical deficiencies. In addition, Cu and Zn interact with each other, affecting the bioavailability of nutrients in the soil. The pathway and mechanism of action of heavy metals, starting from accumulation in the soil passing through the plant uptake reaching different parts of the plant, are shown in Figure 2.4. Heavy metals produce free radicals, resulting in elevation of intracellular levels of ROS, causing oxidative stress, which causes damage to the biological molecules (e.g., proteins, nucleic acids, lipids, and enzymes). Defects in all these biological molecules cause many physiological problems, such as DNA, cell damage, and the inhibition of enzymatic activities, which may ultimately lead to the death of the entire plant.

2.6.6.1 Effect of Cadmium Toxicity on Plants

Cd is considered a non-essential and dangerous heavy metal, as it is found in the environment through anthropogenic sources, causing risks to the whole ecosystem [2]. According to mobility, bioavailability, and concentration, most Cd ions are absorbed by plant roots; however, the

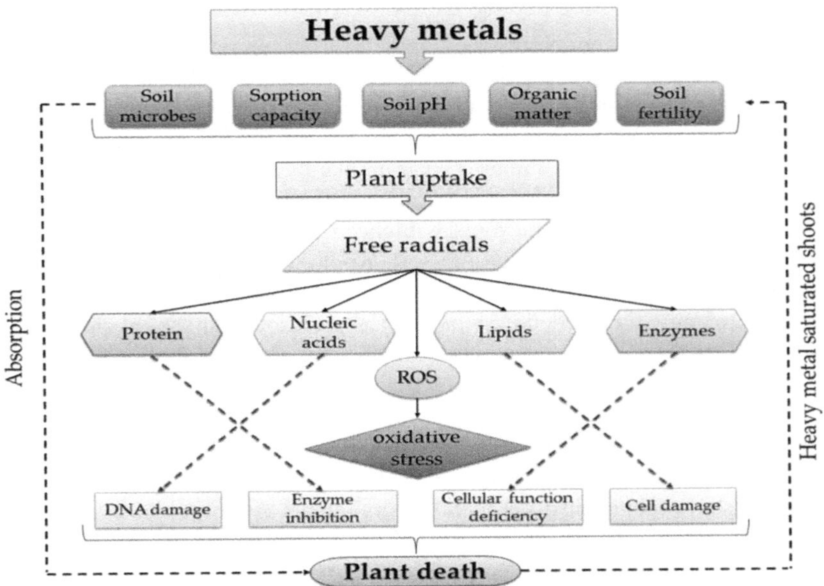

FIGURE 2.4 Mechanism of action and pathway of heavy metal toxicity in soil and plants.

remaining amount can be absorbed directly from the atmosphere. In addition, Cd enters plant cells via other transporters such as calcium (Ca) channels and accumulates in roots, shoots, and edible parts. A high concentration of Cd in plants causes many physiological and biochemical changes. Furthermore, Cd accumulation in plants results in toxic effects such as inhibition of some processes (minerals transportation, photosynthetic apparatus, and nutrient uptake). In addition, it can inhibit the transportation of Fe into plant shoots. Cd has a toxic influence on plant phenotype (e.g., reducing plant weight and length of roots and shoots), cytotoxicity (e.g., reducing chlorophyll content and inhibiting photosynthetic performance), and metabolic processes (e.g., chlorosis and cell damage). Seregin et al. [38] reported that xylem parenchyma was responsible for heavy metal translocation into conducting tissues; however, the transportation of Cd through the xylem was limited. Gratão et al. [39] stated that Cd toxicity could decrease the roots' dry mass and length. Recently, a study done by Vardhan et al. [40] investigated the ecological effect of Cd ions on *Lactuca sativa* seeds using *Pichia* sp. and *Azotobacterchroococcum*. The results revealed that Cd (II) inhibited seed germination, reduced the biomass, and hindered the growth and development of roots. In addition, the Cd (II) had a negative impact on microbial growth.

2.6.6.2 Effect of Lead Toxicity on Plants

Like Cd and mercury (Hg), Pb is not essential for plant growth. Pb is considered a useful and toxic metal at the same time. It has been categorized as a major pollutant due to its high toxicity [2]. Usually, Pb ions are transported from the soil to the plant via roots through the xylem. Pb toxicity is serious to plants even at low concentrations, as it obstructs healthy plant growth and reduces crop yield and productivity. The risk of Pb toxicity in plants is evident in reducing the nutrient uptake and deactivating the permeability of the cell membrane. Pb accumulation in plants causes physiological problems, such as DNA damage and destroying root and shoot systems, and affects the enzymatic activities. The effect of Pb toxicity on plants was studied by Ali et al. [41]. The results indicated that a high concentration of Pb affected the fresh biomass and plant growth. A study on soybean crops done by Hamid et al. [42] investigated the toxic effect of Pb on crop growth; the results obtained demonstrated a decrease in the chlorophyll content in the plant. Kushwaha et al. [43] demonstrated that Pb inhibited seed germination and decreased the protein content. They also noted that if the Pb level exceeds the critical threshold, morphological and physiological processes would be severely affected.

2.6.6.3 Effect of Copper Toxicity on Plants

In terms of worldwide consumption, Cu is ranked third after steel and aluminum. A small amount of Cu is a necessary element for plant nutrition and seed production. However, at high concentrations, Cu is considered a very toxic metal [2]. Cu uptake by the plant depends on several factors, such as physicochemical characteristics of the soil and other physiological parameters of the plant. Naturally, the Cu concentration in the roots is higher than in shoots because the root system is responsible for the Cu ions' uptake from the soil. Kopittke et al. [44] reported that the highest concentration of Cu in the roots was found in the root epidermis. A review article by Adrees et al. [45] stated that Cu toxicity led to a decrease in the crop yield, biosynthesis of chlorophyll, and plant productivity by modification of photosynthesis and nutrients. A concentration of 5 mg Kg^{-1} of Cu is sufficient to harm the plant, reducing plant growth and productivity.

2.6.6.4 Effect of Zinc Toxicity on Plants

Zn is a crucial micronutrient for all living organisms, including plants [2]. Zn is considered the second most readily available transition metal in organisms after Fe, and it has a strong relation to all enzymatic activities. Usually, Zn is transmitted from the soil as Zn^{2+} and enters the plant via roots. Zn plays a critical role in photosynthetic redox reactions. Accumulation of Zn in plant roots or shoots causes severe damage. Excessive Zn in plant cells causes high disruption in physiological processes in plants, followed by plant death. A previous study by Ebbs et al. [46] stated

TABLE 2.3
Heavy Metal Toxicity Forms and Their Toxic Effects on Soil and Plants

Heavy Metals	Toxicity Form	Toxic Effects	
		Soil	Plant
Cd	Cd^{+2}	Kill microorganisms, absorb organic matter, and change soil physicochemical characteristics.	Reduce biomass and root length, inhibit seed germination, and reduce stem conductivity.
Pb	Pb^{+2}	Change soil pH, affect soil sorption capacity, and reduce soil fertility.	DNA damage, decrease chlorophyll content, decrease protein content, and cause stunted foliage.
Cu	Cu salts	Change urease activity, affect microbial communities, and decrease oxidation potential.	Root deformation, decrease shoot length, reduce polypeptides, and change in lipid content.
Zn	Zn^{+2}	Change bicarbonate and organic matter content, inhibit enzymatic activity, and affect soil pH.	Variation in enzymatic activity, obstruction of elements transmission, and cause interveinal chlorosis.

that Zn toxicity led to chlorosis for younger leaves at the early stage of exposure, then reached the old leaves. Another study was done by Hammerschmitt et al. [47] on different types of plants to investigate the Zn toxicity on the young peach tree. The results demonstrated that the accumulation of Zn in the root system prevented the elements from transportation to the leaves. Moreover, the dry matter productivity was also negatively affected. A critical review done by Balafrej et al. [48] summarized the effect of Zn hyperaccumulation in plants and noted that all the physiological and biochemical mechanisms in plants were affected, indicating the harmful effect of Zn accumulation in plants. As a summary of the previously mentioned negative effects of heavy metals, the most common toxic effects of heavy metals (Cd, Pb, Cu, and Zn) on soil and plants are listed in Table 2.3.

REFERENCES

1. Hill MK. *Understanding Environmental Pollution.* Cambridge University Press; 2020.
2. Singh RP, Singh P, Srivastava A, editors. *Heavy Metal Toxicity: Environmental Concerns, Remediation and Opportunities.* Springer; 2023.
3. Xiao R, Wang S, Li R, Wang JJ, Zhang Z. Soil heavy metal contamination and health risks associated with artisanal gold mining in Tongguan, Shaanxi, China. *Ecotoxicol. Environ. Saf.* 2017;141:17–24. doi: 10.1016/j.ecoenv.2017.03.002. [PubMed] [CrossRef] [Google Scholar]
4. Tong S, Li H, Wang L, Tudi M, Yang L. Concentration, spatial distribution, contamination degree and human health risk assessment of heavy metals in urban soils across China between 2003 and 2019—A systematic review. *Int. J. Environ. Res. Public Health.* 2020;17:3099. doi: 10.3390/ijerph17093099. [PMC free article] [PubMed] [CrossRef] [Google Scholar]
5. Sharma RK, Agrawal M. Biological effects of heavy metals: An overview. *J. Environ. Biol.* 2005;26:301–313. [PubMed] [Google Scholar]
6. Filimon MN, Voia SO, Popescu ROXANA, Dumitrescu GABRIEL, Ciochina LP, Mituletu M, et al. The effect of some insecticides on soil microorganisms based on enzymatic and bacteriological analyses. *Romanian Biotechnological Letters.* 2015;20(3):10439–10447.
7. Al-Ani MA, Hmoshi RM, Kanaan IA, Thanoon AA. Effect of pesticides on soil microorganisms. *J Phys Conf Ser.* 2019 Sep;1294(7):072007. IOP Publishing.
8. Yousaf S, Yadav IC, Devi NL. Pesticides classification and its impact on human and environment. *Environ. Sci. Eng.* 2017;6:140–158. [Google Scholar]
9. Goswami RC, Huffling K, Sattler BP. Pesticides and health risks. *J. Obstet. Gynecol. Neonatal Nurs.* 2010;39:103–110. doi: 10.1111/j.1552-6909.2009.01092.x. [PubMed] [CrossRef] [Google Scholar]

10. Wang X, Liu W, Li Z, Teng Y, Christie P, Luo Y. Effects of long-term fertilizer applications on peanut yield and quality and plant and soil heavy metal accumulation. *Pedosphere*. 2020;30:555–562. doi: 10.1016/S1002-0160(17)60457-0. [CrossRef] [Google Scholar]
11. Arif M. Exploring different weeds management practices in wheat. *Pak J Weed Sci Res*. 30 June 2021;27(2):213–225.
12. Baćmaga M, Kucharski J, Wyszkowska J. Microbial and enzymatic activity of soil contaminated with azoxystrobin. *Environ Monit Assess*. 2015;187:1–15.
13. Saha, A. Dissipation and safety evaluation of tebuconazole residues in peanut-field ecosystem. *Proc Natl Acad Sci India Sect B Biol Sci*. 2017;87:753–760.
14. Baćmaga M, Kucharski J, Wyszkowska J. Microbial and enzymatic activity of soil contaminated with azoxystrobin. *Environ Monit Assess*. 2015;187:1–15.
15. Parween S, Creus A, Parrón T, Cebulska-Wasilewska A, Siffel C, Piperakis S, Marcos R. Biomonitoring of four European populations occupationally exposed to pesticides: Use of micronuclei as biomarkers. *Mutagenesis*.2003;18:249–258. doi: 10.1093/mutage/18.3.249. [PubMed] [CrossRef] [Google Scholar]
16. Sharma A, Kumar V, Shahzad B, Tanveer M, Sidhu GPS, Handa N, Kohli SK, Yadav P, Bali AS, Parihar RD. Worldwide pesticide usage and its impacts on ecosystem. *SN Appl. Sci*. 2019;1:1446. doi: 10.1007/s42452-019-1485-1. [CrossRef] [Google Scholar]
17. Kaya A, Doganlar ZB. Exogenous jasmonic acid induces stress tolerance in tobacco (Nicotiana tabacum) exposed to imazapic. *Ecotoxicol Environ Saf*. 2016;124:470–479.
18. Fernandes B, Soares C, Braga C, Rebotim A, Ferreira R, Ferreira J, et al. Ecotoxicological assessment of a glyphosate-based herbicide in cover plants: Medicago sativa L. as a model species. *Appl Sci*. 2020;10(15):5098.
19. Xia XJ, Wang YJ, Zhou YH, Tao Y, Mao WH, Shi K, et al. Reactive oxygen species are involved in brassinosteroid-induced stress tolerance in cucumber. *Plant Physiol*. 2009;150(2):801–814.
20. Ijaz M, Mahmood K, Honermeier B. Interactive role of fungicides and plant growth regulator (Trinexapac) on seed yield and oil quality of winter rapeseed. *Agronomy*. 2015;5(3):435–446.
21. World Health Organization (WHO). *Permissible Limits of Heavy Metals in Soil and Plants*. World Health Organization; 1996. [Google Scholar
22. Chrastný V, Vaněk A, Teper L, Cabala J, Procházka J, Pechar L, Drahota P, Penížek V, Komárek M, Novák M. Geochemical position of Pb, Zn and Cd in soils near the Olkusz mine/smelter, South Poland: Effects of land use, type of contamination and distance from pollution source. *Environ. Monit. Assess*. 2012;184:2517–2536. doi: 10.1007/s10661-011-2135-2. [PubMed] [CrossRef] [Google Scholar]
23. Raţiu IA, Beldean-Galea MS, Bocoş-Binţinţan V, Costea DD. Priority pollutants present in the tisza river hydrographic basin and their effects on living organisms. *Jordan J. Chem*. 2018;13:15–33. doi: 10.47014/13.1.28. [CrossRef] [Google Scholar]
24. Sutkowska K, Teper L, Czech T, Hulok T, Olszak M, Zogala J. Quality of peri-urban soil developed from ore-bearing carbonates: Heavy metal levels and source apportionment assessed using pollution indices. *Minerals*. 2020;10:1140. doi: 10.3390/min10121140. [CrossRef] [Google Scholar]
25. Zeng F, Wei W, Li M, Huang R, Yang F, Duan Y. Heavy metal contamination in rice-producing soils of Hunan province, China and potential health risks. *Int. J. Environ. Res. Public Health*. 2015;12:15584–15593. doi: 10.3390/ijerph121215005. [PMC free article] [PubMed] [CrossRef] [Google Scholar]
26. Chrastný V, Čadková E, Vaněk A, Teper L, Cabala J, Komárek M. Cadmium isotope fractionation within the soil profile complicates source identification in relation to Pb-Zn mining and smelting processes. *Chem. Geol*. 2015;405:1–9. doi: 10.1016/j.chemgeo.2015.04.002. [CrossRef] [Google Scholar]
27. Raiesi F, Sadeghi E. Interactive effect of salinity and cadmium toxicity on soil microbial properties and enzyme activities. *Ecotoxicol. Environ. Saf*. 2019;168:221–229. doi: 10.1016/j.ecoenv.2018.10.079. [PubMed] [CrossRef] [Google Scholar]
28. Oumenskou H, El Baghdadi M, Barakat A, Aquit M, Ennaji W, Karroum LA, Aadraoui M. Assessment of the heavy metal contamination using GIS-based approach and pollution indices in agricultural soils from Beni Amir irrigated perimeter, Tadla plain, Morocco. *Arab. J. Geosci*. 2018;11:692. doi: 10.1007/s12517-018-4021-5. [CrossRef] [Google Scholar]
29. Ijaz T, Ahmad F, Atif M, Ullah S, Waqar AB. Analyses of trace elements, arsenic (As), boron (B), lead (Pb), strontium (Sr), and zinc (Zn), in human body and their correlation with immune status against viral infections. *Pak J Med Health Sci*. 2023;17(04):332–332.
30. Kumar A, Kumar A, Cabral-Pinto M, Chaturvedi AK, Shabnam AA, Subrahmanyam G, Mondal R, Gupta DK, Malyan SK, Kumar SS, et al. Lead toxicity: Health hazards, influence on food Chain, and sustainable remediation approaches. *Int. J. Environ. Res. Public Health*. 2020;17:2179. doi: 10.3390/ijerph17072179. [PMC free article] [PubMed] [CrossRef] [Google Scholar]

31. Khan S, Hesham AEL, Qiao M, Rehman S, He JZ. Effects of Cd and Pb on soil microbial community structure and activities. *Environ. Sci. Pollut. Res.* 2010;17:288–296. doi: 10.1007/s11356-009-0134-4. [PubMed] [CrossRef] [Google Scholar]
32. Caetano AL, Marques CR, Gonçalves F, da Silva EF, Pereira R. Copper toxicity in a natural reference soil: Ecotoxicological data for the derivation of preliminary soil screening values. *Ecotoxicology.* 2016;25:163–177. doi: 10.1007/s10646-015-1577-7. [PubMed] [CrossRef] [Google Scholar]
33. Gülser F, Erdoğan E. The effects of heavy metal pollution on enzyme activities and basal soil respiration of roadside soils. *Environ. Monit. Assess.* 2008;145:127–133. doi: 10.1007/s10661-007-0022-7. [PubMed] [CrossRef] [Google Scholar]
34. Wang L, Xia X, Zhang W, Wang J, Zhu L, Wang J, Wei Z, Ahmad Z. Separate and joint eco-toxicological effects of sulfadimidine and copper on soil microbial biomasses and ammoxidation microorganisms abundances. *Chemosphere.* 2019;228:556–564. doi: 10.1016/j.chemosphere.2019.04.165. [PubMed] [CrossRef] [Google Scholar]
35. Frenk S, Ben-Moshe T, Dror I, Berkowitz B, Minz D. Effect of metal oxide nanoparticles on microbial community structure and function in two different soil types. *PLoS One.* 2013;8:e84441. doi: 10.1371/journal.pone.0084441. [PMC free article] [PubMed] [CrossRef] [Google Scholar]
36. Barman H, Das SK, Roy A. Zinc in soil environment for plant health and management strategy. *Univers. J. Agric. Res.* 2018;6:149–154. doi: 10.13189/ujar.2018.060501. [CrossRef] [Google Scholar]
37. Cabala J, Teper L. Metalliferous constituents of rhizosphere soils contaminated by Zn-Pb mining in southern Poland. *Water. Air. Soil Pollut.* 2007;178:351–362. doi: 10.1007/s11270-006-9203-1. [CrossRef] [Google Scholar]
38. Seregin IV, Kozhevnikova AD. Roles of root and shoot tissues in transport and accumulation of cadmium, lead, nickel, and strontium. *Russ. J. Plant Physiol.* 2008;55:1–22. doi: 10.1134/S1021443708010019. [CrossRef] [Google Scholar]
39. Gratão PL, Monteiro CC, Rossi ML, Martinelli AP, Peres LEP, Medici LO, Lea PJ, Azevedo RA. Differential ultrastructural changes in tomato hormonal mutants exposed to cadmium. *Environ. Exp. Bot.* 2009;67:387–394. doi: 10.1016/j.envexpbot.2009.06.017. [CrossRef] [Google Scholar]
40. Vardhan KH, Kumar PS, Panda RC. A review on heavy metal pollution, toxicity and remedial measures: Current trends and future perspectives. *J. Mol. Liq.* 2019;290:111197. doi: 10.1016/j.molliq.2019.111197. [CrossRef] [Google Scholar]
41. Ali M, Nas FS. The effect of lead on plants in terms of growing and biochemical parameters: A review. *MOJ Ecol. Environ. Sci.* 2018;3:265–268. doi: 10.15406/mojes.2018.03.00098. [CrossRef] [Google Scholar]
42. Hamid N, Bukhari N, Jawaid F. Physiological responses of Phaseolus vulgaris to different lead concentrations. *Pakistan J. Bot.* 2010;42:239–246. [Google Scholar]
43. Kushwaha A, Hans N, Kumar S, Rani R. A critical review on speciation, mobilization and toxicity of lead in soil-microbe-plant system and bioremediation strategies. *Ecotoxicol. Environ. Saf.* 2018;147:1035–1045. doi: 10.1016/j.ecoenv.2017.09.049. [PubMed] [CrossRef] [Google Scholar]
44. Kopittke PM, Menzies NW, de Jonge MD, McKenna BA, Donner E, Webb RI, Paterson DJ, Howard DL, Ryan CG, Glover CJ, et al. In situ distribution and speciation of toxic copper, nickel, and zinc in hydrated roots of cowpea. *Plant Physiol.* 2011;156:663–673. doi: 10.1104/pp.111.173716. [PMC free article] [PubMed] [CrossRef] [Google Scholar]
45. Adrees M, Ali S, Rizwan M, Ibrahim M, Abbas F, Farid M, Zia-ur-Rehman M, Irshad MK, Bharwana SA. The effect of excess copper on growth and physiology of important food crops: A review. *Environ. Sci. Pollut. Res.* 2015;22:8148–8162. doi: 10.1007/s11356-015-4496-5. [PubMed] [CrossRef] [Google Scholar]
46. Ebbs SD, Kochian LV. Toxicity of zinc and copper to brassica species: Implications for phytoremediation. *J. Environ. Qual.* 1997;26:776–781. doi: 10.2134/jeq1997.00472425002600030026x. [CrossRef] [Google Scholar]
47. Hammerschmitt RK, Tiecher TL, Facco DB, Silva LOS, Schwalbert R, Drescher GL, Trentin E, Somavilla LM, Kulmann MSS, Silva ICB, et al. Copper and zinc distribution and toxicity in 'Jade'/'Genovesa' young peach tree. *Sci. Hortic.* 2020;259:108763. doi: 10.1016/j.scienta.2019.108763. [CrossRef] [Google Scholar]
48. Balafrej H, Bogusz D, Triqui ZE, Guedira A, Bendaou N, Smouni A, Fahr M. Zinc hyperaccumulation in plants: A review. *Plants.* 2020;9(5):562.

3 Effect of Pesticides on Environment and Health

Sushila Arya, Sonu Kumar Mahawar, Himani Karakoti, Mozaniel Santana de Oliveira, Ravendra Kumar, and Om Prakash

3.1 INTRODUCTION

Pesticides have a significant role in crop production and the prevention of a variety of human diseases. Pesticides are not only beneficial, but they also pose a health risk to humans and animals due to their toxicity. These poisonous compounds can be found in soil, water, air, plants, food, and livestock feed. Pesticides break down into metabolites once they have been discharged into the environment. These metabolites enter the food chain through plants and animal products, eventually accumulating in humans and animals [1]. Toxic chemicals are ingested through water, air, and agricultural products, as well as their surroundings. They are stored in the tissues of humans and animals, suppressing the action of the brain, kidneys, skin, gastrointestinal, liver, lungs, and spleen, and excreted in various ways. They induce a variety of illnesses, tumors, mutations, and death. They affect human beings, animals, pollinators, honeybees, and soil microorganisms. Birds, wildlife, and soil microorganisms are also severely harmed by the excessive use of pesticides. Organochlorine, organophosphates, carbamate, pyrethroids, and neonicotinoids are some synthetic pesticides that are hazardous to beneficial organisms [2]. Pesticides are an important part of agricultural management since they help increase crop productivity and quality. Because these pesticides are easier, cheaper, and more successful at controlling pests, diseases, and weeds, they are widely utilized around the world. However, while pesticide use has enhanced crop yields, the widespread, nonselective, excessive, and improper use of these chemicals has negatively impacted the ecosystem, increasing toxicity and pollution.

3.2 EFFECTS OF PESTICIDES ON THE ENVIRONMENT

Pesticides have several impacts on environment, which are discussed in the following.

3.2.1 Soil Contamination

The use of pesticides in agriculture crop production for a long time may have adverse effects on soil microbiota and soil processes, which affects the soil nutrient cycles and crop production. Soil health is interlinked with soil biochemical health and soil enzyme activities, and they can be the most important quality indicators in pesticide-contaminated soils. The activity of soil enzymes, namely acid phosphatase, phosphatase, dehydrogenase, fluorescein diacetate hydrolase, β-glucosidase, urease, aryl-sulfatase, and alkaline phosphatase, is adversely affected by pesticidal actions. These enzymes play an important role in enhancing soil quality, managing nutrient cycles of carbon, nitrogen, sulfur and phosphorus (Table 3.1).

Beside the adverse effect of pesticide on soil enzymes, pesticides get converted into different transformation products when applied in soil, which can reduce the population of beneficial microorganisms in soil. For example, 2,4-D inhibits the growth of *Rhizobium*, which is an important bacterium responsible for fixing atmospheric nitrogen, Metsulfuron-methyl inhibits heterotrophic

DOI: 10.1201/9781003463429-5

TABLE 3.1
Effect of Different Classes of Pesticides on Soil Enzyme Activities [3]

Target Enzyme	Class	Pesticide	Enzyme Function
Dehydrogenase	Fungicide	Azoxystrobin, Benomyl, Captan, Carbendazim, Chlorothalonil, Mancozeb, Metalaxyl, Tebuconazole	Indicator of overall microbial activity of soils.
	Herbicide	Atrazine, Butachlor, Dinoterb, Diuron, Glyphosate, Metazachlor, Paraquat	
	Insecticide	Acetamiprid, Chlorpyrifos, Endosulfan, Fenamiphos, Quinalphos, Methyl parathion	
Fluoresceine di-acetate hydrolase	Herbicide	Dichlorophenoxyacetic acid, Glyphosate, Imazethapyr, Metsulfuron-methyl	Commonly used as an indicator of biological activity in soils
	Insecticide	Chlorpyrifos, Endosulfan, Ethion, Novaluron	
Acid and alkaline phosphatase	Fungicide	Mefenoxam, Metalaxyl, Pentachlorophenol, Validamycin	Critical role in phosphorus cycle
	Herbicide	Bromoxynil, Imazethapyr, Rimsulfuron, Aurora	
	Insecticide	Acetamiprid, Chlorpyrifos, Endosulfan	
β-glucosidase	Fungicide	Fluazinam, Metalaxyl, Procloraz, Trichlorophenol	Involved in the decomposition of organic matter in soil. The final product is glucose, an important carbon energy source for soil microorganisms
	Insecticide	Cadusaphos, Endosulfan, Ethion	
	Herbicide	Diflufenican, Glyphosate, Linuron, Metribuzin	
Cellulase	Fungicide	Fluaziram, Propiconazole	Involved in the decomposition of organic matter in soil.
	Herbicide	Bromoxynil, Linuron, Metribuzin	
	Insecticide	Monocrotophos, Profenofos, Endosulfan, Quinalfos, Cypermethrin	
Urease	Fungicide	Carbendazim, Chloramphenacol, Mancozeb, Metalaxyl, Validamycin	Catalyzes the hydrolysis of urea into carbon dioxide and ammonia and is a key component in the nitrogen cycle in soils.
	Herbicide	Aurora, Butachlor, Diuron, Diflufenican, Glyphosate	
	Insecticide	Fenamiphos	
Arylsulfatase	Fungicide	Fluaziram, Metalaxyl	Responsible for the hydrolysis of sulfate esters in the soil.
	Herbicide	Diflufenican, Glyphosate, Metribuzin	
	Insecticide	Cadusaphos, Endosulfan, Selectron	

S-oxidizing and S-reducing bacteria, and Diazinon and imidacloprid inhibit the growth of *Proteus vulgaris*, a urease-producing bacterium [4]. Microbial biomass is an important component of soil organic matter and has an important role in soil nutrient element cycle; thus, a declining population of beneficial microorganisms results in low fertility of soil and can also be responsible for poor production and productivity of crops.

3.2.2 Water Contamination

Pesticides can enter water bodies through surface runoff, draining, drift, and leaching. There are several factors which are responsible for pesticides contaminating water, such as [5]:

- The ability of the pesticide to dissolve in water
- Environmental factors such as soil, weather, season, and distance to water sources
- Application methods

Water-soluble pesticides dissolve in water and enter water bodies, causing harm to the untargeted species there. On the other hand, fat-soluble pesticides like DDT enter animal bodies via fatty tissue by a process known as bio-amplification and accumulate there. It is also known as bio-magnification. When an organism in the higher food chain consumes a lower organism containing such chemicals, the chemicals can accumulate in the higher organism. Pesticides are absorbed by the fatty tissues of animals, which results in the persistence of pesticides in food chains for long periods of time [6].

3.2.3 Air Contamination

Pesticide sprays can cause air pollution when suspended pesticide particles are carried by wind to other areas and potentially contaminate them. They can directly affect the non-target vegetation or can drift or volatilize from the treated area and contaminate air, soil, and non-targeted plants. Drift and volatilization of applied chemicals accounts for 2 to 25% loss of chemicals, which then spread from distances of a few yards to several hundred miles. This contamination has negative impacts on the non-targeted fauna and flora and disturbs the ecosystem [7].

3.3 EFFECT OF PESTICIDES ON BIODIVERSITY

3.3.1 Threats to Aquatic Biodiversity

Aquatic life faces significant threats from water contaminated with pesticides, posing risks to various species. The repercussions include adverse effects on aquatic plants, reduction of dissolved oxygen levels in the water, and the induction of physiological and behavioral changes in fish populations. The persistent and excessive application of insecticides and herbicides in water bodies has led to a notable decline in the populations of diverse aquatic organisms [8]. These contaminants enter the aquatic ecosystem through three primary pathways, whereby aquatic animals may absorb pesticides.

- Dermally: Direct absorption via skin
- Breathing: Uptake via gills during breathing
- Orally: Entry via drinking contaminated water

Aquatic lives mainly depend on aquatic plants for oxygen, as about 80% of total dissolved oxygen is provided by aquatic plants. Killing aquatic plants by the use of herbicides results in drastic reductions in O_2 levels, which in turn leads to suffocation of fish and ultimately to their death. A number of pesticides are responsible for affecting aquatic biodiversity; some are listed in Table 3.2.

TABLE 3.2
Pesticides Responsible for Threatening Aquatic Life

Pesticide	Threats to Aquatic Forms	References
Atrazine	Toxic to some fish species and also indirectly affects the immune system of some amphibians	[9]
Carbaryl	Toxic for several amphibian species	[10]
Glyphosate	Causes high mortality of tadpoles and juvenile frogs	[10]
Malathion	Changes the abundance and composition of plankton and periphyton population that consequently affects the growth of frog tadpoles	[11]
Chlorpyrifos and Endosulfan	Serious damage to amphibians	[12]

3.3.2 Threats to Terrestrial Biodiversity

In addition to destroying harmful pests and diseases, pesticides adversely affect non-targeted plants and beneficial insects. Beneficial insects, such as bees and beetles, that play a significant role in pollination are severely affected by the use of insecticides such as carbamates, organophosphates, pyrethroids, and neonicotinoids. These groups are harmful and toxic to honeybees, significantly reducing their population [13, 14]. Pesticide accumulation in the tissues of bird species can lead to their death. Fungicides, by killing earthworms, indirectly affect bird and mammal populations, as they feed on earthworms. Organophosphate insecticides are highly toxic to birds, and they are known to have poisoned raptors in the fields. Herbicides like glyphosate and chlorpyrifos have deleterious effects on earthworms at the cellular level, causing DNA damage [15]. Earthworms play an important role in the soil ecosystem. Earthworms act as bioindicators of soil contamination and as models for soil toxicity testing, contributing to soil fertility. Many other pesticides produce neurotoxic effects in earthworms, and after long-term exposure, they are physiologically damaged. Thus, these processes disrupt the whole ecosystem by eliminating any member of the food chain.

3.4 EFFECT OF PESTICIDES ON HUMAN HEALTH

Pesticides are ingested by humans through food, air, water, soil, flora, and fauna. Pesticides enter the human body through the dermal, oral, ocular, and respiratory systems. Pesticides are hazardous to all living species at specific concentrations. When they enter the human body, they inhibit enzyme function and interrupt the normal reactions in the body that are required for metabolism. The epidermis, gastrointestinal tract, central nervous system, respiratory, and reproductive systems are all affected by pesticides. Common pesticide effects are revealed in the skin, gastrointestinal tract, central nervous system, and respiratory and reproductive systems. Furthermore, significant pesticide exposure, whether occupational, accidental, or intentional, can cause severe intoxication, resulting in hospitalization and death [16]. Pesticide exposure has been associated with a variety of pathological conditions, including metabolic illnesses, neurotoxicity, immunological toxicity, endocrine disruption, reproductive problems, and cancers. Pesticide exposure can cause both acute and chronic disease, as shown in Table 3.3.

3.4.1 Chronic Toxicity

Pesticide chronic effects are typically lethal and have long-term consequences that affect multiple body organs. The mechanisms of action thought to produce chronic toxic effects comprise genotoxicity, endocrine disruption, and immunotoxicity. Chronic toxicity symptoms include:

- Neurological health effects (loss of coordination and memory, reduced visual ability, and reduced motor signaling) [18].
- Long-term pesticide exposure damages the immune system and causes hypersensitivity, asthma, and allergies.
- Leukemia, brain cancer, lymphoma, breast, prostate, ovaries, and testes cancer (2,4-D).
- Affects reproductive capabilities, for example, birth defects and infertility (Chlorpyrifos, DDE and DDD).
- Damages liver, lungs, and kidneys and may cause blood diseases and hormone disruptions (2, 4-D, lindane and atrazine).

3.5 EFFECT OF PESTICIDES ON VEGETATION

Pesticides are not only exposed to target plants, but they also contact non-target plant species. Pesticides influence sensitive plants at four stages of development: seedling, vegetative phase, seed

TABLE 3.3
Acute Toxicity of Pesticides [17]

Pesticide	Acute toxicity	Diagnosis	Treatment
Organochlorines (e.g. Lindane, Endosulfan)	GABA blockade: tremors, dizziness, seizures	Detectable in blood	Supportive care - Symptomatic Decontamination
Organophosphorus (Clorpyriphos, Diazinon, Azinphos, Parathion)	Irreversible cholinesterase inhibition	Low cholinesterase levels in red blood cells (RBCs)	- Supportive care - Atropine - Oximes Decontamination
Carbamates (Carbaryl, Aldicarb)	Reversible cholinesterase inhibition	Low cholinesterase levels (in RBC)	- Supportive care - Atropine - Decontamination
Pyrethroids (Allethrin, Permethrin, Cypermethrin, Fenvalerate)	Ataxia, irritability, temporary paresthesia, seizures	Urinary 3-phenoxybenzoic acid	Supportive care - Symptomatic Decontamination
Herbicides (2,4-D, paraquat)	Vomiting, corrosive lesions, hepatotoxicity, acute tubular necrosis	Detectable in urine and blood; dithionite test in urine	Decontamination and urine alkalinization, decontamination; avoid O_2 hemoperfusion
Rodenticides (Warfarin, Brodifacoum Diphacinone)	Hemorrhage (vit. K antagonism)	Elevated prothrombin time (PT)	Vitamin K1 (fitomenadione)

formation, and F1 generation stage [19]. Pesticides have a variety of sites of action and activities in plants. The following are some common pesticide modes of action:

- Photosynthesis, respiration, pigment production, and cell growth inhibitors, cell membrane disruptors, and fatty acid and amino acid biosynthesis disruptors.
- Epinasty, leaf mottling, withering, yellowing, leaf and stem twisting, necrosis, and bud deformation are all signs of epinasty.

3.6 EFFECT OF PESTICIDES ON ANIMALS

Pesticide toxicity has a significant impact on the decrease in the population of animals such as bees, birds, amphibians, fish, and small mammals in our environment. Animals' immune systems are weakened by these chemicals, which accumulate in fat. Animals exposed to atrazine have shown reproductive toxicity and sexual maturation delays. The toxicity of pesticides in various species is summarized in the following.

3.6.1 Birds

When an animal consumes food with DDT residue, the DDT accumulates in the tissues of the animal. The transfer of DDT from microscopic aquatic organisms to small fish occurs, which are later consumed by birds and other higher animals. The higher an animal is on the food chain, the greater the concentration of DDT in their body due to a process called biomagnification. The US National Academy of Sciences stated that DDT and its metabolite DDE cause eggshell thinning, leading to a decline in the bald eagle population in the United States. Rachel Carson's book *Silent Spring*

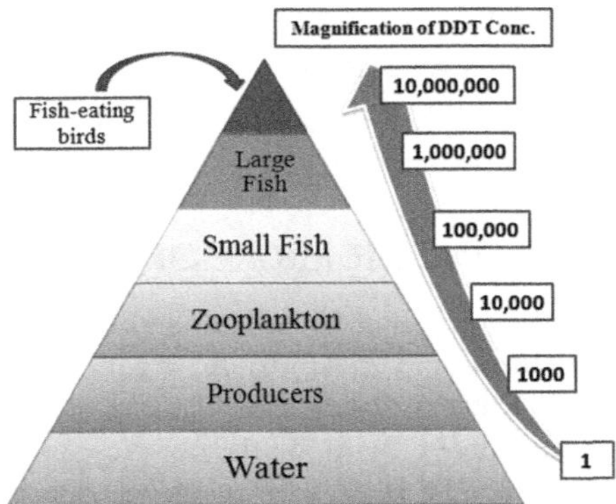

FIGURE 3.1 Biomagnification pyramid [20].

addressed the damage to bird species resulting from pesticide biomagnification. The use of secondary poisoning pesticides can make birds sick after feeding on an animal that is dead or dying from acute exposure to a pesticide. Herbicides may endanger bird populations by reducing the habitat, food, cover, and nesting sites necessary for bird populations, while insecticides may diminish insect populations that bird species feed on. The biological pyramid is shown in Figure 3.1.

3.6.2 Aquatic Biota

Pesticide-contaminated water may affect fish and other aquatic biota. Pesticides used in water bodies can kill vegetation, reducing oxygen levels in the water and suffocating fish. Some pesticides can cause physiological and behavioral changes in aquatic organisms that diminish population size, such as nest abandonment, disease immunity, and increased failure to evade predators, when exposed repeatedly. Pesticides can accumulate in water bodies to the extent that they destroy zooplankton and insects, which are the primary food source for young fish. The faster a given pesticide breaks down in the environment, the less threat it poses to aquatic life. Herbicides and fungicides are usually less hazardous to aquatic life than insecticides.

3.6.3 Amphibians

Pesticide mixtures appear to have a cumulative toxic effect on amphibians (e.g. frogs, toads, salamanders). Tadpoles in pesticide-contaminated ponds take longer to metamorphose, reducing their capacity to catch prey and avoid predators. When aquatic species are exposed to the organochloride insecticide (endosulfan) at high levels, it kills them and causes behavioral and growth problems. In aquatic reptiles and amphibians, the pesticide atrazine has both reproductive and nonreproductive effects. Crocodiles, several turtle species, and some lizards, depending on temperature, lack sex-distinct chromosomes until after fertilization during organogenesis. Pesticide exposure causes sex reversal, reduced hatching success, feminization, skin lesions, and other developmental problems in amphibians.

3.6.4 Pollinators

Pollinators are biotic agents that help in the pollination process of plants. The use of various pesticides in plants reduces pollinator colonies and contributes to toxicity, putting these valuable insects at risk. Bees are killed by neonicotinoids, which causes colony collapse disorder and damages the immune system of the insects [21]. Pesticides excite the neurological system, causing bees to lack coordination, become paralyzed, and eventually die. In honeybees, the fungicide captan can cause larval death and developmental deformity. Herbicides may interfere with pollinators' metabolic and reproductive functions.

3.7 EFFECT OF PESTICIDES ON SOIL MICROORGANISMS

Pesticides are transformed in soil by physical, chemical, and biological processes to control pests. Some pesticides are particularly hazardous to soil living organisms and have a variety of negative impacts on soil microbiological characteristics, nutrient cycling, enzymatic activities, and soil fertility. Buprofezin and Pentachlorophenol (PCP) inhibit the invertase enzyme and reduce soil microbial abundance [22]. These pollutants persist in soil for a long period and have a continual harmful impact on soil microbial flora. Earthworms and other soil microorganisms play an important role in the formation of soil. They have a major role in the breakdown of organic matter in the soil, as well as increasing soil fertility and contributing to soil formation. Pesticides in the soil are extremely toxic to them. These pesticides accumulate in food and increase the risk of toxicity for earthworms and soil-dwelling microorganisms.

3.8 EFFECT OF PESTICIDES ON AGRICULTURE (BENEFITS AND HAZARDS)

The use of pesticides in agriculture has several benefits, along with various consequences and hazardous impacts. According to Paracelsus, "All things are poison, and nothing is without poison; the dosage alone makes it so a thing is not a poison".

3.8.1 Benefits of Pesticide Use in Agriculture

The primary and foremost benefit of pesticide use is the improvement in yield and quality of agricultural produces. The major benefits of pesticide use in agriculture are as follows.

3.8.1.1 Productivity Improvement

As a result of green revaluation in India, food grain production was very high (198 MT) in 1996–97, which is almost four times higher than in 1948–49 (50 MT), from the established permanent cropped land area (196 Mha) [23]. This was achieved by a combination of factors, mainly high use of high-yielding varieties, advanced water management techniques, and increased use of agricultural chemicals, such as chemical fertilizers and pesticides [24]. Along with inputs, a drastic increase in agricultural outputs and productivity was also recorded in most other countries.

3.8.1.2 Quality Improvement

Management of insect pests and diseases at the right time helps in quality enhancement of agricultural products. Damaged food products have lower market values as compared to fresh quality products. Similarly, agricultural products from weed-free fields have better vigor and quality as compared to produce from weed-infested fields. Pesticides used in stored grain pest control also retain quality and market value by protecting the food grain from storage pests.

3.8.2 Consequences of Pesticide Use in Agriculture

Besides the benefits of pesticides, a number of consequences have been reported with respect to agriculture. In general, the persistent and toxic nature of pesticides along with injudicious use of them makes them responsible for their adverse impacts. The major consequences of chemical pesticides in agriculture are as follows.

3.8.2.1 Increased Production Cost

To achieve the expected profit and production from a given area, essential inputs play an important role. The application of pesticides not only adds the cost of pesticide formulations but also adds the cost of application (for manual operations as well as fuel for motor-operated applications). These costs are considered "internal costs" because these costs directly influence the benefit cost ratio to the farmers [25].

3.8.2.2 Ground Water Contamination

From the field, pesticides reach the ground water as well as other water resources via leaching, run-off seepage, and other mechanisms. Slow buildup of pesticide concentrations in water bodies is one of the major concerns related to pesticide use [26]. It was reported that injudicious use of pesticides was found to be responsible for the contamination of water resources in Italian forests, which led to several undesired harmful impacts on aquatic life [27].

3.8.2.3 Alteration in Physical, Chemical, and Biological Properties of Soil

Soil physical properties; chemical properties like soil pH, acidity, alkalinity, cation exchange capacity (CEC); and biological properties such as microbial population and activities and enzymatic activities are directly and indirectly influenced by long-term use of chemical pesticides. Several physical properties of soils are affected by the chemical and biological properties of soils, which are directly altered by the continuous use of pesticides with a similar chemical nature. Pesticides that are acidic in nature cause acidity in the soil, whereas alkaline pesticides lead to alkalinization of soils by alteration of soil pH. Also, the degradation of pesticides generates ionic metabolites in several cases and affects the CEC of soils depending on the nature of pesticides and their metabolites. Similarly, soil acidity and alkalinity also have an adverse impact on different types of microbial and enzymatic activities in soils. Soil microbes respond in different ways depending upon the type of chemical pesticide used in agricultural soils, based on a variety of parameters such as the pesticide's nature, soil qualities, and existing microbe groups in the soil. The total number of bacteria, fungi, protozoa, and algae can rise or decrease depending on the nature of the pesticide, such as its toxicity and potential as a nutrition or energy source. Pesticide use, on the other hand, has the potential to affect the overall structural and functional diversity of soil microbial communities [28]. For example, decreased microbial diversity and enzymatic activities (soil dehydrogenase, urease, β-glucosidase, phosphatase, and arylsulfatase) were reported by increasing doses of cypermethrin insecticide [29].

3.8.2.4 Problem of Resistance Development

According to IRAC (2013), "Resistance can be described as a heritable shift in a pest population's sensitivity, as evidenced by a product's repeated failure to achieve the expected degree of control when used according to the label guideline for that pest species". In a normal population, resistant individuals are uncommon, but non-judicious chemical usage can destroy typical vulnerable populations, giving resistant individuals a selective advantage in the presence of a pesticide. In the absence of competition, resistant individuals continue to grow and eventually form the majority of the population over generations. When the majority of a population is resistant to an insecticide, the pesticide loses its effectiveness, resulting in the emergence or development of insecticide resistance.

3.9 MANAGEMENT

The toughest task in the management of adverse impacts of chemical pesticides is not the lack of suitable approaches but ensuring their acceptance and implementation by farming communities and other pest control agencies and operators [30, 31]. Therefore, the foremost requirement in this respect is creating awareness and providing suitable training, knowledge, and demonstrations to growing communities and pest management operators. This can be done effectively by the involvement of government research originations, state agriculture universities (SAUs), line departments, non-governmental organizations, and other private organizations involved directly and indirectly in pest management strategies. Some of the important available and suitable methods and techniques or strategies for the management of adverse impacts of chemical pesticides are described here.

3.9.1 Optimum Use of Chemical Pesticides

The most important requirement in reduction of environmental contamination is the optimum use of pesticides. In this regard, awareness about the harmful impacts of chemical pesticides is needed in farming communities. Producers should apply optimum doses of chemical pesticides at the right time when needed. Apart from the time and doses of pesticides, their selection is also important. For example, producers should emphasize target-specific insecticides that are more toxic to target species and less toxic or nontoxic to natural enemies and other non-target organisms [31].

3.9.2 Adoption of Integrated Pest Management

The National Academy of Science defined integrated pest management (IPM) as "an ecological approach in pest management in which all available necessary techniques are consolidated in a unified program so that populations can be managed in such a manner that economic damage is avoided and adverse side effects are minimized" [32]. The goal of IPM is not only to mitigate the non-judicious use of chemical pesticides but also to replace unsafe chemicals with safe chemistries and techniques. IPM is a method of attaining long-term, environmentally sound pest control by the combination of a wide variety of techniques and other potential pest management strategies and methods. The components of IPM are as follows.

3.9.2.1 Cultural Methods

Cultural methods for pest control include crop rotation, intercropping, trap crops, refugee crops, border crops, soil manuring and nutrient management, resistant crop varieties, manipulation of sowing and harvesting dates, tillage operations, irrigation, and sanitation [33, 34]. These techniques help in manipulation of environmental conditions in such a way that the survival, growth and development of insect pests become less favorable.

3.9.2.2 Mechanical Methods

Mechanical methods of pest control include management of pests by direct removal with the help of mechanical means (manual devices or labor). Hand picking, hoeing and weeding, the use of different type of traps (such as banding, sticky trapes, pit fall traps), pruning or shaking of crop plants to remove or destroy insect pests and their egg masses, and the use of different types of barriers (such as screens, greenhouses, mulches, trenching, bunding, etc.) are different tools of mechanical methods of pest control in IPM [35].

3.9.2.3 Physical Methods

Physical treatments are used to manipulate pest populations by taking advantage of a physical property of the environment. Mechanical intervention, as well as varying temperatures, humidity levels, and even the environment, can be utilized to control pests. Physical interventions can be the most essential techniques of IPM in instances where the farmer has a high degree of control over the physical environment, such as greenhouses. These methods include soil solarization, burning, flaming, flooding, high levels of nitrogen and CO_2 in storage structures, and the use of light traps.

3.9.2.4 Biological Methods

Biological control helps in controlling pest species through natural control at low density over a long time span. In these methods, the action of biocontrol agents is explored and utilized in pest management such as the use of parasitoids, predators, or pathogens on a pest population. These options are highly selective and have rare negative impacts [36].

3.9.2.5 Chemical Methods

Among chemical methods, both natural and synthetic pesticides are used as a final tool in IPM to get immediate results. Uses of pesticides are considered the most effective and most reliable tool, with a broad spectrum of action [37]. It is important to know the pesticides should be used at right time, of right tyoe, at right dose and finally by the right method of application.

3.10 ALTERNATIVES TO CHEMICAL PESTICIDES

Many eco-friendly pest control agents such as insect growth regulators (IGRs), natural products, semiochemicals, and transgenic plants have been getting attention over chemical pesticides by viewing the negative impacts of chemical pesticides. IGRs are a novel class of chemicals with diverse modes of action such as juvenile hormone mimics, ecdysones, anti-hormones, and chitin inhibitors. Several such compounds are commercially available and are in use widely in the field as well as at the laboratory level against a variety of insect species. Natural products are chemical compounds obtained from natural sources such as animals, plants, and microorganisms such as phenolics and essential oils. Semiochemicals are also a kind of natural product such as allelochemicals and insect pheromones used to manipulate the behavior of insect pests and to manage their populations. The effect of pesticides on the environment and human health is shown in Figure 3.2.

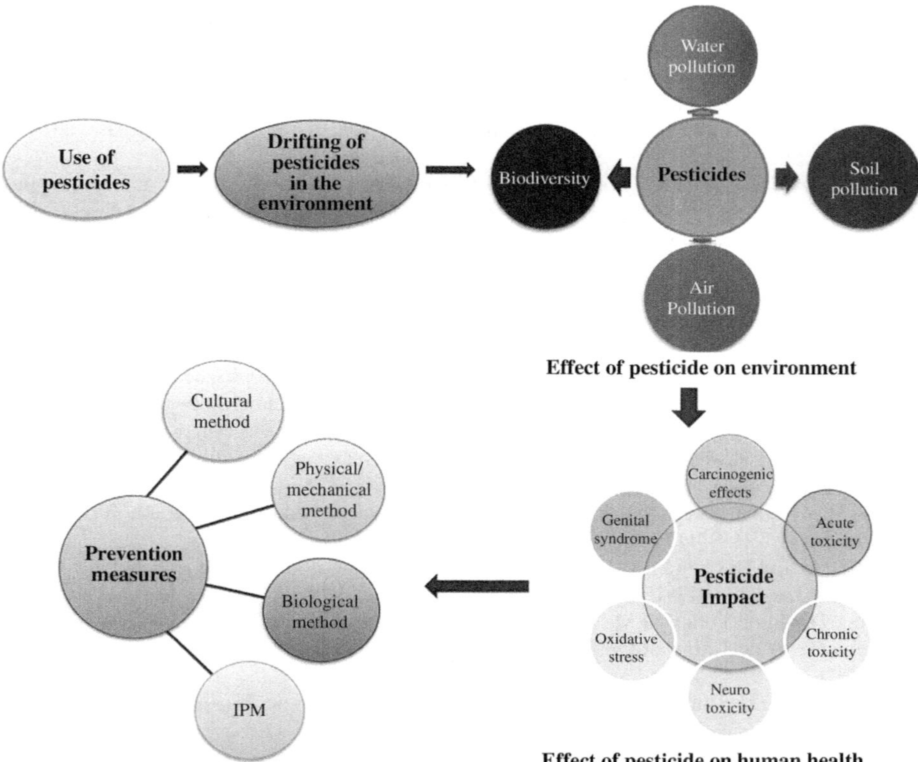

FIGURE 3.2 Effect of pesticides on environment and human health [17].

3.11 CONCLUSION

Pesticides serve the purpose of safeguarding plants from pests in agriculture and are frequently employed in farming practices, with their residues extending through the food chain from one organism to the next. However, these substances adversely impact beneficial organisms such as insects, birds, microbes, animals, and fish, leading to their mortality, slowed growth, and the onset of various diseases. Upon entering the human body, these toxicants affect vital organs such as the kidneys, lungs, skin, liver, cardiovascular system, and nervous system. Pesticide exposures, both acute and chronic, result in a range of poisoning effects. It is imperative that pesticide use be executed with care and guidance followed during application to mitigate the toxic effects on our environment. Adopting alternative cropping methods; maintaining spraying equipment diligently; promoting the adoption of innovative integrated pest management practices among farmers; and developing safer, environmentally friendly pesticide formulations are effective strategies to minimize the harmful consequences and undesirable effects of pesticide exposure. To reduce pesticide hazards, it is crucial to apply these substances in appropriate quantities and only when absolutely necessary. Farmers should be educated on the significance of reducing the use of harmful pesticides. Furthermore, there is a need to explore and develop new types of pesticides that do not pose serious harm to our environment. Overall, a cautious and informed approach to pesticide use is essential for safeguarding both agricultural productivity and environmental health.

REFERENCES

1. Qian, J., Shi, C. Wang, S. Song, Y. Fan, B. and Wu, X., 2018. Cloud-based system for rational use of pesticide to guarantee the source safety of traceable vegetables, *Food Control*, 87, 192–202.
2 .Miller, G.T. and Spoolman, S., 2016. Environmental science. (15th edition). *Cengage Learning*, 171, 232.
3. Riah, W., Laval, K., Laroche-Ajzenberg, E., Mougin, C., Latour, X. and Trinsoutrot-Gattin, I., 2014. Effects of pesticides on soil enzymes: A review. *Environmental Chemical Letters*, 12(2), 257–273.
4. Mandal, A., Sarkar, B., Mandal, S., Vithanage, M., Patra, A.K. and Manna, M.C., 2020. Impact of agrochemicals on soil health. In *Agrochemicals detection, treatment and remediation*. Elsevier, Oxford, pp. 161–187.
5. Srivastava, A., Jangid, N.K., Srivastava, M. and Rawat, V., 2019. Pesticides as water pollutants. In *Handbook of research on the adverse effects of pesticide pollution in aquatic ecosystems*. IGI Global, Hershey, Pennsylvania, pp. 1–19.
6. Mahmood, I., Imadi, S.R., Shazadi, K., Gul, A. and Hakeem, K.R., 2016. Effects of pesticides on environment. In *Plant, soil and microbes*. Springer, Cham, pp. 253–269.
7. Gyawali, K., 2018. Pesticide uses and its effects on public health and environment. *Journal of Health Promotion*, 6, 28–36.
8. Scholz, N.L., Fleishman, E., Brown, L., Werner, I., Johnson, M.L., Brooks, M.L. and Mitchelmore, C.L., 2012. A perspective on modern pesticides, pelagic fi sh declines, and unknown ecological resilience in highly managed ecosystems. *Biosciences*, 62(4), 428–434.
9. Forson, D.D. and Storfer, A., 2006. Atrazine increases Ranavirus susceptibility in the tiger salamander (*Ambystoma tigrinum*). *Ecological Applications*, 16, 2325–2332.
10. Relyea, R.A., 2005. The lethal impact of roundup on aquatic and terrestrial amphibians. *Ecological Applications*, 15, 1118–1124.
11. Relyea, R.A. and Hoverman, J.T., 2008. Interactive effects of predators and a pesticide on aquatic communities. *Oikos*, 117,1647–1658.
12. Sparling, D.W. and Feller, G.M., 2009. Toxicity of two insecticides to California, USA, anurans and its relevance to declining amphibian populations. *Environmental Toxicology and Chemistry*, 28(8), 1696–1703.
13. Pilling, E.D. and Jepson, P.C., 2006. Synergism between EBI fungicides and a pyrethroid insecticide in the honeybee (*Apis mellifera*). *Pesticide Science*, 39, 293–297.
14. Decourtye, A., Lacassie, E. and Pham-Delègue, M.H., 2003. Learning performances of honeybees (*Apis mellifera* L.) are differentially affected by imidacloprid according to the season. *Pest Management Science*, 59, 269–278.

15. Casabe, N., Piola, L., Fuchs, J., Oneto. M.L., Pamparato, L., Basack, S., Giménez, R., Massaro, R., Papa, J.C. and Kesten, E., 2007. Ecotoxicological assessment of the effects of glyphosate and chlorpyrifos in an Argentine soya field. *Journal of Soils and Sediments*, 7, 232–239.
16. Saeed, M.F., Shaheen, M., Ahmad, I., Zakir, A., Nadeem, M., Chishti, A.A, Shahid, M., Bakhsh, K. and Damalas, C.A., 2017. Pesticide exposure in the local community of Vehari District in Pakistan: An assessment of knowledge and residues in human blood. *Science of the Total Environment*, 587, 137–144.
17. Rani, L., Thapa, K., Kanojia, N., Sharma, N., Singh, S., Grewal, A.S. and Kaushal, J., 2021. An extensive review on the consequences of chemical pesticides on human health and environment. *Journal of Cleaner Production*, 283, 124657.
18. Culliney, T.W., Pimentel, D. and Pimentel, M.H., 1992. Pesticides and natural toxicants in foods. *Agricultural Ecosystem Environmental*, 41, 297–320.
19. Isenring, R., 2010. *Pesticides and the loss of biodiversity: How intensive pesticide use affects wildlife populations and species diversity.* PAN Europe, London.
20. Liroff, R.A., 2000. Balancing risks of DDT and malaria in the global POPs treaty. *Pesticide Safety News*, 4, 3.
21. Martin, C., 2015. A re-examination of the pollinator crisis. Elsevier Ltd. *Current Biology*, 25, 811–815.
22. El-Naas, M.H., Mousa, H.A. and El Gamal, M., 2017. Microbial degradation of chlorophenols. In *Microbe-induced degradation of pesticides*. Springer, Cham, pp. 23–58.
23. Aktar, M.W., Sengupta, D. and Chowdhury, A., 2009. Impact of pesticides use in agriculture: their benefits and hazards. *Interdisciplinary Toxicology*, 2(1), 1.
24. Employment Information, 1996. *Indian labour statistics*. Labour Bureau, Ministry of Labour, Chandigarh.
25. Bourguet, D. and Guillemaud, T., 2016. The hidden and external costs of pesticide use. *Sustainable Agriculture Reviews*, 35–120.
26. Pérez-Lucas, G., Vela, N., El Aatik, A. and Navarro, S., 2019. Environmental risk of groundwater pollution by pesticide leaching through the soil profile. In *Pesticides-use and misuse and their impact in the environment*, IntechOpen, pp. 1–28.
27. Trevisan, M., Montepiani, C., Ragozza, L. and Bartoletti, C., Ioannilli, E., Del Re, A.A., 1993. Pesticides in rainfall and air in Italy. *Environmental Pollution*, 80(1), 31–39.
28. Tripathi, S., Srivastava, P., Devi, R.S. and Bhadouria, R., 2020. Influence of synthetic fertilizers and pesticides on soil health and soil microbiology. In *Agrochemicals detection, treatment and remediation*, Elsevier, Oxford, pp. 25–54.
29. Tejada, M., García, C., Hernández, T. and Gómez, I., 2015. Response of soil microbial activity and biodiversity in soils polluted with different concentrations of cypermethrin insecticide. Arch. Environ. *Contamination Toxicology*, 69(1), 8–19.
30. Denholm, I., Birnie, L.C., Kennedy, P.J., Shaw, K.E., Perry, J.N. and Powell, W., 1998. *The complementary roles of laboratory and field testing in ecotoxicological risk assessment. The 1998Brighton Conferences: Pest and Diseases Volume 2*. British Crop Protection Council, Farham, UK, pp. 583–590.
31. Dhaliwal, G.S., Singh, R. and Chhillar, B.S., 2006. *Essentials of agricultural entomology.* KalyaniPublishers, New Delhi, India.
32. NAS, 1969. *Principles of plant and animal pest control, Vol. 3. Insect management and control.* National Academy of Sciences, Washington, DC.
33. Bazdyrev, G. I., 2000. Crop rotations and intercropping as a way of weed control. *Zashchitai Karantin Rasteniĭ*, 10.
34. Gahukar, R.T., 2009. Pest management in cotton: strategy and tools of IPM. *International Journal of Agricultural Sciences*, 5, 307–310.
35. Karuppuchamy, P. and Venugopal, S., 2016. Integrated pest management. In *Ecofriendly pest management for food security.* Academic Press, Cambridge, MA, pp. 651–684.
36. Arya, S., Kumar, R., Prakash, O., Rawat, A. and Pant, A. K., 2022. Impact of insecticides on soil and environment and their management strategies. In *Agrochemicals in soil and environment: Impacts and remediation*. Springer Nature, Singapore, pp. 213–230.
37. Bale, J.S. and Van Lenteren, J.C., 2008. Bigler F. Biological control and sustainable food production. *Philosophical Transactions of the Royal Society B, Biological Sciences*, 363(1492), 761–776.

4 Natural and Synthetic Chemical Products

Protection and Safety, Toxicity in Crops

Marcilene Paiva da Silva, Karyme do Socorro de Souza Vilhena, Fábio José Bonfim Cardoso, Oberdan Oliveira Ferreira, Leonardo Souza da Costa, Fernanda Wariss Figueiredo Bezerra, Mozaniel Santana de Oliveira, and Eloisa Helena de Aguiar Andrade

4.1 INTRODUCTION

In order to maximize food production in response to population demand, there has been notable growth in the agrochemical market worldwide. This growth has resulted in the development and widespread acceptance of synthetic agrochemicals for pest management in food crops. At the same time, increasingly stringent environmental, toxicological, and regulatory factors are further limiting the types of products suitable as pest control options for crop practices. To overcome these challenges and meet the needs of food production, productivity and sustainability of agricultural practices must be improved by introducing new strategies such as the use of natural products in traditional crop management to reduce current impacts (Sparks, Hahn, and Garizi 2017; Kumar et al. 2021).

Just as natural products can be used to control plant cultivation practices, they can also be used to produce natural sources of medicine and health treatments. Among the synthetic products derived from plants is the discovery of the antimalarial compound quinine obtained from the bark of the cinchona tree (*Cinchona officinalis*) (Dagen 2020). Morphine obtained from the species *Paper sommiferum* has been used as an analgesic, replacing synthetic medicine that has serious side effects (Alamgir 2017a).

Other natural products can be obtained from microorganisms, such as bacteria, which are used to control agricultural pests and pathogens. Products such as genes or metabolites of these agents can be used to prevent damage to crop management. The use of natural pesticides is more advantageous than the use of traditional chemical pesticides, as they are environmentally friendly and their application protects cultivated plants from invasive pests that most often infect the production of an agricultural product (Kumar et al. 2021; Lengai and Muthomi 2018).

Several important natural products such as biopesticides are classified according to their sources of extraction and the type of molecule/compound used for their preparation (Kumar et al. 2021; Ruiu 2018). Active molecules/compounds isolated of microorganisms, including bacteria, fungi, and viruses attack specific pest species. The use of the bacterium *Bacillus thuringiensis* has led to the discovery of toxins and virulence factors that can serve as commercial biopesticide products (Ruiu 2018). The formulation of microbial pesticides in the study carried out by Adan et al. (2015) showed that the formulation of *Trichoderma harzianum* had a high level of activity against damping off of eggplant seedlings caused by the fungus *Sclerotium rolfsii*.

DOI: 10.1201/9781003463429-6

Just as microorganisms are used as bioinsecticides in pest control, essential oils can also be attractive alternatives to synthetic insecticides for controlling insect pests. Essential oils are derived from aromatic plants and contain a variety of bioactive chemicals that can act as natural insecticides, reducing the action of various insects in different ways, depending on the physiological characteristics of the insect species as well as the type of plant (Magierowicz, Górska-Drabik, and Golan 2020). *Melaleuca alternifolia* essential oil has the potential to combat early blight, one of the main diseases caused by the fungus *Alternaria solani* that affects tomato leaves (Hendges et al. 2021); *Helianthus anuus* (sunflower) essential oil, rich in α-pinene, also demonstrates antimicrobial potential (Liu et al. 2020). *d*-Limonene and myrcene, the main constituents of the natural product *Citrus sinensis* essential oil, were tested against a cereal leaf beetle (*Oulema melanopus*) on wheat, and a mortality of up to 85% was reported in larvae observed within 48 hours (Zarubova et al. 2015).

In this context, the objective of this research was to gather information on natural products that can be a better alternative to synthetic products because they are chemically viable and inexpensive. In addition, they can contribute through formulations in the production of bioactive products.

4.1.1 Chemical Products of Natural Origin

The human population has been growing steadily around the world, and with it, concerns about the protection and safety of food production have also increased. Billions of dollars are spent annually to try to protect crops from pests, especially insects. The cost of producing agricultural products has been increasing due to the high cost of pesticides, and one of the main associated problems is the loss of productivity. Furthermore, the unrestricted use of chemical herbicides and pesticides, such as glyphosate, in plantations has been generating serious environmental problems, among which the emergence of pests resistant to these products stands out, in addition to the contamination of soil and water with residues from these compounds (Chattopadhyay, Banerjee, and Mukherjee 2017; Sun et al. 2021).

In recent decades, research into the application of natural products as alternatives to synthetic pesticides and agrochemicals has been gaining prominence due to their lower contamination and sustainability. Botanical products such as extracts, essential oils, molecules of natural origin, and nanoformulations, among others, have proven to be environmentally friendly alternatives for controlling pests, insects, fungi, and also for maintaining food quality. Plants and microorganisms are among the main sources of secondary metabolites applicable in agriculture (Ogunnupebi et al. 2020).

Natural products, derived from plants and microorganisms, are an important source for the development of new substances that can be used to control insects, weeds, and pathogens. Natural products, as well as their semi-synthetic derivatives, and compounds inspired by natural products, which have the ability to imitate natural products and semi-synthetic derivatives, account for 17% of the substances applied for crop protection (Sparks and Bryant 2022). Sparks et al. (2023) report that 60% of the products applied in crop protection come from plants, followed by those from bacterial sources with 18%, those of fungal origin corresponding to 16%, and, finally, animal sources contributing with 6% of the natural products applied.

Products containing extracts, essential oils, molecules, and other inputs from plants or microorganisms have also been marketed as biofertilizers or biostimulants, helping plants grow. These include products such as Biosept 33SL (a product based on grapefruit extract), Alga-Fert (a product containing seaweed extract), Bioczos BR (a product containing garlic extract), and Vaxiplant SL (a product containing laminarins—polysaccharides belonging to the glucan class found in brown algae that have glucose units in their structure with β(1-3) and β(1-6) bonds in a linear chain and a branched chain) (Xuan Cuong 2020; Jamiołkowska 2020).

The United States Environmental Protection Agency classifies biopesticides derived from natural products into three classes of compounds, (1) biocontrol organisms, which include bacteria, fungi, viruses, or protozoa as active ingredients; (2) plant-incorporated-protectants (PIPs) derived from transgenes that act in the production of substances that aid in pest control, such as the transgenic Bt toxin; and (3) biochemical pesticides that act through non-toxic mechanisms to control pests (Liu et al. 2021).

This section presents scientific information on natural products obtained from bacteria, fungi, and plants that are applied for the protection, safety, and toxicity of crops. The aim is to provide a general overview of the subject, including information on biological properties related to the safety and protection of crops reported in the literature.

4.1.1.1 Products from Bacteria

Bacterial insecticides are an alternative to chemical pesticides due to their potential to reduce the risks associated with them in agricultural crops. Bacterial insecticides have the advantage of being less toxic; they are also designed to target only pests, small quantities are required, and they decompose quickly in the environment and help reduce chemical pesticide residues in the soil (Chattopadhyay, Banerjee, and Mukherjee 2017).

Microorganisms considered beneficial are those that have the ability to improve plant growth and health and are also used as biological control agents. Among the bacteria that have these characteristics are those belonging to the genera *Pseudomonas*, *Burkholderia*, and *Bacillus*. Among the species of bacteria that have been studied for having the potential to be used in the biological control of weeds are *Pseudomonas fluorescens* and *Xanthomonas campestris*, whose studies indicate their use as bioherbicides (Tulu 2022).

Castaldi et al. (2021) report the potential as a biopesticide and the inhibitory activity of the substance phenanzine (Figure 4.1), which is an organic compound obtained from *Pseudomonas fluorescens*, which demonstrates fungicidal activity against *Macrophomina phaseolina*, a pathogen dangerous to soybeans and other crops. The study reports that, in an *in vitro* assay, the compound phenazine (**1**) presented, at a concentration of 25 μg/mL, inhibitory antifungal activity against the growth of pathogenic fungi isolated from infected soybeans. The growth inhibition potential observed for the compound was 40% for *M. phaseolina*, 50% for *Cercospora nicotianae*, and, finally, 100% for *Colletotrichum truncatum*. Furthermore, the author reports that the variability observed between the inhibitory activity observed for the raw organic extract (75%) and the pure compound (40%) against *M. phaseolina* is probably related to the presence of other metabolites in the raw organic extract.

Another study involving strains of *Pseudomonas fluorescens* ZX by Wang et al. (2021) evaluated the activity of the volatile organic compounds (VOCs) produced and the pure individual components for controlling green mold in citrus fruits infected by *Penicillium digitatum*. The research revealed that at a concentration of 1×10^{10} cfu/mL, the VOCs produced in nutrient broth and agar inhibited conidial germination and mycelial growth of *P. digitatum* by 60%. In addition, the VOCs obtained from the bacterial fluid demonstrated an inhibitory effect on germination and mycelial growth of 75%. The author relates the observed activity to the presence of organosulfur compounds produced by *P. fluorescens* and further considers that at adequately low concentrations, organic acids appear to be promising in controlling green mold, preventing physiological diseases in the fruits. The author reports that among the VOCs produced by *Pseudomonas fluorescens* ZX, the

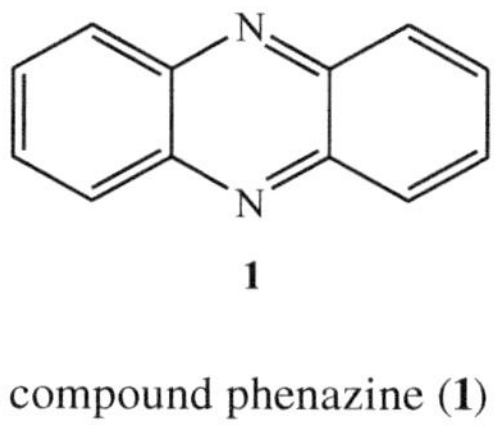

N
N
1

FIGURE 4.1 Molecular structure of the compound phenazine (**1**) obtained from *Pseudomonas fluorescens*. Prepared by the author.

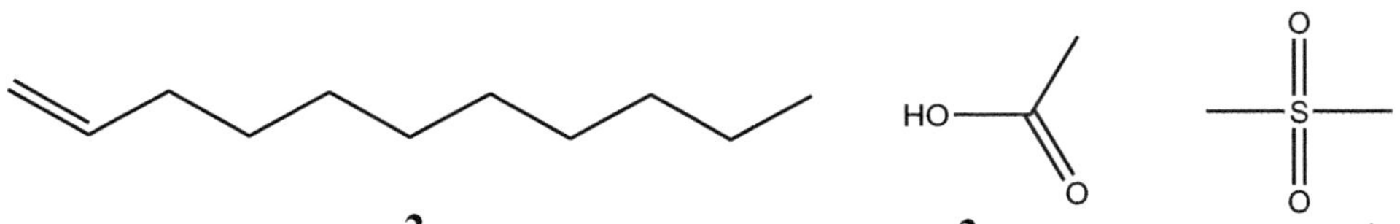

FIGURE 4.2 Molecular structure of the main volatile organic compounds produced by *Pseudomonas fluorescens* ZX: 1-undecene (**2**), methyl sulfone (**3**), and acetic acid (**4**). Prepared by the author.

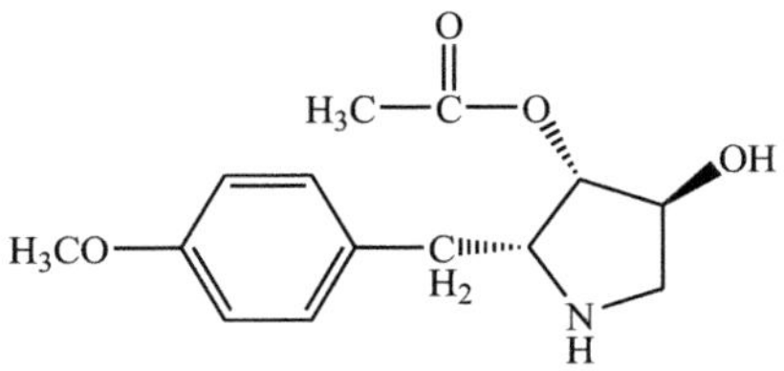

FIGURE 4.3 Molecular structure of the compound anisomycin (**5**) produced by *Streptomyces* sp. Prepared by the author.

main constituents are **1**-undecene (**2**–55.7%), methyl sulfone (**3**–16.6%), and acetic acid (**4**–14.1%). The molecular structure of the reported compounds is shown in Figure 4.2.

Another bacterium of interest for plant protection is from the genus *Streptomyces* sp., which produces a secondary metabolite known as anisomycin (**5**)—see Figure 4.3 (Ocán-Torres et al. 2024). Studies indicate that this compound has herbicidal activity against *Digitaria sanguinalis*, which is a competitive and annual weed that develops rapidly, is highly fecund, produces a large quantity of seeds, and infests the main crops in the world, and also against *Echinochloa crusgalli*, a harmful weed that infests rice fields and causes great losses in agricultural production (Yang et al. 2021; Zhao et al. 2023).

4.1.1.2 Products from Fungi

Among the strategies used in organic farming to control pests and insects is the development and use of mycoacaricide and mycoinsecticide products based on fungi, which appear as an alternative to toxic chemical insecticides. Studies have been carried out based especially on the genera *Beauveria* (Hypocreales: Cordycipitaceae) and *Metarhizium* (Hypocreales: Clavicipitaceae). The fungi of these genera are entomopathogenic; that is, they are parasitic microorganisms capable of infecting and killing arthropods. The filamentous fungus *Beauveria bassiana* (Balsamo-Crivelli) Vuillemin is an insect pathogen (especially in soil) that has a broad spectrum of arthropod hosts, is capable of infecting almost 1,000 species of insects, and is commonly found in arctic and tropical regions around the world. The primary function of these entomopathogenic fungi is the biocontrol of insect populations in the environment (Litwin, Nowak, and Różalska 2020; Pedrini 2022; Iida et al. 2023).

An et al. (2021) report that three main groups of metabolites can be found from *Beauveria* sp. strains, which are alkaloids (e.g. tenellin, bassiatin, pyridovercine, and pyridomacrolidine), cyclodepsipeptides (e.g. beauvericins, alobeauvericins, bassianolides, and beauveriolides), and benzoquinones (oosporein). Several of these compounds have proven biological activities such as insecticidal, anthelmintic, antifungal, antibacterial, and cytotoxic activity (An et al. 2021). Sabbour (2020) reports the insecticidal effect of beauvericin nanogel (2%) against the species *Sitophilus oryzae* and *Sitophilus granarius*, which are harmful insects that cause damage and loss of grain

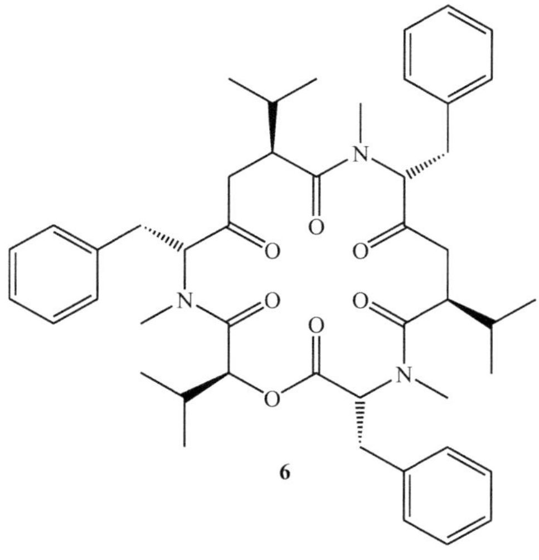

6

FIGURE 4.4 Molecular structure of the compound beauvericin (**6**) produced by *Beauveria* sp. Prepared by the author.

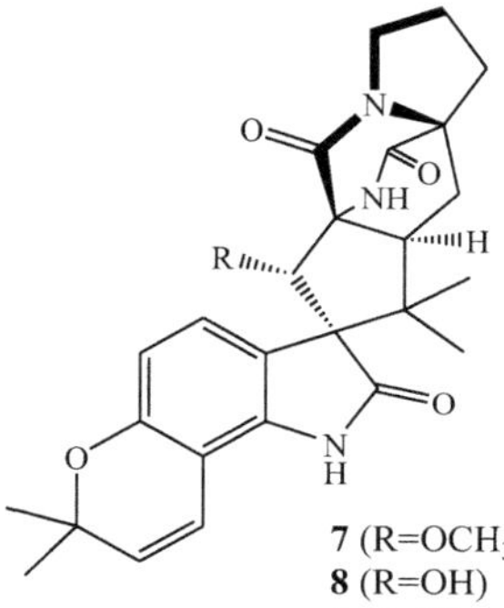

7 (R=OCH_3)
8 (R=OH)
9 (R=H)

FIGURE 4.5 Molecular structures of the compounds sclerotiamide B (**7**), sclerotiamide (**8**), and notoamide B (**9**) produced by *Fusarium sambucinum*. Prepared by the author.

storage. The activity of beauvericin nanogel (2%) was evaluated through persistence and repellency tests. The activity observed for beauvericin (**6**) nanogel demonstrated a cumulative effect dependent on the exposure time. The highest cumulative mortality rate was observed after 48 hours of exposure, and the LC_{50} value was 38 ppm. Furthermore, for the repellency test, beauvericin nanogel demonstrated superior activity to that of essential oils obtained from *Cinnamomum zeylancium* and *Cuminum cyminum*.

A study developed by Zhang et al. (2019) analyzed the larvicidal activity of prenylated indole alkaloids with pyrano [2,3-g] indole moieties obtained from *Fusarium sambucium* TE-6L hosted in *Nicotiana tabacum* L. The compounds sclerotiamide B (**7**), sclerotiamide (**8**) and notoamide B (**9**), whose structures are highlighted in Figure 4.5, demonstrated significant insecticidal potential with mortality rates against first instar larvae of *Helicoverpa armigera*, which is a cosmopolitan and polyphagous insect pest capable of causing great damage to various types of crops such as cotton, corn, sorghum, and soybean. The study used the substance matrine as a positive control, which obtained a mortality rate of 87.4% (Ahmed et al. 2023; Yuan et al. 2020).

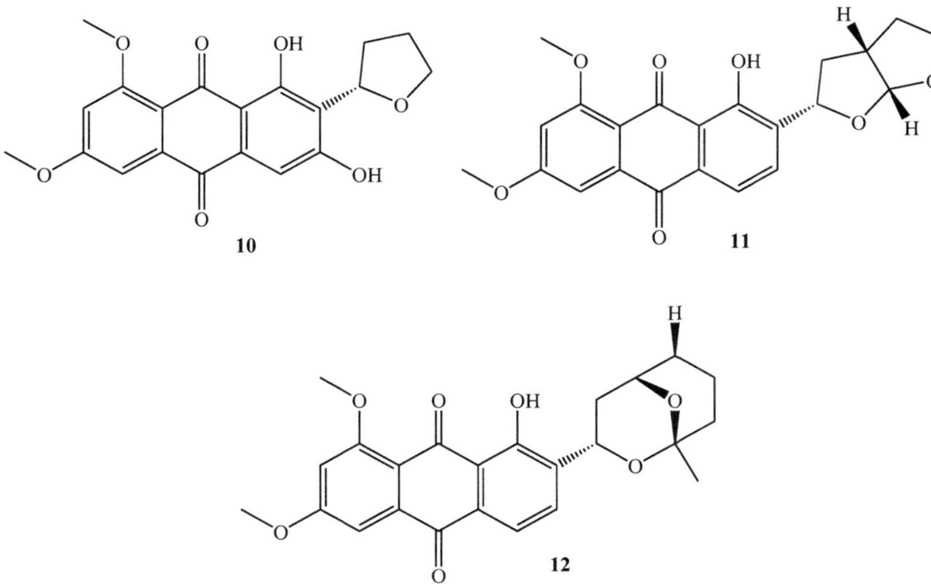

FIGURE 4.6 Molecular structures of the compounds 6,8-di-O-methylbipolarin (**10**), aversin (**11**) and 6,8-di-O-methylaverufin (**12**). Prepared by the author.

Anthraquinone derivatives also show insecticidal activity against *H. armigera*. Yuan et al. (2020) highlight the activity observed for three anthraquinone derivatives isolated from the endophytic fungus *Acremonium vitellinum*. The insecticidal assay was performed by applying the mixed-drug method to artificial insects. The formulations containing the tested compounds were added to the artificial diets, and the toxic food, at concentrations of 0.1, 0.2, 0.3 0.4, 0.6, 0.8, and 1 mg/mL, was placed in 24-well plates together with third-instar larvae of *H. armigera*. Among the three compounds evaluated (**10-12**), the compound 6,8-di-O-methylbipolarin (**10**) demonstrated the most pronounced larvicidal activity against third-instar larvae of *H. armigera* with an LC_{50} of 0.72 mg/mL, followed by the compound aversin (**11**) with an LC_{50} of 0.78 mg/mL and, finally, 6,8-di-O-methylaverufin (**12**), which demonstrated an LC_{50} of 0.87 mg/mL. The substance matrine, whose LC_{50} was 0.29 mg/mL, was used as a positive control.

Wang et al. (2022) studied the inhibitory activity of the ethanolic extract and compounds from the endophytic fungus *Penicillium oxalicum* HLLG-13 against the growth of newly hatched *H. armigera* larvae. The compounds tested were isolated from the ethanolic extract from the fermentation broth obtained from the fungus *P. oxalicum* HLLG-13 derived from *Lumnitzera littorea*. Among the compounds tested are two cyclopian diterpenes, the steroid andrastin H, a butanoate alkaloid, an aliphatic acid, and other compounds. Among the chemical compounds tested, the most significant results against the growth of newly hatched *H. armigera* larvae were demonstrated by the steroid andrastin H (**13**) with IC_{50} of 50 μg/mL, the diterpenes conidiogenone D (**14**) and conidiogenone C (**15**), and the steroids ergosterol (**16**) and β-sitosterol (**17**), which both presented IC_{50} values of 100 μg/mL. Azadirachtin with IC_{50} of 50 μg/mL was used as a positive control. The structures of the compounds mentioned are shown in Figure 4.7.

4.1.1.3 Products from Plants

Plant species are capable of producing and emitting compounds known as allelochemicals. These compounds have the ability to inhibit the reproduction, growth, and development of surrounding flora, acting especially against species considered weeds. Allelochemicals include substances such

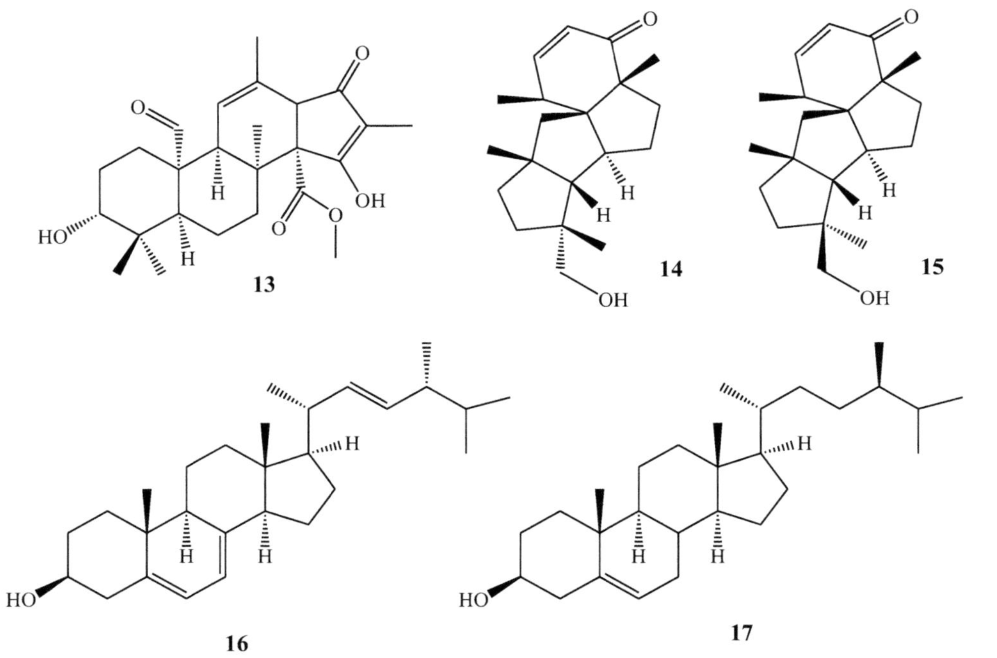

FIGURE 4.7 Molecular structures of the compounds andrastin H (**13**), conidiogenone D (**14**), conidiogenone C (**15**), ergosterol (**16**), and β-sitosterol (**17**) from the endophytic fungus *Penicillium oxalicum* HLLG-13. Prepared by the author.

as alcohols, fatty acids, phenolic compounds, flavonoids, terpenoids, and steroids. Bioherbicides produced from plant extracts and essential oils that contain these classes of substances have been shown to inhibit the growth of target weeds without harming crops (Ammar et al. 2023).

Among the biomolecules originating from plants that are used as a basis for the development of new biopesticides, with multiple forms of action against target pathogens, are naringin, naringenin, and hesperidin. Naringenin is a flavanone, while naringin and hesperidin are glycosidic flavanones, and these compounds are commonly found in *Citrus* species, especially in the epidermis of fruits such as lemon and orange (*Citrus* L.). Naringin is found mainly in the seeds and pulp of grapefruit—*Citrus paradisi* (Jamiołkowska 2020; Franceschini Sarria et al. 2022). In a study developed by Franceschini Sarria et al. (2022), the insecticidal activity of the substances naringin (**18**), hesperidin (**19**), and naringenin (**20**), whose structures are shown in Figure 4.8, as well as of the molecules complexed with copper (II), was evaluated against larvae of *Spodoptera frugiperda*, known as the fall armyworm. The results indicated that naringin and hesperidin, both in free forms, caused the shortening of the larval stage, when compared to the control. Complexed hesperidin demonstrated the greatest toxicity, causing 96.66% mortality of the larval stage of *S. frugiperda*.

Silva et al. (2020) evaluated the bioherbicidal potential of the essential oil obtained from *Eucalyptus saligna* Hook, as well as the fractions and major compounds 1,8-cineole and α-pinene. The inhibitory allelopathic effect on germination and seedling development was analyzed for the following species, some of which are considered invasive weeds in Algerian fields: *Lactuca sativa* (reference), *Amaranthus viridis* (invasive), *Eragrostis plana* (invasive), and *Paspalum notatum* (forage).

The essential oil was obtained from the leaves of *E. saligna* by hydrodistillation and presented as major constituents the oxygenated monoterpene 1,8-cineole (**21**, 32.5%) and α-pinene (**22**, 27.4%), a monoterpene hydrocarbon. The species most affected by the phytotoxic action of the essential

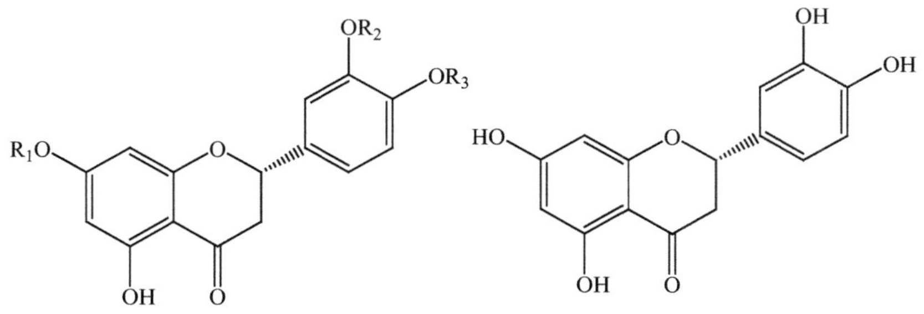

R_1 = Ramnosyl-α–1 → 2-glucose; $R_2 = R_3$ = H, **18**

R_1 = Ramnosyl-α–1 → 6-glucose; $R_2 = R_3$ = H, **19**

20

FIGURE 4.8 Molecular structures of the compounds naringin (**18**), hesperidin (**19**), and naringenin (**20**). Prepared by the author.

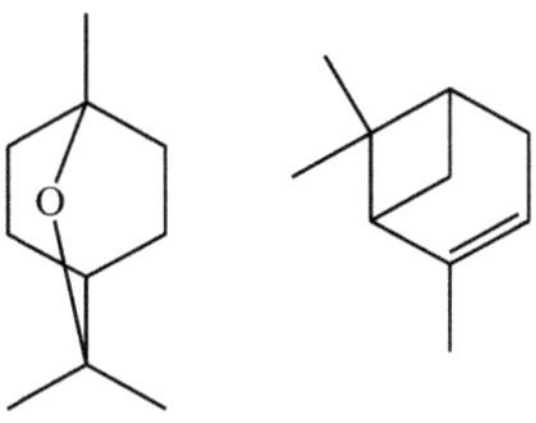

21 **22**

FIGURE 4.9 Molecular structures of the compounds 1,8-cineole (**21**) and α-pinene (**22**), major constituents of the essential oil of *E. saligna*. Prepared by the author.

oil of *E. saligna* were *L. sativa* and the invasive *A. viridis*, followed by the species *E. plana*, also invasive. Both the essential oil and the α-pinene compound were able to inhibit the development of the species *A. viridis*. The results indicate that the phytotoxic effect of the essential oil of *E. saligna* is specific to the tested species and indicates its potential for the control of the invasive species *A. viridis* and *E. plana*. Furthermore, it was observed that phytotoxicity is not directly associated with the presence of the major constituents but occurs due to factors of synergism and/or antagonism between the constituents of the essential oil.

The species *Albizia richardiana* (Voigt.) King & Prain has also been extensively studied due to the phytotoxic potential associated with extracts and compounds obtained from this species (Hossen, Iwasaki et al. 2021b, 2021a; Hossen, Ozaki et al. 2021). Hossen, Ozaki et al. (2021) isolated and evaluated the phytotoxic activity of three compounds obtained from the aqueous methanolic extract of *A. richardiana* against the growth of *Lepidium sativum* (watercress). Three compounds were isolated and tested, two norisoprenoids (**23** and **24**) and one homomonoterpene (**25**), Figure 4.10. The results observed indicated that the activity against seedling growth suppression was concentration dependent. Compound **1**, at a concentration of 0.452 mg/mL, showed a maximum inhibitory potential of 55.3% for the shoot and 81.4% for the root. Compound **2**, at 0.448 mg/mL, demonstrated a maximum inhibition potential of 76.9% and 83.2% for the shoot and root, respectively. Compound **3** showed maximum inhibition rates of 74.3% and 80.0% also for the shoot and root, consecutively, at a concentration of 0.365 mg/mL. Compound **3** was considered the most active among the three analyzed, and it significantly inhibited the growth of *L. sativum*. Therefore, the results suggest that the compounds may contribute to the phytotoxic and allelopathic potential of *A. ricardiana*.

As shown, compounds obtained from microorganisms such as fungi and bacteria, as well as those originating from plants, can present phytotoxic, insecticidal, or allelopathic activity and can

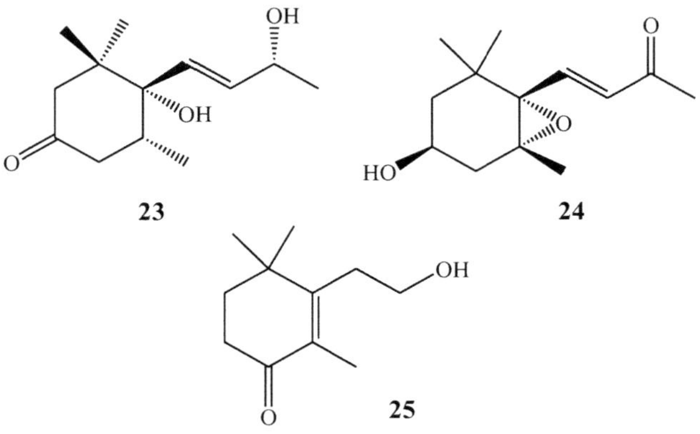

FIGURE 4.10 Molecular structures of the compounds 4,5-dihydrovomifoliol (**23**), 3-hydroxy-5α,6α-epoxy-β-ionone (**24**), and 3-(2-hydroxyethyl)-2,4,4-trimethyl-2cyclohexen-1-one (**25**) obtained from the species *Albizia richardiana*. Prepared by the author.

be applied in the protection and safety of crops without affecting or causing damage to the environment. The use of bioherbicides based on natural compounds emerges as a strategy for sustainable development, since the use of these compounds of natural and biodegradable origin helps in the elimination of weeds without harming the environment or crops. In addition, they target specific species, thus avoiding contamination of the environment by synthetic materials (Ammar et al. 2023).

4.1.2 Synthetic Products Derived from Plant Substances

Medicinal and aromatic plants have been used as a natural source of remedies and health treatments, especially those for ethnopharmocological purposes. This use has been practiced for many years through primitive formulations such as powders, tinctures, macerations, teas, infusions, percolation products, poultices, decoctions, tinctures, inhalations, and other herbal preparations. However, it is known that with the advent of new technologies developed, plants can be transformed into natural medical compounds for use in the production of pharmaceutical products, medicines, and biotechnological treatments (Chaachouay and Zidane 2024).

Among the advances made in synthetic products derived from plants is the discovery of the antimalarial compound quinine obtained from the bark of the cinchona tree (*Cinchona officinalis*). The discovery of this bioactive compound represented a major advance in the treatment of the disease; quinine remained an essential antimalarial drug for centuries until more contemporary therapies were developed (Dagen 2020).

Morphine is an alkaloid obtained from *Papaver somniferum* (Alamgir 2017b), which has been used as an analgesic drug that has serious side effects due to its dependence due to excessive use. This active compound has in its structure a pentacyclic nucleus with five chirals. In this sense, for modern organic chemistry, the synthesis of morphine and its related compounds continues to be an important task, as there is a need for the development of a new and effective method for the synthesis of this alkaloid (Rinner and Hudlicky 2012).

Another bioactive compound from the alkaloid class is berberine, a natural isoquinoline alkaloid isolated from a Chinese herb called *Rhizoma coptidis*. It has been widely used in Chinese medicine due to its great pharmacological effects as an antioxidant and acetylcholinesterase and butyrylcholinesterase inhibitor, it also inhibits monoamine oxidase and reduces cholesterol levels (Ji and Shen 2011).

In addition to alkaloids, there are other chemical classes that have been the target of research and synthesis of new products from plants, such as terpenes, which are compounds that have biological properties such as antiviral, antibacterial, antiseptic, and anti-inflammatory and contain active principles such as caryophyllene and valencene, farnesol, chamazulene, farnesene, limonene, pinene, camphene, cadinene, cedrene, dipentene, phellandrene, terpinene, sabinene, and myrcene (Alamgir 2017b).

Among these active principles, the natural bicyclic sesquiterpenes β-caryophyllene (BCP) and β-caryophyllene oxide (BCPO) stand out, which are present in many aromatic plants worldwide. β-Caryophyllene is a plant compound that occurs in nature mainly as trans-caryophyllene ((E)-BCP) mixed with small amounts of its isomers, (Z)-β-caryophyllene (iso-caryophyllene) and α-humulene (α-caryophyllene), as well as its oxidation derivative, β-caryophyllene oxide. These compounds are used to enhance the efficacy of classical drugs by increasing their concentrations inside cells, mainly in the treatment of cancer cells (Fidyt et al. 2016).

Another compound from the terpene class that stands out is eugenol, which is a compound belonging to the phenylpropanoid family with a pale-yellow coloration that is responsible for giving cloves their characteristic aroma and flavor. In addition to being a natural product, eugenol is also classified as an aromatic compound (Bisergaeva, Takaeva, and Sirieva 2021). In addition, studies show that the compound diaza-bicyclo-naphthalen-oxiranyl-methanone used in the treatment of prostate cancer is obtained from the synthesis of eugenol (Abdou et al. 2021).

The monoterpene compound α-pinene is found in plants of the genus *Juniper* ssp. and *Cannabis* ssp. This compound is a bicyclic hydrocarbon consisting of two isoprene units that is present in the essential oils of aromatic coniferous plants. Furthermore, this monoterpene is used for synthesis (Allenspach and Steuer 2021). Studies report that this compound is used in preclinical pharmacological activities as an antibiotic resistance modulator for *Campylobacter jejuni*, in which it acts in the modulation of antibacterial resistance and in the prevention of antimicrobial efflux (detected by the insertional mutagenesis method) (Salehi et al. 2019).

4.1.3 Natural Products Used as Active Ingredients for Protection and Safety in the Preparation of New Products

In recent decades, concerns about the harmful effects of agrochemicals on human health and the environment have attracted the attention of consumers, leading to a significant increase in the search for alternative products that are sustainable and can be used in crop protection (Ogunnupebi et al. 2020; Sparks, Sparks, and Duke 2023). In this context, research on natural products stands out, especially active compounds derived from plants with potential for biological control (Ogunnupebi et al. 2020).

Table 4.1 presents compounds of natural origin that have potential for the development of new crop protection products. Mohsen et al. (2018) tested lectins isolated and purified from seed extracts of the species *Phaseoulus vulgaris* and *Glycine max* against the phytopathogenic fungi *Fusarium oxysporum* and *Rhizoctonia solani*, with good results in the mycelial growth inhibition test. The fungus *Colletotrichum gloeosporioides* is one of the main causes of diseases in mango crops, including anthracnose; Zhang et al. (2014) evaluated eight compounds isolated from the root bark extract of *Pseudolarix amabilis* against the fungus, obtaining significant results in the mycelial inhibition test with the compounds Pseudolaric acid B and Pseudolaric acid A with EC_{50} (mg/L) of 1.07 and 1.62, respectively, values higher than the result found for the commercial fungicide carbendazim ($EC_{50} = 2.37$ mg/L).

Due to their chemical complexity, essential oils are valuable sources of active compounds that can be used due to their biological activities. The essential oil of *Melaleuca alternifolia* has the potential to combat early blight, one of the main diseases caused by the fungus *Alternaria solani* that affects tomato leaves (Hendges et al. 2021); the essential oil of *Helianthus anuus* (sunflower), rich in α-pinene, also demonstrates antimicrobial potential (Liu et al. 2020). In addition to plants, some microorganisms are also used as an alternative to the use of pesticides and synthetic fertilizers,

TABLE 4.1
Natural Products with Potential for use in Crop Protection

Product	Origin	Results	Reference
Lectins	Plants	Lectins isolated from *Phaseoulus vulgaries* and glycine max seeds demonstrated high inhibition capacity against the phytopathogenic fungi *Fusarium oxysporum* and *Rhizoctonia solani*, with EC_{50} (mg/L) of 5635 and 7781 for *G. max*, respectively; and 3671 and 5465 for *P. vulgaries*, respectively.	(Mohsen et al. 2018)
Pseudolaric acid B and pseudolaric acid A	Plants	Pseudolaric acid B and pseudolaric acid A isolated from the root bark extract of *Pseudolarix amabilis* against the fungus *Colletotrichum gloeosporioides* showed EC_{50} (mg/L) of 1.07 and 1.62, respectively.	(Zhang et al. 2014)
Garlic water extract (main component: hydroxymethylfurfural)	Plants	Twenty-two extracts from different plants were tested against phytopathogens that infest potato crops (*Solanum tuberosum*), with significant results obtained for the aqueous extract of *Allium sativum* (garlic) against *Alternaria solani* (MIC = 6.3 mg/mL) and *Phoma exigua* (MIC = 6.3 mg/mL).	(Steglińska et al. 2022)
Sunflower (*Helianthus anuus* L.)	Plants	Twenty-two extracts from different plants were tested against phytopathogens that infest potato crops (*Solanum tuberosum*), with significant results obtained for the aqueous extract of *Allium sativum* (garlic) against *Alternaria solani* (MIC = 6.3 mg/mL) and *Phoma exigua* (MIC = 6.3 mg/mL).	(Liu et al. 2020)
Tanacetum annuum	Plants	The methanolic extract and essential oil showed antifungal effect against *Fusarium oxysporum*, the fungus that causes Bayoud disease, with MIC (μL/mL) of 3.33 for both substances.	(Ettakifi et al. 2023)
Bacillus velezensis XT1	Microorganisms	The application of the bacteria resulted in the reduction of parameters associated with the infection of tomato and strawberry crops with *Botrytis cinerea*, with a 50% reduction in the incidence rate of the disease and 60% in the severity of the infection in both crops.	(Toral et al. 2020)
Bacillus thuringiensis	Microorganisms	Three bacteria of the species *Bacillus thuringiensis* (Bt) were tested against *Euprosterna elaeasa*, with the following results for the median lethal concentration (LC_{50}): ABTS-1857 var. *aizawai* (LC_{50} = 0.84 mg/mL), GC-91 var. *aizawai* (LC_{50} = 1.13 mg/mL), and HD-1 var. *kurstaki* (LC_{50} = 1.25 mg/mL).	(Plata-Rueda et al. 2020)

as they are capable of inhibiting the action of pathogens, especially bacteria of the genus *Bacillus* (Toral et al. 2020).

4.1.4 Toxicological Properties of Plant Sources

Plants have the natural ability to produce a variety of molecules, especially secondary metabolites that are known to play a role in protecting plants against pathogens due to their biological properties. Among these molecules, more than 3,000 are essential oils, which are complex mixtures consisting mainly of secondary metabolites (Raveau, Fontaine, and Lounès-Hadj Sahraoui

2020). These metabolites predominantly include monoterpenes and sesquiterpenes. Their chemical structures can be modified by oxidation, rearrangement of the chemical structure or by biosynthesis. In this case, specific subunits (alcohols, aldehydes, phenols, ethers, and ketones) or functional groups (sulfur or nitrogen) are attached or integrated into their structures (Cagáň et al. 2022).

To maximize food production to meet population demand, there has been a remarkable growth in the agrochemical market worldwide. This increase in demand has resulted in the development and widespread acceptance of synthetic agrochemicals for pest and weed management in crops. Plant protection products such as synthetic insecticides and herbicides have helped maintain and increase agricultural yields for a long time (Acheuk et al. 2022). However, pesticides can be toxic to other organisms, including birds, fish, beneficial insects, and non-target plants, as well as to air, water, soil, and crops. In addition, pesticide contamination travels away from target plants, resulting in environmental pollution. Thus, these chemical residues impact human health through environmental and food contamination (Tudi et al. 2021).

Compounds with insecticidal, repellent, and antifeedant properties derived from plants can be used as biopesticides and be an alternative to the use of synthetic pesticides. These biopesticides contain several secondary metabolites involved in plant-plant and plant-insect interactions, an action known as allelopathy that can affect the target organism in a positive or negative way. Plant-based products that have a negative effect on plant species or pests can be used to eliminate weeds and insects, but the stimulating effect of plant products on other plant species can be used as growth stimulants (Mežaka et al. 2023). In this scenario, essential oils (EOs) have attracted interest for use as bioinsecticides because they are renewable, natural, biodegradable, non-persistent in the environment, and safe for organisms that are not targets of their action and humans (Garrido-Miranda, Giraldo, and Schoebitz 2022).

Studies have shown that traditional communities in Plateau State, Nigeria, use at least 45 plant species to manage 15 different pests. The most common species are Euphorbiaceae, Fabaceae, and Malvaceae. The study revealed that 33 plant species are used as repellents, 15 as insecticides, 12 as fungicides, and 9 to control rodents. It was also observed that 11 plant species are used as insecticides and repellents and another seven as avicides and rodenticides. Thus, these results provide guidance for the development of stable biopesticides (Ali et al. 2022).

The toxicity of essential oils from the plant species *Annona neolaurifolia*, *Duguetia lanceolata*, and *Xylopia brasiliensis* were evaluated against the fall armyworm *Spodoptera frugiperda*, which is a polyphagous insect that is difficult to control due to its resistance to several active ingredients. The results showed that the most active essential oils against *S. frugiperda* are those extracted from the leaves and stem bark of *D. lanceolata*, with the major compounds being *β*-caryophyllene and caryophyllene oxide; see Figure 4.11. In the toxicity test of the essential oils, it was observed that the exempt oils presented rapid death after application. Thus, the essential oils of *D. lanceolata*,

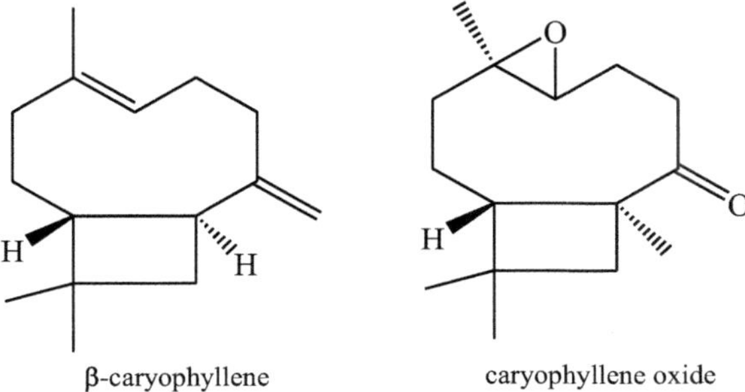

FIGURE 4.11 Structure of the substances *β*-caryophyllene and caryophyllene oxide found in the essential oils of the leaves and stem of *D. lanceolata*. Prepared by the author.

leaves and bark, may be potentially useful for the development of new products for the control of *S. frugiperda* (Paula Rosetti et al. 2023).

Antifungal properties of essential oils against phytopathogens have been reported in several studies; for example, the *in vitro* antifungal activities of vapors of cinnamon, fennel, oregano, and thyme essential oils were analyzed during *in vitro* conidial germination and mycelial growth of the fungus *Alternaria alternata*, a causative agent of tomato leaf spots. The results demonstrated that the vapors of cinnamon essential oil were more effective in suppressing conidial germination. It was also observed that the four oils presented similar antifungal activities against in vitro mycelial growth of *A. alternata*. There was a delay in conidial germination and in the elongation of germ tubes of *A. alternata* on tomato leaves when treated with vapors of cinnamon oil and oregano oil. Thus, these results indicate that essential oils can be used during the production of tomatoes in an ecological way (Hong et al. 2018).

In vitro tests demonstrated that mycelial growth and spore germination of *Botrytis cinerea* were strongly inhibited by *Origanum vulgare* essential oil, as well as by two of its main components, thymol and carvacrol, Figure 4.12. In *in vitro* vapor contact tests of the essential oil, it was observed that the antifungal activity of the oil at 250 mg/L was able to suppress the rotting of cherry tomatoes by 96.4%. In addition, thymol and carvacol, at 125 mg/L concentration, completely eliminated gray mold. According to these results, *O. vulgare* essential oil can be used as a non-toxic and environmentally friendly botanical fungicide for the control of postharvest gray mold (Zhao et al. 2021).

Essential oils are also used as insect attractants in pest control programs. Studies have shown that essential oils from the plant species *Coriandrum sativum* and *Nerium indicum* have a high capacity to attract adults and nymphs of *Cyrtorhinus lividipennis*, which is the main predator of the insect *Nilaparvata lugens*, which is a pest in rice cultivation. Olfactometer assays have identified that the compounds isocaryophyllene and trans-2-dodecenol, Figure 4.13, from essential oils, were the most effective attractants for *C. lividipennis* under laboratory conditions. Thus, these compounds have the potential to attract the predator (*C. lividipennis*) effectively, allowing it to locate the target rice fields during its artificial release and thus increase the efficiency of *N. lugens* biocontrol (Zhao et al. 2021).

In the field of sustainable agriculture, biocontrol has attracted considerable attention due to its focus on developing safe, biodegradable, and environmentally friendly alternatives to traditional biocides. In this sense, essential oils from Moroccan aromatic and medicinal plants (*Ammodaucus leucotrichus*, *Mentha spicata*, *Origanum majorana*, *Rosmarinus officinalis*, *Salvia officinalis*, and *Thymus satureioides*) were analyzed. Antifungal activity bioassays revealed that all essential oils tested reduced the mycelial growth of *Verticillium dahliae* strains, which is the causative agent of verticillium wilt, a disease that affects olive cultivation. These results open up new possibilities for exploiting these plant species for the development of natural biofungicides against *V. dahliae* in olive trees and other crops (Azenzem, Koussa, and Alfeddy 2024).

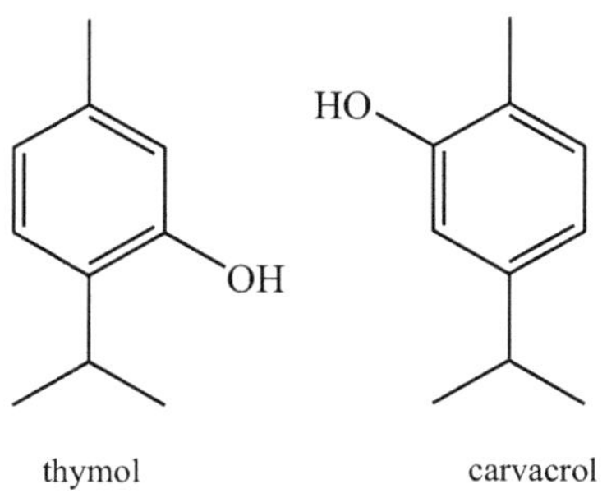

OH HO

thymol carvacrol

FIGURE 4.12 Structure of the substances thymol and carvacrol, oxygenated monoterpenes present in the essential oil of *Origanum vulgare*. Prepared by the author.

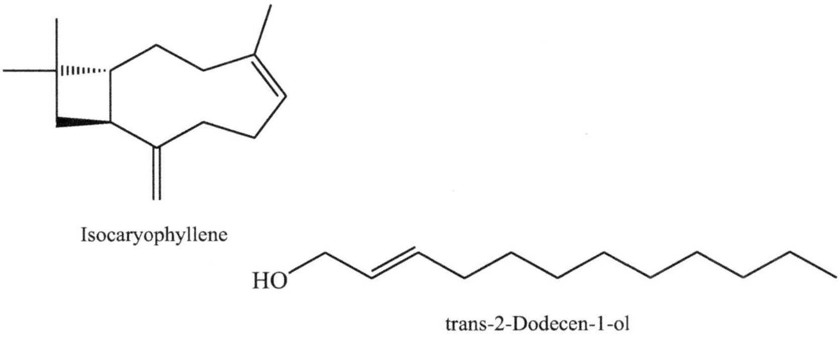

Isocaryophyllene

HO

trans-2-Dodecen-1-ol

FIGURE 4.13 Structure of the substance isocaryophyllene and trans-2-dodecen-1-ol, attractive to *Cyrtorhinus lividipennis*. Prepared by the author.

Studies of the *in vitro* biological activity of essential oils extracted from the plant species *Thymus vulgaris*, *Coriandrum sativum*, and *Cymbopogon martin*, originating from southwestern Ethiopia, demonstrated that all oils presented a high antifungal effect against *Aspergillus flavus*, *A. niger*, *Fusarium graminearum*, and *F. verticillioides*, inhibiting the growth and production of spores, which are pathogens that contaminate corn grains stored after harvest. Thus, oils from these plants may be an attractive option for use as biocides for the management of mycotoxigenic fungi associated with corn and thus for producing grains in a sustainable and environmentally friendly manner (Atnafu et al. 2024).

REFERENCES

Abdou, A., A. Elmakssoudi, A. El Amrani, J. JamalEddine, and M. Dakir. 2021. "Recent Advances in Chemical Reactivity and Biological Activities of Eugenol Derivatives." *Medicinal Chemistry Research* 30 (5): 1011–30. https://doi.org/10.1007/s00044-021-02712-x.

Acheuk, Fatma, Shereen Basiouni, Awad A Shehata, Katie Dick, Haifa Hajri, Salma Lasram, Mete Yilmaz, et al. 2022. "Status and Prospects of Botanical Biopesticides in Europe and Mediterranean Countries." *Biomolecules* 12 (2): 311. https://doi.org/10.3390/biom12020311.

Adan, Md. Jannatul, Md. Abdullahil Baque, Md. Mahfuzar Rahman, Md. Rafiqul Islam, and Afsana Jahan. 2015. "Formulation of Trichoderma Based Biopesticide for Controlling Damping off Pathogen of Eggplant Seedling." *Universal Journal of Agricultural Research* 3 (3): 106–13. https://doi.org/10.13189/ujar.2015.030305.

Ahmed, Arwa Mortada, Basma Khalaf Mahmoud, Natalie Millán-Aguiñaga, Usama Ramadan Abdelmohsen, and Mostafa Ahmed Fouad. 2023. "The Endophytic Fusarium Strains: A Treasure Trove of Natural Products." *RSC Advances* 13 (2): 1339–69. https://doi.org/10.1039/D2RA04126J.

Alamgir, A. N. M. 2017a. "Drugs: Their Natural, Synthetic, and Biosynthetic Sources." In, 105–23. https://doi.org/10.1007/978-3-319-63862-1_4.

Alamgir, A N M. 2017b. "Drugs: Their Natural, Synthetic, and Biosynthetic Sources BT—Therapeutic Use of Medicinal Plants and Their Extracts: Volume 1: Pharmacognosy." In, edited by A N M Alamgir, 105–23. Cham: Springer International Publishing. https://doi.org/10.1007/978-3-319-63862-1_4.

Ali, Ahmed Difa, Lydia Doosuur Ior, Goni Abraham Dogo, John Israila Joshua, and John Stephen Gushit. 2022. "Ethnobotanical Survey of Plants Used as Biopesticides by Indigenous People of Plateau State, Nigeria." *Diversity* 14 (10): 851. https://doi.org/10.3390/d14100851.

Allenspach, Martina, and Christian Steuer. 2021. "α-Pinene: A Never-Ending Story." *Phytochemistry* 190: 112857. https://doi.org/10.1016/j.phytochem.2021.112857.

Ammar, Esraa E., Sohaila A. Elmasry, Soumya Ghosh, Ammar Al-Farga, Youssef K. Ghallab, Abrar M. Fkr Eldeen, Nouran A. El-Shershaby, and Ahmed A. A. Aioub. 2023. "Insightful Review Of Bioherbicides Derived From Plants (Phyto-Herbicides)." *Journal of the Chilean Chemical Society* 68 (2): 5847–52. https://doi.org/10.4067/s0717-97072023000205847.

An, Ran, Maqsood Ahmed, Haiyan Li, Yanbin Wang, Aimin Zhang, Yuhui Bi, and Zhiguo Yu. 2021. "Isolation, Purification and Identification of Biological Compounds from Beauveria Sp. and Their Evaluation as Insecticidal Effectiveness against Bemisia Tabaci." *Scientific Reports* 11 (1): 12020. https://doi.org/10.1038/s41598-021-91574-9.

Atnafu, Birhane, Chemeda Abedeta, Fikre Lemessa, Abdi Mohammed, Safa Oufensou, and Alemayehu Chala. 2024. "Chemical Composition of Selected Aromatic Plant Essential Oils and Their Antifungal Efficacy against Toxigenic Fungi Associated with Maize (Zea Mays L)." *Cogent Food & Agriculture* 10 (1): 2329116. https://doi.org/10.1080/23311932.2024.2329116.

Azenzem, Rachid, Tayeb Koussa, and Mohamed Najib Alfeddy. 2024. "Biocontrol Potential of Essential Oils from Six Moroccan Plants against the Causal Agent of Verticillium Wilt of Olives." *Journal of Natural Pesticide Research* 9 (September): 100085. https://doi.org/10.1016/j.napere.2024.100085.

Bisergaeva, R A., M A. Takaeva, and Y. N. Sirieva. 2021. "Extraction of Eugenol, a Natural Product, and the Preparation of Eugenol Benzoate." *Journal of Physics: Conference Series* 1889 (2): 22085. https://doi.org/10.1088/1742-6596/1889/2/022085.

Cagáň, Ľudovít, Miroslava Apacsová Fusková, Daniela Hlávková, and Oxana Skoková Habuštová. 2022. "Essential Oils: Useful Tools in Storage-Pest Management." *Plants* 11 (22): 3077. https://doi.org/10.3390/plants11223077.

Castaldi, Stefany, Marco Masi, Francisco Sautua, Alessio Cimmino, Rachele Isticato, Marcelo Carmona, Angela Tuzi, and Antonio Evidente. 2021. "Pseudomonas Fluorescens Showing Antifungal Activity against Macrophomina Phaseolina, a Severe Pathogenic Fungus of Soybean, Produces Phenazine as the Main Active Metabolite." *Biomolecules* 11 (11): 1728. https://doi.org/10.3390/biom11111728.

Chaachouay, Noureddine, and Lahcen Zidane. 2024. "Plant-Derived Natural Products: A Source for Drug Discovery and Development." *Drugs and Drug Candidates*. https://doi.org/10.3390/ddc3010011.

Chattopadhyay, Pritam, Goutam Banerjee, and Sayantan Mukherjee. 2017. "Recent Trends of Modern Bacterial Insecticides for Pest Control Practice in Integrated Crop Management System." *3 Biotech* 7 (1): 60. https://doi.org/10.1007/s13205-017-0717-6.

Dagen, Morag. 2020. "History of Malaria and Its Treatment." In Graham L. Patrick (Ed.), *Antimalarial Agents*, 1–48. Amsterdam: Elsevier.

Ettakifi, Hajar, Kaoutar Abbassi, Safae Maouni, El Hadi Erbiai, Abderrahmane Rahmouni, Mounir Legssyer, Rabah Saidi, et al. 2023. "Chemical Characterization and Antifungal Activity of Blue Tansy (Tanacetum Annuum) Essential Oil and Crude Extracts against Fusarium Oxysporum f. Sp. Albedinis, an Agent Causing Bayoud Disease of Date Palm." *Antibiotics* 12 (9). https://doi.org/10.3390/antibiotics12091451.

Fidyt, Klaudyna, Anna Fiedorowicz, Leon Strządała, and Antoni Szumny. 2016. "Β-Caryophyllene and Β-Caryophyllene Oxide—Natural Compounds of Anticancer and Analgesic Properties." *Cancer Medicine* 5 (10): 3007–17. https://doi.org/10.1002/cam4.816.

Franceschini Sarria, André Lucio, Andréia Pereira Matos, Ana Carolina Volante, Antônio Rogério Bernardo, Gracielle Oliveira Sabbag Cunha, João Batista Fernandes, Moacir Rossi Forim, Paulo Cezar Vieira, and Maria Fátima das Graças Fernandes da Silva. 2022. "Insecticidal Activity of Copper (II) Complexes with Flavanone Derivatives." *Natural Product Research* 36 (5): 1342–45. https://doi.org/10.1080/14786419.2020.1868465.

Garrido-Miranda, Karla A, Juan D Giraldo, and Mauricio Schoebitz. 2022. "Essential Oils and Their Formulations for the Control of Curculionidae Pests." *Frontiers in Agronomy* 4 (June). https://doi.org/10.3389/fagro.2022.876687.

Hendges, Camila, José Renato Stangarlin, Vanessa Cristina Zamban, Márcia de Holanda Nozaki Mascaro, and Donizete Batista Carmelo. 2021. "Antifungal Activity and Control of the Early Blight in Tomato through Tea Tree Essential Oil." *Crop Protection* 148 (October): 105728. https://doi.org/10.1016/j.cropro.2021.105728.

Hong, Jeum Kyu, Yeon Sook Jo, Dong Hyun Ryoo, Ji Hwan Jung, Hyun Ji Kwon, Young Hee Lee, Seog Won Chang, and Chang-Jin Park. 2018. "Alternaria Spots in Tomato Leaves Differently Delayed by Four Plant Essential Oil Vapours." *Research in Plant Disease* 24 (4): 292–301. https://doi.org/10.5423/RPD.2018.24.4.292.

Hossen, Kawsar, Arihiro Iwasaki, Kiyotake Suenaga, and Hisashi Kato-Noguchi. 2021a. "Phytotoxic Activity and Growth Inhibitory Substances from Albizia Richardiana (Voigt.) King & Prain." *Applied Sciences* 11 (4): 1455. https://doi.org/10.3390/app11041455.

Hossen, Kawsar, Arihiro Iwasaki, Kiyotake Suenaga, and Hisashi Kato-Noguchi. 2021b. "Phytotoxicity of the Novel Compound 3-hydroxy-4-oxo-β-dehydroionol and Compound 3-oxo-α-ionone from Albizia Richardiana (Voigt.) King & Prain." *Environmental Technology & Innovation* 23 (August): 101779. https://doi.org/10.1016/j.eti.2021.101779.

Hossen, Kawsar, Kaori Ozaki, Toshiaki Teruya, and Hisashi Kato-Noguchi. 2021. "Three Active Phytotoxic Compounds from the Leaves of Albizia Richardiana (Voigt.) King and Prain for the Development of Bioherbicides to Control Weeds." *Cells* 10 (9): 2385. https://doi.org/10.3390/cells10092385.

Iida, Yuichiro, Yumiko Higashi, Oumi Nishi, Mariko Kouda, Kazuya Maeda, Kandai Yoshida, Shunsuke Asano, et al. 2023. "Entomopathogenic Fungus Beauveria Bassiana–Based Bioinsecticide Suppresses Severity of Powdery Mildews of Vegetables by Inducing the Plant Defense Responses." *Frontiers in Plant Science* 14 (August). https://doi.org/10.3389/fpls.2023.1211825.

Jamiołkowska, Agnieszka. 2020. "Natural Compounds as Elicitors of Plant Resistance Against Diseases and New Biocontrol Strategies." *Agronomy* 10 (2): 173. https://doi.org/10.3390/agronomy10020173.

Ji, Hong-Fang, and Liang Shen. 2011. "Berberine: A Potential Multipotent Natural Product to Combat Alzheimer's Disease." *Molecules*. https://doi.org/10.3390/molecules16086732.

Kumar, Jitendra, Ayyagari Ramlal, Dharmendra Mallick, and Vachaspati Mishra. 2021. "An Overview of Some Biopesticides and Their Importance in Plant Protection for Commercial Acceptance." *Plants* 10 (6): 1185. https://doi.org/10.3390/plants10061185.

Lengai, Geraldin M. W., and James W. Muthomi. 2018. "Biopesticides and Their Role in Sustainable Agricultural Production." *Journal of Biosciences and Medicines* 06 (06): 7–41. https://doi.org/10.4236/jbm.2018.66002.

Litwin, Anna, Monika Nowak, and Sylwia Różalska. 2020. "Entomopathogenic Fungi: Unconventional Applications." *Reviews in Environmental Science and Bio/Technology* 19 (1): 23–42. https://doi.org/10.1007/s11157-020-09525-1.

Liu, Xiaoman, Aocheng Cao, Dongdong Yan, Canbin Ouyang, Qiuxia Wang, and Yuan Li. 2021. "Overview of Mechanisms and Uses of Biopesticides." *International Journal of Pest Management* 67 (1): 65–72. https://doi.org/10.1080/09670874.2019.1664789.

Liu, Xin-Sheng, Bo Gao, Xin-Lu Li, Wan-Nan Li, Zi-An Qiao, and Lu Han. 2020. "Chemical Composition and Antimicrobial and Antioxidant Activities of Essential Oil of Sunflower (Helianthus Annuus L.) Receptacle." *Molecules* 25 (22): 5244. https://doi.org/10.3390/molecules25225244.

Magierowicz, Klaudia, Edyta Górska-Drabik, and Katarzyna Golan. 2020. "Effects of Plant Extracts and Essential Oils on the Behavior of Acrobasis Advenella (Zinck.) Caterpillars and Females." *Journal of Plant Diseases and Protection* 127 (1): 63–71. https://doi.org/10.1007/s41348-019-00275-z.

Mežaka, Ieva, Arta Kronberga, Marta Berga, Laura Kaļāne, Laura Pastare, Gundars Skudriņš, and Ilva Nakurte. 2023. "Biochemical and Physiological Responses of Cucumis Sativus L. to Application of Potential Bioinsecticides—Aqueous Carum Carvi L. Seed Distillation By-Product Based Extracts." *Agriculture* 13 (5): 1019. https://doi.org/10.3390/agriculture13051019.

Mohsen, Soad F. E., Moustafa A. Abbassy, Entsar I. Rabea, and Hamdy K. Abou-Taleb. 2018. "Isolation and Antifungal Activity of Plant Lectins against Some Plant Pathogenic Fungi." *Alexandria Science Exchange Journal* 39 (1): 161–67. https://doi.org/10.21608/asejaiqjsae.2018.5872.

Ocán-Torres, Diego, Walter José Martínez-Burgos, Maria Clara Manzoki, Vanete Thomaz Soccol, Carlos José Dalmas Neto, and Carlos Ricardo Soccol. 2024. "Microbial Bioherbicides Based on Cell-Free Phytotoxic Metabolites: Analysis and Perspectives on Their Application in Weed Control as an Innovative Sustainable Solution." *Plants* 13 (14): 1996. https://doi.org/10.3390/plants13141996.

Ogunnupebi, Temitope A., Abimbola P. Oluyori, Adewumi O. Dada, Oluwole S. Oladeji, Adejumoke A. Inyinbor, and Godshelp O. Egharevba. 2020. "Promising Natural Products in Crop Protection and Food Preservation: Basis, Advances, and Future Prospects." *International Journal of Agronomy* 2020 (July): 1–28. https://doi.org/10.1155/2020/8840046.

Paula Rosetti, Mayara Ketllyn de, Dejane Santos Alves, Isabela Caroline Luft, Katiane Pompermayer, Andressa Soares Scolari, Gabriela Trindade de Souza e Silva, Murilo Silva de Oliveira, et al. 2023. "Duguetia Lanceolata A. St.-Hil. (Annonaceae) Essential Oil: Toxicity against Spodoptera Frugiperda (J. E. Smith) (Lepidoptera: Noctuidae) and Selectivity for the Parasitoid Trichogramma Pretiosum Riley (Hymenoptera: Trichogrammatidae)." *Agriculture* 13 (2): 488. https://doi.org/10.3390/agriculture13020488.

Pedrini, Nicolás. 2022. "The Entomopathogenic Fungus Beauveria Bassiana Shows Its Toxic Side within Insects: Expression of Genes Encoding Secondary Metabolites during Pathogenesis." *Journal of Fungi* 8 (5): 488. https://doi.org/10.3390/jof8050488.

Plata-Rueda, Angelica, Hughes Antonio Quintero, José Eduardo Serrão, and Luis Carlos Martínez. 2020. "Insecticidal Activity of Bacillus Thuringiensis Strains on the Nettle Caterpillar, Euprosterna Elaeasa (Lepidoptera: Limacodidae)." *Insects* 11 (5): 1–10. https://doi.org/10.3390/insects11050310.

Raveau, Robin, Joël Fontaine, and Anissa Lounès-Hadj Sahraoui. 2020. "Essential Oils as Potential Alternative Biocontrol Products against Plant Pathogens and Weeds: A Review." *Foods* 9 (3): 365. https://doi.org/10.3390/foods9030365.

Rinner, Uwe, and Tomas Hudlicky. 2012. "Synthesis of Morphine Alkaloids and Derivatives BT—Alkaloid Synthesis." In, edited by Hans-Joachim Knölker, 33–66. Berlin, Heidelberg: Springer Berlin Heidelberg. https://doi.org/10.1007/128_2011_133.

Ruiu, Luca. 2018. "Microbial Biopesticides in Agroecosystems." *Agronomy* 8 (11): 235. https://doi.org/10.3390/agronomy8110235.

Sabbour, Magda M. A. 2020. "Insecticidal Effect of Three Essential Oils and Beauvericin Nano Gel on Sitophilus Oryzae and Sitophilus Granarius (Coleoptera:Curculionidae)." *Journal of Biopesticides* 13(2): 127–34.

Salehi, Bahare, Shashi Upadhyay, Ilkay Erdogan Orhan, Arun Kumar Jugran, Sumali LD Jayaweera, Daniel A. Dias, Farukh Sharopov, Yasaman Taheri, Natália Martins, and Navid Baghalpour. 2019. "Therapeutic Potential of α-and β-Pinene: A Miracle Gift of Nature." *Biomolecules* 9 (11): 738.

Silva, Eliane R., José M. Igartuburu, Gerhard E. Overbeck, Geraldo L. G. Soares, and Francisco A. Macías. 2020. "Bioherbicide Potential of Eucalyptus Saligna Leaf Litter Essential Oil." *Chemistry & Biodiversity* 17 (9). https://doi.org/10.1002/cbdv.202000407.

Sparks, Thomas C., and Robert J. Bryant. 2022. "Impact of Natural Products on Discovery of, and Innovation in, Crop Protection Compounds." *Pest Management Science* 78 (2): 399–408. https://doi.org/10.1002/ps.6653.

Sparks, Thomas C., Donald R. Hahn, and Negar V. Garizi. 2017. "Natural Products, Their Derivatives, Mimics and Synthetic Equivalents: Role in Agrochemical Discovery." *Pest Management Science* 73 (4): 700–15. https://doi.org/10.1002/ps.4458.

Sparks, Thomas C., Janine M. Sparks, and Stephen O. Duke. 2023. "Natural Product-Based Crop Protection Compounds—Origins and Future Prospects." *Journal of Agricultural and Food Chemistry* 71 (5): 2259–69. https://doi.org/10.1021/acs.jafc.2c06938.

Steglińska, Aleksandra, Anastasiia Bekhter, Paweł Wawrzyniak, Alina Kunicka-Styczyńska, Konrad Jastrząbek, Michał Fidler, Krzysztof Śmigielski, and Beata Gutarowska. 2022. "Antimicrobial Activities of Plant Extracts against Solanum Tuberosum L. Phytopathogens." *Molecules* 27 (5): 1579–99. https://doi.org/10.3390/molecules27051579.

Sun, Haishu, Shanxue Jiang, Cancan Jiang, Chuanfu Wu, Ming Gao, and Qunhui Wang. 2021. "A Review of Root Exudates and Rhizosphere Microbiome for Crop Production." *Environmental Science and Pollution Research* 28 (39): 54497–510. https://doi.org/10.1007/s11356-021-15838-7.

Toral, Laura, Miguel Rodríguez, Victoria Béjar, and Inmaculada Sampedro. 2020. "Crop Protection against Botrytis Cinerea by Rhizhosphere Biological Control Agent Bacillus Velezensis XT1." *Microorganisms* 8 (7): 1–17. https://doi.org/10.3390/microorganisms8070992.

Tudi, Muyesaier, Huada Daniel Ruan, Li Wang, Jia Lyu, Ross Sadler, Des Connell, Cordia Chu, and Dung Tri Phung. 2021. "Agriculture Development, Pesticide Application and Its Impact on the Environment." *International Journal of Environmental Research and Public Health* 18 (3): 1112. https://doi.org/10.3390/ijerph18031112.

Tulu, Urgesa Tsega. 2022. "Soil Microbes as Bioherbicides: An Eco-Friendly Approach to Control Striga." *International Journal of Agriculture and Biosciences* 11 (1): 22–28. https://doi.org/10.47278/journal.ijab/2022.003.

Wang, Yue, Wenhao Chen, Zhefei Xu, Qiqi Bai, Xueming Zhou, Caijuan Zheng, Meng Bai, and Guangying Chen. 2022. "Biological Secondary Metabolites from the Lumnitzera Littorea-Derived Fungus Penicillium Oxalicum HLLG-13." *Marine Drugs* 21 (1): 22. https://doi.org/10.3390/md21010022.

Wang, Zhirong, Tao Zhong, Xuhui Chen, Bing Yang, Muying Du, Kaituo Wang, Zsolt Zalán, and Jianquan Kan. 2021. "Potential of Volatile Organic Compounds Emitted by Pseudomonas Fluorescens ZX as Biological Fumigants to Control Citrus Green Mold Decay at Postharvest." *Journal of Agricultural and Food Chemistry* 69 (7): 2087–98. https://doi.org/10.1021/acs.jafc.0c07375.

Xuan Cuong, Dang. 2020. "Laminarin (Beta-Glucan) of Brown Algae Sargassum Mcclurei: Extraction, Antioxidant Activity, Lipoxygenase Inhibition Activity, and Physicochemistry Properties." *World Journal of Food Science and Technology* 4 (1): 31. https://doi.org/10.11648/j.wjfst.20200401.15.

Yang, Qian, Xia Yang, Zichang Zhang, Jieping Wang, Weiguo Fu, and Yongfeng Li. 2021. "Investigating the Resistance Levels and Mechanisms to Penoxsulam and Cyhalofop-Butyl in Barnyardgrass (Echinochloa Crus-Galli) from Ningxia Province, China." *Weed Science* 69 (4): 422–29. https://doi.org/10.1017/wsc.2021.37.

Yuan, Xiao-Long, Xiu-Fang Wang, Kuo Xu, Wei Li, Dan Chen, and Peng Zhang. 2020. "Characterization of a New Insecticidal Anthraquinone Derivative from an Endophyte of Acremonium Vitellinum against Helicoverpa Armigera." *Journal of Agricultural and Food Chemistry* 68 (41): 11480–87. https://doi.org/10.1021/acs.jafc.0c05680.

Zarubova, Lenka, Lenka Kourimska, Miloslav Zouhar, Pavel Novy, Ondrej Douda, and Jiri Skuhrovec. 2015. "Botanical Pesticides and Their Human Health Safety on the Example of Citrus Sinensis Essential Oil and Oulema Melanopus under Laboratory Conditions." *Acta Agriculturae Scandinavica, Section B—Soil & Plant Science* 65 (1): 89–93. https://doi.org/10.1080/09064710.2014.959556.

Zhang, Jing, Li Ting Yan, En Lin Yuan, Hai Xin Ding, Huo Chun Ye, Zheng Ke Zhang, Chao Yan, Ying Qian Liu, and Gang Feng. 2014. "Antifungal Activity of Compounds Extracted from Cortex Pseudolaricis against Colletotrichum Gloeosporioides." *Journal of Agricultural and Food Chemistry* 62 (21): 4905–10. https://doi.org/10.1021/jf500968b.

Zhang, Peng, Xiao-Long Yuan, Yong-Mei Du, Huai-Bao Zhang, Guo-Ming Shen, Zhong-Feng Zhang, Yu-Jun Liang, Dong-Lin Zhao, and Kuo Xu. 2019. "Angularly Prenylated Indole Alkaloids with Antimicrobial and Insecticidal Activities from an Endophytic Fungus Fusarium Sambucinum TE-6L." *Journal of Agricultural and Food Chemistry* 67 (43): 11994–2001. https://doi.org/10.1021/acs.jafc.9b05827.

Zhao, Bochui, Xian Xu, Binghua Li, Zhizun Qi, Jinan Huang, Ali Hu, Guiqi Wang, and Xiaomin Liu. 2023. "Target-Site Mutation and Enhanced Metabolism Endow Resistance to Nicosulfuron in a Digitaria Sanguinalis Population." *Pesticide Biochemistry and Physiology* 194 (August): 105488. https://doi.org/10.1016/j.pestbp.2023.105488.

Zhao, Yun, Yun-Hai Yang, Min Ye, Kai-Bo Wang, Li-Ming Fan, and Fa-Wu Su. 2021. "Chemical Composition and Antifungal Activity of Essential Oil from Origanum Vulgare against Botrytis Cinerea." *Food Chemistry* 365 (December): 130506. https://doi.org/10.1016/j.foodchem.2021.130506.

5 Allelochemicals and Signaling Compounds in the Plant Kingdom
Interactions and Functions

Showket Ahmad Dar, Ayah Talal Abdullah Al-Kilani, Ammar Al-Farga, Kounser Javeed, Wajid Hasan, Yendrembam K. Devi, Dev Karan Bairwa, Manisha Sharma, Dahiphale Kalyan Devidas, I Yimjenjang Longkumer, Kangjam Bumpy, Guneshori Maisnam, Syed Naseem Zaffar G, Sumit Saini, and Iram Khurshid

5.1 INTRODUCTION

Once germinated, plants establish a permanent presence, commencing their life cycle in settings that are influenced by a multitude of biotic and abiotic factors, either directly or indirectly. In contrast to the assumption that they are less effective and passive in response to environmental conditions, particularly stress, sedentary plants employ a range of sophisticated and diverse strategies to enhance their chances of survival and adaptation. A variety of methods have been observed to involve the production, synthesis, vaporization, and release of species-specific chemical molecules [1]. In their natural environment, plants produce and emit a vast array of compounds, including numerous secondary metabolites comprising proteins, carbohydrates, vitamins, and other low-molecular-weight molecules. It has been demonstrated that plants are capable of fixing almost 5 to 30% of CO_2 through the process of photosynthesis, which is released from the breakdown of various metabolites in the environment. It has been demonstrated that plants are the source of more than 80% of natural products and secondary metabolites to date. Consequently, plants serve as a vital and diverse source of active metabolites, which offer numerous benefits to human health and the environment. In essence, the plant-based active metabolites and other ingredients serve to protect the host plant against herbivore predation, microbial invasion, and/or inter-competition by other plants [2]. Plants have developed sophisticated reaction mechanisms to defend themselves against competitors and to safeguard themselves from diseases and herbivorous insects. These mechanisms can be broadly classified into two categories: constitutive and inducible defenses. These defense mechanisms have been shown to be highly effective in preventing crop loss [2].

The two defensive systems of plants are influenced by a multitude of chemical metabolites that are synthesized by plants throughout their developmental stages. Notwithstanding, the significant chemical metabolites, including cyanogenic glycosides (derived from amino acids), alkaloids (plant-based compounds comprising less than one nitrogen molecule), terpenoids (derivatives of isoprene), and flavonoids (polyphenolic secondary metabolites), are among the secondary metabolites that primarily mediate the inducible defense mechanism. These defensive mechanisms play a crucial role in the interactions between host plants and other organisms. Another significant mechanism employed by plants to influence the behavior of surrounding organisms is allelopathy, a biological

DOI: 10.1201/9781003463429-7

phenomenon whereby plants produce compounds that have a detrimental impact on other living organisms. By employing these biochemical processes, plants engage in a phenomenon known as allelopathy, which involves the synthesis and release of defense metabolites, or allelochemicals, that exert an adverse influence on both conspecific (same species) and heterospecific (other species) plants [1]. The surrounding plant development and establishment is significantly impacted by allelochemicals, and their production can be modified by the influence of nearby plants on the defense biochemistry of the plant. In addition to their ability to repel herbivores, directly inhibit pathogens, and suppress plant competitors, the chemicals derived from plants can also function as alarm signals, alerting nearby plants to impending insect herbivorous and microbial attacks [3, 4]. An intra- and inter-plant signal typically enables the detection of nearby plants, comprising genetically related or unrelated plants from the same genus or even between species [5]. The sensing and communication of plants is mediated by these signaling molecules. Plants employ the production and release of allelochemicals and signaling chemicals to defend themselves against herbivores, suppress competing plant species, and attack microbes. Additionally, they regulate microbial populations in the soil, foster beneficial symbiotic relationships, and alter and enhance the below-surface relationships and interactions. Consequently, signaling chemicals and allelochemicals derived from plants are gaining increasing popularity as valuable tools for the management of agricultural pests and diseases. The development of novel ecological techniques for sustainable agriculture is contingent upon a comprehensive understanding of the implications and applications of signaling chemicals and allelochemicals originating from plants.

5.2 ALLELOCHEMICALS

In allelopathic conditions, which are competitive situations in which chemicals cease to impede the growth and spread of plants, the release of chemicals into the environment is commonly known as allelopathy. This is a distinct phenomenon from resource competition [2]. To gain an understanding of plant-plant allelopathic interactions, it is essential to identify the allelochemicals present in plants and their surrounding environment. To date, numerous allelochemicals have been identified in diverse plant species. These allelochemicals belong to various biochemical groups, including alkaloids, N-containing metabolites (such as cyanogenic glycosides, benzoxazinoids, and non-protein components), terpenoids (such as steroids, diterpenes, sesquiterpenes, and monoterpenes), and simple phenolics.

5.3 POTENTIAL ALLELOCHEMICAL-PRODUCING PLANT SPECIES

Juglone was the primary and hypothetical allelochemical known to exist in walnuts. Chemically, it is known to be 1,4-naphthoquinone (a quinone derivative, para-naphthoquinone) produced from the black walnut (*Juglans nigra*), which is responsible for suppressing and hindering the growth of understory vegetation. The recent isolation of allelochemical juglone from various plant parts represents a significant advancement in the field, despite the allelopathic phenomenon having been identified nearly a century ago. The chemical compound juglone is an environmentally friendly and safe metabolite, a naphthol O-glycoside biosynthesized by the walnut plant and emitted into the environment from the plant's leaves, bark, fruit hull, and roots. The phytotoxic juglone is produced by oxidizing the less phytotoxic naphthol aglycone, which is rapidly synthesized from the naphthol O-glycoside through hydrolysis or interactions with soil bacteria. The compound naphthol O-glycoside is non-toxic; however, it undergoes a transformation into the phytotoxic juglone when discharged into the environment.

The occurrence of allelopathic interactions is significantly dependent on the dynamics of plant metabolites (allelochemicals) and their role in the surrounding environment. This is exemplified by the phenomenon of juglone. The efficacy of allelochemicals is contingent upon their proximity to target plants at concentrations that elicit phytotoxicity. Once released from plant tissues, allelochemicals engage in both biotic and abiotic interactions. The majority of allelochemicals undergo a series of fundamental chemical processes, including transformation, modification, signaling, transport, and retention [6]. In plants, allelochemicals interact with soil microbes at a high rate, affecting the deterioration, feeding, utilization, consumption, and breakdown of metabolites from root systems. These interactions result in the production of compounds, namely allelochemicals, that influence the growth and development of plants. The microbial populations present in soil play a significant role in triggering and intensifying plant allelochemical action, thereby impacting the root zone. In the past decade, a multitude of both ancient and novel allelochemicals have been identified and studied. A significant number of the recently discovered plant metabolites that function as allelochemicals have been identified based on their chemical structure, which has been determined from a variety of plant species. In particular, the grass species *Festuca* spp. were identified as a crucial resource, demonstrating an essential ecological function in alpine and subalpine rangeland stress situations. It has been demonstrated that grasses that employ allelopathic effects can displace neighboring plant species through the release of root exudates into the soil, which exert a strong phytotoxic influence. In grasses, the most common and prevalent allelochemical present in roots with phytotoxic actions is m-tyrosine, 4-hydroxyphenylalanine, a non-protein amino acid (i.e., an amino acid that is not involved in protein synthesis and differs from the 22 proteinogenic amino acids occurring in eukaryotic cells). In fescue grasses, the allelopathic effect is attributed to m-tyrosine, which inhibits root growth in other plants and even causes them to become displaced. In grasses, the numerous amino acids that are not proteins were identified as playing a role in protecting plants from diseases, herbivore attack, and other plant rivals and pathogens. While a minority of non-protein amino acids are highly toxic to microorganisms and plants, a minority of them function as deterrents to predators. In addition to their essential functions in food absorption, signaling, and the stress response, the non-protein amino acids present in grasses have also been identified as playing a pivotal role in plant-soil and plant-plant interactions within the ecosystem.

A notable coniferous tree species utilized for the production of industrial timber is the Chinese fir (*Cunninghamia lanceolata*). Conversely, the prevalence of monocultures has exacerbated the issue of its inability to regenerate and decline in productivity [7]. The Chinese fir releases a distinctive allelochemical compound, namely cyclic dipeptide (6-hydroxy-1,3-dimethyl-8-nonadecyl-[1,4]-diazocane-2,5-diketone), which is present in the roots, leaves, and bark. The soil is penetrated by metabolites, including the cyclic dipeptide, which is emitted by the roots of the Chinese fir (*Cunninghamia lanceolate*). As the quantities of this allelochemical increase in subsequent monocultures, self-inhibition (also known as intraspecific allelopathy) occurs. This significant phytotoxic action of the allelochemical cyclic dipeptide limits the ability of Chinese fir plantations to produce offspring. The cultivation of *C. lanceolata* in proximity to *M. macclurei* has been observed to accelerate the decomposition of the metabolic compound cyclic dipeptide in the soil, thereby reducing its formation. This ultimately results in a transition from facilitation to self-inhibition. Moreover, it has been demonstrated that the presence of *M. macclurei* results in a reduction in the chemical compounds released into the environment by an allelopathic species. A plant's defensive strategy may vary based on the identity of its neighbors, as allelochemical synthesis and release are dependent on coexisting plant species [8]. The identification of plant neighbors and allelochemical reactions represents a significant challenge in the context of plant coexistence within any plantation community.

The coexistence of certain weedy species and crop plants has been observed in cropping systems. Crop plants are subject to the effects of allelochemicals, which are emitted by a variety of allelopathic weeds. The allelopathic weed *Ambrosia trifida*, commonly referred to as gigantic ragweed, typically flourishes in agricultural soil, thereby reducing crop productivity. The sesquiterpenes of the carotene class have been observed to impede the development of wheat at low concentrations. Moreover, 1α-angeloyloxycarotol was identified as a prominent constituent of residue-treated soils exhibiting *A. trifida* infestation. Therefore, 1α-angeloyloxycarotol, which *A. trifida* releases into the soil, serves as a crucial allelochemical in wheat fields infested with the plant. The majority of active ingredients found in plants are classified as phytochemicals, and those comprising terpenoids and carotenoid skeletons are particularly bioactive.

5.4 ALLELOCHEMICALS RELEASED FROM VARIOUS PLANTS

The allelochemicals produced and released by allelopathic weeds exert an influence on crop plants. Fortunately, some crop varieties have been observed to exhibit allelopathic properties, which have been demonstrated to inhibit weed growth and development. The allelopathic plant varieties were examined through the analysis of their germplasm collections for the presence of various allelochemicals, with a particular focus on weed species associated with crops such as rice and wheat. The capacity and potential of diverse plant species to synthesize and emit a range of compounds into the surrounding environment. The development of allelochemicals with herbicide-like effects has the potential to facilitate the implementation of environmentally friendly and sustainable weed management strategies. Consequently, there is a significant emphasis on identifying and researching the diverse crops for their allelochemicals and their impact on weeds. The principal allelochemicals derived from diverse plant cultivars of significant economic value have been identified.

The chemical compound sorgoleone, derived from the plant species *Sorghum bicolor*, belongs to the benzoxazinoid hydroxamic acid family. It is found in various species of the genus Zea, including *Zea mays* and *Triticum aestivum*. Additionally, tricin (an O-methylated flavone) and momilactone-B (a lactone with three chiral centers) are two other notable compounds. The 10cβ)-8-Ethenyl-3a,5-a,7,8,9,10,10a,10c-octahydro-3a,8-dimethyl-,4H-3,10b-ethano-1H,3H-benzo-furo[4,3,2-de]-2-benzopyran-4-one, a compound obtained from rice [9], along with its chemical structure. Research has revealed that allelopathic *Oryza* spp. produce a range of significant allelochemicals, including momilactone B and tricin (5,7,4J-trihydroxy-3J,5J-dimethoxyflavone) [9]. Additionally, rice seedlings have been observed to produce 5,4J-dihydroxy-3J,5J-dimethoxy-7-O-β-glucopyranosylflavone, which plays a crucial role in overall crop yield and protection from various adverse factors. Flavone compounds include 7,4J-dihydroxy-3J,5J-dimethoxy-5-O-β-glucopyranosyl. Additionally, rice roots discharge two flavone O-glycosides that are subsequently converted into their aglycone compounds, including 5,7,4J-trihydroxy-3J and 5J-dimethoxyflavone. These compounds exhibit pronounced allelopathic effects on rice weeds and pathogens. Additionally, wheat plants synthesize 2-(2,4-dihydroxy-7-methoxy-1,4-benzoxazin-3-one)-β-D-glucopyranose (DIMBOA-Glc), which is assimilated in cellular vacuoles and exerts a strong allelopathic effect. Once activated by insect feeding, the compound DIMBOA-Glc (hydroxamic acid) is promptly converted into a new derivative aglycone, which is released from roots directly into soil. This process produces compounds including 2,4-dihydroxy-7-methoxy-1,4-benzoxazin-3-one (DIMBOA) and 6-methoxy-benzoxazolin-2-one (MBOA). The two most significant allelochemicals in wheat-weed allelopathic reactions are DIMBOA and MBOA, which have been demonstrated to be highly effective in controlling weeds that have become associated with wheat crops. In comparison to DIMBOA, MBOA displays a more extensive spectrum of bioactivity and exhibits greater resilience in the presence of weak soils with a deficient nutrient composition.

The crucial metabolite, DIMBOA (2,4-dihydroxy-7-methoxy-1,4-benzoxazin-3-one), is a member of the benzoxazinoid hydroxamic acid family, and its structure is derived from 2-hydroxy-2H-1,4-benzoxazin-3(4H)-one. Additionally, other cereal crops, including rice and sorghum, do not produce benzoxazinoids or their chemical skeletons, with the exception of maize and rye. In recent decades, researchers have investigated the significance of benzoxazinoids (a monomeric natural compound with antimicrobial potency but strong allelopathic influence) in plant defense against various species. A correlation has been identified between fungal pathogenic resistance and benzoxazinoid concentrations in cereals. A shift in fungal populations in the wheat rhizosphere was found to benefit DIMBOA and MBOA, which modified the organization of the soil microbial community. Specifically, benzoxazinoids secreted by cereal roots may modify the microbiota from the rhizosphere area, thereby slowing plant development and improving plant defense, while also limiting insect herbivore damage in the subsequent plant life cycle. By modifying root-associated microbial populations, the root-emitted compound DIMBOA is subsequently converted into MBOA, influencing the interaction between the plant and the soil and also generating signaling responses as feedback on defense and developmental fronts [10].

Furthermore, an enhanced comprehension of the interrelationships between allelochemical and soil-dwelling microbes, particularly pathogenic fungi that are soil-borne, may pave the way for the introduction and implementation of disease control measures in agricultural settings. Sorghum is a common cereal crop in semiarid tropical regions. Roots of members of the sorghum genus produce an exudate, sorgoleone, which is a chemical compound with the formula 2-hydroxy-5-methoxy-3-[(8Z,11Z)-8Z,11Z,14Z-pentadecatriene]-p-benzoquinone. The sorgoleone content of sorghum root exudates is typically 85-90% pure. However, significant inter-genotype variation has been observed [10]. It has been established that sorgoleone is an allelochemical with the capacity to inhibit a diverse array of weeds. In comparison to numerous phenolic and terpenoid compounds, as well as the conventional allelochemical juglone, sorgoleone has been observed to exhibit heightened activity. The process of photosynthesis is disrupted by the action of hydroxy-phenyl pyruvate dioxygenase (HPPD), which is significantly inhibited by sorgoleone. Sorgoleone, for instance, has an ongoing impact on soil, and it does not break down rapidly when soil microflora is present. The formulation of sorgoleone as a wettable powder has been demonstrated to enhance the efficacy of weed suppression by tolerant plants, when compared to the use of sorgoleone alone. The use of allelochemicals may be particularly advantageous for the management of agricultural pests, especially those compounds which have been identified as having the capacity to suppress weeds, such as tricin (an O-methylated flavone), momilactone B, benzoxazinoid hydroxamic acids (a secondary metabolite which plays an important role in plant defense), and sorgoleone (a metabolite which has been observed to exert an inhibitory effect on ammonia-oxidizing microbes) which are released from the sorghum grains. These studies collectively provide a comprehensive insight into the mechanisms by which allelopathic crops manage pests through the use of their intrinsic insecticides, or allelochemicals, in cropping systems.

5.5 SIGNALING CHEMICALS

Plants possess both constitutive and inducible chemical defense mechanisms. In this context, biotic and abiotic occurrences may result in the synthesis and emission of active and dynamic plant defense metabolites, which frequently function as signaling molecules that induce defensive responses in plants at both the local and systemic levels. The majority of research on biotic (insect pest)-induced plant stimuli has been conducted in response to pathogen or insect attack. For further details, please see references. The production of allelochemicals is becoming increasingly influenced by the effects of competing neighbors on plant defensive biochemistry. It is widely acknowledged that the presence of weeds may result in an increase in allelochemical production by allelopathic crop plants. This assertion is supported by research findings from studies. The latest research indicates that plants that are allelopathic recognize weeds and respond by elevating their levels of allelochemicals. Prior to initiating allelopathic activity to regulate intra- or inter-specific interactions in intra-species coexistence systems, it has been demonstrated that plants first recognize, detect, and identify their neighbors and surroundings. Two closely related mechanisms that emerge when two or more plants interact are plant neighbor identification and allelochemical reaction. Allelochemicals exhibit considerable diversity in their composition, structure, actions, applications, and impacts. A substantial body of literature exists on intra-specific as well as inter-specific allelopathic interactions. However, the signaling chemicals released by various plant species remain largely unstudied. This has given rise to a great deal of interest and curiosity in recent years regarding the identification, categorization, determination, applications, impacts, and utilization of signaling chemicals obtained from various plant species.

5.6 AIRBORNE PLANT-DERIVED SIGNALING CHEMICALS

Plants and microbes engage in interactions through a range of signaling processes occurring above and below ground [1]. These interactions can be either detrimental or beneficial. In the plant kingdom, species recognize or detect surrounding organisms through the use of plant volatiles, which act as airborne signals [10, 11]. These chemicals function differently than root exudates, which are used for belowground communication linkages. In plants, well-established and developed aboveground signaling linkages are mediated by chemicals transported by wind. In contrast, the signaling chemicals delivered by the soil in belowground plant-organism interactions remain unknown. Consequently, airborne chemical signals, including ethylene-based methyl jasmonate (a principal signaling molecule) and salicylate (a phytohormone instrumental in growth and development), indole (an aromatic heterocyclic compound), and diverse volatile terpenoids, have been the subject of considerable research attention with respect to plant-derived signaling compounds.

5.7 PLANT-DERIVED SIGNALING CHEMICALS

In the event of a pathogenic infection or herbivore attack on a plant, or the incursion of an external agent, the plant in question will release ethylene into the surrounding atmosphere. Ethylene is initially released as a gaseous phytohormone, functioning as a signaling chemical. However, plants produce a diverse array of airborne metabolites that serve as signaling molecules in response to diseases and herbivores. Among the various airborne, fast-disseminating signaling compounds studied in plants are volatile organic compounds such as methyl jasmonate and salicylate, which are defensive molecules that are acidic in nature and commonly produced by plants as secondary metabolites. These include salicylic acid and jasmine acid. Moreover, salicylic acid is a non-volatile, liquid-based jasmonate molecule that functions as a methylated elicitor, facilitating the production of volatile methyl jasmonate, which plays a crucial role in plant defense mechanisms. The compound salicylate is produced in response to an attack or infection by a biotic agent [10, 11].

Methy-jasmonate and salicylate are signaling molecules that are released by plants in response to a perceived threat, including infection, herbivory, competition from other plants, and the presence of invasive species. Ethylene, the first gaseous plant hormone, is among the compounds released into the environment during this process. However, in response to illnesses and herbivores, plants release a diverse range of airborne signaling molecules. The methyl jasmonate and salicylate are among the most extensively studied airborne signaling chemicals. Both salicylic acid and jasmonic acid, which plants produce as secondary metabolites, are highly structured and play a crucial role in plant signaling and defense reactions. In response to an attack or infection, salicylic acid and non-volatile jasmonate are methylated to produce the volatile methyl jasmonate (which functions in defense) and salicylate (which plays a role in signaling and growth) [12]. The odorant 2(E)-hexenal is produced in response to bacterial attack, and a detailed description of this molecule indicates that it functions as a diffused volatile chemical signal which initiates a defense reaction in plants [2]. A recent study has identified the airborne signaling chemical indole as being triggered by the presence of herbivores. The indole emitted by herbivores prompts neighboring plants to synthesize defensive chemicals. With surrounding plants and systemic tissues prepared for potential attacks, indole functions as a reliable and effective aerial signaling agent. A variety of fruits contain 2(*E*)-hexenal, an odorant produced in response to bacterial infection. As elucidated in a comprehensive account of this chemical, it functions as a diffused volatile signal that elicits defensive responses in plants, particularly in relation to the accumulation of sesquiterpenoid phytoalexins-a non-toxic compound that is detoxified. A recent study has demonstrated that herbivores activate the molecule indole, which is an airborne signaling agent. Herbivores release indole, which elicits defensive chemical responses in proximate plants. The aerial signaling agent indole serves as a reliable and efficient means of preparing the surrounding plants and systemic tissues for possible attacks [12]. Furthermore, airborne signaling chemical signals are utilized by parasitic plants to identify their hosts, locate them, and trigger their defense mechanisms [13]

5.8 SOILBORNE SIGNALING CHEMICALS

Moreover, species belonging to the kingdom plantae that are parasitic on terrestrial organisms utilize signaling molecules present in the soil to locate their hosts. The mechanism by which signaling chemicals derived from hosts assist parasitic plants in locating their subterranean hosts has already been elucidated. The germination, sprouting, growth, and plant development of plant-parasitic species that adhere to their hosts, as well as the activation and development of haustorial systems in these parasites, are all influenced by signaling molecules produced by host roots [14]. The discovery of strigol was made several decades ago and established it as the first known stimulant to encourage the development of seeds for the parasitic witchweed (*Striga lutea*). It is therefore postulated that strigol derivatives (orobanchol orobanchyl acetate) act as chemical messengers across and within host plants that are parasitic, such as *Orobanche* spp. and *Striga* spp. The four rinses in the backbone of these molecules resulted in their final categorization as a family of plant hormones called strigolactones. It has been demonstrated that strigolactones are synthesized by both host and non-host plant roots and released into the rhizosphere, which is the area of soil microenvironment impacted with root exudates. It has recently been demonstrated that strigolactones and strigolactone-like compounds are released by roots of multiple field crops into the rhizosphere [15].

Strolactones, which facilitate seed germination, are released into the soil by parasitic plants on roots, resulting in a symbiotic relationship with arbuscular mycorrhizal (AM) fungi. This relationship is aimed at the absorption of atmospheric nitrogen into molecular nitrogen [16]. Arbuscular mycorrhizal fungi facilitate the uptake of nutrients from the soil in their molecular form, which subsequently act as signaling molecules. 5-deoxy-strigol is an essential chemical found in the roots of parasitic plants and fungal symbionts. Secreted signaling chemicals from the roots facilitate interactions between microbes and plants. It is plausible that phenolic flavones serve as chemical signals in the interactions between microbes and plants. Twenty years ago [3, 4, 15], a typical flavone, was identified as a crucial signaling molecule in plants of the Fabaceae family and symbiotic rhizobia. Both rhizobia and mycorrhizal fungi have been observed to stimulate root growth, which in turn promotes shoot growth. It has been demonstrated that the mutual regulation of mycorrhizal fungi and rhizobia by strigolactones and flavones is a crucial factor in coordinating the growth of plants at both the above- and belowground levels.

The interconnection between plant species or plants that have symbiotic microbes is facilitated by belowground signaling, which is mediated by the production of flavones and strigolactones from their roots [17]. Nevertheless, the disparate roles that strigolactones and flavones appear to play in belowground signaling interactions underscore the necessity for further research into the signaling compounds present in soil. In response to the ubiquitous signaling molecule jasmonic acid, plants secrete defensive metabolites. A number of studies have investigated the synthesis of jasmonic acid in the root zone and its role in belowground chemical signaling pathways and actions. A recent study of plant biology has identified the most notable compound, lactone (-)-loliolide, as a key component of belowground signaling pathways and communications. Moreover, the allelopathic wheat crop is capable of detecting, differentiating, and distinguishing between all root-secreted (-)-loliolide compounds while interacting with surrounding plant species. In response, the concentration of DIMBOA is increased. Moreover, the compound (-)-loliolide was identified in the rhizosphere soils of over 100 plant species, functioning as a soil-borne signaling molecule. (-)-Loliolide exhibits high soil mobility and appears to be a shared trait among all species. It is postulated that the allelochemical response in subterranean contacts and plant recognition represent its principal uses. Conversely, (-)-loliolide has been observed to stimulate the accumulation of defensive mechanisms in plants, thereby enhancing their ability to resist disease and herbivory [18]. It can be concluded that (-)-loliolide may be a common signaling molecule transported by the soil in subsurface molecular interactions and cooperation across different plants and other organisms in the field. It is established that jasmonic acid plays a role in regulating many plant defense systems. The regulatory network of (-)-loliolide as a signaling chemical compound important for defensive actions remains unknown to science. Given the role of (-)-loliolide as a crucial soil-borne chemical signaling molecule with high importance and significance in belowground signaling mechanisms, this discovery offers a novel perspective and insights into plant defense pathways and strategies.

In order for insects to adapt to changing environmental conditions, plants have evolved a complex chemical communication network [19]. Despite the existence of a natural link between plant-derived signaling molecules and their interactions with insects, microbes, plants, and soil. Consequently, plants possess a chemical defense advantage against insect pests, which is frequently modified, particularly when secondary metabolites with diverse functions are involved in the production of defense-linked chemical responses resulting from biotic stressors, especially insects, microbes, or plant signaling. In this scenario, signaling molecules originating from plants may serve as the primary regulators in a cellular communications network for plants against various biotic agents. The most significant and intriguing implications and functions observed in both natural and controlled environments, elicited by signaling molecules or biochemical entities, are well documented and of paramount importance for understanding interactions among diverse species in the environment.

5.9 ROLES OF ALLELOCHEMICALS IN SUSTAINABLE AGRICULTURE

It has been demonstrated that the extensive utilization of synthetic pesticides in integrated pest management (IPM) has given rise to concerns pertaining to environmental and public health issues, which have in turn resulted in a number of fatalities. In light of these concerns, it is imperative to explore alternative strategies that can replace the overuse of pesticides in agriculture. One promising approach is to leverage the potential of host plants to defend themselves against various biotic factors in cropping systems. This strategy is particularly promising when considering the significant influence of allelochemical and signaling processes produced by plants. In natural conditions, the management of insect pests is enabled by plant-based allelochemicals, signaling chemicals, and cross-talk, which collectively aim to mitigate the adverse effects of synthetic pesticides on ecosystems. The diverse array of plant-derived allelochemical compounds and signaling molecules provides a sustainable alternative to conventional chemical-based farming practices.

5.10 DEVELOPMENT OF HERBICIDES (ALLELOCHEMICALS)

In the natural environment, plant metabolites (allelochemicals) present a novel approach to herbicide development, offering potential for effective weed management in agricultural systems [20]. In recent decades, numerous allelochemicals have been synthesized and subjected to substantial modifications, including tricin and sorgoleone. In recent years, there has been a significant emphasis on the role of the metabolite tricin (an O-methylated flavone flavonoid, occurring in rice bran, sugarcane, and other cereals). In the development of an herbicide based on allelochemicals, tricine can be generated through the aldol condensation mechanism in an aqueous medium following a chemical reaction between α-chloro-2-hydroxy acetophenone and an aromatic aldehyde. The tricin isomer aurone was, unexpectedly, the principal outcome of this synthesis technique. As the aurone isomer proved to be significantly more effective than tricin in controlling weeds and infections, it was identified as a highly promising chemical for the development of novel herbicides. Significant research has been conducted into the development of hormone-based herbicides. A number of compounds have been synthesized from aurones, including derivatives of benzothiazine and substituted aurones. These compounds demonstrated remarkable bioactivity against weeds. The compound 1,2-benzothiazine and its derivatives, in particular 2-benzoylethen-1-ol, demonstrated a satisfactory pre-emergent weed seed germination influence, thereby preventing HPPD interference in the process of photosynthesis. Moreover, even the concentrations below the limit of detection were found to completely suppress the germination, growth, and development of barnyard grass [16, 19].

In order to launch an allelochemical-based herbicide discovery project, the benzothiazine (a heterocyclic compound) derivative was selected as the target entity. The application of the benzothiazine-derived compound to the rice crop at the recommended dosage levels resulted in the effective reduction of target weeds without any adverse effects on the rice itself [11]. The findings revealed that benzothiazine proved to be an effective method for controlling rice weeds. Additionally, the benzothiazine compounds have been demonstrated to be effective in suppressing a considerable number of weeds in a variety of crops, including wheat, maize, and soybean [12]. Nevertheless, the benzothiazine compounds and their various derivatives exhibit multiple crop selectivities. While they demonstrated strong selectivity for maize, they were found to be unsafe for soybeans. Furthermore, herbicides based on allelochemicals must be environmentally benign and possess low toxicity. Prior to the commercialization of the aromatic derivative of 2,1-benzothiazine, an evaluation of its ecological safety is necessary. The results of the toxicity study indicate that the benzothiazine derivative does not have a detrimental impact on soil earthworms, aquatic zebrafish, or soil microbes. Therefore, 3-(2-chloro-4-methanesulfonyl)-benzoyl-4-hydroxy-2-methyl-2H-1,2-benzothiazine-1,1-dioxide, also known as the benzothiazine derivative is an ecologically safe herbicide based on allelochemicals [9].

The compound is analogous to allelochemicals and various signaling compounds derived from plants were utilized as activators or elicitors to sensitize and initiate defense reactions and responses in plants. The plant-based signaling molecules can protect plants from herbivores, diseases, and weeds without directly exerting insecticidal, antimicrobial, or herbicidal effects by triggering and activating self-defense systems. It has been demonstrated that certain synthetic or naturally occurring chemicals function as plant activators or compounds that resemble plant activators. Specific compounds were employed to produce a range of commercial compounds that were utilized in the generation and expression of high resistance in agricultural crops. The most notable of these included Fytosave and Vacciplant Fruits et Légumes, which both exhibited a high abundance of molecular elicitors of laminarin and chitosan. Notwithstanding the considerable progress made in the synthesis of plant-based secondary metabolites and other signaling compounds, there remain numerous lucrative applications for these compounds, particularly in the context of conventional intercropping techniques and the breeding of economically viable crop varieties that exhibit allelopathic properties [21].

5.11 CONVENTIONAL METHODS OF INTERCROPPING

In the context of integrated pest management, the potential of conventional allelopathic plant intercropping approaches has been explored in agricultural systems [22]. The allelopathic weed *Ageratum conyzoides* is found in various regions of the world, and its allelopathic approach has been employed in conventional and traditional intercropping agricultural systems. In citrus farms globally, the *A. conyzoides* is frequently interplanted as an understory plant species, playing a pivotal role in pest management strategies. The allelochemicals released by *A. conyzoides* into the soil in citrus orchards release biochemicals such as ageratochromene and three flavones, which effectively suppress major soil microbes as well as other weeds. Additionally, *A. conyzoides* plants emit volatile signaling chemicals, including α-bisabolene, E-β-farnesene, and β-caryophyllene, which are released into the atmosphere and exert a significant allelopathic influence [23]. The most efficient natural predators of the citrus mite are predatory mites of the species *Panonychus citri*. The environment in citrus orchards became more favorable for these predatory mites as a result of *A. conyzoides* intercropping, which led to a significant drop in the density of citrus red mites at non-harmful levels. Consequently, the intercropping system of citrus and *A. conyzoides* facilitates the occurrence of natural chemical processes that regulate insect-plant host interactions, thereby ensuring the sustainable management of insect pests.

It has been demonstrated that analogous natural chemical processes are efficacious in mixed-species arboreal ecosystems. It was discovered that the release of allelochemicals from monoculture plantations, such as those containing Chinese fir, a prominent timber tree species, may be the source of replant problems that are frequently observed in managed tree plantations. However, the failure of Chinese fir plantations to regenerate after transplantation and their subsequent loss of yield have been predominantly attributed to the presence of allelochemicals [24]. It is noteworthy that mixed species can help to mitigate the issue. One such species is *Magnolia macclurei, a* non-nitrogen-fixing broadleaf tree species. The mixed-species *M. macclurei* modifies the underground ecological interactions among organisms, thereby encouraging the growth and development of autotoxic *Cunninghamia lanceolata*. Specifically, *M. macclurei* facilitated soil degradation and reduced the emission of various allelochemicals that are designed to inhibit self-growth. In such circumstances, the belowground ecological interactions are mediated by a naturally occurring chemical process in mixed-species plantations. A system of this nature could prove instrumental in the development of strategies aimed at reducing or even eliminating the growth and development of new saplings and plants in managed tree plantations across a range of ecological settings [18].

5.12 BREEDING AND CULTIVATION OF ALLELOPATHIC CULTIVARS

The objective is to enhance the potential of various plants to safeguard themselves from various biotic stresses through breeding procedures. This would be achieved by breeding resistant plants with the aim of offering greater weed suppressive actions using secondary metabolites and their allelopathic effects. In light of the urgent need for economically resistant or defensive plant varieties in the context of climate change, the production of such varieties is of paramount importance [1, 8]. China and Sweden have been instrumental in the development of allelopathic agricultural varieties that are commercially viable. The implementation of breeding programs over the past decade has resulted in the development of numerous allelopathic rice and wheat cultivars. The latest cultivar of rice, designated Huagan-3, represents a significant and pioneering allelopathic cultivar approved for commercial use in China, introduced in 2009. Since 2009, a number of additional varieties have been made available. In South China, a number of allelopathic rice varieties have been successfully cultivated. It has been demonstrated that an overreliance on herbicides should be replaced with a viable alternative that has achieved notable success in developing economically viable allelopathic crop varieties and implementing them in a plant breeding system. To effectively manage weeds in paddy fields in China and the United States, considerable effort has been dedicated to the utilization of allelopathic rice varieties. The objective of allelopathic crop cultivar breeding is to produce high-yield, high-quality crops while simultaneously suppressing weeds. The cultivation of allelopathic crops has the potential to enhance agricultural production, as they are capable of suppressing weeds.

In contrast, allelopathic crop cultivars may generate and release allelochemicals that alter crop allocation strategy, shifting the emphasis from growth to defense [25]. Consequently, the majority of allelopathic crop cultivars are unable to meet the commercial requirements. Any improvement in weed-suppressive ability through crop selection, breeding, and desired crop multiplication is likely to result in a reduction in grain production and quality. This is analogous to the breeding of insect and plant disease-resistant varieties to enhance output. A trade-off was identified between competitive potential and yield performance, with both parameters serving as the basis for increases in grain production. The formation of competitive organs for enhancing reproductive allocation may be inhibited by kin cooperation within a plant species. It can be reasonably deduced that by mitigating the effects of competition, genetic kin and non-kin recognition, and plant incorporation, selection, and cooperation in crop species should result in an increase in grain output [9, 10]. The limitation of high crop quality and bumper yields may not, in fact, affect kin recognition in allelopathic crop cultivars, suggesting a genetic basis for this phenomenon. A recent study has identified allelopathic rice cultivars that are capable of kin identification. The higher grain yields observed in these allelopathic rice cultivars can be attributed to the presence of kin cultivars. This finding is supported by evidence from studies [25]. A novel strategy to enhance the productivity of major grain crops may entail the recognition of kin in mixtures of crop cultivars. The issue of low productivity in the development of commercially viable biotic-tolerant and resistant cultivars may be resolved by increasing grain yield through kin recognition.

5.13 ADVANCES AND TRENDS

In recent years, there has been a notable increase in the recognition of the importance of signaling chemicals and allelochemicals derived from plants, particularly in the context of their potential use in insect pest control [3, 4]. Nevertheless, the biochemical interactions between plants and other microorganisms represent a vast and complex field of study, as evidenced by the extensive literature on the subject. The ability to recognize and detect signaling molecules and allelochemicals from plants and their surrounding environment is a crucial aspect of this field of study. To further facilitate the development of novel pesticides, a comprehensive account of the structural features of diverse allelochemicals is essential. However, the majority of researchers in this field are not familiar with organic chemistry, which results in the omission of crucial processes. In contrast to the general pathways for the synthesis of phytochemicals, metabolites, and allelochemicals, as well as the fate and origin of signaling molecules, it is essential to gather and confirm this information from intact, living organisms, plants, and organic by-products. This can be done either in vivo, in situ, or even in real time [26]. The question of how to extract and characterize the signaling and allelochemicals from plants and their surroundings remains unanswered. Tandem mass spectrometry, in conjunction with GCMS or LC, is typically employed to identify and determine the allelochemical components and signaling compounds from various plant resources and in their surroundings. In particular, the aforementioned chromatographic techniques and tools, which are designed to identify the components of combinations, have proven to be the most effective for the analysis of volatile plant substances. Volatile organic compounds (VOCs) have been the subject of a multitude of biochemical interactions across a diverse range of plant and other organisms. The reliability of GC-MS/MS techniques is the primary factor contributing to this increased attention. However, the majority of signaling molecules and allelochemicals produced and secreted by plants are not volatile. Furthermore, elevated temperatures are necessary for the examination of specimens utilizing GC-MS/MS. GC-MS/MS techniques were frequently inadequate for identifying numerous signaling compounds and other non-volatile biochemicals that play a pivotal role in biochemical interactions across organisms. The majority of investigations examining non-volatile allelochemicals and signaling molecules are based on techniques such as LC-MS/MS analysis. LC-MS/MS is capable of covering a wide range of compounds, including those with a high molecular weight, and exhibits excellent resolution selectivity. The LC-MS/MS method is particularly effective when analyzing watery samples. However, the lack of reliable databases containing chemical structures and the fact that LC-MS/MS analysis is a targeted, high-accuracy method [27]. Consequently, the identification of unknown chemical compounds in plants and their surrounding environment is not feasible through targeted analysis. Instead, it is constrained to the observation of diverse chemical modifications and alterations in the exudation pathways of identified chemical compounds [28]. It is essential to separate each chemical component from the plant and soil samples in order to facilitate the identification of unknown compounds. To ascertain which specific elements may be responsible for the self-defense and biochemical signaling relationships, a bioassay-guided fractionation technique is typically employed. Ultimately, spectroscopic investigations, most notably mass spectrometry (MS) and nuclear magnetic resonance (NMR), were found to be effective in identifying the composition and structure of unknown compounds separated from multiple plant species and their environments. This was achieved through a methodical approach involving the analysis of each compound individually. Such non-targeted investigations may also examine the compounds present in growing soils, plant exudates, and plant roots. Consequently, an increasing number of researchers are developing and employing non-targeted chemical compound analysis of soil and plant samples associated with belowground interactions between plants and organisms. A significant challenge in the study of belowground interactions is the lack of reliable and efficient techniques for obtaining precise data on the quantity, quality, and spatiotemporal dynamics of these interactions. The complexity of the samples presents a significant analytical challenge. In contrast to collections made from live plants in situ, collections of allelochemicals and signaling chemicals from plant root exudates and soils are often made under random conditions. In particular, when considering the soil-borne biochemical, allelochemical, enzyme, and signaling chemical compounds that support various interactions, the design of analytical methods requires a comprehensive investigation of the actual field situations of a plant-soil system [29]. Significant effort has been made to develop a variety of analytical procedures for examining a range of biochemical substances generated from roots. These procedures have been designed to avoid any negative impact on visible root systems or jeopardize the wellbeing of bacteria linked to roots in plant-soil ecosystems. It is noteworthy that the most recent research and development efforts have concentrated on the quantity, quality, availability, spatiotemporal dynamics, and production of a range of metabolites, allelochemicals, and signaling compounds derived from soil. A number of innovative techniques are being employed with the objective of establishing the actual concentrations of the substances released by roots in the soil in their native habitat. Researchers devised an experimental system to ascertain root exudates from plants in their natural habitat. Moreover, the spatiotemporal dynamics and functions of various biochemical substances from forest soils and root exudates have been elucidated [6]. Weidenhamer and colleagues employed silicone tubing micro-extraction to fabricate sampling instruments, thereby facilitating the monitoring of biochemical compound release and their interactions between soil and plant [22]. One promising strategy is the use of a micro-dialysis-based analytical instrument to dynamically sample and quantify compounds from plants grown in soil [13]. The dynamic fluctuations of biochemical compounds released into soil from roots are continually observed and identified using a micro-dialysis probe positioned in soil microsites. This approach facilitates a deeper comprehension of the activities and functions of soil-borne metabolites and signaling molecules under soil conditions between plants and organisms. Moreover, a comprehensive grasp of biochemical interactions and processes occurring below ground hinges on precise data concerning soil-borne signaling compounds and their surrounding environment [16]. The intricate nature of plant-soil interactions presents a significant challenge in accurately identifying and quantifying soil-dwelling metabolites and signaling chemicals [26]. The study of plant-derived biochemical and signaling molecules has made significant theoretical and practical advances [14]. These molecules have been shown to play a crucial role in plant defense, enabling plants to detect and identify herbivores, competitors, and pathogens. Subsequently, allelochemicals are employed for the purpose of safeguarding the identified targets. The relationships between plants and insects are facilitated by the presence of specific metabolites and signaling molecules, which exert a significant influence on both natural and managed ecosystems [15]. One particularly relevant and significant context in which these interactions are observed is that of agro-ecosystems. In order to ensure the future sustainability of agriculture, the field of agricultural pest management is undergoing significant development [17]. The interactions between crops and pests may be regulated by signaling molecules and allelochemicals produced by plants, which could consequently enhance crop productivity. Following the infestation of crop plants by herbivores, pathogens, or weeds, the plants are able to identify and detect the pests, subsequently releasing defensive allelochemicals to counteract them, thereby conferring an advantage to their own growth [10]. The most readily transferable organic defense methods for agricultural integrated pest control and broad spectrum management are those used in organic crop farming systems and small-farm intensive agricultural systems [19]. As a consequence of this significant advancement, current agricultural production systems will witness considerable enhancements in the reduction of heavy pesticide use. Momilactones A and B have been identified as two allelochemicals that have been shown to play a significant role in conferring tolerance to submergence [15] and drought and salinity [28]. Consequently, further research is required to elucidate the mechanisms through which allelochemicals and signaling molecules influence crop protection and yield.

REFERENCES

1. Kildisheva, O. A.; Dixon, K. W.; Silveira, F. A.; Chapman, T.; Di Sacco, A.; Mondoni, A.; Cross, A. T. Dormancy and germination: Making every seed count in restoration. *Restor. Ecol.* **2020**, 28, S256–S265.
2. Böttger, A.; Vothknecht, U.; Bolle, C.; Wolf, A. Plant secondary metabolites and their general function in plants. In *Lessons on Caffeine, Cannabis & Co; Learning Materials in Biosciences*. Springer: Cham, Switzerland, 2018; pp. 3–17.
3. Kong, C.H.; Zhang, S.Z.; Li, Y.H.; Xia, Z.C.; Yang, X.F.; Meiners, S.J.; Wang, P. Plant neighbor detection and allelochemical response are driven by root-secreted signaling chemicals. *Nat. Commun.* **2018**, *9*, 3867.
4. Yang, X.F.; Li, L.L.; Xu, Y.; Kong, C.H. Kin recognition in rice (*Oryza sativa* L.) lines. *New Phytol.* **2018**, *220*, 567–578.
5. Macías, F.A.; Mejías, F.J.R.; Molinillo, J.M.G. Recent advances in allelopathy for weed control: from knowledge to applications. *Pest Manag. Sci.* **2019**, *75*(9), 2413–2436.
6. Xia, Z.C.; Kong, C.H.; Chen, L.C.; Wang, P.; Wang, S.L. A broadleaf species enhances an autotoxic conifers growth through belowground chemical interactions. *Ecology* **2016**, *97*, 2283–2292.
7. Quan, N.V.; Tran, H.D.; Xuan, T.D.; Ahmad, A.; Dat, T.D.; Khanh, T.D.; Teschke, R. Momilactones A and B are alpha-amylase and alpha-glucosidase inhibitors. *Molecules* **2019**, *24*, 482.
8. Berens, M.L.; Berry, H.M.; Mine, A.; Argueso, C.T.; Tsuda, K. Evolution of hormone signaling networks in plant defense. *Annu. Rev. Phytopathol.* **2017**, *55*, 401–425.
9. Hu, L.F.; Robert, C.A.M.; Cadot, S.; Zhang, X.; Ye, M.; Li, B.B.; Manzo, D.; Chervet, N.; Steinger, T.; van der Heijden, M.G.A.; et al. Root exudate metabolites drive plant-soil feedbacks on growth and defense by shaping the rhizosphere microbiota. *Nat. Commun.* **2018**, *9*, 2738.
10. Gfeller, A.; Glauser, G.; Etter, C.; Signarbieux, C.; Wirth, J. *Fagopyrum esculentum* alters its root exudation after *Amaranthus retroflexus* recognition and suppresses weed growth. *Front. Plant Sci.* **2018**, *9*, 50.
11. Martinez-Medina, A.; Fernandez, I.; Lok, G.B.; Pozo, M.J.; Pieterse, C.M.; Van Wees, S.C. Shifting from priming of salicylic acid- to jasmonic acid-regulated defences by Trichoderma protects tomato against the root knot nematode *Meloidogyne incognita*. *New Phytol.* **2017**, *213*, 1363–1377.
12. Murata, M.; Nakai, Y.; Kawazu, K.; Ishizaka, M.; Kajiwara, H.; Abe, H.; Takeuchi, K.; Ichinose, Y.; Mitsuhara, I.; Mochizuki, A.; et al. Loliolide, a carotenoid metabolite, is a potential endogenous inducer of herbivore resistance. *Plant Physiol.* **2019**, *179*, 1822–1833.
13. Yang, X.F.; Lei, K.; Kong, C.H.; Xu, X.H. Effect of allelochemical tricin and its related benzothiazine derivative on photosynthetic performance of herbicide-resistant barnyardgrass. *Pestic. Biochem. Physiol.* **2017**, *143*, 224–230.
14. Zhao, H.H.; Kong, C.H.; Xu, X.H. Herbicidal efficacy and ecological safety of an allelochemical-based benzothiazine derivative. *Pest Manag. Sci.* **2019**, *75*(10), 2690–2697.
15. Lucini, L.; Baccolo, G.; Rouphael, Y.; Colla, G.; Bavaresco, L.; Trevisan, M. Chitosan treatment elicited defence mechanisms, pentacyclic triterpenoids and stilbene accumulation in grape (*Vitis vinifera* L.) bunches. *Phytochemistry* **2018**, *156*, 8.
16. Pétriacq, P.; Williams, A.; Cotton, A.; McFarlane, A.E.; Rolfe, S.A.; Ton, J. Metabolite profiling of non-sterile rhizosphere soil. *Plant J.* **2017**, *92*, 147–162.
17. Quan, N.T.; Xuan, T.D. Foliar application of vanillic and p-hydroxybenozic acids enhanced drought tolerance and formation of phytoalexin momilactones in rice. *Archiv. Agron. Soil Sci.* **2018**, *64*, 1831–1846.
18. Xuan, T.D.; Khang, D.T. Effects of exogenous application of protocatechuic acid and vanillic acid to chlorophylls, phenolics and antioxidant enzymes of rice (*Oryza sativa* L.) in submergence. *Molecules* **2018**, *23*, 620.
19. Tlak Gajger, I.; Svečnjak, L.; Bubalo, D.; Žorat, T. Control of *Varroa destructor* mite infestations at experimental apiaries situated in croatia. *Diversity* **2020**, *12*, 12.
20. Tlak Gajger, I.; Ribarić, J.; Smodiš Škerl, M.; Vlainić, J.; Sikirić, P. Stable gastric pentadecapeptide BPC 157 in honeybee (Apis mellifera) therapy, to control Nosema ceranae invasions in apiary conditions. *J Vet Pharmacol Ther.* **2018** August, *41*(4), 614–621. doi: 10.1111/jvp.12509. Epub 2018 Apr 23. PMID: 29682749.

21. Tlak Gajger, I.; Vlainić, J.; Šoštarić, P.; Prešern, J.; Bubnič, J.; Smodiš Škerl, M.I. Effects onsome therapeutical, biochemical, and immunologicalparameters of honey bee (Apis mellifera) exposed toprobiotic treatments, in field and laboratory conditions. *Insects* **2020**, *11*(9), 638. doi:10.3390/insects11090638.36.
22. Tlak Gajger, I.; Sušec, P. Efficacy of varroacidalfood additive appliance during summer treatment ofhoneybee colonies (Apis mellifera). *Vet. Arhiv.* **2019**, *89*(1), 87–96. doi:10.24099/vet.arhiv.0441
23. Ullah, A.; Tlak Gajger, I.; Majoros, A.; Dar, S.A.; Khan, S.; Kalimullah, H.S.A.; Nasir Khabir, M.; Hussain, R.; Khan, H.U.; Hameed, M.; Anjum, S.I. Viral impacts on honey bee populations: A review. *Saudi J Biol Sci.* **2021** January, *28*(1), 523–530. doi:10.1016/j.sjbs.2020.10.037. Epub 2020 Oct 28. PMID: 33424335; PMCID: PMC7783639.
24. Tlak Gajger, I.; Sakač, M.; Gregorc, A. Impact of thiamethoxam on honey bee queen (*Apis mellifera carnica*) reproductive morphology and physiology. *Bull Environ Contam Toxicol.* **2017**, *99*, 297–302.
25. Zepeda, V.; Martorell, C. Seed mass equalises the strength of positive and negative plant-plant interactions in a semi-arid grassland. *Oecologia* **2019**, *190*, 287–296.
26. Wuest, S.E.; Peter, R.; Niklaus, P.A. Ecological and evolutionary approaches to improving crop variety mixtures. *Nat. Ecol. Evol.* **2021**, *5*, 1068–1077.
27. Zepeda, V.; Martorell, C. Fluctuation-independent niche differentiation and relative non-linearity drive coexistence in a species-rich grassland. *Ecology* **2019**, *100*, e02726.
28. Fréville, H.; Roumet, P.; Rode, N.O.; Rocher, A.; Latreille, M.; Muller, M.H. Preferential helping to relatives: A potential mechanism responsible for lower yield of crop variety mixtures? *Evol. Appl.* **2019**, *12*, 1837–1849.
29. Karban, R. Plant communication. *Ann. Rev. Ecol. Evol. Syst.* **2021**, *52*, 1–24.

6 Pest Management in Agriculture
Chemical and Biological Methods

Shivang Joshi, Sakshi Negi, Pooja Bargali, Himani Karakoti, Mozaniel Santana de Oliveira, and Ravendra Kumar

6.1 INTRODUCTION

A pesticide is any substance or mixture of substances or micro-organisms, including viruses, intended for repelling, destroying, or controlling any pest. Pesticides fight also vectors of human or animal disease, nuisance pests, unwanted species of plants or animals causing harm. This damage may happen during or otherwise interfering with the production, processing, storage, transport, or marketing of food, agricultural commodities, wood and wood products or animal feeding stuffs. The term includes substances intended for use as insect or plant growth regulators; defoliants; desiccants; agents for setting, thinning, or preventing the premature fall of fruit; and substances applied to crops either before or after harvest to protect the commodity from deterioration during storage and transport. The term also includes pesticide synergists and safeners, where they are integral to the satisfactory performance of the pesticide [1, 2].

The term of "chemical pesticide" is considered to embrace inorganic and organic synthetical active ingredients in any form, irrespective of whether, or to what extent, they have been formulated for application. The term is usually associated with materials intended to kill or control pests (insecticides, fungicides, herbicides, etc.), but for the present purposes, it also includes certain materials used to modify the behavior or physiology of pests (e.g. insect repellents and synergists) or of crops during production or storage (herbicide safeners, germination inhibitors) [2].

Biological pesticides, or biopesticides, are pesticides made from biological sources, that is, from toxins which occur naturally—naturally occurring biological agents used to kill pests by causing specific biological effects rather than by inducing chemical poisoning. The active ingredient is a virus, fungus, or bacteria or a natural product derived from a plant source. Biopesticides include naturally occurring substances that control pests (biochemical pesticides), microorganisms that control pests (microbial pesticides), and pesticidal substances produced by plants containing added genetic material (plant-incorporated protectants; PIPs) [3], 4].

Novel pest control methods include nanoencapsulated pesticides with gene-silencing pesticides that are based on RNA interference (RNAi) gene silencing.

Pesticides are crucial in modern agriculture, serving as a primary line of defense against pests, diseases, and weeds that threaten crop yields. The necessity of pesticides stems from the significant losses that can occur without effective pest control measures. It is estimated that pests can cause crop losses of up to 50% if left unchecked, which poses a substantial risk to food security as the global population continues to grow. Furthermore, pesticides play a vital role not only in enhancing agricultural productivity but also in preserving crops during storage and transportation, thereby reducing post-harvest losses. This protective function is essential in maintaining the supply chain and ensuring that food reaches consumers in good condition [3–5].

Current trends in pesticide use reflect a growing awareness of the need for sustainable agricultural practices. The global pesticide market is projected to expand significantly, driven by

DOI: 10.1201/9781003463429-8

innovations in pesticide formulations and application technologies aimed at maximizing effectiveness while minimizing environmental impact [4]. For instance, precision agriculture techniques, including AI-driven spraying technologies, are emerging as transformative tools that can reduce pesticide application by up to 90%. These advancements allow for targeted treatments that focus on specific pest populations while minimizing exposure to non-target organisms and reducing overall chemical usage. This shift towards precision application not only enhances the efficiency of pest control but also aligns with the increasing demand for environmentally responsible farming practices [4–6].

Moreover, there is a notable trend towards integrated pest management (IPM) strategies that combine chemical and biological methods for pest control. IPM emphasizes a holistic approach that reduces reliance on synthetic pesticides by incorporating biological control agents, such as natural predators and biopesticides derived from plants or microorganisms. This method not only addresses pest problems more sustainably but also enhances biodiversity and ecosystem health [6, 7]. Research has shown that implementing IPM can lead to reduced pesticide residues in food and lower health risks for farm workers [7]. As consumer preferences shift towards organic and sustainably produced foods, the adoption of IPM practices is likely to increase, further underscoring the importance of balancing effective pest management with ecological considerations [6, 7].

6.2 HISTORY

The use of pesticides can be traced back thousands of years, as humans have sought ways to protect their crops, livestock, and living spaces from pests, diseases, and invasive species. A brief timeline is shown in the following.

Ancient times: Ancient civilizations, such as the Sumerians and Egyptians, used various natural substances, including sulfur, arsenic compounds, and plant extracts, to control pests and preserve food.

Middle Ages and Renaissance: In Europe, the Middle Ages saw the use of toxic substances like mercury and lead arsenate to combat agricultural pests. During the Renaissance, botanical insecticides derived from plants, such as nicotine from tobacco and pyrethrum from chrysanthemums, gained popularity.

Industrial Revolution Era: The Industrial Revolution in the 18th and 19th centuries brought advancements in chemistry and the development of synthetic pesticides. In 1867, Swiss chemist Jean-François Lallemand discovered the insecticidal properties of the organochlorine compound dieldrin. In the late 19th century, the development of insecticides such as Paris green (copper (II) acetoarsenite) and lead arsenate revolutionized pest control in agriculture.

Post-World War II: The post-World War II period witnessed the rapid growth of synthetic pesticides, particularly organophosphates and organochlorines, such as DDT (dichloro-diphenyl-trichloroethane). DDT became widely used for mosquito control and agricultural purposes due to its effectiveness against pests and vectors of disease. However, its persistence in the environment and harmful effects on wildlife led to its eventual ban in many countries.

Modern Era: Concerns about the environmental and health impacts of conventional pesticides led to the development of more targeted and environmentally friendly alternatives. Integrated pest management approaches gained popularity, emphasizing the integration of various pest control methods, including biological control, cultural practices, and the judicious use of pesticides. Biopesticides, derived from natural sources such as bacteria, fungi, and botanical extracts, have gained prominence as safer alternatives to conventional pesticides. Genetically modified (GM) crops with built-in resistance to pests, known as Bt crops, have reduced the need for external pesticide applications in some cases. Figure 6.1 shows a historical overview of pest-control strategies, from traditional cultural practices and natural remedies to the introduction of synthetic chemical pesticides.

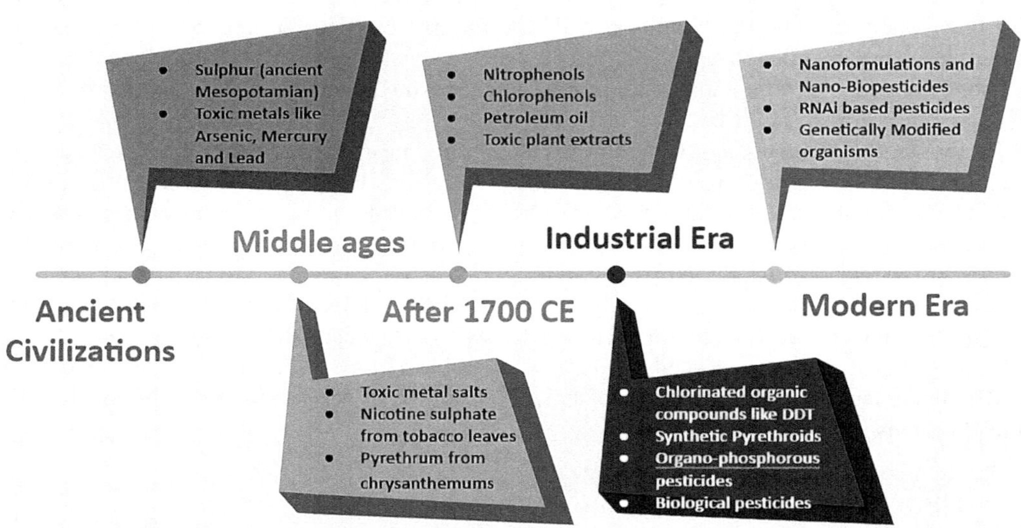

FIGURE 6.1 Timeline of pest-control methods used in agriculture.

6.3 CLASSIFICATION OF PESTICIDES

6.3.1 Classification Based on Pest Targeted and Mechanism of Action

A pest is defined as any species, strain or biotype of plant, animal, or pathogenic agent injurious to plants and plant products, materials, or environments and includes vectors of parasites, pathogens of human and animal disease, and animals causing public health nuisances.

Understanding the classification of pesticides based on the targeted pest is vital for selecting appropriate pest management strategies. By utilizing the right pesticide for the specific pest problem, farmers, gardeners, and pest control professionals can optimize effectiveness while minimizing environmental impacts and potential risks to non-target organisms; classification by target is shown in Table 6.1.

Pesticides play a crucial role in modern agriculture and pest management practices by targeting specific pests that can damage crops, plants, and public health. Effective pest control requires the use of pesticides tailored to the specific pest species or groups. The classification of pesticides based on the targeted pest encompasses various categories.

Insecticides are designed to control and eliminate insect pests such as mosquitoes, flies, termites, ants, beetles, and caterpillars. They utilize different mechanisms of action to disrupt the physiology or nervous system of insects, effectively managing infestations.

Herbicides target and control unwanted plants, commonly referred to as weeds. They inhibit the growth and development of these plants, preventing competition with desired crops or vegetation. Herbicides can be further categorized based on their mode of action, selectivity, and application methods.

Fungicides are specifically formulated to combat fungal diseases that can affect crops, trees, and other plants. They target and kill or inhibit the growth of fungi, preventing the spread of infections and reducing crop losses. Fungicides can be classified based on their spectrum of activity, mode of action, and chemical composition.

TABLE 6.1
Classification of Pesticides Based on Target Pest and Mechanism of Action [4, 5, 9]

Sr. No.	Class of Pesticide	Targeted Pest	Examples and Mechanism of Action
1.	Algaecides	Algae	• Bordeaux and burgundy mixture and other copper chelates: *(Cu2+ and Ca2+ ions inhibit spore germination by affecting enzymes in fungal spores)* • Diuron and simazine: *(inhibits photosynthesis)* • Endothal: *(inhibits phosphotase 2A)*
2.	Avicides	Birds	• Strychnine: *(glycine and acetylcholine receptor antagonist)* • Starlicide or gull toxicant: *(uremic poisoning due to irreversible necrosis of kidney)*
3.	Bactericides	Bacteria	• Streptomycin: *(inhibits bacterial protein synthesis by binding to the 30S ribosomal subunit)* • Copper sulfate: *(disrupts bacterial cell membranes and inhibits enzyme activity)* • Oxytetracycline: *(inhibits protein synthesis in bacteria by binding to the 30S ribosomal subunit)* • Bacillus thuringiensis (Bt): *(produces toxins that disrupt the gut cells of specific bacteria, leading to cell lysis)* • Chloramphenicol: *(inhibits bacterial protein synthesis by binding to the 50S ribosomal subunit)*
4.	Fungicides	Fungi	• Azoxystrobin: *(inhibits mitochondrial respiration in fungi, disrupting energy production)* • Propiconazole: *(inhibits ergosterol biosynthesis, affecting cell membrane integrity in fungi)* • Chlorothaloni: *(disrupts fungal cellular respiration and inhibits spore germination)* • Triazoles (e.g., Tebuconazole): *(inhibits the enzyme lanosterol demethylase, disrupting ergosterol synthesis in fungi)* • Mancozeb: *(multi-site inhibitor that disrupts various metabolic processes in fungi)* • Sulfur: *(disrupts fungal cell membranes and inhibits spore germination through oxidative stress)* • Neem oil: *(contains azadirachtin that disrupts fungal growth and reproduction)* • Benzimidazoles: *(inhibit mitosis in fungi by disrupting microtubule formation)* • Dithiocarbamates: *(interfere with cellular metabolism and inhibit spore germination in fungi)* • Strobilurins: *(inhibit mitochondrial respiration and energy production in fungi)*
5.	Herbicides	Plants	• Sethoxyim, Diclofop-methyl: *(targets lipid synthesis by inhibition of acetyl coenzyme A carboxylase)* • Glyphosate: *(targets protein synthesis by enolpyruvylshikimate 3-phosphate or EPSP synthase enzyme inhibition)* • Sulfonylureas: (such as flazasulfuron, imidazolinones, triazolopyrimidines, pyrimidinyl oxybenzoates, and sulfonylamino carbonyl triazolinones: *(targets amino acid synthesis by inhibition of acetolactate synthase or ACLs* • Phosphinic acid: *(glutamine synthase inhibitor; thus inhibition of protein synthesis)* • Carbamates such as asulam: *(inhibition of dihydropteroate or DHP synthase)* • Pyridynes such as dithiopyr, dinitoanilines such as benefin and butralin: *(inhibition of cell division by disrupting microtubule assembly)* • Carbamates such as propham: *(inhibition of mitosis)* • Tricarbamates and benzofurans: *(target synthesis of very long chain fatty acids)* • Benzamides such as isoxaben, triazolocarboxamides such as flupoxam: *(inhibition of cellulose synthesis)*

(Continued)

TABLE 6.1 ***(Continued)***
Classification of Pesticides Based on Target Pest and Mechanism of Action [4, 5, 9]

Sr. No.	Class of Pesticide	Targeted Pest	Examples and Mechanism of Action
6.	Insecticides	Insects	• Imidacloprid: (*neonicotinoid that binds to nicotinic acetylcholine receptors, causing paralysis*) • Bifenthrin: (*pyrethroid that disrupts sodium channel function, leading to hyperactivity and death*) • Spinosad: (*acts on the insect nervous system by targeting nicotinic acetylcholine receptors*) • Indoxacarb: (*sodium channel blocker that paralyzes insects after metabolic activation*)
7.	Miticides/ Acaricides	Mites	• Bifenazate: (*inhibits mitochondrial respiration, effective against spider mites*) • Abamectin: (*disrupts neurotransmission by binding to glutamate-gated chloride channels*) • Etoxazole: (*inhibits chitin synthesis in mites, disrupting their growth and reproduction*)
8.	Molluscicides	Snails	• Metaldehyde: (*causes hyperactivity and death in slugs and snails by disrupting their nervous system*) • Iron phosphate: (*disrupts feeding behavior in slugs and snails, leading to starvation*)
9.	Nematicides	Nematodes	• Oxamyl: (*carbamate that inhibits acetylcholinesterase, causing paralysis in nematodes*) • 1,3-Dichloropropene: (*disrupts nematode metabolism by acting as a soil fumigant, affecting respiration*)
10.	Rodenticides	Rodents	• Bromadiolone: (*anticoagulant that inhibits vitamin K epoxide reductase, leading to internal bleeding*) • Warfarin: (*disrupts blood clotting processes*)
11.	Slimicides	Slime molds, fungi, algae	• Benzalkonium chloride: (*disrupts cell membranes of slime-producing bacteria, leading to cell lysis*) • Copper sulfate: (*acts as a toxic agent on microbial cell walls, inhibiting growth of slime-forming bacteria*)
12.	Virucides	Viruses	• Hydrogen peroxide: (*oxidizes viral proteins and nucleic acids, leading to viral inactivation*) • Sodium hypochlorite: (*disrupts viral envelope proteins and nucleic acids, effectively inactivating viruses*)

Rodenticides are used to control and manage rodent populations, such as rats and mice, and sometimes squirrels, beavers, and woodchucks, which can cause damage to crops and stored products and transmit diseases. These pesticides are designed to attract rodents and deliver lethal doses of toxic substances, effectively reducing their numbers and minimizing the associated risks. One option that doesn't cause secondary poisoning and can be used to kill burrowing animals like rats and mice is by suffocation is inert gas killing that involves pumping CO_2 in rat burrows. Rodenticides are shown in Figure 6.2, and a classification of rodenticides is shown in Table 6.2. Other eco-friendly rodenticides include powdered gluten and corn starch, which rely on dehydration and the electrolyte balance of the rodent [8].

Molluscicides target and control mollusks, including snails and slugs, which can cause significant damage to crops and gardens. By employing various mechanisms, molluscicides effectively manage these pests and protect plants from feeding damage.

Rodenticide Classification

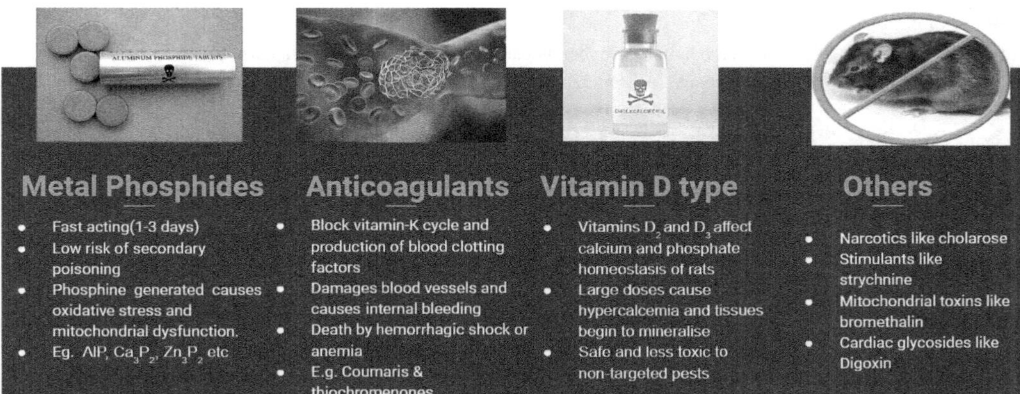

FIGURE 6.2 Chemical classification of rodenticides.

TABLE 6.2
Classification of Rodenticides

Pesticide Class: Rodenticides			
Anticoagulants/vitamin K antagonists	Coumarins/4-hydroxycoumarins	First generation	Warfarin, coumatetralyl, fumarin
		Second generation: (superwarfarins)	Brodifacoum, bromadiolone, difenacoum, difethialone, flocoumafen
	1,3-indandiones		Chlorophacinone, pindone, diphacinone
Convulsants	Crimidine, phenylsilatrane, strychnine, tetramethylenedisulfotetramine, chlorophenylsilatrane, RDX or 1,3,5-trinitro-1,3,5-triazinane		
Calciferols	Cholecalciferol, ergocalciferol		
Inorganic compounds	Aluminum phosphide, arsenic, barium carbonate, calcium phosphide, cyanide, thallium sulfate, zinc phosphide		
Organochlorine	Chloralose, endrin		
Organophosphorus	Phosacetim		
Carbamates	Aldicarb, T-1152, or Stedman's meta compound		
Others	α-naphthylthiourea, bromethalin, fluoroacetamide, flupropadine, 1,3-difluoro-2-propanol (Gliftor), norbormide, pyrinuron, scilliroside, sodium fluoroacetate		

Nematicides are pesticides developed to control nematodes, microscopic worms that can harm plant roots and reduce crop yields. They are designed to eliminate or suppress nematode populations in the soil, mitigating the damage caused by these pests.

Additionally, avicides target birds, acaricides control mites and ticks, and larvicides specifically target the larvae of insects like mosquitoes.

Pest and weed management in agricultural systems entails the utilization of an array of pesticides, encompassing insecticides and herbicides, which are classified according to their chemical properties and modes of action. Among insecticides, the following classes are of particular

TABLE 6.3
Classification of Insecticides

Pesticide Class: Insecticides	
Carbamates	Aldicarb, aminocarb, bendiocarb, butocarboxim, carbaryl, carbofuran, carbosulfan, m-cumenyl, methylcarbamate, ethienocarb, fenobucarb, isoprocarb, methomyl, metolcarb, oxamyl, promecarb, propoxur
Inorganic compounds	Aluminum phosphide, boric acid, chromated copper arsenate, copper (II) arsenate, copper (I) cyanide, cryolite, diatomaceous earth, lead hydrogen arsenate, Paris green, Scheele's green
Insect growth regulators	Benzoylureas, diflubenzuron, flufenoxuron, hydroprene, lufenuron, methoprene, pyriproxyfen
Neonicotinoids	Acetamiprid, clothianidin, dinotefuran, imidacloprid, nitenpyram, nithiazine, thiacloprid, thiamethoxam
Organochlorides	Aldrin, β-hexachlorocyclohexane, carbon tetrachloride, chlordane, cyclodiene, ortho and para dichlorobenzene, 1,1- and 1,2-dichloroethane, dichlorodiphenyldichloroethane (DDD), dichlorodiphenyldichloroethylene (DDE), dichlorodiphenyltrichloroethane (DDT), difluorodiphenyltrichloroethane (DFDT), dicofol, dieldrin, endosulfan, endrin, heptachlor, kepone, lindane, methoxychlor, mirex, tetradifon, toxaphene
Organophosphorus	Acephate, azamethiphos, azinphos-methyl, bensulide, chlorethoxyfos, chlorfenvinphos, chlorpyrifos, chlorpyrifos-methyl, coumaphos, demeton-S-methyl, diazinon, dichlorvos, dicrotophos, diisopropyl fluorophosphates, dimefox, dimethoate, dioxathion, disulfoton, ethion, ethoprop, fenamiphos, fenitrothion, fenthion, fosthiazate, isoxathion, malathion, methamidophos, methidathion, mevinphos, mipafox, monocrotophos, naled, omethoate, oxydemeton-methyl, parathion, parathion-methyl, phenthoate, phorate, phosalone, phosmet, phoxim, pirimiphos-methyl, quinalphos, R-16661 or 2-(methoxy(methylthio) phosphinylimino)-3-ethyl-5-methyl-1,3-oxazolidine, schradan, temefos, tebupirimfos, terbufos, tetrachlorvinphos, tribufos, trichlorfon
Pyrethroids	Acrinathrin, allethrins, bifenthrin, bioallethrin, cyfluthrin, cyhalothrin, cypermethrin, cyphenothrin, deltamethrin, empenthrin, esfenvalerate, etofenprox, fenpropathrin, fenvalerate, flumethrin, fluvalinate, imiprothrin, metofluthrin, permethrin, phenothrin, prallethrin, pyrethrin (I,II), chrysanthemic acid), pyrethrum, resmethrin, silafluofen, tefluthrin, tetramethrin, tralomethrin, transfluthrin
Ryanoids	Chlorantraniliprole, cyantraniliprole, flubendiamide, ryanodine, ryanodol
Other chemicals	Afoxolaner, amitraz, azadirachtin, bensultap, buprofezin, cartap, chlordimeform, chlorfenapyr, cyromazine, fenazaquin, fenoxycarb, fipronil, fluralaner, hydramethylnon, indoxacarb, limonene, lotilaner, pyridaben, pyriprole, sarolaner, sesamex, spinosad, sulfluramid, tebufenozide, tebufenpyrad, veracevine, xanthone, metaflumizone
Metabolites	Oxon, malaoxon, paraoxon, TCPy or 3,5,6-trichloro-2-pyridinol
Biopesticides	*Bacillus thuringiensis, Baculovirus, Beauveria bassiana, Eauveria brongniartii, Isaria fumosorosea, Metarhizium acridum, Metarhizium anisopliae, Nomuraea rileyi, Lecanicillium lecanii, Paenibacillus popilliae, Purpureocillium lilacinum,* Spinosad

note: carbamates, inorganic compounds, insect growth regulators, neonicotinoids, organochlorines, organophosphates, pyrethroids, and ryanoids, as well as other chemical compounds, metabolites, and biopesticides (Table 6.3). Each of these classes possesses distinctive characteristics with regard to toxicity and mode of action and is employed for the specific purpose of pest control in a variety of agricultural contexts.

Herbicides are also classified into several chemical categories, including anilides/anilines, aromatic acids, arsenicals, 4-hydroxyphenylpyruvate dioxygenase (HPPD) inhibitors, organophosphates, phenoxy acids (auxins), acetyl-CoA carboxylase (ACCase) inhibitors, protox inhibitors, pyridines, quaternary ammonium salts (quats), triazines, and ureas. Furthermore, they encompass

TABLE 6.4
Classification of Herbicides

Pesticide Class: Herbicides			
Anilides/anilines	Acetochlor, alachlor, asulam, benfluralin, butachlor, diethatyl, diflufenican, dimethenamid, flamprop, metazachlor, metolachlor, oryzalin, pendimethalin, pretilachlor, propachlor, propanil, trifluralin		
Aromatic acids	Aminopyralid, chloramben, clopyralid, dicamba, picloram, pyrithiobac, quinclorac, quinmerac		
Arsenicals	Cacodylic acid or $(CH_3)_2AsO_2H$, copper arsenate, disodium methyl arsonate or DSMA, monosodium methyl sodium or MSMA		
4-hydroxyphenylpyruvate dioxygenase (HPPD) inhibitors	Mesotrione, sulcotrione, nitisinone, leptospermone, sethoxydim, flurochloridone		
Organophosphorus	Bensulide, bialaphos, ethephon, fosamine, glufosinate, glyphosate, piperophos		
Phenoxys	Auxins	2,4-dichlorophenoxyacetic acid (2,4-D), 4-(2,4-dichlorophenoxy)butyric acid (2,4-DB), dichlorprop, fenoprop, MCPA or (2-methyl-4-chlorophenoxyacetic acid), MCPB or 4-(4-chloro-o-tolyloxy)butyric acid, mecoprop, 2,4,5-T or 2,4,5-trichlorophenoxyacetic acid	
	ACCase or acetyl-CoA carboxylase inhibitors	FOP herbicides	Chlorazifop, cyhalofop, diclofop, fenoxaprop, fluazifop, haloxyfop, quizalofop
		DIM herbicides	Sthoxydim
Protox inhibitors	Nitrophenyl ethers	Acifluorfen, bifenox, fluorodifen, fomesafen, lactofen, nitrofen, oxyfluorfen	
	Pyrimidinediones	Butafenacil, saflufenacil	
	Triazolinones	Carfentrazone, sulfentrazone	
Pyridines	Dithiopyr, fluroxypyr, imazapyr, thiazopyr, triclopyr		
Quaternary ammonium salts or quats	Photosystem I inhibitors	Cyperquat, diquat, paraquat	
Triazines	Photosystem II inhibitors	Ametryn, atrazine, cyanazine, hexazinone, prometon, prometryn, propazine, simazine, simetryn, terbuthylazine, terbutryn	
Ureas	Photosystem II inhibitors	Chlortoluron, (3-(3,4-dichlorophenyl)-1,1-dimethylurea) or DCMU, linuron, monuron, monolinuron, tebuthiuron	
	Acetolactate synthase inhibitors	Chlorsulfuron, flazasulfuron, metsulfuron-methyl, sulfometuron methyl, tribenuron	
Others	3-amino-1,2,4-triazole(3-AT), aclonifen, aminocyclopyrachlor, bentazon, bromoxynil, clomazone, (DCBN or dichlobenil) 2,6-dichlorobenzonitrile, dinoseb, indaziflam, juglone, methazole, metam sodium, metamitron, metribuzin, pyribenzoxim, ziram		

acetolactate synthase (ALS) inhibitors and other substances with particular mechanisms of action, as detailed in Table 6.4. This classification reflects the diversity of compounds available for weed control, with different mechanisms of action to prevent the development of resistance.

This comprehensive examination of the categories of insecticides and herbicides employed in agricultural settings enables the selection of optimal management strategies, with the dual objectives of efficacy and sustainability in the control of pests and weeds.

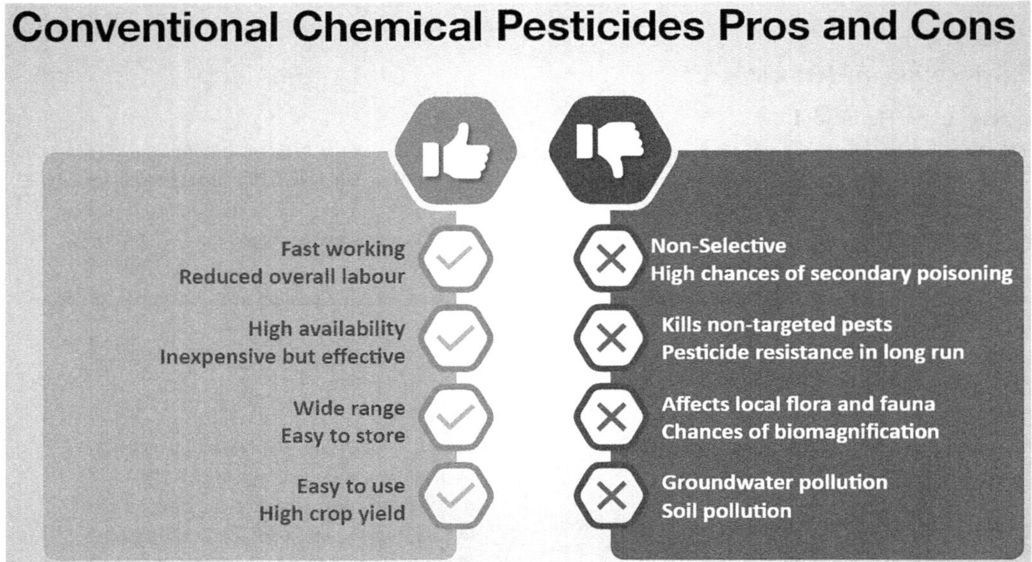

FIGURE 6.3 Pros and cons related to conventional pesticides used in agriculture.

6.3.2 Benefits and Risks Associated with Conventional Pesticides

Conventional pesticides are widely used in agriculture to enhance crop yields and protect food supplies, but their use is accompanied by significant risks and benefits that warrant careful consideration (Figure 6.3). The primary benefit of conventional pesticides lies in their ability to significantly reduce crop losses caused by pests, diseases, and weeds. Research indicates that without pesticides, more than half of the world's crops could be lost, leading to increased food insecurity for millions globally. Pesticides have contributed to a dramatic increase in agricultural productivity since their introduction, with major crops like rice and wheat experiencing yield increases of over 100% due to effective pest management [10]. Additionally, pesticides play a crucial role in public health by controlling disease vectors such as mosquitoes and ticks, thereby preventing the spread of diseases like malaria and dengue fever. Furthermore, they help maintain the quality of harvested crops and extend their shelf life, reducing post-harvest losses and ensuring a steady supply of nutritious food throughout the year [11].

Despite these benefits, the risks associated with conventional pesticide use are substantial [12–14]. Health concerns arise primarily from occupational exposure among agricultural workers who mix and apply these chemicals. Acute exposure can lead to immediate health effects such as respiratory distress, skin irritation, and neurological symptoms; chronic exposure has been linked to more serious conditions, including various cancers, reproductive issues, and developmental disorders in children. The general population is also at risk through pesticide residues found in food and drinking water. Studies have shown widespread contamination of agricultural soils and water sources with pesticide residues, raising concerns about long-term health implications for consumers [15].

Environmental risks are equally concerning. Pesticides can cause significant harm to non-target organisms, including beneficial insects like pollinators, soil microorganisms, and aquatic life [16, 17]. This disruption can lead to declines in biodiversity and the degradation of ecosystem services essential for food production. Moreover, the overuse of pesticides can result in the development of resistant pest populations, necessitating even greater chemical applications and creating a cycle of dependency that undermines sustainable agricultural practices [17].

In developing countries, the risks are exacerbated by the use of banned or restricted chemicals due to inadequate regulation and enforcement. Farmers often lack proper training in safe application techniques, leading to improper handling and increased exposure risks. The cumulative effects of pesticide mixtures—often referred to as the "cocktail effect"—are not well understood, posing additional health risks that current regulatory frameworks may not adequately address [18].

6.4 NOVEL METHODS OF PEST MANAGEMENT

Novel pest-control methods in agriculture are increasingly essential for sustainable farming, addressing the challenges posed by traditional chemical pesticides, a summary can be seen in the Figure 6.4. Biological control leverages natural predators and parasites to manage pest populations, while integrated pest management combines cultural practices, biological control, mechanical methods, and careful pesticide use to minimize environmental impact. Genetic engineering techniques, such as CRISPR, allow for the modification of pest genetics, exemplified by the sterile insect technique (SIT), which reduces pest reproduction. Nanotechnology enhances pesticide efficacy through controlled release mechanisms, reducing overall chemical usage. Smart technologies, including drones and AI, facilitate real-time monitoring and precision application of pest control measures. Additionally, pheromone-based strategies exploit insect communication to disrupt mating behaviors and attract pests away from crops. Together, these innovative approaches promote effective pest management while safeguarding ecological health and ensuring agricultural productivity in an environmentally responsible manner.

6.4.1 Nanopesticides

Nanopesticides are a relatively recent development in the field of crop protection that utilize nanotechnology to enhance the efficiency and effectiveness of pesticide formulations. Nanotechnology involves manipulating materials at the nanoscale, typically below 100 nanometers, to take

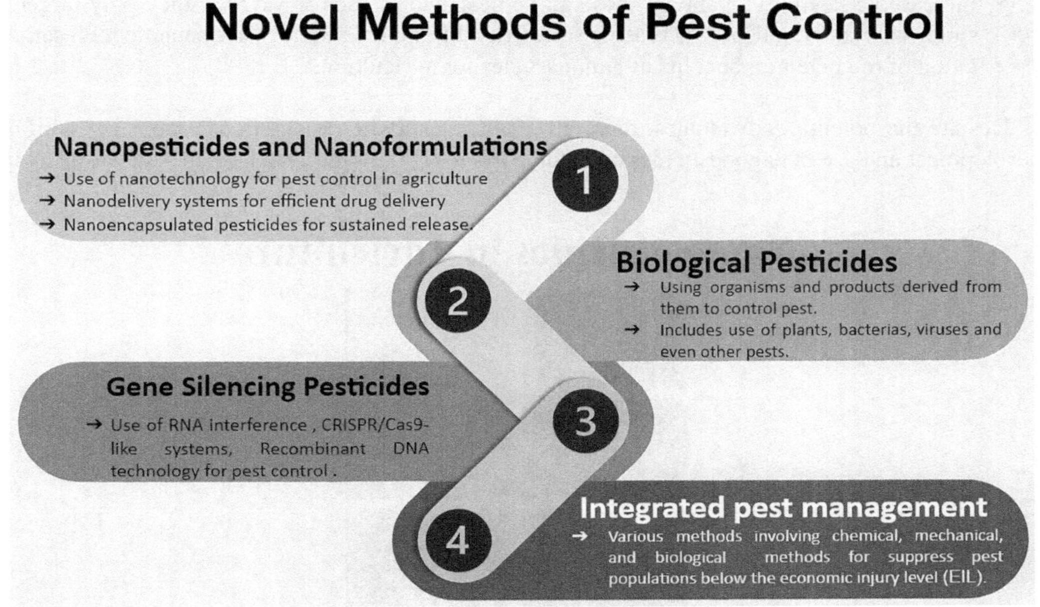

FIGURE 6.4 Novel pest-control methods used in agriculture.

advantage of unique properties and behaviors exhibited at that size range. Nano-pesticides offer several potential benefits compared to traditional pesticide formulations [19, 20]:

1. Increased Surface Area: Nanoparticles of pesticide have a high surface area-to-volume ratio, providing more contact points for interaction with pests or pathogens. This enhanced surface area allows for better adhesion to pest body and improved absorption of the active ingredients into plant or pest.
2. Controlled Release: Nanopesticides can be designed to have controlled-release properties, enabling a gradual and sustained release of the active ingredients. This controlled release mechanism can enhance the longevity of the pesticide's effectiveness, reducing the need for frequent applications.
3. Enhanced Penetration and Targeting: Nanopesticides, due to their very small size, can improve the penetration of active compounds through the protective barriers of pests or plant tissues. They can also be functionalized to specifically target certain pests or pathogens, allowing for more precise and efficient delivery of the pesticide.
4. Increased solubility: Pesticide nanoformulations are designed in such a way that even the components that were poorly soluble in water before can then be dissolved in water to make a stable nanoemulsion, which makes them very easy to use and apply.
5. Reduced Environmental Impact: Nanopesticides have the potential to reduce the overall amount of pesticide required for effective pest control. The targeted delivery and increased efficiency of nanopesticides can minimize off-target effects and reduce the environmental impact associated with conventional pesticide use.
6. Enhanced Stability: Nanoparticles can improve the stability and shelf life of the active ingredients in pesticides, protecting them from degradation or loss of efficacy. This enhanced stability can result in longer product shelf life and improved performance in the field.
7. Enhanced bioavailability and Reduced Overall Cost: Nanopesticides have high bioavailability due to their high selectivity and increased absorption and thus require very low amounts for effective action. This can cut overall cost and labor required in application, required material, and transport.
8. Increased Selectivity: Nanopestides can be designed in such a way that they only target single pest species, hence increasing selectivity. Figure 6.5 provides a schematic representation of the potential benefits of nanoparticles for agriculture.

Despite the potential advantages, there are challenges and considerations associated with the development and use of nanopesticides and nanoparticles in agriculture. These include ensuring the

FIGURE 6.5 Benefits of nanopesticides in agriculture.

safety of nanoparticles to human health and the environment, understanding their long-term effects, optimizing their production methods, and addressing regulatory requirements for their commercialization and use. Ongoing research is focused on exploring different types of nanoformulations and nanopesticides, developing efficient and scalable production techniques, evaluating their efficacy and environmental impact, and addressing any potential risks associated with their use [19–21].

6.4.2 Biological Pesticides

Biological pesticides, often referred to as biopesticides, are natural compounds derived from living organisms such as plants, bacteria, fungi, and other microbes, used to control agricultural pests. Unlike conventional chemical pesticides, biopesticides target specific pests while being environmentally friendly and less toxic to humans and wildlife. They play a crucial role in sustainable agriculture by offering a more ecologically balanced approach to pest control.

With increasing awareness of the adverse effects of synthetic pesticides on the environment and human health, the demand for biological pesticides has grown significantly in recent years. The growing popularity of organic farming, integrated pest management, and stringent regulations on chemical pesticide usage have further driven research and development in this field [22, 23].

6.4.2.1 Classification of Biological Pesticides

Biological pesticides can be broadly classified into three main categories:

6.4.2.1.1 Microbial Pesticides

The utilization of microbial biopesticides has emerged as a promising and sustainable alternative to pest management, capitalizing on the efficacy of specific microorganisms in controlling a range of species. Among the most widely utilized and researched biopesticides are bacteria such as *Bacillus thuringiensis*, which encompasses diverse varieties designed for the control of moth and butterfly caterpillars, mosquito larvae, and beetles. Other bacterial species, including *Bacillus sphaericus*, *Closridium bifermentans*, *Pseudomonas fluorescens*, and *Agrobacterium radiobacter*, have also been shown to be effective in combating insect larvae, various pests, and plant diseases. *Entomopathogenic fungi*, including *Beauveria bassiana*, *Metarhizium anisopliae*, and *Verticillium lecanii*, are extensively utilized for the management of arthropods and insects of agricultural significance. Moreover, organisms such as *Nosema locustae*, *Hetrorhabditis bacteriophora*, and *Steinernema carpocapsae* are employed in the control of grasshoppers, soil nematodes, and soil pests, respectively. Table 6.5 provides an overview of the principal microbial biopesticides currently in use, along with the respective target pests they are designed to control.

6.4.2.1.2 Biochemical Pesticides

Biopesticides, or biochemical pesticides, are defined as naturally occurring substances that are utilized for the management of pests through the deployment of non-toxic mechanisms. Such substances include pheromones, hormones, enzymes, and plant extracts. Notable examples include neem oil, *Azadirachta indica*, which is widely used to repel or kill various pests, and pheromones that interfere with pest mating behaviors, thereby reducing their populations. Insect growth regulators are synthetic analogues of natural hormones that disrupt the development of insect pests.

Natural plant products, including neem, pyrethrum, nicotine, and rotenone, are frequently utilized in pest control due to their bioactive compounds and specific modes of action. The neem tree *Azadirachta indica* produces azadirachtin and other limonoids, which function as insecticides and fungicides through the inhibition of molting and the creation of a repellent effect. Pyrethrum *Tanacetum cinerariifolium* functions as an insecticide by targeting sodium channels through its derivatives of chrysanthemic and pyrethric acids. Nicotine, derived from *Nicotiana spp.*, functions as an insecticide by acting as an acetylcholinesterase agonist, while rotenone, extracted from *Lonchocarpus* and *Tephrosia* species, serves as an insecticide and acaricide through its mitochondrial

TABLE 6.5
Some Microbial Pesticides That Are Currently Being Used and Researched as Biopesticides [24–26]

S. No.	Microbial Pesticide	Target
1.	*Bacillus thuringiensis* (var. kurstaki)	Caterpillars of moths and butterflies
2.	*Bacillus thuringiensis* (var. israelensis)	Mosquito larvae
3.	*Bacillus thuringiensis* (var. san diego)	Larvae of beetles
4.	*Bacillus sphaericus*	Mosquito larvae
5.	*Beauveria bassiana*	Various arthropods
6.	*Metarhizium anisopliae*	Termites
7.	*Nosema locustae*	Grasshoppers
8.	*Bacillus popilliae*	Japanese beetle larvae
9.	*Clostridium bifermentans*	Various insect pests
10.	*Pseudomonas fluorescens*	Various pests
11.	*Agrobacterium radiobacter*	Crown gall disease in plants
12.	*Serratia entomophila*	Various insects
13.	*Streptomyces avermitilis*	Nematodes and insect pests
14.	*Trichoderma harzianum*	Fungal pathogens and soil-borne diseases
15.	*Paecilomyces farinosus*	Various insect pests
16.	*Verticillium lecanii*	Aphids and whiteflies
17.	*Fusarium oxysporum*	Root rot pathogens
18.	*Heterorhabditis bacteriophora*	Soil nematodes
19.	*Steinernema carpocapsae*	Various soil-dwelling insect pests
20.	Nematophagous fungi (e.g., *Arthrobotrys* spp.)	Soil nematodes

TABLE 6.6
Some Commercial Natural Plant Products, Active Compounds, and Modes of Action [27, 28]

Natural Products	Source	Bioactive Compounds	Pesticidal Activity	Mode of Action
Neem	*Azadirachta indica*	Azadirachtin and other limonoids	Insecticide, fungicide	Molting inhibitors/ repellent
Pyrethrum	*Tanacetum cinerariifolium*	Derivatives of chrysanthemic acid and pyrethric acid	Insecticide	Sodium channel agonists
Nicotine	*Nicotiana* spp.	Nicotine	Insecticide	Acetylcholinesterase agonist
Rotenone	*Lonchocarpus* and *Tephrosia* species	Rotenone	Insecticide, acaricide	Mitochondrial cytotoxin
Ryania	*Ryania* spp.	Ryanodine, ryania	Insecticide	Calcium channel agonist
Sabadilla	*Schoenocaulon* spp.	Cevadine, veratridine	Insecticide	Sodium channels agonists
Quassia	Quassia, Aeschrion	Quassin	Insecticide	Unknown

cytotoxic activity. Table 6.6 provides a summary of various commercial natural plant products, their active compounds, their efficacy as pesticides, and the mechanisms of action involved.

6.4.2.1.3 Plant-Incorporated Protectants

PIPs are pesticidal substances produced by plants after specific genes have been introduced through genetic engineering. These substances protect plants from pests by either producing toxins or by strengthening the plant's defenses.

Examples: Bt crops: Genetically engineered plants that produce *Bacillus thuringiensis* toxins, providing built-in pest resistance (e.g., Bt cotton and Bt corn).

Other examples of common biological pesticides include:

Bacillus thuringiensis (Bt): One of the most widely used microbial pesticides, Bt is effective against a variety of insect pests, particularly moth larvae and caterpillars. When ingested by insects, the toxins produced by Bt cause gut paralysis and death. Bt-based products are commonly used in both organic and conventional farming.

Neem Oil: Neem oil, extracted from the neem tree (*Azadirachta indica*), acts as an insect repellent, feeding deterrent, and growth regulator. It is effective against a wide range of pests, including aphids, whiteflies, and caterpillars. Neem oil disrupts the life cycle of pests, reducing their ability to reproduce.

Trichoderma spp.: This genus of fungi is frequently used as a biological control agent for soil-borne diseases. Trichoderma species can parasitize pathogenic fungi and stimulate plant growth and defense mechanisms. They are often used in seed treatments and soil amendments to prevent root rot and other fungal diseases.

Pheromone Traps: Pheromones are chemicals that insects use for communication. Synthetic versions of these pheromones are used in traps to monitor and control insect populations. By attracting insects into traps, farmers can reduce pest numbers and prevent damage to crops.

Beauveria bassiana: This entomopathogenic fungus infects insects by penetrating their cuticle, ultimately killing them. It is effective against a wide range of insect pests, including aphids, whiteflies, and beetles. *Beauveria bassiana* is commonly used in organic farming and greenhouse settings.

6.4.3 Recent Research and Development in Biological Pesticides

Recent advances in biotechnology and microbiology have significantly expanded the potential of biological pesticides. Researchers are exploring new microbial strains, plant-based compounds, and genetically modified crops to enhance pest control efficiency. Some of the recent developments include RNA interference technology, CRISPR and genetic engineering in biopesticides, and nanotechnology in biological pesticides [4].

6.4.3.1 RNA Interference Technology

RNAi technology is being developed as a novel method for pest control. It involves silencing specific genes in pests that are essential for their survival or reproduction. RNAi-based biopesticides target specific pests without affecting non-target organisms, offering a highly selective approach to pest management.

6.4.3.2 CRISPR and Genetic Engineering in Biopesticides

CRISPR technology is being used to create plants with enhanced pest resistance. By editing the genes of plants, scientists can develop crops that produce their own biological pesticides or enhance their natural defenses against pests. For instance, CRISPR has been used to enhance the production of plant secondary metabolites with pesticidal properties.

6.4.3.3 Exploration of Endophytic Fungi and Bacteria

Endophytic microorganisms, which live inside plants without causing harm, are being studied for their potential as biopesticides. These microbes can boost plant health and provide resistance against pests and diseases. Research into identifying and harnessing beneficial endophytes is expanding the scope of biological pest control.

TABLE 6.7
Overview of Nanobiopesticides

S. No.	Nanopesticide	Subtype/Example	Target Pest	References
1.	Nanoemulsions	Insecticide nanoformulation	Aphids	[29]
2.	Nanosuspensions	Fungicide nanoformulation	Fungal pathogens	[29]
3.	Copper oxide nanoparticles	Metal-based nanopesticide	Various insects	[30]
4.	Silver nanoparticles	Metal-based nanopesticide	Bacterial pathogens	[31]
5.	Zinc oxide nanoparticles	Metal-based nanopesticide	Fungal pathogens	[32]
6.	Chitosan nanoparticles	Biopesticide	Nematodes	[29, 33]
7.	Graphene oxide nanoparticles	Carbon-based nanopesticide	Insects	[34]
8.	Polymeric nanoparticles	Biodegradable carrier	Various pests	[29, 35]
9.	Nanoencapsulated essential oils	Biopesticide	Insect pests and fungal pathogens	[36–38]
10.	Nanoscale titanium dioxide	Metal-based nanopesticide	Fungal diseases	[39]
11.	Nanocapsules	Controlled release	Various pests	[29, 40]
12.	Nanoscale iron oxide	Metal-based nanopesticide	Soil-borne pathogens	[41]
13.	Silica nanoparticles	Carrier for various pesticides	Various pests and depends on pesticide	[42]
14.	Bimetallic nanoparticles	Metal-based nanopesticide	Specific insect pests	[43]
15.	Nanoscale calcium carbonate	Mineral-based nanopesticide	Fungal pathogens	[43]
16.	Nanoscale magnesium oxide	Metal-based nanopesticide	Insects	[43, 44]

6.4.3.4 Nanotechnology in Biological Pesticides

The application of nanotechnology to biopesticides has the potential to enhance their stability, delivery, and efficacy. Nanoparticles can be employed to encapsulate biopesticides, thereby safeguarding them from environmental degradation and facilitating controlled release. This results in more effective pest control while reducing the quantity of pesticide necessary. Nano-biological pesticides represent a sustainable and environmentally friendly alternative to synthetic chemical pesticides. Their capacity to target specific pests, reduce environmental contamination, and promote biodiversity renders them an indispensable component of contemporary pest management practices. As research and technological advancements continue, it is probable that the scope and effectiveness of nano-biopesticides will expand, thereby contributing to a more sustainable future for agriculture.

The growing emphasis on precision agriculture, biotechnology, and environmentally conscious practices underscores the significance of biological pesticides in addressing the mounting challenges of food security, pest resistance, and environmental stewardship. The integration of biological pesticides with other pest management strategies enables farmers to safeguard their crops, enhance productivity, and preserve ecosystems for future generations. Table 6.7 provides a comprehensive overview of various nanobiopesticides, including their subtypes, target pests, and the corresponding references. This table illustrates the considerable diversity of nanoformulations, which encompass a range of approaches, including nanoemulsions, nanosuspensions, metal-based carriers, and biodegradable formulations. Each of these has been developed to target specific pest types.

6.5 CONCLUSION

In the realm of agriculture, pest management is a critical area of study, particularly as it pertains to the integration of chemical and biological methods. The current landscape of pest management strategies reflects a growing recognition of the limitations associated with synthetic pesticides, which have historically dominated agricultural practices. Recent research emphasizes the necessity for sustainable alternatives that mitigate the adverse effects of chemical use on human health and the environment. Integrated pest management has emerged as a foundational framework, advocating for a balanced approach that prioritizes biological control methods while employing chemical solutions only as a last resort. This paradigm shift is driven by an increasing body of evidence highlighting the efficacy of biocontrol agents and the importance of ecological principles in establishing resilient agricultural systems.

Recent advancements in biological pest control have underscored its potential to revolutionize pest management practices. Studies indicate that biopesticides, derived from natural materials such as microorganisms and plant extracts, offer promising alternatives to traditional synthetic pesticides. These biopesticides not only demonstrate specificity in targeting pests but also exhibit lower toxicity to non-target organisms, including beneficial insects and humans. Furthermore, innovations in nanobiotechnology have led to the development of nanobiopesticides, which enhance the delivery and effectiveness of biological agents in agricultural settings. Current research is focused on optimizing these biocontrol methods, exploring their mechanisms of action, and addressing challenges related to their commercial viability and shelf life. The integration of these biological methods into existing IPM frameworks represents a significant step toward sustainable agriculture.

As we look toward the future of pest management in agriculture, it is imperative to continue exploring the intersection between chemical and biological methods. Ongoing research aims to refine these strategies through interdisciplinary collaboration, combining insights from entomology, ecology, and molecular biology. The goal is to develop comprehensive pest management systems that not only enhance crop yields but also promote environmental stewardship. By leveraging current scientific findings and fostering innovation in pest control technologies, the agricultural sector can work toward achieving food security while minimizing ecological impacts. The evolution of pest management practices will undoubtedly hinge on our ability to balance effective pest control with sustainable agricultural practices that safeguard both human health and the environment.

REFERENCES

1. *Plant production and protection division: Manual on development and use of FAO and Who specifications for pesticides.* (n.d.). Retrieved January 3, 2024, from https://www.fao.org/agriculture/crops/thematic-sitemap/theme/pests
2. Organization, W. H., & Nations, F. and A. O. of the U. (2010). *International code of conduct on the distribution and use of pesticides: Guidelines for the registration of pesticides.* https://iris.who.int/handle/10665/70293
3. *Guidelines on data requirements for the registration of pesticides. International code of conduct on the distribution and use of pesticides.* (n.d.). Retrieved January 3, 2024, from https://www.who.int/publications/i/item/WHO-HTM-NTD-WHOPES-2013.7
4. Krieger, R. I., Doull, J., & Vega, H. (2010). *Hayes' handbook of pesticide toxicology* (Vol. 1, 3rd ed.). Elsevier.
5. Pohanish, R. P. (2015). *Sittig's handbook of pesticides and agricultural chemicals* (2nd ed.). William Andrew Publishing.
6. Zaman, Q. (Ed.). (2023). *Precision agriculture: Evolution, insights and emerging trends.* Academic Press.
7. Baker, B. P., Green, T. A., & Loker, A. J. (2020). Biological control and integrated pest management in organic and conventional systems. *Biological Control, 140,*
8. Turner, P. V., Hickman, D. L., van Luijk, J., Ritskes-Hoitinga, M., Sargeant, J. M., Kurosawa, T. M., Agui, T., Baumans, V., Choi, W. S., Choi, Y.-K., Flecknell, P. A., Lee, B. H., Otaegui, P. J., Pritchett-Corning, K. R., & Shimada, K. (2020). Welfare impact of carbon dioxide euthanasia on laboratory mice and rats: A systematic review. *Frontiers in Veterinary Science, 7.*
9. Abubakar, Y., Tijjani, H., Egbuna, C., Adetunji, C. O., Kala, S., Kryeziu, T. L., Ifemeje, J. C., & Patrick-Iwuanyanwu, K. C. (2020). Chapter 3—Pesticides, history, and classification. In C. Egbuna & B. Sawicka (Eds.), *Natural remedies for pest, disease and weed control* (pp. 29–42). Academic Press.

10. Aktar, W., Sengupta, D., & Chowdhury, A. (2009). Impact of pesticides use in agriculture: Their benefits and hazards. *Interdisciplinary Toxicology*, *2*(1), 1–12.
11. Tudi, M., Daniel Ruan, H., Wang, L., Lyu, J., Sadler, R., Connell, D., Chu, C., & Phung, D. T. (2021). Agriculture development, pesticide application and its impact on the environment. *International Journal of Environmental Research and Public Health*, *18*(3), 1112.
12. Kaur, R., Choudhary, D., Bali, S., Bandral, S. S., Singh, V., Ahmad, M. A., Rani, N., Singh, T. G., & Chandrasekaran, B. (2024). Pesticides: An alarming detrimental to health and environment. *Science of the Total Environment*, *915*, 170113.
13. Pathak, V. M., Verma, V. K., Rawat, B. S., Kaur, B., Babu, N., Sharma, A., Dewali, S., Yadav, M., Kumari, R., Singh, S., Mohapatra, A., Pandey, V., Rana, N., & Cunill, J. M. (2022). Current status of pesticide effects on environment, human health and it's eco-friendly management as bioremediation: A comprehensive review. *Frontiers in Microbiology*, *13*, 962619.
14. Tang, F. H. M., Lenzen, M., McBratney, A., & Maggi, F. (2021). Risk of pesticide pollution at the global scale. *Nature Geoscience*, *14*(4), 206–210.
15. Hu, R., Huang, X., Huang, J., Li, Y., Zhang, C., Yin, Y., Chen, Z., Jin, Y., Cai, J., & Cui, F. (2015). Long- and short-term health effects of pesticide exposure: A cohort study from china. *PLoS One*, *10*(6), e0128766.
16. Serrão, J. E., Plata-Rueda, A., Martínez, L. C., & Zanuncio, J. C. (2022). Side-effects of pesticides on non-target insects in agriculture: A mini-review. *The Science of Nature*, *109*(2), 17.
17. Trends and challenges in pesticide resistance detection. (2016). *Trends in Plant Science*, *21*(10), 834–853.
18. Göbölös, B., Sebők, R. E., Szabó, G., Tóth, G., Szoboszlay, S., Kriszt, B., Kaszab, E., & Háhn, J. (2024). The cocktail effects on the acute cytotoxicity of pesticides and pharmaceuticals frequently detected in the environment. *Toxics*, *12*(3), 189.
19. Jampílek, J., & Kráľová, K. (2017). 3—Nanopesticides: Preparation, targeting, and controlled release. In A. M. Grumezescu (Ed.), *New pesticides and soil sensors* (pp. 81–127). Academic Press.
20. Yadav, J., Jasrotia, P., Kashyap, P. L., Bhardwaj, A. K., Kumar, S., Singh, M., & Singh, G. P. (2021). Nanopesticides: Current status and scope for their application in agriculture. *Plant Protection Science*, *58*(1), 1–17.
21. Cătălin Balaure, P., Gudovan, D., & Gudovan, I. (2017). 4—Nanopesticides: A new paradigm in crop protection. In A. M. Grumezescu (Ed.), *New pesticides and soil sensors* (pp. 129–192). Academic Press.
22. Ayilara, M. S., Adeleke, B. S., Akinola, S. A., Fayose, C. A., Adeyemi, U. T., Gbadegesin, L. A., Omole, R. K., Johnson, R. M., Uthman, Q. O., & Babalola, O. O. (2023). Biopesticides as a promising alternative to synthetic pesticides: A case for microbial pesticides, phytopesticides, and nanobiopesticides. *Frontiers in Microbiology*, *14*.
23. Dhawan, A. K., & Peshin, R. (2009). Integrated pest management: Concept, opportunities and challenges. In R. Peshin & A. K. Dhawan (Eds.), *Integrated pest management: Innovation-development process* (Vol. 1, pp. 51–81). Springer Netherlands.
24. Hezakiel, H. E., Thampi, M., Rebello, S., & Sheikhmoideen, J. M. (2024). Biopesticides: A green approach towards agricultural pests. *Applied Biochemistry and Biotechnology*, *196*(8), 5533–5562.
25. Koul, O., & Dhaliwal, G. S. (Eds.). (2001). *Microbial biopesticides*. CRC Press.
26. Kumar, P., Kamle, M., Borah, R., Mahato, D. K., & Sharma, B. (2021). Bacillus thuringiensis as microbial biopesticide: Uses and application for sustainable agriculture. *Egyptian Journal of Biological Pest Control*, *31*(1), 95.
27. Khursheed, A., Rather, M. A., Jain, V., Wani, A. R., Rasool, S., Nazir, R., Malik, N. A., & Majid, S. A. (2022). Plant based natural products as potential ecofriendly and safer biopesticides: A comprehensive overview of their advantages over conventional pesticides, limitations and regulatory aspects. *Microbial Pathogenesis*, *173*, 105854.
28. Umetsu, N., & Shirai, Y. (2020). Development of novel pesticides in the 21st century. *Journal of Pesticide Science*, *45*(2), 54–74.
29. Gupta, R., Malik, P., Rani, R., Solanki, R., Ameta, R. K., Malik, V., & Mukherjee, T. K. (2024). Recent progress on nanoemulsions mediated pesticides delivery: Insights for agricultural sustainability. *Plant Nano Biology*, *8*, 100073.
30. Manzoor, M. A., Shah, I. H., Ali Sabir, I., Ahmad, A., Albasher, G., Dar, A. A., Altaf, M. A., & Shakoor, A. (2023). Environmental sustainable: Biogenic copper oxide nanoparticles as nano-pesticides for investigating bioactivities against phytopathogens. *Environmental Research*, *231*, 115941.
31. Bapat, M. S., Singh, H., Shukla, S. K., Singh, P. P., Vo, D.-V. N., Yadav, A., Goyal, A., Sharma, A., & Kumar, D. (2022). Evaluating green silver nanoparticles as prospective biopesticides: An environmental standpoint. *Chemosphere*, *286*, 131761.

32. Agredo-Gomez, A. D., Molano-Molano, J. A., Portela-Patiño, M. C., & Rodríguez-Páez, J. E. (2024). Use of ZnO nanoparticles as a pesticide: In vitro evaluation of their effect on the phytophagous *Puto barberi* (Mealybug). *Nano-Structures & Nano-Objects*, *37*, 101095.
33. Abenaim, L., & Conti, B. (2023). Chitosan as a control tool for insect pest management: A review. *Insects*, *14*(12), 949.
34. Sharma, S., Singh, S., Ganguli, A. K., & Shanmugam, V. (2017). Anti-drift nano-stickers made of graphene oxide for targeted pesticide delivery and crop pest control. *Carbon*, *115*, 781–790.
35. Dasauni, K., Divya, Mathpal, P., & Nailwal, T. K. (2022). Chapter 26—Polymeric nanoparticle-based insecticide: A critical review of agriculture production. In M. Ghorbanpour & M. A. Shahid (Eds.), *Nano-enabled agrochemicals in agriculture* (pp. 445–466). Academic Press.
36. Albuquerque, P. M., Azevedo, S. G., de Andrade, C. P., D'Ambros, N. C. de S., Pérez, M. T. M., & Manzato, L. (2022). Biotechnological applications of nanoencapsulated essential oils: A review. *Polymers*, *14*(24), 5495.
37. Das, S., Singh, V. K., Dwivedy, A. K., Chaudhari, A. K., & Dubey, N. K. (2024). Insecticidal and fungicidal efficacy of essential oils and nanoencapsulation approaches for the development of next generation ecofriendly green preservatives for management of stored food commodities: An overview. *International Journal of Pest Management*, *70*(3), 235–266.
38. Maes, C., Bouquillon, S., & Fauconnier, M.-L. (2019). Encapsulation of essential oils for the development of biosourced pesticides with controlled release: A review. *Molecules*, *24*(14), 2539.
39. Gutiérrez-Ramírez, J. A., Betancourt-Galindo, R., Aguirre-Uribe, L. A., Cerna-Chávez, E., Sandoval-Rangel, A., Ángel, E. C., Chacón-Hernández, J. C., García-López, J. I., & Hernández-Juárez, A. (2021). Insecticidal effect of zinc oxide and titanium dioxide nanoparticles against bactericera cockerelli sulc. (Hemiptera: Triozidae)On tomato solanum lycopersicum. *Agronomy*, *11*(8), 1460.
40. Nuruzzaman, Md., Rahman, M. M., Liu, Y., & Naidu, R. (2016). Nanoencapsulation, nano-guard for pesticides: A new window for safe application. *Journal of Agricultural and Food Chemistry*, *64*(7), 1447–1483.
41. Nadeem, F., Fozia, F., Aslam, M., Ahmad, I., Ahmad, S., Ullah, R., Almutairi, M. H., Aleya, L., & Abdel-Daim, M. M. (2022). Characterization, antiplasmodial and cytotoxic activities of green synthesized iron oxide nanoparticles using nephrolepis exaltata aqueous extract. *Molecules*, *27*(15), 4931.
42. Saw, G., Nagdev, P., Jeer, M., & Murali-Baskaran, R. K. (2023). Silica nanoparticles mediated insect pest management. *Pesticide Biochemistry and Physiology*, *194*, 105524.
43. Pathania, D., Kumar, S., Thakur, P., Chaudhary, V., Kaushik, A., Varma, R. S., Furukawa, H., Sharma, M., & Khosla, A. (2022). Essential oil-mediated biocompatible magnesium nanoparticles with enhanced antibacterial, antifungal, and photocatalytic efficacies. *Scientific Reports*, *12*(1), 11431.
44. Khan, M. R., Siddiqui, Z. A., & Fang, X. (2022). Potential of metal and metal oxide nanoparticles in plant disease diagnostics and management: Recent advances and challenges. *Chemosphere*, *297*, 134114.

Section 2

Essential Oils and Allelochemicals

7 Allelochemical Approaches and Phytotoxic Potential of Essential Oils

Oberdan Oliveira Ferreira, Celeste de Jesus Pereira Franco, Márcia Moraes Cascaes, Karyme do Socorro de Souza Vilhena, Marcilene Paiva da Silva, Mozaniel Santana de Oliveira, and Eloisa Helena de Aguiar Andrade

7.1 INTRODUCTION

Essential oils are derived from the secondary metabolism of aromatic plants. In nature, they perform numerous functions in protecting plants, such as resisting pathogens, functioning as antimicrobial agents, and hindering herbivore attacks [1, 2]. These functions may be associated with a volatile compound or with a combination of various chemical constituents via synergism [1, 3].

The biological activities of volatile oils have attracted considerable interest in various branches and market sectors, primarily for their potential applications in the development of new plant-based products, with a low impact on environmental and human health [4].

To meet market demands, the agricultural sector excessively uses chemical pesticides to control pests. However, this unrestricted use may have negative effects on consumers [5]. Certain commercially available synthetic products severely harm human health, resulting in cancer, allergies, and hypersensitivity [6].

Biomes contain a wide range of aromatic plants. Many of these species produce essential oils, which have a significant potential for use in natural pesticides based on their phytotoxic properties reported in previous studies [7, 8]. Therefore, this chapter aims at reviewing the literature on the phytotoxicity of essential oils.

7.2 ALLELOCHEMICAL APPROACH TO ESSENTIAL OILS

Invasive plants cause serious problems in ecosystems because they compete with the primary crops for water and nutrient absorption, thereby resulting in major losses in productivity [9]. The invasion mechanism of these plants involves contact with plant chemicals, known as allelochemicals [10].

Allelochemicals can affect plant development, including hormonal balance, protein synthesis, respiration, photosynthesis, chlorophyll formation, and the plant–water relationship [11]. Consequently, essential oils with allelochemical functions can inhibit the herbicidal effect of these plants via the defense system [12]. This defense mechanism is attributed to volatile oils and has prompted several studies on natural products because invasive plants have developed a slight resistance to specific synthetic herbicides [13].

Therefore, the allelochemical potential of essential oils is associated with their chemical profile. Table 7.1 characterizes the primary components of volatile oils in aromatic species that were prepared using different extraction techniques. In addition, the table highlights a high variability of chemical compounds. This chemical profile variability is determined by factors such as climate change, light exposure, vegetative stage (flowering and fruiting), aerial parts, environmental

DOI: 10.1201/9781003463429-10

TABLE 7.1
Primary Volatile Compounds of the Essential Oils in Various Aromatic Plants

Species	Collection Site	Plant Part	Extraction Type	Major Components	Reference
Araucaria heterophylla	Egypt	Resin	MAE	α-pinene (62.57%), β-pinene (6.60%), germacrene D (5.88%), and β-caryophyllene (3.56%)	[23]
Calycolpus goetheanus specimen A	Magalhães Barata, Pará, Brazil	Leaves	HD	β-eudesmol (22.83%), (E)-caryophyllene (14.61%), and γ-eudesmol (13.87%)	[16]
C. goetheanus specimen B				1,8-cineole (8.64%), (E)-caryophyllene (5.86%), δ-cadinene (5.78%), and palustrol (4.97%)	
C. goetheanus specimen C				α-cadinol (9.03%), δ-cadinene (8.01%), and (E)-caryophyllene (6.74%)	
Citharexylum spinosum	Monastir (Tunisia)	Roots	HD	α-phellandrene (30.8%), dill ether (20.1%), dillapiol (19.9%), and limonene (15.2%)	[17]
		Leaves		hexahydrofarnesyl acetone (26%), nonadecane (14.5%), linalool (7.5%), and tetradecanoic acid (6.6%).	
		Flowers		2-phenylethyl benzoate (33.5%), linalool (13.4%), benzyl benzoate (13.4%), phenylethyl salicylate (9.7%), and 1-octen-3-ol (5.6%).	
		Stems		Cuparene (16.4%), 1-octen-3-ol (8.6%), dillapiol (7.7%), α-phellandrene (7.5%), 2-heptadecanone (6.9%), hexadecane (6.8%), hinesol (5.8%), and 1-undecanol (5.6%).	
Citrus bergamia	Mugla (Turkey)	Aerial parts	HD	Linalool (41.62%) and limonene (28.02%)	[18]
Cleome droserifolia	Wadi Hagul (Egypt)	Aerial parts (stems, leaves, and flowers)	HD	cis-nerolidol (37.56%), α-cadinol (9.29%), δ-cadinene (7.65%), and γ-muurolene (4.10%)	[15]
Curcuma longa	India	Rhizome and root	Commercial samples	α-zingiberene (24.9 ± 0.8%), β-sesquiphelladrene (11.7 ± 0.3%), ar-curcumene (10.7 ± 0.2%), and β-bisabolene (10.5 ± 0.3%)	[21]
Eremanthus erythropappus	Ouro Preto, Brazil	Leaves	SD	(β)-caryophyllene (29.3%), caryophyllene oxide (22.1%), and β-pinene (12.8%)	[22]
Eucalyptus bicostata	Bizerte (Tunisia)	Leaves	HD	Eucalyptol (59.3%), spathulenol (11.8%), and α-terpineol (4.5%).	[19]
E. brockwayi	New South Wales (Australia)	Fresh leaves	SD	α-pinene (31.1%), isopentyl isovalerate (20.2%), and 1,8-cineole (16.9%)	[20]
E. dundasii				1,8-cineole (65.5%) and α-pinene (19.9%)	
E. salubris				1,8-cineole (57.6%), α-pinene (10.9%), and p-cymene (8.3%)	

(Continued)

TABLE 7.1 ***(Continued)***
Primary Volatile Compounds of the Essential Oils in Various Aromatic Plants

Species	Collection Site	Plant Part	Extraction Type	Major Components	Reference
E. striaticalyx	Zaghouan (Tunisia)	Leaves	HD	eucalyptol (29.6%), *p*-cymene (19.4%), and spathulenol (8.7%)	[24]
Guatteria schomburgkiana	Magalhães Barata, Pará, Brazil	Leaves	HD	spathulenol (22.40%) and caryophyllene oxide (14.70%)	[7]
Xylopia frutescens	Magalhães Barata (Pará—Brazil)	Leaves	HD	α-pinene (35.73%) and β-pinene (18.90%)	[7]

HD: Hydrodistillation; MAE: Microwave-assisted extraction; SD: Steam distillation.

conditions, soil type, extraction method, amount of rainfall, and plant–insect, plant–plant, and plant–microorganism interactions [14].

Monoterpenes are the major components of the essential oils in *Araucaria heterophylla* (Araucariaceae), *Calycolpus goetheanus* specimen B (Myrtaceae), *Citharexylum spinosum* (Verbenaceae), *Citrus bergamia* (Rutaceae), *Eucalyptus bicostata* (Myrtaceae), *E. brockwayi* (Myrtaceae), *E. dundasii* (Myrtaceae), *E. salubris* (Myrtaceae), and *E. striaticalyx* (Myrtaceae) [15–20].

Conversely, sesquiterpenes characterize the profiles of the essential oils in *Calycolpus goetheanus* specimens A and C (Myrtaceae), *Cleome droserifolia* (Cleomaceae), *Curcuma longa* (Zingiberaceae), *Eremanthus erythropappus* (Asteraceae), and *Guatteria schomburgkiana* (Annonaceae) [7, 10, 16, 21, 22].

7.3 PHYTOTOXICITY

The excessive use of chemical pesticides threatens both environmental and human health due to their toxicity and low biodegradability [7, 25]. To mitigate these impacts, the agricultural industry has been recently seeking sustainable alternatives, such as pesticides based on plant essential oils. Essential oils can be used to develop new biopesticides or low-impact synthetic pesticides that behave similarly to that of natural pesticides [26, 27].

Phytotoxicity is defined as a negative effect on plant growth or fitness [25]. This biological effect can affect plant physiology and biochemistry processes, including respiration, photosynthesis, cell division and differentiation, plant–water relationship, enzyme activity, DNA replication, signal transduction, and gene expression [28].

Table 7.2 presents studies on the phytotoxic potential of essential oils, which depends on their chemical composition, primarily oxygenated monoterpenes [29]. For example, 1,8-cineol [30], linalool [31], carvacrol [32], and terpineol [33] are monoterpenes known for their herbicidal effects. However, the phytotoxicity of essential oils is typically related to the synergistic action of several compounds [33] and not just to the biological activity of a single chemical component.

Carvacrol was the primary constituent identified in the essential oil of *Thymus decussatus* studied by Saleh et al. [34]. This oil inhibited *Lactuca sativa* seed germination and both shoot and root growth. Consequently, the essential oil of *Salvia sclarea* exhibited high concentrations of oxygenated monoterpenes, primarily linalool, with phytotoxic potential in *L. sativa*, *Lepidium sativum*,

TABLE 7.2
Phytotoxic Potential of Various Aromatic Species

Species	Collection Site	Plant Part	Results	Reference
Araucaria heterophylla	Egypt	Resin	Seed germination (*Chenopodium murale*)—IC_{50} = 304.0 mg/L Seedling shoot growth (*C. murale*)—C_{50} = 230.1 mg/L Seedling root growth (*C. murale*)—IC_{50} = 147.1 mg/L Seed germination (*Sonchus oleraceus*)—IC_{50} = 295.7 mg/L Seedling shoot growth (*S. oleraceus*)—IC_{50} = 224.5 mg/L Seedling root growth (*S. oleraceus*)—IC_{50} = 106.1 mg/L	[23]
Calycolpus goetheanus	Magalhães Barata (Brazil)	Aerial parts	Germination inhibition (*Mimosa pudica*) = 30.0–33.33% Radicle elongation inhibition (*M. pudica*) = 43.24–57.83% Hypocotyl inhibition (*M. pudica*) = 71.78–80.87% Radicle elongation inhibition (*Senna obtusifolia*) = 57.24–62.57% Hypocotyl inhibition (*S. obtusifolia*) = 34.63–42.79%,	[36]
Citharexylum spinosum	Monastir (Tunisia)	Flowers	Seed germination inhibition (*Lactuca sativa*) = 50% and 100.00% (doses of 0.4 and 1 mg/mL, respectively) Root length inhibition (*L. sativa*) = 72.4% and 100% (doses of 0.4 and 1 mg/mL, respectively) Shoot length inhibition (*L. sativa*) = 100.0% (dose of 1 mg/mL)	[35]
C. spinosum	Monastir (Tunisia)	Root	Seed germination inhibition (*Lactuca sativa*) = 50.0% (dose of 1 mg/mL) Root length inhibition (*L. sativa*) = 60.0% (dose of 1 mg/mL)	
C. spinosum	Monastir (Tunisia)	Leaves	Shoot length inhibition (*Lactuca sativa*) = 100.0% (dose of 1 mg/mL) Root length inhibition (*L. sativa*) = 100.0% (dose of 1 mg/mL)	
C. spinosum	Monastir (Tunisia)	Stem	Seed germination inhibition (*Lactuca sativa*) = 54.3% (dose of 1 mg/mL) Root length inhibition (*L. sativa*) = 60.0% (dose of 1 mg/mL)	
Citrus bergamia	Mugla (Turkey)	Aerial parts	Germination inhibition (*Amaranthus retroflexus*) = 100.0% (dose of 20 µL/mL) Root length inhibition (*A. retroflexus*) = 84.2%, 86.9%, and 100.0% (doses of 10, 15, and 20 µL/mL, respectively) Radicle length inhibition (*A. retroflexus*) = 86.4%, 90.8%, 92.9%, and 100.0% (doses of 5, 10, 15, and 20 µL/mL, respectively)	[18]
Cleome droserifolia	Wadi Hagul (Egypt)	Aerial parts	Speed of germination index (*Chenopodium murale*) = 79.0% (dose of 200 µL/L) Root length inhibition (*C. murale*) = 79.9% (dose of 200 µL/L) Speed of germination index (*Cuscuta trifolii*) = 59.2% (dose of 200 µL/L) Root length inhibition (*C. trifolii*) = 54.2% (dose of 200 µL/L) Shoot length inhibition (*C. trifolii*) = 61.5% (dose of 200 µL/L) Speed of germination index (*Melilotus indicus*) = 73.0% (dose of 200 µL/L)	

			Root length inhibition (*M. indicus*) = 75.9% (dose of 200 µL/L) Shoot length inhibition (*M. indicus*) = 52.9% (dose of 200 µL/L) Speed of germination index (*Trifolium repens*) = 55.6% (dose of 200 µL/L)	[28]
Curcuma longa	India	Root	Seed germination inhibition (*Cortaderia selloana*) = 60.97% and 81.71% (doses of 0.5 and 1 µL/mL, respectively) Seed hypocotyl length inhibition (*C. selloana*) = 83.33% and 97.82% (doses of 0.5 and 1 µL/mL, respectively) Seed radicle length inhibition (*C. selloana*) = 86.08% and 99.74% (doses of 0.5 and 1 µL/mL, respectively) Seed hypocotyl length inhibition (*Portulaca oleracea*) = 56.43% (dose of 1 µL/mL) Seed hypocotyl length inhibition (*Nicotiana glauca*) = 86.22% (dose of 1 µL/mL) Seed radicle length inhibition (*N. glauca*) = 51.42% (dose of 1 µL/mL)	[21]
Eremanthus erythropappus	Ouro Preto (Brazil)	Leaves	Seed germination (*Lactuca sativa*)—IC_{50} = 97.1 µL/mL Radicle length of seedlings (*L. sativa*)—IC_{50} = 35.1 µL/mL Seed germination (*Lycopersicum esculentum*)—IC_{50} = > 200 µL/mL Radicle length of seedlings (*L. esculentum*)—IC_{50} = 130.1 µL/mL Seed germination (*Brassica rapa*)—IC_{50} = 26.5 µL/mL Seed germination (*B. rapa*)—IC_{50} = 54.9 µL/mL Radicle length of seedlings (*B. rapa*)—IC_{50} = 24.5 µL/mL Radicle length of seedlings (*Bidens pilosa*)—IC_{50} = 16.3 µL/mL	[22]
Eucalyptus bicostata	Bizerte (Tunisia)	Leaves	Seed germination inhibition (*Raphanus sativus*) = 92.3% (dose of 1000 µg/mL) Radical length inhibition (*R. sativus*) = 77.14% (dose of 1000 µg/mL) Seed germination inhibition (*Sinapis arvensis*) = 100.0% (doses of 500 and 1000 µg/mL) Radical length inhibition (*S. arvensis*) = 100.0% (doses of 500 and 1000 µg/mL) Radical length inhibition (*Lolium multiflorum*) = 61.1% (dose of 1000 µg/mL)	[19]
E. brockwayi	New South Wales (Australia)	Leaves	Germination inhibition index (So*lanum elaeagnifolium*) = 64.5% Root length index (S. *elaeagnifolium*) = 77.7% Shoot length index (S. *elaeagnifolium*) = 74.3%	[20]
E. dundasii	New South Wales (Australia)	Leaves	Germination inhibition index (So*lanum elaeagnifolium*) = 58.9% Root length index (S. *elaeagnifolium*) = 73.9% Shoot length index (S. *elaeagnifolium*) = 73.2%	[20]
E. gigantea	Bizerte (Tunisia)	Leaves	Seed germination inhibition (*Raphanus sativus*) = 92.78% (dose of 1000 µg/mL) Radical length inhibition (*R. sativus*) = 67.74% (dose of 1000 µg/mL) Seed germination inhibition (*Sinapis arvensis*) = 96.77% (dose of 1000 µg/mL) Radical length inhibition (*S. arvensis*) = 98.21% (dose of 1000 µg/mL) Radical length inhibition (*Lolium multiflorum*) = 88.23% (dose of 1000 µg/mL)	[19]

(*Continued*)

TABLE 7.2 *(Continued)*
Phytotoxic Potential of Various Aromatic Species

Species	Collection Site	Plant Part	Results	Reference
E. intertexta	Zaghouen (Tunisia)	Leaves	Seed germination inhibition (*Lolium multiflorum*) = 50.0% (dose of 1000 μg/mL) Radical length inhibition (*L. multiflorum*) = 88.63% (dose of 1000 μg/mL) Seed germination inhibition (*Raphanus sativus*) = 63.85% and 100.0% (doses of 500 and 1000 μg/mL, respectively) Radical length inhibition (*R. sativus*) = 80% and 100.0% (doses of 500 and 1000 μg/mL, respectively) Seed germination inhibition (*Sinapis arvensis*) = 69.07% and 100.0% (doses of 500 and 1000 μg/mL, respectively) Radical length inhibition (*S. arvensis*) = 81.81% and 100.0% (doses of 500 and 1000 μg/mL, respectively)	[19]
E. obliqua	Siliana (Tunisia)	Leaves	Seed germination inhibition (*Sinapis arvensis*) = 100.0% (doses of 500 and 1000 μg/mL) Radical length inhibition (*S. arvensis*) = 100.0% (doses of 500 and 1000 μg/mL) Radical length inhibition (*Lolium multiflorum*) = 86.66% (dose of 1000 μg/mL) Radical length inhibition (*Raphanus sativus*) = 83.3% (dose of 1000 μg/mL)	[19]
E. occidentalis	Zaghouan (Tunisia)	Leaves	Germination inhibition (*Loliun rigidum)* = 61.57% (doses of 2.25 and 3.00 μL/mL) Shoot growth inhibition (*L. rigidum)* = 1.4 cm (doses of 2.25 and 3.00 μL/mL) Radicle inhibition (*L. rigidum)* = 0.5 cm (dose of 3.00 μL/mL) Germination inhibition (*Sinapis arvensis)* = 75.86% and 100% (doses of 2.25 and 3.00 μL/mL, respectively) Shoot growth inhibition (*S. arvensis)* = 0.1 cm (dose of 3.00 μL/mL) Radicle inhibition (*S. arvensis)* = 0.1 cm (dose of 3.00 μL/mL) Germination inhibition (*Trifolium campestre*) = 78.64% (dose of 3.00 μL/mL) Shoot growth inhibition (*T. campestre*) = 1.3 cm (dose of 3.00 μL/mL) Radicle inhibition (*T. campestre*) = 1.4 cm (dose of 3.00 μL/mL)	[30]
E. pauciflora	Bizerte (Tunisia)	Leaves	Seed germination inhibition (*Lolium multiflorum*) = 100.0% (dose of 1000 μg/mL) Radical length inhibition (*L. multiflorum*) = 85.41% (dose of 1000 μg/mL) Seed germination inhibition (*Raphanus sativus*) = 86.02% and 100.0% (doses of 500 and 1000 μg/mL, respectively) Radical length inhibition (*R. sativus*) = 100.0% (dose of 1000 μg/mL) Seed germination inhibition (*Sinapis arvensis*) = 86.59% and 89.69% (doses of 500 and 1000 μg/mL, respectively) Radical length inhibition (*S. arvensis*) = 81.08% (dose of 1000 μg/mL)	[19]
E. salubris	New South Wales (Australia)	Leaves	Germination inhibition index (So*lanum elaeagnifolium*) = 73.0% Root length index (S. *elaeagnifolium*) = 82.0% Shoot length index (S. *elaeagnifolium*) = 75.7%	[20]

E. striaticalyx	Zaghouan (Tunisia)	Leaves	Germination inhibition (*Loliun rigidum*) = 88.46% (dose of 3.00 µL/mL) Shoot growth inhibition (*L. rigidum*) = 1.5 cm (dose of 3.00 µL/mL) Root inhibition (*L. rigidum*) = 0.3 cm (dose of 3.00 µL/mL) Germination inhibition (*Sinapis arvensis*) = 92.26% and 100% (doses of 2.25 and 3.00 µL/mL, respectively) Shoot growth inhibition (*S. arvensis*) = 0.0 cm (dose of 3.00 µL/mL) Root growth inhibition (*S. arvensis*) = 0.0 cm (dose of 3.00 µL/mL) Germination inhibition (*Trifolium campestre*) = 100% (dose of 3.00 µL/mL) Shoot growth inhibition (*T. campestre*) = 0.0 cm (dose of 3.00 µL/mL) Root growth inhibition (*T. campestre*) = 0.0 cm (dose of 3.00 µL/mL)	[30]
E. stricklandii	Zaghouan (Tunisia)	Leaves	Germination inhibition (*Lolium rigidum*) = 92.26% (dose of 3.00 µL/mL) Shoot growth inhibition (*L. rigidum*) = 1.8 cm (dose of 3.00 µL/mL) Root growth inhibition (*L. rigidum*) = 1.4 cm (dose of 3.00 µL/mL) Germination inhibition (*Sinapis arvensis*) = 100% (dose of 3.00 µL/mL) Shoot growth inhibition (*S. arvensis*) = 0.0 cm (dose of 3.00 µL/mL) Root growth inhibition (*S. arvensis*) = 0.0 cm (dose of 3.00 µL/mL) Germination inhibition (*Trifolium campestre*) = 100% (dose of 3.00 µL/mL) Shoot growth inhibition (*T. campestre*) = 0.0 cm (dose of 3.00 µL/mL) Root growth inhibition (*T. campestre*) = 0.0 cm (dose of 3.00 µL/mL)	[30]
E. tereticornis	Bizerte (Tunisia)	Leaves	Seed germination inhibition (*Lolium multiflorum*) = 100.0% (dose of 1000 µg/mL) Radical length inhibition (*L. multiflorum*) = 100.0% (dose of 1000 µg/mL) Seed germination inhibition (*Raphanus sativus*) = 100.0% (dose of 1000 µg/mL) Radical length inhibition (*R. sativus*) = 100.0% (dose of 1000 µg/mL) Seed germination inhibition (*Sinapis arvensis*) = 100% (doses of 500 and 1000 µg/mL) Radical length inhibition (*S. arvensis*) = 100.0% (doses of 500 and 1000 µg/mL)	[19]
Guatteria schomburgkiana	Magalhães Barata (Brazil)	Leaves	Germination inhibition (*Mimosa pudica*) = 86.67% Radicle elongation inhibition (*M. pudica*) = 50.00% Hypocotyl inhibition (*M. pudica*) = 70.95% Germination inhibition (*Senna obtusifolia*) = 70.95% Radicle elongation inhibition (*S. obtusifolia*) = 46.05% Hypocotyl inhibition (*S. obtusifolia*) = 51.13%	[7]
Lavandula dentata	Diamante (Argentina)	Aerial parts	Radicle growth inhibition (*Lactuca sativa*) = 92.5% (dose of 12.5 µL/mL) Radicle growth inhibition (*L. sativa*) = 100.0% (dose of 25.0 µL/mL)	[37]

(*Continued*)

TABLE 7.2 ***(Continued)***
Phytotoxic Potential of Various Aromatic Species

Species	Collection Site	Plant Part	Results	Reference
Mentha piperita	Valencia (Spain)	-	Seed germination inhibition (*Lolium multiflorum*) = 100.00% (doses of 0.125, 0.25, 0.5, and 1 μL/mL) Seedling growth hypocotyl inhibition (*Echinochloa crus-galli*) = 80.72% (dose of 0.25 μL/mL) Seedling growth hypocotyl inhibition (*E. crus-galli*) = 84.95% (dose of 0.5 μL/mL) Seedling growth hypocotyl inhibition (*E. crus-galli*) = 86.64% (dose of 1 μL/mL) Seedling growth radicle inhibition (*E. crus-galli*) = 81.42% (dose of 0.25 μL/mL) Seedling growth radicle inhibition (*E. crus-galli*) = 82.09% (dose of 1 μL/mL)	[38]
Pimpinella anisum	Valencia (Spain)	-	Seedling growth hypocotyl inhibition (*Lolium multiflorum*) = 73.89% (dose of 1 μL/mL) Seedling growth radicle inhibition (*Echinochloa crus-galli*) = 65.39% (dose of 1 μL/mL) Seedling growth radicle inhibition (*E. crus-galli*) = 63.81% (dose of 1 μL/mL)	[38]
Pine pinaster	Ain Draham (Tunisia)	Branches	Shoot growth inhibition (*Sinapis arvensis*) = 79.67% (dose of 3 μL/mL) Radical elongation inhibition (*S. arvensis*) = 87.78% (dose of 3 μL/mL) Shoot growth inhibition (*Phalaris canariensis*) = 78.28% (dose of 3 μL/mL) Radical elongation inhibition (*P. canariensis*) = 64.82% (dose of 3 μL/mL) Shoot growth inhibition (*Triticum durum*) = 77.21% (dose of 3 μL/mL) Radical elongation inhibition (*T. durum*) = 78.37% (dose of 3 μL/mL)	[39]
P. pinaster	Ain Draham (Tunisia)	Needles	Germination inhibition (*Sinapis arvensis*) = 96.67% and 100% (doses of 2 and 3 μL/mL, respectively) Shoot growth inhibition (*S. arvensis*) = 100% (dose of 3 μL/mL) Radical elongation inhibition (*S. arvensis*) = 100% (dose of 3 μL/mL) Shoot growth inhibition (*Phalaris canariensis*) = 88.76% (dose of 3 μL/mL) Radical elongation inhibition (*P. canariensis*) = 89.06% (dose of 3 μL/mL) Shoot growth inhibition (*Triticum durum*) = 68.55% (dose of 3 μL/mL) Radical elongation inhibition (*T. durum*) = 79.76% (dose of 3 μL/mL)	[39]
P. pinaster	Ain Draham (Tunisia)	Cones	Germination inhibition (*Sinapis arvensis*) = 100% (dose of 3 μL/mL) Shoot growth inhibition (*S. arvensis*) = 100% (dose of 3 μL/mL) Radical elongation inhibition (*S. arvensis*) = 100% (dose of 3 μL/mL) Germination inhibition (*Phalaris canariensis*) = 92.31% (dose of 3 μL/mL) Shoot growth inhibition (*P. canariensis*) = 85.21% (dose of 3 μL/mL) Radical elongation inhibition (*P. canariensis*) = 89.06% (dose of 3 μL/mL) Germination inhibition (*Triticum durum*) = 70% (dose of 3 μL/mL) Shoot growth inhibition (*T. durum*) = 74.21% (dose of 3 μL/mL) Radical elongation inhibition (*T. durum*) = 79.93% (dose of 3 μL/mL)	[39]

Pinus nigra	Ain Draham (Turkey)	Aerial parts	Germination inhibition (*Sinapis arvensis*) = 100.00% (dose of 4 µg/mL) Germination inhibition (*S. arvensis*) = 100.00% (dose of 5 µg/mL) Germination inhibition (*Trifolium campestre*) = 100.0% (dose of 5 µg/mL) Germination inhibition (*Phalaris canariensis*) = 100.0% (dose of 5 µg/mL)	[40]
Psidium cattleianum	Espírito Santo (Brazil)	Leaves	Germination inhibition (*Lactuca sativa*) = 74.6% (dose of 3000 µL/mL) Germination inhibition (*Sorghum bicolor*) = 92.6% (dose of 3000 µL/mL)	[41]
P. myrtoides	Espírito Santo (Brazil)	Leaves	Germination inhibition (*Lactuca sativa*) = 47.7% (dose of 3000 µL/mL) Germination inhibition (*Sorghum bicolor*) = 90.4% (dose of 3000 µL/mL)	[41]
P. friedrichsthalianum	Espírito Santo (Brazil)	Leaves	Germination inhibition (*Sorghum bicolor*) = 93.1% (dose of 1500 µL/mL)	[41]
P. gaudichaudianum	Espírito Santo (Brazil)	Leaves	Germination inhibition (*Lactuca sativa*) = 90.7% (dose of 1500 µL/mL) Germination inhibition (*Sorghum bicolor*) = 97.1% (dose of 1500 µL/mL)	[41]
Pulicaria undulata	Wadi Hagul (Egypt)	Aerial parts	Germination inhibition (*Bidens pilosa*) = 86.67% (dose of 100 µL/L) Shoot inhibition (*B. pilosa*) = 79.23% (dose of 100 µL/L) Root inhibition (*B. pilosa*) = 94.17% (dose of 100 µL/L) Germination inhibition (*Dactyloctenium aegyptium*) = 96.67% (dose of 100 µL/L) Shoot inhibition (*D. aegyptium*) = 100.0% (dose of 100 µL/L) Root inhibition (*D. aegyptium*) = 100.0% (dose of 100 µL/L)	[42]
Pulicaria undulata	Riyadh (Saudi Arabia)	Aerial parts	Germination inhibition (*Bidens pilosa*) = 66.67% (dose of 100 µL/L) Shoot inhibition (*B. pilosa*) = 74.59% (dose of 100 µL/L) Root inhibition (*B. pilosa*) = 83.47% (dose of 100 µL/L) Germination inhibition (*Dactyloctenium aegyptium*) = 93.33% (dose of 100 µL/L) Shoot inhibition (*D. aegyptium*) = 100.0% (dose of 100 µL/L) Root inhibition (*D. aegyptium*) = 100.0% (dose of 100 µL/L)	[42]
Ruta graveolens	Mugla (Turkey)	Aerial parts	Germination inhibition (*Amaranthus retroflexus*) = 100.0% (dose of 20 µL/mL) Root length inhibition (*A. retroflexus*) = 89.5% and 100.0% (doses of 15 and 20 µL/mL, respectively) Radicle length inhibition (*A. retroflexus*) = 89.7%, 94.0%, 90.8%, and 100.0% (doses of 5, 10, 15, and 20 µL/mL, respectively)	[18]
Salvia sclarea	Turkey	Aerial parts	Germination inhibition (*Lactuta sativa*) = 94.00% (dose of 1.24 µg/mL) Radicle length inhibition (*L. sativa*) = 96.00% (dose of 1.24 µg/mL) Plumule length inhibition (*L. sativa*) = 96.00% (dose of 1.24 µg/mL) Germination inhibition (*Lepidium sativum*) = 100.0% (dose of 1.24 µg/mL) Radicle length inhibition (*L. sativum*) = 100.0% (dose of 1.24 µg/mL) Plumule length inhibition (*L. sativum*) =100.0% (dose of 1.24 µg/mL)	[32]

(Continued)

TABLE 7.2 (Continued)
Phytotoxic Potential of Various Aromatic Species

Species	Collection Site	Plant Part	Results	Reference
			Germination inhibition (*Portulaca. oleracea*) = 50.0% (dose of 1.24 μg/mL)	
			Radicle length inhibition (*P. oleracea*) = 81.0% (dose of 1.24 μg/mL)	
			Plumule length inhibition (*P. oleracea*) = 81.0% (dose of 1.24 μg/mL)	
Satureja calamintha subsp. *nepeta*	Algeria	Aerial parts	Seed germination inhibition (*Sinapis arvensis*) = 50.0% (dose of 0.03%)	[43]
			Seed germination inhibition (*Avena fatua*) = 90.1% (dose of 0.03%)	
			Seed germination inhibition (*Sonchus oleraceus*) = 93.8% (dose of 0.03%)	
			Seed germination inhibition (*Xanthium strumarium*) = 81.6% (dose of 0.03%)	
			Seed germination inhibition (*Cyperus rotundus*) = 93.1% (dose of 0.03%)	
Satureja montana	Valencia (Spain)	-	Seed germination inhibition (*Lolium multiflorum*) = 100.0% (doses of 0.125, 0.25, 0.5, and 1 μL/mL)	[38]
			Seed germination inhibition (*Portulaca oleracea*) = 100.0% (doses of 0.125, 0.25, 0.5, and 1 μL/mL)	
			Seed germination inhibition (*Echinochloa crus-galli*) = 100.0% (doses of 0.125, 0.25, 0.5, and 1 μL/mL)	
Teucrium polium	St. Katherine Protectorate (Egypt)	Aerial parts	Germination inhibition (*Lactuta sativa*) = 51.9% (dose of 1000 μL/L)	[34]
			Root growth inhibition (*L. sativa*) = 54.4% (dose of 1000 μL/L)	
T. polium	St. Katherine Protectorate (Egypt)	Aerial parts	Germination inhibition (*Lactuta sativa*) = 50.9% (dose of 1000 μL/mL)	[34]
			Root growth inhibition (*L. sativa*) = 42.3.4% (dose of 1000 μL/L)	
Thymbra capitata	Enna province (Sicily, Italy)	Aerial parts (blooming)	Seed germination inhibition (*Portulaca oleracea*) = 91.95% (dose of 0.250 μL/mL)	[31]
			Seed germination inhibition (*P. oleracea*) = 100.0% (doses of 0.5 and 1 μL/mL)	
			Seed germination inhibition (*Conyza canadensis*) = 100.0% (doses of 0.125, 0.250, 0.5, and 1 μL/mL)	
T. capitata	Caltanissetta province (Sicily, Italy)	Aerial parts (vegetative stage)	Seed germination inhibition (*Portulaca oleracea*) = 95.40% (dose of 1 μL/mL)	
			Seed germination inhibition (*P. oleracea*) = 100.0% (doses of 0.250 and 0.5 μL/mL)	
			Seed germination inhibition (*Conyza canadensis*) = 100.0% (doses of 0.125, 0.250, 0.5, and 1 μL/mL)	
T. capitata	Seville province (Spain)	Aerial parts	Seed germination inhibition (*Solanum nigrum*) = 73.7% (dose of 0.125 μL/mL)	
			Seed germination inhibition (*S. nigrum*) = 75.7% (dose of 0.250 μL/mL)	
			Seed germination inhibition (*S. nigrum*) = 100.0% (doses of 0.5 and 1 μL/mL)	
			Seed germination inhibition (*Setaria verticillata*) = 80.0% (dose of 0.125 μL/mL)	
			Seed germination inhibition (*S. verticillata*) = 86.3% (dose of 0.250 μL/mL)	
			Seed germination inhibition (*S. verticillate*) = 100.0% (doses of 0.250 and 0.5 μL/mL)	
			Seed germination inhibition (*Chenopodium album*) = 100.0% (doses of 0.5 and 1 μL/mL)	
			Seed germination inhibition (*Sonchus oleraceus*) = 100.0% (doses of 0.125, 0.250, 0.5, and 1 μL/mL)	

Thymus decussatus	St. Katherine Protectorate (Egypt)	Aerial parts	Germination inhibition (*Lactuta sativa*) = 77.7% (dose of 100 μL/L)	[34]
			Shoot growth inhibition (*L. sativa*) = 85.8% (dose of 100 μL/L)	
			Root growth inhibition (*L. sativa*) = 84.6% (dose of 100 μL/L)	
T. decussatus	St. Katherine Protectorate (Egypt)	Aerial parts	Germination inhibition (*Lactuta sativa*) = 86.6% (dose of 100 μL/L)	[34]
			Shoot growth inhibition (*L. sativa*) = 87.4% (dose of 100 μL/L)	
			Root growth inhibition (*L. sativa*) = 89.9% (dose of 100 μL/L)	
T. fontanesii	Algeria	Aerial parts	Seed germination inhibition (*Sinapis arvensis*) = 100.0% (dose of 0.03%)	[43]
			Seed germination inhibition (*Avena fatua*) = 100.0% (dose of 0.03%)	
			Seed germination inhibition (*Sonchus oleraceus*) = 100.0% (dose of 0.03%)	
			Seed germination inhibition (*Xanthium strumarium*) = 84.5% (dose of 0.03%)	
			Seed germination inhibition (*Cyperus rotundus*) = 100.0% (dose of 0.03%)	
Xylopia frutescens	Magalhães Barata (Pará—Brazil)	Leaves	Germination inhibition (*Mimosa pudica*) = 86.67 ± 5.77%	[7]
			Radicle elongation inhibition (*M. pudica*) = 55.22 ± 2.72%	
			Hypocotyl inhibition (*M. pudica*) = 71.12 ± 3.80%	
			Germination inhibition (*Senna obtusifolia*) = 13.33 ± 5.77%	
			Radicle elongation inhibition (*S. obtusifolia*) = 60.43 ± 4.63%	
			Hypocotyl inhibition (*S. obtusifolia*) = 51.38 ± 1.05%	
Zingiber officinale	Indonesia	Rhizome	Seed germination inhibition (*Portulaca oleracea*) = 45.34% (dose of 1 μL/mL)	[21]
			Seed hypocotyl length inhibition (*P. oleracea*) = 82.73% (dose of 1 μL/mL)	
			Seed radicle length inhibition (*P. oleracea*) = 86.05% (dose of 1 μL/mL)	
			Seed germination inhibition (*Lolium multiflorum*) = 46.67% (dose of 1 μL/mL)	
			Seed hypocotyl length inhibition (*L. multiflorum*) = 66.85% (dose of 1 μL/mL)	
			Seed radicle length inhibition (*L. multiflorum*) = 72.35% (dose of 1 μL/mL)	
			Seed germination inhibition (*Cortaderia selloana*) = 43.90% (dose of 1 μL/mL)	
			Seed hypocotyl length inhibition (*C. selloana*) = 73.67% (dose of 1 μL/mL)	
			Seed radicle length inhibition (*C. selloana*) = 75.25% (dose of 1 μL/mL)	
			Seed hypocotyl length inhibition (*Nicotiana glauca*) = 63.77% (dose of 1 μL/mL)	
			Seed radicle length inhibition (*N. glauca*) = 75.25% (dose of 1 μL/mL)	

and *Portulaca oleracea*. This essential oil was particularly effective in *L. sativum*, which exhibited a 100% inhibition of seed germination and reduced radicle and plumule lengths [32].

1,8-cineol is found in essential oils of various species of the genus *Eucalyptus*. Khammassi et al. [30] evaluated the essential oils of *Eucalyptus occidentalis, E. striaticalyx*, and *E. stricklandii* and demonstrated that these oils particularly affected the germination of the weed *Sinapis arvensis*. The essential oils of these three species had the strongest inhibitory effect on the shoot and radicle growth of the weeds *Lolium rigidum, S. arvensis,* and *Trifolium campestre*.

According to de Moraes et al. [7], the essential oils of *G. schomburgkiana* and *Xylopia frutescens* exhibit phytotoxic potential at different response intensities depending on the weed species and growth parameter. Regarding seed germination, the weed *Senna obtusifolia* (13.33% inhibition) was more resistant than *Mimosa pudica* (86.67% inhibition). Regarding radicle and hypocotyl elongation, *S. obtusifolia* and *M. pudica* were more sensitive to the essential oil of *X. frutescens*. This result was possibly due to oxygenated monoterpene and hydrocarbon sesquiterpene compounds that characterize the chemical profile of the samples.

Treatment with the essential oil of *E. erythropappus* significantly decreased *Brassica rapa* and *Bidens pilosa* seed germination and radicle length [22]. These authors observed the strongest inhibitory effects on seed germination and radicle length in *B. rapa* (IC_{50} = 26.5 μL/mL) and *B. pilosa* seedlings (IC_{50} = 16.3 μL/mL), respectively.

El Ayeb-Zakhama et al. [35] assessed the essential oils of the different plant parts (flowers, stem, root, and leaves) of *C. spinosum* and showed that the essential oil extracted from flowers significantly affected *L. sativa* seed germination, causing 100% inhibition at a dose of 1 mg/mL. In addition, this essential oil completely inhibited root growth, thus matching the effect of the essential oil of leaves, while the essential oil of roots reached 60% root growth inhibition. This study demonstrated that the phytotoxic potential of an essential oil depends on the plant part that is used to extract the oil.

7.4 CONCLUSION

Essential oils contain a wide range of volatile compounds, which explain their high biological activities.

As shown in this bibliographic study, the phytotoxic profile of several essential oils enables the development of biopesticides and insecticides that are less toxic to humans.

REFERENCES

1. Ferreira, O.O.; Franco, C. de J.P.; Varela, E.L.P.; Silva, S.G.; Cascaes, M.M.; Percário, S.; de Oliveira, M.S.; Andrade, E.H. de A. Chemical Composition and Antioxidant Activity of Essential Oils from Leaves of Two Specimens of Eugenia Florida DC. *Molecules* **2021**, *26*, 5848, doi:10.3390/molecules26195848.
2. Cascaes, M.M.; Marques da Silva, S.H.; de Oliveira, M.S.; Cruz, J.N.; de Moraes, Â.A.B.; do Nascimento, L.D.; Ferreira, O.O.; Guilhon, G.M.S.P.; Andrade, E.H. de A. Exploring the Chemical Composition, in Vitro and in Silico Study of the Anticandidal Properties of Annonaceae Species Essential Oils from the Amazon. *PLoS One* **2023**, *18*, e0289991.
3. Cruz, J.N.; Oliveira, M.S.; Cascaes, M.; Mali, S.N.; Tambe, S.; Santos, C.B.; Zoghbi, M.D.; Andrade, E.H. Variation in the Chemical Composition of Endemic Specimens of Hedychium Coronarium J. Koenig from the Amazon and In Silico Investigation of the ADME/Tox Properties of the Major Compounds. *Plants* **2023**, *12*.
4. Isman, M.B.; Grieneisen, M.L. Botanical Insecticide Research: Many Publications, Limited Useful Data. *Trends Plant Sci.* **2014**, *19*, 140–145, doi:10.1016/j.tplants.2013.11.005.
5. Alavanja, M.C.R.; Bonner, M.R. Occupational Pesticide Exposures and Cancer Risk: A Review. *J. Toxicol. Environ. Heal. Part B* **2012**, *15*, 238–263, doi:10.1080/10937404.2012.632358.
6. Anand, S.P.; Sati, N. Artificial Preservatives and Their Harmful Effects: Looking toward Nature for Safer Alternatives. *Int. J. Pharm. Sci. Res.* **2013**, *4*, 2496–2501.
7. de Moraes, Â.A.B.; Cascaes, M.M.; do Nascimento, L.D.; de Jesus Pereira Franco, C.; Ferreira, O.O.; Anjos, T.O. dos; Karakoti, H.; Kumar, R.; da Silva Souza-Filho, A.P.; de Oliveira, M.S.; et al. Chemical

Evaluation, Phytotoxic Potential, and In Silico Study of Essential Oils from Leaves of Guatteria Schomburgkiana Mart. and Xylopia Frutescens Aubl. (Annonaceae) from the Brazilian Amazon. *Molecules* **2023**, *28*, 2633, doi:10.3390/molecules28062633.

8. de Oliveira, M.S.; da Costa, W.A.; Pereira, D.S.; Botelho, J.R.S.; de Alencar Menezes, T.O.; de Aguiar Andrade, E.H.; da Silva, S.H.M.; da Silva Sousa Filho, A.P.; de Carvalho, R.N. Chemical Composition and Phytotoxic Activity of Clove (Syzygium Aromaticum) Essential Oil Obtained with Supercritical CO2. *J. Supercrit. Fluids* **2016**, *118*, 185–193, doi:10.1016/j.supflu.2016.08.010.
9. El Sawi, S.A.; Ibrahim, M.E.; El-Rokiek, K.G.; El-Din, S.A.S. Allelopathic Potential of Essential Oils Isolated from Peels of Three Citrus Species. *Ann. Agric. Sci.* **2019**, *64*, 89–94, doi:10.1016/j.aoas.2019.04.003.
10. Abd-ElGawad, A.M.; Elshamy, A.I.; Al-Rowaily, S.L.; El-Amier, Y.A. Habitat Affects the Chemical Profile, Allelopathy, and Antioxidant Properties of Essential Oils and Phenolic Enriched Extracts of the Invasive Plant Heliotropium Curassavicum. *Plants* **2019**, *8*.
11. Gharibvandi, A.; Karimmojeni, H.; Ehsanzadeh, P.; Rahimmalek, M.; Mastinu, A. Weed Management by Allelopathic Activity of Foeniculum Vulgare Essential Oil. *Plant Biosyst.—An Int. J. Deal. with all Asp. Plant Biol.* **2022**, *156*, 1298–1306, doi:10.1080/11263504.2022.2036848.
12. Abd El-Gawad, A.M. Chemical Constituents, Antioxidant and Potential Allelopathic Effect of the Essential Oil from the Aerial Parts of Cullen Plicata. *Ind. Crops Prod.* **2016**, *80*, 36–41, doi:10.1016/j.indcrop.2015.10.054.
13. Taban, A.; Saharkhiz, M.J.; Hadian, J. Allelopathic Potential of Essential Oils from Four Satureja Spp. *Biol. Agric. Hortic.* **2013**, *29*, 244–257, doi:10.1080/01448765.2013.830275.
14. Feitosa, B.D.; Ferreira, O.O.; Mali, S.N.; Anand, A.; Cruz, J.N.; Franco, C.D.; Mahawer, S.K.; Kumar, R.; Cascaes, M.M.; Oliveira, M.S.; et al. Chemical Composition, Preliminary Toxicity, and Antioxidant Potential of Piper Marginatum Sensu Lato Essential Oils and Molecular Modeling Study. *Molecules* **2023**, *28*.
15. Abd El-Gawad, A.M.; El-Amier, Y.A.; Bonanomi, G. Essential Oil Composition, Antioxidant and Allelopathic Activities of Cleome Droserifolia (Forssk.) Delile. *Chem. Biodivers.* **2018**, *15*, e1800392.
16. Franco, C.D.; Ferreira, O.O.; Cruz, J.N.; Varela, E.L.; de Moraes, Â.A.; Nascimento, L.D.; Cascaes, M.M.; Souza Filho, A.P.; Lima, R.R.; Percário, S.; et al. Phytochemical Profile and Herbicidal (Phytotoxic), Antioxidants Potential of Essential Oils from Calycolpus Goteanus (Myrtaceae) Specimens, and in Silico Study. *Molecules* **2022**, *27*.
17. El Ayeb-Zakhama, A.; Sakka-Rouis, L.; Flamini, G.; Ben Jannet, H.; Harzallah-Skhiri, F. Chemical Composition and Allelopathic Potential of Essential Oils from Citharexylum Spinosum L. Grown in Tunisia. *Chem. Biodivers.* **2017**, *14*, e1600225.
18. Bozhuyuk, A.U. Herbicidal Activity and Chemical Composition of Two Essential Oils on Seed Germinations and Seedling Growths of Three Weed Species. *J. Essent. Oil-Bearing Plants* **2020**, *23*, 821–831, doi:10.1080/0972060X.2020.1828178.
19. Polito, F.; Kouki, H.; Khedhri, S.; Hamrouni, L.; Mabrouk, Y.; Amri, I.; Nazzaro, F.; Fratianni, F.; De Feo, V. Chemical Composition and Phytotoxic and Antibiofilm Activity of the Essential Oils of Eucalyptus Bicostata, E. Gigantea, E. Intertexta, E. Obliqua, E. Pauciflora and E. Tereticornis. *Plants* **2022**, *11*, 3017, doi:10.3390/plants11223017.
20. Zhang, J.; An, M.; Wu, H.; Liu, D.L.; Stanton, R. Chemical Composition of Essential Oils of Four Eucalyptus Species and Their Phytotoxicity on Silverleaf Nightshade (Solanum Elaeagnifolium Cav.) in Australia. *Plant Growth Regul.* **2012**, *68*, 231–237, doi:10.1007/s10725-012-9711-5.
21. Ibáñez, M.; Blázquez, M. Ginger and Turmeric Essential Oils for Weed Control and Food Crop Protection. *Plants* **2019**, *8*, 59, doi:10.3390/plants8030059.
22. Pinto, A.P.R.; Seibert, J.B.; dos Santos, O.D.H.; Filho, S.A.V.; do Nascimento, A.M. Chemical Constituents and Allelopathic Activity of the Essential Oil from Leaves of Eremanthus Erythropappus. *Aust. J. Bot.* **2018**, *66*, 601, doi:10.1071/BT18138.
23. Abd-ElGawad, A.M.; Saleh, I.; El-Razek, M.H.A.; Elkarim, A.S.A.; El-Amier, Y.A.; Mohamed, T.A.; El Gendy, A.E.N.G.; Afifi, S.M.; Esatbeyoglu, T.; Elshamy, A.I. Chemical Profiling of Significant Antioxidant and Phytotoxic Microwave-Extracted Essential Oil from Araucaria Heterophylla Resin. *Separations* **2023**, *10*, 141, doi:10.3390/separations10020141.
24. Khammassi, M.; Polito, F.; Amri, I.; Khedhri, S.; Hamrouni, L.; Nazzaro, F.; Fratianni, F.; De Feo, V. Chemical Composition and Phytotoxic, Antibacterial and Antibiofilm Activity of the Essential Oils of Eucalyptus & Occidentalis, E. Striaticalyx and E. Stricklandii. *Molecules* **2022**, *27*.
25. Werrie, P.-Y.; Durenne, B.; Delaplace, P.; Fauconnier, M.-L. Phytotoxicity of Essential Oils: Opportunities and Constraints for the Development of Biopesticides. A Review. *Foods* **2020**, *9*, 1291, doi:10.3390/foods9091291.

26. Verdeguer, M.; Sánchez-Moreiras, A.M.; Araniti, F. Phytotoxic Effects and Mechanism of Action of Essential Oils and Terpenoids. *Plants* **2020**, *9*, 1–48, doi:10.3390/plants9111571.
27. Cascaes, M.M.; De Moraes, Â.A.B.; Cruz, J.N.; Franco, C. de J.P.; E Silva, R.C.; Nascimento, L.D. do; Ferreira, O.O.; Anjos, T.O. dos; de Oliveira, M.S.; Guilhon, G.M.S.P. Phytochemical Profile, Antioxidant Potential and Toxicity Evaluation of the Essential Oils from Duguetia and Xylopia Species (Annonaceae) from the Brazilian Amazon. *Antioxidants* **2022**, *11*, 1709.
28. Abd El-Gawad, A.M.; El-Amier, Y.A.; Bonanomi, G. Essential Oil Composition, Antioxidant and Allelopathic Activities of Cleome Droserifolia (Forssk.) Delile. *Chem. Biodivers.* **2018**, *15*, e1800392, doi:10.1002/cbdv.201800392.
29. Dhifi, W.; Bellili, S.; Jazi, S.; Bahloul, N.; Mnif, W. Essential Oils' Chemical Characterization and Investigation of Some Biological Activities: A Critical Review. *Medicines* **2016**, *3*, 25, doi:10.3390/medicines3040025.
30. Khammassi, M.; Polito, F.; Amri, I.; Khedhri, S.; Hamrouni, L.; Nazzaro, F.; Fratianni, F.; De Feo, V. Chemical Composition and Phytotoxic, Antibacterial and Antibiofilm Activity of the Essential Oils of Eucalyptus Occidentalis, E. Striaticalyx and E. Stricklandii. *Molecules* **2022**, *27*, 5820, doi:10.3390/molecules27185820.
31. Verdeguer, M.; Torres-Pagan, N.; Muñoz, M.; Jouini, A.; García-Plasencia, S.; Chinchilla, P.; Berbegal, M.; Salamone, A.; Agnello, S.; Carrubba, A.; et al. Herbicidal Activity of Thymbra Capitata (L.) Cav. Essential Oil. *Molecules* **2020**, *25*, 2832, doi:10.3390/molecules25122832.
32. Bozok, F.; Ulukanli, Z. Volatiles from the Aerial Parts of East Mediterranean Clary Sage: Phytotoxic Activity. *J. Essent. Oil-Bearing Plants* **2016**, *19*, 1192–1198, doi:10.1080/0972060X.2015.1119066.
33. Zhou, L.; Li, J.; Kong, Q.; Luo, S.; Wang, J.; Feng, S.; Yuan, M.; Chen, T.; Yuan, S.; Ding, C. Chemical Composition, Antioxidant, Antimicrobial, and Phytotoxic Potential of Eucalyptus Grandis × E. Urophylla Leaves Essential Oils. *Molecules* **2021**, *26*, 1450, doi:10.3390/molecules26051450.
34. Saleh, I.; Abd-ElGawad, A.; El Gendy, A.E.-N.; Abd El Aty, A.; Mohamed, T.; Kassem, H.; Aldosri, F.; Elshamy, A.; Hegazy, M.-E.F. Phytotoxic and Antimicrobial Activities of Teucrium Polium and Thymus Decussatus Essential Oils Extracted Using Hydrodistillation and Microwave-Assisted Techniques. *Plants* **2020**, *9*, 716, doi:10.3390/plants9060716.
35. El Ayeb-Zakhama, A.; Sakka-Rouis, L.; Flamini, G.; Ben Jannet, H.; Harzallah-Skhiri, F. Chemical Composition and Allelopathic Potential of Essential Oils from Citharexylum Spinosum L. Grown in Tunisia. *Chem. Biodivers.* **2017**, *14*, e1600225, doi:10.1002/cbdv.201600225.
36. Franco, C. de J.P.; Ferreira, O.O.; Cruz, J.N.; Varela, E.L.P.; de Moraes, Â.A.B.; Nascimento, L.D. do; Cascaes, M.M.; Souza Filho, A.P. da S.; Lima, R.R.; Percário, S.; et al. Phytochemical Profile and Herbicidal (Phytotoxic), Antioxidants Potential of Essential Oils from Calycolpus Goetheanus (Myrtaceae) Specimens, and in Silico Study. *Molecules* **2022**, *27*, 4678, doi:10.3390/molecules27154678.
37. Wagner, L.S.; Sequin, C.J.; Foti, N.; Campos-Soldini, M.P. Insecticidal, Fungicidal, Phytotoxic Activity and Chemical Composition of Lavandula Dentata Essential Oil. *Biocatal. Agric. Biotechnol.* **2021**, *35*, 102092, doi:10.1016/j.bcab.2021.102092.
38. Ibáñez, M.; Blázquez, M. Phytotoxicity of Essential Oils on Selected Weeds: Potential Hazard on Food Crops. *Plants* **2018**, *7*, 79, doi:10.3390/plants7040079.
39. Sana, K.; Marwa, K.; Ismail, A.; Yassine, M.; Dhaouadi, F.; Samia, G. Phytochemical Studies on Essential Oils OfPinus Pinaster Aiton and Evaluation of Their Biological Activities. *Arab. J. Med. Aromat. Plants* **2022**, *8*, 75–98, doi:10.48347/IMIST.PRSM/ajmap-v8i2.30781.
40. Amri, I.; Hanana, M.; Jamoussi, B.; Hamrouni, L. Essential Oils of Pinus Nigra J.F. Arnold Subsp. Laricio Maire: Chemical Composition and Study of Their Herbicidal Potential. *Arab. J. Chem.* **2017**, *10*, S3877–S3882, doi:10.1016/j.arabjc.2014.05.026.
41. Vasconcelos, L.C.; de Souza Santos, E.; de Oliveira Bernardes, C.; da Silva Ferreira, M.F.; Ferreira, A.; Tuler, A.C.; Carvalho, J.A.M.; Pinheiro, P.F.; Praça-Fontes, M.M. Phytochemical Analysis and Effect of the Essential Oil of Psidium L. Species on the Initial Development and Mitotic Activity of Plants. *Environ. Sci. Pollut. Res.* **2019**, *26*, 26216–26228, doi:10.1007/s11356-019-05912-6.
42. Abd-ELGawad, A.M.; Al-Rowaily, S.L.; Assaeed, A.M.; EI-Amier, Y.A.; El Gendy, A.E.-N.G.; Omer, E.; Al-Dosari, D.H.; Bonanomi, G.; Kassem, H.S.; Elshamy, A.I. Comparative Chemical Profiles and Phytotoxic Activity of Essential Oils of Two Ecospecies of Pulicaria Undulata (L.) C.A. Mey. *Plants* **2021**, *10*, 2366, doi:10.3390/plants10112366.
43. Benchaa, S.; Hazzit, M.; Zermane, N.; Abdelkrim, H. Chemical Composition and Herbicidal Activity of Essential Oils from Two Labiatae Species from Algeria. *J. Essent. Oil Res.* **2019**, *31*, 335–346.

8 Essential Oils as Potential Integrated Pest Management Strategies

Marcilene Paiva da Silva, Karyme do Socorro de Souza Vilhena, Eliza de Jesus Barros dos Santos, Oberdan Oliveira Ferreira, Leonardo Souza da Costa, Fernanda Wariss Figueiredo Bezerra, Mozaniel Santana de Oliveira, and Eloisa Helena de Aguiar Andrade

8.1 INTRODUCTION

Plant substances from aromatic plants are well recognized as a rich source of novel chemistry. Most of these chemical compounds are known as secondary compounds that probably exhibit defensive agents against insects and plant pathogens. These "plant secondary metabolites" have some demonstrated bioactivity in insects (at least in laboratory tests) (Isman 2017, 2020a).

The complex mixture of secondary metabolites of essential oils (EOs) presents bioactivity due to the result of the synergy between their major constituents. In the research, rosemary (*Rosemarinus officinalis*) and lemongrass (*Cymbopogon citratus*) oils developed a synergy between the main constituents, resulting in crop protection by increasing resistance against plant pathogens (Tak and Isman 2017; Tak and Isman 2016; Tak, Jovel, and Isman 2017).

The use of conventional pesticides requires safety measures to protect their handling, especially for agricultural workers, due to the frequent occurrence of poisoning caused by these synthetic products. Therefore, this aggravating factor provides an impetus for the development and use of plant-based pesticides. Pyrethrum of the genus *Pyrethrum*, an oleoresin obtained from the flowers of *Tanacetum cinerariefolium* (Asteraceae), has been used for insect control for a long time, continues to be a natural insecticide, and has been used to this day (Isman 2017). *Zingiber officinale*, a species native to Southeast Asia and the Pacific Islands, produces essential oil with antifungal properties described against *Aspergillus flavus*, used as a natural pesticide (Nerilo et al. 2016). The chemical composition of EO from Algerian *Eucalyptus citriodora* was evaluated as a potential bioherbicide to control some of the most problematic weeds in Algerian crop fields (*Sinapis arvensis* L., *Sonchus oleraceus* L., *Xanthium strumarium* L., and *Avena fatua* L.) (Benchaa, Hazzit, and Abdelkrim 2018).

Although the application of EOs and their bioactive compounds are promising against pests, their use remains quite limited due to their high volatility, low water solubility, and susceptibility to degradation. Therefore, some EO-based products are currently being developed to improve their stability, thus expanding their applicability. Nanoformulations/nanoemulsions of EOs are being considered a promising tool to increase their bioefficacy (Chaudhari et al. 2021). The chemical composition eugenol, the major compound obtained from the EO of clove flowers (*Syzygium aromaticum*), showed insecticidal activity against the larvae of the potato moth—*Phthorimaea operculella*. The formulation of the encapsulated essential oil of *P. operculella* demonstrated a prolonged action in the mortality of the larvae of the potato moth of the same species (Milićević et al. 2022).

DOI: 10.1201/9781003463429-11

Therefore, the aim of this study was to gather information on aromatic plant species as pest control products, as well as the application of EOs and their bioactive compounds as green pesticides. In addition, information on the advantages and challenges of applying EOs to overcome the problems of using conventional chemical pesticides is reported.

8.1.1 Aromatic Plants Considered Ecological Products for Pest Control

Aromatic plants produce EOs with great diversity and chemical complexity and may present properties of interest for pest management, especially due to the presence of compounds that exhibit biological activities, which can be used as alternatives in the production of synthetic pesticides (Jouini et al. 2020).

Table 8.1 describes aromatic plants that demonstrate potential for use in pest management.

Zingiber officinale, a species native to Southeast Asia and the Pacific Islands, produces an essential oil with antifungal properties described against *Aspergillus flavus* (Nerilo et al. 2016) and *Phytophthora colocasiae* (Kalhoro et al. 2022). One of the chemotypes of *Origanum vulgare* produces an essential oil rich in thymol; in the study by Gruľová et al. (2020), the oil was able to inhibit the growth of four species of phytopathogenic fungi (*Monilinia fructicola*, *Aspergillus niger*, *Pinicillium expansum*, and *Botrytis cinerea*).

TABLE 8.1
Aromatic Plants with Biopesticide Potential

Specie	Results	Reference
Allium sativum	The aqueous extract of garlic (*Allium sativum*) was tested against larvae and adults of *Tribolum castaneum*, obtaining mean lethality (LC_{50}) of 267.37 and 12.90 μL/L, respectively, after 24 h.	Mobki et al. (2014)
Ocimum basilicum	The methanolic extract completely inhibited the growth of the fungi *Bipolaris hawaiensis*, *B. spicifera*, and *Cochliobulus cynodontis* at a concentration of 16 mg/mL.	Elsherbiny, Safwat, and Elaasser (2017)
Origanum onites	The organic extract of the plant presented MIC of 40 μg/mL against *Clavibacter michiganensis* ssp. *michiganensis*, *Erwinia amylovora*, *E. carotovora* ssp. *atroceptica*, *Pseudomonas syringae* pv. *Syringae*, *Xanthomonas axonopodis* pv. *malvacearum*, *X. axonopodis* pv. *vsicatoria*, and *X. axonopodis* pv. *campestres*, while the essential oil presented MIC of 7.81 μL/mL against *X. campestris* pv. *raphanin*.	Kotan et al. (2014)
Origanum vulgare	*Origanum vulgare* essential oil was able to inhibit the growth of four species of phytopathogenic fungi (*Monilinia fructicola*, *Aspergillus niger*, *Pinicillium expansum*, and *Botrytis cinerea*) at concentrations of 1000 and 500 ppm.	Gruľová et al. (2020)
Tanacetum annuum	The methanolic extract and essential oil of *Tanacetum annuum* showed significant antifungal action against *Fusarium oxysporum*, with MIC of 3.33 μL/mL.	Ettakifi et al. (2023)
Zingiber officinale	The mycelial growth of *Aspergillus flavus* was significantly reduced at a concentration of 150 μg/mL of the essential oil.	Nerilo et al. (2016)
Zingiber officinale	The minimum inhibitory concentration for 100% inhibition of the growth of *Phytophthora colocasiae* was 1250 ppm.	Kalhoro et al. (2022)
Ziziphus spina-christi	The methanolic extract of the leaves and fruits of *Ziziphus spina-christi* was able to inhibit the growth of Altemaria alternata with inhibition percentages in the range between 37.04 and 69.96% and 42.59 and 71.48%, respectively.	El-Shahir et al. (2022)

Kotan et al. (2014) tested the hexane extract and essential oil of *Origanum onites* against several phytopathogenic bacteria, obtaining significant results. The methanolic extract and essential oil of *Tanacetum annuum* showed antifungal effect against *Fusarium oxysporum*, which causes Bayoud disease (Ettakifi et al. 2023). Elsherbiny, Safwat, and Elaasser (2017) evaluated the methanolic extract of *Ocimum basilicum* against some species of phytopathogenic fungi, obtaining complete inhibition of the growth of the fungi *Bipolaris hawaiensis*, *B. spicifera*, and *Cochliobulus cynodontis*.

8.2 ESSENTIAL OILS AND THEIR BIOACTIVE COMPOUNDS AS ENVIRONMENTALLY FRIENDLY GREEN PESTICIDES IN PEST CONTROL

EOs are products of the secondary metabolism of plants, in which they act as important defense and signaling mechanisms for plants (KUMAR et al. 2022). In addition, these volatile oils have a complex mixture of bioactive compounds that are obtained by steam distillation, hydrodistillation, and other extraction methods, whether conventional or traditional (Mossa 2016).

These extraction methods (solvent-free) ensure that the process of obtaining EOs is considered safe and free from harm to the environment; that is, they are considered environmentally friendly agents as a safe and viable alternative in combating pests, because they have low toxicity, large-scale availability, renewable nature of the source materials, and minimal chances of generating resistance due to the mixture of multiple components that cause toxicity by interfering in many aspects of the physiology and biochemistry of insects (Chaudhari et al. 2021).

In this sense, EOs are natural products that represent an excellent alternative to synthetic pesticides as a means of reducing the negative impacts on human health and the environment as potential natural pesticides (Bajpai 2019). From this information, it can be seen that Table 8.2 presents the results and major advances in research with EOs as potential natural pesticides.

8.3 ESSENTIAL OIL-BASED PRODUCTS EFFECTIVE IN PEST CONTROL

Pesticides are formulations or agents formed by different bioactive compounds, which have diverse actions such as herbicides, insecticides, nematicides, rodenticides, avicides, algaecides, fungicides, and bactericides (Garrido-Miranda, Giraldo, and Schoebitz 2022). Natural pesticides, also called biopesticides, are formulations capable of efficiently reducing the attack of pests or insects, with minimal toxicity to organisms for which they are not intended, including humans, animals, and natural enemies, as well as to the external environment (Chaudhari et al. 2021; Garrido-Miranda, Giraldo, and Schoebitz 2022). Natural pesticides can be classified into four categories: mineral, botanical, microbial, and essential oil derivatives (Assadpour et al. 2024; Mfarrej and Rara 2019).

EOs obtained from plant species of families such as Asteraceae, Myrtaceae, Apiaceae, Lamiaceae, and Rutaceae have been used as natural biopesticides due to their insecticidal activity. Oils obtained from plant species of these families are active as repellents, fumigants, larvicides, and adulticides against insects of the orders Lepidoptera, Diptera, Coleopterans, Hemipterans, and Isopterans (Mossa 2016). The use of EOs in bioinsecticide formulations began in the 2000s in the United States of America and was favored by the fact that some main components and even certain EOs used as active ingredients in pesticides are exempt from the normal regulatory process. As a result of this exemption from regulation, the products are marketed in less time and at lower costs when compared to conventional products (Isman 2020b).

The advantage of using EOs from plants in the formulation of biopesticides or bioinsecticides is that they have properties and characteristics that can influence the metabolic, physiological, and biochemical functions of insects and fungi, thus modifying the biological activities of the target sites of these organisms. These plant-derived formulations can have antifungal and antipest properties that demonstrate broad or narrow spectrum. In the first case, the formulations are effective

TABLE 8.2
Main Insecticidal Activities Found in Volatile Oils

Essential Oil	Insects	Results	Reference
Allium sativum (Amaryllidaceae)	*Spodoptera littoralis*	The study explored a garlic essential oil-based nanoemulsion as an alternative control tool. The results demonstrated that the garlic-EO nanoemulsion effectively killed larvae and reduced feeding activity in laboratory tests. The authors emphasize that these findings highlight the potential of botanical insecticides for sustainable pest management.	(Giuliano et al. 2024)
Blumea balsamifera (Asteraceae)	*Tribolium castaneum, Lasioderma serricorne,* and *Sitophilus oryzae*	The compounds borneol (23.3%), β-caryophyllene (20.9%), and camphor (11.8%) were the main compounds found in EO. The results demonstrated that EO had no fumigant toxicity to the three target insects, but exhibited significant contact activity against *L. serricorne* (LD_{50} = 12.4 µg/adult) and *S. oryzae* (LD_{50} = 44.4 µg/adult). Meanwhile, EO showed a remarkable repellent effect on *T. castaneum* at all test concentrations and a general repellent effect on *S. oryzae* at high concentrations (78.63 nL/cm^2). Finally, β-caryophyllene showed the best performance in the contact toxicity bioassays against the three insects.	(Qi et al. 2024)
Cymbogon citratus (Poaceae)	*Sitophilus zeamais* and *Tribolium castaneum*	Geraniol (20.86%) was identified as the main constituent of the essential oil. This tested EO showed pronounced insecticidal activity against the pest species in relation to the applied doses.	(Moutassem et al. 2024)
Lantana camara (Verbenaceae)	*Tribolium castaneum, Lasioderma serricorne,* and *Callosobruchus chinensis*	The major EO compounds identified were caryophyllene (69.96%), isoledene (12%), and *a*-copaene (4.11%). In this study, the essential oil exhibited excellent fumigant toxicity (LC_{50} of 16.70 mg/L air for *T. castaneum*, 4.141 mg/L air for *L. serricorne*, and 6.245 mg/L air for *C. chinensis* in 24 h).	(Aisha et al. 2024)
Lavandula angustifolia (Lamiaceae)	*Plodia interpunctella*	The essential oil presented limonene (66.12%), eucalyptol (66.12%), and camphor (14.31%) as major components. This EO presented the highest mortality percentage (90%) with the highest treatment of 0.50 µL/cm^2 in the first 3 hours.	(Moullamri et al. 2024)
Mentha pulegium (Lamiaceae)		The pulegone compound (83.06%) was predominant in the EO. The insecticidal potential in this oil was potent in a short period of exposure (3h), in which 92% and 100% mortality were observed with concentrations of 0.25 and 0.50 µL/cm^2, respectively.	
Ocimum basilicum (Lamiaceae)	*Drosophila suzukii* flies	The compounds (E)-anethole, linalool, and 1,8-cineole were major in the EO. Regarding its insecticidal potential, the essential oil showed toxicity against *Drosophila suzukii* flies (LC_{50} = 1.19 mg mL^{-1})	(Oliveira et al. 2024)
Salvia officinalis (Lamiaceae)	*Plodia interpunctella*	The essential oil was characterized by the major components α-thujone (26.99%) and camphor (25.50%). This EO showed the highest mortality percentage (90%) with the highest treatment of 0.50 µL/cm^2 in the first 3 hours.	(Moullamri et al. 2024)

(*Continued*)

TABLE 8.2 ***(Continued)***
Main Insecticidal Activities Found in Volatile Oils

Essential Oil	Insects	Results	Reference
	Drosophila suzukii flies	The compounds camphor, β-pinene, and 1,8-cineole were major in the EO. Regarding its insecticidal potential, the essential oil showed toxicity against *Drosophila suzukii* flies (LC_{50} = 3.51 mg mL^{-1}).	(Oliveira et al. 2024)
Thymus pallescens (Lamiaceae)	*Sitophilus zeamais* and *Tribolium castaneum*	Carvacrol (56.04%) was identified as the major constituent of the essential oil. The study revealed that EO effected corrected mortality rates ranging from 42.5–100% to 25–100% in *S. zeamais* with corresponding lethal concentration (LC_{50}) values of 17.7 µL/ml and 15 µL/L air, respectively.	(Moutassem et al. 2024)
Thymus vulgaris (Lamiaceae)	*Drosophila suzukii* flies	The compounds thymol, ρ-cymene, and carvacrol were major components of the EO. Regarding its insecticidal potential, the essential oil showed toxicity against *Drosophila suzukii* flies (LC_{50} = 1.24 mg mL^{-1}).	(Oliveira et al. 2024)

against a range of insect and fungal pests. In the second, they affect only a specific species of fungus or insect (Dassanayake et al. 2021; Mfarrej and Rara 2019).

EOs have already been tested as bioinsecticides against various pests that attack crops such as corn, coffee, lettuce, vegetables, soybeans, grains, and cereals (Menossi et al. 2021; Garrido-Miranda, Giraldo, and Schoebitz 2022). Bioinsecticides based on EOs obtained from plants can be developed considering different formulations. These include products developed from active ingredients that come from (1) a mixture of EOs, (2) only from a single type of essential oil or major component present in the essential oil used, (3) from a synthetic mixture of terpenoids that are similar to the essential oil produced by a plant, and (4) a new mixture of terpenoids (not natural) from different plant sources (Isman 2020a, 2020b).

8.3.1 Essential Oils in Nanoformulations and Emulsions

The use of encapsulated EOs or in nanoformulations has proven to be an advantageous technique for preserving bioactive substances. In encapsulation, the bioactive compound is reserved, surrounded by a carrier matrix, which allows its release to be regulated (Milićević et al. 2022). The encapsulation of bioactive oils becomes an advantageous alternative since, in addition to being poorly soluble in water and unstable, encapsulation allows the oil to be protected from light, heat, oxygen, and moisture. Among the polymeric carriers most used for encapsulating EOs are: natural polymers, alginate, cellulose derivatives, chitosan, starch and maltodextrin-based systems, whey protein, and gelatin (Chiriac et al. 2021).

A study carried out by Milićević et al. (2022) demonstrated that the EO obtained from clove flowers (*Syzygium aromaticum*) has insecticidal activity against the larvae of the potato moth—*Phthorimaea operculella*. The EO was analyzed for chemical composition, with eugenol (79.70%) as the major component. The EO at a concentration of 0.768 µL.L^{-1} air killed 99% of *P. operculella* adults after 24 hours. The encapsulated formulations (synthetic zeolite, natural zeolite, and bovine gelatin) demonstrated a longer-lasting action in adult mortality. The most efficient formulation was the one in which the essential oil was encapsulated with synthetic zeolite, whose activity lasted for 14 days of exposure, while the others (natural zeolite and bovine gelatin) lasted only 10 days (Milićević et al. 2022).

Fumigant activity is related to the ability of substances present in plant EOs to pass into the vapor phase at certain temperatures. The inhalation of some substances by insects promotes, due to toxicity, an insecticidal effect. Among the components of EOs that exhibit pronounced fumigant toxicity in stored product pests are camphene, carvacrol, carveol, caryophyllene oxide, cinnamaldehyde, citronellal, eugenol, eucalyptol, isoeugenol, geraniol, limonene, and linalool, among others (Karabörklü and Ayvaz 2023).

Research developed by Rajkumar et al. (2020) evaluated the fumigant toxicity of pure essential oil of *Piper nigrum* and the nanoformulation encapsulated in chitosan against the storage pests *Tribolium castaneum* and *Sitophilus oryzae*. The observed LC_{50} and LC_{90} values for the nanoformulation were 25.03 and 44.31 μL/L air for *S. oryzae*, respectively. In relation to the pest *T. castaneum*, the values were 29.02 and 59.13 μL/L air for LC_{50} and LC_{90}. The results observed were more satisfactory than those demonstrated by the pure oil for *S. oryzae* (48.97 and 85.77 μL/L air—LC_{50} and LC_{90}) and *T. castaneum* (55.77 and 97.93 μL/L air—LC_{50} and LC_{90}). The nanoformulation was also more active against the mortality of *S. oryzae* and *T. castaneum*, reaching an index of 100% at a concentration of 75.0 μL/L air, while, for the pure oil, the value was 100 μL/L air (Rajkumar et al. 2020). Studies with some information on nanoformulations of EOs with activity against pests are presented in Table 8.3.

Studies applying essential oil encapsulation techniques and testing their activities as insecticides, fumigants, fungicides, among others, reveal that these techniques are promising, as they considerably increase the bioactivity of EOs while maintaining the active constituents. The toxicity associated with essential oil nanoformulations can be attributed to their nanometric scale. This factor increases the bioactivity and bioavailability of the active substances present in EOs (Assadpour et al. 2024; Chiriac et al. 2021; Palermo et al. 2021).

8.3.2 Essential Oils in Commercial Pesticides

Recently, the use of EOs in insecticides has intensified. The production of this type of product uses oils from various species such as rosemary, peppermint, cinnamon, eucalyptus, clove and thyme, and these oils are used in mixtures with other active compounds or isolated. The EOs most commonly used in commercial products are those that contain eugenol in their composition, such as the EOs of bay, clove, anise, pine and eucalyptus. An example of a biopesticide commercially approved in Europe is the product Prev-Am (Oro Agri, Cape Town, and South Africa), which contains orange essential oil as a bioactive ingredient in a proportion of 5 to 6%. Orange oil is composed mainly of *d*-limonene, in a percentage above 90%. EOs such as thyme and lemongrass have proven insecticidal activity attributed to the main constituent, in the case of the former it would be thymol and for lemongrass citral (Isman, 2020a, 2020b; Dunan et al. 2023).

In the United States, products based on EOs and their constituents are available and approved for sale as insecticides. Among these, the EcoTrol product is composed of 2% peppermint essential oil, 10% rosemary essential oil, and 5% geraniol. The Requiem product is also available in the American market as an insecticide and has 17.9% *d*-limonene, 22.4% *p*-cymene, and 59.7% α-terpinene in its composition, characterized by a synthetic mixture of terpenoids that are similar to the essential oil produced by a plant. Another product that is also sold as an insecticide is TetraCurb, which is composed of a mixture of EOs that are 1.95% peppermint essential oil, 3% clove essential oil, and 50% rosemary essential oil (Kheloul et al. 2023).

The marketing of some oils with fungicidal and bactericidal action is permitted in European countries. The product Bioxeda, which contains clove essential oil, is applied as a fungicide against the action of storage pathogens in apple and pear trees. Another example is the product Biox-M, which contains essential oil of *Mentha spicata* L., popularly known as mint, which helps control growth in potatoes. Other products that act as insecticides against potato leafhoppers and whiteflies are Limocide, Orocide, Prev-Am, and Essen'ciel with orange essential oil (Raveau, Fontaine, and Lounès-Hadj Sahraoui 2020; Kheloul et al. 2023).

TABLE 8.3
Nanoformulations of Essential Oils with Activity Against Pests

Essential oil (EO)	Method of obtaining	Formulation	Insect/Fungus	Activity	Major components of EO	Reference
Allium sativum	Essential oil purchased commercially (Esperis S.p.A.—Milan, Italy)	Essential oil nanoemulsion (Tween 80: distilled water 80% and essential oil 15%)	*Tribolium confusum*	Repellent and acute toxicity (fumigant)—The formulation demonstrated repellent activity (area preference bioassay). Nanoemulsion 10 times more toxic (compared to citrus formulations)	Diallyl disulfide (30.09%), diallyl trisulfide (22.70%), diallyl tetrasulfide (13.17%), and diallyl sulfide (11.09%)	Palermo et al. (2021)
Artemesia herba-alba (plant)	Hydrodistillation in a Clevenger type apparatus	Pure essential oil encapsulated in β-cyclodextrin modified with succinic acid	*Botrytis cinerea*	Fungicide—encapsulated essential oil demonstrated greater ability to inhibit *B. cinerea* when compared to pure essential oil	α-Thujone (65.0%)	Ez-Zoubi et al. (2023)
Artemisia vulgaris	Commercially purchased essential oil (Esperis S.p.A.—Milan, Italy).	Essential oil nanoemulsion (Tween 80: distilled water 80% and essential oil 15%)	*Tribolium confusum*	Repellent and acute toxicity—The tested formulation demonstrated repellent activity in an area preference bioassay	α-Thujone (28.42%), camphor (22.21%), β-thujone (13.35%), chrysanthenone (5.97%), and chrysantenyl acetate (4.97%)	Palermo et al. (2021)
Foeniculum vulgare Mill. (in flowering phase)	Hydrodistillation in a Clevenger type apparatus	Pure EOs encapsulated in β-cyclodextrin	*Leptinotarsa decemlineata* Say	Larvicide/insecticide—It was observed that the inclusion complexes of EOs reduced the body mass of the larvae	Anethole (37.77%), α-pinene (23.51%), and α-phellandrene (11.98%)	Devrnja et al. (2024)
Foenicum vulgare	Commercially purchased essential oil (Esperis S.p.A.—Milan, Italy).	Essential oil nanoemulsion (Tween 80: distilled water 80% and essential oil 15%)	*Tribolium confusum*	Repellent and acute toxicity—The tested formulation demonstrated repellent activity in an area preference bioassay	(E)-Anethole (42.13%) and estragol (2.10%)	Palermo et al. (2021)
Lavandula angustifolia	Commercially purchased essential oil (Esperis S.p.A.—Milan, Italy).	Essential oil nanoemulsion (Tween 80: distilled water 80% and essential oil 15%)	*Tribolium confusum*	Repellent and acute toxicity—The tested formulation demonstrated repellent activity in an area preference bioassay	Linalool (38.12%), linalyl acetate (36.31%), camphor (7.01%), and eucalyptol (5.01%)	Palermo et al. (2021)

(*Continued*)

TABLE 8.3 *(Continued)*
Nanoformulations of Essential Oils with Activity Against Pests

Essential oil (EO)	Method of obtaining	Formulation	Insect/Fungus	Activity	Major components of EO	Reference
Mentha piperita	Commercially purchased essential oil (Esperis S.p.A.—Milan, Italy).	Essential oil nanoemulsion (Tween 80: distilled water 80% and essential oil 15%)	*Tribolium confusum*	Repellent and acute toxicity—The tested formulation demonstrated repellent activity in an area preference bioassay	(-)-Menthol (37.67%), menthone (22.10%), isomenthone (8.75%), limonene (7.73%), and (-)-neomenthol (6.32%)	Palermo et al. (2021)
Mentha pulegium L. (leaves and flowers)	Hydrodistillation in a Clevenger type apparatus	Essential oil encapsulated in yeast cells	*Myzus persicae*	Insecticide—The insecticidal effect of the encapsulated oil was more intense (72 hours after treatment) when compared to the pure oil	Pulegone (43.07%) and piperitenone (37.45%)	Kavetsou et al. (2019)
Ocimum basillicum L.	Hydrodistillation in a Clevenger type apparatus	Pure EOs encapsulated in β-cyclodextrin	*Leptinotarsa decemlineata* Say	Larvicide/Insecticide—It was observed that the inclusion complexes of EOs reduced the body mass of the larvae	Linalool (32.6%) and methyl chavicol (23.2%)	Palermo et al. (2021)
Piper nigrum (fresh and dried fruit)	Hydrodistillation in a Clevenger type apparatus	Pure and encapsulated essential oil (chitosan nanoparticles)	*Tribolium castaneum* and *Sitophylus oryzae*	Fumigant—encapsulated essential oil was more efficient than pure against *T. castaneum* and *S. oryzae*	B—Caryophyllene (29.49%), 3-carene (19.20%), and limonene (18.68%)	Rajkumar et al. (2020)
Pimpinella anisum	Commercially purchased essential oil (Esperis S.p.A.—Milan, Italy).	Essential oil nanoemulsion (Tween 80: distilled water 80% and essential oil 15%)	*Tribolium confusum*	Repellent and acute toxicity—The tested formulation demonstrated repellent activity in an area preference bioassay	(E)-Anethole (87.11%), estragol (3.98%), foeniculin (2.01%), and linalool (1.94%)	Palermo et al. (2021)
Rosmarinus officinalis	Óleo essencial adquirido comercialmente (Esperis S.p.A.—Milan, Italy).	Essential oil nanoemulsion (Tween 80: distilled water 80% and essential oil 15%)	*Tribolium confusum*	Repellent and acute toxicity—The tested formulation demonstrated repellent activity in an area preference bioassay	Eucalyptol (52.72%), camphor (12.04%), α-pinene (11.28%), ortho-cymene (4.01%), and β-pinene (4.01%)	Palermo et al. (2021)
Salvia officinalis	Óleo essencial adquirido comercialmente (Esperis S.p.A.—Milan, Italy).	Essential oil nanoemulsion (Tween 80: distilled water 80% and essential oil 15%)	*Tribolium confusum*	Repellent and acute toxicity—The tested formulation demonstrated repellent activity in an area preference bioassay	α-Pinene (21.20%), eucalyptol (20.07%), α-thujone (16.23%), camphor (12.85%), and camphene (4.67%)	Palermo et al. (2021)
Syzygium aromaticum (flowers)	Destilação a vapor	Pure and encapsulated essential oil (synthetic, natural zeolite and bovine gelatin)	*Phthorimaea operculella*	Insecticide—encapsulated essential oil demonstrated efficacy against *P. operculella*, providing prolonged oil efficacy for up to 14 days	Eugenol (79.70%), eugenol acetate (11.83%), and caryophyllene (E-) (4.51%)	Milićević et al. (2022)

8.4 ADVANTAGES, APPLICATIONS, AND CHALLENGES OF ESSENTIAL OILS IN PEST MANAGEMENT

Historically, there has been a reliance on synthetic pesticides for pest control and with the increase in food production, the overuse of these chemicals has become a significant problem due to the indiscriminate application of agrochemicals to crops, resulting in soil and water contamination (Kamanula et al. 2010). This not only pollutes the environment but also presents high toxicity to humans and animals. These practices raise serious concerns about the safety and sustainability of agricultural systems, threatening public health and ecological integrity (Aktar, Sengupta, and Chowdhury 2009; Kim, Kabir, and Jahan 2017).

The application of EOs as biopesticides (bioherbicides, bioinsecticides, and biofungicides) in pest management has emerged as a sustainable and effective alternative to synthetic pesticides for various crops (Fierascu et al. 2020). The diversity of chemical compounds present in EOs makes it difficult for pests to develop in crops. While synthetic insecticides have a single mechanism of action, EOs act through multiple simultaneous modes of action, making it more difficult for pests to develop resistance, promoting long-lasting effectiveness in controlling infestations. Additionally, EOs are natural and, when used properly, tend to be less toxic than many synthetic insecticides, and their rapid degradation in the environment reduces human exposure to potentially harmful residues, decreasing the risk of adverse health effects (Gonzales Correa et al. 2015; Pavela and Benelli 2016; Campos et al. 2019).

EOs are composed of a variety of natural chemicals such as monoterpenes, sesquiterpenes and phenylpropanoids that have the ability to interact with the biochemical, physiological and metabolic systems of insects and fungi. These interactions can directly affect the biological processes of these organisms, such as respiration, reproduction and communication, leading to effects that can be detrimental to their survival and reproduction (Mossa 2016; Fierascu et al. 2020).

A study developed by Trombin de Souza and collaborators (2022) demonstrated that three specimens of *Rosmarinus officinalis* presented in their essential oil, camphor (12.1–20.5%), 1,8-cineole (13.0–49.1%), and α-pinene (9.3–20.1%), as common major compounds showed neurotoxic effect in *Drosophila suzukii*. The neurotoxicity of these compounds interferes with the nervous system of the pest, leading to its immobilization and death. In addition, these compounds exhibit a deterrent effect, repelling flies and reducing oviposition in the treated areas. These compounds also influenced the larvicidal effect that occurs when larvae came into direct contact with the compounds, resulting in high larval mortality. The methods of evaluating efficacy in the study included topical application bioassays, where the essential oil is applied directly to the pest, and ingestion bioassays, where the pest ingests the food treated with the essential oil. These assays provide a measure of the efficacy of the active compounds against *Drosophila suzukii*.

In a study involving basil essential oil (*Ocimum basilicum*) and its effects on the corn weevil (*Sitophilus zeamais*), the main components of the essential oil, estragole (85%) and linalool (12%), and their mechanism of action by fumigation in *S. zeamais* were evaluated. Fumigation toxicity tests showed that the essential oil was lethal to *S. zeamais*, with a LC_{50} of 25.4 μL L^{-1} of air (Moura et al. 2021). Holy basil (*Ocimum tenuiflorum*) essential oil has also been shown to be a natural insecticide against *Sitophilus oryzae*. Gas chromatography-mass spectrometry (GC-MS) analysis identified eugenol (50.4%) and caryophyllene (20.1%) as the main chemical components of the essential oil. Toxicity tests showed high levels of insect mortality, with LC_{50} of 0.376 μL/cm^2 for contact and 963.3 μL/L for fumigation. The efficacy of the oil is a result of the inhibition of acetylcholinesterase, resulting in paralysis and death of the insects.

In addition to insects, fungi also pose a significant threat to agriculture, causing major losses in agricultural production (Liu et al. 2016). One example is *Penicillium digitatum*, a phytopathogenic fungus that causes postharvest decay of Mexican lemons (*Citrus aurantiifolia*). This fungus is the causative agent of green rot, a disease that mainly affects citrus fruits during storage and transportation. Especially in tropical regions where high temperature and relative humidity (RH) provide suitable conditions for the development of fungi on the fruit (Smilanick et al. 2006; Li et al. 2022).

The study by Atrash et al. (2018) investigated the biological potential of *Satureja hortensis* essential oil in the post-harvest preservation of Mexican lemon against the fungus *Penicillium digitatum*. The essential oil presented as main chemical compounds carvacrol (55.67%) and γ-terpinene (31.98%), showing strong antifungal action and reduction of the degrading enzymatic activity of fungi in the fruits.

Ben Jabeur, Somai-Jemmali, and Hamada (2017) also evaluated the biofungicidal potential of the essential oil, offering a promising alternative to traditional chemical fungicides. In the study with thyme (*Thymus vulgaris*), they obtained the essential oil by hydrodistillation and subsequently analyzed the chemical composition of thyme, highlighting the main chemical constituents thymol (76.96%), ρ-cymene (9.89%), γ-terpinene (1.92%), and caryophyllene oxide (1.69%). The antifungal activity of the essential oil and its major component, thymol, was evaluated *in vitro* against *Mycosphaerella graminicola*. The essential oil demonstrated greater antifungal activity compared to pure thymol. This superiority of the essential oil can be attributed to the synergistic effect of the various chemical components present, which together exert a more potent antifungal action than thymol alone.

The mechanisms of action were investigated by analyzing gene expression in *M. graminicola*. Treatment with thyme essential oil resulted in the overexpression of the MgMfs1, MgAtr4, MgBcy1, and MgHog1 genes, and in the repression of the MgAtr4, MgSlt2, and MgBcy1 genes. In contrast, pure thymol repressed only the MgAtr4 and MgSlt2 genes. These data highlight the ability of thyme essential oil to affect multiple genes involved in the development and virulence of the fungus *M. graminicola*, explaining its greater antifungal activity compared to thymol alone.

Bioherbicides produced from EOs have also emerged as promising alternatives for pest management. Plant-derived compounds, such as EOs, have the potential to effectively and sustainably control weed growth (Cai and Gu 2016). These bioherbicides work through allelopathic mechanisms, interfering with the growth and development of invasive plants.

In the study by Benchaa, Hazzit, and Abdelkrim (2018) the potential of *Eucalyptus citriodora* essential oil as a bioherbicide to control weeds such as *Sinapis arvensis*, *Sonchus oleraceus*, *Xanthium strumarium*, and *Avena fatua* was evaluated. Using analytical techniques such as gas chromatography-flame ionization detection (GC-FID) and GC-mass spectrometry (GC-MS), the researchers identified the main compounds in the oil, citronellal (64.7%) and citronellol (10.9%). Tests carried out in the laboratory and in the greenhouse with different concentrations of the oil (0.01%, 0.02%, and 0.03% in the laboratory and 1%, 2%, and 3% in the greenhouse) demonstrated a significant reduction in the germination and growth of the weeds, with 0.03% of the oil completely suppressing the germination of *S. arvensis*. In addition, treatment with the essential oil also affected plant chlorophyll pigmentation and plant membrane integrity. The results indicate that *E. citriodora* essential oil has natural bioherbicidal potential. Table 8.4 presents a summary of the applicability of EOs in pest management, including information on the plant, active compound, target pest, mechanism of action, and efficacy assessment method.

The effective implementation of EOs as biopesticides faces several challenges that need to be overcome, such as the chemical instability of EOs when exposed to environmental factors such as air, light, and humidity. Exposure to air can lead to oxidation of volatile compounds, especially to UV light, which causes photodegradation resulting in loss of biological activities and alteration of organoleptic properties (Kumar et al. 2019).

In addition to chemical instability, phytotoxicity is another significant challenge to overcome when implementing EOs as biopesticide. These compounds not only cause damage to weeds but can also affect crop plants and other non-target plants. The active compounds present in EOs, such as monoterpenes and phenylpropanoids, can interfere with fundamental cellular processes, such as photosynthesis and the integrity of cell membranes. This interference can result in adverse effects, compromising plant health and development, and consequently, affecting agricultural productivity (Singh et al. 2020).

TABLE 8.4
Applicability of Essential Oils in Pest Management

Species	Active Compound	Target Pest	Mechanism of Action	Method of Evaluating Effectiveness	Reference
Rosmarinus fficinalis	Camphor (12.1–20.5%) 1,8-cynole (13.0–49.1%) α-pinene (9.3–20.1%)	*Drosophila suzukei*	Neurotoxic; deterrent effect; larvicidal	Topical application bioassay; ingestion bioassay	Trombin de Souza et al. (2022)
Ocimum basilicum	Estragole (85%) linalool (12%)	*Sitophilus zeamais*	Fumigation	Fumigation toxicity test	Moura et al. (2021)
Ocimum tenuiflorum	Eugenol (50.4%) β-caryophyllene (20.1%)	*Sitophilus orzyae*	Inhibition of acetylcholinesterase activity	*In vivo* tests; toxicity tests	Bhavya, Chandu, and Devi (2018)
Satureja hortensis	Carvacrol (55.67%) γ-terpinene (31.98%)	*Penicillium digitatum*	Fungicidal action	*In vivo* tests	(Atrash et al. 2018)
Thymus vulgaris	Thymol (76.96%) ρ-cymene (9.89%) γ-terpinene (1.92%) caryophyllene (1.69%)	*Mycosphaerella graminicola*	Overexpression and repression of genes involved in fungal development and detoxification	*In vitro* assay evaluating the expression of specific genes	(Ben Jabeur, Somai-Jemmali, and Hamada 2017)
Eucalyptus citriodora	Citronellal (64.7%) citronellol (10.9%)	*Sinapis arvensis, Sonchus oleraceus, Xanthium strumarium*, and *Avena fatua*	Herbicidal action	Seed germination test; evaluation of allelopathic effect under greenhouse conditions	Benchaa, Hazzit, and Abdelkrim (2018)

REFERENCES

Aisha, Kolapparamban, Naduvilthara U Visakh, Berin Pathrose, Nicola Mori, Rowida S Baeshen, and Rady Shawer. 2024. "Extraction, Chemical Composition and Insecticidal Activities of Lantana Camara Linn. Leaf Essential Oils against Tribolium Castaneum, Lasioderma Serricorne and Callosobruchus Chinensis." *Molecules*. https://doi.org/10.3390/molecules29020344.

Aktar, Wasim, Dwaipayan Sengupta, and Ashim Chowdhury. 2009. "Impact of Pesticides Use in Agriculture: Their Benefits and Hazards." *Interdisciplinary Toxicology* 2 (1): 1–12. https://doi.org/10.2478/v10102-009-0001-7.

Assadpour, Elham, Aslı Can Karaça, Mahdis Fasamanesh, Sahar Akhavan Mahdavi, Mahya Shariat-Alavi, Jianguo Feng, Mohammad Saeed Kharazmi, Abdur Rehman, and Seid Mahdi Jafari. 2024. "Application of Essential Oils as Natural Biopesticides; Recent Advances." *Critical Reviews in Food Science and Nutrition* 64 (19): 6477–97. https://doi.org/10.1080/10408398.2023.2170317.

Atrash, Sara, Asghar Ramezanian, Majid Rahemi, Reza Mostofizadeh Ghalamfarsa, and Elhadi Yahia. 2018. "Antifungal Effects of Savory Essential Oil, Gum Arabic, and Hot Water in Mexican Lime Fruits." *HortScience* 53 (4): 524–30. https://doi.org/10.21273/HORTSCI12736-17.

Bajpai, Susmita. 2019. "Essential Oils as Green Pesticides." *Think India Journal* 22 (14): 11957–76.

Benchaa, Sara, Mohamed Hazzit, and Hacène Abdelkrim. 2018. "Allelopathic Effect of *Eucalyptus Citriodora* Essential Oil and Its Potential Use as Bioherbicide." *Chemistry & Biodiversity* 15 (8). https://doi.org/10.1002/cbdv.201800202.

Cai, Xiaoya, and Mengmeng Gu. 2016. "Bioherbicides in Organic Horticulture." *Horticulturae* 2 (2): 3. https://doi.org/10.3390/horticulturae2020003.

Campos, Estefânia V. R., Patrícia L. F. Proença, Jhones L. Oliveira, Mansi Bakshi, P. C. Abhilash, and Leonardo F. Fraceto. 2019. "Use of Botanical Insecticides for Sustainable Agriculture: Future Perspectives." *Ecological Indicators* 105 (October): 483–95. https://doi.org/10.1016/j.ecolind.2018.04.038.

Chaudhari, Anand Kumar, Vipin Kumar Singh, Akash Kedia, Somenath Das, and Nawal Kishore Dubey. 2021. "Essential Oils and Their Bioactive Compounds as Eco-Friendly Novel Green Pesticides for Management of Storage Insect Pests: Prospects and Retrospects." *Environmental Science and Pollution Research* 28 (15): 18918–40. https://doi.org/10.1007/s11356-021-12841-w.

Chiriac, Aurica P., Alina G. Rusu, Loredana E. Nita, Vlad M. Chiriac, Iordana Neamtu, and Alina Sandu. 2021. "Polymeric Carriers Designed for Encapsulation of Essential Oils with Biological Activity." *Pharmaceutics* 13 (5): 631. https://doi.org/10.3390/pharmaceutics13050631.

Dassanayake, Mackingsley Kushan, Chien Hwa Chong, Teng-Jin Khoo, Adam Figiel, Antoni Szumny, and Chee Ming Choo. 2021. "Synergistic Field Crop Pest Management Properties of Plant-Derived Essential Oils in Combination with Synthetic Pesticides and Bioactive Molecules: A Review." *Foods* 10 (9): 2016. https://doi.org/10.3390/foods10092016.

Devrnja, Nina, Boban Anđelković, Jovana Ljujić, Tatjana Ćosić, Sofija Stupar, Milica Milutinović, and Jelena Savić. 2024. "Encapsulation of Fennel and Basil Essential Oils in β-Cyclodextrin for Novel Biopesticide Formulation." *Biomolecules* 14 (3): 353. https://doi.org/10.3390/biom14030353.

Dunan, Lana, Tara Malanga, Sylvain Benhamou, Nicolas Papaiconomou, Nicolas Desneux, Anne-Violette Lavoir, and Thomas Michel. 2023. "Effects of Essential Oil-Based Formulation on Biopesticide Activity." *Industrial Crops and Products* 202 (October): 117006. https://doi.org/10.1016/j.indcrop.2023.117006.

El-Shahir, Amany A., Deiaa A. El-Wakil, Arafat Abdel Hamed Abdel Latef, and Nora H. Youssef. 2022. "Bioactive Compounds and Antifungal Activity of Leaves and Fruits Methanolic Extracts of Ziziphus Spina-Christi L." *Plants* 11 (6): 1–18. https://doi.org/10.3390/plants11060746.

Elsherbiny, E. A., N. A. Safwat, and M. M. Elaasser. 2017. "Fungitoxicity of Organic Extracts of Ocimum Basilicum on Growth and Morphogenesis of Bipolaris Species (Teleomorph Cochliobolus)." *Journal of Applied Microbiology* 123 (4): 841–52. https://doi.org/10.1111/jam.13543.

Ettakifi, Hajar, Kaoutar Abbassi, Safae Maouni, El Hadi Erbiai, Abderrahmane Rahmouni, Mounir Legssyer, Rabah Saidi, et al. 2023. "Chemical Characterization and Antifungal Activity of Blue Tansy (Tanacetum Annuum) Essential Oil and Crude Extracts against Fusarium Oxysporum f. Sp. Albedinis, an Agent Causing Bayoud Disease of Date Palm." *Antibiotics* 12 (9). https://doi.org/10.3390/antibiotics12091451.

Ez-Zoubi, Amine, Yassine Ez Zoubi, Fatiha Bentata, Ayoub El-Mrabet, Chaymae Ben Tahir, Mustapha Labhilili, and Abdellah Farah. 2023. "Preparation and Characterization of a Biopesticide Based on Artemisia Herba-Alba Essential Oil Encapsulated with Succinic Acid-Modified Beta-Cyclodextrin." Edited by Jo o Paulo Leal. *Journal of Chemistry* 2023 (May): 1–8. https://doi.org/10.1155/2023/3830819.

Fierascu, Radu Claudiu, Ioana Catalina Fierascu, Cristina Elena Dinu-Pirvu, Irina Fierascu, and Alina Paunescu. 2020. "The Application of Essential Oils as a Next-Generation of Pesticides: Recent Developments and Future Perspectives." *Zeitschrift Für Naturforschung C* 75 (7–8): 183–204. https://doi.org/10.1515/znc-2019-0160.

Garrido-Miranda, Karla A., Juan D. Giraldo, and Mauricio Schoebitz. 2022. "Essential Oils and Their Formulations for the Control of Curculionidae Pests." *Frontiers in Agronomy* 4 (June). https://doi.org/10.3389/fagro.2022.876687.

Giuliano, Gaetano, Orlando Campolo, Giuseppe Forte, Alberto Urbaneja, Meritxell Pérez-Hedo, Ilaria Latella, Vincenzo Palmeri, and Giulia Giunti. 2024. "Insecticidal Activity of Allium Sativum Essential Oil-Based Nanoemulsion against Spodoptera Littoralis." *Insects*. https://doi.org/10.3390/insects15070476.

Gonzales Correa, Yenis Del Carmen, Lêda R. A. Faroni, Khalid Haddi, Eugênio E. Oliveira, and Eliseu José G. Pereira. 2015. "Locomotory and Physiological Responses Induced by Clove and Cinnamon Essential Oils in the Maize Weevil Sitophilus Zeamais." *Pesticide Biochemistry and Physiology* 125 (November): 31–37. https://doi.org/10.1016/j.pestbp.2015.06.005.

Gruľová, Daniela, Lucia Caputo, Hazem S. Elshafie, Beáta Baranová, Laura De Martino, Vincent Sedlák, Zuzana Gogaľová, Vincenzo De Feo, Janka Poráčová, and Ippolito Camele. 2020. "Thymol Chemotype Origanum Vulgare L. Essential Oil as a Potential Selective Bio-Based Herbicide on Monocot Plant Species." *Molecules* 25 (3): 595.

Isman, Murray B. 2017. "Bridging the Gap: Moving Botanical Insecticides from the Laboratory to the Farm." *Industrial Crops and Products* 110 (December): 10–14. https://doi.org/10.1016/j.indcrop.2017.07.012.

Isman, Murray B. 2020a. "Commercial Development of Plant Essential Oils and Their Constituents as Active Ingredients in Bioinsecticides." *Phytochemistry Reviews* 19 (2): 235–41. https://doi.org/10.1007/s11101-019-09653-9.

Isman, Murray B. 2020b. "Bioinsecticides Based on Plant Essential Oils: A Short Overview." *Zeitschrift Für Naturforschung C* 75 (7–8): 179–82. https://doi.org/10.1515/znc-2020-0038.

Jabeur, M. Ben, L. Somai-Jemmali, and W. Hamada. 2017. "Thyme Essential Oil as an Alternative Mechanism: Biofungicide-Causing Sensitivity of *Mycosphaerella Graminicola*." *Journal of Applied Microbiology* 122 (4): 932–39. https://doi.org/10.1111/jam.13408.

Jouini, Amira, Mercedes Verdeguer, Samuele Pinton, Fabrizio Araniti, Eristanna Palazzolo, Luigi Badalucco, and Vito Armando Laudicina. 2020. "Potential Effects of Essential Oils Extracted from Mediterranean Aromatic Plants on Target Weeds and Soil Microorganisms." *Plants* 9 (10): 1–24. https://doi.org/10.3390/plants9101289.

Kalhoro, Muhammad Talib, Hong Zhang, Ghulam Mujtaba Kalhoro, Fukai Wang, Tianhong Chen, Yahya Faqir, and Farhan Nabi. 2022. "Fungicidal Properties of Ginger (Zingiber Officinale) Essential Oils against Phytophthora Colocasiae." *Scientific Reports* 12 (1): 1–10. https://doi.org/10.1038/s41598-022-06321-5.

Kamanula, John, Gudeta W. Sileshi, Steven R. Belmain, Phosiso Sola, Brighton M. Mvumi, Greenwell K. C. Nyirenda, Stephen P. Nyirenda, and Philip C. Stevenson. 2010. "Farmers' Insect Pest Management Practices and Pesticidal Plant Use in the Protection of Stored Maize and Beans in Southern Africa." *International Journal of Pest Management* 57 (1): 41–49. https://doi.org/10.1080/09670874.2010.522264.

Karabörklü, Salih, and Abdurrahman Ayvaz. 2023. "A Comprehensive Review of Effective Essential Oil Components in Stored-Product Pest Management." *Journal of Plant Diseases and Protection* 130 (3): 449–81. https://doi.org/10.1007/s41348-023-00712-0.

Kavetsou, Eleni, Spyridon Koutsoukos, Dimitra Daferera, Moschos G. Polissiou, Dimitrakis Karagiannis, Dionysios Ch. Perdikis, and Anastasia Detsi. 2019. "Encapsulation of Mentha Pulegium Essential Oil in Yeast Cell Microcarriers: An Approach to Environmentally Friendly Pesticides." *Journal of Agricultural and Food Chemistry* 67 (17): 4746–53. https://doi.org/10.1021/acs.jafc.8b05149.

Kheloul, Lynda, Sylvia Anton, Dimitri Bréard, and Abdellah Kellouche. 2023. "Fumigant Toxicity of Some Essential Oils and Eucalyptol on Different Life Stages of Tribolium Confusum (Coleoptera: Tenebrionidae)." *Botany Letters* 170 (1): 3–14. https://doi.org/10.1080/23818107.2021.1982767.

Kim, Ki-Hyun, Ehsanul Kabir, and Shamin Ara Jahan. 2017. "Exposure to Pesticides and the Associated Human Health Effects." *Science of the Total Environment* 575 (January): 525–35. https://doi.org/10.1016/j.scitotenv.2016.09.009.

Kotan, Recep, Ahmet Cakir, Hakan Ozer, Saban Kordali, Ramazan Cakmakci, Fatih Dadasoglu, Neslihan Dikbas, Tuba Aydin, and Cavit Kazaz. 2014. "Antibacterial Effects of Origanum Onites against Phytopathogenic Bacteria: Possible Use of the Extracts from Protection of Disease Caused by Some Phytopathogenic Bacteria." *Scientia Horticulturae* 172: 210–20. https://doi.org/10.1016/j.scienta.2014.03.016.

Kumar, Sandeep, Gagana Kuamr Mahapatro, Dinesh Kumar Yadav, Kailashpati Tripathi, Pushpendra Koli, Parshant Kaushik, Kuldeep Sharma, and Suresh Nebapure. 2022. "Essential Oils as Green Pesticides: An Overview." *The Indian Journal of Agricultural Sciences* 92 (11 SE-Review Article): 1298–305. https://doi.org/10.56093/ijas.v92i11.122746.

Kumar, Sandeep, Monika Nehra, Neeraj Dilbaghi, Giovanna Marrazza, Ashraf Aly Hassan, and Ki-Hyun Kim. 2019. "Nano-Based Smart Pesticide Formulations: Emerging Opportunities for Agriculture." *Journal of Controlled Release* 294 (January): 131–53. https://doi.org/10.1016/j.jconrel.2018.12.012.

Li, Yongmei, Mengyuan Xia, Pengbo He, Qiaoming Yang, Yixin Wu, Pengfei He, Ayesha Ahmed, et al. 2022. "Developing Penicillium Digitatum Management Strategies on Post-Harvest Citrus Fruits with Metabolic Components and Colonization of Bacillus Subtilis L1-21." *Journal of Fungi* 8 (1): 80. https://doi.org/10.3390/jof8010080.

Liu, Yihua, Shiliang Li, Zhanglin Ni, Minghua Qu, Donglian Zhong, Caifen Ye, and Fubin Tang. 2016. "Pesticides in Persimmons, Jujubes and Soil from China: Residue Levels, Risk Assessment and Relationship between Fruits and Soils." *Science of the Total Environment* 542 (January): 620–28. https://doi.org/10.1016/j.scitotenv.2015.10.148.

Menossi, Matías, Romina P. Ollier, Claudia A. Casalongué, and Vera A. Alvarez. 2021. "Essential Oil-loaded Bio-nanomaterials for Sustainable Agricultural Applications." *Journal of Chemical Technology & Biotechnology* 96 (8): 2109–22. https://doi.org/10.1002/jctb.6705.

Mfarrej, Manar Fawzi Bani, and Fatimetou Mohamed Rara. 2019. "Competitive, Sustainable Natural Pesticides." *Acta Ecologica Sinica* 39 (2): 145–51. https://doi.org/10.1016/j.chnaes.2018.08.005.

Milićević, Zoran, Slobodan Krnjajić, Milan Stević, Jovana Ćirković, Aleksandra Jelušić, Mira Pucarević, and Tatjana Popović. 2022. "Encapsulated Clove Bud Essential Oil: A New Perspective as an Eco-Friendly Biopesticide." *Agriculture* 12 (3): 338. https://doi.org/10.3390/agriculture12030338.

Mobki, Marzieh, Seyed Ali Safavi, Mohammad Hasan Safaralizadeh, and Omid Panahi. 2014. "Toxicity and Repellency of Garlic (Allium Sativum L.) Extract Grown in Iran against Tribolium Castaneum (Herbst) Larvae and Adults." *Archives of Phytopathology and Plant Protection* 47 (1): 59–68. https://doi.org/10.1080/03235408.2013.802896.

Mossa, Abdel-Tawab H. 2016. "Green Pesticides: Essential Oils as Biopesticides in Insect-Pest Management." *Journal of Environmental Science and Technology* 9 (5): 354–78. https://doi.org/10.3923/jest.2016.354.378.

Moullamri, Mouna, Kacem Rharrabe, Houssam Annaz, Amin Laglaoui, Filippo Alibrando, Francesco Cacciola, Luigi Mondello, Noureddin Bouayad, Saïd Zantar, and Mohammed Bakkali. 2024. "Salvia Officinalis,

Lavandula Angustifolia, and Mentha Pulegium Essential Oils: Insecticidal Activities and Feeding Deterrence against Plodia Interpunctella (Lepidoptera: Pyralidae)." *Journal of Essential Oil Bearing Plants* 27 (1): 16–33. https://doi.org/10.1080/0972060X.2023.2301699.

Moura, Eridiane da Silva, Lêda Rita D'Antonino Faroni, Fernanda Fernandes Heleno, and Alessandra Aparecida Zinato Rodrigues. 2021. "Toxicological Stability of Ocimum Basilicum Essential Oil and Its Major Components in the Control of Sitophilus Zeamais." *Molecules* 26 (21): 6483. https://doi.org/10.3390/molecules26216483.

Moutassem, Dahou, Tahar Boubellouta, Yuva Bellik, Zyed Rouis, Dmitry E. Kucher, Aleksandra O. Utkina, Olga D. Kucher, Olga A. Mironova, Nyasha J. Kavhiza, and Nazih Y. Rebouh. 2024. "Insecticidal Activity of Thymus Pallescens de Noë and Cymbogon Citratus Essential Oils against Sitophilus Zeamais and Tribolium Castaneum." *Scientific Reports* 14 (1): 13951. https://doi.org/10.1038/s41598-024-64757-3.

Nerilo, Samuel B., Gustavo Henrique O. Rocha, Caroline Tomoike, Simone A. G. Mossini, Renata Grespan, Jane M. G. Mikcha, and Miguel Machinski. 2016. "Antifungal Properties and Inhibitory Effects upon Aflatoxin Production by Zingiber Officinale Essential Oil in Aspergillus Flavus." *International Journal of Food Science and Technology* 51 (2): 286–92. https://doi.org/10.1111/ijfs.12950.

Oliveira, Cassia Duarte, Maria das Graças Cardoso, Luis Roberto Batista, Eduardo Alves, Maria Beatriz Pereira Rosa, Vanuzia Rodrigues Fernandes Ferreira, Luciano de Souza, Maria Pineda, Antonia Isadora Fernandes, and David Lee Nelson. 2024. "The Antibacterial, Antioxidant, and Insecticidal Activities of Essential Oils from Thymus Vulgaris L., Salvia Officinalis L., and Ocimum Basilicum L." *Journal of Food Safety* 44 (3): e13145.

Palermo, Davide, Giulia Giunti, Francesca Laudani, Vincenzo Palmeri, and Orlando Campolo. 2021. "Essential Oil-Based Nano-Biopesticides: Formulation and Bioactivity against the Confused Flour Beetle Tribolium Confusum." *Sustainability* 13 (17): 9746. https://doi.org/10.3390/su13179746.

Pavela, Roman, and Giovanni Benelli. 2016. "Essential Oils as Ecofriendly Biopesticides? Challenges and Constraints." *Trends in Plant Science* 21 (12): 1000–1007. https://doi.org/10.1016/j.tplants.2016.10.005.

Qi, Yuan-Tong, Jia-Zhu Wang, Yu Zheng, Jia-Wei Zhang, and Shu-Shan Du. 2024. "Chemical Composition and Insecticidal Activities of Blumea Balsamifera (Sambong) Essential Oil Against Three Stored Product Insects." *Journal of Food Protection* 87 (1): 100205. https://doi.org/10.1016/j.jfp.2023.100205.

Rajkumar, Vallavan, Chinappan Gunasekaran, Jayaraman Dharmaraj, Panneerselvam Chinnaraj, Cheruvathur Amita Paul, and Inbaraj Kanithachristy. 2020. "Structural Characterization of Chitosan Nanoparticle Loaded with Piper Nigrum Essential Oil for Biological Efficacy against the Stored Grain Pest Control." *Pesticide Biochemistry and Physiology* 166 (June): 104566. https://doi.org/10.1016/j.pestbp.2020.104566.

Raveau, Robin, Joël Fontaine, and Anissa Lounès-Hadj Sahraoui. 2020. "Essential Oils as Potential Alternative Biocontrol Products against Plant Pathogens and Weeds: A Review." *Foods* 9 (3): 365. https://doi.org/10.3390/foods9030365.

Singh, Narayan, Harminder Pal Singh, Daizy Rani Batish, Ravinder Kumar Kohli, and Surender Singh Yadav. 2020. "Chemical Characterization, Phytotoxic, and Cytotoxic Activities of Essential Oil of Mentha Longifolia." *Environmental Science and Pollution Research* 27 (12): 13512–23. https://doi.org/10.1007/s11356-020-07823-3.

Smilanick, J. L., M. F. Mansour, F. Mlikota Gabler, and W. R. Goodwine. 2006. "The Effectiveness of Pyrimethanil to Inhibit Germination of Penicillium Digitatum and to Control Citrus Green Mold after Harvest." *Postharvest Biology and Technology* 42 (1): 75–85. https://doi.org/10.1016/j.postharvbio.2006.05.008.

Tak, Jun-Hyung, and Murray B. Isman. 2016. "Metabolism of Citral, the Major Constituent of Lemongrass Oil, in the Cabbage Looper, Trichoplusia Ni, and Effects of Enzyme Inhibitors on Toxicity and Metabolism." *Pesticide Biochemistry and Physiology* 133 (October): 20–25. https://doi.org/10.1016/j.pestbp.2016.03.009.

Tak, Jun-Hyung, and Murray B. Isman. 2017. "Acaricidal and Repellent Activity of Plant Essential Oil-Derived Terpenes and the Effect of Binary Mixtures against Tetranychus Urticae Koch (Acari: Tetranychidae)." *Industrial Crops and Products* 108 (December): 786–92. https://doi.org/10.1016/j.indcrop.2017.08.003.

Tak, Jun-Hyung, Eduardo Jovel, and Murray B. Isman. 2017. "Effects of Rosemary, Thyme and Lemongrass Oils and Their Major Constituents on Detoxifying Enzyme Activity and Insecticidal Activity in Trichoplusia Ni." *Pesticide Biochemistry and Physiology* 140 (August): 9–16. https://doi.org/10.1016/j.pestbp.2017.01.012.

Trombin de Souza, Michele, Mireli Trombin de Souza, Daniel Bernardi, Daiana da Costa Oliveira, Maíra Chagas Morais, Douglas José de Melo, Vinicius Sobrinho Richardi, Paulo Henrique Gorgatti Zarbin, and Maria Aparecida Cassilha Zawadneak. 2022. "Essential Oil of *Rosmarinus Officinalis* Ecotypes and Their Major Compounds: Insecticidal and Histological Assessment Against *Drosophila Suzukii* and Their Impact on a Nontarget Parasitoid." *Journal of Economic Entomology* 115 (4): 955–66. https://doi.org/10.1093/jee/toab230.

9 Antimicrobial Activity of Essential Oils

Uses and Applications Against Plant Pathogens

Arley Rey Páez and João Paulo Viana Leite

9.1 INTRODUCTION

Plants, as autotrophic organisms, use their own biochemical machinery to meet all of their nutritional needs. Primary metabolism, which encompasses processes common to practically all living beings, including animals, plants, and microorganisms, refers to the set of all these metabolic reactions related to the plant's survival and the generation of chemical energy. Through their catabolic pathways, primary metabolites make biosynthetic building blocks for a huge variety of biomolecules that are called secondary metabolites or natural products. These compounds in plants, including phenolics, terpenoids, alkaloids, and other natural compounds, contribute to distinctive traits such as aroma, flavor, and color while also playing a crucial role in the plant's adaptation to its environment.

Various secondary metabolites, including those belonging to the classes of terpenoids and phenylpropanoids, play a fundamental role in the defense of plants against phytopathogens. These bioactive compounds are present in essential oils (EOs). The biosynthesis of terpenoid secondary metabolites in plants, such as those found in EOs, begins with simple precursors like mevalonic acid (through the mevalonate pathway—MEV) or pyruvate and glyceraldehyde-3-phosphate (through the MEP pathway). These precursors form isoprene units, which condense to generate different types of terpenes, such as monoterpenes and sesquiterpenes. Specific enzymes catalyze these biosynthetic reactions, and the combination of these isoprene units determines the structure and function of terpenes, many of which are key components of essential oils (Leite, 2008). Many of these compounds exhibit antimicrobial, antifungal, and insecticidal properties that protect plants against phytopathogens, such as fungi, bacteria, and herbivorous insects.

Essential oils serve as a potent defense mechanism for plants against phytopathogens. The antimicrobial properties of essential oils disrupt the biological processes of these harmful microorganisms. Specific components within the essential oils can destabilize the cell membranes of pathogens, leading to cell lysis and death. Other compounds interfere with the reproduction and proliferation of the pathogens by inhibiting enzymes critical for their metabolism and survival. Furthermore, essential oils can have the added benefit of attracting pollinators and other beneficial organisms to the plant, creating an unfavorable environment for the pathogens and promoting the plant's overall health and resilience. This multifaceted mode of action makes essential oils a valuable natural defense system for plants against a wide range of fungal, bacterial, and insect-borne diseases.

Recognizing the potent defensive properties of EOs, researchers and agriculturalists have explored ways to harness these natural compounds as sustainable alternatives to synthetic pesticides. EOs can be extracted directly from plants and applied as standalone biopesticides on agricultural crops, or they can be incorporated into more complex formulations. The application of EOs has proven effective in controlling a wide range of plant pests and diseases, providing a more environmentally friendly solution compared to traditional chemical pesticides. Importantly, EOs

DOI: 10.1201/9781003463429-12

are biodegradable, and when used correctly, they are less likely to contribute to the development of pathogen resistance—a major challenge with synthetic pesticides (Raveau *et al.*, 2020). This makes essential oils a promising natural tool for integrated pest management programs aimed at sustainable agriculture and reduced environmental impact.

This chapter provides a review of the effects of various essential oils against phytopathogens, such as bacteria, fungi, and nematodes, indicating possible mechanisms of action. This understanding makes it possible to identify the most effective components and direct their use toward specific pests. This information is crucial for the creation of innovative, effective, and safe biopesticides that can contribute to more sustainable agriculture.

9.2 EOS IN THE CONTROL OF PLANT PATHOGEN BACTERIA

EOs have acquired recognition in agriculture as friendly and sustainable options to replace traditional methods for fighting harmful plant bacteria (Raveau *et al.*, 2020). These bacteria cause diseases such as galls, wilts, leaf spots, specks soft rots, scabs, cankers and blights, which reduce farm output and crop quality (Mansfield *et al.*, 2012). This discussion looks at some general aspects of the specific uses and advantages of EOs in dealing with plant pathogen bacteria.

While EOs have been extensively studied for their use in medical settings, their potential applications in plant health are still relatively underexplored. However, several studies have demonstrated that EOs exhibit strong antibacterial properties against a range of plant-harming bacteria (Raveau *et al.*, 2020). For instance, oregano (*Origanum* spp.) EO, which is rich in compounds like carvacrol (20–70%) and thymol (5–20%), has proven effective against *Xanthomonas campestris*, the bacterium responsible for black spot disease in cabbage and cauliflower (Simirgiotis *et al.*, 2020). Although the precise mechanisms by which oregano EO inhibits bacterial growth are not fully understood, early research suggests that it disrupts the bacterial cell membrane, leading to the leakage of cellular contents and ultimately causing cell death (Wijesundara & Rupasinghe, 2018).

The EOs contain a diverse array of chemical compounds that have demonstrated significant antibacterial properties against a range of plant pathogens. Key active components include resorcyl acid, carvacrol, cinnamaldehyde, eugenol, trans-cinnamaldehyde, thymol, cineol, menthol, α-pinene, p-cymene, γ-terpinene, linalool, geraniol, and vanillin (Figure 9.1). These compounds

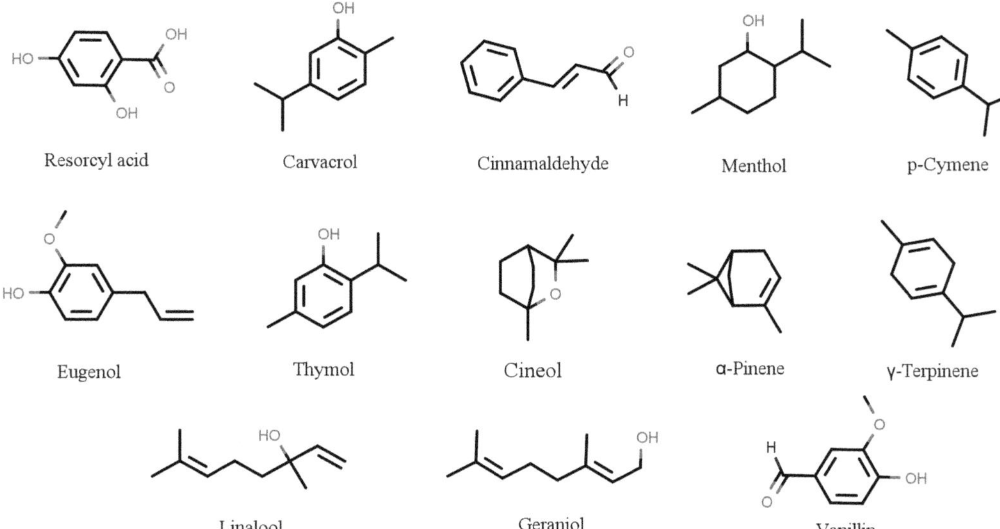

FIGURE 9.1 Chemical structures of some compounds present in the EOs with antimicrobial activity.

have been found effective against harmful bacteria such as *X. campestris*, *Pseudomonas syringae*, *Ralstonia solanacearum*, *Clavibacter michiganensis* subsp. *michiganensis*, *Erwinia amylovora*, *Agrobacterium tumefaciens*, *Xylella fastidiosa*, *Pectobacterium carotovorum*, *Xanthomonas axonopodis* pv. *phaseoli*, and *Pseudomonas syringae* pv. *tomato*, among others (Mansfield *et al.*, 2012; Sousa *et al.*, 2022).

For instance, cinnamon (*Cinnamomum* spp.) EO, which contains high levels of cinnamaldehyde (60–90%), eugenol (20–60%), and cinnamyl acetate (5–20%), has shown potent activity against *E. amylovora*, the causative agent of fire blight, a serious disease affecting fruit trees like apples and pears. Laboratory tests revealed that cinnamon EO produced an inhibition zone of 37.7 mm and had a minimum bactericidal concentration (MBC) of 1.04%, indicating its strong antibacterial potential (Doukkali *et al.*, 2022). Furthermore, cinnamon bark essential oil effectively reduced populations of *X. campestris*, a bacterium responsible for black spot disease in cabbages and other cruciferous crops. Under controlled conditions of humidity and temperature, the bacterial count decreased by approximately 0.3 to 2.6 log CFU/g after 48 hours (Chung *et al.*, 2022).

Electron microscopy studies have shown that exposure to cinnamon EO leads to significant alterations in the structure of bacterial cell walls and membranes. These observations suggest that the antibacterial action of the oil may stem from its ability to disrupt the integrity of the cell membrane by destabilizing the phospholipid bilayer, ultimately causing the cell wall to break down (Nowotarska *et al.*, 2017). Furthermore, cinnamon EO appears to interfere with key bacterial enzymes, such as glucosamine-6-phosphate deaminase (YvcK), which is crucial for the synthesis of UDP-N-acetylglucosamine (UDP-GlcNAc). This compound is essential for the production of peptidoglycan, the fundamental component of bacterial cell walls, as it provides N-acetylglucosamine units that combine with N-acetylmuramic acid (MurNAc) to form the repeating sugar units in the peptidoglycan structure (Sun *et al.*, 2021). Similarly, clove (*Syzygium* spp.) EO, which is rich in eugenol (70–90%), eugenol acetate (5–20%), and beta-caryophyllene (5–15%), has demonstrated strong antibacterial effects against several plant-harming bacteria. Notably, it is effective against *P. syringae* pv. *tomato*, which causes bacterial speck in tomatoes, and *E. carotovora* subsp. *atroseptica*, responsible for soft rot in tubers and black stem rot in potatoes (Silva *et al.*, 2014).

Research on the minimum inhibitory concentration (MIC) of oregano (*Origanum vulgare*) EO has highlighted its potent antibacterial properties against various plant-harming bacteria, including *X. campestris*, *P. syringae*, and *E. carotovora*. The EO, rich in compounds such as carvacrol (40–90%), thymol (10–30%), γ-terpinene (5–10%), and p-cymene (5–15%), exhibits inhibitory effects at concentrations ranging from 2 to 200 mg/mL (Carezzano *et al.*, 2017). In addition to its well-known ability to destabilize bacterial cell membranes, *O. vulgare* EO has been shown to interfere with the tricarboxylic acid cycle (TCA) by inhibiting key enzymes like citrate synthase (CS) and mitochondrial isocitrate dehydrogenase (ICDHm), disrupting essential metabolic processes (Cui *et al.*, 2019). Moreover, carvacrol, a primary component of oregano EO, has been found to interact directly with bacterial DNA. Studies on methicillin-resistant *Staphylococcus aureus* (MRSA) indicate that carvacrol can form complexes with DNA, leading to structural alterations. This interaction is evidenced by changes in the UV absorption spectrum, where the carvacrol absorption peak shifts and weakens upon binding to DNA, suggesting conformational changes (Cui *et al.*, 2019; Nieto *et al.*, 2018). These structural modifications can impede critical cellular functions, including replication, transcription, and translation, by disrupting DNA topoisomerases, which are essential enzymes for these processes.

Similar to *O. vulgare* EO, eucalyptus (*Eucalyptus* spp.) EO has demonstrated strong antibacterial activity against plant-harming bacteria such as *P. syringae*, *X. campestris*, *E. carotovora*, and *R. solanacearum*. This oil is primarily composed of 1,8-cineole (60–85%), α-pinene (2–14%), p-cymene (1–3%), and limonene (1–5%) (Mulyaningsih *et al.*, 2011). *In vitro* experiments have shown that eucalyptus EO can inhibit *P. syringae* growth, producing inhibition zones up to 33 mm, with a minimum inhibitory concentration (MIC) of 300 μL/mL and a bactericidal minimum concentration (BMC) of 60 μL/mL (Sabir *et al.*, 2017). Additionally, studies on *Eucalyptus torelliana* EO have

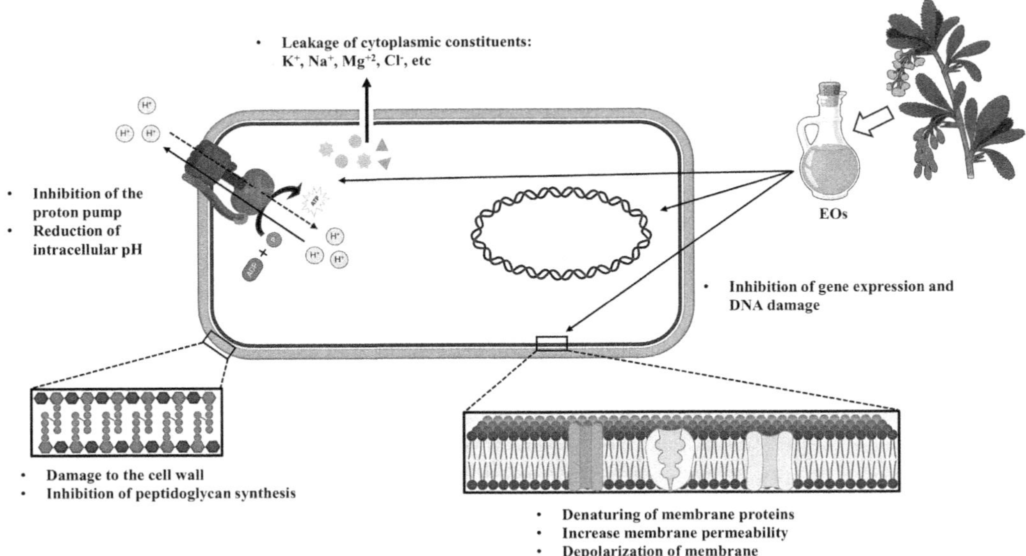

FIGURE 9.2 Possible mode of action of EOs on phytopathogenic bacteria.

shown its effectiveness against *X. campestris* and *E. carotovora*, resulting in inhibition zones ranging from 6 to 22 mm at concentrations above 100 μL/mL (Adekoya & Akeem, 2021). The primary component, 1,8-cineole, is believed to disrupt the cell membranes of both Gram-positive and Gram-negative bacteria, leading to cell lysis. Meanwhile, other components such as p-cymene and limonene contribute to this membrane-disrupting effect, enhancing the oil's antimicrobial properties (Čmiková *et al.*, 2023; Sewanu *et al.*, 2015). Additionally, α-pinene has been found to interfere with the synthesis of nucleic acids and proteins, thereby inhibiting bacterial growth and reproduction.

The EOs are derived from a wide range of plant families and possess a complex chemical composition, making it difficult to attribute their antibacterial properties to a single compound or mechanism of action (Swamy *et al.*, 2016). The effectiveness of EOs largely stems from the synergistic interactions among their diverse constituents, which enhance their overall antimicrobial activity by targeting multiple sites within bacterial cells (Figure 9.2). This synergy manifests through several mechanisms, such as increasing the potency of chemical compounds, reducing bacterial resistance, and attacking different structural and functional parts of the bacterial cell. When these mechanisms operate simultaneously, it becomes challenging for bacteria to adapt and develop resistance, thereby enhancing the oils' efficacy. Research by Skandamis and Nychas, 2001; Carson *et al.*, 2002 underscores the importance of these synergistic effects, demonstrating that the combined action of EO components significantly contributes to their antimicrobial potential. Consequently, these findings suggest that a holistic approach, focusing on the complete blend of EO components rather than isolating individual compounds, is more effective in harnessing their antibacterial properties.

The chemical structure of EO components plays a critical role in determining their antibacterial activity and mechanisms of action. Phenolic compounds like carvacrol and thymol illustrate the importance of the hydroxyl group within their molecular framework. Research has shown that this group is essential for their antimicrobial efficacy (Ultee *et al.*, 2002). Interestingly, the position of the hydroxyl group on the phenolic ring does not significantly alter the bactericidal or bacteriostatic effects of these compounds (Nazzaro *et al.*, 2013). For instance, thymol, which has a meta-positioned hydroxyl group, demonstrates similar effectiveness against *Bacillus cereus*, *S. aureus*, and *Pseudomonas aeruginosa* as carvacrol, where the hydroxyl group is in the ortho position (Ultee *et al.*, 2002). Despite being structural isomers, studies suggest that thymol and carvacrol may impact Gram-positive and Gram-negative bacteria through distinct pathways, indicating potential differences in their mechanisms of action (Kachur & Suntres, 2020). The significance of the phenolic ring structure becomes even clearer when comparing the antibacterial potency of menthol, which, despite being similar to thymol and carvacrol, exhibits lower antimicrobial activity. This reduced efficacy can be attributed to its lack of the unstable electron configuration present in the benzene ring of carvacrol and thymol, which is crucial for disrupting bacterial cells (Ultee *et al.*, 2002). Furthermore, modifications to the molecular structures of EOs, such as the addition of organic substituents, can markedly enhance or reduce their antimicrobial potency. For example, the addition of an acetate group, as seen in geranyl acetate compared to geraniol, significantly boosts its effectiveness against a broad range of Gram-positive and Gram-negative bacteria (Celuppi *et al.*, 2022). These insights underscore the critical role of chemical structure in the development and application of essential oils for antimicrobial purposes, highlighting that even small structural changes can lead to notable differences in antibacterial activity.

The antibacterial activity of EOs is primarily attributed to the hydrophobic nature of their molecules, which allows them to easily penetrate bacterial cell membranes. This interaction disrupts the membrane's structural integrity, leading to ion leakage, osmotic imbalance, and ultimately, cell lysis and bacterial death (Rath & Priyadarshanee, 2017). This mechanism, which combines bactericidal and bacteriostatic effects, also minimizes the likelihood of bacteria developing resistance compared to antibiotics that act through a single, specific mode of action (Visan & Neguț, 2024). However, the antibacterial effectiveness of EOs can vary significantly depending on their chemotype (Vidic *et al.*, 2010). Factors such as the geographical origin of the plant, harvest time, genetic variations, drying techniques, and the specific parts of the plant used for extraction all influence the chemical composition of EOs (Zgheib *et al.*, 2016). Additionally, extraction and storage processes can further impact the concentration and stability of the active components, affecting the oil's overall effectiveness. Given this variability, standardization of production methods and comprehensive chemical analyses are essential to maintain the consistent efficacy, safety, and quality of EOs, especially in therapeutic, food, and agricultural applications. Ensuring a standardized composition is crucial not only for optimizing their antimicrobial properties but also for providing reliable and reproducible results across different uses. This consistency is vital for the broader adoption of EOs, allowing them to be more effectively integrated into various industries.

9.3 EOS AND THEIR ANTIFUNGAL ACTIVITY AGAINST PLANT PATHOGEN FUNGI

Molecular techniques have revealed that the early association between plants and fungi began around 400 million years ago, coinciding with the period when plants were first establishing themselves on land (Krings *et al.*, 2007). Since this initial interaction, a complex process of co-evolution has taken place, resulting in the development of mutually beneficial symbiotic relationships (Strack *et al.*, 2003). These associations have played a crucial role in the origin and evolution of terrestrial ecosystems, helping plants adapt to and thrive in various environments. Fungi have successfully colonized both the root tissues and aerial parts of plants, forming diverse symbiotic relationships. While the benefits of interactions between plant roots and fungi (mycorrhizae) are well documented—such as enhanced nutrient uptake and improved stress tolerance—there is still much to learn about the range of plant-fungal interactions. Not all fungal associations are advantageous for plant physiology; some can lead to disease or competition, highlighting the complexity and variability of these relationships in natural ecosystems.

Fungal colonization or infection of plant tissues can occur through three main modes: biotrophic, necrotrophic, and hemibiotrophic. Biotrophic fungi extract nutrients exclusively from living cells, necrotrophs feed on dead tissue, and hemibiotrophs exhibit both biotrophic and necrotrophic phases during their life cycle (Perfect & Green, 2001). Biotrophic plant pathogenic fungi, such as powdery mildew and rust fungi, are responsible for a wide range of plant diseases. Their rapid proliferation within plant tissues can lead to abnormal cell growth, resulting in hyperplasia or hypertrophy of organs, which eventually manifests as pustules and spots on the affected plant surfaces (Micali *et al.*, 2008). In contrast, necrotrophic fungi cause decay by breaking down plant tissues, leading to symptoms such as rotting trunks, roots, and the presence of lesions. These fungi often attack the plant's vascular system, specifically targeting xylem and phloem vessels, causing tracheomycosis, which can severely impair the plant's ability to transport water and nutrients (Shao *et al.*, 2021).

Approximately 10% of all known fungi are capable of becoming plant pathogens, posing a threat to over 10,000 plant species worldwide (Raza *et al.*, 2022). Common fungal pathogens, such as rust, downy mildew, and *Fusarium*, have the potential to decrease agricultural yields by up to 50%, posing a significant risk to global food security and leading to economic losses that amount to millions of dollars annually (Fones *et al.*, 2020). Moreover, the financial burden of managing these diseases is exacerbated by the high costs associated with fungicide treatments and other control measures, increasing production expenses. This economic strain affects both smallholder farmers and large agricultural enterprises, highlighting the urgent need for sustainable and effective disease management strategies.

Among the numerous fungi capable of acting as plant pathogens, a select few are particularly significant for agriculture due to their widespread impact on key crops. These include *Magnaporthe oryzae*, *Botrytis cinerea*, *Puccinia* spp., *Fusarium graminearum*, *Fusarium oxysporum*, *Blumeria graminis*, *Mycosphaerella graminicola*, *Colletotrichum* spp., *Ustilago maydis*, and *Melampsora lini* (Dean *et al.*, 2012). The economic implications of these pathogens are substantial. For instance, *M. oryzae*, responsible for rice blast disease, causes global losses exceeding US$5 billion annually, severely impacting rice production (Li *et al.*, 2020). *B. cinerea*, which affects fruit and vegetable crops, leads to over US$1 billion in losses each year due to gray mold infections (Hua *et al.*, 2018). Additionally, *Puccinia* spp., known for causing rust diseases in cereal crops, account for approximately US$3 billion in economic losses annually, underlining the importance of effective management strategies to mitigate these threats (Degete & Chala, 2019).

Currently, particularly in developing countries, the management of plant pathogenic fungi relies heavily on the use of synthetic fungicides (McLaughlin *et al.*, 2023). However, this strategy presents several critical challenges. For example, the overuse of strobilurin-based fungicides has led to the emergence of *B. cinerea* strains resistant to these treatments, especially in strawberry crops (Habib *et al.*, 2020). Additionally, residues from fungicides like chlorothalonil can persist in soil and water, posing risks to human health and disrupting ecosystems by harming beneficial fauna (Meftaul *et al.*, 2021). In the case of *Phytophthora infestans*, the pathogen responsible for potato late blight, resistance has developed against fungicides such as metalaxyl, further complicating disease management (Bhatt & Sharma, 2020). Moreover, the prolonged use of copper-based fungicides in grapevine cultivation has resulted in the toxic accumulation of copper in the soil, negatively affecting soil microbiota and reducing soil fertility (Mackie *et al.*, 2013).

Plant-derived fungicides, especially EOs, are increasingly recognized as promising alternatives for the integrated management of fungal plant pathogens. These bioactive compounds have naturally evolved to defend plants against a broad spectrum of pathogenic microorganisms, making them effective in agricultural disease control (Allagui *et al.*, 2023). Their effectiveness stems from their chemical diversity and multiple modes of action, including the inhibition of mycelial growth, disruption of cell membrane integrity, and activation of plant defense mechanisms (Basak & Guha, 2018). Moreover, EOs offer key advantages over synthetic fungicides due to their low toxicity and biodegradability, which make them more sustainable and environmentally friendly options for long-term use (Pavela & Benelli, 2016).

The antifungal properties of EOs are largely attributed to the presence of phenolic compounds, such as carvacrol, thymol, safrole, and eugenol, as well as terpenes like pinene, limonene, myrcene, terpinene, p-cymene, citronellal, linalool, and nerolidol. These bioactive compounds have demonstrated significant efficacy in inhibiting the growth and proliferation of various plant pathogenic fungi (Barros *et al.*, 2023). They work through multiple mechanisms, including disrupting the permeability of fungal cell membranes, denaturing proteins, and inhibiting key enzymes involved in fungal metabolism (Bakkali *et al.*, 2008). Additionally, terpenoids can interfere with the biosynthesis of ergosterol, an essential component of fungal cell membranes, leading to membrane destabilization and the eventual death of the pathogen (OuYang *et al.*, 2021). Phenolic compounds like eugenol and carvacrol also possess antioxidant properties that protect plants from oxidative stress induced by fungal infections, further enhancing plant resilience (Didehdar *et al.*, 2022).

EOs can exhibit either fungistatic (growth-inhibiting) or fungicidal (lethal) effects against a wide range of fungal species, making them valuable for both agricultural and medical applications (Okorska *et al.*, 2023). Their efficacy in agriculture is primarily determined by their chemical composition, which can vary significantly depending on factors such as plant species, geographic origin, environmental conditions, and cultivation practices. Additionally, the effectiveness of EOs as fungicides depends on the susceptibility of the target fungal species, with different pathogens requiring tailored approaches to optimize control strategies (Basak & Guha, 2018). The method of application, whether through foliar sprays, soil treatments, or seed coatings, also plays a critical role in their effectiveness, influencing how well the EOs interact with the fungal pathogens and plant tissues (Jugreet *et al.*, 2020). Thus, a strategic approach that considers the specific EO composition, the target pathogen, and the application method is essential for maximizing the benefits of EOs in plant disease management.

The MIC, which measures the effectiveness of EOs in inhibiting fungal growth, can be influenced by several variables, including the volume and concentration of the fungal sample, the type of culture medium, pH level, and incubation time (Ćosić *et al.*, 2014). Additionally, interactions between EO components and factors such as nutrients and pH during antifungal testing can reduce the fungicidal activity, thereby impacting their stability and overall efficacy. The antimicrobial effectiveness of EOs can also decline due to oxidation and degradation of active compounds during storage or when exposed to light and air. Furthermore, the low solubility of some EO components in the test medium can limit their interaction with fungal pathogens, reducing their antifungal performance. Environmental conditions like temperature and humidity can alter the volatility and bioavailability of EOs, further influencing their efficacy in antimicrobial assays (Tserennadmid *et al.*, 2010).

Early research on the antifungal properties of EOs highlighted the efficacy of *Pepper betel* leaf EO in combating the plant pathogenic fungus *Aspergillus flavus* (Dubey & Tripathi, 1987). Subsequent studies revealed that EOs from *Melaleuca alternifolia*, which are rich in terpinen-4-ol (37–45%), α-terpinene (4–12%), and α-terpineol (3–7%), were effective antifungal agents against *B. cinerea*, the pathogen responsible for grey mold disease in various plants (Bishop & Reagan, 1998). Further research by Nguefack *et al.* (2004) demonstrated that EOs from *Ocimum gratissimum*, *Thymus vulgaris*, and *Cymbopogon citratus* exhibited antifungal properties against *Fusarium moniliforme*, *A. flavus*, and *A. fumigatus* when tested on cornmeal agar. The effective concentrations for these EOs were 800, 1000, and 1200 μg/mL, respectively, showcasing their potential as natural antifungal agents across different fungal species.

Although there is extensive literature on the antifungal properties of EOs, this review will concentrate on those that have demonstrated widespread use and proven efficacy. Specifically, it will cover EOs from *O. vulgare* (oregano), *S. aromaticum* (clove), *Lavandula angustifolia* (lavender), *T. vulgaris* (thyme), and *C. verum* (cinnamon), among others. To maintain clarity and conciseness, the selection of oils discussed has been intentionally limited to these well-documented examples. However, we recognize that many other plant species also produce EOs with potent antifungal activity, showing potential against a diverse range of plant pathogenic fungi.

O. vulgare EO is widely recognized for its antibacterial, antioxidant, and anti-inflammatory properties (Walasek-Janusz *et al.*, 2024). One of the earliest studies to demonstrate its antifungal potential was conducted by Vági *et al.* (2005), who found that *O. majorana* EO effectively inhibited the growth of foodborne fungi such as *A. flavus*, *Trichoderma viride*, and *Penicillium cyclopium*.

Further research by Soylu *et al.* (2006) highlighted the antifungal effects of *O. syriacum* cv. Bevanii, showing that a concentration of 6.4 μg/mL effectively suppressed the growth of *P. infestans*, a major plant pathogen. Recent studies have confirmed the efficacy of *O. vulgare* EO against *B. cinerea*, the pathogen responsible for gray mold in grape and strawberry plants, with notable growth inhibition achieved at 500 μg/mL (Álvarez-García *et al.*, 2023). Additionally, the EO has been shown to combat *C. gloeosporioides*, which causes anthracnose in tropical fruits, with substantial growth suppression observed at concentrations ranging from 250 to 1000 μg/mL (Yilmaz *et al.*, 2016).

Several plant species within the Lamiaceae family, beyond oregano and thyme, have shown notable antifungal properties against plant pathogenic fungi. For example, *Mentha pulegium* EO has demonstrated efficacy against *Mauginiella scaettae*, the causative agent of inflorescence rot in date palms, effectively inhibiting the pathogen at a concentration of 200 μg/mL (Hadjra *et al.*, 2023). Likewise, *M. piperita* EO has proven to be highly effective against *F. moniliforme*, a fungus known for causing stem rot and vascular wilt in key crops such as corn, cotton, tomato, wheat, and rice (Goudjil *et al.*, 2016). Additionally, *Salvia officinalis* EO has exhibited potent antifungal activity against *Sclerotinia sclerotiorum*, a significant pathogen that affects a broad spectrum of vegetable and field crops, underscoring its potential for agricultural disease management (Pansera *et al.*, 2013).

The EO of *L. angustifolia*, a member of the Lamiaceae family, is notably rich in linalool (20–50%), linalyl acetate (15–30%), and terpinen-4-ol (2–10%). It has demonstrated strong antifungal properties, effectively inhibiting the growth of *B. cinerea* and various *Fusarium* species at concentrations of 223 μg/mL and 520 μg/mL, respectively (Caprari *et al.*, 2023). Similarly, *Rosmarinus officinalis* (rosemary), another prominent Lamiaceae plant, has been widely studied for its well-established antifungal effects. Rosemary EO, which primarily consists of 1,8-cineole (20–50%), α-pinene (10–20%), and camphor (4–25%), has proven effective against a diverse range of plant pathogenic fungi. These include *Fusarium verticillioides*, *B. cinerea*, *F. oxysporum*, *F. proliferatum*, *Mucor pusillus*, *A. oryzae*, and *Alternaria alternata*, with effective concentrations ranging from 50 to 500 μg/mL (Barakat & Ghazal, 2016). However, the sensitivity of these pathogens to rosemary EO varies significantly depending on the specific microorganism and disease. For instance, *A. oryzae* exhibited up to 93% growth inhibition at a concentration of 50 μg/mL, whereas *F. verticillioides* showed only 67% inhibition at a higher concentration of 500 μg/mL (Ferdes *et al.*, 2017). Additionally, *B. cinerea* growth was completely inhibited at concentrations below 50 μg/mL, highlighting how the antimicrobial efficacy of EOs can vary significantly based on conditions and the specific pathogens being studied (Mutlu-Ingok *et al.*, 2020). These findings underscore the need for tailored approaches when utilizing EOs for plant disease management, as effectiveness can depend on both the EO composition, and the pathogen targeted.

Eucalyptus (*Eucalyptus* spp.) and clove (*S. aromaticum*), both belonging to the Myrtaceae family, are renowned for their rich source of bioactive compounds with strong antifungal properties. EOs derived from these plants have been effectively used to manage various plant pathogenic fungi, including *F. oxysporum*, *B. cinerea*, and *A. alternata* (Batish *et al.*, 2008). Research conducted by Abdallah *et al.* (2015) demonstrated the fungicidal properties of *E. globulus* EO, which is primarily composed of 1,8-cineole (70–90%), α-pinene (5–15%), and limonene (2–10%). At a concentration of 500 μg/mL, the oil was able to inhibit 90% of *F. oxysporum* growth, highlighting its potent antifungal activity. Similarly, the EO of *S. aromaticum*, rich in eugenol (75–90%) and β-caryophyllene (15–25%), has shown strong antifungal effects against *B. cinerea*. Studies have found that this EO can effectively suppress the growth of the pathogen at concentrations as low as 250 μg/mL (Morjane *et al.*, 2024).

Several other botanical families have also shown significant antifungal properties against plant pathogenic fungi, including Rutaceae, Asteraceae, Zingiberaceae, Apiaceae, Lauraceae, Piperaceae, and Brassicaceae (Daradka *et al.*, 2021; Tripathi, 2023). The Rutaceae family, in particular, stands out for its EOs derived from lemon (*Citrus limon*) and orange (*Citrus sinensis*), which are rich in limonene (65–90%), β-pinene (5–25%), γ-terpinene (2–15%), and citral (2–10%) (Bhandari *et al.*,

2021). These compounds have been shown to effectively inhibit the growth of several plant pathogens. Within the Asteraceae family, essential oils from plants like chrysanthemum (*Chrysanthemum indicum*) and wormwood (*Artemisia absinthium*) have also demonstrated notable antifungal effects. These oils contain a diverse array of bioactive compounds, including camphor (12–35%), 1,8-cineole (3–15%), cis-sabinol (5–20%), trans-thujone (40–60%), chamazulene (5–15%), and sabinene (2–20%). Research has shown that these essential oils can effectively inhibit the growth and mycelial development of a variety of fungal species, such as *Colletotrichum* spp., *B. cinerea*, *Fusarium* spp., and *A. solani* (Zeng *et al.*, 2020).

The antifungal efficacy of EOs derived from plants in the Zingiberaceae family, such as ginger (*Zingiber officinale*) and turmeric (*Curcuma longa*), is primarily attributed to the presence of active compounds including zingiberene (20–30%), β-sesquiphellandrene (10–20%), camphene (7–25%), β-phellandrene (5–15%), curcumene (2–10%), and sabinene (1–8%) (Gakuubi *et al.*, 2017; Sánchez-Hernández *et al.*, 2024). Similarly, essential oils from the Lauraceae family, particularly those extracted from cinnamon (*Cinnamomum* spp.), exhibit potent antifungal properties, primarily due to their high cinnamaldehyde content. This compound is highly effective against several fungal species, including *F. oxysporum* and *A. niger*, making cinnamon EO a valuable natural alternative for managing fungal infections in agricultural settings (Zhou *et al.*, 2024).

Piper nigrum (black pepper), belonging to the Piperaceae family, produces an EO that is rich in bioactive compounds, including monoterpenes such as sabinene (20–40%), β-pinene (15–25%), and limonene (5–15%), as well as sesquiterpenes like β-caryophyllene (10–25%) and γ-elemene (7–20%). These components exhibit potent antifungal properties, effectively inhibiting the growth of plant pathogenic fungi such as *Fusarium* spp. and *Rhizoctonia solani* (Kumar *et al.*, 2021). In addition, plants from the Brassicaceae family, including mustard (*Brassica nigra*) and radish (*Raphanus sativus*), have demonstrated significant antifungal activity against pathogens like *A. brassicae* and *Sclerotinia sclerotiorum*. This effectiveness is mainly attributed to the presence of glucosinolates and their degradation products, which possess strong bioactive properties (Calmes *et al.*, 2015). These findings highlight the potential of EOs and natural compounds from the Piperaceae and Brassicaceae families as eco-friendly alternatives for managing fungal infections in crops.

The antifungal mode of action of EOs shares some similarities with their antibacterial effects; however, fungi present unique challenges as a target. Unlike bacteria, fungi are eukaryotic organisms with complex cellular structures that closely resemble those of plant cells. This structural and functional resemblance, a result of evolutionary convergence, can complicate the diagnosis and treatment of fungal infections, making it more difficult to distinguish from diseases caused by nutritional deficiencies or abiotic stress. One key difference is that while plant cell walls are composed of cellulose, fungal cell walls primarily consist of chitin. This distinction makes chitin a specific target for bioactive compounds present in EOs, providing an avenue for selective antifungal action (Ibe & Munro, 2021). Additionally, the plasma membranes of fungi contain ergosterol, whereas plant cell membranes are composed of other sterols like sitosterol, stigmasterol, and campesterol (Khan & Quinn, 2017). Moreover, fungi have specific nutritional needs, including vitamins and amino acids, which are essential for their complex metabolic pathways (Choi *et al.*, 2024). These structural and nutritional distinctions between fungi and plants can be exploited to create more selective and effective fungicides, reducing the risk of collateral damage to non-target organisms and improving overall disease management strategies.

EOs demonstrate both fungicidal (lethal) and fungistatic (growth-inhibiting) effects through multiple mechanisms, including disrupting the fungal cell wall, destabilizing the cytoplasmic membrane, and interfering with ATP synthesis (Figure 9.3). These actions lead to significant cell membrane damage and the leakage of essential macromolecules, ultimately resulting in cell death. The lipophilic nature of the compounds in EOs allows them to easily penetrate the cell wall, where they compromise the integrity of the cytoplasmic membrane by disrupting its layers of proteins, phospholipids, fatty acids, and polysaccharides. This disruption weakens the structural integrity of the cell, causing membrane disintegration and affecting critical cellular processes (Hou *et al.*, 2022).

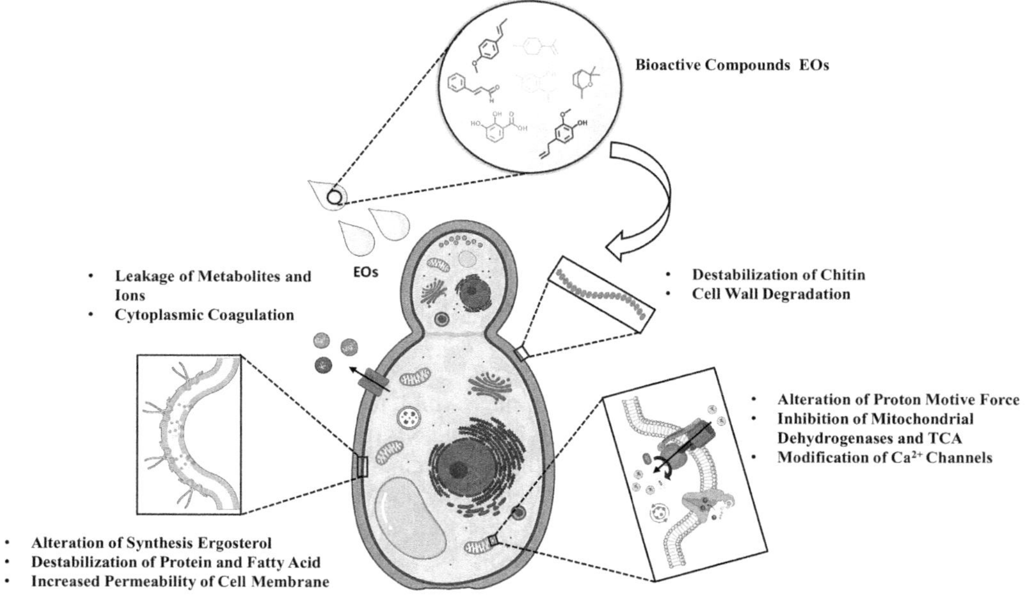

FIGURE 9.3 Mode of action of EO against fungal species and their constituents.

Certain compounds in EOs, including anethole, limonene, carvacrol, thymol, eugenol, and thymoquinone, have been found to disrupt the process of chitin polymerization in fungal cells. Additionally, they can form DNA complexes similar to those created by 4,6-diamidino-2-phenylindole, though distinct from the binding mechanism of ethidium bromide (Shariati *et al.*, 2022; Yin *et al.*, 2024). These disruptions affect key cellular processes such as cell division, cell wall maturation, and the formation of septa and bud rings. For instance, trans-anethole, a major component of *Pimpinella anisum* and *Foeniculum vulgare* EOs, has demonstrated inhibitory effects on hyphal growth in *Mucor mucedo*, a fungus responsible for the spoilage of fruits and vegetables during storage (Vieira *et al.*, 2019; Yutani *et al.*, 2011). This effect is attributed to the inhibition of chitin synthase, an enzyme crucial for fungal cell wall formation and growth. Similarly, compounds like limonene and thymoquinone have shown the ability to hinder the growth of *A. niger* and *F. moniliforme* (Al-Qurashi *et al.*, 2007). These substances cause the fungal hyphal walls to become thinner and deformed, eventually leading to wall rupture and leakage of cytoplasmic contents. Through these mechanisms, EOs present a promising natural approach for controlling fungal growth and mitigating plant pathogens.

Tatsadjieu *et al.* (2009) demonstrated that *Lippia rugosa* EO could completely inhibit the growth of *A. flavus* at a concentration of 1000 μg/mL, primarily by targeting the H^+-ATPase enzyme. This enzyme is crucial for maintaining the electrochemical gradient of the fungal cell membrane, which is essential for nutrient absorption, pH regulation, and pathogenicity (Vázquez-Carrada *et al.*, 2022). The fungicidal effects of *L. rugosa* EO are largely attributed to its main components, eugenol and thymol, which disrupt the function of this proton pump, leading to cell death (Monk & Perlin, 2008). Additionally, thymol and carvacrol have been found to enhance the efficacy of azole antifungals, such as fluconazole, by inhibiting the overexpression of efflux pump genes like *CDR1* and *MDR1* in *Candida albicans* (Nazzaro *et al.*, 2017). This synergistic interaction not only improves the effectiveness of conventional antifungal treatments but also helps to combat resistance in fungal pathogens.

EOs significantly affect mitochondrial function by disrupting the membrane potential, altering calcium channels, and inhibiting key dehydrogenase enzymes, such as lactate, malate, and succinate dehydrogenases (Zeng *et al.*, 2015). Research has shown that EOs from *Anethum graveolens* exert antifungal effects by interfering with the TCA cycle and impairing ATP production within the mitochondria (Chen *et al.*, 2013). When *Saccharomyces cerevisiae* cells are exposed to EOs from *O. compactum*, *A. herba-alba*, and *C. camphora*, noticeable cytoplasmic changes occur, indicative of mitochondrial damage (Bakkali *et al.*, 2006). Furthermore, terpenoids in EOs have been observed to reduce mitochondrial populations significantly, impacting ATP synthesis and increasing reactive oxygen species (ROS) levels (Haque *et al.*, 2016). Interestingly, EOs have also been shown to decrease nitric oxide (NO) levels, which in turn reduces the production of hydrogen peroxide (H_2O_2) and NO synthase. This reduction is particularly relevant because NO serves as a defense mechanism, protecting fungal mycelium from oxidative damage caused by thermal stress (Shen *et al.*, 2016). These findings highlight the diverse ways in which EOs disrupt mitochondrial function, ultimately inhibiting fungal growth and survival.

Ergosterol, the primary sterol in fungal cell membranes, is crucial for maintaining membrane fluidity and structural integrity. This makes it an ideal target for several bioactive compounds found in EOs. Notably, compounds such as carvacrol, thymol, methyl chavicol, linalool, and geraniol exhibit antifungal properties by interacting with ergosterol, thereby destabilizing the fungal membrane (Shariati *et al.*, 2022). Studies have demonstrated that EOs from *Ocimum sanctum*, which are rich in linalool (20–40%) and methyl chavicol (10–30%), can reduce ergosterol synthesis in *C. albicans* strains. However, other findings suggest that these compounds may not significantly inhibit ergosterol production, even at higher concentrations exceeding 20 mg/mL (Khan *et al.*, 2010). Despite this variability, the ability of EO components to disrupt fungal membranes presents an opportunity to enhance the effectiveness of azole antifungals, which also target ergosterol biosynthesis (Cardoso *et al.*, 2016). Combining EOs with azoles could help overcome issues related to azole resistance, providing a synergistic approach that improves the overall efficacy of antifungal therapies.

9.4 EOS AS NEMATICIDES IN PLANT PROTECTION

Like bacteria and fungi, plant parasitic nematodes (PPNs) are another important group of microorganisms that negatively impact agricultural productivity. Unlike these microorganisms, whose pathologies can be observed in the different aerial parts of plants, pathologies generated by PPNs often escape the human eye due to their small size (200 μm to 10 mm in length) and preferential location in plant roots (Bernard *et al.*, 2017). There, they pierce and kill root cells through their stylet, which can be hollow, as in the case of sedentary PPNs (*Meloidogyne* spp., *Heterodera* spp., *Globodera* spp., etc.), or solid, as in those PPNs of a migratory nature (*Pratylenchus* spp., *Helicotylenchus* spp., *Aphelenchoides* spp., etc.). The activity of this stylet, together with the effector proteins secreted by PPNs, leads to root lesions resulting in symptoms such as yellowing, dwarfing, wilting and reduced crop yield (Kumar & Yadav, 2020).

Global agricultural losses caused by PPNs are estimated to be between 10–20% of world production, which in economic terms represents more than US$100 billion per year (Cao *et al.*, 2023). Although there are more than 4000 species of PPNs, only a few are of importance in agriculture, due to their complex and specific interaction with host plants, their ability to infect a wide range of crops and the considerable damage they can cause to water and nutrient uptake (Abd-Elgawad & Askary, 2015). Among the 10 most important PPNs of agricultural importance are the root-knot nematodes (*Meloidogyne* spp.), cyst nematodes (*Heterodera* and *Globodera* spp.), root lesion nematode (*Pratylenchus* spp.), burrowing nematode (*Radopholus similis*), *Ditylenchus dipsaci*, pine disease nematode (*Bursaphelenchus xylophilus*), reniform nematode (*Rotylenchulus reniformis*), *Xiphinema index*, *Nacobbus aberrans*, and *Aphelenchoides besseyi* (Jones & Fosu-Nyarko, 2014).

Currently, there are different alternatives for the management and control of PPNs, such as crop rotation, genetically modified crops, organic amendments, biological control, and chemical control (Pan *et al.*, 2016). Control using fumigant (e.g., metam sodium, 1,3-dichloropropene, chloropicrin methyl bromide, etc.) and non-fumigant (e.g., aldicarb, carbofuran, oxamyl, fenamiphos, etc.)

nematicides is still the most popular method for the management and control of PPNs. This is primarily due to its short-term efficiency, ease of application, and commercial availability (Dawabah *et al.*, 2019). As with the chemical management of other plant pathogens, fumigant and non-fumigant nematicides are no longer accepted by environmental and animal protection agencies (Ganguly *et al.*, 2021), as in addition to their proven harmful effects on human and environmental health, they also have a negative impact on soil microbiota, which are essential for the biodegradation processes of organic and inorganic "residues", as well as for soil fertility and structure (Castellano-Hinojosa *et al.*, 2022).

Despite restrictions by the US Environmental Protection Agency (EPA), the European Food Safety Authority (EFSA), and the World Health Organization (WHO), the nematicide market continues to grow, from generating dividends of US$1 billion in 2011 to more than US$1.78 billion in 2022 (Request free Sample: Global Nematicides Market Share, Size, Trends, n.d.; Oka, 2020). Thus, there is growing interest in the search for and development of new nematicides that are more selective, photodegradable, have lower ecotoxicological profiles, and have improved efficiency to help prevent the emergence of resistance in PPNs populations (Desaeger *et al.*, 2020).

In contrast to synthetic nematicides, the degradation products of EOs are not toxic to non-target organisms, which means that they can specifically target nematodes without causing harm to other beneficial micro-organisms in the soil. EOs from plants have many bioactive compounds that can interfere with nematodes' biological functions, like stopping their metabolic and reproductive processes, without having a big negative impact on the good microbiota in the soil (Catani *et al.*, 2023). Being compounds of natural origin and not containing toxic elements such as chlorine, bromine, fluorine, or sulfur, they decompose rapidly in the environment, thus reducing the risk of soil and water contamination (El-Ansary *et al.*, 2020). In addition, as we saw previously, EOs have multiple mechanisms of action on target organisms, which decreases the likelihood of resistance development by PPNs. Thus, EOs constitute a promising alternative for sustainable agricultural practices in the management and control of PPNs.

Several *in vitro* studies have assessed at the nematicidal potential of EOs from many different plant families, such as Lamiaceae, Asteraceae, Myrtaceae, Rutaceae, Lauraceae, Poaceae, and Apiaceae, against the 10 most important PPNs for agriculture. Notably, the root-knot nematodes, *Meloidogyne* spp., and the pinewood nematode, *B. xylophilus*, have been the subjects of extensive research (Catani *et al.*, 2023). EOs from aromatic plants such as those belonging to the genera *Artemisia*, *Cymbopogon*, *Lavandula*, *Allium*, *Mentha*, *Origanum*, *Ocimum*, *Rosmarinus*, *Thymus*, *Pimpinella*, *Piper*, and *Syzygium*, as well as those from aromatic trees of the genera *Citrus*, *Eucalyptus*, *Eugenia*, and *Melaleuca*, have demonstrated significant in vitro activity against theses PPNs (D'Addabbo *et al.*, 2021; Laquale *et al.*, 2015).

Although EOs from various *Artemisia* species have been shown to be effective against human parasites such as *Ascaris suum*, *Plasmodium* spp., *Leishmania* spp., *Giardia lamblia*, and *Toxoplasma gondii*, as well as animal parasites like *Haemonchus contortus*, *Schistosoma mansoni*, and *Eimeria* spp. (Băieş *et al.*, 2022;), relatively few studies have investigated their efficacy against PPNs. Some research suggests that EOs from *Artemisia absinthium*, commonly known as wormwood, may exhibit nematostatic effects on second-stage juveniles (J2) of certain *Meloidogyne* species, as well as reduce root gall formation in tomato and carrot plants (Ozdemir & Gozel, 2018). However, other studies have found that *A. absinthium* EOs do not exhibit nematicidal or nematostatic effects against *M. incognita* and *M. javanica* and, in some cases, may even have adverse effects, increasing the number of eggs and galls in tomato roots (García-Rodríguez *et al.*, 2015). These findings highlight the need for further research to better understand the factors influencing EOs' effectiveness under different conditions, as the various compounds present in EOs can interact with environmental components, potentially reducing their efficacy.

There is substantial evidence supporting the nematicidal activity of *O. vulgare* and *T. vulgaris* EOs against PPNs of the genus *Meloidogyne spp.* Oka *et al.* (2000) was among the first to evaluate the nematicidal effects of *T. vulgaris* EO against *M. incognita* and *M. javanica*, reporting significant nematicidal activity against J2 at a concentration of 800 µl/L. Additionally, the EO of *O. vulgare* has been shown to be highly toxic to J2 of *M. incognita* at concentrations ranging from 2 to 4 mg/L, with complete inhibition of hatching observed at 4 mg/L (Ibrahim *et al.*, 2006). Another study determined that the lethal dose (LD_{50}) for *M. incognita* mortality after 72 hours of exposure to *O. glandulosum* EO was 89.7 µL/L, demonstrating higher effectiveness compared to other plants such as *S. officinalis* and *A. herba-alba* (Sellami *et al.*, 2010). More recent research corroborates these findings; for instance, Uludamar (2023) evaluated the nematicidal activity of various EOs, reporting a 100% mortality rate of *M. chitwoodi* after 24 hours of exposure to all tested concentrations (25, 250, 500, 10,000, 20,000, and 30,000 µg/mL) of thyme EO. The nematicidal effects of these EOs are likely attributable to the high concentrations of carvacrol and thymol, which destabilize nematode cell membranes and disrupt nerve functions by interfering with acetylcholine's nicotinic receptors (Marjanović *et al.*, 2020).

Recent studies by Jardim *et al.* (2018, 2020) have demonstrated the potent nematicidal activity of EOs from *Cinnamomum cassia* and *Allium sativum* against J2 of *M. incognita. In vitro* experiments revealed that *C. cassia* EO was lethal at concentrations of 62, 125, and 250 µg/mL. In contrast, the nematode population exhibited slightly more resistance to *A. sativum* EO, with significant mortality observed only at concentrations exceeding 500 µg/mL. The nematicidal effects of some *Cinnamomum* species EOs are believed to stem from the presence of key chemical compounds such as (E)-cinnamaldehyde, o-methoxycinnamaldehyde, and benzaldehyde, whose efficacy is thought to be on par with that of the commercial nematicide carbofuran (Barros *et al.*, 2021). Hyphenated chromatographic and spectrometric analyses (GC-MS) indicate that *A. sativum* EO is primarily composed of organosulfur compounds like diallyl trisulfide and diallyl disulfide, along with cinnamyl acetate and cinnamaldehyde. These compounds likely disrupt the organization of fibrous proteins and collagen in the nematode cuticle, leading to osmotic imbalance and eventual death (Arbach, 2014). Moreover, thiols and low molecular weight cinnamaldehyde are known to induce redox imbalance, resulting in DNA damage and cell apoptosis (Eder *et al.*, 2021).

Among the various EOs tested against *B. xylophilus*, those from cinnamon (*Cinnamomum verum*), red thyme (*Thymbra capitata*), garlic (*Allium sativum*), oregano (*Origanum vulgare*), clove (*Syzygium aromaticum*), winter savory (*Satureja montana*), and lemongrass (*Cymbopogon citratus*) have demonstrated the most potent nematicidal activity. This nematode is a major cause of forest devastation in many parts of the world, particularly in Asia and Europe. Specifically, the EOs of *S. montana* and *T. capitata* achieved corrected mortality rates of ≥96% at a concentration of 2 µL/mL, with lethal concentrations (LC_{100}) below 0.4 µL/mL (Faria *et al.*, 2013). Additionally, the EOs of cinnamon and garlic exhibited significant nematicidal activity, with LC_{50} values for diallyl trisulfide, diallyl disulfide, and cinnamyl acetate measured at 2.79, 37.06, and 32.81 µL/mL, respectively (Park *et al.*, 2005).

Recently, Faria and Vicente (2021) reported strong nematicidal activity of EOs from *Allium sativum*, *Eucalyptus globulus*, *Salvia officinalis*, *Hyssopus cuspidatus*, *Kaempferia galanga*, *Mentha canadensis*, *Ocimum basilicum*, and *Valeriana amurensis* against J2 of various *Heterodera* and *Globodera* species. The EOs of *A. sativum*, *E. globulus*, and *S. officinalis* demonstrated significant activity against the hatching of *G. rostochiensis* eggs, while the EOs of *H. cuspidatus*, *K. galanga*, *M. canadensis*, *O. basilicum*, and *V. amurensis* were most effective against J2 of *H. avenae*, *H. cajani*, and *H. schachtii* at concentrations ranging from 0.09 to 0.48 mg/mL. The compound with the highest nematicidal activity was ethyl p-methoxy cinnamate, a phenylpropanoid ester found in high concentrations in *K. galanga* and *O. basilicum* (El-Soud *et al.*, 2015; Umar *et al.*, 2012). In another study, Sosa *et al.* (2020) evaluated the nematicidal effects of EOs from *Pimpinella anisum* (anise) and *O. vulgare* against the false root-knot nematode *Nacobbus aberrans*. Fisher's LSD tests revealed significant nematicidal effects ($p \leq 0.05$) with LD_{100} values of 200 and 600 µL/L for anise and oregano, respectively. Finally, Avato *et al.* (2016) reported the nematicidal activity of EOs from *A. herba-alba*, *Citrus sinensis*, *Rosmarinus officinalis*, and *Thymus satureioides* against the nematodes *Pratylenchus vulnus* and *Xiphinema index*. A concentration of 50 µg/mL of EOs from

A. herba-alba, *R. officinalis*, and *T. satureioides* was sufficient to achieve 100% mortality of *X. index* after 24 hours of exposure, whereas 75% mortality of *P. vulnus* was observed after 96 hours of exposure to 15 μg/mL of *R. officinalis* EO.

9.5 CONCLUSIONS

Unlike synthetic pesticides used for managing and controlling phytopathogens, EOs—due to their natural origin—are more easily biodegradable and do not leave toxic residues in food or the environment. This makes them a promising alternative for disease control in crops intended for human and animal consumption. EOs have a lower impact on beneficial flora and fauna, including pollinating insects and soil microorganisms. Additionally, because they contain compounds with various modes of action, they reduce the selection pressure on phytopathogenic microorganisms, promoting more sustainable agricultural disease management. Beyond their antimicrobial properties, certain essential oils can stimulate plant growth, enhancing overall plant health and disease resistance. They are highly adaptable to different production systems and cultivation needs, with applications ranging from fumigants on diseased crops and seeds to coatings on fruits and vegetables during storage, among other uses. Furthermore, they can be combined with other control methods, such as cultural practices or biological control agents, to create a synergistic effect in disease management. Lastly, nanobiotechnological formulations, including emulsions, microencapsulates, and nanoemulsions, can enhance the bioavailability, absorption, stability, and shelf life of essential oils, making them more effective in the control and management of phytopathogens.

ACKNOWLEDGMENTS

The authors would like to express their sincere gratitude to Chemaxon Ltd. for providing the license to use the MarvinSketch software (v.24.1.2). This software was instrumental in the design of the chemical structures of some compounds present in the EOs (https://www.chemaxon.com).

REFERENCES

Abdallah, M. E., Ali, H. B., Mohamed, A. E., & Mohamed, A. A. (2015). *In vitro* antifungal activity of some plant essential oils. *International Journal of Pharmacology*, 11(1), 56–61. https://doi.org/10.3923/ijp.2015.56.61.

Abd-Elgawad, M. M. M., & Askary, T. H. (2015). Impact of phytonematodes on agriculture economy. In *CABI eBooks* (pp. 3–49). https://doi.org/10.1079/9781780643755.0003.

Adekoya, M. A., & Akeem, N. J. (2021). Phytochemical analysis and antimicrobial properties of *Eucalyptus torelliana* oils. *FUOYE Journal of Engineering and Technology*, 6(2). https://doi.org/10.46792/fuoyejet.v6i2.644.

Allagui, M. B., Moumni, M., & Romanazzi, G. (2023). Antifungal activity of thirty essential oils to control pathogenic fungi of postharvest decay. *Antibiotics*, 13(1), 28. https://doi.org/10.3390/antibiotics13010028.

Al-Qurashi, A. R., Akhtar, N., Al-Jabre, S., AL-Akloby, O., & Randhawa, M. A. (2007). Anti-fungal activity of thymoquinone and amphotericin B against *Aspergillus niger*. *Scientific Journal of King Faisal University* (Basic and Applied Sciences), 8(1), 143. https://www.researchgate.net/publication/268001401_Anti-Fungal_Activity_of_Thymoquinone_and_Amphotericine_B_Against_Aspergillus_Niger.

Álvarez-García, S., Moumni, M., & Romanazzi, G. (2023). Antifungal activity of volatile organic compounds from essential oils against the postharvest pathogens Botrytis cinerea, Monilinia fructicola, Monilinia fructigena, and Monilinia laxa. *Frontiers in Plant Science*, 14, 1274770. https://doi.org/10.3389/fpls.2023.1274770.

Arbach, M. (2014, July 1). *Diallyl polysulfides from garlic: Mode of action and applications in agriculture*. https://ethos.bl.uk/OrderDetails.do?uin=uk.bl.ethos.620813.

Avato, P., Laquale, S., Argentieri, M. P., Lamiri, A., Radicci, V., & D'Addabbo, T. (2016). Nematicidal activity of essential oils from aromatic plants of Morocco. *Journal of Pest Science*, 90(2), 711–722. https://doi.org/10.1007/s10340-016-0805-0.

Băieş, M. H., Gherman, C., Boros, Z., Olah, D., Vlase, A. M., Cozma-Petruţ, A., Györke, A., Miere, D., Vlase, L., Crişan, G., Spînu, M., & Cozma, V. (2022). The effects of *Allium sativum* L., *Artemisia absinthium* L., *Cucurbita pepo* L., *Coriandrum sativum* L., *Satureja hortensis* L. and *Calendula officinalis* L. on the Embryogenesis of *Ascaris suum* eggs during an *in vitro* experimental study. *Pathogens*, 11(9), 1065. https://doi.org/10.3390/pathogens11091065.

Bakkali, F., Averbeck, S., Averbeck, D., & Idaomar, M. (2008). Biological effects of essential oils—A review. *Food and Chemical Toxicology*, 46(2), 446-475. https://doi.org/10.1016/j.fct.2007.09.106.

Bakkali, F., Averbeck, S., Averbeck, D., Zhiri, A., Baudoux, D., & Idaomar, M. (2006). Antigenotoxic effects of three essential oils in diploid yeast (*Saccharomyces cerevisiae*) after treatments with UVC radiation, 8-MOP plus UVA and MMS. *Mutation Research. Genetic Toxicology and Environmental Mutagenesis*, 606(1–2), 27–38. https://doi.org/10.1016/j.mrgentox.2006.02.005.

Barakat, H., & Ghazal, G. (2016). Antifungal and antioxidant activities of rosemary (*Rosmarinus officinalis* L.) essential oil. *Journal of Food and Dairy Sciences*, 7(5), 273–282. https://doi.org/10.21608/jfds.2016.43002.

Barros, A. F., Campos, V. P., De Paula, L. L., Pedroso, L. A., De Jesus Silva, F., Da Silva, J. C. P., De Oliveira, D. F., & Silva, G. H. (2021). The role of *Cinnamomum zeylanicum* essential oil, (E)-cinnamaldehyde and (E)-cinnamaldehyde oxime in the control of *Meloidogyne incognita*. *Journal of Phytopathology*, 169(4), 229–238. https://doi.org/10.1111/jph.12979.

Barros, D. B., Nascimento, N. S., Sousa, A. P., Barros, A. V., Borges, Y. W. B., Silva, W. M. N., Motta, A. B. S., Pinto, J. E. L., Sampaio, M. G. V., Barbosa, M. F. S., Fonseca, M. C., Silva, L. A., Lima, L. O., Borges, M. G. S. A., Oliveira, M. B. M., Correia, M. T. S., Castellano, L. R. C., Guerra, F. Q. S., & Silva, M. V. (2023). Antifungal activity of terpenes isolated from the Brazilian Caatinga: A review. *Brazilian Journal of Biology*, 83. https://doi.org/10.1590/1519-6984.270966.

Basak, S., & Guha, P. (2018). A review on antifungal activity and mode of action of essential oils and their delivery as nano-sized oil droplets in food system. *Journal of Food Science and Technology*, 55(12), 4701–4710. https://doi.org/10.1007/s13197-018-3414-8.

Batish, D. R., Singh, H. P., Kohli, R. K., & Kaur, S. (2008). Eucalyptus essential oil as a natural pesticide. *Forest Ecology and Management*, 256(12), 2166–2174. https://doi.org/10.1016/j.foreco.2008.08.008.

Bernard, G. C., Egnin, M., & Bonsi, C. (2017). The impact of plant-parasitic nematodes on agriculture and methods of control. *In Nematology—Concepts, Diagnosis and Control*. InTech eBooks. https://doi.org/10.5772/intechopen.68958.

Bhandari, D. P., Poudel, D. K., Satyal, P., Khadayat, K., Dhami, S., Aryal, D., Chaudhary, P., Ghimire, A., & Parajuli, N. (2021). Volatile compounds and antioxidant and antimicrobial activities of selected citrus essential oils originated from Nepal. *Molecules*, 26(21), 6683.

Bhatt, B., & Sharma, G. (2020). Metalaxyl resistance in *Phytophthora infestans*: An overview. *International Journal of Chemical Studies*, 8(4), 266–271. https://doi.org/10.22271/chemi.2020.v8.i4ak.10114.

Bishop, C. D., & Reagan, J. (1998). Control of the storage pathogen *Botrytis cinerea* on Dutch white cabbage (*Brassica oleracea* var. capitata) by the essential oil of *Melaleuca alternifolia*. *Journal of Essential Oil Research*, 10(1), 57–60. https://doi.org/10.1080/10412905.1998.9700838.

Calmes, B., N'Guyen, G., Dumur, J., Brisach, C. A., Campion, C., Iacomi, B., & Simoneau, P. (2015). Glucosinolate-derived isothiocyanates impact mitochondrial function in fungal cells and elicit an oxidative stress response necessary for growth recovery. *Frontiers in Plant Science*, 6, 414. https://doi.org/10.3389/fpls.2015.00414.

Cao, Y., Ikram, A. U., Chen, J., Sun, Z., & Chen, J. (2023). The marksman: Bioactivated nematicides selectively kill plant-parasitic nematodes. *Journal of Integrative Plant Biology*, 65(10), 2239–2241. https://doi.org/10.1111/jipb.13546.

Caprari, C., Fantasma, F., Monaco, P. J., Divino, F., Iorizzi, M., Ranalli, G., Fasano, F., & Saviano, G. (2023). Chemical profiles, *in vitro* antioxidant and antifungal activity of four different Lavandula angustifolia L. EOs. *Molecules*, 28(1), 392. https://doi.org/10.3390/molecules28010392.

Cardoso, N. N. R., Alviano, C. S., Blank, A. F., Romanos, M. T. V., Fonseca, B. B., Rozental, S., Rodrigues, I. A., & Alviano, D. S. (2016). Synergism effect of the essential oil from *Ocimum basilicum* var. Maria Bonita and its major components with fluconazole and its influence on ergosterol biosynthesis. *Evidence-Based Complementary and Alternative Medicine: eCAM*, 2016, 1–12. https://doi.org/10.1155/2016/5647182.

Carezzano, M., Sotelo, J., Primo, E., Reinoso, E., Rovey, M., Demo, M., Giordano, W., & Oliva, M. (2017). Inhibitory effect of *Thymus vulgaris* and *Origanum vulgare* essential oils on virulence factors of phytopathogenic *Pseudomonas syringae* strains. *Plant Biology*, 19(4), 599–607. https://doi.org/10.1111/plb.12572.

Carson, C. F., Hammer, K. A., & Riley, T. V. (2002). *Melaleuca alternifolia* (Tea Tree) oil: A review of antimicrobial and other medicinal properties. *Clinical Microbiology Reviews*, 15(2), 235–238. https://doi.org/10.1128/CMR.15.2.235-238.2002.

Castellano-Hinojosa, A., Boyd, N. S., & Strauss, S. L. (2022). Impact of fumigants on non-target soil microorganisms: A review. *Journal of Hazardous Materials*, 427, 128149. https://doi.org/10.1016/j.jhazmat.2021.128149.

Catani, L., Manachini, B., Grassi, E., Guidi, L., & Semprucci, F. (2023). Essential oils as nematicides in plant protection—A review. *Plants*, 12(6), 1418. https://doi.org/10.3390/plants12061418.

Celuppi, L., Capelezzo, A. P., Cima, L. B., Zeferino, R., Zanetti, M., Riella, H., & Fiori, M. (2022). Antimicrobial cellulose acetate films by incorporation of geranyl acetate for active food packaging application. *Research, Society and Development*, 11(1), e25141. https://doi.org/10.33448/rsd-v11i1.25141.

Chen, Y., Zeng, H., Tian, J., Ban, X., Ma, B., & Wang, Y. (2013). Antifungal mechanism of essential oil from *Anethum graveolens* seeds against *Candida albicans*. *Journal of Medical Microbiology*, 62(8), 1175–1183. https://doi.org/10.1099/jmm.0.055467-0.

Choi, J.-Y., Gihaz, S., Munshi, M., Singh, P., Vydyam, P., Hamel, P., Adams, E. M., Sun, X., Khalimonchuk, O., Fuller, K. K., & Ben Mamoun, C. (2024). Vitamin B5 metabolism is essential for vacuolar and mitochondrial functions and drug detoxification in fungi. *Communications Biology*. https://doi.org/10.1038/s42003-024-06595-7.

Chung, M. Y., Kim, H., Beuchat, L. R., & Ryu, J.-H. (2022). Antimicrobial activities of plant essential oil vapours against *Acidovorax citrulli* and *Xanthomonas campestris* on Cucurbitaceae, Brassicaceae and Solanaceae seeds. *Journal of Applied Microbiology*, 132(3), 2189–2202. https://doi.org/10.1111/jam.15352.

Čmiková, N., GaloviČová, L., Schwarzová, M., Vukic, M. D., Vukovic, N. L., Kowalczewski, P. Ł., Bakay, L., Kluz, M. I., Puchalski, C., & KaČániová, M. (2023). Chemical composition and biological activities of *Eucalyptus globulus* essential oil. *Plants*, 12(5), 1076. https://doi.org/10.3390/plants12051076.

Ćosić, J., VrandeČić, K., & Jurkovic, D. (2014). The effect of essential oils on the development of phytopathogenic fungi. In N. Sharma (Ed.), *Biological controls for preventing food deterioration: Strategies for pre-and postharvest management* (pp. 273–291). Hoboken, NJ: John Wiley and Sons, Ltd. https://doi.org/10.1002/9781118533024.ch12.

Cui, H., Zhang, C., Li, C., & Lin, L. (2019). Antibacterial mechanism of oregano essential oil. *Industrial Crops and Products*, 139, 111498. https://doi.org/10.1016/j.indcrop.2019.111498.

D'Addabbo, T., Laquale, S., Argentieri, M. P., Bellardi, M. G., & Avato, P. (2021). Nematicidal Activity of essential oil from lavandin (*Lavandula × intermedia* Emeric ex Loisel.) as related to chemical profile. *Molecules*, 26(21), 6448. https://doi.org/10.3390/molecules26216448.

Daradka, H., Saleem, A., & Obaid, W. A. (2021). Antifungal effect of different plant extracts against phytopathogenic fungi *Alternaria alternata* and *Fusarium oxysporum* isolated from tomato plant. *Journal of Pharmaceutical Research International*, 33(31A), 188–197. https://doi.org/10.9734/jpri/2021/v33i31A31681.

Dawabah, A. A. M., Al-Yahya, F. A., & Lafi, H. A. (2019). Integrated management of plant-parasitic nematodes on guava and fig trees under tropical field conditions. *Egyptian Journal of Biological Pest Control*, 29(1). https://doi.org/10.1186/s41938-019-0133-9.

Dean, R., Van Kan, J. A. L., Pretorius, Z. A., Hammond-Kosack, K. E., Di Pietro, A., Spanu, P. D., Rudd, J. J., Dickman, M., Kahmann, R., Ellis, J., & Foster, G. D. (2012). The top 10 fungal pathogens in molecular plant pathology. *Molecular Plant Pathology*, 13(4), 414–430. https://doi.org/10.1111/j.1364-3703.2011.00783.x.

Degete, A. G., & Chala, A. (2019). Effects of stem rust (*Puccinia graminis* f.sp. tritici) on yield, physical and chemical quality of durum wheat varieties in East Shoa Zone, Ethiopia. *American Journal of Agriculture and Forestry*, 7(2), 95–101. https://doi.org/10.11648/j.ajaf.20190702.15.

Desaeger, J., Wram, C., & Zasada, I. (2020). New reduced-risk agricultural nematicides—rationale and review. *Journal of Nematology*, 52(1), 1–16. https://doi.org/10.21307/jofnem-2020-091.

Didehdar, M., Chegini, Z., & Shariati, A. (2022). Eugenol: A novel therapeutic agent for the inhibition of *Candida* species infection. *Frontiers in Pharmacology*, 13, 872127. https://doi.org/10.3389/fphar.2022.872127.

Doukkali, E., Radouane, N., Tahiri, A., Tazi, B., Guenoun, F., & Lahlali, R. (2022). Chemical composition and antibacterial activity of essential oils of *Cinnamomum cassia* and *Syzygium aromaticum* plants and their nanoparticles against *Erwinia amylovora*. *Journal of Plant Pathology*, 101(1), 123–130. https://doi.org/10.1080/03235408.2021.2015865.

Dubey, P., & Tripathi, S. (1987). Studies on antifungal, physico-chemical and phytotoxic properties of the essential oil of *Piper betle*. *Proceedings of the National Academy of Sciences, India Section B: Biological Sciences*, 57(3), 237–241. https://www.jstor.org/stable/43383242.

Eder, R., Consoli, E., Krauss, J., & Dahlin, P. (2021). Polysulfides applied as formulated garlic extract to protect tomato plants against the root-knot nematode *Meloidogyne incognita*. *Plants*, 10(2), 394. https://doi.org/10.3390/plants10020394.

El-Ansary, M. S. M., Hamouda, R. a. F., & Ahmed-Farid, O. A. (2020). Bioremediation of oxamyl compounds by algae: description and traits of root-knot nematode control. *Waste and Biomass Valorization*, 12(1), 251–261. https://doi.org/10.1007/s12649-020-00950-5.

El-Soud, N. H. A., Deabes, M., El-Kassem, L. A., & Khalil, M. (2015). Chemical composition and antifungal activity of *Ocimum basilicum* L. Essential Oil. Open Access *Macedonian Journal of Medical Sciences*, 3(3), 374–379. https://doi.org/10.3889/oamjms.2015.082.

Faria, J. M., Barbosa, P., Bennett, R. N., Mota, M., & Figueiredo, A. C. (2013). Bioactivity against *Bursaphelenchus xylophilus*: Nematotoxics from essential oils, essential oils fractions and decoction waters. *Phytochemistry*, 94, 220–228. https://doi.org/10.1016/j.phytochem.2013.06.005.

Faria, J. M., & Vicente, C. (2021). Essential oils and volatiles as nematodicides against the cyst nematodes globodera and heterodera. *Biology and Life Sciences Forum*. https://doi.org/10.3390/iecag2021-09689.

Ferdes, M., Al Juhaimi, F., Özcan, M. M., & Ghafoor, K. (2017). Inhibitory effect of some plant essential oils on growth of *Aspergillus niger*, *Aspergillus oryzae*, *Mucor pusillus* and *Fusarium oxysporum*. *South African Journal of Botany*, 113, 457–460. https://doi.org/10.1016/j.sajb.2017.09.015.

Fones, H., Bebber, D., Chaloner, T. M., Kay, W., Steinberg, G., & Gurr, S. (2020). Threats to global food security from emerging fungal and oomycete crop pathogens. *Nature Food*, 1(6), 332–342. https://doi.org/10.1038/s43016-020-0075-0.

Gakuubi, M. M., Maina, A. W., & Wagacha, J. M. (2017). Antifungal activity of essential oil of Eucalyptus camaldulensis Dehnh. against selected Fusarium spp. *International Journal of Microbiology*, 2017(1), 8761610.

Ganguly, R. K., Mukherjee, A., Chakraborty, S. K., & Verma, J. P. (2021). Impact of agrochemical application in sustainable agriculture. In *Elsevier eBooks* (pp. 15–24). https://doi.org/10.1016/b978-0-444-64325-4.00002-x.

García-Rodríguez, J. J., Andrés, M. F., Ibañez-Escribano, A., Julio, L. F., Burillo, J., Bolás-Fernández, F., & González-Coloma, A. (2015). Selective nematocidal effects of essential oils from two cultivated *Artemisia absinthium* populations. *Zeitschrift Für Naturforschung C*, 70(9–10), 275–280. https://doi.org/10.1515/znc-2015-0109.

Goudjil, M. B., Ladjel, S., Zighmi, S., Hammoya, F., Bensaci, M. B., Mehani, M., & Bencheikh, S. (2016). Bioactivity of *Laurus nobilis* and *Mentha piperita* essential oils on some phytopathogenic fungi (*in vitro* assay). *Journal of Materials and Environmental Science*, 7(12), 4525–4533.

Habib, W., Saab, C., Malek, R., Kattoura, L., Rotolo, C., Gerges, E., Baroudy, F., Pollastro, S., Faretra, F., & De Miccolis Angelini, R. M. (2020). Resistance profiles of *Botrytis cinerea* populations to several fungicide classes on greenhouse tomato and strawberry in Lebanon. *Plant Pathology*, 69(8), 1683–1695. https://doi.org/10.1111/ppa.13228.

Hadjra, H., Hammia, H., & Boufadi, Y. (2023). Evaluation of antifungal potential of *Mentha pulegium* essential oil in biological control against the pathogen of inflorescence rot disease of date palm (*Mauginiella scaettae*). *Jordan Journal of Biological Sciences*, 16(4), 411–419. https://doi.org/10.54319/jjbs/160412.

Haque, E., Irfan, S., Kamil, M., Sheikh, S., Hasan, A., Ahmad, A., Lakshmi, V., Nazir, A., & Mir, S. S. (2016). Terpenoids with antifungal activity trigger mitochondrial dysfunction in *Saccharomyces cerevisiae*. *Microbiology*, 85(4), 436–443. https://doi.org/10.1134/s0026261716040093.

Hou, T., Sana, S. S., Li, H., Xing, Y., Nanda, A., Netala, V. R., & Zhang, Z. (2022). Essential oils and its antibacterial, antifungal and anti-oxidant activity applications: A review. *Food Bioscience*, 47(101716), 101716. https://doi.org/10.1016/j.fbio.2022.101716.

Hua, L., Yong, C., Zhang, Z., Li, B., Qin, G., & Tian, S. (2018). Pathogenic mechanisms and control strategies of *Botrytis cinerea* causing post-harvest decay in fruits and vegetables. *Food Quality and Safety*, 2(3), 111–119. https://doi.org/10.1093/fqsafe/fyy016.

Ibe, C., & Munro, C. A. (2021). Fungal cell wall: An underexploited target for antifungal therapies. *PLoS Pathogens*, 17(4), e1009470. https://doi.org/10.1371/journal.ppat.1009470.

Ibrahim, S. K., Traboulsi, A. F., & El-Haj, S. (2006). Effects of essential oils and plant extracts on hatching, migration and mortality of "*Meloidogyne incognita*." *Phytopathologia Mediterranea*, 45(3), 238–246. https://doi.org/10.1400/56486.

Jardim, I. N., Oliveira, D. F., Campos, V. P., Silva, G. H., & Souza, P. E. (2020). Garlic essential oil reduces the population of *Meloidogyne incognita* in tomato plants. *European Journal of Plant Pathology*, 157(1), 197–209. https://doi.org/10.1007/s10658-020-02000-1.

Jardim, I. N., Oliveira, D. F., Silva, G. H., Campos, V. P., & De Souza, P. E. (2018). (E)-cinnamaldehyde from the essential oil of *Cinnamomum cassia* controls *Meloidogyne incognita* in soybean plants. *Journal of Pest Science*, 91(1), 479–487. https://doi.org/10.1007/s10340-017-0850-3.

Jones, M. G. K., & Fosu-Nyarko, J. (2014). Molecular biology of root lesion nematodes (Pratylenchus spp.) and their interaction with host plants. *Annals of Applied Biology*, 164(2), 163–181.

Jugreet, B. S., Suroowan, S., Rengasamy, R., & Mahomoodally, M. (2020). Chemistry, bioactivities, mode of action and industrial applications of essential oils. *Trends in Food Science & Technology*, 101, 89–105. https://doi.org/10.1016/j.tifs.2020.04.025.

Kachur, K., & Suntres, Z. (2020). The antibacterial properties of phenolic isomers, carvacrol and thymol. *Critical Reviews in Food Science and Nutrition*, 60(18), 3042–3053. https://doi.org/10.1080/10408398.2019.1675585.

Khan, A., Ahmad, A., Akhtar, F., Yousuf, S., Xess, I., Khan, L. A., & Manzoor, N. (2010). Ocimum sanctum essential oil and its active principles exert their antifungal activity by disrupting ergosterol biosynthesis and membrane integrity. *Research in Microbiology*, 161(10), 816–823.

Khan, J. A., & Quinn, J. (2017). Targeting pathogen sterols: Defence and counterdefence? *PLoS Pathogens*, 13(3), e1006297. https://doi.org/10.1371/journal.ppat.1006297.

Krings, M., Taylor, T. N., Hass, H., Kerp, H., Dotzler, N., & Hermsen, E. J. (2007). Fungal endophytes in a 400-million-yr-old land plant: infection pathways, spatial distribution, and host responses. *New Phytologist*, 174(3), 648–657. https://doi.org/10.1111/j.1469-8137.2007.02008.x.

Kumar, B. M., Sasikumar, B., & Kunhamu, T. K. (2021). Agroecological aspects of black pepper (Piper nigrum L.) cultivation in Kerala: A review. *AGRIVITA Journal of Agricultural Science*, 43(3), 648–664.

Kumar, Y., & Yadav, B. C. (2020). Plant-parasitic nematodes: nature's most successful plant parasite. *International Journal of Research*, 7(3), 379–386. https://www.ijrrjournal.com/IJRR_Vol.7_Issue.3_March2020/IJRR0050.pdf.

Laquale, S., Candido, V., Avato, P., Argentieri, M., & D'Addabbo, T. (2015). Essential oils as soil biofumigants for the control of the root-knot nematode *Meloidogyne incognita* on tomato. *Annals of Applied Biology*, 167(2), 217–224. https://doi.org/10.1111/aab.12221.

Leite, J. P. V. *Fitoterapia: bases científicas e tecnológicas*. São Paulo: Atheneu, 2008. 328p.

Li, X.-P., Ma, X.-C., Wang, H., Zhu, Y.-X., Liu, X., Li, T., Zheng, Y.-P., Zhao, J.-Q., Zhang, J., Huang, Y.-Y., Pu, M., Feng, H., Fan, J., Li, Y., & Wang, W.-M. (2020). Osa-miR162a fine-tunes rice resistance to *Magnaporthe oryzae* and yield. *Rice*, 13, 32. https://doi.org/10.1186/s12284-020-00396-2.

Mackie, K. A., Müller, T., Zikeli, S., & Kandeler, E. (2013). Long-term copper application in an organic vineyard modifies spatial distribution of soil micro-organisms. *Soil Biology and Biochemistry*, 65, 245–253. https://doi.org/10.1016/j.soilbio.2013.06.003.

Mansfield, J., Genin, S., Magori, S., Citovsky, V., Sriariyanum, M., Ronald, P., Dow, M., Verdier, V., Beer, S. V., Macho, A. P., Tardieu, F., & Royale, J. (2012). Top 10 plant pathogenic bacteria in molecular plant pathology. *Molecular Plant Pathology*, 13(6), 614–629. https://doi.org/10.1111/j.1364-3703.2012.00804.x

Marjanović, D. S., Zdravković, N., Milovanović, M., Trailović, J. N., Robertson, A. P., Todorović, Z., & Trailović, S. M. (2020). Carvacrol acts as a potent selective antagonist of different types of nicotinic acetylcholine receptors and enhances the effect of monepantel in the parasitic nematode *Ascaris suum*. *Veterinary Parasitology*, 278, 109031. https://doi.org/10.1016/j.vetpar.2020.109031.

McLaughlin, M. S., Roy, M., Abbasi, P., Carisse, O., Yurgel, S. N., & Ali, S. (2023). Why do we need alternative methods for fungal disease management in plants? *Plants*, 12(22), 3822. https://doi.org/10.3390/plants12223822.

Meftaul, I. M., Venkateswarlu, K., Annamalai, P., Parven, A., Sobhani, Z., & Megharaj, M. (2021). Behavior and fate of fungicide chlorothalonil in urban landscape soils and associated environmental concern. *Journal of Environmental Science and Health Part B*, 56(12), 1066–1077. https://doi.org/10.1080/03601234.2021.2014255.

Micali, C., Göllner, K., Humphry, M., Consonni, C., & Panstruga, R. (2008). The Powdery Mildew Disease of Arabidopsis: A Paradigm for the Interaction between Plants and Biotrophic Fungi. *The Arabidopsis Book*, 6, e0115. https://doi.org/10.1199/tab.0115.

Monk, B. C., & Perlin, D. S. (2008). Fungal plasma membrane proton pumps as promising new antifungal targets. *Critical Reviews in Microbiology*, 20(3), 209–223. https://doi.org/10.3109/10408419409114555.

Morjane, A., Oubihi, A., Bouigua, H., & Ouhssine, M. (2024). Antifungal activity of clove (*Syzygium aromaticum*) essential oil. *International Journal of Chemical and Biological Sciences*, 25(19). https://doi.org/10.62877/26-ijcbs-24-25-19-26.

Mulyaningsih, S., Sporer, F., Reichling, J., & Wink, M. (2011). Antibacterial activity of essential oils from Eucalyptus and of selected components against multidrug-resistant bacterial pathogens. *Pharmaceutical Biology*, 49(9), 893–899. https://doi.org/10.3109/13880209.2011.553625.

Mutlu-Ingok, A., Devecioglu, D., Dikmetas, D. N., Karbancioglu-Guler, F., & Capanoglu, E. (2020). Antibacterial, antifungal, antimycotoxigenic, and antioxidant activities of essential oils: An updated review. *Molecules*, 25(20), 4711. https://doi.org/10.3390/molecules25204711.

Nazzaro, F., Fratianni, F., Coppola, R., & De Feo, V. (2017). Essential oils and antifungal activity. *Pharmaceuticals*, 10(4), 86.

Nazzaro, F., Fratianni, F., De Martino, L., Coppola, R., & De Feo, V. (2013). Effect of essential oils on pathogenic bacteria. *Pharmaceuticals*, 6(12), 1451–1474. https://doi.org/10.3390/ph6121451.

Nguefack, J., Leth, V., Amvam Zollo, P. H., & Mathur, S. B. (2004). Evaluation of five essential oils from aromatic plants of Cameroon for controlling food spoilage and mycotoxin producing fungi. *International Journal of Food Microbiology*, 94(3), 329-334. https://doi.org/10.1016/j.ijfoodmicro.2004.02.017.

Nieto, G., Ros, G., & Castillo, J. (2018). Antioxidant and antimicrobial properties of rosemary (*Rosmarinus officinalis*, L.): A review. *Medicines*, 5(3), 98. https://doi.org/10.3390/medicines5030098.

Nowotarska, S., Nowotarski, K., Grant, I., Elliott, C., Friedman, M., & Situ, C. (2017). Mechanisms of antimicrobial action of cinnamon and oregano oils, cinnamaldehyde, carvacrol, 2,5-dihydroxybenzaldehyde, and 2-hydroxy-5-methoxybenzaldehyde against *Mycobacterium avium* subsp. paratuberculosis (Map). *Foods*, 6. https://doi.org/10.3390/foods6090072.

Oka, Y. (2020). From old-generation to next-generation nematicides. *Agronomy*, 10(9), 1387. https://doi.org/10.3390/agronomy10091387.

Oka, Y., Nacar, S., Putievsky, E., Ravid, U., Yaniv, Z., & Spiegel, Y. (2000). Nematicidal activity of essential oils and their components against the root-knot nematode. *Phytopathology*, 90(7), 710–715. https://doi.org/10.1094/phyto.2000.90.7.710.

Okorska, S., Głowacka, K., Pszczółkowska, A., Jankowski, K. J., Jastrzębski, J. P., Oszako, T., & Okorski, A. (2023). The fungicidal effect of essential oils of fennel and hops against Fusarium disease of pea. *Applied Sciences*, 13(10), 6282. https://doi.org/10.3390/app13106282.

OuYang, Q., Liu, Y., Oketch, O. R., Zhang, M., Shao, X., & Tao, N. (2021). Citronellal exerts its antifungal activity by targeting ergosterol biosynthesis in *Penicillium digitatum*. *Journal of Fungi*, 7(6), 432. https://doi.org/10.3390/JOF7060432.

Ozdemİr, E., & Gozel, U. (2018). Nematicidal activities of essential oils against *Meloidogyne incognita* on tomato plant. *Fresenius Environmental Bulletin*, 27(6), 4511–4517. https://www.cabdirect.org/cabdirect/abstract/20183386844.

Pan, L., Li, X. Z., Sun, D. A., Jin, H., Guo, H. R., & Qin, B. (2016). Design and synthesis of novel coumarin analogs and their nematicidal activity against five phytonematodes. *Chinese Chemical Letters*, 27(3), 375–379. https://doi.org/10.1016/j.cclet.2016.01.029.

Pansera, M. R., Pauletti, M., Fedrig, C. P., & Sartori, V. C. (2013). Utilization of essential oil and vegetable extracts of *Salvia officinalis* L. in the control of rot sclerotinia in lettuce. *Applied Research & Agrotechnology*, 6(2), 83–88. https://doi.org/10.5935/PAeT.V6.N2.09.

Park, I. K., Kim, K. H., Choi, K. S., Kim, C. S., Choi, I. H., Park, J. Y., & Shin, S. C. (2005). Nematicidal activity of plant essential oils and components from garlic (*Allium sativum*) and cinnamon (*Cinnamomum verum*) oils against the pine wood nematode (*Bursaphelenchus xylophilus*). *Nematology*, 7(5), 767–774. https://doi.org/10.1163/156854105775142946

Pavela, R., & Benelli, G. (2016). Essential oils as ecofriendly biopesticides? Challenges and constraints. *Trends in Plant Science*, 21(12), 1000–1006. https://doi.org/10.1016/j.tplants.2016.10.005.

Perfect, S. E., & Green, J. R. (2001). Infection structures of biotrophic and hemibiotrophic fungal plant pathogens. *Molecular Plant Pathology*, 2(2), 101–108. https://doi.org/10.1046/j.1364-3703.2001.00055.x.

Rath, C. C., & Priyadarshanee, M. (2017). Evaluation of *in-vitro* antibacterial activity of selected essential oils. *Journal of Essential Oil-Bearing Plants*, 20(2), 359–367. https://doi.org/10.1080/0972060x.2017.1326321.

Raveau, R., Fontaine, J., & Lounès-Hadj Sahraoui, A. (2020). Essential oils as potential alternative biocontrol products against plant pathogens and weeds: A review. *Foods*, 9(3), 365. https://doi.org/10.3390/foods9030365.

Raza, A., Hairs, M., Riaz, A., Rafiq, R., Basharat, M., Shaheen, M., Goshi, S., Asghar, S., Khan, A., & Rehman, Z. (2022). Trending killers of plants; An overview of the most prevailing pathogenic fungi. *Journal of Microbiology and Genetics*, 3(3), 73. https://doi.org/10.52700/jmmg.v3i3.73.

Request free Sample—Global Nematicides Market Share, Size, Trends. (n.d.). *Spherical insights*. https://www.sphericalinsights.com/%20request-sample/1768.

Sabir, Khalfi B, E., Errachidi, Chemsi, Serrano, & Soukri. (2017). Evaluation of the potential of some essential oils in biological control against phytopathogenic agent *Pseudomonas syringae* pv. Tomato DC3000 responsible for the tomatoes speck. *Journal of Plant Pathology & Microbiology*, 08(09). https://doi.org/10.4172/2157-7471.1000420.

Sánchez-Hernández, E., Teixeira, A., Martín-Ramos, P., Cunha, A., & Oliveira, R. (2024). Antifungal and anti-oomycete activities of a *Curcuma longa* L. Hydroethanolic extract rich in bisabolene sesquiterpenoids. *Horticulturae*, 10(2), 124. https://doi.org/10.3390/horticulturae10020124.

Sellami, S., Mezrket, A., & Dahmane, T. (2010). Activité nématicide de quelques huiles essentielles contre *Meloidogyne incognita* nematicidal activity of some essential oils against Meloidogyne incognita. *Nematologia Mediterranea*, 38(2). http://journals.fcla.edu/nemamedi/article/view/87037.

Sewanu, S., Bongekile, M., Folusho, O., Adejumobi, L., & Rowland, O. (2015). Antimicrobial and efflux pumps inhibitory activities of *Eucalyptus grandis* essential oil against respiratory tract infectious bacteria. *Journal of Medicinal Plants Research*, 9, 343-348. https://doi.org/10.5897/JMPR2015.

Shao, D., Smith, D. L., Kabbage, M., & Roth, M. (2021). Effectors of plant necrotrophic fungi. *Frontiers in Plant Science*, 12, 687713. https://doi.org/10.3389/fpls.2021.687713.

Shariati, A., Didehdar, M., Razavi, S., Heidary, M., Soroush, F., & Chegini, Z. (2022). Natural compounds: A hopeful promise as an antibiofilm agent against Candida species. *Frontiers in Pharmacology*, 13, 917787. https://doi.org/10.3389/fphar.2022.917787.

Shen, Q., Zhou, W., Li, H., Hu, L., & Mo, H. (2016). ROS involves the fungicidal actions of thymol against spores of *Aspergillus flavus* via the induction of nitric oxide. *PloS One*, 11(5), e0155647. https://doi.org/10.1371/journal.pone.0155647.

Silva, É., Martins, S., & Alves, E. (2014). Essential oils for the control of bacterial speck in tomato crop. *African Journal of Agricultural Research*, 9(34), 2628–2634. https://doi.org/10.5897/ajar2014.8918.

Simirgiotis, M. J., Burton, D., Parra, F., López, J., Muñoz, P., Escobar, H., & Parra, C. (2020). Antioxidant and antibacterial capacities of *Origanum vulgare* L. essential oil from the arid Andean region of Chile and its chemical characterization by GC-MS. *Metabolites*, 10(10), 414. https://doi.org/10.3390/metabo10100414.

Skandamis, P. N., & Nychas, G. J. (2001). Effect of oregano essential oil on microbiological and physico-chemical attributes of minced meat stored in air and modified atmospheres. *Journal of Applied Microbiology*, 91(6), 1011–1022.

Sosa, A. L., Girardi, N. S., Rosso, L. C., Salusso, F., Etcheverry, M. G., & Passone, M. A. (2020). *In vitro* compatibility of *Pimpinella anisum* and *Origanum vulgare* essential oils with nematophagous fungi and their effects against *Nacobbus aberrans*. *Journal of Pest Science*, 93(4), 1381–1395. https://doi.org/10.1007/s10340-020-01252-4.

Sousa, L. G. V., Castro, J., Cavaleiro, C., Salgueiro, L., Tomás, M., Palmeira-Oliveira, R., Martinez-Oliveira, J., & Cerca, N. (2022). Synergistic effects of carvacrol, α-terpinene, γ-terpinene, p-cymene and linalool against Gardnerella species. *Scientific Reports*, 12(1), 3429. https://doi.org/10.1038/s41598-022-08217-w.

Soylu, E. M., Soylu, S., & Kurt, S. (2006). Antimicrobial activities of the essential oils of various plants against tomato late blight disease agent *Phytophthora infestans*. *Mycopathologia*, 161(2), 119–128. https://doi.org/10.1007/s11046-005-0206-z.

Strack, D., Fester, T., Hause, B., Schliemann, W., & Walter, M. H. (2003). Arbuscular mycorrhiza: Biological, chemical, and molecular aspects. *Journal of Chemical Ecology*, 29(9), 1955–1979. https://doi.org/10.1023/a:1025695032113.

Sun, L., Rogiers, G., & Michiels, C. W. (2021). The natural antimicrobial trans-cinnamaldehyde interferes with UDP-N-acetylglucosamine biosynthesis and cell wall homeostasis in *Listeria monocytogenes*. *Foods* (Basel, Switzerland), 10(7), 1666. https://doi.org/10.3390/foods10071666.

Swamy, M. K., Akhtar, M. S., & Sinniah, U. R. (2016). Antimicrobial properties of plant essential oils against human pathogens and their mode of action: An updated review. *Evidence-Based Complementary and Alternative Medicine: eCAM*, 2016(1). https://doi.org/10.1155/2016/3012462.

Tatsadjieu, L. N., Dongmo, P. M. J., Ngassoum, B. M., Etoa, F. X., & Mbofung, F. M. C. (2009). Investigations on the essential oil of *Lippia rugosa* from Cameroon for its potential use as antifungal agent against *Aspergillus flavus* Link ex. Fries. *Food Control*, 20(2), 161–166. https://doi.org/10.1016/j.foodcont.2008.03.008.

Tripathi, S. (2023). Antifungal Activity of Apiaceae family plants against *Aspergillus fumigates* and *Fusorium solani*. *Journal of Plant Pathology*, 50(3), 123–130. https://doi.org/10.21203/rs.3.rs-1902271/v1.

Tserennadmid, R., Takó, M., Galgóczy, L., Papp, T., Vágvölgyi, C., Gerő, L., & Krisch, J. (2010). Antibacterial effect of essential oils and interaction with food components. *Central European Journal of Biology*, 5(5), 641–648. https://doi.org/10.2478/S11535-010-0058-5.

Ultee, A., Bennik, M. H., & Moezelaar, R. (2002). The phenolic hydroxyl group of carvacrol is essential for action against the food-borne pathogen *Bacillus cereus*. *Applied and Environmental Microbiology*, 68(4), 1561–1568. https://doi.org/10.1128/AEM.68.4.1561-1568.2002.

Uludamar, E. B. K. (2023). Screening of the nematicidal potential of some essential oils against the columbia root-knot nematode, *Meloidogyne chitwoodi*. ÇUkurova ÜNiversitesi Ziraat FaküLtesi Dergisi/ÇUkurova Tarım Ve Gıda Bilimleri Dergisi:/ÇUkurova Tarım Ve Gıda Bilimleri Dergisi. https://doi.org/10.36846/cjafs.2023.123.

Umar, M. I., Asmawi, M. Z., Sadikun, A., Atangwho, I. J., Yam, M. F., Altaf, R., & Ahmed, A. (2012). Bioactivity-guided isolation of ethyl-p-methoxycinnamate, an anti-inflammatory constituent, from Kaempferia galanga L. Extracts. *Molecules*, 17(7), 8720–8734. https://doi.org/10.3390/molecules17078720.

Vági, E., Simándi, B., Suhajda, Á., & Héthelyi, É. (2005). Essential oil composition and antimicrobial activity of *Origanum majorana* L. extracts obtained with ethyl alcohol and supercritical carbon dioxide. *Food Research International*, 38(1), 51–57. https://doi.org/10.1016/j.foodres.2004.07.006.

Vázquez-Carrada, M., Feldbrügge, M., Olicón-Hernández, D. R., Guerra-Sánchez, G., & Pardo, J. P. (2022). Functional Analysis of the Plasma Membrane H+-ATPases of *Ustilago maydis*. *Journal of Fungi*, 8(6), 550. https://doi.org/10.3390/jof8060550.

Vidic, D., Maksimović, M., Ćavar, S., & Siljak-Yakovlev, S. (2010). Influence of the continental climatic conditions on the essential-oil composition of *Salvia brachyodon* Vandas transferred from Adriatic coast. *Chemistry & Biodiversity*, 7(5), 1200–1210. https://doi.org/10.1002/cbdv.200900126.

Vieira, J. N., Gonçalves, C. L., Villarreal, J. P. V., Gonçalves, V. M., Lund, R. G., Freitag, R. A., Silva, A. F., & Nascente, P. S. (2019). Chemical composition of essential oils from the Apiaceae family, cytotoxicity, and their antifungal activity in vitro against candida species from oral cavity. *Brazilian Journal of Biology*, 79(3), 432–437. https://doi.org/10.1590/1519-6984.182206.

Visan, A., & Neguț, I. (2024). Coatings based on essential oils for combating antibiotic resistance. *Antibiotics*, 13(7), 625. https://doi.org/10.3390/antibiotics13070625.

Walasek-Janusz, M., Grzegorczyk, A., Malm, A., Nurzyńska-Wierdak, R., & Zalewski, D. P. (2024). Chemical composition, and antioxidant and antimicrobial activity of oregano essential oil. *Molecules*, 29(2), 435. https://doi.org/10.3390/molecules29020435.

Wijesundara, N. M., & Rupasinghe, H. (2018). Essential oils from *Origanum vulgare* and *Salvia officinalis* exhibit antibacterial and anti-biofilm activities against *Streptococcus* pyogenes. *Microbial Pathogenesis*, 118, 238–246. https://doi.org/10.1016/j.micpath.2018.02.026.

Yilmaz, A., Ermis, E., & Boyraz, N. (2016). Investigation of *in vitro* and *in vivo* anti-fungal activities of different plant essential oils against postharvest apple rot diseases—*Colletotrichum gloeosporioides*, *Botrytis cinerea* and *Penicillium expansum*. *Journal of Food Safety and Food Quality*, 67(5), 122–131. https://doi.org/10.2376/0003-925X-67-122.

Yin, J., Wu, S., Yang, Y., Wang, D., Ma, Y., Zhao, Y., Sheth, S., Huang, H., Song, B., & Chen, Z. (2024). In addition to damaging the plasma membrane, phenolic monoterpenoid carvacrol can bind to the minor groove of DNA of phytopathogenic fungi to potentially control tea leaf spot caused by *Lasiodiplodia theobromae*. *Phytopathology*, 114(4), 700–716. https://doi.org/10.1094/phyto-07-23-0263-r.

Yutani, M., Hashimoto, Y., Ogita, A., Kubo, I., Tanaka, T., & Fujita, K. (2011). Morphological changes of the filamentous fungus *Mucor mucedo* and inhibition of chitin synthase activity induced by anethole. *Phytotherapy Research*, 25(11), 1707–1713. https://doi.org/10.1002/ptr.3579.

Zeng, H., Chen, X., & Liang, J. (2015). *In vitro* antifungal activity and mechanism of essential oil from fennel (*Foeniculum vulgare* L.) on dermatophyte species. *Journal of Medical Microbiology*, 64(1), 93–103. https://doi.org/10.1099/jmm.0.077768-0.

Zeng, Z., Li, Q., Huo, Y., Chen, C., Duan, S., Xu, F., Cheng, Y., & Dong, X. (2020). Inhibitory effects of essential oils from Asteraceae plant against pathogenic fungi of *Panax notoginseng*. *Journal of Applied Microbiology*, 130(2), 592–603. https://doi.org/10.1111/jam.14606.

Zgheib, R., Chaillou, S., Ouaini, N., Kassouf, A., Rutledge, D., El Azzi, D., & El Beyrouthy, M. (2016). Chemometric tools to highlight the variability of the chemical composition and yield of Lebanese *Origanum syriacum* L. essential oil. *Chemistry & Biodiversity*, 13(10), 1480–1489. https://doi.org/10.1002/cbdv.201600061.

Zhou, L.-R., Hu, H.-J., Wang, J., Zhu, Y.-X., Zhu, X.-D., Ma, J.-W., & Liu, Y.-Q. (2024). Cinnamaldehyde acts as a fungistat by disrupting the integrity of *Fusarium oxysporum* Fox-1 cell membranes. *Horticulturae*, 10(1), 48. https://doi.org/10.3390/horticulturae10010048.

Section 3

Pesticides Based on Natural Products

10 High-Throughput Screening to Detect Natural Compounds with Pesticidal Properties

Luis C. Vesga, Jonathan Lara-Sanchez, Gloria Stefany Avendaño-Mora, Stelia Mendez-Sanchez, and Jonny E Duque

10.1 INTRODUCTION

Developing new, biocompatible, and biosafe pesticides is essential due to the harmful effects of traditional pesticides on human health and the environment. These synthetic chemicals can cause various health issues, from mild irritations to severe conditions like respiratory problems, neuro-psychiatric disorders, cancer, and reproductive harm (Kumar & Kumar, 2019; Ren & Jiang, 2022). Conventional pesticides also negatively affect animal health, altering their microbiome, causing mitochondrial DNA damage, or acting as endocrine disruptors (Alimova et al., 2022; Pinos et al., 2021). Resistance to chemically synthesized pesticides among pests and parasites is a concerning issue that has been extensively researched across various species. For instance, the black cutworm *Agrotis ipsilon* (Lepidoptera: Noctuidae) is a common agricultural pest that infests a wide range of crops, leading to significant yield losses. The repeated use of traditional insecticides has resulted in *A. ipsilon* developing resistance to organophosphates, carbamates, and pyrethroids (Moustafa et al., 2022). Another example is the tomato moth, *Tuta absoluta* (Lepidoptera: Gelechiidae), which is a major threat to tomato crops. The larvae of this pest are known to damage leaves, shoots, and immature fruit calyx during feeding, impacting photosynthesis and causing substantial economic losses (Castresana & Puhl, 2017; Salas Gervassio, 2019). In recent years, resistance of *T. absoluta* to diamide-type insecticides has emerged (García Vidal et al., 2016). Similarly, in the livestock industry, the cattle tick *Rhipicephalus microplus* (Ixodida: Ixodidae) has developed resistance mutations against various acaricides like ethion, fipronil, and amitraz (Gasparotto et al., 2020; Nava et al., 2018; Torres-Santos et al., 2021). Resistance mechanisms arising from mutations at the target site have been documented, manifesting through various factors including insensitivity and metabolic resistance. Research indicates that metabolic resistance in ticks may correlate with the activity of specific enzymes, such as multifunctional oxidases, glutathione-S-transferases, and esterases, with the latter being particularly significant in relation to certain acaricides; Esteve-Gasent et al., 2020). Regarding insensitivity, some arthropods have exhibited resistance to ivermectin linked to the GluCl receptor (Rodriguez-Vivas et al., 2018).

Various research studies have indicated that arthropods like the mite *Tetranychus urticae* (Trombidiformes: Tetranychidae) have developed resistance to different categories of pesticides, such as organophosphorus compounds, pyrethroids, and abamectin, among others, which poses a significant threat to the agricultural industry (Tian et al., 2022; Dermauw et al., 2013). The resistance to insecticides and herbicides has resulted in substantial economic costs in agriculture and could potentially lead to fatalities due to insect-borne diseases (Gould et al., 2018). It is important to recognize that several mechanisms contribute to insecticide resistance, including behavioral changes, altered cuticular penetration, modification of the site of action, and metabolic resistance (Avilés et al., 2021). The development of biocompatible pesticides can help address the issue of pesticide residues in human and animal plasma, as well as in water effluents, underscoring the

DOI: 10.1201/9781003463429-14

significance of safer alternatives (Jean et al., 2019). Furthermore, multiple studies have emphasized implementing training, education, and awareness programs to enhance farmers' understanding of the risks associated with pesticides and promote their proper use (Sumudumali et al., 2021).

In this chapter, we present a high-throughput approach for identifying natural compounds with potent pesticide properties, combining virtual and electrophysiological techniques. In this sense, to create libraries of compounds with specific pesticidal properties, various methods can be employed in virtual screening, such as molecular docking, pharmacophore modeling, and structure-based approaches, to forecast the biological activities of compounds against particular targets (Ding et al., 2023; Xu et al., 2023). These techniques entail screening extensive compound databases, predicting compound activities, and executing molecular docking simulations to develop potential pesticides or inhibitors based on desired activity profiles. Furthermore, the exploration of novel pharmacological targets within mitochondria may yield unidentified molecules with pesticidal activity, which, upon traversing the cell membrane with the assistance of a potential transporter, could exert a significant impact on this organelle (Alarcón-Aguilar et al., 2019). The method of electroantennography (EAG) enables the assessment of pesticides' influence on the olfactory systems of insects or arthropods, aiding in the discovery of substances that may serve as repellents or disruptors for pest management. Its capacity to detect and assess the effects of pesticides on insect behavior and sensory perception has been emphasized (Stanczyk et al., 2019; Wu et al., 2023).

10.2 SCREENING METHODS FOR NATURAL COMPOUND IDENTIFICATION AS NEW PESTICIDES

High-throughput *in silico* screening is a valuable and efficient bioinformatics tool that enhances optimization and experimental design for new drug development, complementing traditional pharmacology. This method facilitates identifying and verifying compounds for specific research goals by utilizing innovative search techniques within databases, repositories, and digital libraries. Additionally, it employs visualization and simulation techniques to predict the outcomes of interaction detection assays (Cavasotto, 2015). Starting with an *in silico* screening before the bioassay is highly effective, allowing researchers to save time and resources while achieving precise results. Its advantages have led to widespread acceptance, and use of this technique originated in the 1970s; the search and discovery of drugs or compounds by their primary function have increased significantly in recent years. This surge is due to enhanced computational screening capabilities, including improved hardware; advanced algorithms like neural networks and artificial intelligence; a plethora of open-source, commercial, and free software; and frequent updates to specialized databases (Shoichet, 2004); McInnes, 2007).

The initial approach to screening new pesticides was centered on high-throughput screening (HTS) of biochemical compounds. This process was accompanied by implementing high-throughput docking (HTD) and pharmacophore-based search algorithms. Both of these methods increase the efficiency of compound search by utilizing three-dimensional structures, e-format compounds, software with docking scoring and classification algorithms, and virtual inspection of the complexes formed (Phatak et al., 2009a). One important facet of this advanced screening methodology is the ligand-based approach.

10.2.1 Ligand-Based Approach

Ligand-based virtual screening (LBVS) involves examining various techniques to identify and elucidate potential ligands. This process considers factors such as flexibility, interaction with active sites in low-energy bioactive configurations, chemical characteristics, and known activities based on structural similarity and molecular fields. By integrating LBVS with HTS and HTD, researchers can further enhance the precision and efficiency of compound discovery and optimization (Pitt et al., 2013). These methods have been developed using both 2D and 3D pharmacophores. The 2D approach relies on molecular fingerprints to identify the presence or absence of specific molecular fragments (Halder et al., 2013). In contrast, the 3D screening approach allows for the evaluation of millions of compounds in a database, identifying the essential molecular features necessary for evaluating new compounds (Sun, 2008).

The 3D methodology is often used with artificial intelligence and machine learning. This combination has been successfully employed to identify interleukin-6 inhibitors in coronavirus patients (Tran et al., 2022), anticancer agents, antituberculosis agents, anticonvulsants (Andrade et al., 2010), and candidates for insect repellents based on three-dimensional quantitative structure-activity relationship (QSAR) studies (Bhattacharjee et al., 2005).

Various platforms are available for ligand design, employing different detection approaches and utilizing databases that include drugs, bioactive molecules, and commercial compounds. These platforms include BIOVA Discovery Studio 15 (Jejurikar & Rohane, 2021), CATALYST (Krovat et al., 2004; Jejurikar & Rohane, 2021), MOE (Roy & Luck, 2007), Phase module from Schrödinger (Dixon et al., 2006), LigandScout (Dixon et al., 2006; Wolber & Langer, 2004), Flexophore (von Korff et al., 2008), and GALAHAD (Richmond et al., 2006; von Korff et al., 2008). The selection of open-source and commercial software depends to a large extent on the detection and interpretation capabilities of each user ("Strategies for 3D pharmacophore-based virtual screening", 2010).

Included within ligand-based approaches, quantitative structure-activity relationship models connect the structures of molecules with biological activity to their physicochemical properties, allowing for the prediction of a molecule's activity (Guha, 2008). Machine learning algorithms have been used to develop 2D and 3D field-based models for ligand structures, which enable the determination of the activity of compounds based on their electrostatic, hydrophobic, and steric fields. This approach offers a significant advantage over traditional methods due to the unknown nature of ligands and the difficulty of optimizing their structures for analysis (Cappel et al., 2015). Various algorithmic variants of this method, such as Q-RASAR, have been developed to correlate physicochemical values and predict the toxicity of compounds ("Machine-learning-based similarity meets traditional QSAR: 'q-RASAR' for the enhancement of the external predictivity and detection of prediction confidence outliers in an hERG toxicity dataset", 2023).

Algorithms have been developed to expand the scope of possible new molecule designs in scientific research. One such tool is the automated fragment-based quantum-assisted structure generator (GFASG), which uses artificial intelligence to design small molecules tailored to a specific protein from libraries. These algorithms have the potential to open up exciting new avenues of research, motivating us to explore the unknown (Evteev et al., 2024).

The majority of the techniques analyzed are in conjunction with virtual screening (VS), molecular docking, and molecular dynamics methods (Pal et al., 2019), which incorporate pharmacophores to generate a virtual activity profile of each compound (Pal et al., 2019; Steindl et al., 2006). The researchers utilized these tools t to identify new agonists of nicotinic acetylcholine receptors by using the structure of first-, second-, and third-generation imidacloprid analogs, providing a theoretical basis for the designation and development of higher active insecticides (Crisan et al., 2022; Webb et al., 2022).

10.2.2 Structure-Based Method

The method is based on the receptor's attributes, which include high-performance docking, the search for protein cavities, the identification of allosteric and covalent binding sites, and the ligand's pharmacological potential. The preprocess allows for the evaluation of protein plasticity, ligand-receptor docking, analysis of the complexes formed, and classification of the results (Zoete et al., 2009).

The approach involves a series of methodologies to develop these simulations, which include identifying the molecular determinants of the protein structure to be studied and examining the ligands or ligand databases to be assessed. Consequently, the diversity of structures presents a challenge for analysis (Velazquez-Campoy & Freire, 2001).

One of the structure-based methods involves the analysis of the homology modeling of the protein sequence of interest, by aligning the amino acid sequence with crystallized proteins, it can be modeled, as facilitated by services like SWISS-MODEL, a pioneer in this type of service (Schwede et al., 2003). Another service, MODELLER, also supports this type of modeling (Bitencourt-Ferreira & de Azevedo, 2019b), while the RaptorX server allows for the modeling and probabilistic alignment of sequences with those reported in the database (Källberg et al., 2014) contributing to structural biology by predicting protein structures and annotating functions (Mohseni Moghadam et al., 2019; Yang & Zhang, 2015). Additionally, services such as Phyre2 and Robetta enable the prediction and analysis of protein functions based on homology-modeled structures (Chivian et al., 2003; Kelley et al., 2015). Alternatively, artificial intelligence algorithms have been developed that have advanced in the field of structure elucidation (Mirdita et al., 2022), like AlphaFold, developed by a Google Startup that revolutionized the prediction of protein structures through the use of neural networks and automated training of the algorithm (Jumper et al., 2021). RoseTTAFold, another algorithm developed for structure prediction by deep learning, has been updated with the help of AI (Watson et al., 2023), tools that permit more analysis of structures modeled both de novo and by homology, but which entails changes for structural biologists (Baek & Baker, 2022).

Besides homology modeling, one advantage of structure-based methods is the use of databases containing crystallized proteins by X-ray diffraction or crystallography, which allows for greater fidelity of the structures and aids in identifying compounds as targets within these protein structures (Muhammed & Aki-Yalcin, 2019). The preparation of these protein structures depends on the characteristics of the study, including selection of the crystal structure in databases such as RCSB PDB (https://www.rcsb.org/) and INTERPRO (https://www.ebi.ac.uk/interpro/) (Blum et al., 2021; Kouranov et al., 2006). Once the protein structure is obtained, the interaction of ligands within the binding pocket can be assessed through molecular docking analyses.

10.2.3 Molecular Docking Characteristics and Software

Molecular docking is a powerful tool for discovering new ligands capable of eliciting specific protein actions. This target molecule interrelation has led to the development of multiple algorithms for high-throughput docking (Moitessier et al., 2008), which predict the binding structure and affinity energies of compounds and optimize ligands based on the protein structure (Paggi et al., 2024). The basic principle behind molecular docking is to assess the binding strength of small molecules by positioning them within the active cavity of the receptor. This process utilizes geometric, energetic, and chemical environment complementarity to identify the optimal binding mode of small molecules (Chen et al., 2020). This approach also helps in understanding the intramolecular interactions that occur within the molecular context of the generated ligands (Ferreira et al., 2015). Depending on the selected method, molecular docking can be categorized as rigid, semi-rigid, or flexible. These categories differ in terms of the conformational freedom of the receptor and the ligand. The accuracy of the results can vary significantly based on the chosen method, with flexible docking generally providing the most accurate results. However, this increased accuracy comes at the cost of higher computational expense (Hou et al., 2023; Roy et al., 2018).

Since the development of the first docking software in 1982, numerous molecular docking programs have been created and continue to be updated (Tao et al., 2020). Today, there are over 60 different molecular docking software options, with AutoDock, AutoDock Vina, and Glide being the most frequently used. It is important to note that the popularity of these software programs does not necessarily indicate greater accuracy, and their higher citation rates may be attributed to the fact that AutoDock and AutoDock Vina are free and were developed earlier than many other programs (Sabe et al., 2021; Tao et al., 2020).

10.2.4 Molecular Docking Applications for Pesticide Molecule Identification Targeting Mitochondria

Molecular docking is a valuable approach for virtual screening when combined with pharmacophore design, homology modeling, and molecular dynamics simulations (Guedes et al., 2018). In the discovery of new pesticide molecules, molecular docking can enhance the identification of active compounds while reducing screening costs and improving the feasibility of drug screening. Thus, it has become a relevant method for virtual screening in pesticide research (Hou et al., 2023). Duque et al. (2023) recently investigated the larvicidal effects of over 20 essential oils derived from American native plants on *Aedes aegypti* larvae. Several of these oils demonstrated promising lethal concentration 50 (LC50) values of less than 100 ppm, indicating high larvicidal activity. The research focused on mitochondrial proteins as potential targets to elucidate the possible mechanisms of action of these essential oils. This choice was driven by mitochondria's critical role in cell death induction processes, intracellular calcium regulation, reactive oxygen species production, and various other physiological functions. Homology modeling was employed to build the 3D structures of mitochondrial proteins in *Ae. aegypti*. This computational approach allowed the researchers to generate accurate models of the target proteins, facilitating further analysis. Subsequently, virtual screening of the secondary metabolites of the essential oils was performed using molecular docking with Glide (Duque et al., 2023).

This approach identified active compounds of secondary metabolites such as carvacrol, thymol, *p*-pinene, carvone, and limonene. It also explored the possible mechanism of action, suggesting that these compounds inhibit electron transport through the mitochondrial respiratory chain. These compounds are currently being tested independently and in combination to design a new, effective pesticide (Duque et al., 2023; Vanegas-Estévez et al., 2024). Similarly, J.-Y. Wang, Gao, et al. (2022) screened an extensive database to establish a pharmacophore model, followed by a docking strategy. This process resulted in the identification of five compounds that matched well with the pharmacophore model and exhibited favorable properties for the development of new 4-hydroxyphenylpyruvate dioxygenase (HPPD) herbicides (J.-Y. Wang, Zhao, et al., 2022). Later, the same authors screened an extensive database of over 100,000 compounds, followed by molecular dynamics (MD) simulations. They analyzed the interaction patterns with key residues within the binding pocket of HPPD. Their findings suggested strong π–π interactions with Phe 381 and Phe 424, which could contribute to the potential inhibition of the enzyme (J.-Y. Wang, Gao et al., 2022). Gang et al., 2020 constructed a homology model of subunit H of ATPase and screened a database of compounds using LibDock, identifying two lead compounds with the highest scores. Subsequently, they synthesized over 20 different analogs and tested their insecticidal activity. The results showed promising insecticidal effects, identifying potential candidates for insecticide formulation. However, the two lead sulfonamide compounds, 2-(4-chloro-2-methyl-5-(N-(p-tolyl)sulfamoyl)phenoxy)acetamide and 2-(4-chloro-5-(N-(4-ethoxyphenyl)sulfamoyl)-2-methylphenoxy)acetamide, have not been further studied due to their activity against human homologous targets (Gang et al., 2020).

Structure-based and ligand-based approaches are utilized not only for insecticide development but also to identify compounds with repellent effects. For example, Portilla-Pulido et al., 2020 applied *in silico* techniques, such as molecular docking and homology modeling, to identify geranyl acetate, nerolidol, α-bisabolol, and nerol, which exhibited high affinity to AaegOBP1 in *Ae. aegypti*. These compounds demonstrated an extended repellent effect compared to the commercial repellent DEET. The study also considered ADME properties and various descriptors, including human cytotoxicity and mutagenicity. The principal insight from this research was that a simple mixture of these compounds achieved a 100% repellent effect for more than three hours (Portilla-Pulido et al., 2020).

10.3 WORKFLOW TO IDENTIFY NATURAL COMPOUNDS WITH PESTICIDAL ACTIVITY

In silico screening methods are employed to acquire precise information regarding the search, selection, design, and optimization of potential pharmacophores or ligands that function as substrates or inhibitors of specific proteins. These methods rely on structural data, affinity energy calculations, and pose information obtained from the analysis of various configurations. As shown in Figure 10.1, the strategy for processing this information and selecting lead compounds for further experiments includes the following.

10.3.1 Search and Selection of the Protein of Interest

Proteins utilized for ligand screening analysis in protein-ligand interactions are typically sourced from crystallographic or X-ray elucidated structures available in databases such as RCSB PDB (Burley et al., 2023). The best alignment score is determined based on the database's conserved protein families and motif. Protein structures can also be modeled de novo using tools like AlphaFold3 (Jumper et al., 2021) or RoseTTAFold (Baek et al., 2021) or through homology, as previously described.

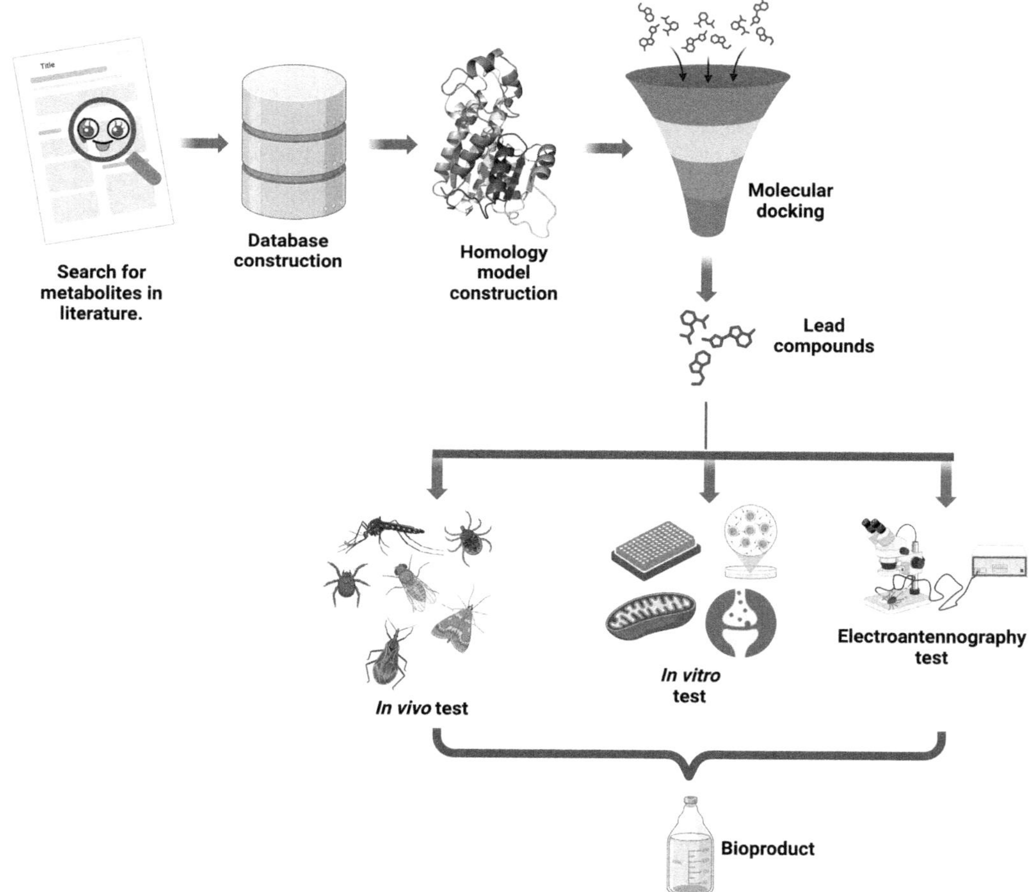

FIGURE 10.1 A schematic workflow using in silico approaches and testing the biological activity by in vitro, in vivo, and EAG methods.

10.3.2 Structure Preparation, Optimization, and Detection of Protein Binding Sites

Once the optimal crystal structure or ideal homology model is selected, the protein of interest can be prepared using commercial software such as Maestro Schrödinger (via the Protein Preparation Wizard) (Bell et al., 2012), MOE (Roy & Luck, 2007), or Discovery studio (Jejurikar & Rohane, 2021). Open-source options like AutoDock Vina can also be employed (Trott & Olson, 2010).

During this preparation, water molecules are removed, ion binding sites are detected, and residues are protonated at a specified pH (using Epik in Maestro). Additional steps include bond assignment, optimization of bond assignments, and hydrogen replacement. The correct tautomeric states of histidine (HIS) residues, as well as the rotameric states of asparagine (ASN) and glutamine (GLN), are selected. The protein is then minimized using the force field of interest (Sastry et al., 2013). Additionally, detecting ligand-binding sites is crucial for determining the characteristics enhancing docking performance. This involves assessing the cavity volume and spatial arrangement to understand how interactions with the ligand or its conformations develop (Sun, 2008).

10.3.3 Library Selection, Preparation, Ligand Optimization

Regarding ligands, a database of natural compounds, designed drugs, elucidated structures, or structures created by the user and commercially available should be selected according to the interest (Phatak et al., 2009b). The ligands used must be available in an electronic format for later handling; there is also software for designing of these molecules. The pre-filtering of the databases is mainly focused on the analysis of ADMET properties, pharmacophore models, racemic chiral centers in the ligand structure, physicochemical properties, hydrogen acceptor and donor elements, ionizable, aromatic, hydrophilic and hydrophobic characteristics, metal binding, exclusion, and inclusion sphere volumes (Martínez-Archundia et al., 2018).

Ligand preparation involves transforming structures into three-dimensional formats that can be analyzed and utilized in molecular docking programs. Software programs that facilitate this process include the LigPrep module from Maestro Schrödinger LCC, Corina Classic (*3D Structures for the Compounds Published in Molbank*, n.d.), ConfGen (Watts et al., 2010), and OMEGA (OpenEye) (Hawkins et al., 2010), although the free version of the software includes Balloon, Confab, Frog2, and RDKit. Each of these tools has specific functionalities that make it suitable for different purposes based on the type of ligand (Ebejer et al., 2012). The quality of the software and algorithms used to generate these conformations and establish conformational energies (Δew) plays a crucial role in the process (Foloppe & Chen, 2019). Energy fields for ligand minimization, such as OPLS (Jorgensen & Tirado-Rives, 1988), MMFFs (Watts et al., 2014), MECBM (Bai et al., 2010), GAFF (Sprenger et al., 2015), and ECEPP (Arnautova et al., 2006), are commonly used in protein minimization.

10.3.4 Molecular Docking for Screening Insecticide Compounds

As previously mentioned, molecular docking is a valuable tool for screening extensive libraries of compounds to identify lead candidates for insecticide analysis. Among the most commonly used software for molecular docking is AutoDock Vina, an open-source platform that allows for the docking of multiple ligands. It features enhancements such as the Broyden-Fletcher-Goldfarb-Shannon (BFGS) algorithm for local simulations, increasing its efficiency (Eberhardt et al., 2021). On the other hand, Glide within the Maestro Schrödinger platform offers a user-friendly interface and supports multiple sequence workflows. It classifies ligands with high-performance virtual detection using its XP and SP GlideScore (Vass et al., 2012). Other software, such as Dockey, provides a flexible graphical interface and can dock thousands of ligands simultaneously (Du et al., 2023), MetaDOCK combines algorithms from AutoDock, Ledock, and rDOCK and is freely available (Kamal & Chakrabarti, 2023). FlexAID and MDock are other software that accompanies the algorithms used to perform molecular docking (Ma & Zou, 2021), as well as platforms such as

SwissDock (Bitencourt-Ferreira & de Azevedo, 2019a), Udock (Levieux et al., 2014), CrystalDock (Bitencourt-Ferreira & de Azevedo, 2019a).

While the high availability of docking algorithms offers significant advantages, it also presents limitations. Developing robust strategies is essential for obtaining accurate results, as factors such as the flexibility of the protein and ligand must be considered to account for their mobility rather than treating them as rigid structures. Accurate docking scores and binding entropy estimates are crucial for reliable results (Huang & Zou, 2010).

10.3.5 Molecular Dynamics Simulations

This process, which is relevant for interaction studies, has been implemented to improve the scores under simulations that are solved under a system encompassing the Newtonian movement of each atom present in the system ("Molecular dynamics simulations of protein dynamics and their relevance to drug discovery", 2010), Unlike docking, force fields such as CHARMM (Brooks et al., 2009), AMBER, and GROMACS are used (Oostenbrink et al., 2004), and although the simulation suites are the same, NAMD (Phillips et al., 2020) and Desmond (Grossman et al., 2008) can be added.

The relevance of using of molecular dynamics is based on studying the performance of a molecular system as a function of time, including conformational changes in the protein and ligand-receptor interactions preceding the molecular interactions presented in the simulation and the dynamism of the molecules in the system (Pieroni et al., 2023). The analysis of the simulations is measured by the simulation itself and the equilibrium time of the system, that is, the interactions that are presented in a determined time concerning the ligand, which is based on the calculation of the root mean square deviation (RMSD), in addition to the calculations of the temperature, pressure, and fluctuation of the simulations, if they are physically realistic (Salsbury, 2010).

10.4 ELECTROANTENNOGRAPHY AS AN ALTERNATIVE METHOD FOR TESTING NEW NATURAL COMPOUND-BASED REPELLENTS

As mentioned, identifying active compounds by using computational approaches is an efficient alternative for repellent products. Portilla-Pulido et al. (2020) used molecular docking assays to explain the possible repellent effect of some natural compounds in *A. aegypti*. For this, the AaegOBP1 protein was used as a target, showing interesting results about their interactions with this protein involved in the odorant system. As a relevant insight, a formulation of secondary plant metabolites was used with a 100% repellent effectivity.

Once lead compounds have been identified by virtual screening, some techniques like electroantennography and *in vitro* assays have emerged as an alternative for testing these compounds in the insect's odorant system. Electroantennography (EAG) is a method for exploring the olfactory mechanisms of insects and arthropods in response to semiochemicals, pheromones, and volatile compounds. This technique enables the measurement of the electrical reactions of olfactory receptor neurons in the antennae of insects (Yang et al., 2019) or olfactory organs of arthropods (Krieger et al., 2010), offering insights into sensory reception, communication, and reactions to environmental cues, as well as foraging or reproductive behaviors (Jacob, 2018; Perera et al., 2022). EAG has been widely employed in studies concerning insect behavior and the creation of artificial repellents. For instance, it has been used to distinguish between the modes of action of natural and synthetic repellents in mosquitoes like *Ae. aegypti (Portilla Pulido et al., 2022)*. Furthermore, the response of *Drosophila melanogaster* to various volatile compounds has been assessed to determine the range of olfactory sensory neurons in their sensillae (Wang et al., 2016). This technique has enhanced our comprehension of how these insects detect and react to chemical stimuli in their surroundings (Batra et al., 2019).

This method demonstrates a certain level of resilience, as it can identify biologically active compounds, aid in the purification of extracts to isolate active components, select synthetic compounds with activity, measure concentrations in the field, and even function as a detector in conjunction with gas chromatography (Jacob, 2018; Miano et al., 2022; Xue et al., 2022). Additionally, it is a relatively straightforward technique that does not necessitate specialized or excessively expensive equipment; instead, it requires creativity and improvisation on the part of the researcher to adapt the technique to the insect or individual based on the morphology of the olfactory or chemoreceptor system.

10.4.1 Instrumentation for Electroantennography

As shown in Figure 10.2, the necessary equipment for performing an electroantennographic test includes an airflow pump (Synthec stimulus controller CS-55 or an upgraded version), an IDAC intelligent data controller, conductive metal electrodes, a signal amplifier, an anti-vibration table, a data acquisition system for recording and analyzing electrophysiological responses, and a stereomicroscope (Velásquez & Portilla, 2024). In this instance, tungsten electrodes were employed, following the methodology established by Velasquez et al. in 2024 in their research involving triatomines. Panel a) displays the electrode both before sharpening (upper) and following electrolytic sharpening (lower) using a 1 M NaOH solution. Panel b) showcases the electroantennography equipment, which

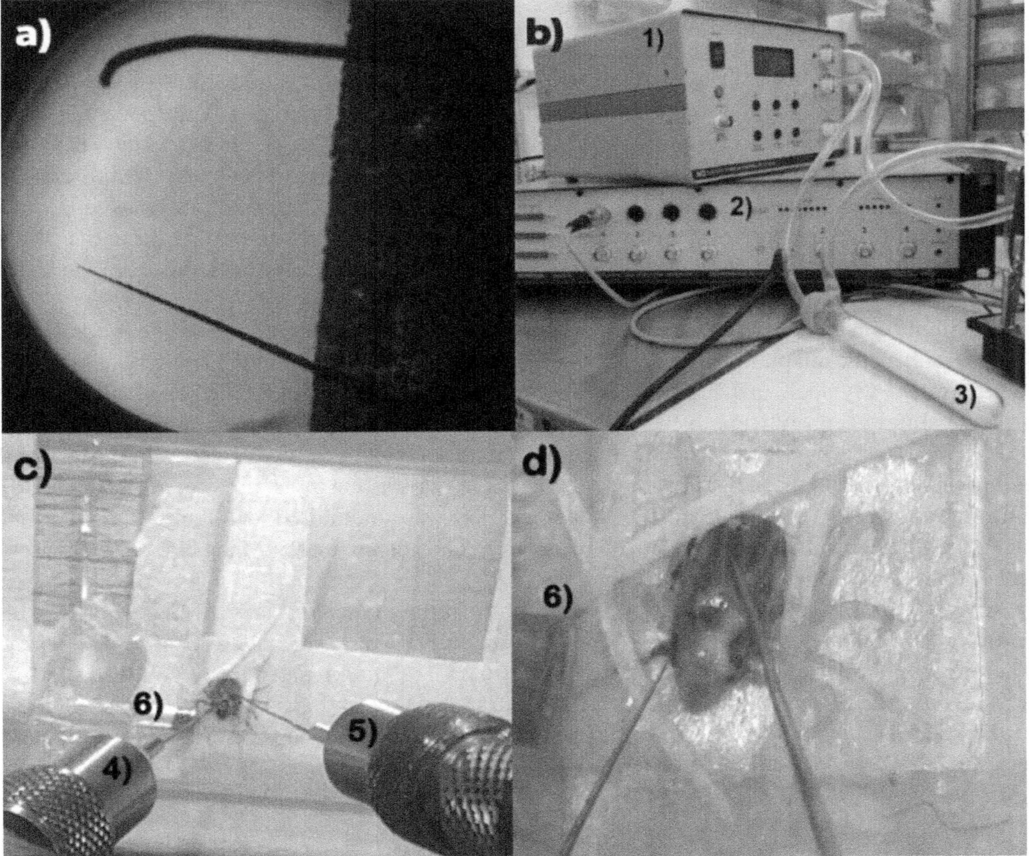

FIGURE 10.2 Equipment used in an electroantennography test.

Note: Photo taken from (Velásquez & Portilla, 2024) with some modifications.

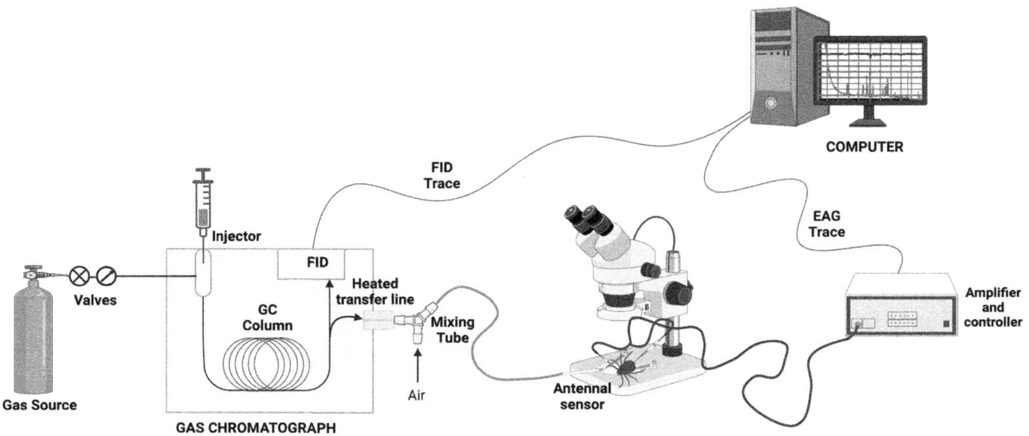

FIGURE 10.3 Gas chromatography coupled to an electroantennography equipment (GC-EAD).

includes an airflow controller system for delivering volatile compounds to the insect or mite (1) and an intelligent data acquisition system (IDAC, Synthetic) (2). Additionally, a tube with a side outlet containing cotton, where filter paper saturated with the compounds of interest is positioned (3), is connected. The airflow passes through this tube, carrying the substances to the tick. In panel c), the tick is secured in place using double-sided tape in a ventral position, with masking tape binding its four pairs of legs on each side. This immobilization method is employed to prevent interference with the emitted signal and facilitate the insertion of the electrodes (Velásquez & Portilla, 2024). The observation of two probes with measurement electrodes (4) and reference electrodes (5) inserted in coxa I and the lower part of the dorsal area, respectively, is noted. The utilization of micromanipulators (Syntech) is necessary for the insertion of the electrodes. Lastly, using a stereomicroscope, the adjustment of the electrodes to the insect and the capillary tube (6) to the first leg where Haller's organ is situated is carried out to guarantee the arrival of volatiles from the airflow system (Velásquez & Portilla, 2024).

It is important to highlight that there are opportunities to conduct more specialized experimental setups, such as the adaptation of electroantennography to gas chromatographs (GCs), as illustrated in Figure 10.3 (Syntech, 2015). In this configuration, the effluent from the GC capillary column is directed simultaneously to both the flame ionization detector (FID) and the individual's antenna through a transfer line in the side hole of a tube, which is typically heated. To prevent the loss of the compound during its journey to EAG, it is essential to introduce a moist air stream (Syntech, 2015; Tamiru et al., 2017). The outputs from EAG and FID were monitored and analyzed concurrently using the Syntech software package. GC peaks were considered active if they elicited responses in three or more of the five connected runs.

An illustration of the outcomes obtained from this type of analysis is presented in Figure 10.4, focusing on a female parasitic vespa of the *Cotesia sesamiae* (Hymenoptera: Braconidae) species. The upper line displays the EAG responses from the *C. sesamiae* antenna. In contrast, the lower lines exhibit the responses from the flame ionization detector and gas chromatography (GC-FID) of the volatile sample. The identified compounds that triggered responses in three or more antennas include: (1) (R)-linalool, (2) (E)-4,8-dimethyl-1,3,7-nonatriene (DMNT), (3) decanal, (4) (E)-caryophyllene, (5) (E,E)-4,8,12-trimethyl-1,3,7,11-tridecatetraene (TMTT) (Tamiru et al., 2017).

10.4.2 Sensils and the Electrical Principles of the EAG Technique

The insects' antennae and palps contain sensilla structures (Figure 10.5) that allow volatile compounds to enter the insect's olfactory and nervous system (Liu et al., 2020). These structures vary

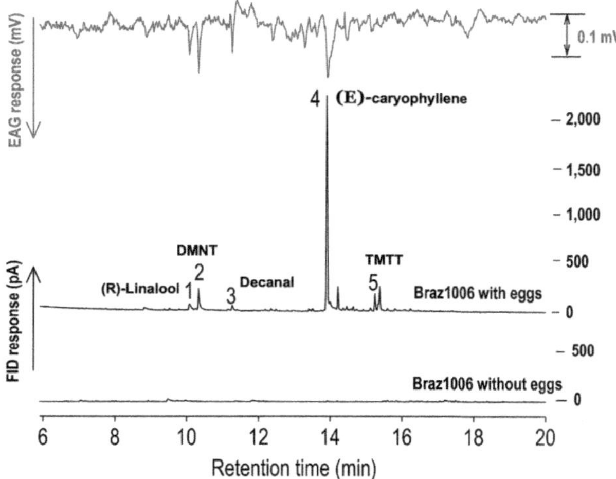

FIGURE 10.4 Chromatogram and electroantennogram obtained from a combination of GC-EAD (Tamiru et al., 2017).

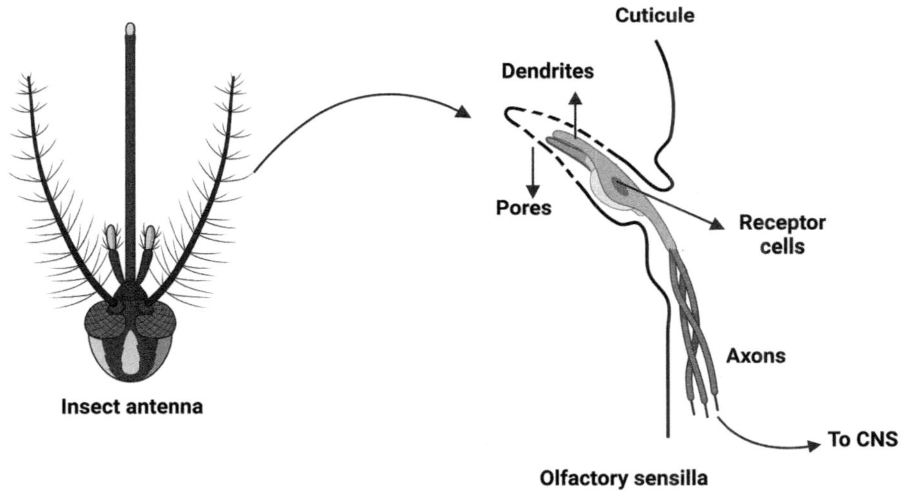

FIGURE 10.5 Description of the sensilla in an antenna and their parts.

in shape and distribution and are categorized into various types based on their form and function. Some of the most common types of sensilla found in insects include trichoid, basiconic, aurichilic, chaetic, coeloconic, campaniform, and styloconic sensilla, among others (Diakova & Polilov, 2020; Sevarika et al., 2021). Each sensilla consists of a group of neurons that facilitate the transmission of signals along the nervous system. These sensilla are a collection of cuticular or epidermal cells that resemble hairs or rod-like structures and are innervated by dendrites of sensory neurons (Figure 10.5). Additionally, they are typically surrounded by three or four accessory cells that encase the sensory neurons and associated cavities. The morphological diversity of sensilla reflects the adaptation of insects to different environmental stimuli, such as odor detection, moisture sensitivity, thermoreception, and mechanoreception (Freelance et al., 2019; Khalil et al., 2009; Wang et al., 2023).

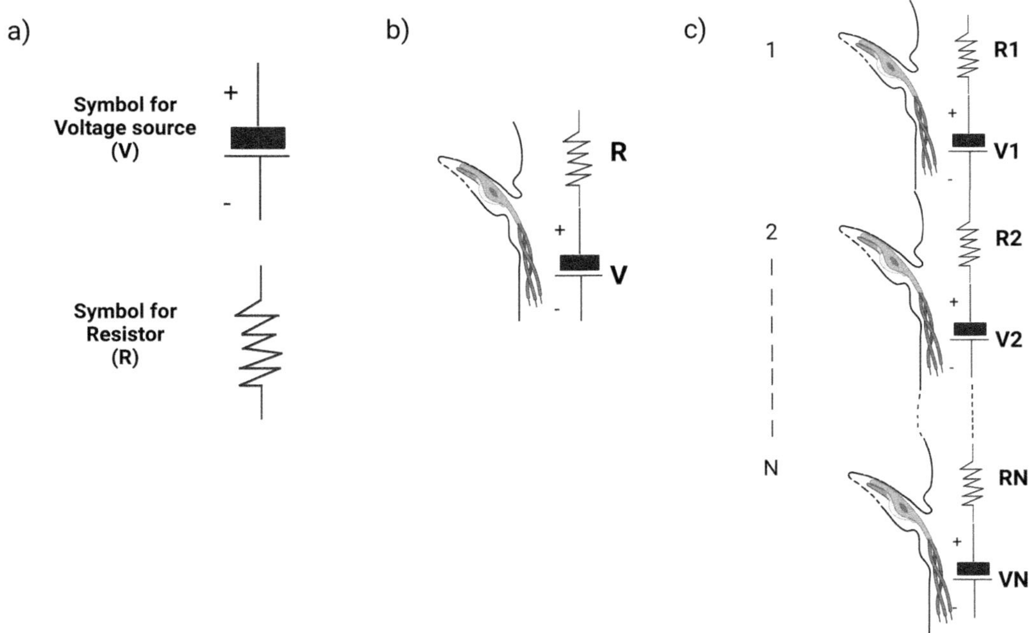

FIGURE 10.6 Analog representation of the sensilla with a voltage source and resistor in a series connected circuit.

If sensilla are conceptualized using an electrical circuit perspective, they can be illustrated as a circuit linked sequentially or alternatively, in a series (Wang et al., 2015). In Figure 10.6, standard symbols are depicted for the voltage source (V) and a resistor (R) in a), each receiver cell in b) mimics the resistor connected to the voltage source, and last, the numerous sensillae that can be organized in a single antenna are connected in the shape of a series circuit with N resistors and N voltage sources, as illustrated in c) (Syntech, 2015).

The insect's antennae or palps emit voltage signals that are very low, typically on the order of microvolts (1uV = 10^{-6} V) or millivolts (1mV=10^{-3} V). To amplify these signals without disrupting the insect's physiological processes, a specialized system is required. This system ensures that the physiological process of the antenna is not interfered with during signal amplification. The antenna is considered like a voltage source (V) with a resistance (R), while the amplifier input consists of a constant voltage source (Vi) and a resistance (Ri). Applying Ohm's Law when closing the circuit shows that V is proportional to the ratio of resistances R and Ri. This relationship demonstrates that a higher voltage response in electroantennography (EAG) is achieved when Ri > R (Shuttleworth & Johnson, 2020; Syntech, 2015). Currently, electroantennography equipment utilizes amplifiers with resistances that are 10 to 100 times greater than those of the insect's antennae, enabling accurate measurement of electrical signals (Syntech, 2015).

The signal-to-noise ratio (S/N ratio) quality is contingent upon reducing potential noise sources within the recording system. This can be achieved by mitigating noise factors such as inadequate antenna preparation, disturbances from insect movements during live insect experiments, improper amplifier utilization, and external noises stemming from suboptimal placement of the EAG equipment (Syntech, 2015). Additionally, it is imperative to consider the necessity of shielding the recording site from electromagnetic interference, which can be accomplished through the utilization of a Faraday cage. Furthermore, other external noise sources include electrical system fluctuations, abrupt changes in humidity levels, and researcher movements (Glasscott et al., 2022; Syntech, 2015).

Conversely, it is important to consider that the antennal response of a particular insect species and sex may be influenced by various factors, including the type and strength of the stimulus (concentration), the condition of the antennae, the duration of the preparation, the insect's previous exposure to stimuli, the quality of the amplifier input, temperature, humidity, and the insect's physiological state (Gadenne et al., 2016). Furthermore, environmental factors such as temperature and humidity, as well as individual physiological conditions, can also impact the antennal response, highlighting the intricate nature of sensory perception and the ability of insects to adapt to their surroundings (Sane et al., 2010; Szyszka et al., 2014).

Briefly, electroantennography presents a promising alternative to traditional methods for evaluating repellent compounds, particularly those identified by *in silico* approaches. While computational techniques offer a rapid and cost-effective way to screen potential repellent candidates, their predictive accuracy often requires empirical validation, and here EAG provides a robust, biological assessment of a compound's efficacy by directly measuring the olfactory response of insect antennae to chemical stimuli. This technique enables researchers to bridge the gap between theoretical predictions and practical application, ensuring that selected compounds elicit the desired repellent effects in target insect species. By incorporating EAG into the development pipeline, formulators can enhance the reliability and effectiveness of new repellents, ultimately leading to more successful and sustainable pest management solutions.

10.5 CONCLUSIONS AND FUTURE PERSPECTIVES

Developing new, biocompatible, and safe pesticides is essential due to the harmful effects of traditional pesticides on human health and the environment. These synthetic chemicals can cause various health issues, from mild irritations to severe conditions like respiratory problems, neuropsychiatric disorders, cancer, and reproductive harm. Consequently, researchers have been focusing on formulating new and efficient biopesticides.

Computational approaches have become valuable tools in identifying new bioactive compounds that target novel and interesting molecular targets such as mitochondrial proteins. Mitochondria are not only the cell's powerhouse but also play crucial roles in cell death processes and serve as intracellular calcium reservoirs, making them key organelles for pesticide development. Identifying bioactive compounds using computational approaches such as homology modeling, ligand-based screening, structure-based screening, molecular docking, and molecular dynamics simulations can save both time and cost.

Once lead compounds are identified, in vitro and in vivo assays can measure their biological activity. Electroantennography is emerging as a vital tool for developing new repellents. Combining computational approaches with electroantennography can provide valuable insights into animal and human health and for controlling pests in the agricultural industry.

ACKNOWLEDGMENTS

The authors thank the "Vicerrectoria de Investigación y Extensión VIE" for the support of the project; grant number 3739.

REFERENCES

Alarcón-Aguilar, A., Maycotte-González, P., Cortés-Hernández, P., López-Diazguerrero, N. E., & Königsberg, M. (2019). Dinámica mitocondrial en las enfermedades neurodegenerativas. *Gaceta Medica de Mexico*, *155*(3), 276–283. https://doi.org/10.24875/GMM.18004337

Alimova, A. A., Sitnikov, V. V., Pogorelov, D. I., Boyko, O. N., Vitkalova, I. Y., Gureev, A. P., & Popov, V. N. (2022). High doses of pesticides induce mtDNA damage in intact mitochondria of potato in vitro and do not impact on mtDNA integrity of mitochondria of shoots and tubers under in Vivo exposure. *International Journal of Molecular Sciences*, *23*(6). https://doi.org/10.3390/ijms23062970

Andrade, C. H., Pasqualoto, K. F. M., Ferreira, E. I., & Hopfinger, A. J. (2010). 3D-Pharmacophore mapping of thymidine-based inhibitors of TMPK as potential antituberculosis agents. *Journal of Computer-Aided Molecular Design*, *24*(2), 157–172. https://doi.org/10.1007/s10822-010-9323-y

Arnautova, Y. A., Jagielska, A., & Scheraga, H. A. (2006). A new force field (ECEPP-05) for peptides, proteins, and organic molecules. *The Journal of Physical Chemistry. B*, *110*(10), 5025–5044. https://doi.org/10.1021/jp054994x

Avilés, L. C., Chávez, E. C., Rodríguez, J. F. R., Beache, M. B., Fuentes, Y. M. O., & Félix, S. V. (2021). Cuantificación de enzimas relacionadas a la resistencia de insecticidas en Bemisia tabaci del estado de Sinaloa. *Revista Mexicana de Ciencias Agrícolas*, *12*(1), 77–88. https://doi.org/10.29312/remexca.v12i1.2504

Baek, M., & Baker, D. (2022). Deep learning and protein structure modeling. *Nature Methods*, *19*(1), 13–14. https://doi.org/10.1038/s41592-021-01360-8

Baek, M., DiMaio, F., Anishchenko, I., Dauparas, J., Ovchinnikov, S., Lee, G. R., Wang, J., Cong, Q., Kinch, L. N., Schaeffer, R. D., Millán, C., Park, H., Adams, C., Glassman, C. R., DeGiovanni, A., Pereira, J. H., Rodrigues, A. V., van Dijk, A. A., Ebrecht, A. C., . . . Baker, D. (2021). Accurate prediction of protein structures and interactions using a three-track neural network. *Science*, *373*(6557), 871–876. https://doi.org/10.1126/science.abj8754

Bai, F., Liu, X., Li, J., Zhang, H., Jiang, H., Wang, X., & Li, H. (2010). Bioactive conformational generation of small molecules: A comparative analysis between force-field and multiple empirical criteria based methods. *BMC Bioinformatics*, *11*, 545. https://doi.org/10.1186/1471-2105-11-545

Batra, S., Corcoran, J., Zhang, D.-D., Pal, P., K.p., U., Kulkarni, R., Löfstedt, C., Sowdhamini, R., & Olsson, S. B. (2019). A functional agonist of insect olfactory receptors: Behavior, physiology and structure. *Frontiers in Cellular Neuroscience*, *13*. https://doi.org/10.3389/fncel.2019.00134

Bell, J. A., Cao, Y., Gunn, J. R., Day, T., Gallicchio, E., Zhou, Z., Levy, R., & Farid, R. (2012). *Abstract. F*, 534–538. https://doi.org/10.1107/97809553602060000864

Bhattacharjee, A. K., Dheranetra, W., Nichols, D. A., & Gupta, R. K. (2005). 3D Pharmacophore model for insect repellent activity and discovery of new repellent candidates. *QSAR & Combinatorial Science*, *24*(5), 593–602. https://doi.org/10.1002/qsar.200430914

Bitencourt-Ferreira, G., & de Azevedo, W. F. (2019a). Docking with SwissDock. *Docking Screens for Drug Discovery*, 189–202. https://doi.org/10.1007/978-1-4939-9752-7_12

Bitencourt-Ferreira, G., & de Azevedo, W. F. (2019b). Homology modeling of protein targets with modeller. *Docking Screens for Drug Discovery*, 231–249. https://doi.org/10.1007/978-1-4939-9752-7_15

Blum, M., Chang, H.-Y., Chuguransky, S., Grego, T., Kandasaamy, S., Mitchell, A., Nuka, G., Paysan-Lafosse, T., Qureshi, M., Raj, S., Richardson, L., Salazar, G. A., Williams, L., Bork, P., Bridge, A., Gough, J., Haft, D. H., Letunic, I., Marchler-Bauer, A., . . . Finn, R. D. (2021). The InterPro protein families and domains database: 20 years on. *Nucleic Acids Research*, *49*(D1), D344–D354. https://doi.org/10.1093/nar/gkaa977

Brooks, B. R., Brooks III, C. L., Mackerell Jr, A. D., Nilsson, L., Petrella, R. J., Roux, B., . . . Karplus, M. (2009). CHARMM: The biomolecular simulation program. *Journal of Computational Chemistry*, *30*(10), 1545–1614.

Burley, S. K., Bhikadiya, C., Bi, C., Bittrich, S., Chao, H., Chen, L., Craig, P. A., Crichlow, G. V., Dalenberg, K., Duarte, J. M., Dutta, S., Fayazi, M., Feng, Z., Flatt, J. W., Ganesan, S., Ghosh, S., Goodsell, D. S., Green, R. K., Guranovic, V., . . . Zardecki, C. (2023). RCSB Protein Data Bank (RCSB.org): Delivery of experimentally-determined PDB structures alongside one million computed structure models of proteins from artificial intelligence/machine learning. *Nucleic Acids Research*, *51*(D1), D488–D508. https://doi.org/10.1093/nar/gkac1077

Cappel, D., Dixon, S. L., Sherman, W., & Duan, J. (2015). Exploring conformational search protocols for ligand-based virtual screening and 3-D QSAR modeling. *Journal of Computer-Aided Molecular Design*, *29*(2), 165–182. https://doi.org/10.1007/s10822-014-9813-4

Castresana, J., & Puhl, L. (2017). Comparative study among a variety of solar-powered LED traps to capture tomato leafminers Tuta absoluta adults by mass trapping in tomato greenhouses in the Province of Entre Ríos, Argentina. *Idesia (Arica)*, *35*(4), 87–95. https://doi.org/10.4067/S0718-34292017000400087

Cavasotto, C. N. (2015). *In silico drug discovery and design: Theory, methods, challenges, and applications*. CRC Press. https://play.google.com/store/books/details?id=oV1ECgAAQBAJ

Chen, G., Seukep, A. J., & Guo, M. (2020). Recent advances in molecular docking for the research and discovery of potential marine drugs. *Marine Drugs*, *18*(11), 545. https://doi.org/10.3390/md18110545

Chivian, D., Kim, D. E., Malmström, L., Bradley, P., Robertson, T., Murphy, P., Strauss, C. E. M., Bonneau, R., Rohl, C. A., & Baker, D. (2003). Automated prediction of CASP-5 structures using the Robetta server. *Proteins*, *53 Suppl 6*, 524–533. https://doi.org/10.1002/prot.10529

Crisan, L., Borota, A., Bora, A., Suzuki, T., & Funar-Timofei, S. (2022). Application of molecular docking, homology modeling, and chemometric approaches to neonicotinoid toxicity against aphis craccivora. *Molecular Informatics*, *41*(3), e2100058. https://doi.org/10.1002/minf.202100058

Dermauw, W., Wybouw, N., Rombauts, S., Menten, B., Vontas, J., Grbic, M., Clark, R. M., Feyereisen, R., & Van Leeuwen, T. (2013). A link between host plant adaptation and pesticide resistance in the polyphagous spider mite *Tetranychus urticae*. *Proceedings of the National Academy of Sciences of the United States of America*, *110*(2), E113–E122. https://doi.org/10.1073/pnas.1213214110

Diakova, A. V., & Polilov, A. A. (2020). Sensation of the tiniest kind: The antennal sensilla of the smallest free-living insect Scydosella musawasensis (Coleoptera: Ptiliidae). *PeerJ*, *8*, e10401. https://doi.org/10.7717/peerj.10401

Ding, Y., Chen, S., Liu, H., Liu, T., & Yang, Q. (2023). Discovery of multitarget inhibitors against insect chitinolytic enzymes via machine learning-based virtual screening. *Journal of Agricultural and Food Chemistry*, *71*(23), 8769–8777. https://doi.org/10.1021/acs.jafc.3c00633

Dixon, S. L., Smondyrev, A. M., & Rao, S. N. (2006). PHASE: A novel approach to pharmacophore modeling and 3D database searching. *Chemical Biology & Drug Design*, *67*(5), 370–372. https://doi.org/10.1111/j.1747-0285.2006.00384.x

Du, L., Geng, C., Zeng, Q., Huang, T., Tang, J., Chu, Y., & Zhao, K. (2023). Dockey: A modern integrated tool for large-scale molecular docking and virtual screening. *Briefings in Bioinformatics*, *24*(2), bbad047. https://doi.org/10.1093/bib/bbad047

Duque, J. E., Urbina, D. L., Vesga, L. C., Ortiz-Rodríguez, L. A., Vanegas, T. S., Stashenko, E. E., & Mendez-Sanchez, S. C. (2023). Insecticidal activity of essential oils from American native plants against Aedes aegypti (Diptera: Culicidae): An introduction to their possible mechanism of action. *Scientific Reports*, *13*(1), 2989. https://doi.org/10.1038/s41598-023-30046-8

Ebejer, J.-P., Morris, G. M., & Deane, C. M. (2012). Freely available conformer generation methods: How good are they? *Journal of Chemical Information and Modeling*, *52*(5), 1146–1158. https://doi.org/10.1021/ci2004658

Eberhardt, J., Santos-Martins, D., Tillack, A. F., & Forli, S. (2021). AutoDock Vina 1.2.0: New docking methods, expanded force field, and python bindings. *Journal of Chemical Information and Modeling*. https://doi.org/10.1021/acs.jcim.1c00203

Esteve-Gasent, M. D., Rodríguez-Vivas, R. I., Medina, R. F., Ellis, D., Schwartz, A., Cortés Garcia, B., Hunt, C., Tietjen, M., Bonilla, D., Thomas, D., Logan, L. L., Hasel, H., Alvarez Martínez, J. A., Hernández-Escareño, J. J., Mosqueda Gualito, J., Alonso Díaz, M. A., Rosario-Cruz, R., Soberanes Céspedes, N., Merino Charrez, O., . . . Pérez de León, A. A. (2020). Research on integrated management for cattle fever ticks and bovine babesiosis in the United States and Mexico: Current status and opportunities for binational coordination. *Pathogens*, *9*(11). https://doi.org/10.3390/pathogens9110871

Evteev, S., Ivanenkov, Y., Semenov, I., Malkov, M., Mazaleva, O., Bodunov, A., Bezrukov, D., Sidorenko, D., Terentiev, V., Malyshev, A., Zagribelnyy, B., Korzhenevskaya, A., Aliper, A., & Zhavoronkov, A. (2024). Quantum-assisted fragment-based automated structure generator (QFASG) for small molecule design: An in vitro study. *Frontiers in Chemistry*, *12*, 1382512. https://doi.org/10.3389/fchem.2024.1382512

Ferreira, L. G., Dos Santos, R. N., Oliva, G., & Andricopulo, A. D. (2015). Molecular docking and structure-based drug design strategies. *Molecules*, *20*(7), 13384–13421. https://doi.org/10.3390/molecules200713384

Foloppe, N., & Chen, I.-J. (2019). Energy windows for computed compound conformers: Covering artefacts or truly large reorganization energies? *Future Medicinal Chemistry*, *11*(2), 97–118. https://doi.org/10.4155/fmc-2018-0400

Freelance, C. B., Majoe, M., Tierney, S. M., & Elgar, M. A. (2019). Antennal asymmetry is not associated with social behaviour in Australian Hymenoptera. *Australian Entomologist*, *58*(3), 589–594. https://doi.org/10.1111/aen.12368

Gadenne, C., Barrozo, R. B., & Anton, S. (2016). Plasticity in insect olfaction: To smell or not to smell? *Annual Review of Entomology*, *61*, 317–333. https://doi.org/10.1146/annurev-ento-010715-023523

Gang, F., Li, X., Yang, C., Han, L., Qian, H., Wei, S., Wu, W., & Zhang, J. (2020). Synthesis and insecticidal activity evaluation of virtually screened phenylsulfonamides. *Journal of Agricultural and Food Chemistry*. https://doi.org/10.1021/acs.jafc.0c02153

García Vidal, L., Martínez Aguirre, M. del R., & Bielza Lino, P. (2016). *Evolution of Tuta absoluta (Meyrick) (Lepidoptera: Gelechiidae) resistance to diamide insecticides over the last 5 years*. Universidad Politécnica de Cartagena. https://repositorio.upct.es/handle/10317/10918

Gasparotto, P. H. G., Santos, C. A. F. dos, Dantas Filho, J. V., Ferraz, R. C. S., Silva, F. R. C. da, & Daudt, C. (2020). Resistance of rhipicephalus (Boophilus) microplus (Canestrini, 1888) to acaricides used in dairy cattle of Teixeirópolis, Rondônia, Brazil. *Acta Veterinaria Brasilica*, *14*(2), 99–105. https://doi.org/10.21708/avb.2020.14.2.9123

Glasscott, M. W., Brown, E. W., Dorsey, K., Laber, C. H., Conley, K., Ray, J. D., Moores, L. C., & Netchaev, A. (2022). Selecting an optimal faraday cage to minimize noise in electrochemical experiments. *Analytical Chemistry*, *94*(35), 11983–11989. https://doi.org/10.1021/acs.analchem.2c02347

Gould, F., Brown, Z. S., & Kuzma, J. (2018). Wicked evolution: Can we address the sociobiological dilemma of pesticide resistance? *Science*, *360*(6390), 728–732. https://doi.org/10.1126/science.aar3780

Grossman, J. P., Richard, H., & Kolossváry, I. (2008). Anton, a special-purpose machine for molecular dynamics simulation. *Communications of the ACM*. https://doi.org/10.1145/1364782.1364802

Guedes, I. A., Pereira, F. S. S., & Dardenne, L. E. (2018). Empirical scoring functions for structure-based virtual screening: Applications, critical aspects, and challenges. *Frontiers in Pharmacology*, *9*, 1089. https://doi.org/10.3389/fphar.2018.01089

Guha, R. (2008). On the interpretation and interpretability of quantitative structure–activity relationship models. *Journal of Computer-Aided Molecular Design*, *22*(12), 857–871. https://doi.org/10.1007/s10822-008-9240-5

Halder, A. K., Saha, A., & Jha, T. (2013). The role of 3D pharmacophore mapping based virtual screening for identification of novel anticancer agents: An overview. *Current Topics in Medicinal Chemistry*, *13*(9), 1098–1126. https://doi.org/10.2174/1568026611313090009

Hawkins, P. C. D., Geoffrey Skillman, A., Warren, G. L., Ellingson, B. A., & Stahl, M. T. (2010). *Conformer generation with OMEGA: Algorithm and validation using high quality structures from the protein databank and Cambridge structural database*. https://doi.org/10.1021/ci100031x

Hou, Y., Bai, Y., Lu, C., Wang, Q., Wang, Z., Gao, J., & Xu, H. (2023). Applying molecular docking to pesticides. *Pest Management Science*, *79*(11), 4140–4152. https://doi.org/10.1002/ps.7700

Huang, S.-Y., & Zou, X. (2010). Advances and challenges in protein-ligand docking. *International Journal of Molecular Sciences*, *11*(8), 3016–3034. https://doi.org/10.3390/ijms11083016

Jacob, V. E. J. M. (2018). Current source density analysis of electroantennogram recordings: A tool for mapping the olfactory response in an insect antenna. *Frontiers in Cellular Neuroscience*, *12*, 287. https://doi.org/10.3389/fncel.2018.00287

Jean, S., Emmanuel, F., Edouard, A. N., & Brownlinda, S. (2019). Farmers' knowledge, attitude and practices on pesticide safety: A case study of vegetable farmers in mount-bamboutos agricultural area, Cameroon. *Agricultural Sciences in China/Sponsored by the Chinese Academy of Agricultural Sciences*, *10*(08), 1039–1055. https://doi.org/10.4236/as.2019.108079

Jejurikar, B. L., & Rohane, S. H. (2021). Drug designing in discovery studio. *Asian Journal of Research in Chemistry*, *14*(2), 135–138. https://doi.org/10.5958/0974-4150.2021.00025.0

Jorgensen, W. L., & Tirado-Rives, J. (1988). The OPLS [optimized potentials for liquid simulations] potential functions for proteins, energy minimizations for crystals of cyclic peptides and crambin. *Journal of the American Chemical Society*, *110*(6), 1657–1666. https://doi.org/10.1021/ja00214a001

Jumper, J., Evans, R., Pritzel, A., Green, T., Figurnov, M., Ronneberger, O., Tunyasuvunakool, K., Bates, R., Žídek, A., Potapenko, A., Bridgland, A., Meyer, C., Kohl, S. A. A., Ballard, A. J., Cowie, A., Romera-Paredes, B., Nikolov, S., Jain, R., Adler, J., . . . Hassabis, D. (2021). Highly accurate protein structure prediction with AlphaFold. *Nature*, *596*(7873), 583–589. https://doi.org/10.1038/s41586-021-03819-2

Källberg, M., Margaryan, G., Wang, S., Ma, J., & Xu, J. (2014). RaptorX server: A resource for template-based protein structure modeling. *Protein Structure Prediction*, 17–27. https://doi.org/10.1007/978-1-4939-0366-5_2

Kamal, I. M., & Chakrabarti, S. (2023). MetaDOCK: A combinatorial molecular docking approach. *ACS Omega*, *8*(6), 5850–5860. https://doi.org/10.1021/acsomega.2c07619

Kelley, L. A., Mezulis, S., Yates, C. M., Wass, M. N., & Sternberg, M. J. E. (2015). The Phyre2 web portal for protein modeling, prediction and analysis. *Nature Protocols*, *10*(6), 845–858. https://doi.org/10.1038/nprot.2015.053

Khalil, M., Abo-zeid, N., Hafez, S., Hamed, R., & Hamouda, L. (2009). Morphological changes induced in the antenna of cowpea beetle, Callosobruchus maculatus (Fabricius) (Coleoptera: Bruchidae) after treatment with lufenuron. *Egyptian Academic Journal of Biological Sciences*, *2*(1), 207–218. https://doi.org/10.21608/eajbsa.2009.15615

Kouranov, A., Xie, L., de la Cruz, J., Chen, L., Westbrook, J., Bourne, P. E., & Berman, H. M. (2006). The RCSB PDB information portal for structural genomics. *Nucleic Acids Research*, *34*(Database issue), D302–D305. https://doi.org/10.1093/nar/gkj120

Krieger, J., Sandeman, R. E., Sandeman, D. C., Hansson, B. S., & Harzsch, S. (2010). Brain architecture of the largest living land arthropod, the giant robber crab birgus latro (Crustacea, Anomura, Coenobitidae): Evidence for a prominent central olfactory pathway? *Frontiers in Zoology*, *7*, 25. https://doi.org/10.1186/1742-9994-7-25

Krovat, E. M., Frühwirth, K. H., & Langer, T. (2004). *Pharmacophore identification, in silico screening, and virtual library design for inhibitors of the human factor Xa.* https://doi.org/10.1021/ci049778k

Kumar, V., & Kumar, P. (2019). Pesticides in agriculture and environment: Impacts on human health. In *Contaminants in agriculture and environment: Health risks and remediation* (pp. 76–95). Agro Environ Media—Agriculture and Ennvironmental Science Academy. https://doi.org/10.26832/aesa-2019-cae-0160-07

Levieux, G., Tiger, G., Mader, S., Zagury, J.-F., Natkin, S., & Montes, M. (2014). Udock, the interactive docking entertainment system. *Faraday Discussions*, *169*(0), 425–441. https://doi.org/10.1039/C3FD00147D

Liu, Y., Cui, Z., Wang, G., Zhou, Q., & Liu, Y. (2020). Cloning and functional characterization of three odorant receptors from the chinese citrus fly bactrocera minax (Diptera: Tephritidae). *Frontiers in Physiology*, *11*, 246. https://doi.org/10.3389/fphys.2020.00246

Machine-learning-based similarity meets traditional QSAR: "q-RASAR" for the enhancement of the external predictivity and detection of prediction confidence outliers in an hERG toxicity dataset. (2023). *Chemometrics and Intelligent Laboratory Systems*, *237*, 104829. https://doi.org/10.1016/j.chemolab.2023.104829

Martínez-Archundia, M., Bello, M., & Correa-Basurto, J. (2018). Design of drugs by filtering through ADMET, physicochemical and ligand-target flexibility properties. *Methods in Molecular Biology*, *1824*, 403–416. https://doi.org/10.1007/978-1-4939-8630-9_24

Ma, Z., & Zou, X. (2021). MDock: A suite for molecular inverse docking and target prediction. *Protein-Ligand Interactions and Drug Design*, 313–322. https://doi.org/10.1007/978-1-0716-1209-5_18

Miano, R. N., Ayelo, P. M., Musau, R., Hassanali, A., & Mohamed, S. A. (2022). Electroantennogram and machine learning reveal a volatile blend mediating avoidance behavior by Tuta absoluta females to a wild tomato plant. *Scientific Reports*, *12*(1), 8965. https://doi.org/10.1038/s41598-022-13125-0

Mirdita, M., Schütze, K., Moriwaki, Y., Heo, L., Ovchinnikov, S., & Steinegger, M. (2022). ColabFold: Making protein folding accessible to all. *Nature Methods*, *19*(6), 679–682. https://doi.org/10.1038/s41592-022-01488-1

Mohseni Moghadam, Z., Halabian, R., Sedighian, H., Behzadi, E., Amani, J., & Imani Fooladi, A. A. (2019). Designing and analyzing the structure of DT-STXB fusion protein as an anti-tumor agent: An approach. *Iranian Journal of Pathology*, *14*(4), 305–312. https://doi.org/10.30699/ijp.2019.101200.2004

Moitessier, N., Englebienne, P., Lee, D., Lawandi, J., & Corbeil, C. R. (2008). Towards the development of universal, fast and highly accurate docking/scoring methods: A long way to go. *British Journal of Pharmacology*, *153*(S1), S7–S26. https://doi.org/10.1038/sj.bjp.0707515

Molecular dynamics simulations of protein dynamics and their relevance to drug discovery. (2010). *Current Opinion in Pharmacology*, *10*(6), 738–744. https://doi.org/10.1016/j.coph.2010.09.016

Moustafa, M. A. M., Elmenofy, W. H., Osman, E. A., El-Said, N. A., & Awad, M. (2022). Biological impact, oxidative stress and adipokinetic hormone activities of Agrotis ipsilon in response to bioinsecticides. *Plant Protection Science = Ochrana Rostlin/Czech Academy of Agricultural Sciences*, *58*(4), 326–337. https://doi.org/10.17221/46/2022-pps

Muhammed, M. T., & Aki-Yalcin, E. (2019). Homology modeling in drug discovery: Overview, current applications, and future perspectives. *Chemical Biology & Drug Design*, *93*(1), 12–20. https://doi.org/10.1111/cbdd.13388

Nava, S., Morel, N., Mangold, A. J., & Guglielmone, A. A. (2018). Un caso de resistencia de Rhipicephalus microplus (Acari: Ixodidae) al fipronil detectado en pruebas de campo en el este de Santiago del Estero, Argentina. *FAVE sección Ciencias Veterinarias*, *17*(1), 1–5. https://doi.org/10.14409/favecv.v17i1.7158

Oostenbrink, C., Villa, A., Mark, A. E., & van Gunsteren, W. F. (2004). A biomolecular force field based on the free enthalpy of hydration and solvation: The GROMOS force-field parameter sets 53A5 and 53A6. *Journal of Computational Chemistry*, *25*(13), 1656–1676. https://doi.org/10.1002/jcc.20090

Paggi, J. M., Pandit, A., & Dror, R. O. (2024). *The Art and Science of Molecular Docking*. https://doi.org/10.1146/annurev-biochem-030222-120000

Pal, S., Kumar, V., Kundu, B., Bhattacharya, D., Preethy, N., Reddy, M. P., & Talukdar, A. (2019). Ligand-based pharmacophore modeling, virtual screening and molecular docking studies for discovery of potential topoisomerase I inhibitors. *Computational and Structural Biotechnology Journal*, *17*, 291–310. https://doi.org/10.1016/j.csbj.2019.02.006

Perera, N. N., Weston, P. A., Barrow, R. A., Weston, L. A., & Gurr, G. M. (2022). Contrasting volatilomes of livestock dung drive preference of the dung beetle bubas bison (Coleoptera: Scarabaeidae). *Molecules*, *27*(13). https://doi.org/10.3390/molecules27134152

Phatak, S. S., Stephan, C. C., & Cavasotto, C. N. (2009a). High-throughput and in silico screenings in drug discovery. *Expert Opinion on Drug Discovery*. https://doi.org/10.1517/17460440903190961

Phatak, S. S., Stephan, C. C., & Cavasotto, C. N. (2009b). High-throughput and in silico screenings in drug discovery. *Expert Opinion on Drug Discovery*, *4*(9), 947–959. https://doi.org/10.1517/17460440903190961

Phillips, J. C., Hardy, D. J., Maia, J. D. C., Stone, J. E., Ribeiro, J. V., Bernardi, R. C., Buch, R., Fiorin, G., Hénin, J., Jiang, W., McGreevy, R., Melo, M. C. R., Radak, B. K., Skeel, R. D., Singharoy, A., Wang, Y., Roux, B., Aksimentiev, A., Luthey-Schulten, Z., . . . Tajkhorshid, E. (2020). Scalable molecular dynamics on CPU and GPU architectures with NAMD. *The Journal of Chemical Physics*, *153*(4), 044130. https://doi.org/10.1063/5.0014475

Pieroni, M., Madeddu, F., Di Martino, J., Arcieri, M., Parisi, V., Bottoni, P., & Castrignanò, T. (2023). MD-ligand-receptor: A high-performance computing tool for characterizing ligand-receptor binding interactions in molecular dynamics trajectories. *International Journal of Molecular Sciences*, *24*(14). https://doi.org/10.3390/ijms241411671

Pinos, H., Carrillo, B., Merchán, A., Biosca-Brull, J., Pérez-Fernández, C., Colomina, M. T., Sánchez-Santed, F., Martín-Sánchez, F., Collado, P., Arias, J. L., & Conejo, N. M. (2021). Relationship between prenatal or postnatal exposure to pesticides and obesity: A systematic review. *International Journal of Environmental Research and Public Health*, *18*(13). https://doi.org/10.3390/ijerph18137170

Pitt, W. R., Calmiano, M. D., Kroeplien, B., Taylor, R. D., Turner, J. P., & King, M. A. (2013). Structure-based virtual screening for novel ligands. *Protein-Ligand Interactions*, 501–519. https://doi.org/10.1007/978-1-62703-398-5_19

Portilla-Pulido, J. S., Castillo-Morales, R. M., Barón-Rodríguez, M. A., Duque, J. E., & Mendez-Sanchez, S. C. (2020). Design of a repellent against aedes aegypti (Diptera: Culicidae) using in silico simulations with AaegOBP1 protein. *Journal of Medical Entomology*, *57*(2), 463–476. https://doi.org/10.1093/jme/tjz171

Portilla Pulido, J. S., Urbina Duitama, D. L., Velasquez-Martinez, M. C., Mendez-Sanchez, S. C., & Duque, J. E. (2022). Differentiation of action mechanisms between natural and synthetic repellents through neuronal electroantennogram and proteomic in Aedes aegypti (Diptera: Culicidae). *Scientific Reports*, *12*(1), 20397. https://doi.org/10.1038/s41598-022-24923-x

Ren, Z., & Jiang, H. (2022). Risk cognition, agricultural cooperatives training, and farmers' pesticide overuse: Evidence from Shandong province, China. *Frontiers in Public Health*, *10*, 1032862. https://doi.org/10.3389/fpubh.2022.1032862

Richmond, N. J., Abrams, C. A., Wolohan, P. R. N., Abrahamian, E., Willett, P., & Clark, R. D. (2006). GALAHAD: 1. pharmacophore identification by hypermolecular alignment of ligands in 3D. *Journal of Computer-Aided Molecular Design*, *20*(9), 567–587. https://doi.org/10.1007/s10822-006-9082-y

Rodriguez-Vivas, R. I., Jonsson, N. N., & Bhushan, C. (2018). Strategies for the control of Rhipicephalus microplus ticks in a world of conventional acaricide and macrocyclic lactone resistance. *Parasitology Research*, *117*(1), 3–29. https://doi.org/10.1007/s00436-017-5677-6

Roy, S., Narang, B. K., Gupta, M. K., Abbot, V., Singh, V., & Rawal, R. K. (2018). Molecular docking studies on isocytosine analogues as xanthine oxidase inhibitors. *Drug Research*, *68*(7), 395–402. https://doi.org/10.1055/s-0043-125210

Roy, U., & Luck, L. A. (2007). Molecular modeling of estrogen receptor using molecular operating environment. *Biochemistry and Molecular Biology Education: A Bimonthly Publication of the International Union of Biochemistry and Molecular Biology*, *35*(4), 238–243. https://doi.org/10.1002/bmb.65

Sabe, V. T., Ntombela, T., Jhamba, L. A., & Kruger, H. G. (2021). Current trends in computer aided drug design and a highlight of drugs discovered via computational techniques: A review. *European Journal of Medicinal Chemistry*, *224*, 113705. https://doi.org/10.1016/j.ejmech.2021.113705

Salas Gervassio, N. G. (2019). *Perspectivas del uso del endoparasitoide nativo Pseudapanteles dignus (Muesebeck) (Hymenoptera: Braconidae) para el control biológico de la polilla del tomate Tuta absoluta (Meyrick) (Lepidoptera: Gelechidae)*. Universidad Nacional de La Plata. https://doi.org/10.35537/10915/59264

Salsbury Jr, F. R. (2010). Molecular dynamics simulations of protein dynamics and their relevance to drug discovery. *Current Opinion in Pharmacology*, *10*(6), 738–744.

Sane, S. P., Srygley, R. B., & Dudley, R. (2010). Antennal regulation of migratory flight in the neotropical moth Urania fulgens. *Biology Letters*, *6*(3), 406–409. https://doi.org/10.1098/rsbl.2009.1073

Sastry, G. M., Adzhigirey, M., Day, T., Annabhimoju, R., & Sherman, W. (2013). Protein and ligand preparation: Parameters, protocols, and influence on virtual screening enrichments. *Journal of Computer-Aided Molecular Design*, *27*(3), 221–234. https://doi.org/10.1007/s10822-013-9644-8

Schwede, T., Kopp, J., Guex, N., & Peitsch, M. C. (2003). SWISS-MODEL: An automated protein homology-modeling server. *Nucleic Acids Research*, *31*(13), 3381–3385. https://doi.org/10.1093/nar/gkg520

Sevarika, M., Rossi Stacconi, M. V., & Romani, R. (2021). Fine morphology of antennal and ovipositor sensory structures of the gall chestnut wasp, dryocosmus kuriphilus. *Insects*, *12*(3). https://doi.org/10.3390/insects12030231

Shoichet, B. K. (2004, December 15). *Virtual screening of chemical libraries*. Nature Publishing Group UK. https://doi.org/10.1038/nature03197

Shuttleworth, A., & Johnson, S. D. (2020). Using two confluent capillary columns for improved gas chromatography-electroantennographic detection (GC-EAD). *Entomologia Experimentalis et Applicata*, *168*(2), 191–197. https://doi.org/10.1111/eea.12873

Sprenger, K. G., Jaeger, V. W., & Pfaendtner, J. (2015). The general AMBER force field (GAFF) can accurately predict thermodynamic and transport properties of many ionic liquids. *The Journal of Physical Chemistry. B*, *119*(18), 5882–5895. https://doi.org/10.1021/acs.jpcb.5b00689

Stanczyk, N. M., Brugman, V. A., Austin, V., Sanchez-Roman Teran, F., Gezan, S. A., Emery, M., Visser, T. M., Dessens, J. T., Stevens, W., Smallegange, R. C., Takken, W., Hurd, H., Caulfield, J., Birkett, M., Pickett, J., & Logan, J. G. (2019). Species-specific alterations in Anopheles mosquito olfactory responses caused by Plasmodium infection. *Scientific Reports*, *9*(1), 3396. https://doi.org/10.1038/s41598-019-40074-y

Steindl, T. M., Schuster, D., Laggner, C., & Langer, T. (2006). Parallel screening: A novel concept in pharmacophore modeling and virtual screening. *Journal of Chemical Information and Modeling*, *46*(5), 2146–2157. https://doi.org/10.1021/ci6002043

Strategies for 3D pharmacophore-based virtual screening. (2010). *Drug Discovery Today. Technologies*, *7*(4), e221–e228. https://doi.org/10.1016/j.ddtec.2010.11.004

Sumudumali, R. G. I., Jayawardana, J. M. C. K., Piyathilake, I. D. U. H., Randika, J. L. P. C., Udayakumara, E. P. N., Gunatilake, S. K., & Malavipathirana, S. (2021). What drives the pesticide user practices among farmers in tropical regions? A case study in Sri Lanka. *Environmental Monitoring and Assessment*, *193*(12), 860. https://doi.org/10.1007/s10661-021-09611-z

Sun, H. (2008). Pharmacophore-based virtual screening. *Current Medicinal Chemistry*, *15*(10), 1018–1024. https://doi.org/10.2174/092986708784049630

Syntech. (2015). *Electroantennography a practical introduction* (Version Syntech 2015). Syntech. https://www.ockenfels-syntech.com/wp-content/uploads/EAGpract_man_fin

Szyszka, P., Gerkin, R. C., Galizia, C. G., & Smith, B. H. (2014). High-speed odor transduction and pulse tracking by insect olfactory receptor neurons. *Proceedings of the National Academy of Sciences of the United States of America*, *111*(47), 16925–16930. https://doi.org/10.1073/pnas.1412051111

Tamiru, A., Bruce, T. J. A., Richter, A., Woodcock, C. M., Midega, C. A. O., Degenhardt, J., Kelemu, S., Pickett, J. A., & Khan, Z. R. (2017). A maize landrace that emits defense volatiles in response to herbivore eggs possesses a strongly inducible terpene synthase gene. *Ecology and Evolution*, *7*(8), 2835–2845. https://doi.org/10.1002/ece3.2893

Tao, X., Huang, Y., Wang, C., Chen, F., Yang, L., Ling, L., Che, Z., & Chen, X. (2020). Recent developments in molecular docking technology applied in food science: A review. *International Journal of Food Science & Technology*, *55*(1), 33–45. https://doi.org/10.1111/ijfs.14325

Tian, T., Wu, M., Zhang, Y., Xu, D., Wu, M., Xie, W., Su, Q., & Wang, S. (2022). Pesticide resistance and related mutation frequencies of tetranychus urticae in Hainan, China. *Horticulturae*, *8*(7), 590. https://doi.org/10.3390/horticulturae8070590

Torres-Santos, P. T., Farias, I. F., Passos, G. dos S., Almeida, M. D., & Horta, M. C. (2021). Avaliação In Vitro Da Resistência Do Carrapato Rhipicephalus microplus a diferentes carrapaticidas. *Veterinaria E Zootecnia*, *28*. https://doi.org/10.35172/rvz.2021.v28.550

Tran, Q.-H., Nguyen, Q.-T., Vo, N.-Q.-H., Mai, T. T., Tran, T.-T.-N., Tran, T.-D., Le, M.-T., Trinh, D.-T. T., & Thai, K.-M. (2022). Structure-based 3D-Pharmacophore modeling to discover novel interleukin 6 inhibitors: An in silico screening, molecular dynamics simulations and binding free energy calculations. *PloS One*, *17*(4), e0266632. https://doi.org/10.1371/journal.pone.0266632

Trott, O., & Olson, A. J. (2010). AutoDock Vina: Improving the speed and accuracy of docking with a new scoring function, efficient optimization, and multithreading. *Journal of Computational Chemistry*, *31*(2), 455–461. https://doi.org/10.1002/jcc.21334

Vanegas-Estévez, T., Duque, F. M., Urbina, D. L., Vesga, L. C., Mendez-Sanchez, S. C., & Duque, J. E. (2024). Design and elucidation of an insecticide from natural compounds targeting mitochondrial proteins of Aedes aegypti. *Pesticide Biochemistry and Physiology*, *198*, 105721. https://doi.org/10.1016/j.pestbp.2023.105721

Vass, M., Tarcsay, Á., & Keserű, G. M. (2012). Multiple ligand docking by Glide: Implications for virtual second-site screening. *Journal of Computer-Aided Molecular Design*, *26*(7), 821–834. https://doi.org/10.1007/s10822-012-9578-6

Velásquez, M. C., & Portilla, J. S. (2024). Selección de compuestos con actividad biológica repelente, atrayente o insecticida por medio de análisis con electroantenografía. In J. E. Duque (Ed.), *Vectores de chagas y su control. Hemiptera: Reduviidae: Triatominae* (pp. 107–117). División de publicaciones UIS.

Velazquez-Campoy, A., & Freire, E. (2001). Incorporating target heterogeneity in drug design. *Journal of Cellular Biochemistry*, *84*(S37), 82–88. https://doi.org/10.1002/jcb.10068

Virtual screening strategies in drug discovery. (2007). *Current Opinion in Chemical Biology*, *11*(5), 494–502. https://doi.org/10.1016/j.cbpa.2007.08.033

von Korff, M., Freyss, J., & Sander, T. (2008). Flexophore, a new versatile 3D pharmacophore descriptor that considers molecular flexibility. *Journal of Chemical Information and Modeling*, *48*(4), 797–810. https://doi.org/10.1021/ci700359j

Wang, B., Liu, Y., He, K., & Wang, G. (2016). Comparison of research methods for functional characterization of insect olfactory receptors. *Scientific Reports*, *6*, 32806. https://doi.org/10.1038/srep32806

Wang, J.-Y., Gao, S., Shi, J., Cao, H.-F., Ye, T., Yue, M.-L., Ye, F., & Fu, Y. (2022). Virtual screening based on pharmacophore model for developing novel HPPD inhibitors. *Pesticide Biochemistry and Physiology*, *184*, 105109. https://doi.org/10.1016/j.pestbp.2022.105109

Wang, J.-Y., Zhao, L.-X., Shi, J., Gao, S., Ye, F., & Fu. (2022). Discovery of novel HPPD inhibitors based on a combination strategy of pharmacophore, consensus docking and molecular dynamics. *Journal of Molecular Liquids*, *362*, 119683. https://doi.org/10.1016/j.molliq.2022.119683

Wang, W.-W., Peng-Yang, H. E., Tong-Xian, L. I. U., Jing, X.-F., & Zhang, S.-Z. (2023). Morphology and distribution of antennal sensilla on Spodoptera frugiperda (Lepidoptera: Noctuidae) larvae and adults. *Authorea Preprints*. https://doi.org/10.22541/au.167715708.87484000/v1

Wang, Z., Yang, P., Chen, D., Jiang, F., Li, Y., Wang, X., & Kang, L. (2015). Identification and functional analysis of olfactory receptor family reveal unusual characteristics of the olfactory system in the migratory locust. *Cellular and Molecular Life Sciences: CMLS*, *72*(22), 4429–4443. https://doi.org/10.1007/s00018-015-2009-9

Watson, J. L., Juergens, D., Bennett, N. R., Trippe, B. L., Yim, J., Eisenach, H. E., Ahern, W., Borst, A. J., Ragotte, R. J., Milles, L. F., Wicky, B. I. M., Hanikel, N., Pellock, S. J., Courbet, A., Sheffler, W., Wang, J., Venkatesh, P., Sappington, I., Torres, S. V., . . . Baker, D. (2023). De novo design of protein structure and function with RFdiffusion. *Nature*, *620*(7976), 1089–1100. https://doi.org/10.1038/s41586-023-06415-8

Watts, K. S., Dalal, P., Murphy, R. B., Sherman, W., Friesner, R. A., & Shelley, J. C. (2010). ConfGen: A conformational search method for efficient generation of bioactive conformers. *Journal of Chemical Information and Modeling*, *50*(4), 534–546. https://doi.org/10.1021/ci100015j

Watts, K. S., Dalal, P., Tebben, A. J., Cheney, D. L., & Shelley, J. C. (2014). Macrocycle conformational sampling with MacroModel. *Journal of Chemical Information and Modeling*, *54*(10), 2680–2696. https://doi.org/10.1021/ci5001696

Webb, D. T., Nagorzanski, M. R., Cwiertny, D. M., & LeFevre, G. H. (2022). Combining experimental sorption parameters with QSAR to predict neonicotinoid and transformation product sorption to carbon nanotubes and granular activated carbon. *ACS ES&T Water*, *2*(1), 247–258. https://doi.org/10.1021/acsestwater.1c00492

Wolber, G., & Langer, T. (2004). *LigandScout: 3-D pharmacophores derived from protein-bound ligands and their use as virtual screening filters*. https://doi.org/10.1021/ci049885e

Wu, W.-Y., Liao, L.-H., Lin, C.-H., Johnson, R. M., & Berenbaum, M. R. (2023). Effects of pesticide-adjuvant combinations used in almond orchards on olfactory responses to social signals in honey bees (Apis mellifera). *Scientific Reports*, *13*(1), 15577. https://doi.org/10.1038/s41598-023-41818-7

Xue, J., Ai, D., Xu, X., Wang, C., Jiang, X., Han, T., & Er, D. (2022). Isolation and identification of volatile substances with attractive effects on wohlfahrtia magnifica from Vagina of Bactrian Camel. *Veterinary Science in China*, *9*(11). https://doi.org/10.3390/vetsci9110637

Xu, L., Sun, L., Su, P., Ma, T., Yu, Y., Liu, H., & Huang, X. (2023). Identification of potential inhibitors of PDE5 based on structure-based virtual screening approaches. *Current Computer-Aided Drug Design*, *19*(3), 234–242. https://doi.org/10.2174/1573409919666221208143327

Yang, J., & Zhang, Y. (2015). I-TASSER server: New development for protein structure and function predictions. *Nucleic Acids Research*, *43*(W1), W174–W181. https://doi.org/10.1093/nar/gkv342

Yang, L., Liu, Y., Richoux, G. M., Bernier, U. R., Linthicum, K. J., & Bloomquist, J. R. (2019). Induction coil heating improves the efficiency of insect olfactory studies. *Frontiers in Ecology and Evolution*, *7*. https://doi.org/10.3389/fevo.2019.00247

Zoete, V., Grosdidier, A., & Michielin, O. (2009). Docking, virtual high throughput screening and in silico fragment-based drug design. *Journal of Cellular and Molecular Medicine*, *13*(2), 238–248. https://doi.org/10.1111/j.1582-4934.2008.00665.x

11 Microbe-Based Pesticides
Promising Alternative to Chemical Pesticides

Saranya R, Amrutha Lakshmi M, and Ritu Mawar

11.1 INTRODUCTION

In the new era of agriculture, pesticides have always played important roles in ensuring agricultural production as a necessity to increase grain production, regulate crop growth, and control plant diseases and insect pests. Following economic globalization, the problems of chemical pesticide residue have attracted the attention of many European countries. Many of them have increasingly strict trade standards for pesticide residue in agricultural products. Ideally, pesticides must be lethal to target pests but not to non-target species, including humans. Biopesticides can be applied as an alternative to the use of chemical pesticides, as they have been shown to be effective for pest management and the generation of sustainable agricultural products (Prabha et al., 2016). However, technical obstacles and challenges have limited their effectiveness to date (Nielsen et al., 2008). In the European Union, the impact of integrated pest management (IPM) policies provides the incentive for novel pest management strategies, especially the use of biological-based pesticides, including living microorganisms or natural products (Chandler et al., 2011). Biological pesticides, mainly produced with naturally occurring bacteria, fungi (including some protozoa and yeasts), and viruses, have been attracting widespread attention due to their advantages in target specificity, environmental safety, efficacy, biodegradability, and applicability in integrated disease management programs. Biopesticides can be classified into different categories, such as microbial pesticides, plant-incorporated protectants, and biochemicals. Biopesticides are a crucial component of integrated pest management programs for pest control, which lead to more natural alternatives to chemical pesticides that are eco-friendly and safer. Since the emergence of biopesticides for potential pest management, numerous products have been released, and some of them dominate the market. This chapter will discuss the current status and substitutes for chemical fungicides and future prospects and challenges associated with the use of biopesticides in plant disease management.

11.2 CAN MICROBIAL PESTICIDES BE AN ALTERNATIVE TO CHEMICAL PESTICIDES?

Several factors indicate that microbe-based pesticides are excellent alternatives to synthetic pesticides. Specifically, they are highly effective and target specific and have fewer environmental risks. Microbial pesticides can come from a wide diversity of organisms, and many products have been released and registered in the agromarket. Microbial pesticides allow for a sustainable approach for improved crop production, which should increase their use and popularity in the coming years (Mishra et al., 2015). Moreover, opinions about microbial pesticide use have begun to change because of the recent recognition of the environmental consequences of chemical pesticides. Other factors contributing to their increased use are advances in activity spectra spots and improved options for delivery and application (Glare et al., 2012). Current research into microbial pesticides focuses on the improvement of their mode of action, including mechanisms to replace the

DOI: 10.1201/9781003463429-15

use of chemical pesticides in integrated disease management (IDM) programs (Nawaz et al., 2016). IDM is a method to incorporate chemical, biological, and physical methods for disease control (Ghewande and Nandagopal, 1997). The main challenges of new biological pesticides in growth and utilization are how to market or promote them (Tripathi et al., 2020) and how to enhance the stability and residual action of microbial pesticides (Damalas and Koutroubas, 2018). Previously, there were several methods applied for handling pests, such as traditional methods, chemical methods, and biological methods (Lindsey et al., 2020).

11.3 MICROBIAL PESTICIDES

Microbial pesticides are compounds used to control plant diseases that are derived from beneficial microorganisms or the metabolic products they produce. The microbiome or microorganisms as a biological pesticide is another option that has been shown to be effective against insect pests and diseases. Therefore, switching to biopesticides is absolutely necessary to replace these chemical pesticides (Kumari et al., 2014). Microbial pesticides and biopesticides (plant growth–promoting rhizobacteria (PGPR), biostimulants) based on living microbes and their bioactive components have been intensively studied, published, and promoted in order to efficiently manage plant diseases and stimulate plant growth. This is done by applying large varieties of microorganisms, including fungi, bacteria, viruses, protozoans, and nematodes, as microbial pesticides (Singh, 2022). *Pseudomonas fluorescens, Bacillus subtilis, Beauvaria bassiana, Gliocladium* spp., *Trichoderma*, and *Metarrhizium anisopliae* are among the nine microbes listed in a schedule published in the *Indian Gazette* on March 26, 1999, as an amendment to the Insecticides Act, 1968, for the commercial production of biopesticides. The Insecticide Act of 1968 included 26 additional microorganisms on its list of targets for the development of microbial biopesticides (Singh, 2022).

11.4 WHY MICROBIAL PESTICIDES?

Microbial pesticides are species specific and do not harm other beneficial organisms (Kumari et al., 2014). They may occur naturally in the environment or be genetically modified for the production of toxicants, which exhibit species-specific reactions. Today's commonly used microbial pesticides are microorganisms that have a harmful effect on pest populations. These biopesticides are also effective in small amounts and dissolve quickly and more environmentally friendly technique (Thakur et al., 2020). Despite traditional agrochemical insecticides being efficient at controlling pests for many years, their continued usage has put their efficacy in jeopardy. It involves the emergence of pesticide resistance and the suspension or deregistration of insecticide usage as a result of environmental and human health issues. As a result, an eco-friendly substitute is urgently needed. The higher quality and increased quantity of agricultural products are examples of pest management strategy advancements. As a result, there is a critical need for the development of beneficial, sustainable, and non-toxic biopesticides. To avoid the negative effects of chemical insecticides, priority should be given to the use of biopesticides against pests (Mazid et al., 2011; Pathak et al., 2017).

11.5 TYPES AND FUNCTION OF MICROBIAL PESTICIDES

11.5.1 Bacteria

Bacteria have been introduced into soil, seeds, roots, and other plant structures for many decades in order to boost plant advancement and development. The purpose of utilizing bacteria is to promote the beneficial relationship of nitrogen fixation, hazardous chemical degradation, plant growth promotion, and biological control of pathogenic microorganisms. It has been reported that a large number of bacterial genera, including *Acinetobacter, Agrobacterium, Alcaligenes, Arthrobacter, Azospirillum, Azotobacter, Bacillus, Bradyrhizobium, Frankia, Pantoea, Pseudomonas, Rhizobium, Serratia,*

TABLE 11.1
Commercially Available Bacterial Bioformulations for Disease Management

Product Name	Bacteria	Phytopathogen/Targeted Disease
Biosave	*Pseudomonas syringae*	*Botrytis cinera, Mucor piriformis, Geotrichum, Penicillium*
Biosave	*P. syringae*	*Botrytis, Penicillium, Mucor, Geotrichum*
Blight Ban A506	*P. fluorescens*	*Erwinia amylovora*
Epic	*Bacillus subtilis*	Phytopathogenic fungi
Sonata	*Bacillus pumilus* QST 2808	Powdery mildew, downey mildew, and rust
Galltrol-A	*Agrobacterium radiobacter*	*Agrobacterium tumefaciens*
Cedomon	*P. chlororaphis*	*Sclerotinia*
Spot-Less	*P. auerofaciens*	*Gaeumannomyces graminis* var. *Tritici*
Kodiac	*Bacillus* sp.	*Podosphaera fusca*
Biotilis	*Bacillus subtilis*	*Pythium, Phytopthora, Botrytis, Alterneria, Phakopsora, Peranospora, Oidiopsis, Leveillula, Rhizoctonia, Sclerotium, Sclerotinia,* and *Xanthomonas*
Serenade	*Bacillus subtilis*	*Botrytis, Alterneria, Colletotrichum graminicola, Rhizoctonia spp., Sclerotinia, Magnaporthe poae, Fusarium nivale, Pythium* spp., and *Phytophthora*
Proradix	*Pseudomonas* spp.	*Rhizoctonia* on potato

Stenotrophomonas, Streptomyces, Enterobacter, Erwinia, Thiobacillus, and *Xanthomonas*, have plant disease protection activity against fungal and bacterial pathogens (Whipps, 2001; Montesinos et al., 2015). Among these, the bacteria belonging to the genera *Pseudomonas* spp. and *Bacillus* spp. have been studied intensively and registered as commercial products (Table 11.1). Numerous studies on bacteria have shown that they can prevent a variety of plant diseases, including fungal diseases and bacterial diseases (Table 11.2) (Tariq et al., 2020).

In order to shield plants against pathogen infections, bacterial biocontrol agents employ a wide range of strategies. They may engage directly or inadvertently with the pathogen to use one or a combination of strategies to prevent or minimize plant disease (Köhl et al., 2019). They can compete for resources and space, interfere with the pathogen's pathogenicity, and directly interact with the pathogen by secreting antimicrobial compounds. Many bacterial biocontrol agents synthesize and release antimicrobial metabolites like lipopeptides, bacteriocins, antibiotics, biosurfactants, cell-wall disintegrating enzymes, or microbial volatile compounds. By enzymatically destroying or preventing the synthesis of signal molecules that are used to start infections, it may also interfere with the quorum sensing (QS) system of the pathogens (Kalia et al., 2019; Bonaterra et al., 2022).

11.6 FUNGI

Fungi have systematically and biologically separated a large number of their parts in order to limit the growth of other pathogenic fungi that are harmful to plant development and advancement. The development of fungal strains to serve as biocontrol agents for plant diseases has received a lot of attention. *Trichoderma* sp. are among the ones that are most frequently studied. Because of their ability to minimize abiotic and biotic stress on host plants, such as by suppressing a variety of plant diseases and nematodes, members of the rhizospheric and epiphytic *Trichoderma* spp. are employed in various products. *Trichoderma viride* is deadly to soilborne diseases belonging to the genera *Fusarium, Sclerotium, Rhizoctonia*, and *Pythium*. *T. harzianum* has the same ability to antagonize fungal pathogens as *T. viride*, along with the pathogen species of *Botrytis, Gaeumannomyces,*

TABLE 11.2
Bacterial Biocontrol Agents and Their Target Pathogen/Disease

Bacterial Biocontrol Agent	Test Plant/Disease	Target Pathogen	References
Azospirillum brasilense	Strawberry/anthracnose	*Colletotrichum acutatum*	Tortora et al. (2011)
Azotobacter chrooccum	Cotton and rice	*Rhizoctonia solani*	Chauhan et al. (2012)
B. subtilis BY-2	Oil seed rape	*S. sclerotium*	Hu et al. (2014)
B. licheniformis BLO6	Pepper	*Phytophthora capsici*	La et al. (2020)
B. megaterium	Curus fruit	Blue mold	Mohammadi et al. (2017)
B. methylotrophicus	Maize/stalk rot	*Fusarium graminearum*	Cheng et al. (2019)
B. subtilis 26DCryChs	Wheat	*Stagonospora nodorum Berk.*	Maksimov et al. (2020)
B. thuringiensis	Sclerotiniose/*Brassica campestris* L.	*Sclerotinia sclerotiorum*	Wang et al. (2020)
Brevibacillus brevis	Strawberry/grey mold	*Botrytis cinerea*	Haggag Wafaa (2008)
Burkholderia cepacia strain BY	Tomato/damping off	*Rhizoctonia solani*	Szczech and Shoda (2004)
Ochrobactrum anthropi BMO-111	Tea/blister blight	*Exobasidium vexans*	Sowndhararajan et al. (2013)
P. chlororaphis	Canola plant	*S. sclerotiorum*	Savchuk and Fernando (2004)
Paenibacillus alvei K-165	Cotton/black root rot	*Thielaviopsis basicola*	Schoina et al. (2011)
Paenibacillus polymyxa BRF-1	Soybean/root rot	*Phialophora gregata*	Zhou et al. (2008)
Pantoea agglomerrans	Wheat root rot	*Rhizoctonia solani* AG-B	Barnett et al. (2006)
Pantoea agglomerrans	Banana/crown rot	*Colletotrichum musae* and *Lasiodiplodia theobromae*	Gunasinghe and Karunaratne (2009)
Pseudomonas and *Burkholderia*	NA	*Phytophthora capsici*	Khatun et al. (2018)
Pseudomonas fluorescens	Apple/mucor rot	*Mucor piriformis*	Wallace et al. (2018)
Pseudomonas parafulva JBCS 1880	Soybean/bacterial pustule rice/panicle blight	*Xanthomonas axonopodis* pv. *Glycines, Burkholderia glumae*	Kakembo and Lee (2019)
Pseudomonas putida BP 25	Blast disease/rice	*Magnaporthe oryzae*	Ashajyothia et al. (2020)
Rhizobium japonicum	Soybean/root rot	*Fusarium solani; Macrophomina phaseolina*	Al-Ani et al. (2012)

Adapted from Tariq et al. (2020).

Sclerotinia, Verticillium, and wood-rot fungi (Vosátka et al., 2012). Other fungal species from genera, *Gliocladium, Aspergillus*, and so on have been reported to have antagonistic activities against plant fungal pathogens, including *Alternaria, Pythium, Aspergillus, Fusarium, Rhizoctonia, Phytophthora, Botrytis, Pyricularia*, and *Gaeumannomyces* (Pal and Gardener, 2006).

Nematophagous fungus have long been thought to be a promising biological agent for nematode management. These fungi have been categorized into several genera based on the shape of their conidia and conidiophores (Senthilkumar et al., 2020). According to Dasgupta and Khan (2015), some of the most investigated genera and species of nematophagous fungus are *Arthrobotrys* spp., *Duddingtonia flagrans, Monacrosporium* spp., and *Dactylaria* spp. In the tropics and subtropics of the world, as well as in the majority of agricultural soils, *Paecilomyces lilacinus* is a soilborne fungus that has a higher potential for use as a biological control agent against plant-parasitic nematodes (Table 11.3).

Entomopathogenic fungi are a type of fungus that lives in soil and infects insects by breaching their cuticle and then penetrating their bodies, killing and feeding on them (Dara, 2017).

TABLE 11.3
Fungal Biocontrol Agents and their Target Pathogen/Disease

Fungal Strains	Test Plant/Disease	Target Pathogen	References
Aspergillus fumigatus	Cocoa/black pod	*Phytophthora palmivora*	Adebola and Amadi (2010)
Paecilomyces lilacinus	Tomato/root-knot disease	*Meloidogyne javanica*	Hanawi (2016)
Penicillium oxalicum	Tomato/wilt	*Fusarium oxyysporum* f. sp. *lycopersici*	Sabuquillo et al. (2006)
Penicillium sp. EU0013	Tomato and cabbage/wilt	*Fusarium oxysporum*	Alam et al. (2010)
Pochonia chlamydosporia	Carrot/root-knot disease	*Meloidogyne incognita*	Bontempo et al. (2017)
Purpureocilium lilacinum	Vigna radiata/root-knot disease	*Meloidogyne incognita*	Khan et al. (2019)
Purpureocilium lilacinum	Pineapple/root knot disease	*Meloidogyne javanica*	Kiriga et al. (2018)
Trichoderma hamatum	Cabbage	*S. sclerotiorum*	Jones et al. (2014)
Trichoderma asperellum	Beans	*S. sclerotiorum*	Geraldine et al. (2013)
Trichoderma asperellum	Onion	*Sclerotium cepivorum*	Rivera-Mendez et al. (2020)
Trichoderma asperellum T8a	Mango/anthracnose	*C. gleosporiodes*	Santos-Villalobos et al. (2013)
Trichoderma atroviridae	Beans	*Botrytis cinerea*	Brunner et al. (2005)
Trichoderma harzianum	Rice/brown spot	*Bipolaris oryzae*	Brunner et al. (2005)
Trichoderma harzianum T-22	Soya bean	*S. sclerotiorum*	Zeng et al. (2012a)
Trichoderma spp.	Tobacco/root rot	*Rhizoctonia solani*	Gveroska and Ziberoski (2011)
Trichoderma virens	Okra/root-knot disease	*Meloidogyne incognita*	Tariq et al. (2018)

Adapted from Tariq et al. (2020).

Entomopathogenic fungi are divided into the Ascomycota, Zygomycota, Deuteromycota, Oomycota, and Chytridiomycota divisions (Mora et al., 2017). However, the majority of these products are based on only the entomopathogenic fungal genera/species *Beauveria bassiana* and *Metharhizum.*

According to Brunner-Mendoza et al. (2019), entomopathogenic fungi produce conidia that adhere to the insect cuticle and undergo an enzymatic production process, primarily through the production of fungal protease. These enzymes, combined with a mechanical process through the development of appressoria, break the insect cuticle and penetrate the body insect to invade and kill the insect while obtaining nutrients from the insect tissues.

11.7 ACTINOMYCETES

Actinomycetes can be used as a biocontrol agent on phytopathogenic fungus as an alternative to applying synthetic fungicides in recent years (Boukaew et al., 2022). Actinomycetes are Gram-positive bacteria that can live in a range of environments with varying humidity, pH levels, and temperatures. Actinomycetes have been isolated from a variety of habitats, including wetlands, terrestrial, marine, and hypersaline conditions (Jagannathan et al., 2021).

Competence for space and nutrients, production of antibiotics, siderophores, lytic enzymes, volatile organic compounds (VOCs), and host resistance induction are the principal antagonistic methods used by actinomycetes to suppress phytopathogenic fungi. Actinomycetes also aid in the production of phytohormones, atmospheric nitrogen fixation, mineral solubilization, and other processes that aid in the growth and development of plants. Actinomycetes have been the subject of several research; however, due to their utility as biocontrol agents in agriculture and their antagonistic processes to phytopathogenic fungi, few of these studies have concentrated on actinomycetes isolated from different environments (Table 11.5).

TABLE 11.4
Commercially Available Fungal Bioformulations for Disease Management

Product	Organism	Target Pathogen/Disease
AQ10 Biofungicide	*Ampelomyces quisqualis*	Powdery mildew
Aspire	*Candida oleophila*	*Penicillium, Botrytis*
Binab-T	*Trichoderma harzianum, T. polysporll*	Take all, wilt, root rot
BiofoxC	*Fusarium oxysporium*	*F. oxysporium, F. moniliforme*
Bio-Fungus	*Trichoderma* spp.	*Sclerotinia, Rhizoctonia solani, Pythium, Phytophthora, Fusarium, Verticillium*
Contans	*Coniothyrium minitans*	*Sclerotinia sclerotiorum* and *S. minor*
DiTera	*Myrothecium verrucaria*	Root-knot, citrus cyst, stubby root, sting, lesion, and burrowing nematodes
Fusaclean	*Fusarium oxysporium* (non pathogenic)	*F. oxysporium*
Root stop	*Phlebia gigantean*	*Heterobasidion annosum*
Root shield	*T. harzianum* starin T-22	*Pythium, R. solani, Fusarium* spp.
Soil guard/GlioGard	*Gliocladium virens*	*Rhizoctonia solani, Pythium*
Supresivit	*T. harzianum*	Broad spectrum
T-22G 19, T-22 HB	*T. harzianum*and *T. viride*	*Fusarium, Rhizoctonia, Sclerotinia homeocarpa, Pythium*
Trichodex	*T. harzianum*	*Colletotrichum, Botrytis, Plasmopara viticola*
Trichopel, trichoject, Trichodowel, Trichoseal	*T. harzianum* and *T. viride*	Broad spectrum
Trichoderma 2000	*Trichoderma* sp.	*R. solani, S. rolfsii, Pythium*
Tricha	*T. harzianum*NBRI-1055	*R. solani, S. rolfsii, Pythium*
Sardar Eco, Green Biofungicide	*T. harzianum*NBRI-1055	*R. solani, S. rolfsii, Pythium*
Asperello T34 Biocontrol	*Trichoderma asperellum*	*Pythium, Phytophthora, Rhizoctonia,* and *Fusarium*
Afla-guard	*Aspergillus flavus*	Aflatoxin-producing fungi of corn and peanuts
Trichox	*Trichoderma harizanum*	*Rhizoctonia solani* and *Fusarium oxzsporium* f. sp. *Dianthi*
BIO-TAM 2.0	1–3% *Trichoderma asperellum* + 1–3% *Trichoderma gamsii*	*Pythium, Phytophthora, Rhizoctonia, Fusarium, Sclerotium rolfsii, Verticillium* spp., and *Thielaviopsis basicola*

11.8 ACTINOMYCETES AS BIOCONTROL AGENTS

The significance of using actinomycetes as biocontrol agents is explained by their inherent beneficial traits, which include the following: (1) they don't affect human or animal health, (2) they're not poisonous to plants, (3) they increase plant yield, and (4) they reduce the need for synthetic fungicides. Streptomyces has received the most research attention among the several genera since it is simple to isolate. Bacteria grow faster than actinomycetes. To acquire desired actinomycetes in culture media, growth enhancement approaches should be used. These methods are based on the pretreatment of samples using wet, chemical, and other pretreatment methods as well as selected isolation media and soil with calcium carbonate. By incorporating biostimulants and organic fertilizers like compost and vermicompost, one can encourage the populations of actinomycete in the soil. The maximum antagonistic activity was demonstrated by *S. sampsonii* and *S. flavovariabilis* isolates from soil enriched with vermicompost towards *Rhizoctonia solani, Alternaria tenuissima, Aspergillus*

TABLE 11.5
Registered Actinomycete Bioformulations for Disease Management

Product Name	Actinomycete	Phytopathogen/Targeted Disease	Registered Countries
Mycostop	*S. griseoviride*	*Alternaria* sp., *R. solani, Phytophthora, Botrytis, Fusarium* and *Pythium*	Canada, USA and UE countries
Actinovate	*S. lydicus*	*Pythium, Fusarium, Phytophthora, Verticillium, Botrytis, Alternaria, Sclerotinia, Geotrichum,* and powdery and downy mildew	USA and Canada
Mycocide KIBC	*S. colombiensis*	Grey mold, brown patch, and powdery mildews	South Korea
Kasugamycin (Kasumin)	*S. kasugaensis*	Root rot, scab, and leaf spot	Ukraine
Safegrow KIBC	*S. kasugaensis*	Sheath blight	South Korea
Agrimycin, Paushak, Cuprimicin 17, Astrepto 17	*S. griseus*	*Xanthomonas* and *Pseudomonas*	Indea, Canada, China, New Zealand, USA and Ukraine
Polyoxorim (Polyoxin Z), Endorse, and Spotil	*S. cacaoi* var. *asoensis*	*Alternaria, Sclerotium, Corynespora, Cochilobolus, Helminthosporium, Botrytis,* powdery mildew, and sheath blight	UE countries
Validamycin, Valium	*S. hygroscopicus*	*Rhizoctonia*	-

niger, and *Penicillium expansum*. Actinomycete populations in the soil were also increased by adding *Brassica napus* and *Brassica rapa* leaf residue to the soil. The decline in the *R. solani* wilt disease was strongly correlated with the rise in the actinomycete population (Ascencion et al., 2015).

11.9 MYCOVIRUS

Fungal viruses, also known as mycoviruses, are not known to kill or infect their hosts, but they are known to alter gene expression, morphology, growth, sporulation, pathogenicity, and interactions with other species. The possibility that mycoviruses could be a useful tool for biocontrol of fungi has sparked research into mycoviruses. Mycoviruses are known to slow down the development rate and/or virulence of their hosts.

11.10 MYCOVIRUSES AS BIOCONTROL AGENTS

The majority of plant diseases are brought on by fungi. Fungicide application is a typical technique for managing fungal phytopathogenic diseases. Continuous use of fungicides results in the development of fungicide-resistant strains, which has potential negative impacts on both humans and the environment. The use of biocontrol agents as an alternative to chemical control has grown significantly in relevance in recent years. Mycoviruses could also be employed as biocontrol agents (Table 11.6). Nuss (1992) was the first to demonstrate the use of a dsRNA virus as a biocontrol agent against chestnut blight. The production of a cDNA copy of the viral RNA associated with the hypovirus and its transformation into a virulent fungal strain turned the compatible, virulent strain into a hypovirulent strain. Chestnut fungus was successfully managed in Europe with the use of CHV-1 cDNA, but in North America, the fungus had a wider genetic range and was more difficult

TABLE 11.6
Details of Mycoviruses and Their Fungal Plant Hosts

Mycoviruses	Fungi	Fungal Host	Viral Genome Type
FgV1-DK21	*F. graminearum* Isolate DK21	Barley, maize, (Korea)	dsRNA (1 segment)
FgV2	*F. graminearum* Isolate 98-8-60	Barley (Korea)	dsRNA1, dsRNA2, dsRNA3, dsRNA4, dsRNA5
FgV3	*F. graminearum* DK3		dsRNA
FgV4	*F. graminearum* Isolate DK3		dsRNA1, dsRNA2
FgV-ch9	*F. graminearum* Isolate china9	Cereals (China)	dsRNA1, dsRNA2, dsRNA3, dsRNA4, dsRNA5
FpV1	*F. poae*	Wheat	dsRNA1, dsRNA2 (VLP)
FoV1	*F. oxysporum* f.sp. *melonis*		dsRNA1, dsRNA2, dsRNA3
FsV1	*F. solani* f. sp. *robiniae*		dsRNA1, dsRNA2
PdV1	*P. digitatum* strain HS-RH1	Citrus fruit	dsRNA
VdPV1	*Verticillium dahlia* Vd08284	Cotton, potato, soybean, tomato, and beet	dsRNA1, dsRNA2
NoNRV1	*Nigrospora oryzae*	Rice, maize, sorghum, cotton, weeds, Kentucky blue grass, brown mustard, aloe vera	Non-segmented RNA virus
SsHADV-1	*Sclerotinia sclerotiorum* (hypovirulent strain DT-8)	Brown mustard, aloe vera, rapeseed, soybean, sunflower, and numerous others	ssDNA

Modified from the paper Abid et al. (2018)

to manage (Nuss, 1992). There are barriers to the viruses' successful transmission when they are transmitted by hyphal contact. The beneficial infection of viruses that decreases the pathogenicity of plant pathogenic fungi is known as hypovirulence. The most prevalent mycovirus trait, hypovirulence, is utilized to biologically control a number of plant diseases, including *Cryphonecteria parasitica* (chestnut blight), woody plant white root rot, rice blast, and a variety of soilborne infections (Zhang and Nuss, 2016).

11.11 MECHANISM

The exact mechanism underlying hypopathogenicity is still unknown, although numerous theories have been offered by various researchers, including signal transduction pathways, RNA silencing of the fungus, and counter-silencing processes exploited by the hypovirus (Nuss, 2011). There are also other documented pathways for low virulence, including nuclear alterations, plasmids, and mutations of the mitochondria. The expansion of the host range of mycoviruses without the challenges of vegetative incompatibility is made possible by the production of cDNA infectious clones and other transformation techniques (Chen et al., 1994; Abid et al., 2018)

11.12 BACTERIOPHAGES

Viruses that infect and multiply in bacteria are known as phages. Temperate and lytic pathways are used in phage replication cycles, with the lytic pathway being the easier and more essential pathway for use in phage biocontrol. The challenges with antibiotic resistance, ineffective cultural practices, copper-based pesticides, and environmental chemical contamination have sparked global interest in Xanthomonas phage research and the use of biocontrol in agriculture (Nakayinga et al., 2021).

Bacteriophage-based biocontrol for the phytopathogen *Xanthomonas* has a long history, going all the way back to the early 19th century, when a filtrate of decaying cabbage prevented the spread of a disease known as *Xanthomonas campestris* pv. *campestris*. A few decades later, similar biocontrol efficacy was noted with phage-containing lysates that prevented *Xanthomonas campestris* pv. *pruni*-caused bacterial spot disease in peaches. Phage applications have advanced from in vitro studies to field testing in a number of instances. Against bacterial leaf blight of rice (*X. oryzae* pv. *oryzae*), leaf blight of onion (*X. axonopodis* pv. *allii*), bacterial spots of tomato (*X. campestris* pv. *vesicatoria*), citrus cankers (*X. axonopodis* pv. *citri*) and citrus bacterial spots (*X. axonopodis* pv. *citrumelo*), a number of bacteriophages were isolated and evaluated. It has been demonstrated that two Xanthomonas phage products made by AgriPhage are effective in managing the pathogens responsible for citrus cankers and tomato and pepper spot disease (Nakayinga et al., 2021).

11.13 APPLICATION OF MICROBIAL PESTICIDES IN PLANT DISEASE MANAGEMENT

The production of microbial pesticides, which are used to manage a variety of diseases in agricultural crops all over the world, involves the use of fungi, bacteria, actinomycetes, bacteriophages, mycoviruses, and nematodes (Yadav and Devi, 2017). They are applied by spraying, soaking, or directly applying them to the roots in the form of a tablet, coating the seeds or treating the roots before planting, as well as by various insects (Montesinos and Bonaterra, 2009)

A variety of microbial pesticide formulations are available, including wettable powders, suspensions, oil suspensions, water-dispersible granules, and suspension seed coatings (Table 11.7). However, microbial pesticide manufacturing is more difficult than chemical pesticide production (Lin et al., 2023). The microorganisms in pesticides are more susceptible to outside environmental conditions during preparation, such as UV radiation, temperature, and humidity.

11.14 FORMULATION

The term "formulation" describes the process of creating a product from an active ingredient by adding specific active (functional) and inert (inactive) substances.

11.15 PRINCIPLES OF FORMULATION

In order to maximize lifespan in storage, optimize application to the target, and protect the organisms after application, prepared organisms are suspended in a suitable carrier that is supplemented by additives. They differ from chemical active components in that they are particulate, living, or proteinous in nature, making them more susceptible to environmental factors and storage conditions.

There are two types of biopesticide formulations based on their physical states.

- Liquid formulations
- Dry formulations

Liquid formulations may be water, oil or polymer based or the combination of all. Water based formulations like suspension concentrate (SC), capsule suspension (CS), and suspo-emulsions (SE) require the addition of inert materials like surfactants, stickers, stabilizers, antifreezing agents, coloring agents, and nutritive materials.

TABLE 11.7
Types of Biopesticide Formulations and Their Application

Formulation Type and Example	Formulation Technique (Processing)	Application/Remarks
Liquid formulation i) Emulsifiable concentrate	Biopesticides are dissolved in a solvent with emulsifiers and surfactants to create ECs	Low cost, more stable, prepared by simple mixing
ii) Suspo-emulsions (SEs)	A combination of water-insoluble active substances dispersed in an aqueous solution, one (or more) of which is in suspension form and one (or more) of which is in emulsion form	Increases stability
iii) O/W emulsions (EWs)	In an EW formulation, a solid active is dissolved in a water-immiscible solvent, disseminated into a continuous water phase, and mixed at high shear. It has particles between 5 and 6 μm	Solvent free, no phytotoxic effects, and safer during transport and storage
iv) Suspension concentrates (SCs)	A finely ground solid particle that has a combination of wetting/dispersing, antifoaming, and thickening ingredients distributed in water. Sizes of the particles range from 1 to 10 μm	Easy to use and environmentally safe and secure
v) Oil dispersions (ODs)	The solid active components are disseminated in a non-aqueous liquid, typically oil. Usually it needs to be diluted before usage	Enhanced retention, spreading, and penetration
vi) Capsule suspensions (CSs)	A method that is reliable for encapsulating active components (organisms) in substances like gelatin or starch	Highly effective biopesticide formulation that is more stable in use due to the controlled release of active components
Dry formulation i) Dusts (DP)	Formulated by sorption of microbes onto a fine solid mineral carrier particle (talc, fly ash, clay) ranging from 50 to 100 μm	Directly applied either mechanically or manually
ii) Powders for seed dressing (DS)	For seed coating, the active ingredient (organism) is combined with a powdered carrier material and various adjuvants	It is applied to seeds by tumbling them in a prepared microbe mixed powder formulation
iii) Granules (GRs) and microgranules (MGs)	It is similar to dust, but they are larger in size than dust formulations (granules 100–1000 μ; microgranules 100– 600 μ)	Slow release, easy to manufacture; mostly applied to control nematodes, weeds, and insect pests living in soil
iv) Water dispersible granules (WGs)	It is formulated by using various techniques like extrusion granulation, fluid bed granulation, spray drying. It is developed to solve the problems of dustiness of powder formulation	Dust free and good storage stability with better convenience in application
v) Wettable powders (WPs)	A dry, finely ground formulation that needs to be mixed with water before use and has a particle size of about 5 microns	Long storage stability; good miscibility with water; easy to apply using conventional spraying equipment

Dry formulations like dusts (DP), granules (Gs), micro granules (MGs), wettable powders (WPs), and water dispersible granules (WGs) are prepared by adding wetting agents, binding agents, and dispersants using technologies like spray drying, air drying, or freeze drying (Tadros, 2005; Brar et al., 2006; Knowles, 2008).

Microbial pesticides are applied through various delivery methods based on the type of formulation. The major delivery methods followed in microbial pesticide formulations are presented in Table 11.7.

TABLE 11.8
Different Types of Adjuvants Used in Microbial Pesticide Formulation

Adjuvant type	Example
Carrier	Vegetable oil
*Liquid carriers	Kaolinite clay, diatomaceous earth
*Mineral carriers	Grain flours
*Organic carriers	
Surfactants	Triton X-100, tween-80, sodium dodecyl sulfate, tergitol NP10, brij35, etc.
*Synthetic surfactants	
*Biosurfactants	Glycoproteins, glycolipids, glycopeptides, glycosides, peptidoglycan, and lipopolysaccharides
Dispersants	Microcrystalline cellulose
UV protectants	Oxybenzone, milk, rubus oil, hemp seed oil, kojic acid, etc.
Stabilizer	Lactose, sodium benzoate
Nutritional adjuvant	Molasses, peptone
Binders	Gum arabic, carboxy methyl cellulose (CMC)
Desiccants	Silica gel, anhydrous salts
Thickeners	Xanthan gum
Stickers	Pregelatinized corn flour
Emulsifier	Calcium alkyl dodecyl benzene sulfonate, sodium dodecyl benzene sulfonate, castor oil with 20e60 Eo, alkyl phenols with 3e6 ethylene oxide Eo, Alkyl phenols with 8e30 Eo

11.16 CLASSIFICATIONS AND FUNCTIONS OF ADJUVANTS USED FOR MICROBIAL PESTICIDES

The four basic categories of adjuvants employed in the production of microbial pesticides are surfactants, transporters, protective agents, and nutritional adjuvants (Table 11.8).

11.17 CARRIERS

The basic function of a carrier is to serve as a tiny container or diluent for the active components of microbial pesticides. High-concentration powders, wettable powders, or granules can be produced using carriers with strong adsorption capabilities, such as diatomite, attapulgite, silica, and bentonite. To create low-concentration powders, diluents and fillers such as talc, pyrophyllite, sepiolite, and clay materials with low or moderate adsorption capabilities are typically utilized. Biodegradable and environmentally safe materials such as chitosan, carbon nanotube nanocomposite, biodiesel, and plant-based substances like bagasse, corn bagasse, chaff powder, tobacco powder, and walnut shell powder are used as carriers. Though plant-based carriers have the property of absorbing ultraviolet rays, their use in commercial production is very rare.

11.18 SURFACTANTS

Surfactants are crucial in preserving the physical stability of pesticides over the long term and enhancing their biological capabilities. Generally, microbial pesticides comprise one or more surfactants. This surfactant helps to increase cost effectiveness, save energy and raw materials, and increase processing efficiency.

11.19 PROTECTIVE AGENTS

11.19.1 UV Protection Agents

UV ray absorbers and anti-oxidative UV protection agents are the two categories of UV protection agents utilized in the manufacturing of microbial pesticides. Milk, optical brighteners, lignosulfonates, and locust toxin are examples of UV ray absorbers that may absorb the UV portion of sunlight and fluorescent light sources without transforming themselves. Rubus oil, hemp seed oil, kojic acid, hydroxykynurenic acid, flavonoids, and lecithin are anti-oxidative UV protection agents which have strong antioxidative effects.

11.20 NUTRITIONAL ADJUVANTS

Microorganisms require water, inorganic salts, carbon sources, nitrogen sources, and growth nutrients in order to function normally throughout the growth phase. Nutritional adjuvants can give nutrition to the microorganisms that make up microbial pesticides so they can reproduce more easily and grow and proliferate more quickly in the field.

11.21 CHALLENGES AND OPPORTUNITIES FOR COMMERCIALIZATION OF MICROBIAL PESTICIDES

Synthetic chemical pesticides were formerly the mainstay of crop protection, but their use is now declining due to new rules and restrictions, as well as the evolution of pathogen resistance. As a result, a new pest management strategy is required. A microbial pesticide is an ideal solution for synthetic chemical pesticides based on natural or living microorganisms. Microbial pesticides have acquired importance as an essential part of IPM due to their economic feasibility and sustainability in comparison to chemical synthetic pesticides (Birch, 2011). Microbial pesticides are able to overcome the limitations of conventional fungicides as part of IPM programs since they are host-specific, biodegradable, and less toxic to beneficial microbes (Matyjaszczyk, 2015).

The effectiveness of microbial pesticides in combating plant diseases depends on the microbial agent (mechanism of action, conditioning, dose, methods of application), the host (cultivar type, physical characteristics), the targeted plant pathogens (sensitivity), and environmental factors (biotic and abiotic factors, chemical residues, nutrient availability, temperature, and moisture). Variable factors like location, time period, host plants, and their interactions may alter the effectiveness of microbial agents. Lack of efficacy and varied field performance have frequently been documented. The effectiveness and consistency of biological control must therefore be compared to those of conventional chemical fungicide and bactericide treatments under sufficiently broad production settings in fields representing various environments and agricultural practices (Montesinos 2003; Köhl et al. 2011; Bonaterra et al. 2022).

The demand for safer plant disease control products leads to a preference for microbial biopesticide formulations with effective antagonistic activity and good stability. Biopesticides offer environmentally benign alternatives to synthetic pesticides, but they encounter a number of challenges during their development, formulation, and use. Maximum attention is required to maintain the microbial population and efficacy throughout the entire use of biopesticides because they typically contain living organisms. The commercialization of biopesticides could benefit greatly from research into their manufacture and formulation. As their application methods advance and less expensive inert components are acknowledged for various formulations, it looks like biopesticides will be used more widely in the future.

The introduction of new adjuvants demonstrated a notable increase in microbial activity and suggested a new avenue of research. The stability of the product, its longer shelf life, and the performance of the microorganisms in the field may all be improved by choosing the right formulation.

In the future, formulation products should have a better balance between production cost and efficiency, as is the case with biopesticides, which provide a more balanced plant protection product application (Glare et al. (2012)). The formulation type could evolve from dusts to granules, from suspension concentrates and wettable powders to water-dispersible granules, and from products based on a single bacterium to formulations based on microbial consortia.

11.22 CONCLUSION

The fact that components of microbial pesticides are widely regarded as safe and are target specific means that they generally have no negative effects on the environment and human beings. Compared to chemical pesticides, the use of them produces fewer greenhouse gas emissions. Moreover, a wide range of organisms can be used to develop microbial biopesticides to tackle the issue of resistance and guarantee sustainability. It may be difficult for producers, marketers, and end users to manage their storage and transportation because different microorganisms employed as biopesticides may require varied storage conditions. Therefore, more study is required to ensure that microbial pesticides have a sustainable and extended shelf life.

REFERENCES

Abid M., Khan M., Mushtaq S., Afzaal S. and Haider, M. (2018) A comprehensive review on mycoviruses as biological control agent. *World J. Biol. Biotechnol.*, 3(2): 187–192. https://doi.org/10.33865/wjb.003.02.0146

Adebola M.O. and Amadi J.E. (2010) Screening three *Aspergillus* species for antagonistic activities against the cocoa blacknpod organism (*Phytophthora palmivora*). *Agric. Biol. J. North Am.,* 1: 362–365.

Al-Ani R.A., Adhab M.A., Mahdi M.H. and Abood H.M. (2012) *Rhizobium japonicum* as a biocontrol agent of soybean root rot disease caused by *Fusarium solani* and *Macrophomina phaseolina. Plant Prot. Sci.*, 48:149–155.

Alam S.S., Sakamoto K., Amemiya Y. and Inubushi K. (2010) Biocontrol of soil-borne Fusarium wilts of tomato and cabbage with a root colonizing fungus, *Penicillium* sp. EU0013. *Proc. 19th World Congress of Soil Science, Soil Solutions for a Changing World.* August; Brisbane: Australia, pp. 1–6.

Ascencion L.C., Liang W.J. and Yen T.B. (2015) Control of *Rhizoctonia solani* damping-off disease after soil amendment with dry tissues of *Brassica* results from increase in Actinomycetes population. *Biol. Control* 82: 21–30.

Ashajyothia M., Kumara A., Sheorana N., Ganesana P., Gogoia R., Subbaiyanb G.K. and Bhattacharya R. (2020) Black pepper (*Piper nigrum* L.) associated endophytic *Pseudomonas putida* BP25 alters root phenotype and induces defense in rice (*Oryza sativa* L.) against blast disease incited by *Magnaporthe oryzae*. *Biol Control.*, 143: 104181.

Barnett S.J., Roget D.K. and Ryder M.H. (2006) Suppression of *Rhizoctonia solani* AG-8 induced disease on wheat by the interaction between *Pantoea, Exiguobacterium* and *Microbacteria. Soil Res.*, 44: 331–342.

Birch A.N.E. (2011) How agro-ecological research helps to address food security issues under new IPM and pesticide reduction policies for global crop production systems. *J. Exp. Bot.*, 62: 3251–3261.

Bonaterra A., Badosa E., Daranas N., Francés J., Roselló G. and Montesinos E. (2022) Bacteria as Biological Control Agents of Plant Diseases. *Microorganisms* 10: 1759. https://doi.org/10.3390/microorganisms10091759

Bontempo A.F., Lopes E.A., Fernandes R.H., De Freitas L.G. and Dallemole-Giaretta R. (2017) Dose-response effect of *Pochonia chlamydosporia* against *Meloidogyne incognita* on carrot under field conditions. *Revista Caatinga, Mossoró.*, 30(1): 258–262.

Boukaew S., Yossan S., Cheirsilp B. and Prasertsan P. (2022) Impact of environmental factors on *Streptomyces* spp. metabolites against *Botrytis cinerea. J. Basic Microbiol.*, 62, 611–622.

Brar, S.K., Tyagi V.R.D. and Valero J.R. (2006) Recent advances in downstream processes and formulations of *Bacillus thuringiensis* based biopesticide. *Process Biochem.*, 41: 323–342.

Brunner K., Zeilinger S., Ciliento R., Woo S.L., Lorito M., Kubicek C.P. and Mach R.L. (2005) Improvement of the fungal biocontrol agent Trichoderma atroviride to enhance both antagonism and induction of plant systemic disease resistance. *Appl Environ Microbiol.*, 71: 3959–3965.

Brunner-Mendoza C., Reyes-Montes M.R., Moonjely S., Bidochka M.J. and Toriello C. (2019) A review on the genus *Metarhizium* as an entomopathogenic microbial biocontrol agent with emphasis on its use and utility in Mexico. *Biocontrol. Sci. Tech.*, 29: 83102.

Chandler D., Bailey A.S., Tatchell G.M., Davidson G. and Greaves J. et al. (2011) The development, regulation and use of biopesticides for integrated pest management. *Philosophical Trans. Royal Society B: Biol. Sci.*, 366: 1987–1998. DOI: 10.1098/rstb.2010.0390

Chauhan S., Wadhwa K., Vasudeva M. and Narula N. (2012) Potential of *Azotobacter* spp. As biocontrol agents against *Rhizoctonia solani* and *Fusarium oxysporum* in cotton (*Gossypium hirsutum*), guar (*Cyamopsis tetragonoloba*) and tomato (*Lycopersicum esculentum*). *Arch. Agron. Soil Sci.*, 58: 1365–1385.

Chen B., Choi G.H. and Nuss D.L. (1994) Attenuation of fungal virulence by synthetic infectious hypovirus transcripts. *Science*, 264(5166): 1762–1764.

Cheng X., Ji X., Li J., Qi W., Ge Y. and Qiao K. (2019) Characterization of antagonistic *Bacillus methylotrophicus* isolated from rhizosphere and its biocontrol effects on maize stalk rot. *Phytopathol.*, 109: 571–581.

Damalas C.A. and Koutroubas S.D. (2018) Current status and recent developments in biopesticide use. *Agriculture*, 8: 1–6.

Dara S.K. (2017) Entomopathogenic microorganisms: Modes of action and role in intregrated pest management. *UC ANR e-Journal of Entomology and Biologicals*, 20 May 2017. https://ucanr.edu/blogs/blogcore/postdetail.cfm?postnum 5 24119.

Dasgupta M.K. and Khan M.K. (2015) Nematophagous Fungi: Ecology, diversity and geographical distribution. In: Askary T.H. and Martinelli P.R.P. (eds.), *Biocontrol Agents of Phytonematodes*. CAB International, p. 126.Geraldine A.M., Lopes F.A.C., Carvalho D.D.C., Barbosa E.T., Rodrigues A.F., Brandao R.S., Ulhoa C.J. and Junior M.L. (2013) Cell wall-degrading enzymes and parasitism of sclerotia are key factors on field biocontrol of white mold by *Trichoderma* spp. *Biol Control.*, 67: 308–316.

Ghewande M.P. and Nandagopal V. (1997) Integrated pest management in groundnut (*Arachishypogaea* L.) in India. *Integr. Pest Manag. Rev.*, 2: 1–15. https://doi.org/10.1023/a:1018488326980

Glare T., Caradus J., Gelernter W., Jackson T., Keyhani N., Köhl J., et al. (2012) Have biopesticides come of age? *Trends Biotechnol.*, 30: 250–258. https://doi.org/10.1016/j.tibtech.2012.01.003

Gunasinghe W.K.R.N. and Karunaratne A.M. (2009) Interactions of *Colletotrichum musae* and *Lasiodiplodia theobromae* and their biocontrol by *Pantoea agglomerans* and *Flavobacterium* sp. in expression of crown rot of 'Embul' banana. *Biocontrol.*, 54: 587–596.

Gveroska B. and Ziberoski J. (2011) The influence of *Trichoderma harzianum* on reducing root rot disease in tobacco seedlings caused by *Rhizoctonia solani*. *Int. J. Pure Appl. Sc. Technol.*, 2: 1–11.

Haggag Wafaa M. (2008) Isolation of bioactive antibiotic peptides from *Bacillus brevis* and *Bacillus polymyxa* against Botrytis grey mould in strawberry. *Arch. J. Phytopathol. Plant Prot.*, 41: 477–491.

Hanawi M.J. (2016) Tagetes erecta with native isolates of *Paecilomyces lilacinus* and *Trichoderma hamatum* in controlling root-knot nematode *Meloidogyne javanica* on tomato. *Int. J. Appl. or Innovation Eng. Manage.*, 5(1): 81–88.

Hu X., Roberts D.P., Xie L., Maul J.E., Yu C., Li Y., Jiang M., Liao X., Che Z. and Liao X. (2014) Formulation of *Bacillus subtilis* by BY-2 suppresses *Sclerotinia sclerotiorum* on oilseed rape in the field. *Biol. Control.*, 70: 54–64.

Jagannathan S.V., Manemann E.M., Rowe S.E., Callender M.C. and Soto W. (2021) Marine actinomycetes, new sources of biotechnological products. *Mar. Drugs*, 19: 365.

Jones E.E., Rabeendran N. and Stewart A. (2014) Biocontrol of *Sclerotinia sclerotiorum* infection of cabbage by *Coniothyrium minitans* and *Trichoderma* spp. *Biocontrol Sci. Technol.*, 24: 1363–1382.

Kakembo D. and Lee Y.H. (2019) Analysis of traits for biocontrol performance of *Pseudomonas parafulva* JBCS1880 against bacterial pustule in soybean plants. *Biol. Control.*, 134: 72–81.

Kalia V.C., Patel S.K.S., Kang Y.C. and Lee J.K. (2019) Quorum sensing inhibitors as antipathogens: Biotechnological applications. *Biotechnol. Adv.*, 37: 68–90.

Khan A., Tariq M., Asif M., Khan F., Ansari T. and Siddiqui M.A. (2019) Integrated management of *Meloidogyne incognita* infecting *Vigna radiata* L. using biocontrol agent *Purpureocillium lilacinum*. *Trends Appl. Sci. Res.*, 14: 119–124.

Khatun A., Farhana T., Sabir A.A., Smn I., West H.M., Rahman M. and Islam T. (2018) *Pseudomonas* and *Burkholderia* inhibit growth and asexual development of *Phytophthora capsici*. *Zeitschrift für Naturforschung.*, 73: 123–135.

Kiriga A.W., Haukeland S., Kariuki G.M., Coyne D.L. and Beek N.V. (2018) Effect of *Trichoderma* Spp. and *Purpureocillium lilacinum* on *Meloidogyne javanica* in commercial Pineapple production in Kenya. *Biol. Control.*, 119: 27–32.

Knowles A. (2008) Recent developments of safer formulations of agrochemicals. *Environmentalist*, 28(1):35–44. https://doi.org/10.1007/s10669-007-9045-4

Köhl J., Kolnaar R. and Ravensberg W.J. (2019) Mode of action of microbial biological control agents against plant diseases: Relevance beyond efficacy. *Front. Plant Sci.*, 10: 845.

Köhl J., Postma J., Nicot P., Ruocco M. and Blum B. (2011) Stepwise screening of microorganisms for commercial use in biological control of plant-pathogenic fungi and bacteria. *Biol. Control.*, 57: 1–12.

Kumari S., Kumar S.C., Jha M.N., Kant R., Upendra S. and Kumar P. (2014) Microbial pesticide: A boom for sustainable agriculture. *Int. J. Sci. Eng. Res.*, 5(6): 1394–1397.

La Y., Feng X., Wang X., Zheng L. and Liu H. (2020) Inhibitory effects of *Bacillus licheniformis* BL06 on *Phytophthora capsici* in pepper by multiple modes of action. *Biol. Control.*, 144: 104210.

Lin F., Mao Y., Zhao F., Idris A.L., Liu Q., Zou S., Guan X. and Huang, T. (2023) Towards sustainable green adjuvants for microbial pesticides: Recent progress, upcoming challenges and future perspectives. *Microorganisms*, 11: 364. https://doi.org/10.3390/microorganisms11020364

Lindsey A.P.J., Murugan S. and Renitta R.E. (2020) Biocatalysis and agricultural biotechnology microbial disease management in agriculture: Current status and future prospects. *Biocatalysis Agric. Biotechnol.*, 23: 1–12.

Maksimov I.V., Blagova D.K., Veselova S.V., Sorokan A.V., Burkhanova G.F. and Cherepanova E.A. (2020) Recombinant *Bacillus subtilis* 26DCryChS line with gene Btcry1Ia encoding Cry1Ia toxin from *Bacillus thuringiensis* promotes integrated wheat defense against pathogen *Stagonospora nodorum* Berk and greenbug *Schizaphis graminum* Rond. *Biol. Control.*, 144: 104242.

Matyjaszczyk E. (2015) Products containing microorganisms as a tool in integrated pest management and the rules of their market placement in the European Union. *Pest Manag. Sci.*, 71: 1201–1206.

Mazid S., Kalita J.C. and Rajkhowa R.C. (2011) A review on the use of biopesticides in insect pest management. *Int. J. Sci. Adv. Technol.*, 1: 169–178.

Mishra J., Tewari S., Singh S. and Arora N.K. (2015) Biopesticides: Where we stand? In: N.K. Arora (ed.), *Plant Microbes Symbiosis: Applied Facets*. Springer, pp. 37–75.

Mohammadi P., Tozlu E., Kotan R. and Kotan M.S. (2017) Potential of some bacteria for biological control of postharvest citrus green mold caused by *Penicillum digitatum*. *Plant Prot. Sci.*, 53: 1–14.

Montesinos E. (2003) Development, registration and commercialization of microbial pesticides for plant protection. *Int. Microbiol.*, 6: 245–252.

Montesinos E. and Bonaterra A. (2009) Microbial pesticides. In: Schaechter M. (ed.), *Encyclopedia of Microbiology*, 3rd ed. Elsevier, pp. 110–120.

Montesinos E., Francés J., Badosa E. and Bonaterra A. (2015) Post harvest control. In: Ben Lugtenberg (ed.), *Principles of Plant-Microbe Interactions: Microbes for Sustainable Agriculture*, Springer, Germany, Berlin, pp. 193–202.

Mora M.A.E., Conteiro A.M., Castillo C. and Fraga M.E. (2017) Classification and infection mechanism of entomopathogenic fungi. *Arq. Inst. Biol.*, 84: 110. https://doi.org/10.1590/1808-1657000552015

Nakayinga R., Makumi A., Tumuhaise V. and Tinzaara W. (2021) *Xanthomonas* bacteriophages: a review of their biology and biocontrol applications in agriculture *BMC Microbiol.*, 21: 291 https://doi.org/10.1186/s12866-021-02351-7

Nawaz M., Mabubu J.I. and Hua H. (2016) Current status and advancement of biopesticides: Microbial and botanical pesticides. *J. Entomol. Zool. Stud.*, 4: 241–246.

Nielsen A.L., Spence K.O. and Lewis, E.E. (2008) Efficacy patterns of biopesticides used in potting media. *Biopesticide Int.*, 4: 87–101.

Nuss D.L. (1992) Biological control of chestnut blight: An example of virus-mediated attenuation of fungal pathogenesis. *Microbiol. Rev.*, 56(4): 561–576.

Nuss D.L. (2011) Mycoviruses, RNA silencing, and viral RNArecombination. In: *Advances in Virus Research*. Elsevier, pp. 25–48.

Pal K.K. and Gardener B.M. (2006) Biological control of plant pathogens. *Plant Health Instr.*, 1–25. https://doi.org/10.1094/PHI-A-2006-1117-02.

Pathak D., Yadav R. and Kumar M. (2017) Microbial pesticides: development, prospects and popularization in India. In: Singh D.P. et al. (eds.), *Plant Microbe Interactions in Agro-Ecological Perspectives*. Springer, pp. 455–471.Prabha S., Yadav Ashwani A. Kumar and Yadav et al. (2016) Biopesticides—an alternative and eco-friendly source for the control of pests in agricultural crops. *Plant Archives*, 16: 902–906Rivera-Méndeza W., Obregónc M., Morán-Diezb M.E., Hermosab R. and Monteb E. (2020) *Trichoderma asperellum* biocontrol activity and induction of systemic defenses against *Sclerotium cepivorum* in onion plants under tropical climate conditions. *Biol. Control.*, 141: 104145.

Sabuquillo P., Cal A.D. and Melgarejo P. (2006) Biocontrol of tomato wilt by *Penicillium oxalicum* formulations in different cropconditions. *Biol. Control.*, 37: 256–265.

Santos-Villalobos S.D.L., Guzmán-Ortiz D.A., Gómez-Lim M.A., Délano-Frier J.P., de Folter S., Sánchez-García P. and Peña-Cabriales J.J. (2013) Potential use of *Trichoderma asperellum* (Samuels, Liechfeldt et Nirenberg) T8a as a biological control agent against anthracnose in mango (*Mangifera indica* L.). *Biol. Control.*, 64: 37–44.

Savchuk S. and Fernando W.D.D. (2004) Effect of timing of application and population dynamics on the degree of biological control of *Sclerotinia sclerotiorum* by bacterial antagonists. *FEMS Microbiol. Ecol.*, 49: 379–388.

Schoina C., Stringlis I.A., Pantelides I.S., Tjamos S.E. and Paplomatas E.J. (2011) Evaluation of application methods and biocontrol efûcacy of *Paenibacillus alvei* strain K-165, against the cotton black root rot pathogen *Thielaviopsis basicola. Biol. Control.*, 58: 68–73.

Senthilkumar M., Anandham R. and Krishnamoorthy R. (2020) Chapter 41—Paecilomyces. In: Amaresan N., Senthil Kumar M., Annapurna K. and Sankaranarayanan K.K.A. (eds.), *Beneficial Microbes in Agro-Ecology*. Academic Press, pp. 793–808.

Singh H.B. (2022) Current status and recent developments in microbial pesticide use in India. *Indian J. Plant Genet. Resour.*, 35(3): 369–374. https://doi.org/10.5958/0976-1926.2022.00102.4

Sowndhararajan K., Marimuthu S. and Manian S. (2013) Integrated control of blister blight disease in tea using the biocontrol agent *Ochrobactrum anthropi* strain BMO-111 with chemical fungicides. *J. Appl. Microbiol. PMID.*, 114(5): 1491–1499.

Szczech M. and Shoda M. (2004) Biocontrol of Rhizoctonia dampingoff of tomato by *Bacillus subtilis* combined with *Burkholderia cepacia. J. Phytopathol.,* 152: 549–556.

Tadros F. (2005) *Applied Surfactants, Principles, and Applications*. Wiley-VCH Verlag GmbH and Co. KGaA, pp. 187–256.

Tariq M., Khan A., Asif M., Khan F., Ansari T., Shariq M. and Siddiqui M.A. (2020) Biological control: A sustainable and practical approach for plant disease management. *Acta Agric. Scand.-B Soil Plant Sci.*, 70(6): 507–524. https://doi.org/10.1080/09064710.2020.1784262

Tariq M., Khan A., Asif M. and Siddiqui M.A. (2018) Interactive effect of *Trichoderma virens* and *Meloidogyne incognita* and their influence on plant growth character and nematode multiplication on *Abelmoschus esculentus* (L.) Moench. *Current Nematol.*, 29: 1–9.

Thakur N., Kaur S., Tomar P, Thakur S. and Yadav A.N. (2020) Chapter 15—Microbial biopesticides: Current status and advancement for sustainable agriculture and environment. In: Rastegari A. A., Yadav A. N. and Yadav N. (eds.), *New and Future Developments in Microbial Biotechnology and Bioengineering*. Elsevier, pp. 243–282.

Tortora M.L., Dýaz-Ricci J.C. and Pedraza R.O. (2011) *Azospirillum brasilense* siderophores with antifungal activity against *Colletotrichum acutatum. Arch. Microbiol.*, 193: 275–286.

Tripathi Y.N., Divyanshu K., Kumar S., Jaiswal L.K. and Khan A. et al. (2020) Biopesticides: Current status and future prospects in India. In: Keswani C. (ed.), *Bioeconomy for Sustainable Development*. Springer, pp: 79–109. ISBN-13: 978-981-13-9431-7

Vosátka M., Látr A., Gianinazzi S. and Albrechtová J. (2012) Development of arbuscular mycorrhizal biotechnology and industry: Current achievements and bottlenecks. *Symbiosis*,58: 29–37. https://doi.org/10.1007/s13199-012-0208-9

Wallace R.L., Hirkala D.L. and Nelson L.M. (2018) Efficacy of *Pseudomonas fluorescens* for control of Mucor rot of apple during commercial storage and potential modes of action. *Can. J. Microbiol.*, 64: 420–431.

Wang M., Geng L., Sun X., Shu C., Song F. and Zhang J. (2020) Screening of *Bacillus thuringiensis* strains to identify new potential biocontrol agents against *Sclerotinia sclerotiorum* and *Plutella xylostella* in *Brassica campestris* L. *Biol. Control.*, 145: 1–9.

Whipps J.M. (2001) Microbial interactions and biocontrol in the rhizosphere. *J. Exp. Bot.*, 52: 487–511.

Yadav I.C. and Devi N.L. (2017) Pesticides classification and its impact on human and environment. *Environ Sci Engg* 6: Toxicology.

Zeng W., Kirk W. and Hao J. (2012) Field management of Sclerotinia stem rot of soybean using biological control agents. *Biol. Control.*, 60: 141–147.

Zhang D.X. and Nuss D. L. (2016) Engineering super mycovirus donor strains of chestnut blight fungus by systematic disruption of multilocusvic genes. *Proc. Nat. Aca. Sci*: 201522219.

Zhou K., Yamagishi M. and Osaki M. (2008) Biocontrol of brown stem rot disease in soybean *Paenibacillus* BRF-1 has biocontrol ability against *Phialophora gregata* disease and promotes soybean growth. *Soil Sci. Plant Nutr.,* 54: 870–875.

12 Phenylpropanoids as Potential Natural Pesticides

Karyme do Socorro de Souza Vilhena, Fábio José Bonfim Cardoso, Leonardo Souza da Costa, Anderson de Santana Botelho, Tatiani da Luz Silva Vasconcelos, Oberdan Oliveira Ferreira, Marcilene Paiva da Silva, Mozaniel Santana de Oliveira, and Eloisa Helena de Aguiar Andrade

12.1 INTRODUCTION

Phenylpropanoids are secondary metabolites that form a class of compounds whose structure consists of a phenyl group bonded to a three-carbon atom chain (C_6–C_3). This basic nine-carbon atom unit (C_6–C_3) originates from the enzyme-catalyzed deamination of two aromatic amino acids: L-phenylalanine, and L-tyrosine (Figure 12.1). In terrestrial plants, the first step in the formation of phenylpropanoids involves the removal of the side chain ammonia group, catalyzed by L-phenylalanine ammonia lyase (PAL) to yield cinnamic acid (E or *trans*) if L-phenylalanine is the precursor, or catalyzed by L-tyrosine ammonia lyase (TAL) to yield *p*-coumaric acid if L-tyrosine is the precursor (Vanholme, El Houari, and Boerjan 2019; Dewick 2009; de Vries et al. 2021).

Plant foods, including fruit, coffee, vegetables and cereal grains (for example, potatoes, wheat, rye, oats, barley and rice bran, and mushrooms), and wines are among the main phenylpropanoid sources. Cinnamic acid is the main active compound found in species of the *Cinnamomum* spp. genus, commonly known as cinnamon. Other plant species that produce cinnamic acid include coriander (*Coriandrum sativum* L.), clove (*Eugenia caryophyllata* Thunb.), black pepper (*Piper nigrum* L.), and turmeric (*Curcuma longa* L.). (Neelam, Khatkar, and Sharma 2020).

Phenylpropanoids are economically important metabolites because they are constituents of the human diet, and act as nutraceuticals with antioxidant, chemopreventive, antimycotic, neuroprotective, cardioprotective, and anti-inflammatory activity. (Cesarino et al. 2022). Moreover, the development of antimicrobial drug resistance and the spread of bacteria have become sources of morbidity and mortality worldwide. This compels the development of new, more effective, and safer antimicrobial agents, capable of mitigating the widespread drug resistance. Thus, natural products, including secondary metabolites of medicinal plants, can serve as a foundation for the design of new drugs (Dagne et al. 2023).

The biological activity of essential oils and their main components have been widely described in scientific literature. For example, both cinnamon essential oil and its main component, cinnamaldehyde (a phenylpropanoid), exhibit antifungal activity. The essential oil was able to inhibit the growth of the fungi *Aspergillus niger* and *Alternaria alternata* at a concentration of 66 µg/mL. Cinnamaldehyde also inhibited the growth of both microorganisms at concentrations of 75 and 41 µg/mL, respectively, as well as of two other fungi, *Aspergillus ochraceus* and *Fusarium oxysporum*, at a concentration of 58 µg/mL. Both cinnamon essential oil and cinnamaldehyde exhibited significant antiparasitic activity against *Trypanosoma* (*T.*) *cruzi* and *Leishmania* (*L.*) *mexicana*. Cinnamon essential oil exhibited IC_{50} values of 23 µg/mL and 21 µg/mL against *T. cruzi* and *L. mexicana*, respectively. Whereas cinnamaldehyde showed activity at 10.4 (*T. cruzi*) and 8.6 µg/mL

DOI: 10.1201/9781003463429-16

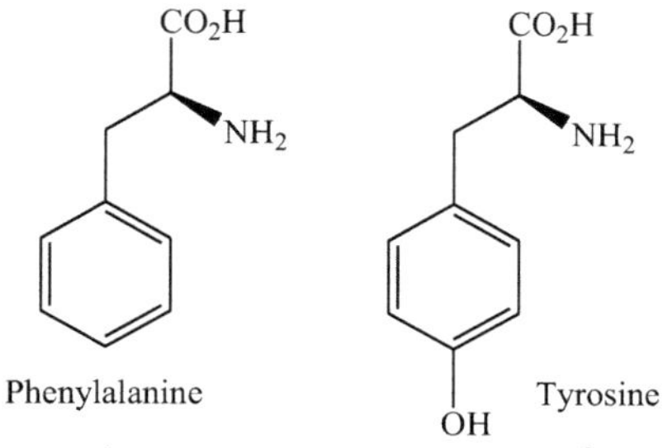

CO_2H NH_2 Phenylalanine 1

CO_2H NH_2 Tyrosine OH 2

FIGURE 12.1 Molecular structures of aromatic amino acids L-phenylalanine (**1**) and L-tyrosine (**2**). Structure prepared by the author.

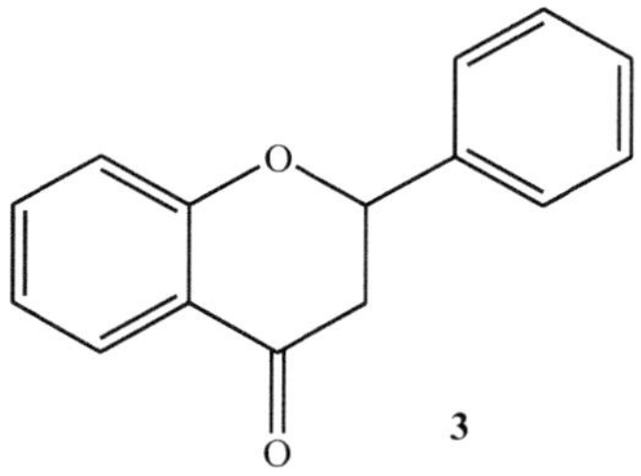

O O 3

FIGURE 12.2 Basic structure of a benzopyrone. Structure elaborated by the author.

(*L. mexicana*), respectively. These results indicate different biological activities of essential oils and their constituents. (Andrade-Ochoa et al. 2021).

Phenylpropanoids also stand out for their plant protection capability. Pathogens interacting with plant cell walls and penetrating the cell wall trigger a series of plant defense reactions in the phenylpropanoid pathway, which bifurcates into the production of a vast array of compounds, including phenylpropanoids and more complex derivatives (flavonoids, lignins, monolignols, stilbenes, and coumarins), based on intermediates of the shikimate pathway. The whole metabolomic pathway constitutes a network regulated by multiple gene families, with regulatory mechanisms at the transcriptional, post-transcriptional, and post-translational levels. The pathway genes are involved in the production of antimicrobial compounds and signaling molecules (Yadav et al. 2020). Given the biochemical importance of this class of compounds, we discuss the main applications of phenylpropanoids as natural pesticides, highlighting their classification, biosynthesis, and their action in protecting plants.

12.2 CLASSIFICATION OF PHENYLPROPANOIDS

Metabolites formed in the phenylpropanoid metabolic pathway can be classified into flavonoids, monolignols, lignins, phenolic acids, stilbenes, or coumarins, based upon their chemical structure (Siebeneichler et al. 2024). The majority of these phenolic compounds are synthesized by the shikimate (shikimic acid) pathway in plants (Metsämuuronen and Sirén 2019).

Flavonoids are polyphenolic compounds with a characteristic central core consisting of a benzopyrone (Figure 12.2, **3**) bonded to a benzene ring (Liu and Ding 2019; Ullah et al. 2020). These compounds are abundant in nature, with many variations in the location of the aromatic rings in the central core, and in the degrees of hydroxylation, oxidation, and saturation of the heterocyclic ring (Shamsudin et al. 2022).

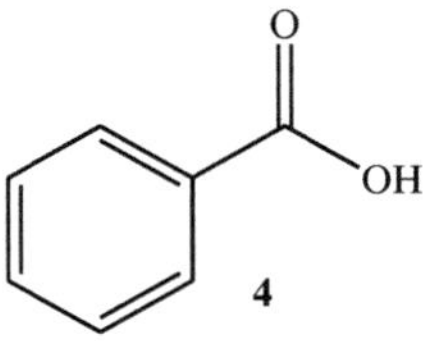

FIGURE 12.3 Basic structure of acids derived from benzoic acid (**4**) and cinnamic acid (**5**). Structure prepared by the author.

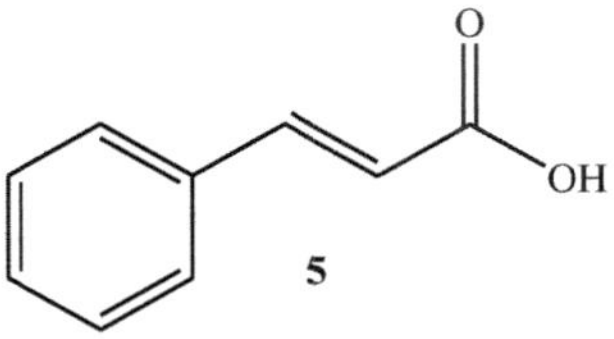

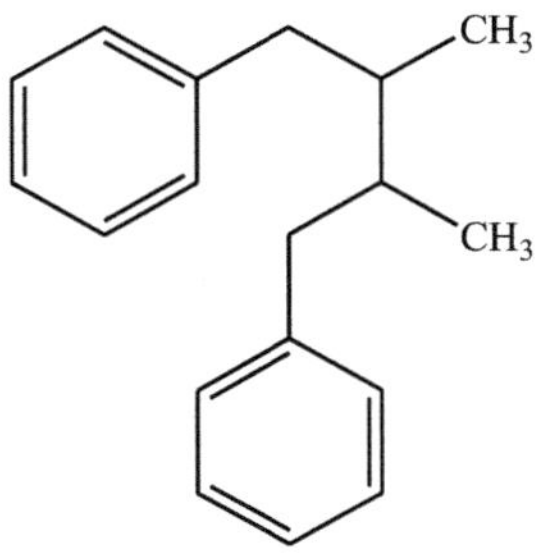

FIGURE 12.4 Basic structure of dibenzylbutane (**6**), responsible for the formation of simple lignans. Structure prepared by the author.

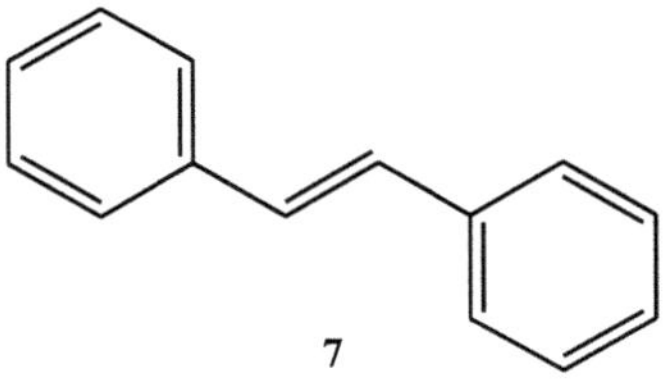

FIGURE 12.5 Common structure of stilbenes: *trans* (**7**) and *cis* isomers (**8**). Structure prepared by the author.

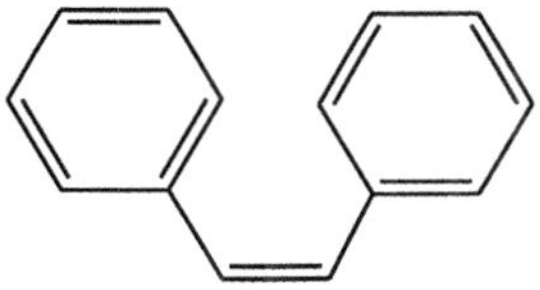

Phenolic acids are commonly divided into two categories: hydroxybenzoic acids, derived from benzoic acid (Figure 12.3, **4**), and hydroxycinnamic acids, derived from cinnamic acid (Figure 12.3, **5**), (Abotaleb et al. 2020; Heleno et al. 2015). The characteristic structure of phenolic acids consists of at least one aromatic ring, wherein at least one of its hydrogen atoms is substituted by a hydroxyl group (Heleno et al. 2015).

The structure of lignans is formed by the combination of two phenylpropanoid units, with coniferyl alcohol as the precursor. The core scaffold of lignan molecules is formed by dibenzylbutane (Figure 12.4 **6**) (Cui et al. 2020; Satake et al. 2015). Lignans can be found in more than 70 families in the plant kingdom, with a high prevalence in several species from various genera of the Lauraceae family (Cui et al. 2020).

Stilbenes are another class of secondary metabolites originating from the shikimic acid pathway (Valletta, Iozia, and Leonelli 2021). These compounds share a common structure comprising two aromatic rings linked by an ethylene moiety (Figure 12.5), the configuration of which can lead to two naturally occurring isomeric forms (Figure 12.5, **7** and **8**): *trans*-stilbenes ((E)-1,2-diphenylethylene), and *cis*-stilbenes ((Z)-1,2-diphenylethylene). (Navarro-Orcajada et al. 2023; Pecyna et al. 2020).

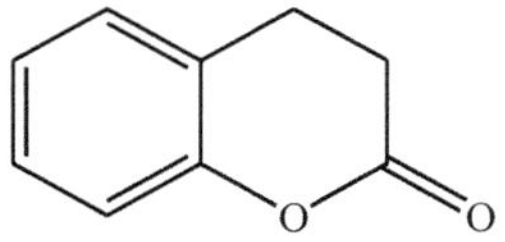

9

FIGURE 12.6 Common structure of coumarins. Structure elaborated by the author.

Finally, coumarins are synthesized in many species of different genera and families, and can be found in roots, seeds, fruits, and leaves (Loncaric et al. 2020). The common coumarin structure consists of the combination between a benzene ring and a lactone ring (Figure 12.6, **9**)—hence making coumarins a part of the benzopyrone family, characterized by being oxygen-containing heterocyclic compounds. Coumarins can be found in many medicinal plants (El-Sawy and Abdelwahab 2021; Keri, Budagumpi, and Balappa Somappa 2022; Sharifi-Rad et al. 2021).

12.3 BIOSYNTHESIS OF PHENYLPROPANOIDS

12.3.1 Shikimate Biosynthetic Pathway

The shikimate biosynthetic pathway is responsible for the biosynthesis of phenylpropanoids, as it generates L-phenylalanine (Figure 12.1), an aromatic amino acid that is the basic unit of production of phenylpropanoids. The biosynthesis of these compounds and their derivatives is characterized by the action of three enzymes which are integral to the central biosynthesis pathway: i) phenylalanine ammonia lyase, which deaminates phenylalanine, yielding cinnamic acid, ii) cinnamate 4-hydroxylase (C4H), which hydroxylates cinnamic acid, yielding *p*-coumaric acid, and iii) 4-coumaroyl-CoA ligase (4CL), which converts *p*-coumaric acid to *p*-coumaroyl-CoA (Biala and Jasienski 2018; Ferreira and Antunes 2021).

The shikimate biosynthetic pathway starts with precursors from the glycolysis and pentose phosphate pathways (Figure 12.7). Initially, there is a condensation between phosphoenolpyruvate (**PEP**) and D-erythrose 4-phosphate (**DE4P**) generating 3-deoxy-D-arabino-heptulosonicacid 7-phosphate, which is subsequently converted into 3-dehydroquinic acid. Through subsequent enzymatic dehydration and reduction reactions, 3-dehydroquinic acid is converted into shikimic acid, which is then phosphorylated by shikimate kinase. The resulting product is condensed with another PEP molecule, undergoing several other enzyme-catalyzed reactions, culminating in conversion into chorismic acid by chorismate synthase. An alternative pathway involves the conversion of the intermediate 3-dehydroshikimic acid into protocatechuic acid via dehydration and enolization, or into gallic acid via oxidation and enolization (Almeida et al. 2024; Dewick 2009; Shende, Bauman, and Moore 2024). Protocatechuic acid and gallic acid are considered simple, yet important, phenolic acids. For instance, the latter is the basic component of tannins, molecules generally found in plant shoots, fruits, seeds, roots, and branches (Tong et al. 2022), which modulate plant-herbivore interactions by protecting against infections caused by microorganisms, or by deterring herbivores (Singh, Kaur, and Kariyat 2021).

The biosynthesis of 3-dehydroshikimic acid involves two different types of enzyme—type I and type II. Only the former is found in bacteria and higher plants, while both types can be found in fungi. After the biosynthesis of chorismic acid, phenylpropanoid generation depends on the production of phenylalanine and tyrosine. In photosynthetic organisms, both amino acids are synthesized in the chloroplasts and plastids of non-green plant cells (Pascual et al. 2016). Humans, however, are unable to synthesize either amino acid and must instead obtain them from their diet, hence the classification as "essential amino acids".

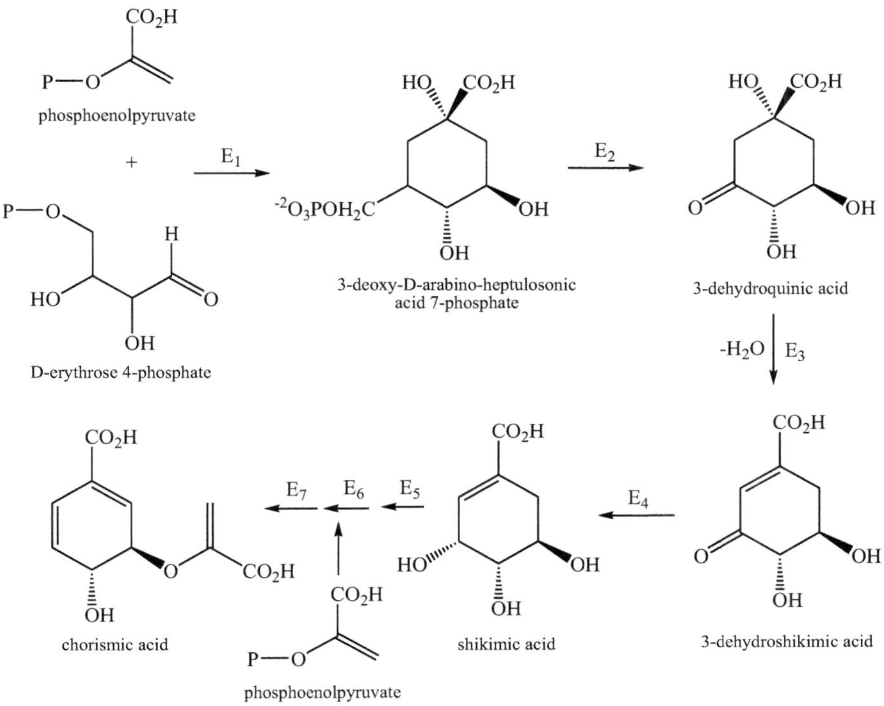

FIGURE 12.7 Biosynthetic pathways for the formation of shikimic acid and chorismic acid. Elaborated by the author.

Key: E_1: 3-deoxy-D-arabino-heptulosonic acid 7-phosphate synthase; E_2: 3-dehydroquinate synthase; E_3: 3-dehydroquinase; E_4: shikimate dehydrogenase; E_5: shikimate kinase; E_6: 5-enolpyruvylshikimic acid 3-phosphate synthase; E_7: chorismate synthase.

Chorismic acid is converted into prephenic acid via a rearrangement wherein the PEP-derived side chain becomes directly bonded to the carbocycle, yielding the basic C_6–C_3 carbon skeleton, which is a precursor to the formation of phenylalanine and tyrosine (Figure 12.8). The use of prephenic acid in the production of these aromatic amino acids depends on the organism; moreover, a particular species may possess more than one route, according to the enzymatic activities which are available (Dewick 2009; Mir, Jallu, and Singh 2015). The conversion of prephenic acid essentially involves three types of reactions: decarboxylative aromatization, transamination and, for the biosynthesis of tyrosine, an oxidation (Dewick 2009).

After the formation of phenylalanine and tyrosine by the shikimate pathway, the general phenylpropanoid pathway (GPP) begins, which consists of three stages (Figure 12.9): first is the PAL-catalyzed deamination of phenylalanine, yielding cinnamic acid, the precursor of phenylpropanoids from secondary metabolism; second is the enzymatic reaction in the conversion of cinnamic acid into *p*-coumaric acid, that is, the C4H-catalyzed introduction of a hydroxyl group in the *para* position of cinnamic acid's phenyl ring; and third is the 4CL-catalyzed conversion of *p*-coumaric acid into *p*-coumaroyl-CoA, an important precursor in the biosynthetic cascade of various phenylpropanoids (Deng and Lu 2017; Dong and Lin 2021; Vanholme, El Houari, and Boerjan 2019).

While it is generally accepted that phenylpropanoid biosynthesis begins with the formation of phenylalanine synthesized via the shikimate pathway, tyrosine has been proposed as the starting point of phenylpropanoid biosynthesis in some monocotyledons, fungi, and bacteria (Deng and Lu

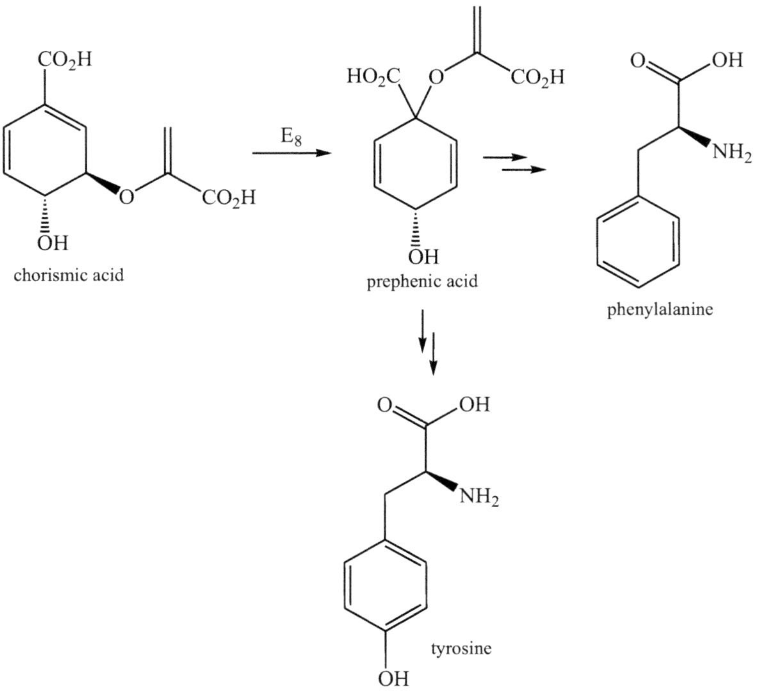

FIGURE 12.8 Biosynthesis of aromatic amino acids phenylalanine and tyrosine from chorismic acid. Elaborated by the author.

Key: E_8: chorismate mutase.

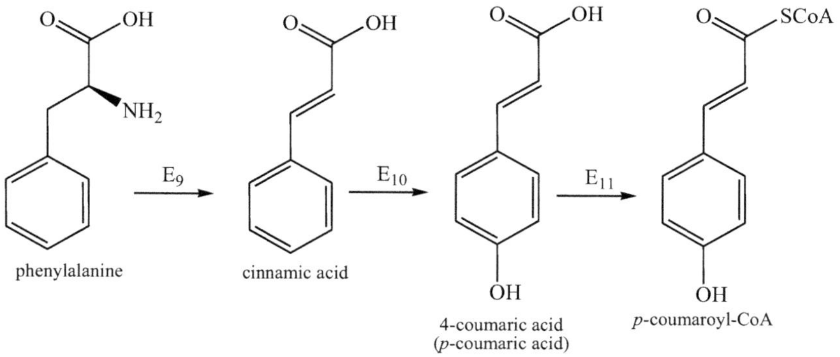

FIGURE 12.9 Biosynthesis of cinnamic acid and *p*-coumaric acid from phenylalanine.

Key: E_9: phenylalanine ammonia lyase (PAL); $E_{10:}$ cinnamic acid 4-hydroxylase (C4H); $E_{11:}$ 4-coumaroyl CoA Ligase (4CL).

2017). In this situation, the enzymes tyrosine ammonia lyase or bifunctional ammonia lyase (PTAL) act by catalyzing the conversion of tyrosine into *p*-coumaric acid, which is subsequently converted into *p*-coumaroyl-CoA by 4CL, Figure 12.10 (Yadav et al. 2020).

Among the main phenylalanine- and tyrosine-derived phenylpropanoid derivatives (Figure 12.11) are aromatic phenolic compounds, such as: monolignols (C_6–C_3), lignans and neolignans $(C_6–C_3)_2$,

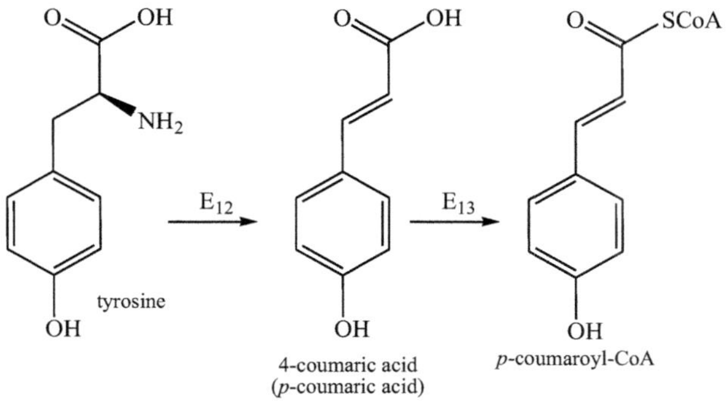

FIGURE 12.10 Biosynthesis of *p*-coumaric acid from tyrosine.

Key: E_{12}: tyrosine ammonia lyase (TAL) or bifunctional ammonia lyase (PTAL); $E_{13:}$ 4-coumaroyl CoA ligase (4CL).

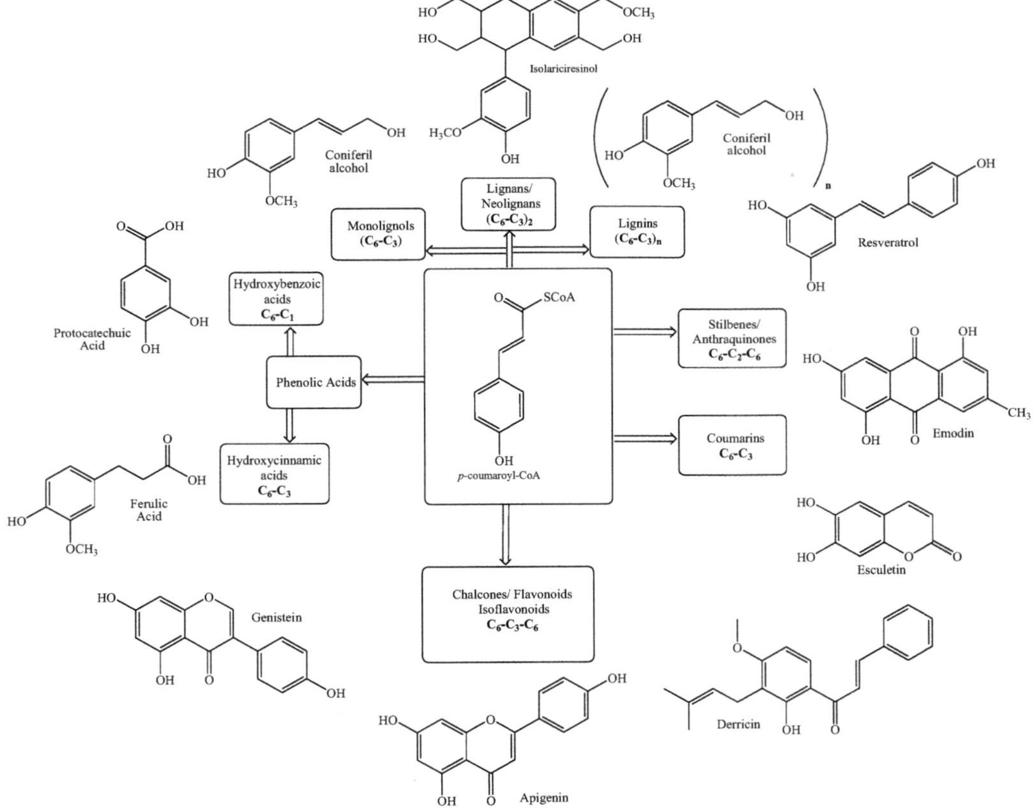

FIGURE 12.11 Schematic representation of the biosynthetic phenylpropanoids derivatives indicating *p*-coumaroyl CoA, in the center, as the precursor of phenylpropanoids. The chemical structures of resveratrol, emodin, esculetin, genistein, apigenin, derricin, ferulic acid, protocatechuic acid, coniferyl alcohol, and isolariciresinol, are highlighted. Elaborated by the author.

lignins $(C_6–C_3)_n$, coumarins $(C_6–C_3)$, phenolic acids—subdivided into hydroxybenzoic acids $(C_6–C_1)$ or hydroxycinnamic acids $(C_6–C_3)$, stilbenes and anthraquinones $(C_6–C_2–C_6)$, and chalcones, flavonoids, isoflavonoids, and neoflavonoids $(C_6–C_2–C_6)$ (Deng and Lu 2017; Norma Francenia et al. 2019).

Not all phenylpropanoid pathway-derived compound classes are present in all plant species. The hydroxycinnamic acid and flavonoid classes are generally found in many higher plants, however some compounds belonging to these classes may only occur in some genera or in certain species, particularly those with specific substitution patterns (Dixon et al. 2002).

12.4 PHENYLPROPANOIDS IN PLANT PROTECTION

12.4.1 Antimicrobial Activity in Plant Protection

Plants are important nutrient sources for various organisms, including insects and vertebrates. Although they do not have defense mechanisms equal to those of animals, they contain in their composition a mixture of complex chemical products called secondary metabolites, that exhibit a defensive action against pathogens. These secondary metabolites have various functions, including the regulation of interactions between plants and the environment. They are generally grouped into three classes: terpenoids, phenolics, and alkaloids. Among these classes, phenylpropanoid compounds have the ability to protect plants against different microbial diseases (Ismail et al. 2022; Tiku 2018).

The first plant component that pathogens encounter when attacking plants is the cell wall, the breach of which activates the phenylpropanoid pathway. Plants develop a multi-layered defense network to combat infection by microbial pathogens in this first contact (Yadav et al. 2020). In addition, these compounds have attracted attention because they are involved in various plant functions, such as providing color to flowers, fruits, grains, and vegetables, and protection against UV radiation and environmental stress (Dwivedi, Upadhyaya, and Chung 2016).

The coumarin class of phenylpropanoids contains the largest number of compounds, and the ability of this class of compounds to suppress the appetite of insects plays an important role in the chemical defense of plants. Despite their pleasant, sweet aroma, coumarins have a bitter taste that animals tend to avoid (Ortiz and Sansinenea 2023). They also act like phytoalexins—terpene derivatives with antimicrobial activity—helping to protect plants against microbial pathogens (Stringlis, de Jonge, and Pieterse 2019; Yang et al. 2016).

Stassen et al. (2021) emphasize the fundamental role of coumarin communication along the microbiome-root-shoot axis, stimulating their selective antimicrobial activity and shaping the formation of the root microbiome, thereby promoting plant growth, health, and immunity. Ramaroson, Koutouan, and Helesbeux (2022) indicate that the phenylpropanoid pathway is triggered towards plant defense through signaling, and that these compounds are regulated by hydroxycinnamic acids and monolignols through complex plant mechanisms. Thus, phenylpropanoid-derived substances such as flavonoids, isoflavonoids, monolignols, phenylpropenes, phenolic acids, stilbenes and stilbenoids, lignin, suberin, and sporopollenin, act as plant defenders against a wide range of microorganisms.

The types of pathogens that phenylpropanoids and their derivatives can protect plants against, and the mechanisms of action, are summarized in Table 12.1.

12.4.2 Allelopathic Activity

The significant biological activity of plants, particularly the volatile compounds present in essential oils that play a crucial role in plant defense against natural enemies, has been examined in many studies. Plants from different genera, and plant species from the same genus, differ in their essential oil composition and concentration. These variations can significantly influence the defensive

TABLE 12.1
Examples of Antimicrobial Activity of Phenylpropanoids and Their Derivatives across Plant Species

Defense Compounds/Plant Species	Pathogens	Antimicrobial Activity	Reference
Hydroxycinnamate/*Citrus sinensis* **(L.) Osbeck leaves grafted onto** *Citrus limonia* **Osbeck**	Bacterial (*Candidatus liberibacter*)	Inoculation of orange leaves with hydroxycinnamate increased the plant's defense against infection by *Candidatus liberibacter.*	Hijaz et al. (2013)
Plant-derived coumarins from *Daphne* **species** (*D. odora*, *D. gnidium*, *D. mezereum*, *D. giraldii*, *D. Koreana* **Nakai**, *D. tangutica* **and** *D. oleoides*).	Bacterial (*Ralstonia solanacearum*)	Daphnetin (hydroxycoumarin) exhibited the best antibacterial activity against *R. solanacearum* in this study.	Yang et al. (2016)
Phenylpropanoid derivatives/*Amorphophallus muelleri* **and** *Amorphophallus konjac*	Bacterial (*Pectobacterium carotovorum*)	The phenylpropanoid biosynthesis pathway is closely related to plant resistance to disease, due to the presence of PAL.	Gao et al. (2024)
Phenylpropanoid derivatives /*Citrus sinensis*	Fungal (*Penicillium digitatum*)	The involvement of the phenylpropanoid pathway in inducing the accumulation of secondary metabolites such as flavanones, flavones, polymethoxylated flavones, and scoparone in citrus fruits happens in response to infection by *Penicillium digitatum*.	Ballester (2013)
Phenylpropanoid derivatives/ soybean cultivars	Fungal (*Corynespora cassiicola*)	The strong presence of lignins and flavonoids in soybean leaves contributed to resistance against the fungus, via a putative activation of the phenylpropanoid pathway to obtain cultivars with high levels of resistance against this important leaf disease.	Fortunato, Araujo, and Rodrigues (2017)
Phenylpropanoid derivatives/*Capsicum annuum*	Fungal (*Verticillium dahliae*)	Pepper developed tolerance against *Verticillium dahliae* through various mechanisms involving phenylpropanoid metabolism, which probably resulted in fungitoxic activity.	Novo et al. (2017)
Phenylpropanoid derivatives/*Linum usitatissimum*	Fungal (*Fusarium oxysporum*)	The involvement of different branches of the shikimic acid/phenylpropanoid pathway in the initial response of flax to *Fusarium oxysporum* attack was evaluated. The metabolic pathway generates a variety of compounds (benzoates, phenolic acids, lignin, flavonoids, etc.), and their interactions were relevant to the initial phases of the plant's antifungal response.	Boba et al. (2017)
Lignin/*Solanum lycopersicum*	Fungal (*Alternaria solani*)	The importance of the phytoalexin and phenylpropanoid biosynthetic pathways, together with lignin accumulation, were emphasized, revealing a possible basis for disease resistance in wild tomatoes	Shinde et al. (2017)
Phenylpropanoid derivatives/ *Malus domestica*	Fungal (*Botrytis cinerea*)	Data indicate that phenylpropanoid metabolism is closely associated with apple resistance to gray mold disease.	Ma et al. (2018)

(*Continued*)

TABLE 12.1 ***(Continued)***
Examples of Antimicrobial Activity of Phenylpropanoids and Their Derivatives across Plant Species

Defense Compounds/Plant Species	Pathogens	Antimicrobial Activity	Reference
Flavonoids/*Daucus carota*	Fungal (*Alternaria dauci*)	High levels of secondary metabolites (feruloylquinic acid, 4-O and 7-O-glucosides of apigenin, luteolin, and chrysoberyl) present in carrot leaves may be linked to resistance to *Alternaria dauci*.	Koutouan et al. (2018)
Phenylpropanoid derivatives/*Sorghum bicolor*	Fungal (*Colletotrichum sublineolum*)	Nine significant metabolic pathways of phenylalanine, stilbenoids and gingerol, flavonoids, tryptophan, riboflavin and tyrosine and the biosynthesis of phenylpropanoids were identified as collectively defining resistance to *Colletotrichum sublineolum*.	Tugizimana et al. (2019)
Phenylalanine/petunia and Arabidopsis	Fungal (*Botrytis cinerea*)	The increased accumulation of phenylalanine and phenylpropanoid-derived metabolites in plant leaves significantly decreased plant susceptibility to *Botrytis cinerea*.	Oliva et al. (2020)
p-coumaric acid/*Cerasus pseudocerasus*	Fungal (*Botrytis cinerea*/*Penicillium expansum*)	*p*-coumaric acid inhibited the mycelial growth of *Botrytis cinerea* and *Penicillium expansum*, and reduced the rate of natural deterioration of the fruit, which was attributed to the activation of the phenylpropanoid metabolic pathway.	X. Liu et al. (2020)
Ferulic acid/*Zea mays*	Fungal (*Fusarium verticillioides*)	*In vitro* assays with ferulic acid extracts in the 0.25-0.50 mM range inhibited the growth of the fungus with a resistance effect against corn ear rot.	Martnez-Fraca et al. (2022)
Phenylalanine/*Mangifera indica*	Fungal (*Colletotrichum gloeosporioides*)	The action of phenylalanine in fruit-fungus interactions activates multi-layered defense mechanisms in the fruit, preventing the development of anthracnose caused by the fungus.	Patel et al. (2023)
Ferulic acid/*Malus domestica*	Fungal (*Penicillium expansum*)	Ferulic acid boosted the biosynthesis of phenylpropanoid secondary metabolites, with greater inhibitory effect against *Penicillium expansum* in *in vitro* assays, at a concentration of 1.0 g L^{-1}.	Guo et al. (2023)
Phenylalanine/*Malus domestica*	Fungal (*Venturia inaequalis*)	The higher activity of PAL and the increased accumulation of phenylpropanoids in *Venturia inaequalis*-infected plants treated with plant hormone methyl jasmonate indicated the tolerance to scab in apples, induced through the exogenous application of an optimized concentration (100 μM) of the hormone and resulting in increased membrane stability and decreased levels of oxidative stress.	(Demiwal et al. 2024)

strength exerted by these compounds. In Brazil, the first chemical substances were only isolated, identified, and tested for their allelopathic potential in the 1990s. Estimates suggest that more than 100 chemical compounds with medium to high polarity have already been isolated, identified and tested, the vast majority of which are from native Brazilian flora species, and from the Amazon region in particular (da Souza Filho 2023).

Phenylpropanoids and their derivatives are among the most common active components of many foods, flavors, fragrances, wines, essential oils, and drinks. Among many studies examining plant allelopathic activity, de Souza Filho et al. (2009) showed that the essential oil of *Ocimum americanum*, in concentrations ranging from 100 to 2,000 mg/L, inhibited the germination and development of the radicle and hypocotyl of seeds of two weed species that invade pastures: *Mimosa pudica* and *Senna obtusifolia*. The effects were attributed to the presence of monoterpenes, oxygenated monoterpenes, sesquiterpenes, aliphatics, and phenylpropanoids in the plant's essential oil, with emphasis on the constituents with proven allelopathic activity, such as limonene, camphor, and linalool.

Tynanthus micranthus Corr. Mello, a species of the Bignoniaceae family popularly known in Brazil as "cipó-cravo" or "craveiro", is geographically distributed in the states of Paraná, São Paulo, and Mato Grosso, Brazil. Compounds such as apigenin, eugenol, and β-sitosterol have been isolated from polar extracts obtained from this species, as well as other compound classes such as coumarins, flavonoids, anthraquinones, steroids, and triterpenoids (Cansian et al. 2015). The allelopathic activity of ethanolic extract fractions obtained from roots, shoots, and leaves of this species was evaluated against *Lactuca sativa* diaspores; the chloroform fraction of shoots inhibited diaspore germination at a concentration of 0.8 mg/L, whereas the hydroalcoholic fraction favored germination across all concentrations analyzed (Cansian et al. 2013).

Eugenol, a phenylpropanoid, is widely used in insect repellent formulations developed against oriental fruit flies (*Bactrocera dorsalis*) and melon flies (*Bactrocera cucurbitae*). Eugenol is not toxic in food but is toxic by inhalation, and high doses of eugenol can cause liver damage. Clove (*Syzygium aromatum*) is the richest source of eugenol, with concentrations between 80 and 90% in clove oil and between 82 and 88% in clove-leaf oil (Zheng and Bossier 2023). Other plants that contain eugenol in large quantities include *Artemisia absinthium*, *Cinnamomum tamala*, *Myristica fragrans*, *Ocimum gratissimum*, and *O. basilicum* (Tripathi and Mishra 2016).

Mazzafera (2003) showed that the ethanolic extract of cloves at concentrations between 31-62 mg/mL, and (pure) eugenol, have an allelopathic effect and strongly inhibit the germination of seeds of several plant species, including: tomato (*Lycopersicum esculentum* Mill.), balsam (*Impatiens balsamina* L.), showy rattlebox (*Crotalaria spectabilis* L.), radish (*Raphanus sativus* L.), wheat (*Triticum aestivum* L.), and lettuce (*Lactuca sativa* L.). Spraying of the ethanolic extract also reduced the growth of these species. Eugenol and *Syzygium aromaticum* extracts also exhibit nematocidal, insecticidal, antiviral, and fungicidal activity. The bactericidal and fungicidal effects of eugenol may explain why spraying clove seeds with fungicide does not improve their germination. Furthermore, as a fungicide, clove essential oil can be used to combat diseases in banana cultivations, and as an alternative in the post-harvest treatment of bananas (Gusman, Vieira, and Vestena 2014).

Lippia thymoides is another species with described allelopathic activity. Its phytotoxic effect was shown in the inhibition of germination of the weeds commonly known in Brazil as malícia (*Mimosa pudica*) and mata-pasto (*Senna obtusifolia*). Gas chromatography coupled with mass spectrometry (GC/MS) analysis of the essential oil, obtained by hydrodistillation, identified thymol as the major constituent (Silva et al. 2019). Miranda et al. (2015) performed a similar study using lettuce seeds as the recipient plant and showed that the extract had no phytotoxic effect on the seeds, but the main chemical constituent, thymol, had a significant inhibitory effect on germination of the lettuce seeds.

The essential oil and extracts of plant species of the *Aristolochia* (Aristolochiaceae) genus display several different biological activities, owing to their diverse chemical composition. The main essential oil components include myristicin, borneol, α-copaene, selin-6-en-4-ol, and cadinene. An evaluation of the chemical composition of the crude ethanolic extract of *Aristolochia* sp. identified tannins, flavanonols, flavanones, terpenoids, and steroids. The ethanolic extract showed low toxicity against *Artemia salina*, whereas the essential oil was considered toxic. The antifungal activity of these samples, tested against *Colletotrichum gloeosporioides*, was considered microbiostatic (Lerma-Herrera et al. 2022). The effect of aqueous extracts from different parts of *Aristolochia esperanzae* on the germination and growth of lettuce and radish have also been reported. The ethanolic extract from leaves exerted the largest effect on percentage germination, with several different concentrations delaying germination. Ethanolic extracts from the aerial parts and roots caused seedling abnormalities. Seedlings grown on filter paper showed greater growth inhibition than those grown on coconut fiber substrate (Laemah and Waewthongrak 2022).

12.5 CONCLUSIONS

It is well established that volatile phenylpropanoids possess biological activities that are beneficial for human health. This metabolic class of compounds can be used to safeguard crops as it exhibits phytotoxic properties that can impede the growth of invasive plants and regulate phytopathogens. Research that further elucidates these activities would be beneficial. This review presents insights into the potential applications of phenylpropanoids, which can be utilized in isolation or in complex mixtures, such as essential oils, thereby contributing to the rational use of natural products in agriculture.

ACKNOWLEDGMENTS

We would like to thank the authors for their contributions and commitment to writing this scientific work, and the editors of this book for the invitation and vote of confidence.

REFERENCES

Abotaleb, Mariam, Alena Liskova, Peter Kubatka, and Dietrich Büsselberg. 2020. "Therapeutic potential of plant phenolic acids in the treatment of cancer." *Biomolecules* 10 (2): 1–23. https://doi.org/10.3390/biom10020221.

Almeida, Aline Marengoni, Rogério Marchiosi, Josielle Abrahão, Rodrigo Polimeni Constantin, Wanderley Dantas dos Santos, and Osvaldo Ferrarese-Filho. 2024. "Revisiting the shikimate pathway and highlighting their enzyme inhibitors." *Phytochemistry Reviews* 23 (2): 421–457. https://doi.org/10.1007/s11101-023-09889-6.

Andrade-Ochoa, Sergio, Karla F. Chacón-Vargas, Luvia E. Sánchez-Torres, Blanca E. Rivera-Chavira, Benjamín Nogueda-Torres, and Guadalupe V. Nevárez-Moorillón. 2021. "Differential antimicrobial effect of essential oils and their main components: insights based on the cell membrane and external structure." *Membranes* 11 (6). https://doi.org/10.3390/membranes11060405.

Ballester, Ana-Rosa and. 2013. "Citrus phenylpropanoids and defence against pathogens. Part II: Gene expression and metabolite accumulation in the response of fruits to Penicillium digitatum infection." *Food Chemistry* 136 (1): 285–291. https://doi.org/10.1016/j.foodchem.2012.08.006. https://linkinghub.elsevier.com/retrieve/pii/S0308814612012897.

Biala, Wanda, and Michal Jasienski. 2018. "The phenylpropanoid case — It is transport that matters." *Frontiers in Plant Science* 9. https://www.frontiersin.org/journals/plant-science/articles/10.3389/fpls.2018.01610.

Boba, Aleksandra, Kamil Kostyn, Anna Kostyn, Wioleta Wojtasik, Mariusz Dziadas, Marta Preisner, Jan Szopa, and Anna Kulma. 2017. "Methyl salicylate level increase in flax after fusarium oxysporum infection is associated with phenylpropanoid pathway activation." *Frontiers in Plant Science* 7. http://journal.frontiersin.org/article/10.3389/fpls.2016.01951/full.

Cansian, Fernanda Colombi, Francis José Zortéa Merino, Josiane de Fátima Gaspari Dias, Sandra Maria Warumby Zanin, Obdulio Gomes Miguel, and Marilis Dallarmi Miguel. 2015. "Chemical review and studies related to species from the genus *Tynanthus* (Bignoniaceae)." *Brazilian Journal of Pharmaceutical Sciences* 51.

Cansian, Fernanda, Cristina Lima, Zortéa Merino, Obdulio Miguel, and Marilis Miguel. 2013. "Allelopathic effects of tynanthus micranthus Corr. Mello ex. Schum. (Bignoniaceae) on diaspores of *Lactuca sativa* L." *Revista de Ciencias Farmaceuticas Basica e Aplicada* 34: 137–140.

Cesarino, Igor, Aymerick Eudes, Breeanna Urbanowicz, and Meng Xie. 2022. "Editorial: Phenylpropanoid systems biology and biotechnology." *Frontiers in Plant Science* 13. https://www.frontiersin.org/journals/plant-science/articles/10.3389/fpls.2022.866164.

Cui, Qinghua, Ruikun Du, Miaomiao Liu, and Lijun Rong. 2020. "Lignans and their derivatives from plants as antivirals." *Molecules* 25 (1): 1–17. https://doi.org/10.3390/molecules25010183.

Dagne, Abebe, Sileshi Degu, Abiy Abebe, and Daniel Bisrat. 2023. "Antibacterial activity of a phenylpropanoid from the root extract of carduus leptacanthus fresen." *Journal of Tropical Medicine* 2023 (1): 4983608. https://onlinelibrary.wiley.com/doi/abs/10.1155/2023/4983608.

Da Souza Filho, A. P. S. 2023. *A história das pesquisas em alelopatia no Brasil*. Brasília, DF: Embrapa.
Da Souza Filho, A. P. S., J. C. Bayma, G. M. S. P. Guilhon, and M. G. B. Zoghbi. 2009. "Atividade potencialmente alelopática do óleo essencial de *Ocimum americanum*." *Planta Daninha* 27.
de Vries, Sophie, Janine Fürst-Jansen, Iker Irisarri, Amra Dhabalia Ashok, Till Ischebeck, Kirstin Feussner, Ilka N. Abreu, Maike Petersen, Ivo Feussner, and Jan de Vries. 2021. "The evolution of the phenylpropanoid pathway entailed pronounced radiations and divergences of enzyme families." *The Plant Journal* 107 (4): 975–1002. https://doi.org/10.1111/tpj.15387.
Demiwal, Pratibha, Sajad Un Nabi, Javid Iqbal Mir, Mahendra K. Verma, Shri Ram Yadav, Partha Roy, and Debabrata Sircar. 2024. "Methyl jasmonate improves resistance in scab-susceptible Red Delicious apple by altering ROS homeostasis and enhancing phenylpropanoid biosynthesis." *Plant Physiology and Biochemistry* 207: 108371. https://doi.org/10.1016/j.plaphy.2024.108371. https://linkinghub.elsevier.com/retrieve/pii/S0981942824000391.
Deng, Yuxing, and Shanfa Lu. 2017. "Biosynthesis and regulation of phenylpropanoids in plants." *Critical Reviews in Plant Sciences* 36 (4): 257–290. https://doi.org/10.1080/07352689.2017.1402852.
Dewick, Paul M. 2009. *Medicinal Natural Products: A Biosynthetic Approach*. Edited by John Wiley & Sons Ltd. 3rd ed. West Sussex, United Kingdom: John Wiley & Sons Ltd.
Dixon, Richard A., Lahoucine Achnine, Parvathi Kota, Chang-Jun Liu, M. S. Srinivasa Reddy, and Liangjiang Wang. 2002. "The phenylpropanoid pathway and plant defence— a genomics perspective." *Molecular Plant Pathology* 3 (5): 371–390. https://doi.org/10.1046/j.1364-3703.2002.00131.x.
Dong, Nai-Qian, and Hong-Xuan Lin. 2021. "Contribution of phenylpropanoid metabolism to plant development and plant—environment interactions." *Journal of Integrative Plant Biology* 63 (1): 180–209. https://doi.org/10.1111/jipb.13054.
Dwivedi, Sangam L., Hari D. Upadhyaya, and Ill-Min and Chung. 2016. "Exploiting phenylpropanoid derivatives to enhance the nutraceutical values of cereals and legumes." *Frontiers in Plant Science* 7. http://journal.frontiersin.org/Article/10.3389/fpls.2016.00763/abstract.
El-Sawy, Eslam Reda, and Ahmed Bakr Abdelwahab. 2021. "Rings with multi heteroatoms." *Molecules* 26 (11).
Ferreira, Savio S., and Mauricio S. Antunes. 2021. "Re-engineering plant phenylpropanoid metabolism with the aid of synthetic biosensors." *Frontiers in Plant Science* 12. https://www.frontiersin.org/journals/plant-science/articles/10.3389/fpls.2021.701385.
Fortunato, Alessandro Antnio, Leonardo Araujo, and Fabrcio vila Rodrigues. 2017. "Association of the production of phenylpropanoid compounds at the infection sites of corynespora cassiicola with soybean resistance against target spot." *Journal of Phytopathology* 165 (2): 131–142. https://onlinelibrary.wiley.com/doi/10.1111/jph.12546.
Gao, Penghua, Ying Qi, Lifang Li, Shaowu Yang, Jianwei Guo, Jiani Liu, Huanyu Wei, Feiyan Huang, and Lei Yu. 2024. "Phenylpropane biosynthesis and alkaloid metabolism pathways involved in resistance of Amorphophallus spp. against soft rot disease." *Frontiers in Plant Science* 15. https://www.frontiersin.org/articles/10.3389/fpls.2024.1334996/full.
Guo, Mi, Canying Li, Rui Huang, Linhong Qu, Jiaxin Liu, Chenyang Zhang, and Yonghong Ge. 2023. "Ferulic acid enhanced resistance against blue mold of Malus domestica by regulating reactive oxygen species and phenylpropanoid metabolism." *Postharvest Biology and Technology* 202: 112378. https://doi.org/10.1016/j.postharvbio.2023.112378. https://linkinghub.elsevier.com/retrieve/pii/S0925521423001394.
Gusman, Grasielle Soares, Licielo Romero Vieira, and Silvane Vestena. 2014. "Atividade alelopática e moluscicida de *Syzygium aromaticum* (L.) Merr. & Perry (Myrtaceae)." *Evidų╙ncia* 14 (2): 113–128.
Heleno, Sandrina A., Anabela Martins, Maria Jоцёо R. P. Queiroz, and Isabel C. F. R. Ferreira. 2015. "Bioactivity of phenolic acids: Metabolites versus parent compounds: A review." *Food Chemistry* 173: 501–513. https://doi.org/10.1016/j.foodchem.2014.10.057.
Hijaz, Faraj M., John A. Manthey, Svetlana Y. Folimonova, Craig L. Davis, Shelley E. Jones, and Jos I. Reyes-De-Corcuera. 2013. "An HPLC-MS characterization of the changes in sweet orange leaf metabolite profile following infection by the bacterial pathogen candidatus liberibacter asiaticus." *PLoS One* 8 (11): e79485. https://dx.plos.org/10.1371/journal.pone.0079485.
Ismail, Muhammad, Salma Javed, Muhammad Kazim, Abdur Razaq, Ejaz Hussain, Wahab Attia tul, Sajjad Ali, and Muhammad Iqbal Choudhary. 2022. "Phenylpropanoids from tanacetum baltistanicum with nematocidal and insecticidal activities." *Chemistry of Natural Compounds* 58 (4): 637–643. https://link.springer.com/10.1007/s10600-022-03759-x.
Keri, Rangappa S., Srinivasa Budagumpi, and Sasidhar Balappa Somappa. 2022. "Synthetic and natural coumarins as potent anticonvulsant agents: A review with structure—activity relationship." *Journal of Clinical Pharmacy and Therapeutics* 47 (7): 915–931. https://doi.org/10.1111/jcpt.13644.

Koutouan, Claude, Valrie Le Clerc, Raymonde Baltenweck, Patricia Claudel, David Halter, Philippe Hugueney, Latifa Hamama, Anita Suel, Sbastien Huet, Marie-Hlne Bouvet Merlet, and Mathilde Briard. 2018. "Link between carrot leaf secondary metabolites and resistance to Alternaria dauci." *Scientific Reports* 8 (1): 13746. https://www.nature.com/articles/s41598-018-31700-2.

Laemah, Arofah, and Waewruedee Waewthongrak. 2022. "Allelopathic effect of eichhornia crassipes aqueous extract against growth of mimosa pudica." *Sains Malaysiana* 51: 3153–3162. https://doi.org/10.17576/jsm-2022-5110-03.

Lerma-Herrera, Martín A., Lidia Beiza-Granados, Alejandra Ochoa-Zarzosa, Joel E. López-Meza, Pedro Navarro-Santos, Rafael Herrera-Bucio, Judit Aviña-Verduzco, and Hugo A. García-Gutiérrez. 2022. "Biological activities of organic extracts of the genus aristolochia: A review from 2005 to 2021." *Molecules* 27 (12). https://doi.org/10.3390/molecules27123937.

Liu, Aimin, and Shifang Ding. 2019. "Anti-inflammatory Effects of dopamine in inhibiting NLRP3 In fl ammasome activation." *Annals of Clinical & Laboratory Science* 49 (3): 353–360.

Liu, Xiaoyun, Dongchao Ji, Xiaomin Cui, Zhanquan Zhang, Boqiang Li, Yong Xu, Tong Chen, and Shiping Tian. 2020. "p-Coumaric acid induces antioxidant capacity and defense responses of sweet cherry fruit to fungal pathogens." *Postharvest Biology and Technology* 169: 111297. https://doi.org/10.1016/j.postharvbio.2020.111297. https://linkinghub.elsevier.com/retrieve/pii/S0925521420308693.

Loncaric, Melita, Dajana Gaso-Sokac, Jokic Stela, and Maja Molnar. 2020. Recent advances in the synthesis of coumarins. *Biomolecules* 10 (1): 151.

Ma, Lijing, Junhua He, Huan Liu, and Huiling Zhou. 2018. "The phenylpropanoid pathway affects apple fruit resistance to Botrytis cinerea." *Journal of Phytopathology* 166 (3): 206–215. https://onlinelibrary.wiley.com/doi/10.1111/jph.12677.

Martnez-Fraca, Javier, M. Eugenia de la Torre-Hernndez, Max Meshoulam-Alamilla, and Javier Plasencia. 2022. "In search of resistance against fusarium ear rot: ferulic acid contents in maize pericarp are associated with antifungal activity and inhibition of fumonisin production." *Frontiers in Plant Science* 13. https://www.frontiersin.org/articles/10.3389/fpls.2022.852257/full.

Mazzafera, Paulo. 2003. "Efeito alelopático do extrato alcoólico do cravo-da-índia e eugenol." *Brazilian Journal of Botany* 26.

Metsämuuronen, Sari, and Heli Sirén. 2019. Bioactive phenolic compounds, metabolism and properties: A review on valuable chemical compounds in Scots Pine and Norway Spruce. *Phytochemistry Reviews* 18, 623–664.

Mir, Rafia, Shais Jallu, and T. P. Singh. 2015. "The shikimate pathway: Review of amino acid sequence, function and three-dimensional structures of the enzymes." *Critical Reviews in Microbiology* 41 (2): 172–189. https://doi.org/10.3109/1040841X.2013.813901.

Miranda, Cíntia Alvarenga Santos Fraga de, Maria das Graças Cardoso, Maria Laene de Moreira Carvalho, Samísia Maria Fernandes Machado, Milene Aparecida Andrade, and Christiane Maria de Oliveira. 2015. "Análise comparativa do potencial alelopático do óleo essencial de *Thymus vulgaris* e seu constituinte majoritário na germinação e vigor de sementes de alface (*Lactuca sativa* L.)."

Navarro-Orcajada, Silvia, Irene Conesa, Francisco José Vidal-Sánchez, Adrián Matencio, Lorena Albaladejo-Marico, Francisco García-Carmona, and José Manuel López-Nicolás. 2023. "Stilbenes: Characterization, bioactivity, encapsulation and structural modifications. A review of their current limitations and promising approaches." *Critical Reviews in Food Science and Nutrition* 63 (25): 7269–7287. https://doi.org/10.1080/10408398.2022.2045558.

Neelam, Anurag Khatkar, and Krishna Kant Sharma. 2020. "Phenylpropanoids and its derivatives: Biological activities and its role in food, pharmaceutical and cosmetic industries." *Critical Reviews in Food Science and Nutrition* 60 (16): 2655–2675. https://doi.org/10.1080/10408398.2019.1653822.

Norma Francenia, Santos-Sánchez, Salas-Coronado Raúl, Hernández-Carlos Beatriz, and Villanueva-Canongo Claudia. 2019. "Shikimic acid pathway in biosynthesis of phenolic compounds." In *Plant Physiological Aspects of Phenolic Compounds*, edited by Soto-Hernández Marcos, García-Mateos Rosario and Palma-Tenango Mariana, Ch. 3. Rijeka: IntechOpen.

Novo, Marta, Cristina Silvar, Fuencisla Merino, Teresa Martnez-Corts, Fachuang Lu, John Ralph, and Federico Pomar. 2017. "Deciphering the role of the phenylpropanoid metabolism in the tolerance of Capsicum annuum L. to Verticillium dahliae Kleb." *Plant Science* 258: 12–20. https://doi.org/10.1016/j.plantsci.2017.01.014. https://linkinghub.elsevier.com/retrieve/pii/S0168945216305349.

Oliva, Moran, Erel Hatan, Varun Kumar, Ortal Galsurker, Ada Nisim-Levi, Rinat Ovadia, Gad Galili, Efraim Lewinsohn, Yigal Elad, Noam Alkan, and Michal Oren-Shamir. 2020. "Increased phenylalanine levels in plant leaves reduces susceptibility to Botrytis cinerea." *Plant Science* 290: 110289. https://doi.org/10.1016/j.plantsci.2019.110289. https://linkinghub.elsevier.com/retrieve/pii/S0168945219310179.

Ortiz, Aurelio, and Estibaliz Sansinenea. 2023. "Phenylpropanoid derivatives and their role in plants' health and as antimicrobials." *Current Microbiology* 80 (12): 380. https://link.springer.com/10.1007/s00284-023-03502-x.

Pascual, María B., Jorge El-Azaz, Fernando N. de la Torre, Rafael A. Cañas, Concepción Avila, and Francisco M. Cánovas. 2016. "Biosynthesis and metabolic fate of phenylalanine in conifers." *Frontiers in Plant Science* 7. https://www.frontiersin.org/journals/plant-science/articles/10.3389/fpls.2016.01030.

Patel, Manish Kumar, Dalia Maurer, Oleg Feyngenberg, Danielle Duanis-Assaf, Noa Sela, Rinat Ovadia, Michal Oren-Shamir, and Noam Alkan. 2023. "Revealing the mode of action of Phenylalanine application in inducing fruit resistance to fungal pathogens." *Postharvest Biology and Technology* 199: 112298. https://doi.org/10.1016/j.postharvbio.2023.112298. https://linkinghub.elsevier.com/retrieve/pii/S0925521423000595.

Pecyna, Paulina, Joanna Wargula, Marek Murias, and Malgorzata Kucinska. 2020. "More than resveratrol: New insights into stilbene-based compounds." *Biomolecules* 10 (8): 1–40. https://doi.org/10.3390/biom10081111.

Ramaroson, Marie-Louisa, Claude Koutouan, and Jean-Jacques and Helesbeux. 2022. "Role of phenylpropanoids and flavonoids in plant resistance to pests and diseases." *Molecules* 27 (23): 8371. https://doi.org/10.3390/molecules27238371. https://www.mdpi.com/1420-3049/27/23/8371.

Satake, Honoo, Tomotsugu Koyama, Sedigheh Esmaeilzadeh Bahabadi, Erika Matsumoto, Eiichiro Ono, and Jun Murata. 2015. "Essences in metabolic engineering of lignan biosynthesis." *Metabolites* 5 (2): 270–290. https://doi.org/10.3390/metabo5020270.

Shamsudin, Nur Farisya, Qamar Uddin Ahmed, Syed Mahmood, Syed Adnan Ali Shah, Alfi Khatib, Sayeed Mukhtar, Meshari A. Alsharif, Humaira Parveen, and Zainul Amiruddin Zakaria. 2022. "Antibacterial effects of flavonoids and their structure-activity relationship study: A comparative interpretation." *Molecules* 27 (4). https://doi.org/10.3390/molecules27041149.

Sharifi-Rad, Javad, Natália Cruz-Martins, Pía López-Jornet, Eduardo Pons Fuster Lopez, Nidaa Harun, Balakyz Yeskaliyeva, Ahmet Beyatli, Oksana Sytar, Shabnum Shaheen, Farukh Sharopov, Yasaman Taheri, Anca Oana Docea, Daniela Calina, and William C. Cho. 2021. "Natural coumarins: Exploring the pharmacological complexity and underlying molecular mechanisms." *Oxidative Medicine and Cellular Longevity* 2021. https://doi.org/10.1155/2021/6492346.

Shende, Vikram V., Katherine D. Bauman, and Bradley S. Moore. 2024. "The shikimate pathway: Gateway to metabolic diversity." *Natural Product Reports* 41 (4): 604–648. https://doi.org/10.1039/d3np00037k. https://www.sciencedirect.com/science/article/pii/S0265056824000266.

Shinde, Balkrishna A., Bhushan B. Dholakia, Khalid Hussain, Sayantan Panda, Sagit Meir, Ilana Rogachev, Asaph Aharoni, Ashok P. Giri, and Avinash C. Kamble. 2017. "Dynamic metabolic reprogramming of steroidal glycol-alkaloid and phenylpropanoid biosynthesis may impart early blight resistance in wild tomato (Solanum arcanum Peralta)." *Plant Molecular Biology* 95 (4–5): 411–423. http://link.springer.com/10.1007/s11103-017-0660-2.

Siebeneichler, Tatiane Jéssica, Rosane Lopes Crizel, Cesar Valmor Rombaldi, and Vanessa Galli. 2024. "Regulation of phenylpropanoid biosynthesis in strawberry ripening: molecular and hormonal mechanisms." *Phytochemistry Reviews* 7. https://doi.org/10.1007/s11101-023-09907-7.

Silva, Sebastião, Renato Costa, Jorddy Cruz, Mozaniel Oliveira, Lidiane Nascimento, Wanessa Almeida, José Costa, Daniel Pereira, Antonio Filho, and Eloisa Andrade. 2019. "Composição química e atividade herbicida (fitotóxica) do óleo essencial de *Lippia thymoides* Mart. & Schauer (Verbenaceae)." 98–107.

Singh, Sukhman, Ishveen Kaur, and Rupesh Kariyat. 2021. "The multifunctional roles of polyphenols in plant-herbivore interactions." *International Journal of Molecular Sciences* 22 (3): 1442. https://www.mdpi.com/1422-0067/22/3/1442.

Stassen, Max J. J., Shu-Hua Hsu, Corn M. J. Pieterse, and Ioannis A. Stringlis. 2021. "Coumarin communication along the microbiome—root—shoot axis." *Trends in Plant Science* 26 (2): 169–183. https://doi.org/10.1016/j.tplants.2020.09.008. https://linkinghub.elsevier.com/retrieve/pii/S1360138520302867.

Stringlis, Ioannis A., Ronnie de Jonge, and Cornï M. J. Pieterse. 2019. "The age of coumarins in plant—microbe interactions." *Plant and Cell Physiology* 60 (7): 1405–1419. https://doi.org/10.1093/pcp/pcz076. https://academic.oup.com/pcp/article/60/7/1405/5488005.

Tiku, Anupama Razdan. 2018. "Antimicrobial compounds and their role in plant defense." 283–307. https://doi.org/10.1007/978-981-10-7371-7_13

Tong, Zhenkai, Wenfeng He, Xiao Fan, and Aiwei Guo. 2022. "Biological function of plant tannin and its application in animal health." *Frontiers in Veterinary Science* 8. https://doi.org/10.3389/fvets.2021.803657. https://www.frontiersin.org/journals/veterinary-science/articles/10.3389/fvets.2021.803657.

Tripathi, Arun K., and Shikha Mishra. 2016. "Chapter 16—Plant monoterpenoids (prospective pesticides)." In *Ecofriendly Pest Management for Food Security*, edited by Omkar, 507–524. San Diego: Academic Press.

Tugizimana, Fidele, Arnaud T. Djami-Tchatchou, Paul A. Steenkamp, Lizelle A. Piater, and Ian A. Dubery. 2019. "Metabolomic analysis of defense-related reprogramming in sorghum bicolor in response to colletotrichum sublineolum infection reveals a functional metabolic web of phenylpropanoid and flavonoid pathways." *Frontiers in Plant Science* 9. https://www.frontiersin.org/article/10.3389/fpls.2018.01840/full.

Ullah, Asad, Sidra Munir, Syed Lal Badshah, Noreen Khan, Lubna Ghani, Benjamin Gabriel Poulson, Abdul Hamid Emwas, and Mariusz Jaremko. 2020. "Important flavonoids and their role as a therapeutic agent." *Molecules* 25 (22): 1–39. https://doi.org/10.3390/molecules25225243.

Valletta, Alessio, Lorenzo Maria Iozia, and Francesca Leonelli. 2021. "Impact of environmental factors on stilbene biosynthesis." *Plants* 10 (1): 1–40. https://doi.org/10.3390/plants10010090.

Vanholme, Bartel, Ilias El Houari, and Wout Boerjan. 2019. "Bioactivity: phenylpropanoids—best kept secret." *Current Opinion in Biotechnology* 56: 156–162. https://doi.org/10.1016/j.copbio.2018.11.012. https://www.sciencedirect.com/science/article/pii/S0958166918301058.

Yadav, Vivek, Zhongyuan Wang, Chunhua Wei, Aduragbemi Amo, Bilal Ahmed, Xiaozhen Yang, and Xian Zhang. 2020. "Phenylpropanoid pathway engineering: An emerging approach towards plant defense." *Pathogens* 9 (4): 312. https://doi.org/10.3390/pathogens9040312. https://www.mdpi.com/2076-0817/9/4/312.

Yang, Liang, Wei Ding, Yuquan Xu, Dousheng Wu, Shili Li, Juanni Chen, and Bing Guo. 2016. "New insights into the antibacterial activity of hydroxycoumarins against ralstonia solanacearum." *Molecules* 21 (4): 468. https://doi.org/10.3390/molecules21040468. https://www.mdpi.com/1420-3049/21/4/468.

Zheng, Xiaoting, and Peter Bossier. 2023. "Toxicity assessment and anti-activity of essential oils: Potential for application in shrimp aquaculture." *Reviews in Aquaculture* 15 (4): 1554–1573. https://doi.org/10.1111/raq.12795.

13 Terpenes and Terpenoids, Potential Alternatives for the Control of Weeds

Hamdy A. Shaaban and Mohamed H. Soliman

13.1 INTRODUCTION

Terpenoids, also known as isoprenoids, are a class of natural products occurring in organic chemicals (any chemical compound that contains carbon–hydrogen or carbon–carbon bonds) derived from the five-carbon compound isoprene and its derivatives called terpenes, diterpenes, and so on. While sometimes used interchangeably with "terpenes", terpenoids contain additional functional groups (a substituent or moiety in a molecule that causes the molecule's characteristic chemical reactions), usually containing oxygen. When combined with hydrocarbon terpenes, terpenoids comprise about 80,000 compounds (1).

Terpenoids are the largest class of plant secondary metabolites, representing about 60% of known natural products (2). Many terpenoids have substantial pharmacological bioactivity and are therefore of interest to medicinal chemists (3).

Plant terpenoids are used for their aromatic qualities and play a role in traditional herbal remedies. Terpenoids contribute to the scent of eucalyptus; the flavors of cinnamon, cloves, and ginger; the yellow color in sunflowers; and the red color in tomatoes (4). Well-known terpenoids include citral, menthol, camphor, and salvinorin A in the plant *Salvia divinorum*; ginkgolide and bilobalide found in *Ginkgo biloba*; and the cannabinoids found in cannabis. The provitamin beta carotene is a terpene derivative called a carotenoid. Steroids and sterols in animals are biologically produced from terpenoid precursors. Sometimes terpenoids are added to proteins, for example, to enhance their attachment to the cell membrane; this is known as isoprenylation. Terpenoids play a role in plant defense as prophylaxis against pathogens and attractants for the predators of herbivores (5).

13.2 STRUCTURE AND CLASSIFICATION

Terpenes are a class of natural products consisting of compounds with the formula $(C_5H_8)n$ for $n \geq 2$. Comprising more than 30,000 compounds, these unsaturated hydrocarbons are produced predominantly by plants, particularly conifers (4). Terpenes or isoprenoids are the major constituents found in essentiel oils (EOs) with molecular structures containing carbon backbones of 2-methylbuta-1,3-diene (isoprene units) which can be rearranged into cyclic structures. The number of isoprene units is primarily responsible for structural diversity of terpenes. Hemiterpenes are formed by one isoprene unit (C5), monoterpenes (C10), sesquiterpenes (C15), diterpenes (C20), triterpenes (C30), and tetraterpenes (C40). Hemiterpenes are a minor part of terpenes found in EOs. An important HT is isoprene, which is emitted from the herbs and leaves of many trees such as conifers, oaks, poplars, and willows. Examples of hemiterpenes include angelic, tiglic, isovaleric, and senecioic acids. Monoterpenes are the predominant components of EOs (90%), followed by sesquiterpenes. Diterpenes, triterpenes, and tetraterpenes with their oxygenated derivatives are also detected in small amounts (6, 7). Examples of terpenes and terpenoids are presented in Figure 13.1.

DOI: 10.1201/9781003463429-17

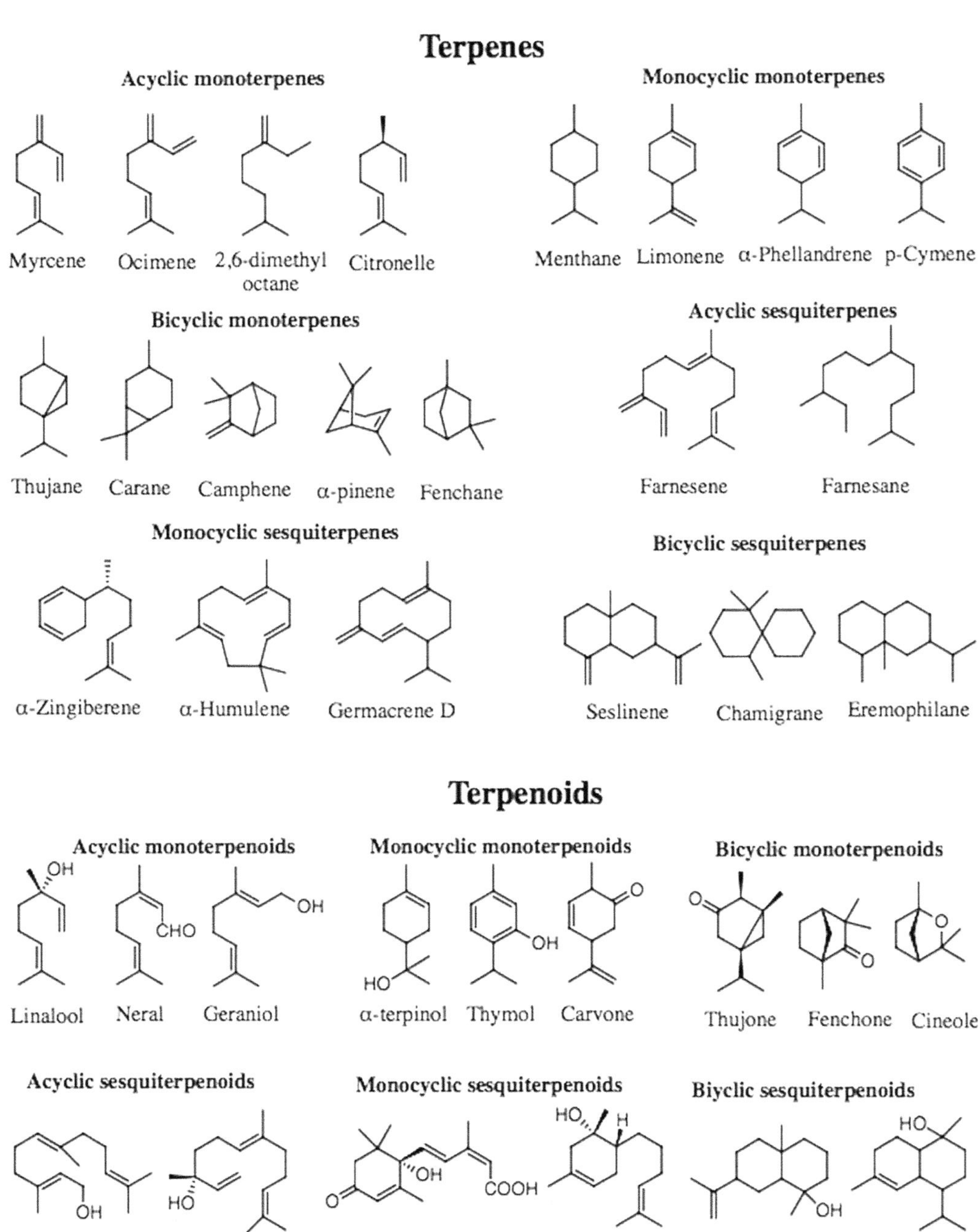

FIGURE 13.1 The chemical structure and classification of terpenes and terpenoids.

Terpenoids can also be classified according to the type and number of cyclic structures they contain: linear, acyclic, monocyclic, bicyclic, tricyclic, tetracyclic, pentacyclic, or macrocyclic (8). Terpenoids are another type of terpenes containing oxygen molecules that are constructed via biochemical modifications (removal or addition of methyl groups) (6). Terpenoids can be divided into alcohols, aldehydes, esters, ether, epoxides, ketones, and phenols. Examples of terpenoids are:

carvacrol, citronellal, geraniol, linalool, linalyl acetate, piperitone, menthol, and thymol. These bioactive compounds confer several biological activities such as natural insecticides, anticancer, antiallergic, antibacterial, and antioxidant (7).

Terpenes and terpenoids are synthesized by the mevalonic acid (MVA) pathway in the cytosol and the 2C-methyl-D-erythritol-4-phosphate (MEP) pathway in the plastid for the formation of precursors: isopentenyl pyrophosphate (IPP) and dimethylallyl pyrophosphate. The Salkowski test can be used to identify the presence of terpenoids (9).

13.3 NATURAL PRODUCTS OF TERPENES

13.3.1 Terpenoids: Natural Products for Cancer Therapy

Terpenoids constitute the largest class of natural products and are a rich reservoir of candidate compounds for drug discovery. Recent efforts into the research and development of anti-cancer drugs derived from natural products have led to the identification of a variety of terpenoids that inhibit cancer cell proliferation and metastasis via various mechanisms. (10).

13.4 USES OF TERPENES AND TERPENOIDS IN PLANT PROTECTING

13.4.1 Biochemistry of Terpenes and Recent Advances in Plant Protection

Biodiversity is adversely affected by the growing levels of synthetic chemicals released into the environment due to agricultural activities. This has been the driving force for embracing sustainable agriculture. Plant secondary metabolites (PSMs) offer promising alternatives for protecting plants against microbes, feeding herbivores, and weeds. Terpenes are the largest among PSMs and have been extensively studied for their potential as antimicrobial, insecticidal, and weed control agents. They also attract natural enemies of pests and beneficial insects, such as pollinators and dispersers. However, most of these research findings are shelved and fail to pass beyond the laboratory and greenhouse stages. This review provides an overview of terpenes, types, biosynthesis, and their roles in protecting plants against microbial pathogens, insect pests, and weeds to rekindle the debate on using terpenes for the development of environmentally friendly biopesticides and herbicides. (11).

13.4.2 Terpenes as Natural Insecticides

The 11 terpene ketones, thymoquinone, (*R*)-carvone, (*S*)-carvone, pulegone, dihydrocarvone, menthone, verbenone, ocimenone, camphor, α-thujone, and piperitenone, were tested as contact and fumigant insecticides against adults of *Sitophilus zeamais* Motschulsky under laboratory conditions. The results show that thymoquinone was found to be more toxic than the other ketones, with lethal doses values being LC_{50} 16.5 μg/cm^2 and LC_{50} 13.8 μL/L air (24 h after treatment). These ketones were also subjected to multiple regression analysis and the results derived from the quantitative structure activity relationship (QSAR) model suggested two topological indicators. The Balaban and Randic indices were the best descriptors of insecticidal activity of the ketones. However, orbital electronegativity of the carbonyl group was the main parameter that connected the inhibition activity of ketones on acetylcholinesterase (AChE). Thus, the insecticidal activity of terpene ketones was primarily explained by the shape of molecules and the branching of the carbon-atom skeleton, while inhibition of activity of AChE was mainly due to the electronic descriptors (12). Essential oils, which are mixtures of terpenes, frequently show stronger insecticide activity, that is, lower lethal dose 50 (LC_{50}), than their most abundant terpenes. Synergy between terpenes provides a plausible explanation, but its demonstration has been elusive. In the present work, we look for an alternative explanation by considering the influence of insect metabolic detoxification. Basically, we propose a model (metabolic model; MM) in which the LC_{50} of the major terpene in a mixture is expected to include a fraction that is detoxified by the insect, whereas a minor terpene would act unimpeded, showing a lower LC_{50} than when acting alone. In order to test this idea, we analyzed the effects of inhibiting the cytochrome P450 detoxification system with piperonyl butoxide (PBO) on the lethal concentration of terpenes as fumigants against *Musca domestica*. We found that, within a group of 10 terpenes [linalool, citronellal, (*R*)-α-pinene, 1,8-cineole, γ-terpinene, limonene, α-terpinene, (*S*)-β-pinene, thymol and (*R*)-pulegone], seven showed an LC_{50}PBO (the lethal concentration for PBO-treated flies) between 1.7 and 12.4 times lower than the corresponding LC_{50} when P450 was not inhibited. Only in one case, that of (*R*)-pulegone, was the LC_{50}PBO greater than the LC_{50}, while two terpenes [(*S*)-β-pinene and thymol] showed no changes in toxicity. The increased activity of most terpenes (particularly linalool and citronellal) in PBO-treated flies supports our hypothesis that normally the LC_{50} includes a fraction of inactive compound, due to detoxification. Having previously determined that *M. domestica* preferentially oxidizes the most abundant terpene in a mixture, while terpenes in smaller proportions are poorly or not detoxified by the P450 system, we assessed whether the toxicity of minority terpenes in a mixture is similar to their activity under P450 inhibition. We chose suitable binary combinations in such a way that one terpene (in greater proportion) should be the target of P450 while the other (in smaller proportion) should intoxicate the fly with LC_{50}PBO or similar. Combinations of 1,8-cineole-citronellal, 1,8-cineole-linalool, linalool-citronellal, (*R*)-pulegone-linalool, (*R*)-pulegone-1,8-cineole, and (*R*)-pulegone-citronellal were assayed against *M. domestica*, and the LC_{50} of each mixture was determined and compared to values predicted by MM (considering the LC_{50}PBO for minor component) or by the classical approach (LC_{50} for both components). The MM showed the best fit to the data, suggesting additive rather than synergistic effects, except for the combination of (*R*)-pulegone-citronellal that was clearly synergistic. Thus, the experimental data indicate that the insect preferentially oxidizes the major component in a mixture, while the terpene in lesser proportion acts as a toxicant, with higher toxicity than when it was assayed alone. These findings contribute to a deeper understanding of the higher toxicity of essential oils compared to their component terpenes and provide important information for the design of effective insecticides based on essential oils or terpenes (13).

13.4.3 Terpene-Based Bio-Pesticides as Potential Alternatives to Synthetic Insecticides for Control of Aphid Pests on Protected Ornamentals

Continued reliance on synthetic pesticides to provide effective control of crop pests, weeds, and diseases is under increasing pressure. The number of active ingredients permitted for use in the European Union (EU) has declined from *c.* 1000 actives in 1993 to around 250 in 2011 (14). This decline is largely the consequence of regulatory changes, e.g. implementation of EU Regulation 1107/2009 and an associated shift from risk to hazard based assessment of pesticide safety in terms of human health and the environment. There are also financial and time constraints on the development of active ingredients. The cost of bringing a new active ingredient to market has increased from $152 million in 1995 to $256 million in 2005 (14), while the time taken to develop and register a new conventional pesticide is now around 10 years (15). As a result, while there were around 70 new active ingredients in the development pipeline in 2000, this number had dropped to 28 in 2012 (14). These challenges facing the industry are compounded by increasing numbers of cases of pesticide resistance, with over 580 arthropod species being recorded as having developed resistance globally (16). Biopesticides offer a promising alternative to the use of conventional synthetic pesticides due to reduced risk of resistance developing in pest populations, lower development costs, higher target specificity, lower environmental persistence, and generally improved compatibility with biological controls (17, 18). Plant essential oils are synthesized through secondary metabolic pathways and have long been recognized to possess insecticidal and/or repellent properties (19). Although plant essential oils are typically complex mixtures, they are often dominated by two or three chemical compounds that can be usually classified into two chemical groups, terpenes or phenylpropanoids (20).

There are many examples of biopesticides based on terpenes, including orange or citrus oils, essential oil derived from *Chenopodium ambrosioides* variety nr. *ambrosioides* (Chenopodiaceae), and neem extracts. Orange or citrus oils include *d*-limonene as the major component and are known to be toxic to a wide range of insect pest species (21). Products based on orange or citrus oils work through contact as well as, in some cases, fumigation action. The essential oil derived from *Chenopodium ambrosioides* variety nr. *ambrosioides* L., consists of a mixture of 14 monoterpenes and is known to be toxic to a range of insect and mite pest species. Finally, oil extracted from *Azadirachta indica* A. Juss. (neem) seeds includes, as the major compound, azadirachtin A (a triterpene), which has been shown to have anti-feedant and repellent properties as well as inducing sterility in insects, such as the peach-potato aphid (*Myzuspersicae* (Sulzer) (22). Although biopesticides based on plant essential oils have previously been shown to have insecticidal effects, their modes of action are poorly understood. Around 100 species of aphid are considered to be agricultural pests of a wide range of crops (23). Two of the most important aphid pest species are *M. persicae* and the melon and cotton aphid (*Aphis gossypii* Glover) due to the wide range of crops that they may infest and their ability to develop resistance to conventional synthetic insecticides, such as carbamates (24), pyrethroids (24) and neonicotinoids (25).

Biopesticides based on plant extracts offer a promising alternative to the use of conventional synthetic pesticides. However, biopesticide products must provide acceptable levels of control. To date, few studies have investigated the efficacy of biopesticide products under conditions that reflect commercial practice. Here we report results from three experiments, one completed under glasshouse conditions in 2014 and two completed under polytunnel conditions in 2015 and 2016, respectively. These experiments tested the efficacy of three terpene based biopesticides used to control two aphid species, peach-potato aphid (*Myzuspersicae*) and melon and cotton aphid (*Aphis gossypii*), on ornamental crops. The three biopesticide products tested were orange oil (60 g active ingredient per liter, formulated as a soluble liquid), the essential oil from *Chenopodium ambrosioides* variety nr. *ambrosioides* (16.75% active ingredient, formulated as an oil dispersion), and neem oil (1% active ingredient, formulated as emulsifiable concentrate). The biopesticides tested were applied as foliar sprays using a water volume of 600 l/Ha, and all experiments were done at Harper Adams University, Shropshire, UK. The biopesticide products tested gave statistically similar levels of control of *M. persicae* populations on pansy plants as the conventional synthetic insecticide flonicamid (150 g/l active ingredient, formulated as a wettable granule) and spirotetramat (150 g/l active ingredient, formulated as an oil dispersion). All products reduced numbers of aphids by at least 85% during the experimental period. Orange oil also gave a similar speed of kill to flonicamid and was faster acting than spirotetramat, two conventional synthetic insecticides that are widely used to control aphid pests of ornamental crops. Against *A. gossypii* on Hebe, orange oil gave similar levels of control (90% reduction in aphid numbers) as flonicamid (98% reduction in aphid numbers) when applied with a spray interval of three days (per label recommendation). The essential oil from *Chenopodium ambrosioides* variety nr. *ambrosioides* was not as effective as flonicamid but significantly reduced (80% reduction in aphid numbers) numbers of *A. gossypii* on *Hebe* compared to a water control when applied with a spray interval of five days. Neem oil was not effective against *A. gossypii*. Importantly, there was little evidence of any phytotoxicity caused by any of the biopesticide products tested. The potential to use these products as part of an insecticide resistance management (IRM) program is discussed (26).

13.4.3.1 Therapeutic and Medicinal Uses of Terpenes

Among the natural products that provide medical benefits for an organism, terpenes play a variety of major roles. The common plant sources of terpenes are tea, thyme, cannabis, Spanish sage, and citrus fruits (e.g., lemon, orange, mandarin). Terpenes have a wide range of medicinal uses, among which antiplasmodial activity is notable, as its mechanism of action is similar to the popular antimalarial drug in use, chloroquine. Monoterpenes specifically are widely studied for their antiviral properties. With growing incidents of cancer and diabetes in the modern world, terpenes also have the potential to serve as anticancer and antidiabetic reagents. Along with these properties, terpenes also allow for flexibility in route of administration and suppression of side effects. Certain terpenes were widely used in natural folk medicine. One such terpene is curcumin, which holds anti-inflammatory, antioxidant, anticancer, antiseptic, antiplasmodial, astringent, digestive, diuretic, and many other properties. Curcumin has also become a recent trend in healthy foods and opens doors for several medical studies. This chapter summarizes the various terpenes, their sources, medicinal properties, mechanism of action, and the recent studies that are underway for designing terpenes as a lead molecule in modern medicine.

13.4.3.2 Properties Associated with Terpene

Important properties associated with terpene are difficult to overstress. There are many important uses with terpene, and these include anti-insect properties, antimicrobial properties, and anti-herbivore properties. Terpene can be extracted through plants and some insects.

13.4.3.3 Anti-Insect

Without using harsh chemicals that could potentially contain side effects, terpene is a healthy alternative to ward off insects. There have been many pesticides made for killing domestic pests like lice or mites. In these cases, it is very important to make sure that these pesticides do not affect humans in harmful ways. There are many options like shampoo, sprays, and lotions that were manufactured against pests that include one or more terpenes that are employed in the instant invention. These naturally occurring terpenes are generally not modified; they were used in their raw form, and the Environmental Protection Agency in the USA classified as them GRAS (generally regarded as safe). Certain terpenes are highly effective against both lice and lice eggs, and there is a less than significant chance of resistance developing against terpene-based pesticides; the reason for this is their observed modes of action. Unlike other types of pediculosis medication, terpene-based instant interventions are not neurotoxins. Terpenes are also used combined with a terpene aldehyde called citral. Citral derives from an essential oil that is extracted from lemongrass. Citral possesses antibacterial and antifungal properties, while lemongrass possesses anti-insect properties. A series of anti-insect formulation contains many terpenes. Most of these pesticides are a mix of terpene and citral (27). Table 13.1 shows what these terpenes include.

TABLE 13.1
Terpenes Added in Anti-Insect Formulations

Terpene type	Function	Features
Limonene	This is strongly preferred. Limonene enhances the properties of other terpenes	Redistilled limonene has less odor and is more stable than D-limonene
Beta-ionone	Antibacterial and antifungal properties	Beta-ionone has prophylactic value
Geraniol	Similar level activity to beta-ionone. Geraniol possesses antibacterial and antifungal properties	Geraniol gives a pleasant fragrance
Eugenol	This is also the active terpene in clove oil. This possesses anesthetic properties, which help with the itching that comes with bug bites. Also has antibacterial and antifungal properties	Contains a distinct fragrance, which is like geraniol
Myrcene	Possesses antifungal, antibacterial properties	Famous for its fragrance properties

13.4.3.4 Anticancer

The medicinal benefits of terpenes are not limited to pathogenic diseases. Terpenes are widely acclaimed for their anticancer activity, too. An early 1997 study concluded that a combination of monoterpenes, diterpenes, and sesquiterpenes can be effectively used to treat cancers that occur in the colon, brain, prostate glands, and bones (28). It also claimed that administration of terpenes in humans inhibited the growth of prostate cancer cells and sensitized the tumor in such a way that it becomes susceptible to radiotherapy.

The major advantage of this treatment is that the drug can be administered through several routes, among which oral and topical were preferred.

Among the different kinds of terpenes, limonene is well recognized as an anticancer agent. Limonene is a bioactive food component found in citrus peels, orange peels, and several other citrus fruits (26). Studies have reported limonene to exhibit strong cancer inhibition activity both in vitro and in vivo. The mechanism behind limonene activity is still under investigation. A study (26) reported that limonene acts through induction of transforming growth factor B-1 and mannose-6-phosphate/insulin-like growth factor II receptors. In contrast, a study (27) suggested that limonene eliminates cancer cells by induction of apoptosis. Structural studies on limonene reported that it is lipophilic and has a tendency to be deposited in fatty tissues when administered orally. This indicates that limonene can act as an excellent chemopreventive drug for cancer, as it can be deposited in the body (29). Another study in 2013 concluded that limonene acts by suppressing the expression of breast tumor cyclin D1 (30). This led to cell cycle arrest and mitigated proliferation of cancer cells in women with early stages of breast cancer. A recent study showed that limonene from pinecones can kill lung cancer cells in vitro by an apoptotic mechanism that is activated through the caspase-3 pathway (31), fighting and preventing cancer. Not just limonene, but also its metabolite, perillyl alcohol, is also said to exhibit antitumor activity in pancreatic cell lines through apoptotic mechanisms (32).

Apart from limonene, the terpene thymoquinone has been widely studied for its chemoprotective and chemotherapeutic activity. Thymoquinone is found to be an active constituent of the volatile oils of an annual herbaceous plant, black cumin (33). The pathways affected by thymoquinone to exert its antitumor properties are the p53, PPARγ, MAPK, NF-κB, PI3K/AKT, and STAT3 signaling pathways. Thymoquinone has proved to be anticancerous against several cancers such as breast cancer, skin cancer, non–small cell lung cancer, bile duct cancer, and brain cancer. The basic mechanisms underlying the cancer inhibition are apoptosis and cell cycle arrest. Most of the cancer-related studies were performed using thermoquinone obtained from *N. sativa* extracts. A 2012 study showed that thermoquinone can be obtained in larger amounts from the mint family, *Monarda didyma* and *Monarda media* (34). Thus, thermoquinone from alternative sources has to be tested for its precious potential in cancer therapy. Other terpenes that have reported cytotoxic effects on cancer cells include alloocimene, camphor, beta-myrcene, pinene, alpha- and gamma-thujaplicin, terpinene, thymohydroquinone, carvone, camphene, and cymene. Terpenes, being natural compounds, are unlikely to affect healthy cells or create side effects, which attracts many researchers to exploit their capability in cancer treatment.

13.5 ANTIDIABETIC

Diabetes is one of the widely prevalent diseases in the world. It affects both children and adults in both developing and developed nations (35, 36). The social and economic burden of diabetes continues to grow, and it is expected to rise rapidly in developing countries (37). In the USA, diabetes is one of the leading causes of visual impairment, limb amputation, renal diseases, heart diseases, and death (38). Diabetes can be of two types—type 1 (where the immune system of the body acts against the insulin-producing organs) and type 2 (where the insulin produced cannot be used by the body or insulin is produced in low amounts) (39). Although there are several medications available, their use is limited due to their adverse effects. Some of the commonly found side effects include low blood sugar, vomiting, nausea, diarrhea, bloating, and weight gain (40). This led to research for natural products to be used as effective antidiabetic medication. Phytochemicals from the medicinal plants have been recommended for treating type 2 diabetes, of which terpene forms a major constituent (41).

Medicinal plants of Morocco were studied for their antidiabetic properties in rats. The report showed that terpenes, terpene diols, and terpene diol glucosides form major components of the extracts of the plants under study (42). This study was focused on non–insulin-dependent diabetes mellitus (type 2), and it proved that terpenes, along with few other secondary metabolites such as alkaloids and flavonoids, exhibit antidiabetic potential (43).

The most promising terpene compound for treating diabetes is called rographolide, which is a diterpenoid lactone (44). This compound forms the major component of the leaves of the small herbaceous plant *A. paniculata*. It is an Asian plant that has already been reported to be used in traditional medicines for its therapeutic nature (44). The terpenoid acts by reducing plasma glucose and increasing the utilization of glucose by the body in diabetes mellitus rats (45). The actual mechanism by which it does this is that it activates the alpha-adrenoreceptors to increase the release of an opioid peptide, beta-endomorphin (44), which is reported to be secreted in low amounts in diabetic rats (46). This increased secretion in turn activates the opioid μ-receptors. These receptors can effectively curb hepatic gluconeogenesis (glucose synthesis from non-carbohydrate precursors) and elevate the utilization of glucose by muscles. Finally, this results in a reduced plasma glucose concentration (44). Andrographolide is also observed to prevent the secondary complications of diabetes such as diabetic retinopathy, a condition that will lead to blindness (44). It significantly weakens retinal angiogenesis and inflammation during the development of the disease (44). Moreover, it can also fix impaired or extended estrous cycle in diabetic rats (46). Andrographolide was orally administered in all the previous studies. This indicates its efficiency for being used as a lead molecule in future drugs designed for treating diabetes mellitus.

Another widely known terpene is curcumin obtained from turmeric (47). It exhibits high antidiabetic properties and acts by quashing oxidative stress and inflammation. By regulating the polyol pathway, it can also reduce plasma glucose and levels of glycosylated hemoglobin (47). Moreover, curcumin is also reported to activate the enzymes present in the liver that are essential for glycolysis, gluconeogenesis, and lipid metabolism (48). Like andrographolide, curcumin is also reported to reduce the complications of diabetes (47), for example, liver disorder, which is a common manifestation of diabetes type 2 (48). Curcumin treats these disorders by reducing the liver weight and lipid peroxidation products. Further, it is also reported to normalize the levels of fetuin-A in a serum that contributes to insulin resistance and fatty liver in diabetic rats (48). Other complications that can be attenuated by curcumin include diabetes-associated retinopathy, microangiopathy, neuropathy, and nephropathy (48). These findings confirm that curcumin is likely to be used in the future for diabetes treatment.

13.6 ANTIDEPRESSANTS

Depression has become a serious health concern by contributing to emerging mental and emotional disorders throughout the world. It is hitting both developed and developing countries. Depression can pave the way to various health issues from alcoholism to heart diseases (49). It is also said to increase the rate of mortality significantly in breast cancer patients (50). Moreover, depression immobilizes its victims, thereby leading to economic loss (51). By analyzing the social and economic burdens caused by depression, researchers have stepped towards finding novel stress-relieving drugs. Synthetic drugs have serious side effects and unintended interactions with the body that negatively affect treatment outcome (52). This necessitated the need for natural drugs. Terpenes serve as one of the most relevant bioactive compounds for treating depression and therefore can open doors for designing natural or synthetic antidepressant drugs (53).

Twenty-five percent of antidepressant drugs prescribed by doctors are obtained from herbs through various extracts (54). To estimate the important compounds contributing to the antidepressant effect, (54) an electronic database-based study was performed. The results revealed that terpenes formed a major part of the extracts of medicinal plants that exerted antidepressant effects (54) in extracts contributing to antistress effects. Different plants had different acting compounds.

Among the several terpenes, linalool and beta-pinene are commonly found to be active principles. They were discovered from the extracts of medicinal plants and flowers of lavender (55). These monoterpenes act by interacting with the 5HT1A receptors of the serotonergic pathway. Serotonins are important in the fact that their release and re-uptake levels can be altered to overcome stress. They also interact with adrenergic receptors of the body that play a major role in stress-induced behavioral changes (56, 57). Another interesting finding is the interaction of beta-pinene with dopaminergic receptors, namely Dl receptors. This is the mechanism followed by most of the antidepressant drugs available in the market (58). A more interesting study would be to examine beta-pinene and linalool efficiency through inhalation tests. This is because these monoterpenes are aromatic compounds that generally have enhanced activity when inhaled, as they can directly hit the central nervous system.

Apart from monoterpenes, sesquiterpenes also exhibit antidepressant effects. One striking example is beta-caryophyllene, which was proved to ameliorate depressive symptoms in mice. The underlying mechanism of this compound is binding to a receptor called CB2 and activating it. CB2 is found in the brain and immune cells and plays a major role in regulating depressive-related disorders. Thus beta-caryophyllene curbs depression by acting as a CB2 receptor agonist (59).

Other terpenes that have effective antidepressant properties include hyperforin (60). It has been shown that the extracts of *H. perforatum* differ in their antidepressant potential with the difference in concentration of hyperforin present in the extract (61). Similar to many other antidepressants, hyperforin acts by inhibiting the neuronal uptake of mood regulators such as serotonin, dopamine, and norepinephrine. In addition, it also has its own unique mechanism of controlling depression by inhibiting the neurotransmitters GABA and L-glutamate uptake (62).

Another fascinating antidepressant plant is *Valeriana,* which is a short perennial herb. This plant not only reduces stress and anxiety levels but also improves the symptoms of depression in humans. The major components of Valeriana extracts are terpenoids called maaliol, patchouli alcohol, and 8-acetoxypatchouli alcohol. The terpenoid-less extract of Valeriana was found to be devoid of antidepressant activity, which indicates that terpenes are the active components involved in reducing depression.

13.7 USES IN FOLK MEDICINE

Folk medicine has always been an inspiration for designing novel drugs for diseases. To be more specific, almost three-fourths of plant-based drugs were created based on the knowledge of folk medicine (Table 13.2) (63). Realizing this fact, the western world is now turning back to old medicines and bioactive plant components to treat modern diseases (63). This has boosted the export rates of Chinese medicinal products (based on traditional Chinese medicine) from China to other developed nations. Plants used in traditional Chinese medicine (TCM) are being extensively studied for their secondary metabolites and their therapeutic properties (64). One of the active principles of TCM products is terpenes (65). Due to their high availability and diversity, terpenes contribute the most to industrial and medicinal applications among all the secondary metabolites of plants (66).

TABLE 13.2
Drugs Created Based on Knowledge of Folk Medicine

Scientific Name	Common Name	Abundant Terpene	Uses in Ayurveda	Uses in TCM
Citrus limon	Lemon	Limonene	Cavities, digestive problems, abdominal colic pain, and cough	Digestive problems and cleansing the body
Citrus reticulata	Orange	Limonene	Digestive disorders, abdominal colic pain, and worm infestation	Stomach ache and cough
Juniperus communis	Juniper	Limonene	Antiseptic, cellulite, pain, and swelling	Colds and urinary problems
Phyllanthus emblica	Indian gooseberry	Phyllaembicilins	Boost immunity Strengthen hair follicles, cure acne and pimples, and improve circulatory system Cure diarrhea	Diarrhea, jaundice, and inflammation
Panax sp.	Ginseng	Humulene	Boosts energy Used to treat musculoskeletal problems such as rheumatism, arthritis, and so on	• Memory booster • Reduce fatigue • Reduce menopause symptoms
Cinnamomum verum	Cinnamon	Alpha-pinene, caryophyllene, linalool, alpha-phelandrene, cymene, humulene	Cold, diabetes, high cholesterol, digestive problems, bronchitis, sinus congestion	Cold, diabetes, high cholesterol, digestive problems; control sweating Chest pain
Lyciumchinense	Goji berry	Beta-carotene	Maintain kidney functions Improve eyesight, fertility, circulation Increase lifetime	Improve eyesight, fertility, circulation; increase lifetime
Zingiber officinale	Ginger	Zingiberene	Digestive problems, Joint pain and air sickness	Cold, cough, wheezing, asthma
Allium sativum	Garlic	Nerolidol, alpha-pinene, and terpinolene	Treat pimples, tumor, snakebites, wounds, headache, heart disease, gastric problems, ulcer, and measles	Food poisoning and digestive problems
Ocimumtenuiflorum	Holy basil/tulsi	Eugenol, β-elemene, β-caryophyllene, and germacrene	• Restore functions of nervous system • Increase fertility • Used to treat asthma and cold	Restore functions of the nervous system

REFERENCES

1. Christianson DW. Structural and chemical biology of terpenoid cyclases. *Chem Rev.* 2017;117(17):11570–11648. doi:10.1021/acs.chemrev.7b00287. PMC 5599884. PMID 28841019.
2. Firn R. *Nature's chemicals.* Oxford: Biology; 2010.
3. Ashour M, Wink M, Gershenzon J. Biochemistry of terpenoids: Monoterpenes, sesquiterpenes and diterpenes. In: *Biochemistry of Plant Secondary Metabolism*; 2020, pp. 258–303. doi: 10.1002/9781444320503.ch5. ISBN 9781444320503.
4. Breitmaier E. *Terpenes: Flavors, Fragrances, Pharmaca, Pheromones.* Wiley-VCH; 2006. doi: 10.1002/9783527609949. ISBN 9783527609949.

5. Singh B, Sharma RA. Plant terpenes: Defense responses, phylogenetic analysis, regulation and clinical applications. *3 Biotech.* 2015;5(2):129–151. doi: 10.1007/s13205-014-0220-2. ISSN 2190-572X. PMC 4362742. PMID 28324581.
6. Davis EM, Croteau R. Cyclization enzymes in the biosynthesis of monoterpenes, sesquiterpenes, and diterpenes. In: *Biosynthesis*; 2000, pp. 53–95. doi: 10.1007/3-540-48146-X_2. ISBN 978-3-540-66573-1. {{cite book}}: |journal= ignored (help)
7. What are Terpenes. rareterpenes.com. 13 April 2021.
8. Ludwiczuk A, Skalicka-Woźniak K, Georgiev MI. Terpenoids. In: *Pharmacognosy*; 2017, pp. 233–266. doi:10.1016/B978-0-12-802104-0.00011-1. ISBN 9780128021040.
9. Ayoola GA. Phytochemical screening and antioxidant activities of some selected medicinal plants used for malaria therapy in Southwestern Nigeria. *Trop J Pharm Res.* 2008;7(3):1019–1024. doi:10.4314/tjpr.v7i3.14686.
10. Huang M, Huang J, Bao X, Wang YT. Terpenoids: Natural products for cancer therapy. *Published Online.* 2012;1801–1818.
11. Zhang L, Yan J, Fu Z, Yang T, Zeng H. Biochemistry of terpenes and recent advances in plant protection. *Int J Mol Sci.* 2021;22(11):10. https://doi.org/10.3390/ijms22115710.
12. Jimena P, Herrera M, Zunino MP, et al. Terpene ketones as natural insecticides against Sitophilus zeamais. *Ind Crops Prod.* 2015;70:435–442.
13. Scalerandi E, Flores GA, Palacio M, et al. Understanding synergistic toxicity of terpenes as insecticides: Contribution of metabolic detoxification in Musca domestica. *Plant Sci.* 2018;9:1579. https://doi.org/10.3389/fpls.2018.01579.
14. Chapman P. Is the regulatory regime for the registration of plant protection products in the EU potentially compromising food security? *Food Energy Secur.* 2014;3(1):1–6.
15. Glare T, et al. Have biopesticides come of age? *Trends Biotechnol.* 2012;30(5):250–258.
16. Sparks TC, et al. IRAC: Mode of action classification and insecticide resistance management. *Pestic Biochem Physiol.* 2015;121:122–128.
17. Copping LG, et al. Biopesticides: A review of their action, applications and efficacy. *Pest Manag Sci.* 2000;56(8):651–676.
18. Pastor S, Creus A, Parrón T, et al. Biomonitoring of four European populations occupationally exposed to pesticides: Use of micronuclei as biomarkers. *Mutagenesis.* 2003;18:249–258. doi: 10.1093/mutage/18.3.249. [PubMed] [CrossRef] [Google Scholar]
19. Isman MB. Plant essential oils for pest and disease management. *Crop Prot.* 2000;19(8–10):603–608.
20. Cooper J, Dobson H. The benefits of pesticides to mankind and the environment. *Crop Prot.* 2007;26:1337–1348. doi: 10.1016/j.cropro.2007.03.022. [CrossRef] [Google Scholar]
21. Rosell G, Quero C, Coll J, Guerrero A. Biorational insecticides in pest management. *J Pestic Sci.* 2008;33:103–121. doi: 10.1584/jpestics.R08-01. [CrossRef] [Google Scholar]
22. Chaudhary C, et al. Progress on Azadirachta indica based biopesticides in replacing synthetic toxic pesticides. *Front Plant Sci.* 2017;6:610.
23. Verger PJ, Boobis AR. Reevaluate pesticides for food security and safety. *Science.* 2013;341:717–718. doi: 10.1126/science.1241572. [PubMed] [CrossRef] [Google Scholar]
24. Kim KH, Kabir E, Jahan SA. Exposure to pesticides and the associated human health effects. *Sci Total Environ.* 2017;575:525–535. doi: 10.1016/j.scitotenv.2016.09.009. [PubMed] [CrossRef] [Google Scholar]
25. Smith GH, Roberts JM, Pope TW. Terpene based biopesticides as potential alternatives to synthetic insecticides for control of aphid pests on protected ornamentals. *Crop Prot.* 2018;110:125–130.
26. Jirtle RL, et al. Increased mannose 6-phosphate/insulin-like growth factor II receptor and transforming growth factor beta 1 levels during monoterpene induced regression of mammary tumors. *Cancer Res.* 1993;53(17):3849–3852. [PubMed] [Google Scholar]
27. Bishayee A, Rabi T. D-Limonene sensitizes docetaxel-induced cytotoxicity in human prostate cancer cells: generation of reactive oxygen species and induction of apoptosis. *J Carcinogen.* 2009;8(1):9. doi: 10.4103/1477-3163.51368.
28. Gould MN. Cancer chemoprevention and therapy by monoterpenes. *Environ. Health Perspect.* 1997;105(suppl 4):977–979.
29. Miller JA, et al. Limonene: A bioactive food component from citrus and evidence for a potential role in breast cancer prevention and treatment. *Oncol Rev.* 2011;5(1):31–42. doi: 10.4081/oncol.2011.31.
30. Miller JA, et al. Human breast tissue disposition and bioactivity of limonene in women with early-stage breast cancer. *Cancer Prev Res.* 2013;6(6):577–584. doi: 10.1158/1940-6207.CAPR-12-0452.

31. Lee TK, et al. Pinecone of Pinus koraiensis inducing apoptosis in human lung cancer cells by activating Caspase-3 and its chemical constituents. *Chem Biodivers.* 2017;14(4):1612–1880. doi: 10.1002/cbdv.201600412.
32. Sobral MV, et al. Antitumor activity of monoterpenes found in essential oils. *Sci World J.* 2014;2014:953451. doi: 10.1155/2014/953451.
33. Majdalawieh AF, et al. Anti-cancer properties and mechanisms of action of thymoquinone, the major active ingredient of Nigella sativa. *Crit Rev Food Sci Nutr.* 2017;57(18):3911–3928. doi: 10.1080/10408398.2016.1277971.
34. Taborsky J, et al. Identification of potential sources of thymoquinone and related compounds in Asteraceae, cupressaceae, lamiaceae, and ranunculaceae families. *Open Chem.* 2012;10(6).
35. You W, Henneberg M. Type 1 diabetes prevalence increasing globally and regionally: the role of natural selection and life expectancy at birth. *BMJ Open Diabetes Res Care.* 2016;4 (1): e000161. doi: 10.1136/bmjdrc-2015-000161. [PMC free article] [PubMed] [CrossRef] [Google Scholar]
36. Narayan V, et al. Diabetes—a common, growing, serious, costly, and potentially preventable public health problem. *Diabetes Res Clin Pract.* 2000;50(2):77–84. doi: 10.1016/S0168-8227(00)00183-2. [PubMed] [CrossRef] [Google Scholar]
37. Sarwar N, et al. Diabetes mellitus, fasting blood glucose concentration, and risk of vascular disease: a collaborative meta-analysis of 102 prospective studies. Emerging risk factors collaboration. *Lancet.* 2010;26(375):2215–2222. [PMC free article] [PubMed] [Google Scholar]
38. Saddinne JB, et al. Prevalence of self-rated visual impairment among adults with diabetes—United States. *Am J Public Health.* 1999;89(8):1200–1205. doi: 10.2105/AJPH.89.8.1200. [PMC free article] [PubMed] [CrossRef] [Google Scholar].
39. www.diabetes.ca/about-diabetes/types-of-diabetes. Accessed 23 Jan 2018.
40. www.diabetes.co.uk/features/diabetes-medication-side-effects.html. Accessed 23 Jan 2018.
41. Jung M, et al. Antidiabetic agents from medicinal plants. *Curr Med Chem.* 2006;13(10):1203–1218. doi: 10.2174/092986706776360860. [PubMed] [CrossRef] [Google Scholar].
42. Bnouham M, et al. Antidiabetic effect of some medicinal plants of oriental Morocco in neonatal non-insulin-dependent diabetes mellitus rats. *Hum Exp Toxicol.* 2010;29(10):865–871. doi: 10.1177/0960327110362704. [PubMed] [CrossRef] [Google Scholar]
43. Jung M, et al. Antidiabetic agents from medicinal plants. *Curr Med Chem.* 2006;13(10):1203–1218. doi: 10.2174/092986706776360860. [PubMed] [CrossRef] [Google Scholar]
44. Brahmachari G. *Discovery and Development of Antidiabetic Agents from Natural Products Natural Product Drug Discovery.* Amsterdam: Elsevier; 2017. [Google Scholar]
45. Gupta R, et al. An overview of Indian novel traditional medicinal plants with anti-diabetic potentials. *African J Tradition Complement Alternat Med.* 2008;51:1–17. [PMC free article] [PubMed] [Google Scholar]
46. Forman LJ, et al. Diabetes induced by streptozocin results in a decrease in immunoreactive beta-endorphin levels in the pituitary and hypothalamus of female rats. *Diabetes.* 1985;34(11):1104–1107. doi: 10.2337/diab.34.11.1104. [PubMed] [CrossRef] [Google Scholar]
47. Nabavi S, et al. Curcumin: A natural product for diabetes and its complications. *Curr Top Med Chem.* 2015;15(23):2445–2455. doi: 10.2174/1568026615666150619142519. [PubMed] [CrossRef] [Google Scholar]
48. Zhang DW, et al. Curcumin and diabetes: A systematic review. *Evid Based Complement Alternat Med.* 2013;2013:636053. [PMC free article] [PubMed] [Google Scholar]
49. Holden C. Mental health: Global survey examines impact of depression. *Science.* 2000;288(5463):39–40. doi: 10.1126/science.288.5463.39. [PubMed] [CrossRef] [Google Scholar]
50. Hjerl K, et al. Depression as a prognostic factor for breast cancer mortality. *Psychosomatics.* 2003;44(1):24–30. doi: 10.1176/appi.psy.44.1.24. [PubMed] [CrossRef] [Google Scholar]
51. Jawaid T, Gupta R, Siddiqui ZA. A review on herbal plants showing antidepressant activity. *Int J Pharm Sci Res.* 2011;90(24):3051–3060. [Google Scholar]
52. Jawaid T, Gupta R, Siddiqui ZA. A review on herbal plants showing antidepressant activity. *Int J Pharm Sci Res.* 2011;90(24):3051–3060. [Google Scholar]
53. Bahramsoltani R, et al. Phytochemical constituents as future antidepressants: A comprehensive review. *Rev Neurosci.* 2015;26(6):699–719. doi: 10.1515/revneuro-2015-0009. [PubMed] [CrossRef] [Google Scholar]
54. Saki K, et al. The effect of most important medicinal plants on two importnt psychiatric disorders (anxiety and depression)-a review. *Asian Pac J Trop Med.* 2014;7:S34–S42. doi: 10.1016/S1995-7645(14)60201-7. [PubMed] [CrossRef] [Google Scholar]

55. Appleton J. Lavender oil for anxiety and depression. *Nat Med J.* 2012;4(2).
56. Pandey SC, et al. β-Adrenergic receptor subtypes in stress-induced behavioral depression. *PharmacolBiochemBehav.* 1995;51(2–3):339–344. doi: 10.1016/0091-3057(94)00392-V. [PubMed] [CrossRef] [Google Scholar
57. Guzmán-Gutiérrez SL, et al. Linalool and β-pinene exert their antidepressant-like activity through the monoaminergic pathway. *Life Sci.* 2015;128:24–29. doi: 10.1016/j.lfs.2015.02.021. [PubMed] [CrossRef] [Google Scholar]
58. Guzmán-Gutiérrez SL, et al. Linalool and β-pinene exert their antidepressant-like activity through the monoaminergic pathway. *Life Sci.* 2015;128:24–29. doi: 10.1016/j.lfs.2015.02.021. [PubMed] [CrossRef] [Google Scholar]
59. Bahi A, et al. β-Caryophyllene, a CB2 receptor agonist produces multiple behavioral changes relevant to anxiety and depression in mice. *PhysiolBehav.* 2014;135:119–124. doi: 10.1016/j.physbeh.2014.06.003. [PubMed] [CrossRef] [Google Scholar]
60. Subhan F, et al. Terpenoid content of Valeriana wallichii extracts and antidepressant-like response profiles. *Phytother Res.* 2010;24:686–691. [PubMed] [Google Scholar]
61. Laakmann G, et al. St. Johns wort in mild to moderate depression: The relevance of hyperforin for the clinical efficacy. *Complement Ther Med.* 1999;7(4):265. [PubMed] [Google Scholar]
62. Müller WE, et al. Hyperforin—antidepressant activity by a novel mechanism of action. *Pharmacopsychiatry.* 2001;34(1):98–102. doi: 10.1055/s-2001-15512. [PubMed] [CrossRef] [Google Scholar]
63. Efferth T, et al. Phytochemistry and pharmacogenomics of natural products derived from traditional Chinese medicine and Chinese Materia Medica with activity against tumor cells. *Mol Cancer, Ther.* 2008;7(1):152–161. doi: 10.1158/1535-7163.MCT-07-0073. [PubMed] [CrossRef] [Google Scholar]
64. Efferth T, et al. Molecular target-guided tumor therapy with natural products derived from traditional Chinese medicine. *Curr Med Chem.* 2007;14 (19):2024–2032. doi: 10.2174/092986707781368441. [PubMed] [CrossRef] [Google Scholar]
65. Liu QM, Jiang JG. Antioxidative activities of medicinal plants from TCM. *Mini-Rev Med Chem.* 2012;12(11):1154–1172. doi: 10.2174/138955712802762239. [PubMed] [CrossRef] [Google Scholar]
66. Zwenger S, Basu C. Plant terpenoids: applications and potentials. *Biotechnol Mol Biol Rev.* 2008;3:1–7. [Google Scholar]

14 Traditional Knowledge and the Development of Botanical Biocides in Brazil for the Control of *Aedes aegypti*

Paula Maria Correa de Oliveira, João Paulo Barreto Sousa, Lorena C. Albernaz, Ellem Suane Ferreira Alves, Laila Salmen Espindola, and Márlia Coelho-Ferreira

14.1 INTRODUCTION

Records from 2000 B.C. on Hindu populations already point to the use of plants to control insects considered pests, either because they act to destroy crops or to transmit diseases. The Egyptians during the time of the Pharaohs and the Chinese around 1200 B.C. also used plants as sources of insecticidal compounds [1]. Travel between the Old and New Worlds allowed for the exchange of plants, animals, and, consequently, insects that caused disease. However, these trips also led to the discovery of new ways to control these organisms, as many native cultures already used plant extracts to control pests [2]. The seeds of the *Sabadilla officinarum* Brandt lily, for example, were used by Indigenous Venezuelans for this purpose in the mid-20th century, while in North America, nicotine extracted from *Nicotiana tabacum* L. was the main insecticidal compound used. Another example is the use of cinchona (*Quassia amara* L.) by Central American indigenous people against pathogenic insects. In China and India, the *Acorus calamus* L. was the main species used in this sense. From the exchange of knowledge during these trips, the chemical control of arthropods, based on plants, spread throughout the world [1].

Despite the advent of organo-synthetic insecticides after World War II, plants remain promising candidates for the formulation of insecticides and biocides. This is due to the increase in problems related to insect resistance to synthetic organo-synthetic insecticides, resurgence and eruption of pests, and the problems arising from the indiscriminate use of these substances in the environment [3].

There are different ways of selecting plants for the discovery and development of new drugs, which can directly influence the time, effectiveness, costs, and other aspects that involve the preparation of a ready-to-use drug. The pathways used can be random selection followed by chemical screening, random selection followed by one or more biological assays, monitoring biological activity reports, use of databases, and ethnobiological studies of plants [4].

This last approach, in general, is based on studies in the area of ethnobotany and ethnopharmacology, which have contributed substantially to the discovery of drugs since the 19th century [5]. Several studies in these areas have investigated the conception of local populations about insecticidal plants, as well as diseases transmitted by mosquitoes, and have shown promising species for the control of vectors such as *Aedes aegypti* [6–9].

Aedes aegypti is one of the most important mosquito species vectors of the viruses that cause dengue, Zika, chikungunya, and urban yellow fever arboviruses, which together cause thousands of deaths annually in most of the world [10]. The control of this mosquito has been a public health

 DOI: 10.1201/9781003463429-18

challenge, since its distribution currently occurs in all continents, except Antarctica. Studies indicate that global climate change can boost the adaptation of this species to habitats that were previously inhospitable and may become favorable [11]. In view of this, this study presents a bibliographic review of plants traditionally documented in Brazil as repellents and/or insecticides in order to highlight potential species to advance research on the control of *Aedes aegypti*.

14.2 MATERIALS AND METHODS

The compiled studies were found in Scopus, Google Academic, Scielo, Pubmed, Science Direct, FioCruz, and CAPES Portal de Periódicos databases, using the keywords: ethnobotanical studies, Ethnopharmacology, traditional medicine, medicinal plants, repellent plants, and plants insecticides.

The selection research followed the criteria: (i) ethnobotanical studies developed in Brazil from 2018 to 2022, published in national and international journals, in Portuguese, English, and Spanish (ii) that presented one or more reports on the use of plants by local and/or traditional populations as repellents, insecticides, or for "insect bites" in (iii) works with presentation of sufficiently clear methodological guidelines regarding the precision in the identification of species.

14.3 RESULTS AND DISCUSSION

Among the 424 works consulted, 45 recorded plant species were traditionally used against mosquitoes (Table 13.1). These studies, developed in the field of ethnobotany and ethnopharmacology, were realized in municipalities from 16 states located in the five regions of Brazil. These represent a heterogeneous sample of the different sociocultural contexts such as indigenous lands, extractive reserves, urban areas, riverside communities, quilombolas, and small farmers.

The higher number of species was recorded in the north region (39%), followed by the northeast (21%), southeast (14%), south (13%), and midwest (12%). In recent years, the north region has been a stage for the endemism of diseases such as dengue, malaria, leishmaniasis, and others classified among the neglected diseases [12]. The epidemiological profile of a given region reflects the use of resources as a control alternative by local populations. In addition, this is an issue that is neglected by health authorities, especially in areas where the most vulnerable populations live [13].

14.3.1 Insecticide and Repellent Plants Traditionally Used in Brazil

We documented the use of the 103 species, belonging to 79 genera and 43 botanical families (Table 14.1). Asteraceae, Fabaceae, and Poaceae are the most representative families (Figure 14.2). These families also stand out in reviews of species traditionally used as repellents in Brazil and other parts of the world [14, 15]. Asteraceae and Fabaceae are among the families whose species provide essential oils that have been extensively tested against *Ae. aegypti*. Asteraceae was the most important for adulticidal, repellent, and oviposition effects [16]. Among the substances of insecticidal importance in this family are the pyrethrins isolated from the dried flowers of *Chrysanthemum cinerariaefolium* Vis. and different monoterpenes, sesquiterpenes and phototoxic thiophenes, such as tertienyl (2,2':5',2"-tertiophene) present in the genera *Tagetes* and *Porophyllum* [17].

Several species of Fabaceae have rotenone, a potent insecticide, which gives them high commercial value [18]. Essential oils and extracts of the plants Poaceae, especially those of the genus *Cymbopogon*, are used as ingredients in herbal mosquito repellents [19]. Citral and geranyl acetate constitute the majority components of the essential oils of the Poaceae plants, to which different insecticidal activities are attributed [20].

Recorded species are predominantly trees (48%) native to Brazil (65%) (Figure 14.1a, and Table 14.1, respectively). Woody plants stand out mainly in the north, northeast and midwest regions. In the south and southeast regions, exotic herbs are more frequent. There are several herbs whose biological potential is known throughout the country. Regardless, the expressiveness of tree and

TABLE 14.1
Plant Species Traditionally Indicated as Repellents or Insecticides in Ethnobotanical Studies Conducted in Brazil

Family/Species	Traditional Name	Use	H	Or	UP	Mode of Preparation and Application	Brazilian States	Reference
Acanthaceae								
Avicennia germinans (L.) L.	siriúba	repellent	tree	nat	–	fumigation/ smoke	PA	[23]
Amaranthaceae								
Amaranthus viridis L.	bredo-de-porco	repellent	herb	ex	–	–	PE	[24]
Amaranthus spinosus L.	bredo-de-espinho	repellent	herb	ex	–	–	PE	[24]
Dysphania ambrosioides (L.) Mosyakin & Clemants	santa maria, herb-de-santa maria	repellent of fleas	herb	ex	–	–	SC	[25]
	mastruz, mastruço	insecticide repellent	shrub	ex	leaves	fumigation/ smoke	AP	[26]
Anacardiaceae								
Anacardium occidentale L.	caju	repellent	tree	nat	fruits	infusion bath	PA	[27]
Annona exsucca DC.	envira-preta	repellent	tree	nat	bark	maceration bath	PA	[9]
Annona squamosa L.	ata	insecticide	tree	ex	leaves	maceration	PE	[28]
Astronium urundeuva (M. Allemão) Engl.	aroeira	insect bites	tree	nat	stem bark	–	PE	[29]
Annonaceae			tree					
Annona coriacea Mart.	araticum	insecticide	tree	nat	seed	*in natura* topical	PI	[30]
Annona crassifla Mart.	araticum	insecticide	tree	nat	–	–	GO	[31]
Apocynaceae			tree					
Aspidosperma nitidum Benth. ex Müll. Arg.	carapanã	repellent	tree	nat	stem bark	maceration bath	PA	[9]
Tabernaemontana catharinensis A.DC.	cobrina	repel insects	tree	nat	leaves	alcoholic infusion oral	RS	[32]
Geissospermum argenteum Woodson	taquarirana	repellent	tree	nat	stem bark	maceration bath	PA	[9]
Araceae								
Montrichardia linifera (Arruda) Schott	aninga	repellent	shrub	nat	leaves	fumigation/ smoke	PA	[9]
Araucariaceae								
Columbea angustifolia Bertol.	pinheiro-do-paraná	repellent	ar	nat	leaves	infusion topical	PR	[33]
Arecaceae								
Astrocaryum vulgare Mart.	tucum	repellent	ar	nat	leaves fruits	fumigation/ smoke	PI	[30]

(*Continued*)

TABLE 14.1 *(Continued)*
Plant Species Traditionally Indicated as Repellents or Insecticides in Ethnobotanical Studies Conducted in Brazil

Family/Species	Traditional Name	Use	H	Or	UP	Mode of Preparation and Application	Brazilian States	Reference
Cocos nucifera L.	coco	repellent	herb	ex	fruits	fumigation/ smoke	PA	[23]
	coco	repellent	herb	ex	fruits	fumigation/ smoke	BA	[34]
	coco	repellent	herb	ex	fruits	fumigation/ smoke	BA	[34]
Copernicia prunifera (Mill.) H.E. Moore	carnaúba	repellent	tree	nat	leaves	fumigation/ smoke	PI	[30]
Astrocaryum aculeatum G. Mey.	tucumã	repellent	tree	nat	–	*in natura*	RO	[35]
		repellent	tree	nat	–	fumigation/ smoke	PA	[23]
Aristolochiaceae								
Aristolochia triangularis Cham. & Schltdl.	cipó-mil-homens	repellent	liana	nat	leaves stem bark	infusion topical	PR	[33]
Asparagaceae								
Aloe arborescens Mill.	–	mosquito bites	herb	ex	–	–	SC	[36]
Aloe vera (L.) Burm.f.	babosa	insecticide, repellent	herb	ex	leaves	–	MG	[37]
Asteraceae								
Artemisia alba Turra	canforinha	repellent	tree	ex	–	–	RJ	[38]
Bidens pilosa L.	picão, pião-branco	repel insects	herb	ex	–	–	SP	[39]
Cabobanthus polysphaerus (Baker) H. Rob.	assa-peixe	repellent	herb	ex	leaves, roots	infusion topical	PR	[33]
Calea unifla Less.	arnica-da-praia, arnica-do campo,	repel insects	herb	nat	flowers	juice topical	SC	[40]
Lychnophora ericoides Mart.	–	repellent	tree	nat	all plant	alcoholic infusion	GO	[41]
Mikania glomerata Spreng.	guaco	repel insects	lia	nat	–	–	SP	[42]
Rolandra fruticosa (L.) Kuntze	carica-á	repellent	tree	nat	roots	infusion bath	PA	[9]
Solidago chilensis Meyen		repellent	shrub	nat	all plant	alcoholic infusion	GO	[41]
	arnica	insect bites	shrub	nat	–	–	SP	[39]
Sphagneticola trilobata (L.) Pruski	–	mosquito bites	herb	nat	–	–	SC	[36]
Symphyopappus casarettoi B.L. Rob.	vassoura-de-bicho, vassourinha, vassourão-braba, vassoura-mansa	repellent	shrub	nat	–	tea oral	SC	[25]

(Continued)

TABLE 14.1 *(Continued)*
Plant Species Traditionally Indicated as Repellents or Insecticides in Ethnobotanical Studies Conducted in Brazil

Family/Species	Traditional Name	Use	H	Or	UP	Mode of Preparation and Application	Brazilian States	Reference
Tagetes erecta L.	cravo-de-defunto	insecticide or repellent	herb	ex	flowers	–	MG	[37]
Crescentia cujete L.	cuieira	repellent	shrub	ex	fruits	*in natura* topical	PA	[43]
		repellent	shrub		–	fumigation/ smoke	PA	[23]
Mansoa alliacea (Lam.) A.H. Gentry	cipó-d'alho	repellent	liana	nat	leaves	juice topical	PA	[43]
Bixaceae								
Bixa orellana L.	coloral	repellent	tree	nat	fruits, seeds	infusion	MT	[44]
Burseraceae								
Protium heptaphyllum (Aubl.) Marchand	breu branco	repellent	tree	nat	stem bark	fumigation/ smoke	PA	[23]
	breu branco	repellent	tree	nat	–	–	PA	[23]
Caryophyllaceae								
Drymaria cordata (L.) Willd. ex Roem. & Schult.	herb-sapinho	insecticide	herb	ex	–	–	RJ	[38]
Calophyllaceae								
Calophyllum brasiliense Cambess.	jacareúba	repellent	tree	nat	resin	–	AM	[45]
Convolvulaceae								
Ipomoea asarifolia (Desr.) Roem. & Schult.	salsa	repellent	liana	nat	leaves	alcoholic infusion bath	PA	[43]
Ipomoea carnea Jacq.	algodão-bravo	repellent	liana	nat	leaves	infusion bath	PA	[43]
Cannabaceae								
Trema micrantha (L.) Blume	grandiuva, periquiteiro	repel insect	tree	nat	–	–	MT	[46]
Combretaceae								
Laguncularia racemosa (L.) C.F. Gaertn.	tinteiro	repellent	tree	nat	–	fumigation/ smoke	PA	[23]
Commelinaceae								
Murdannia nudifla (L.) Brenan	trapoeraba	insect bites	herb	ex	leaves	–	RJ	[47]
Chrysobalanaceae								
Hymenopus heteromorphus (Benth.) Sotherbs & Prance	caripé	repelnte	tree	nat	–	–	PA	[48]

(Continued)

TABLE 14.1 *(Continued)*
Plant Species Traditionally Indicated as Repellents or Insecticides in Ethnobotanical Studies Conducted in Brazil

Family/Species	Traditional Name	Use	H	Or	UP	Mode of Preparation and Application	Brazilian States	Reference
Cucurbitaceae								
Cayaponia botryocarpa C. Jeffrey	cipó-alho-bravo	repellent	liana	nat	leaves	topical	AM	[49]
Momordica charantia L.	melão-de-são-caetano.	repellent	liana	ex	leaves	juice, tea	PA	[43]
	melão-de-são-caetano.	repellent	liana	ex	leaves	decoction fumigation	GO	[41]
	melão-são-caetano	insecticide	liana	ex	leaves	*in natura* topical, spatial	PI	[30]
Ebenaceae								
Diospyros guianensis L. (Aubl.) Gürke	maria preta	repel insects from fungal wounds	tree	nat	fruits	*in natura* topical: minced fruits applied as poultice	PA	[9]
Euphorbiaceae								
Croton campestris A. St.-Hil.	velame	repellent, insecticide	tree	nat	leaves	*in natura* topical	PI	[30]
Croton tricolor Klotzsch ex Baill.	sacatinga; marmeleiro-branco	repellent	tree	nat	–	–	PE	[24]
Jatropha gossypiifolia L.	peão-roxo	repellent	tree	nat	leaves	juice	MA	[50]
Manihot esculenta Crantz	mandioca	insecticide	tree	ex	roots	topical	PI	[30]
Ricinus communis L.	mamona	repellent	tree	ex	seed	fumigation/ smoke	PA	[43]
Sapium marmieri Huber.	burra-leiteira	repellent	tree	nat	látex	fumigation/ smoke	PA	[51]
Fabaceae								
Amburana acreana (Ducke) A.C. Sm.	cumaru-de-cheiro	repellent	tree	nat	stem bark	decoction fumigation	AM	61
	imburana, amburana-decheiro, cerejeira-do-nordeste	repellent	tree	nat	stem bark	–	MT	[52]
Amburana cearensis (Allemão) A.C. Sm.	imburana	repellent	tree	nat	stem bark	fumigation/ smoke	PI	[30]
Copaifera langsdorffii Desf.	copaíba, paud'óleo, podoi	repellent	tree	nat	stem bark oil seed	decocção, maceração, *in natura*	MT	[53]

(*Continued*)

TABLE 14.1 *(Continued)*
Plant Species Traditionally Indicated as Repellents or Insecticides in Ethnobotanical Studies Conducted in Brazil

Family/Species	Traditional Name	Use	H	Or	UP	Mode of Preparation and Application	Brazilian States	Reference
Derris sp.	timbó	repellent	tree	–	roots	–	RO	[35]
Hymenaea courbaril L.	jatobá	repellent	tree	nat	stem bark	decoction fumigation	PI	[30]
	jatobá/breu jutaicíca	repellent	tree	nat	stem bark	maceration/ fumigation	PA	[9]
Libidibia ferrea (Mart. ex Tul.) L.P. Queiroz	jucá	insect bites	tree	nat	fruits	–	PA	[54]
Pterodon emarginatus Vogel	sucupira	repellent	tree	nat	leaves	spatial repellent	PI	[30]
Lamiaceae								
Hyptidendron canum (Pohl ex Benth.) Harley	hortelã-do-campo	repellent	herb	nat	–	–	MT	[55]
Plectranthus barbatus Andrews	boldo	insecticide	herb	ex	leaves	topical	PI	[30]
Lecythidaceae								
Bertholletia excelsa Bonpl.	castanheira	repellent	tree	nat	stem barks, fruits	fumigation/ smoke	AM	[56]
Couratari sp.	tauari	repellent			stem barks	–	PA	[51]
Malvaceae								
Sida spinosa L.	malva-lanceta, malva-relógio	repel insect	shrub	nat	leaves	topical	PB	[57]
Meliaceae								
Carapa guianensis Aubl.	andiroba	repellent	tree	nat	roots	fumigation/ smoke	AM	[56]
		repellent	tree		seeds	fumigation/ smoke	PA	[23]
		repellent	tree		seeds	*in natura*	AM	[49]
		repellent	tree		seeds	*in natura* topical	PA	[58]
		repellent	tree		oil seed	*in natura* topical	AM	[59]
		repellent	tree		oil seed	*in natura* topical	AM	[60]
		repellent and insecticide	tree		oil seed	*in natura* topical	AM	[61]
		repellent	tree		oil seed	*in natura* topical	PA	[43]
		repellent	tree		oil seed	*in natura* topical	AP	[25]

(Continued)

TABLE 14.1 *(Continued)*
Plant Species Traditionally Indicated as Repellents or Insecticides in Ethnobotanical Studies Conducted in Brazil

Family/Species	Traditional Name	Use	H	Or	UP	Mode of Preparation and Application	Brazilian States	Reference
		repellent	tree		oil seed	tea juice topical	PA	[62]
		insect bites	tree		oil seed stem bark	*in natura* topical	PA	[54]
		repellent	tree		–	–	PA	[48]
		repellent	tree		oil seed	*in natura*/ fumigation	PA	[9]
Melia azedarach L.	cinamão	repellent	tree	ex	leaves	infusion topical	Pr	[33]
Azadirachta indica A. Juss.	nin	repellent	tree	ex	leaves	juice topical	PA	[43]
		insecticide and repellent	tree	ex		–	PA	[63]
		repellent	tree	ex	leaves	infusion topical	SP	[64]
		repellent, insecticide			leaves, all plant	spatial	PI	[30]
Moringaceae								
Moringa oleifera Lam.	noni	–	tree	ex	–	–	MT	[53]
Myrtaceae								
Corymbia citriodora (Hook.) K.D. Hill & L.A.S. Johnson	–	repelente	tree	ex	leaves	infusion	GO	[41]
Eucalyptus cinerea F. Muell. ex Benth.	eucalipto	repel insects	tree	ex	leaves	alcoholic infusion	RJ	[65]
Eucalyptus sp.	eucalipto	repelente	tree	ex	leaves	fumigation/ smoke	PI	[30]
Syzygium sp.		repel insects	tree	ex	–	–	SP	[39]
Piperaceae								
Piper callosum Ruiz & Pav.	óleo elétrico	repel insects	shrub	nat	leaves	tea	AM	[66]
	elixir paregórico	repellent	shrub	nat	roots	infusion topical	PA	[27])
Piper klotzschianum (Kunth) C. DC.	jambú do mato	repellent	shrub	nat	leaves	topical: crushed and applied	PA	[9]
Phytolaccaceae					leaves			
Petiveria alliacea L.	mucura-caá	repellent	herb	ex	leaves	juice	PA	[43]
		repellent	herb	ex	roots leaves	juice topical	PA	[62]

(Continued)

TABLE 14.1 *(Continued)*
Plant Species Traditionally Indicated as Repellents or Insecticides in Ethnobotanical Studies Conducted in Brazil

Family/Species	Traditional Name	Use	H	Or	UP	Mode of Preparation and Application	Brazilian States	Reference
		repellent	herb	ex	leaves	fumigation/ smoke infusion/ topical	AP	[25]
	guiné	repellent	herb	ex	leaves	infusion	Pr	[33]
Plantaginaceae								
Stemodia maritima L.	meladinha	repellent	herb	nat	all plant	juice	PB	[57]
Poaceae								
Cymbopogon citratus (DC.) Stapf	capim santo, capim cheiroso, capim marinho, capim limão	repellent	herb	ex	all plant	fumigation/ smoke	AP	[25]
Cymbopogon nardus (L.) Rendle	–	repellent	herb	ex	leaves	alcoholic infusion	GO	[41]
	citronela	repellent	herb	ex	leaves	alcoholic infusion	SP	[67]
	citronela	repellent	herb	ex	leaves	maceration topical	SP	[67]
	citronela	insecticide or repellent	herb	ex	leaves	–	MG	[37]
	capim-citronela	repellent	herb	ex	leaves	–	BA	[34]
Cymbopogon winterianus Jowitt ex Bor	–	repellent	herb	ex	leaves	–	BA	[34]
	citronela	repellent	herb	ex	leaves	alcoholic infusion	MT	[52]
	citronela	repellent	herb	ex	–	–	RJ	[38]
Cymbopogon sp.	citronela	repellent	herb	ex	all plant	spatial	PI	[30]
Guadua superba Huber	taboca	repellent	tree	nat	–	fumigation/ smoke	PA	[23]
Guadua weberbaueri Pilg.	tabuqui	repel insects	herb	nat	leaves	*in natura*	AM	[66]
Melinis minutifla P. Beauv.	capim-gordura	repel insects	herb	ex	all plant	infusion	RJ	[65]
Polygonaceae								
Muehlenbeckia sagittifolia (Ortega) Meisn.	salsaparrilha	repel insects	shrub	nat	–	–	RS	[27]
Rubiaceae								
Hamelia patens Jacq.	falsa-erva-de-rato; amélia	repel insects	tree	nat	–	–	–	[68]
Ruta graveolens L	arruda	insecticide	herb	ex	–	–	PA	[69]

(Continued)

TABLE 14.1 *(Continued)*
Plant Species Traditionally Indicated as Repellents or Insecticides in Ethnobotanical Studies Conducted in Brazil

Family/Species	Traditional Name	Use	H	Or	UP	Mode of Preparation and Application	Brazilian States	Reference
		insecticide and repellent	herb	ex	leaves flower	tea maceration	AP	[25]
	arruda	insecticide and repellent	herb	ex	leaves flower	–	MG	[37]
Ruta graveolens L	arruda	repellent	herb	ex	leaves	alcoholic infusion	MT	[52]
	arruda	insecticide, repellent	herb	ex	–	–	RJ	[38]
	arruda	repellent	herb	ex	leaves roots	infusion	Pr	[33]
	arruda	insecticide	herb	ex	leaves	–	PI	[30]
Rutaceae								
Citrus limon (L.) Osbeck	limão-azedo	insecticide	tree	ex	fruit	–	PI	[30]
Rhamnaceae								
Ziziphus joazeiro Mart	juá	repellent	tree	nat	leaves	juice topical	PA	[43]
Rhizophoraceae								
Rhizophora racemosa G. Mey.	mangue	repellent	tree	nat	–	fumigation/ smoke	PA	[23]
Sapindaceae								
Magonia pubescens A. St.-Hil.	tingui	insecticide, kills lice	tree	nat	–	–	GO	[31]
Sapindus saponária L.	saboneteira	insecticide	tree	nat	seed	*in natura* topical	–	[70]
Simaroubaceae								
Homalolepis maiana (Casar.) Devecchi & Piran	prá-tudo	insecticide	tree	nat	stem bark	–	PI	[30]
Homalolepis cedron (Planch.) Devecchi and Piran	pau-para-tudo	repellent	tree	nat	stem bark	maceration bath	PA	[71]
Siparunaceae								
Siparuna guianensis Aubl.	negramina	repellent	shrub	nat	leaves	maceration	MT	[53]
Solanaceae								
Capsicum frutescens L.	pimenta malagueta	repellent	shrub	ex	–	–	PA	[72]
	pimenta malagueta	repellent	shrub	ex	seeds	fumigation/ smoke	PI	[30]
Nicotiana tabacum L.	tabaco	repellent	shrub	ex	–	–	RO	[35]
	fumo	insecticide	shrub	ex	–	–	MT	[73]
	fumo	repellent	shrub	ex	fruit	infusion topical	PR	[33]

(Continued)

TABLE 14.1 *(Continued)*
Plant Species Traditionally Indicated as Repellents or Insecticides in Ethnobotanical Studies Conducted in Brazil

Family/Species	Traditional Name	Use	H	Or	UP	Mode of Preparation and Application	Brazilian States	Reference
Solanum lycopersicum L.	tomate	insect bites	shrub	ex	fruit	–	RJ	[47]
Verbenaceae								
Lippia alba (Mill.) N.E. Br. ex Britton & P. Wilson	salva	insect bites	shrub	nat	–	–	RS	[27]
Lippia grata Schauer	alecrim	repellent	shrub	nat	leaves	fumigation/ smoke	PI	[30]

H: Habit; Or: origin, nat: native to Brazil, ex: exotic or introduced; UP: used part. PA: Pará, PE: Pernambuco, SC: Santa Catarina, AP: Amapá, PI: Piauí, GO: Goiás, RS: Rio Grande do Sul, PR: Paraná, BA: Bahia, RO: Roraima, MG: Minas Gerais, RJ: Rio de Janeiro, SP: São Paulo, MT: Mato Grosso, AM: Amazonas, MA: Maranhão, PB: Paraíba.

native species may be related to the availability of these resources throughout the year, their versatility, and cultural factors. Tree resources have stood out among ethnobotanical and ethnopharmacological research in the northeast, both because herbs and leaves of deciduous woody plants are only available for a short period (generally 2–4 months) in habitats such as the Brazilian Caatinga and because supply is often unpredictable due to annual variation in climatic conditions. This result points to the importance of the sustainable exploitation of tree resources, since native plants are sources of substances of economic, technological, and social importance, especially for the control of disease-carrying mosquitoes.

14.3.2 Plant Parts Used and Methods of Preparation and Application of Botanical Repellents and Insecticides

Some of the evaluated works did not present information about the plant parts used or details about the methods of preparation and use of the mentioned species as repellents or insecticides. Different plant organs were mentioned in traditional preparations, with leaves (43%) most frequently used (Figure 14.1b). Fumigation and infusion are the most commonly used preparation methods (Figure 14.1c). Concerning application, topical use (50%) and smoking (32%) stood out (Figure 14.1d) (Table 14.1). Topical use often occurs in the application of vegetable oils and juices throughout the body, as well as through maceration or infusion baths. Sample plants with topical application are effective in repelling mosquitoes but little or no toxic to humans [21]. Fumigation occurs from the burning of plant parts so that the resulting smoke is spread around residences in order to ward off mosquitoes. This process mostly involves woody parts of trees or shrubs or even residues resulting from processing or plant use. Examples are the bark of coconut fruits (*Cocus nucifera* L.), residues of andiroba seeds, and the bark of species such as *Bertholletia excelsa* Bonpl. and *Hymenaea courbaril* L. Expulsion of smoke and/or heat stood out among the methods used in the preparation of herbal repellents on the African continent [22]. Dube et al. (2011) evaluated the repellent potential of species traditionally used in Ethiopia to ward off mosquitoes and observed that volatile extracts from the smoke of burning dried leaves were found to be more repellent than those from fresh leaves.

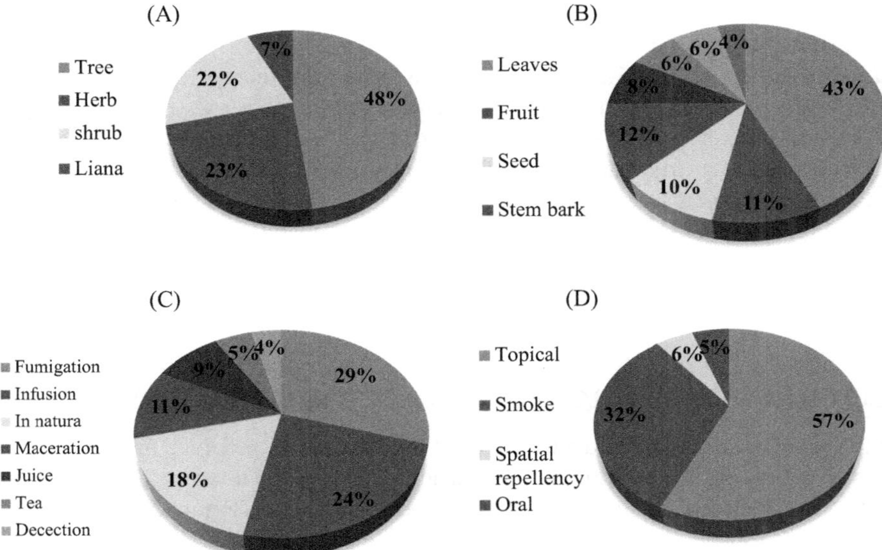

FIGURE 14.1 Ethnobotanical characteristics of the species traditionally used in Brazil as insecticides and repellents: (A) habits; (B) used plant parts; (C) forms of preparation; (D) modes of application.

14.3.3 Species with the Highest Citation Frequency

Carapa guianensis stood out among the species recorded in this review, followed by *Ruta graveolens* L, *Cymbopogon nardus* (L.) Rendle, *Petiveria alliacea* L., *Azadirachta indica* A. Juss., and *Cymbopogon winterianus* Jowitt ex Bor (Table 14.1). These species, except *C. guianensis*, are aromatic, release substances capable of repelling insects, including mosquitoes, due to their neurotoxic activities, through GABA, octopamine synapses, and acetylcholinesterase inhibition [18].

Andiroba is historically used in the north region as a repellent and insecticide. The oil extracted from its seeds is commonly applied as a topical repellent or through of the fumigation with smoke of burn of the seed or stem bark. This oil did not show significant repellency results against *Ae. aegypti* [74]. On the other hand, the hexane:ethyl acetate:dichloromethane seed shell extract showed larvicidal activity, as well as a residual effect of approximately 10 days against L3 larvae of these mosquito [9].

R. graveolens use occurs in all regions of Brazil (Table 14.1). This plant, known for its odorous characteristics, is used as a spatial repellent, topical and hair insecticide. Its essential oil showed significant larvicidal and repellent activity against *Ae. aegypti* [75].

The *Cymbopogon* species mentioned previously are worldwide known as citronella. These herbs are employed as a repellent and/or insecticide throughout the country. Beyond essential oil being traded internationally, Brazilian populations cultivate in them backyards to promote spatial repellency. They also use a maceration or alcoholic infusion of its leaves as a topical repellent. The insecticidal and repellent activities of *C. nardus* and *C. winterianus* have been recorded in different studies against insect pests and mosquito vectors of infectious diseases [76].

Petiveria alliacea, known as mucura-caá, is a naturalized species, widely distributed in Central and South America. Studies report the use of its crushed or burned leaves as repellents. The insecticidal and acaricidal activity of this species was related to the compound dibenzyl-trisulfide (DBTS) isolated from its roots [77]. Later studies showed the significant toxic effect of *P. alliacea* on the whitefly *Trialeurodes vaporariorum* [78], as well as the cytotoxic activity of hydroalcoholic leaf extract against *Ae. aegypti* larvae [79].

Azadirachta indica, a tree historically known as neem, originates in India and is widely cultivated in Brazil, especially for its insecticidal and repellent properties. This plant is cultivated around homes to promote spatial repellency, and the juice and infusions of its leaves are applied as topical repellents. Extracts obtained with different solvents, as well as its essential oil, showed insecticidal and repellent activity against insects, including the various disease vector mosquitoes, and this activity is attributed to the abundant presence of limonoids in this plant, especially azadiactin [80]. Neem derivatives have taken the industrial market by storm, especially for agricultural pest control, with an estimated market value of US$16.2 billion, and are projected to grow to more than US$27 billion by 2025 [81].

14.4 SPECIES EVALUATED AGAINST THE VECTOR *AEDES AEGYPTI*

Crude extracts and essential and fixed oils of 67 of the 103 species recorded have already been submitted to biological assays against immature and adult forms of *Ae. aegypti* (Table 14.2). Larvicidal tests are the most frequent (65 spp.), followed by repellent (12 spp.) (Figure 14.2).

The literature specializing in tests against *Ae. aegypti* considers a $LC_{50} < 50$ µg/mL very active; a $LC_{50} = 50$–100 µg/mL active, and a $LC_{50} > 100$ µg/mL weak/inactive [16]. The repellency results ranged from 47 to 100%, and the protection period was not disclosed for all species.

Crude extracts are obtained by different extraction methods with organic solvents or water. Only two plant extracts show repellency against *Ae. aegypti*. *C. winterianus* extract at a concentration of 75% showed 92.26% protection against this mosquito during 5 h of exposure [113].

The study that evaluated the topical repellency of extracts obtained from the seeds of *Bixa orellana* L. using three different solvents showed that the dry hexane extract resuspended in acetone was the most active, followed by the hydroalcoholic extract, with 90% and 73% (113.8 µg/ mL) of protection, respectively [98].

The extracts recorded in Table 14.2 were mostly submitted to larvicidal activity assays. Of the 39 extracts, 10 showed high larvicidal activity ($LC_{50} < 50$ µg/mL), 4 were active (LC_{50} 50–100 µg/mL), and 18 showed weak activity ($LC_{50} > 100$ µg/mL).

The ethanolic extract of the stem bark of *Annona crassiflora* Mart. showed greater larvicidal activity (LC_{50} 0.71 µg/ml) in a study that evaluated the larvicidal activity of 84 extracts obtained from 51 Brazilian plant species. Ethanol extracts from the fruit and bark of *Anacadium occidentale* also stand out, with LC_{50} values 7.84 and 4.01 µg/mL, respectively. In cashew fruit extract, several phenolic compounds were identified, such as anacardic acids, cardol, and cardanol, which in turn also showed significant larvicidal activity against the same vector (LC_{50} 5.93, 6.14 and 20.32 µg/mL) [83].

The species from which the essential oils were evaluated are predominantly herbaceous and/or aromatic and are the main samples used in repellency studies. Essential oils have been widely investigated for insect control, as they concentrate compounds capable of scaring away different species. They are classified as insecticides, anti-ovipositional, and repellents of great importance. In addition, these oils are composed of volatile substances with fast degradation, thus causing little or no damage to other organisms in the environment.

Cymbopogon species are the most studied regarding repellency. These plants have repellent activity not only against *Ae. aegypti* but also against other vector mosquitoes such as *Anopheles stephensi* and *Anopheles dirus* [146].

Cymbopogon winterianus Jowitt ex Bor essential oil applied at a rate of 0.1 mL/30 cm^2 provided 75.7% protection for 2 h against *Ae. aegypti*. This result was more effective than that observed with the application of standard DEET 20%, a commercial repellent [106]. The main compounds of the essential oil of *C. winterianus*, citronellal, citrolenol, and geraniol, were tested separately and showed repellency >70% for 1 h against *Ae. aegypti* [147]. The essential oil of *C. nardus* showed 47.03% protection for 6 hours against this vector [109].

TABLE 14.2
Plant species Used as Repellents and/or Insecticides, According to the studies in Table 14.1, with Their Respective Biological Activity Information against the *Aedes aegypti* Vector

Plant Species	Part Used	Extraction Solvent	Activity	Result	Reference
Aloe vera (L.) Burm.f.	Leaves	Petroleum ether	Larvicidal	LC_{50} 253 µg/mL	[82]
Amburana cearensis (Allemão) A.C. Sm.	Seed	Water	Larvicidal	LC_{50} 8100 µg/mL	[71]
Anacardium occidentale L.	Fruit	Ethanol	Larvicidal	LC_{50} 7.84 µg/mL	[83]
	Fruit	Ethanol	Pupicidal	LC_{50} 338.28 µg/mL	[83]
	Fruit	Ethanol	Larvicidal	LC_{50} 6.55 µg/mL	[84]
	Oil fruit	-	Larvicidal	LC_{50} 14.5 µg/mL	[85]
	Bark	Hexane	Larvicidal	LC_{50} 4.01 µg/mL	[86]
Annona coriacea Mart.	Seed	Ethanol	Larvicidal	LC_{50} 50 µg/mL	[87]
Annona crassiflora Mart.	Bark root, bark stem	Ethanol	Larvicidal	LC_{50} 0.71 µg/mL	[88]
Annona exsucca DC.	Wood	Hexane:ethyl acetate:dichloromethane (45:45:10)	Larvicidal	Not active	[9]
Annona squamosa L.	Root, leaves, seed	Ethanol	Larvicidal	LC_{50} 31.9 µg/mL	[88]
Aspidosperma nitidum Benth. ex Müll.Arg.	Wood	Hexane:ethyl acetate:dichloromethane	Larvicidal	LC_{50} 137 µg/mL	[9]
Aristolochia triangularis Cham. & Schltdl.	Not given	Methanol; dichloromethane	Larvicidal	LC_{50} 285.5; 188.1 µg/mL	[89]
Astrocaryum aculeatum G. Mey.	Fruit	Methanol	Larvicidal	LC_{50} 341.5 µg/mL	[90]
Astronium urundeuva (M. Allemão) Engl.	Seed	Ethanol	Larvicidal	LC_{50} 15.2 µg/mL	[91]
Astronium urundeuva (M. Allemão) Engl.	Seed	Water	Larvicidal	LC_{50} 370 µg/mL	[92]
	Seed	Sodium phosphate buffer	Oviposition	98.4% [20%]	[92]
	Wood	Nacl	Larvicidal	LC_{50} 40 µg/mL	[93]
Avicennia germinans (L.) L.	Leaves	Water	Larvicidal	LC_{50} 10.465 µg/mL	[94]
Azadirachta indica A. Juss.	Oil fruit	-	Repellent	84.21% [5%]	[95]
	Leaves	Acetone	Larvicidal	LC_{50} 50 µg/mL	[96]
Bidens pilosa L.	Leaves	Water	Oviposition	37. 5% [not informed]	[97]
Bixa orellana L.	Seed	Hexane	Repellent	90% [113.2 µg/mL]	[98]
	Seed	Water	Larvicidal	not active	[99]
Calophyllum brasiliense Cambess.	Leaves	Dichloromethane	Larvicidal	not active	[100]
Capsicum frutescens L.	Fruit	Ethanol	Larvicidal	LC_{50} 141.92 µg/mL	[100]
Carapa guianensis Aubl.	Oil seed	–	Larvicidal	LC_{50} 136 µg/mL	[101]
	Oil seed	–	Repellent	not active	[102]
	Stem wood	Hexane:ethyl acetate:dichloromethane (45:45:10)	Not active	Not active	[9]

(Continued)

TABLE 14.2 *(Continued)*
Plant species Used as Repellents and/or Insecticides, According to the studies in Table 14.1, with Their Respective Biological Activity Information against the *Aedes aegypti* Vector

Plant Species	Part Used	Extraction Solvent	Activity	Result	Reference
	Seed shells	Hexane:ethyl acetate:dichloromethane (45:45:10)	Larvicida	LC_{50} 70 µg/mL	[9]
Citrus limon (L.) Osbeck	Essential oil	–	Larvicidal	LC_{50} 1.48 µg/mL	[103]
Copaifera langsdorffii Desf.	Oil bark	–	Larvicidal	LC_{50} 41 µl/mL	[85]
Corymbia citriodora (Hook.) K.D. Hill & L.A.S. Johnson	Essential oil leaves	–	Larvicidal	LC_{50} 38.7 µg/mL	[104]
	Essential oil	–	Repellent	99.34% [1%]	[105]
	Essential oil	–	Repellent	59.4% [0.1 mL/30 cm²] – 2 h e 30 min	[106]
Crescentia cujete L.	Leaves	Ethanol	Larvicidal	not active	[107]
Cymbopogon citratus (DC.) Stapf	Essential oil bark	–	Adulticidal	87.3% [5%]	[108]
Cymbopogon nardus (L.) Rendle	Essential oil leaves	–	Repellent	47.03%-6 h	[109]
	Essential oil leaves	–	Larvicidal	LC_{50} 35.133 µg/mL	[110]
	Comercial essential oil	–	Repellent	60% [50%] – 2 h	[111]
Cymbopogon winterianus Jowitt ex Bor	Essential oil	–	Larvicidal	LC_{50} 38.8 µg/mL	[112]
	–	Crude extract	Repellent	92.26% [5%]	[113]
	Essential oil	–	Repellent	75.7% [0.1 mL/30 cm²] – 2 h	[106]
Drymaria cordata (L.) Willd. ex Roem. & Schult.	Leaves	Petroleum ether, metanol, acetone, water	Larvicidal	not active	[114]
Dysphania ambrosioides (L.) Mosyakin & LCemants	Aerial parts	N-hexane	Larvicidal	LC_{50} 50 µg/mL	[115]
Diospyros guianensis L. (Aubl.) Gürke	Fruits	Hexane:ethyl acetate:dichloromethane (45:45:10)	Larvicidal	LC_{50} 62 µg/mL	[9]
Eucalyptus cinerea F. Muell. ex Benth.	Essential oil	–	Larvicidal	LC_{50} 0.38 µg/mL	[116]
Geissospermum argenteum Woodson	Stem bark, wood, leaves	Hexane:ethyl acetate:dichloromethane (45:45:10)	Larvicidal	Not active	[9]
Hymenaea courbaril L.	Essential oil fruit	–	Larvicidal	LC_{50} 14.8 µg/mL	[117]
Homalolepis cedron (Planch.) Devecchi and Piran	Stem, leaves, wood	Hexane:ethyl acetate:dichloromethane (45:45:10)	Larvicidal	Not active	[9]
Ipomoea carnea Jacq.	Stem	Ethyl Acetate	Larvicidal	LC_{50} 119.61 µg/mL	[118]

(Continued)

TABLE 14.2 *(Continued)*
Plant species Used as Repellents and/or Insecticides, According to the studies in Table 14.1, with Their Respective Biological Activity Information against the *Aedes aegypti* Vector

Plant Species	Part Used	Extraction Solvent	Activity	Result	Reference
Libidibia ferrea (Mart. ex Tul.) L.P. Queiroz	Fruit	Etanol 70%	Larvicidal	LC_{50} 14900 μg/mL	[119]
	Seed	Sodium phosphate (50 mm,	Larvicidal	LC_{50} 12020 μg/mL	[92]
Lippia alba (Mill.) N.E. Br. ex Britton & P. Wilson	Essential oil aerial parts	–	Larvicidal	LC_{50} 113.99 μg/mL	[120]
	Oil essential leaves	–	Oviposition	50.4% at 1000 μg/mL	[121]
	Oil essential oil	–	Adulticidal	not active	[121]
Lippia alba (Mill.) N.E. Br. ex Britton & P. Wilson	Oil essential	–	Repellent	50% {1 mg/cm²]	[121]
Lippia grata Schauer	Essential oil	–	Larvicidal	LC_{50} 22.79 μg/mL	[122]
Magonia pubescens A. St.-Hil.	Bark stem	Ethanol	Larvicidal	LC_{50} 60000 μg/mL	[123]
Manihot esculenta Crantz	Leaves	Water	Larvicidal	LC_{50} 66.14 μg/mL	[124]
Melia azedarach L.	Leaves	Ethanol	Larvicidal	LC_{50} 760 μg/mL	[125]
	Leaves	Ethanol	Oviposition	1g/L	[125]
	Seed	Ethanol	Larvicidal	LC_{50} 34 μg/mL	[126]
Momordica charantia L.	Fruit	N-Hexane	Larvicidal	LC_{50} 122.45 μg/mL	[127]
	Flower, fruit	Hexane ethylacetate methanol (400 ml; 1:1).	Larvicidal	LC_{50} 37.2 μg/mL	[128]
	Leaves	Water	Oviposition	87.7%	[97]
Moringa oleifera Lam.	Seed	Water	Larvicidal	LC_{50} 1260 μg/mL	[129]
Nicotiana tabacum L.	Leaves	Ethanol	Adulticidal	90% [5%] – 4 h	[130]
	Leaves	Ethanol	Larvicidal	LC_{50} 1022.97 μg/mL	[130]
Petiveria alliacea L.	Leaves	Water	Larvicidal	LC_{50} 25% *v/v*	[79]
Plectranthus barbatus Andrews	Leaves	Acetone	Larvicidal	LC_{50} 610 μg/mL	[131]
Protium heptaphyllum (Aubl.) Marchand	Leaves, bark	Etanol	Larvicidal	not active	[132]
Protium heptaphyllum (Aubl.) Marchand	Essential oil nanoemulsion	–	Adulticidal	200 μg/mL	[133]
	Essential oil nanoemulsion	–	Repellent	77.67%. [200 μg/mL] – 3 h	[133]
Pterodon emarginatus Vogel	Essential oil fruit	–	Larvicidal	LC_{50} 134.90 g/mL	[134]
Pterodon emarginatus Vogel	Fruit	N-Hexane	Larvicidal	LC_{50} 23.99 g/mL.	[134]
Ricinus communis L.	Oil seed	–	Oviposition	LC_{50} 0.26 μL/mL	[135]

(Continued)

TABLE 14.2 *(Continued)*
Plant species Used as Repellents and/or Insecticides, According to the studies in Table 14.1, with Their Respective Biological Activity Information against the *Aedes aegypti* Vector

Plant Species	Part Used	Extraction Solvent	Activity	Result	Reference
	Oil seed	–	Larvicidal	LC_{50} 0.029 µl/mL	[135]
	Oil seed	–	Repellent	86% [100%] 3 h	[136]
	Leaves	Methanol	Larvicidal	LC_{50} 191.54 µl/mL	[137]
	Seed	Methanol	Larvicidal	LC_{50} 15.52 µl/mL	[137]
Rolandra fruticosa (L.) Kuntze	Aerial parts, roots	Hexane:ethyl acetate:dichloromethane (45:45:10)	Larvicidal	Not active	[9]
Ruta graveolens L	Essential oil leaves	–	Larvicidal	LC_{50} 14.37 µl/mL	[75]
	Essential oil leaves	–	Repellent	100% [0.187 mg/cm²]	[75]
	Essential oil leaves	–	Larvicidal	LC_{50} 61.641 µl/mL	[138]
Sapindus saponária L.	Fruit	Metanol	Larvicidal	LC_{50} 601.81 µl/mL	[139]
Siparuna guianensis Aubl.	Essential oil leaves	–	Oviposition	80% [0.5 µg/mL]	[140]
	Essential oil leaves	–	Oviposition	1 µg/mL	[140]
	Stem, leaf, and fruit essential oils		Larvicidal	LC_{50} 1.76, 0.98, 2.46 µg/mL	[140]
	Essential oil leaves	–	Pupicidal	LC_{50} 1.69 µg/mL	[140]
	Essential oil leaves	–	Repellent	100% [0.450 µg/cm²] – 2 h	[140]
Solanum lycopersicum L.	Leaves	Methanol	Larvicidal	LC_{50} 20323 µl/mL	[141]
Solidago chilensis Meyen	Aerial parts	Water	Larvicidal	not active	[142]
Stemodia maritima L.	Essential oil leaves	–	Larvicidal	LC_{50} 55.4 µl/mL	[143]
	Essential oil bark		Larvicidal	LC_{50} 22.9 µl/mL	[143]
Tagetes erecta L.	Essential oil steam		Larvicidal	LC_{50} 79.78 µg/mL	[144]
Ziziphus joazeiro Mart				not active	[145]

The essential oils obtained from leaves, stems, and fruits of the native species *Siparuna guianensis* Aubl. showed significant ovicidal (80% [0.5 µg/mL), larvicidal (LC_{50} 1.76 µg/mL), pupicidal (LC_{50} 1.69 µg/mL), anti-oviposition (1 µg/mL), and repellent (100% [0.450 µg/cm²—2 h]) properties. The main components of the leaf oils were β-myrcene (79.71%) and 2-undecanone (14.58%): from the stem, β-myrcene (26.91%), δ-elemene (20.92%), germacrene D (9.42%), α-limonene (7.91%), and bicyclo-germacrene (7.79%) and from the fruit, 2-tridecanone (38.75%), 2-undecanone (26.5%), and β-myrcene (16.42%) [140].

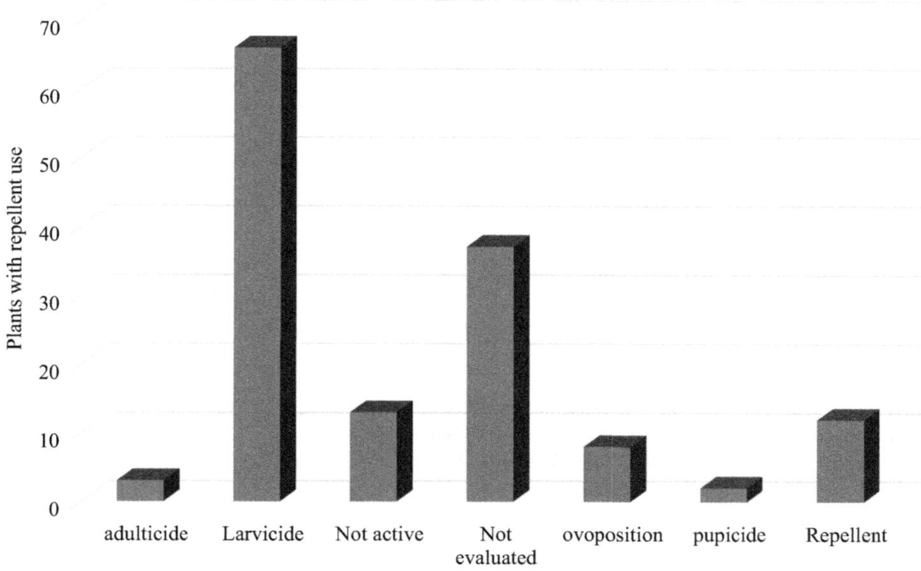

FIGURE 14.2 Evaluation of crude extracts, essential oils, and fixed oils from 67 plant species for larvicidal and repellent activities against *Aedes aegypti.*

14.5 CONCLUSION

This result points to two important issues. The first refers to the progress in research carried out with plants widely used by traditional populations in Brazil, bringing to light species that can act as sources of effective products in the control of *Ae. aegyti.* On the other hand, almost half of the species recorded for their traditional use have not yet been the subject of studies related to the control of this vector, which may direct further research in the area of biopesticides.

Therefore, ethnobotany and ethnopharmacology studies encourage the development of research regarding mosquito vectors using plant species. This parallel between the traditional use of repellent plants and their potential against this mosquito reinforces the importance of developing multidisciplinary studies that seek to value traditional local knowledge, as well as strategies for the development of different plant-based insecticides.

ACKNOWLEDGMENTS

We would like to thank the CNPq—Brasil for financial support.

REFERENCES

1. Thacker JRM. *An introduction to arthropod pest control.* Cambridge University Press. 2002. Tng DY, Apgaua, DMG Lisboa, MM El-Hani, CN. Gender differences in plant use knowledge within a traditional fishing community in northeastern Brazil. *Ethnobotany Research and Applications*, 2016, 21, 1–36.
2. Casida JE, Quistad, GB. Golden age of insecticide research: past, present, or future? *Annual Review of Entomology*, 1998, 43, 1–16.
3. Bellinato DF, Viana-Medeiros PF, Araújo SC, Martins AJ, Lima JBP, Valle D. Resistance Status to the Insecticides Temephos, Deltamethrin, and Diflubenzuron in Brazilian *Aedes aegypti* Populations. *BioMed Research International*, 2016, 1–12.

4. Rout SP, Choudary KA, Kar DM, Das L, Jain A. Plants in traditional medicinal system—future source of new drugs. *International Journal of Pharmacy and Pharmaceutical Sciences*, 2009, 1, 1.
5. Heinrich M, Gibbons S. Ethnopharmacology in drug discovery: an analysis of its role and potential contribution. *Journal of Pharmacy and Pharmacology*, 2001, 53, 425–432.
6. Orozco OL, Lentz DL. Poisonous plants and their uses as insecticides in Cajamarca. *Peru. Economic Botany*, 2005, 59, 2, 166–173.
7. Luna JDS, Dos Santos AF, De Lima MRF, De Omena MC, De Mendonça FAC, Bieber LW. A study of the lavicidal and molluscicidal activities of some medicinal plants from northeast Brazil. *Journal of Ethnopharmacology*, 2005, 97, 199–206.
8. Dos Santos EA, De Carvalho CM, Costa AL, Conceição AS, Moura FB, Santana AE. Bioactivity evaluation of plant extracts used in indigenous medicine against the snail, *Biomphalaria glabrata*, and the Larvae of *Aedes aegypti*. *Evid Based Complement Alternat Medicine*, 2012, 1–9.
9. Oliveira-Melo PMC, Lima PGC, Costa JC, Coelho-Ferreira MR. Ethnobotanical study in a rural settlement in Amazon: contribution of local knowledge to public health policies. *Research, Society and Development*, 2022, 11, 56911125258–1.
10. WHO. Neglected tropical diseases. (Accessed August 20, 2020, at https://www.who.int/neglected-tropical-diseases-progress-dashboard-2011-2020).
11. Laporta GZ, Potter AM, Oliveira JF, Bourke BP, Pecor DB, Linton YM. Global distribution of Aedes aegypti and Aedes albopictus in a climate change scenario of regional rivalry. *Insects*, 2023, 14, 1, 49.
12. Sousa FDCA, Soares HVA, Lemos LEAS, Reis DM, Da Silva WC, De Sousa Rodrigues LA. Perfil epidemiológico de doenças negligenciadas de notificação compulsória no Brasil com análise dos investimentos governamentais nessa área. *Research, Society and Development*, 2020, 9, 1, e62911610-e62911610.
13. Fonseca BDP, Albuquerque PC, Zicker F. Neglected tropical diseases in Brazil: lack of correlation between disease burden, research funding and output. *Tropical Medicine & International Health*, 2020, 25, 11, 1373–1384.
14. Rehman JU, Ali A, Khan IA. Plant based products: Use and development as repellents against mosquitoes: A review. *Fitoterapia*, 2014, 95, 65–74.
15. Pavela R, Benelli G. Ethnobotanical knowledge on botanical repellents employed in the African region against mosquito vectors–a review. *Experimental Parasitology*, 2016, 167, 103–108.
16. Silvério MRS, Espindola LS, Lopes NP, Vieira PC. Plant natural products for the control of *Aedes aegypti*: The main vector of important arboviruses. *Molécules*, 2020, 25.
17. Sukh D, Opender K. *Insecticides of natural origin* (1st ed.). Routledge, 1997. https://doi.org/10.1201/9780203750759.
18. Regnault-Roger C, Philogène BJR. Past and current prospects for the use of botanicals and plant allelochemicals in integrated pest management. *Pharmaceutical Biology*, 2008, 46, 41–52. https://doi.org/10.1080/13880200701729794.
19. Dias CN, Moraes DFC. Essential oils and their compounds as *Aedes aegypti* L. (Diptera: Culicidae) larvicides. *Parasitology Research*, 2014, 113, 2, 565–592.
20. Plata-Rueda A, Martínez LC, Da Silva Rolim G, Coelho RP, Santos MH, De Souza Tavares W. Insecticidal and repellent activities of *Cymbopogon citratus* (Poaceae) essential oil and its terpenoids (citral and geranyl acetate) against Ulomoides dermestoides. *Crop Protection*, 2020, 137, 105299.
21. Wangai LN, Kamau KK, Munyekenye G, Nderu D, Maina E, Gitau, Otieno F. Efficacy of plant-based repellents against Anopheles mosquitoes: a systematic review. *Biomedical Science*, 2020, 6, 44.
22. Pavela R, Benelli G. Ethnobotanical knowledge on botanical repellents employed in the African region against mosquito vectors–a review. *Experimental Parasitology*, 2016, 167, 103–108.
23. Rocha TT, Tavares-Martins ACC, Lucas FCA. Traditional populations in environmentally protected areas: an ethnobotanical study in the soure marine extractive reserve of Brazil. *Boletín Latinoamericano y del Caribe de Plantas Medicinales y Aromáticas*, 2017, 16, 4, 410–427.
24. Albuquerque UP, Andrade LHC. Conhecimento botânico tradicional e conservação em uma área de caatinga no estado de Pernambuco, Nordeste do Brasil. *Acta Botanica Brasilica*, 2002, 16, 3, 273–285.
25. Melo S, Lacerda VD, Hanazaki N. Espécies de restinga conhecidas pela comunidade do Pântano do Sul, Florianópolis, Santa Catarina, Brasil. *Rodriguésia*, 2008, 59, 4, 799–812.
26. Lima RB, Dos Santos JUM, Da Luz Freitas J, Souto RNP. Caracterização agroecológica e socioeconômica dos moradores da comunidade quilombola do Curiaú, Macapá-AP, Brasil. *Biota Amazônia (Biote Amazonie, Biota Amazonia, Amazonian Biota)*, 2013, 3, 3, 113–138.
27. Ritter RA, Monteiro MVB, Monteiro FOB, Rodrigues ST, Soares ML, Silva JCR, . . . Tourinho MM. Ethnoveterinary knowledge and practices at Colares island, Pará state, eastern Amazon, Brazil. *Journal of Ethnopharmacology*, 2012, 144, 2, 346–352.

28. Magalhães SIR, Oliveira SA, Araújo LM, Costa PRC, Araújo Roland I, Borrás MRL. Determination of Cu, Fe, Mn, and Zn in the leaves and tea of *Arrabidaea chica* (Humb. & Bompl.) Verl. *Biological Trace Element Research*, 2019, 132, 1, 239–246.
29. Albuquerque UP, Monteiro JM, Ramos MA, Amorim ELC. Medicinal and magic plants from a public market in northeastern Brazil. *Journal of Ethnopharmacology*, 2007, 110, 1, 76–91.
30. Almeida Neto JR, Dos Santos KPP, Chaves ME, De Morais RF, Neto EMC, Silva PRR, Barros RFM. Conhecimento sobre uso de plantas repelentes e inseticidas em duas comunidades rurais do Complexo vegetacional de Campo Maior, Nordeste do Brasil. *Gaia Scientia*, 2017, 11, 1.
31. Souza CDD, Felfili JM. Uso de plantas medicinais na região de Alto Paraíso de Goiás, GO, Brasil. *Acta Botanica Brasilica*, 2006, 20, 135–142.
32. Garlet TMB, Irgang BE. Plantas medicinais utilizadas na medicina popular por mulheres trabalhadoras rurais de Cruz Alta, Rio Grande do Sul, Brasil. *Revista Brasileira de plantas medicinais*, 2001, 4, 1, 9–18.
33. Burei ST, Santana DAD, Lopez BB, Sotomaior CS, Acra LA, Weber SH, Ollhoff RD. Ethnoveterinary knowledge and practice applied to domestic animals raised in the Ukraine Colonization Community of Palmital, Paraná State, Brazil. *Economic Botany*, 2022, 1–12.
34. Tng DY, Apgaua DMG, Lisboa MM, El-Hani CN. Gender differences in plant use knowledge within a traditional fishing community in northeastern Brazil. *Ethnobotany Research and Applications*, 2016, 21, 1–36.
35. Pinho RCD, Pedreira JL, Rocha JC, Perez IU, Hada AR, Miller RP, Alfaia SS. Agroextrativismo indígena no Lavrado: o caso da Terra Indígena Araçá, Roraima, Amazônia. *Terra e Água-degradação e desenvolvimento sustentável*, 2012, 143–164.
36. Merétika AHC, Peroni N, Hanazaki N. Local knowledge of medicinal plants in three artisanal fishing communities (Itapoá, Southern Brazil), according to gender, age, and urbanization. *Acta Botanica Brasilica*, 2010, 24, 2, 386–394.
37. Prado AC, Rangel EB, Sousa HCD, Messias MCT. Etnobotânica como subsídio à gestão socioambiental de uma unidade de conservação de uso sustentável. *Rodriguésia*, 2019, 70.
38. Horta BO, Neto GM. Levantamento etnobotânico de plantas medicinais em duas comunidades da região serrana do Rio de Janeiro, Brasil. *Revista Fitos*, 2019, 13, 3, 212–231.
39. Pilla MAC, Amorozo, MCDM, Furlan, A. Obtenção e uso das plantas medicinais no distrito de Martim Francisco, Município de Mogi-Mirim, SP, Brasil. *Acta Botanica Brasilica*, 2006, 20, 789–802.
40. Ramos LS, Cardoso PS, Freitas MD, Paghan R, Borges MS, Citadini-Zanette V, . . . Dalbo S. Popular medicinal uses of *Calea uniflora* Less.(Asteraceae) and its contribution to the study of Brazilian medicinal plants. *Anais da Academia Brasileira de Ciências*, 2016, 88, 2319–2330.
41. Guimarães OB, Morais LI, Oliveira PA. Medicinal plants and their popular use in Boa Esperança Settlement, Piracanjuba, Goiás, Brazil. *Boletín Latinoamericano y del Caribe de Plantas Medicinales y Aromáticas*, 2022, 21, 4.
42. Macedo AF, Oshiiwa M, Guarido CF. Ocorrência do uso de plantas medicinais por moradores de um bairro do município de Marília-SP. *Revista de Ciências Farmacêuticas Básica e Aplicada*, 2007, 28, 1.
43. Monteiro MVB, Bevilaqua CML, Palha MDDC, Braga RR, Schwanke K Rodrigues ST, Lameira OA. Ethnoveterinary knowledge of the inhabitants of Marajó Island, eastern Amazonia, Brazil. *Acta Amazonica*, 2011, 41, 2, 233–242.
44. Pinto JDS, Oliveira AKMD, Fernandes V, Matias R. Ethnobotany and popular culture in the use of plants in settlements on the southern edge of southern Pantanal Mato Grosso. *Journal of Biosciences*, 2017, 193–203.
45. Santos J, Pagani E, Ramos J, Rodrigues E. Observations on the therapeutic practices of riverine communities of the Unini River, AM, Brazil. *Journal of Ethnopharmacology*, 2012, 142, 2, 503–515. https://doi.org/10.1016/j.jep.2012.05.027.
46. Bieski IGC, Leonti M, Arnason JT, Ferrier J, Rapinski M, Violante IMP, . . . de Oliveira Martins DT. Ethnobotanical study of medicinal plants by population of Valley of Juruena Region, Legal Amazon, Mato Grosso, Brazil. *Journal of Ethnopharmacology*, 2015, 173, 383–423.
47. Boscolo OH, Senna Valle L. Plantas de uso medicinal em Quissamã, Rio de Janeiro, Brasil. *Iheringia, Série Botânica*, 2008, 63, 2, 263–278.
48. Shanley P, Rosa NA. Eroding knowledge: an ethnobotanical inventory in eastern Amazonia's logging frontier. *Economic Botany*, 2004, 58, 2, 135–160.
49. Pedrollo CT, Kinupp VF, Shepard Jr G, Heinrich M. Medicinal plants at Rio Jauaperi, Brazilian Amazon: Ethnobotanical survey and environmental conservation. *Journal of Ethnopharmacology*, 2016, 186, 111–124.

50. Scudeller VV, Veiga JD, Araújo-Jorge, LD. Etnoconhecimento de plantas de uso medicinal nas comunidades São João do Tupé e Central (Reserva de Desenvolvimento Sustentável do Tupé). *Biotupé: meio físico, diversidade biológica e sociocultural do Baixo Rio Negro, Amazônia Central*, 2009, 2, 185–189.
51. Santos JX, Reis ARS, Parry SM, Leão F.M, Carvalho JC. Caracterização etnobotânica de essências florestais com fins medicinais utilizados pela Etnia Xipaya, no município de Altamira-PA. *Biota Amazônia (Biote Amazonie, Biota Amazonia, Amazonian Biota)*, 2016, 6, 2, 1–8.
52. Cavalheiro L, Guarim-Neto G. Ethnobotany and regional knowledge: combining popular knowledge with the biotechnological potential of plants in the Aldeia Velha community, Chapada dos Guimarães, Mato Grosso, Brazil. *Boletín Latinoamericano y del Caribe de Plantas Medicinales y Aromáticas*, 2018, 17, 2.
53. Ribeiro RV, Bieski IGC, Balogun SO, Oliveira DTM. Ethnobotanical study of medicinal plants used by Ribeirinhos in the North Araguaia microregion, Mato Grosso, Brazil. *Journal of Ethnopharmacology*, 2017, 205, 69–102.
54. Magno-Silva RE, Rocha TT, Tavares-Martins CAC. Ethnobotany and ethnopharmacology of medicinal plants used in communities of the Soure Marine Extractive Reserve, Pará State, Brazil. *Boletín Latinoamericano y del Caribe de Plantas Medicinales y Aromáticas*, 2020, 19, 1.
55. Fiebig DA, Pasa GMC. A Etnobotânica na Comunidade Passagem da Conceição em Várzea Grande, Mato Grosso, Brasil. *Biodiversidade*, 2016, 15, 2.
56. Costa PSP. Estudo Etnobotânico e Farmacognóstico de plantas antimaláricas de uso popular na comunidade Céu do Mapiá, Pauini-AM. *Dissertação (Mestrado em ciências farmacêuticas)—UFAM, Manaus*, 2013.
57. Agra MF, Baracho GS, Nurit K, Basílio IJLD, Coelho VPM. Medicinal and poisonous diversity of the flora of "Cariri Paraibano", Brazil. *Journal of Ethnopharmacology*, 2007, 111, 2, 383–395.
58. Shanley P, Luz L. The impacts of forest degradation on medicinal plant use and implications for health care in eastern Amazonia. *BioScience*, 2003, 53, 6, 573–584.
59. Silva AL, Tamashiro J, Begossi A. Ethnobotany of riverine populations from the Rio Negro, Amazonia (Brazil). *Journal of Ethnobiology*, 2007, 27, 1, 46–72.
60. Frausin G, Hidalgo A, Lima RB, Kinupp VF, Ming LC, Pohlit AM, Milliken W. An ethnobotanical study of anti-malarial plants among indigenous people on the upper Negro River in the Brazilian Amazon. *Journal of Ethnopharmacology*, 2015, 174, 238–252. https://doi.org/10.1016/j.jep.2015.07.033.
61. Mendonça AP, Ferraz IDK. Óleo de andiroba: processo tradicional da extração, uso e aspectos sociais no estado do Amazonas, Brasil. *Acta Amazônica*, 2007, 37, 3, 353–364.
62. Carvalho TLGS. Etnofarmacologia e fisiologia de plantas medicinais do quilombo Tiningú, Santarém, Pará, Brasil. *Dissertação (Mestrado em Ciências Ambientais) UFOPA, Santarém*, 2015.
63. Flor ASSO, Barbosa WLR. Sabedoria popular no uso de plantas medicinais pelos moradores do bairro do sossego no distrito de Marudá-PA. *Revista Brasileira de Plantas Medicinais*, 2015, 17, suppl 1, 757–768.
64. Antonio RL, Souza RM, Furlan MR, Pedro CR, Cassas F, Honda S, Rodrigues E. Investigation of urban ethnoveterinary in three veterinary clinics at east zone of São Paulo city, Brazil. *Journal of Ethnopharmacology*, 2015, 173, 183–190.
65. Parente CET, Da Rosa MMT. Plantas comercializadas como medicinais no Município de Barra do Piraí, RJ. *Rodriguésia*, 2001, 52, 47–59.
66. Filho PGX, Ribeiro AF, Moraes AF, Penha WF, Borges, WL, Santos RHS. Ethnobotanical knowledge on non-conventional food and medicinal plants in Rio Cajari Extractivist Reserve, Amazon, Brazil. *Research Square*, 2020.
67. Sauini T, Fonseca-Kruel VS, Baptistela Yazbek P, Matta P, Cassas F, Da Cruz C, . . . Rodrigues E. Participatory methods on the recording of traditional knowledge about medicinal plants in Atlantic forest, Ubatuba, São Paulo, Brazil. *Plos One*, 2020, 15, 5, e0232288.
68. Dias SRK, Alcantara MMAC, Pessoa SMA. Aspectos etnobotánicos, fitoquímicos y farmacológicos de especies de Rubiaceae en Brasil. *Revista Cubana de Plantas Medicinales*, 2013, 18, 1, 140–156.
69. Martins AG, Rosário DLD, Barros MND, Jardim MAG. Levantamento etnobotânico de plantas medicinais, alimentares e tóxicas da Ilha do Combu, Município de Belém, Estado do Pará, Brasil. *Revista Brasileira de Farmacologia*, 2005, 86, 1, 21–30.
70. Guarin Neto G, Santana SR, Silva, J. Notas etnobotânicas de espécies de Sapindaceae jussieu. *Acta Botânica Brasílica*, 2000, 14, 327.
71. Farias DF, Cavalheiro MG, Viana MP, Queiroz VA, Rocha-Bezerra LC, Vasconcelos IM. Water extracts of Brazilian leguminous seeds as rich sources of larvicidal compounds against *Aedes aegypti* L. *Anais da Academia Brasileira de Ciências*, 2010, 82, 3, 585–594.

72. Roman ALC, Ming LC, Carvalho ID, Sablayrolles MDGP. Uso medicinal da pimenta malagueta (*Capsicum frutescens* L.) em uma comunidade de várzea à margem do rio Amazonas, Santarém, Pará, Brasil. Boletim do Museu Paraense Emílio Goeldi. *Ciências Humanas* 2011, 543–557.
73. Carniello MA, Silva RDS, Cruz MA, Guarim Neto G. Quintais urbanos de Mirassol D'Oeste-MT, Brasil: uma abordagem etnobotânica. *Acta amazonica*, 2010, 40, 3, 451–470.
74. Miot HA, Batistella RF, Batista KDA, Volpato DEC, Augusto LST, Madeira NG, . . . Miot LDB. Comparative study of the topical effectiveness of the andiroba oil (*Carapa guianensis*) and DEET 50% as repellent for *Aedes* sp. *Revista do Instituto de Medicina Tropical de São Paulo*, 2004, 46, 253–256.
75. Tabanca N, Demirci B, Kiyan HT, Ali A, Bernier UR, Wedge DE, Başer KHC. Repellent and larvicidal activity of *Ruta graveolens* essential oil and its major individual constituents against *Aedes aegypti*. *Planta Medica*, 2012, 78, 05, 90.
76. Devi MA, Sahoo D, Singh TB, Rajashekar Y. Toxicity, repellency and chemical composition of essential oils from *Cymbopogon* species against red flour beetle *Tribolium castaneum* Herbst (Coleoptera: Tenebrionidae). *Journal of Consumer Protection and Food Safety*, 2020, 15, 2, 181–191.
77. Williams LAD, Rosner H, Levy HG, Barton EN. A critical review of the therapeutic potential of dibenzyl trisulphide isolated from *Petiveria alliacea* L (guinea hen weed, anamu). *West Indian Medical Journal*, 2007, 56(1), 17.
78. García-Mateos MR, Elizalde Sánchez E, Espinosa-Robles P, Álvarez-Sánchez ME. Toxicidad de *Petiveria alliacea* L.: Sobre la mosca blanca (*Trialeurodes vaporariorum* west.). *Interciencia*, 2007, 32, 2, 121–124.
79. Hartmann I, Silva AD, Walter EM, Jeremias WDJ. Investigação do efeito larvicida de *Petiveria alliacea* (Guiné) sobre as larvas de mosquitos da espécie *Aedes aegypti*. *Revista virtual de Química*. 2018, 10, 3.
80. Ferreira LS, Alves SN. Neem (*Azadirachta indica*): larvicidal properties—a review. *Revista Conexão Ciência*, 2021, 16, 1.
81. Globe Newswire. Pest control market projected to reach $27.5 Billion by 2025—Report by Marketsand MarketsTM. Marketsand Markets. 2019. (Accessed Februery 10, 2022, at https://www.globenewswire.com/news-release/2019/10/07/1925966/0/en/Pest-Control-Market-Projected-to-Reach-27-5-Billion-by-2025-Report-by-MarketsandMarkets.html.)
82. Subramaniam J, Kovendan K, Kumar PM, Murugan K, Walton W. Mosquito larvicidal activity of *Aloe vera* (Family: Liliaceae) leaf extract and *Bacillus sphaericus*, against Chikungunya vector, *Aedes aegypti*. *Saudi Journal of Biological Sciences*, 2012, 19, 4, 503–509.
83. Carvalho GHF, Andrade MA, Araújo CN, Santos ML, Castro NA, Charneau S, . . . Bastos IMD. Larvicidal and pupicidal activities of eco-friendly phenolic lipid products from Anacardium occidentale nutshell against arbovirus vectors. *Environmental Science and Pollution Research*, 2019, 26, 6, 5514–5523.
84. Guissoni ACP, Silva IG, Geris R, Cunha LCD, Silva, HHGD. Atividade larvicida de *Anacardium occidentale* como alternativa ao controle de Aedes aegypti e sua toxicidade em *Rattus norvegicus*. *Revista Brasileira de Plantas Medicinais*, 2013, 15, 363–367.
85. Mendonça FA, Da Silva KFS, Dos Santos KK, Júnior KR, Sant'ana AEG. Activities of some Brazilian plants against larvae of the mosquito *Aedes aegypti*. *Fitoterapia*, 2005, 76, 629–636.
86. Torres RC, Garbo AG, Walde RZML. Characterization and bioassay for larvicidal activity of *Anacardium occidentale* (cashew) shell waste fractions against dengue vector *Aedes aegypti*. *Parasitology Research*, 2015, 114, 10, 3699–3702.
87. Dill EM. Pereira MJB, Costa MS. Efeito residual do extrato de *Annona coriacea* sobre *Aedes aegypti*. *Arquivos do Instituto Biológico*, 2012, 79, 595–602.
88. Omena MC, Navarro DMAF, De Paula JE, Luna JS, De Lima M.F, Sant'ana AEG. Larvicidal activities against *Aedes aegypti* of some Brazilian medicinal plants. *Bioresource Technology*, 2007, 98, 2549–2556.
89. Ciccia G, Coussio J, Mongelli E. Insecticidal activity against *Aedes aegypti* larvae of some medicinal South American plants. *Journal of Ethnopharmacology*, 2000, 72, 185–189.
90. Hidalgo PDSP, Nunomura RDCS, Nunomura, SM, Pinto SAC, Tadei WP. Larvicidal Activities against *Aedes aegypti* and *Culex quinquefasciatus* of some extracts from Amazon edible fruits. *Journal of Mosquito Research*, 2017, 6.
91. Souza TM, Farias DV, Soares BM, Viana MP, Lima GPG, Machado LKA. . . Carvalho AFU. Toxicity of Brazilian plant seed extracts to two strains of *Aedes aegypti* (Diptera: Culicidae) and non-target animals. *Journal of Medical Entomology*, 2011, 48, 846–851.
92. Barbosa PBBM, De Oliveira JM, Chagas JM, Rabelo LMA, Medeiros GF, Giodani RB, . . . Ximenes MFFM. Evaluation of seed extracts from plants found in the Caatinga biome for the control of *Aedes aegypti*. *Parasitology Research*, 2014, 113, 10, 3565–3580.

93. Sá RA, De Lima Santos ND, Da Silva CSB, Napoleão TH, Gomes F.S, Cavada BS, . . . Paiva PMG. Larvicidal activity of lectins from Myracrodruon urundeuva on *Aedes aegypti*. *Comparative Biochemistry and Physiology Part C: Toxicology & Pharmacology*, 2009, 149, 3, 300–306.
94. Mohammed A, Chadee DD. An evaluation of some Trinidadian plant extracts against larvae of *Aedes aegypti* mosquitoes. *Journal of the American Mosquito Control Association*, 2007, 23, 2, 172–176.
95. Kiplang'at KP, Mwangi RW. Repellent Activities of *Ocimum basilicum, Azadirachta indica* and *Eucalyptus citriodora* Extracts on Rabbit Skin against *Aedes aegypti*. *Journal of Entomology and Zoology Studies*, 2013, 84.
96. Nour AH, Jessinta D, Nour AH. Larvicidal activity of extracts from different parts of neem (*Azadirachta indica*) against *Aedes Aegypti* mosquitoes larvae. *Scientific Research and Essays*, 2012, 7, 31, 2810–2815.
97. Ajileye AY, Muse WA. Biological effect of some plant extractions on the reproductive development of the yellow fever mosquito *Aedes aegypti* Linn. *Ife Journal of Science*, 2006, 8, 1, 23–27.
98. Giorgi A, De Marinis P, Granelli G, Chiesa LM, Panseri S. Secondary metabolite profile, antioxidant capacity, and mosquito repellent activity of *Bixa orellana* from Brazilian Amazon region. *Journal of Chemistry*, 2013, 1, 409826.
99. Sanabria L, Segovia EA, González N, Alcaraz P, Bilbao VN. Larvicidal activity of aqueous plants extracts on *Aedes aegypti* larva (first trials). *Memorias del Instituto de Investigaciones en Ciencias de la Salud*, 2009, 26–31.
100. Alvarez MRS, Heralde Iii FM. Quiming NS. Potent larvicidal activities of *Capsicum frutescens* (L.) fruit ethanolic and partially purified extracts against *Aedes aegypti* (L.) and *Aedes albopictus* (S.). *Der Pharmacia Lettre*, 2015, 7, 11, 94–99.
101. Prophiro JS. Da Silva MAN, Kanis LA, Da Rocha, LCB, Duque-Luna, JE, Da Silva, O. S. First report on susceptibility of wild *Aedes aegypti* (Diptera: Culicidae) using *Carapa guianensis* (Meliaceae) and *Copaifera* sp. (Leguminosae). *Parasitology Research*, 2012, 110, 699–705.
102. Miot HA, Batistella RF, Batista KDA, Volpato DEC, Augusto LST, Madeira NG, . . . Miot LDB. Comparative study of the topical effectiveness of the andiroba oil (*Carapa guianensis*) and DEET 50% as repellent for *Aedes* sp. *Revista do Instituto de Medicina Tropical de São Paulo*, 2004, 46, 253–256.
103. Gomes PRB, Oliveira MB, De Sousa DA, Silva JC, Fernandes RP, Louzeiro HC, . . . Mouchrek Filho VE. Larvicidal activity, molluscicide and toxicity of the essential oil of *Citrus limon* peels against, respectively, *Aedes aegypti, Biomphalaria glabrata* and *Artemia salina*. *Eclética Química*, 2019, 44, 4, 85–95.
104. Massebo F, Tadesse M, Bekele T, Balkew M, Gebre-Michael T. Evaluation on larvicidal effects of essential oils of some local plants against *Anopheles arabiensis* Patton and *Aedes aegypti* Linnaeus (Diptera, Culicidae) in Ethiopia. *African Journal of Biotechnology*, 2009, 8, 17, 4183–4188,
105. Kiplang'at KP, Mwangi RW. Repellent Activities of *Ocimum basilicum, Azadirachta indica* and *Eucalyptus citriodora* extracts on Rabbit Skin against *Aedes aegypti*. *Journal of Entomology and Zoology Studies*, 2013, 84.
106. Amer A, Mehlhorn, H. Repellency effect of forty-one essential oils against *Aedes*, *Anopheles*, and *Culex* mosquitoes. *Parasitology Research*, 2006, 99, 4, 478–490.
107. Porto KR, Motti PR, Yano M, Roel AR, Cardoso CA, Matias R. Screening of plant extracts and fractions on *Aedes aegypti* larvae found in the state of Mato Grosso do Sul (linnaeus, 1762)(culicidae). *Anais da Academia Brasileira de Ciências*, 2017, 89, 895–906.
108. Soonwera M, Sittichok S. Adulticidal activities of *Cymbopogon citratus* (Stapf.) and *Eucalyptus globulus* (Labill.) essential oils and of their synergistic combinations against *Aedes aegypti* (L.), *Aedes albopictus* (Skuse), and *Musca domestica* (L.). *Environmental Science and Pollution Research*, 2020, 27, 16, 20201–20214.
109. Arpiwi NL, Muksin IK, Kartini NL. Essential oil from *Cymbopogon nardus* and repellant activity against *Aedes aegypti*. *Biodiversitas Journal of Biological Diversity*, 2020, 21, 8.
110. Kurniasih N, Nuryadin W, Harahap MN, Supriadin A, Kinasih I. Toxicity of essential oils from orange (*Citrus sinesis* L. Obbeck) and lemongrass (*Cymbopogon nardus* L. Rendle) on *Aedes aegypti* a vector of Dengue Hemorrhagic Fever (DHF). *Journal of Physics: Conference Series*. IOP Publishing, 2021, 012015.
111. Trongtokit Y, Rongsriyam Y, Komalamisra N, Apiwathnasorn C. Comparative repellency of 38 essential oils against mosquito bites. *Phytotherapy Research: An International Journal Devoted to Pharmacological and Toxicological Evaluation of Natural Product Derivatives*, 2005, 19, 4, 303–309.
112. Manh HD, Hue DT, Hieu NTT, Tuyen DTT, Tuyet OT. The mosquito larvicidal activity of essential oils from *Cymbopogon* and *Eucalyptus* species in vietnam. *Insects*, 2020, 11, 2, 128.

113. Yanti CA, Sari M, Triana ADPSW. (*Cymbopogon winterianus* Jowitt) sebagai Repelen dari Nyamuk Aedes aegypti. *Jurnal Vektor Penyakit*, 2021, 15, 2, 99–106.
114. Anoopkumar AN, Puthur S, Rebello S, Aneesh EM. Screening of a few traditionally used medicinal plants for their larvicidal efficacy against *Aedes aegypti* Linn (Diptera: Culicidae), a dengue fever vector. *Symbiosis Microbiology & Infectious Diseases*, 2017a, 5, 1–5.
115. Pinto LPDAP, da Cruz ILS, Dias TD, Honório, NA, da Cunha Gonçalves SJ, Maleck M. Extratos de Erva-de-Santa-Maria na saúde pública: controle do vetor de arboviroses. *Revista Pró-univerSUS*, 2019, 10, 1, 102–105.
116. Cavalca PAM, Lolis MIGDA, Reis B, Bonato CM. Homeopathic and larvicide effect of *Eucalyptus cinerea* essential oil against *Aedes aegypti. Brazilian Archives of Biology and Technology*, 2010, 53, 835–843.
117. Aguiar JCD, Santiago GM, Lavor PL, Veras HN, Ferreira YS, Lima MA, . . . Braz-Filho R. Chemical constituents and larvicidal activity of *Hymenaea courbaril* fruit peel. *Natural Product Communications*, 2010, 5, 12.
118. Khatiwora E, Adsul V.B, Pawar P, Joseph M, Deshpande NR, Kashalkar RV. Larvicidal activity of *Ipomoea carnea* stem extracts and its active ingredient dibutyl phthalate against *Aedes aegypti* and *Culex quinquefasciatus*. *Der Pharma Chemica*, 2014, 6, 1, 155–161.
119. Andrade TAS, Santana IM, Jimenez GC, Farias ETN, Macedo LO, Alves LC, . . . de Carvalho GA. Larvicidal activity of *Caesalpinia ferrea* Mart. and *Lippia origanoides* Cham against *Aedes aegypti* (LINNAEUS, 1762) (Diptera: Culicidae). *Revista de Patologia Tropical/Journal of Tropical Pathology*, 2020, 49, 1, 45–54.
120. Sobrinho ACN, Morais, SM, Marinho MM, Souza NV, Lima DM. Antiviral activity on the Zika virus and larvicidal activity on the *Aedes* spp. of *Lippia alba* essential oil and β-caryophyllene. *Industrial Crops and Products*, 2021, 162, 113281.
121. Mahanta S, Sarma R, Khanikor, B. The essential oil of *Lippia alba* Mill (Lamiales: Verbenaceae) as mosquitocidal and repellent agent against *Culex quinquefasciatus* Say (Diptera: Culicidae) and *Aedes aegypti* Linn (Diptera: Culicidae). *The Journal of Basic and Applied Zoology*, 2019, 80, 1, 1–7.
122. Felix SF, Rodrigues AM, Rodrigues ALM, Freitas JCC, Alves DR, Silva AA, . . . de Morais SM. Chemical composition, larvicidal activity, and enzyme inhibition of the essential oil of *Lippia grata* Schauer from the Caatinga Biome against dengue vectors. *Pharmaceuticals*, 2021, 14, 3, 250.
123. Silva IG, Guimarães VP, Lima CG, Silva HHG, Elias CN, Mady CM, . . . Isac E. Efeito larvicida e toxicológico do extrato bruto etanólico da casca do caule de *Magonia pubescens* sobre *Aedes aegypti* (Diptera, Culicidae) em criadouros artificiais. *Revista de Patologia Tropical*, 2003, 32, 73–86.
124. Velayutham K, Ramanibai R, Umadevi M. Green synthesis of silver nanoparticles using Manihot esculenta leaves against Aedes aegypti and Culex quinquefasciatus. *The Journal of Basic & Applied Zoology*, 2016, 74, 37–40.
125. Coria C, Almiron W, Valladares G, Carpinella C, Ludueña F, Defago M, Palacios S. Larvicide and oviposition deterrent effects of fruit and leaf extracts from *Melia azedarach* L. on *Aedes aegypti* (L.) (Diptera: Culicidae). *Bioresource Technology*, 2008, 99, 8, 3066–3070.
126. Wandscheer CB, Duque JE, Da Silva MA, Fukuyama Y, Wohlke J.L, Adelmann J, Fontana JD. Larvicidal action of ethanolic extracts from fruit endocarps of *Melia azedarach* and *Azadirachta indica* against the dengue mosquito *Aedes aegypti*. *Toxicon*, 2004, 44, 8, 829–835.
127. Singh RK, Dhiman RC, Mittal PK. Mosquito larvicidal properties of *Momordica charantia* Linn (family: Cucurbitaceae). *Journal of Vector Borne Diseases*, 2006, 43, 2, 88
128. Mituiassu LMP, Serdeiro MT, Vieira RRBT, Oliveira LS, Maleck M. *Momordica charantia* L. extracts against *Aedes aegypti* larvae. *Brazilian Journal of Biology*, 2021, 82.
129. Ferreira PMP, Carvalho AF, Farias DF, Cariolano NG, Melo VM, Queiroz MG. Larvicidal activity of the water extract of *Moringa oleifera* seeds against *Aedes aegypti* and its toxicity upon laboratory animals. *Anais da Academia Brasileira de Ciências*, 2009, 81, 2, 207–216.
130. Sandralintang TC, Fauzantoro A, Hermansyah H, Jufri M, Gozan M. Use of *Nicotiana tabacum* L extract for anti-*Aedes Aegypti* mosquito paint. In: *AIP Conference Proceedings*. AIP Publishing LLC, 2018, 030010.
131. Musau JK, Mbaria JM, Nguta JM, Mathiu M, Kiama SG. Phytochemical composition and larvicidal properties of plants used for mosquito control in Kwale County, Kenya. 2016, 3, 3, 12–17.
132. Oliveira PV, Ferreira JC, Moura FS, Lima GS, Oliveira FM, Oliveira PES, . . . Lemos, RPL. Larvicidal activity of 94 extracts from ten plant species of northeastern of Brazil against *Aedes aegypti* L. (Diptera: Culicidae). *Parasitology Research*, 2010, 107, 2, 403–407.

133. Faustino CG, Medeiros FA, Galardo AKR, Rodrigues ABL, Da Costa ALP, Martins RL, . . . da Silva de Almeida SSM. Biocidal activity of a nanoemulsion containing essential oil from *Protium heptaphyllum* Resin against *Aedes aegypti* (Diptera: Culicidae). *Molecules*, 2021, 26, 21, 6439.
134. Pimenta AT, Santiago GM, Arriaga Â, Menezes GH, Bezerra SB. Estudo fitoquímico e avaliação da atividade larvicida de *Pterodon polygalaeflorus* Benth (Leguminosae) sobre *Aedes aegypti. Revista brasileira de farmacognosia*, 2006, 16, 501–505.
135. Candido LP, Beserra EB. Repellent activity of *Cnidoscolus phyllacanthus* Mart. and *Ricinus communis* L. extracts against *Aedes aegypti* L. oviposition behavior. *Biotemas*, 2015, 28, 4, 105–112.
136. Cerna GO, Rodríguez JR. Actividad repelente y tiempo de protección experimental del aceite del endospermo de Ricinus communis (Euphorbiaceae) en Aedes aegypti. *Revista Rebiolest*, 2014, 2, 2, 48–60.
137. Sogan N, Kapoor N, Singh H, Kala S, Nayak A, Nagpal BN. Larvicidal activity of *Ricinus communis* extract against mosquitoes. *Journal of Vector Borne Diseases*, 2018, 55, 4, 282.
138. Orlanda JFF, Mouchrek AN. Efeito larvicida do óleo essencial das folhas de *Ruta graveolens* L. no controle de *Aedes aegypti* (L, 1762) (Diptera: Culicidae). *Research, Society and Development*, 2021, 10, 12, e115101220028-e115101220028,
139. Amariles-Barrera S, García-Pajón C, Parra-Henao G. Actividad insecticida de extractos vegetales sobre larvas de *Aedes aegypti*, Diptera: Culicidae. *CES Medicina*, 2013, 27, 2, 193–203.
140. Aguiar RWS, Santos SF, Morgado SF, Ascencio SD, Mendonça LM, Viana KF, . . . Ribeiro BM. Insecticidal and repellent activity of *Siparuna Guianensis* Aubl. (Negramina) against *Aedes Aegypti* and *culex quinquefasciatus. Plos One*, 2015, 10, 2, E0116765.
141. Nityasree BR, Chalannavar RK, Ghosh SK, Divakar MS, Sowmyashree K. Effect of *Solanum lycopersicum* leaf extracts against larvicidal activity of *Aedes aegypti* L. *Biomedicine*, 2020, 40, 4, 467–473.
142. Guarda C, Lutinski JA, Roman-Junior WA, Busato MA. Atividade larvicida de produtos naturais e avaliação da susceptibilidade ao inseticida temefós no controle do *Aedes aegypti* (Diptera: culicidae). *Interciência*, 2016, 41, 4, 243–247.
143. Arriaga AM, Rodrigues FE, Lemos TL, Oliveira MDC, Lima JQ, Santiago GM. Composition and larvicidal activity of essential oil from *Stemodia maritima* L. *Natural Product Communications*, 2007, 2, 12, 934578X0700201209.
144. Marques MM, Moraes SM, Vieira ÍG, Vieira MG, Silva ARA, Almeida RR, Guedes MIF. Larvicidal activity of *Tagetes erecta* against *Aedes aegypti. Journal of the American Mosquito Control Association*, 2011, 27, 2, 156–158.
145. Medeiros VF. Potencial larvicida de extratos de plantas regionais no controle de larvas de *Aedes aegypti* (Diptera: Clucidae). Dissertação de Mestrado. Programa de pós-graduação em Ciências Biológicas. *Universidade Federal do Rio Grande do Norte*. 2007.
146. Asadollahi A, Khoobdel M, Zahraei-Ramazani A, Azarmi S, Mosawi SH. Effectiveness of plant-based repellents against different Anopheles species: A systematic review. *Malaria Journal*, 2019, 18, 1, 1–20.
147. Eden WT, Alighiri D, Supardi KI, Cahyono, E. The mosquito repellent activity of the active component of air freshener gel from Java Citronella oil (*Cymbopogon winterianus*). *Journal of Parasitology Research*, 2020, 1, 9053741.

15 Azadirachtin from Neem (*Azadirachta indica*)

Efficacy and Mechanisms Against Insects and Diseases

Abdulrahman A. Almadiy, Mariam S. Al-Ghamdi, Fahd Mohammed Abd Al Galil, and Showket Ahmad Dar

15.1 INTRODUCTION

In recent times, there has been a notable shift in the field of holistic pest management and plant disease management, with an increasing reliance on organic substances that living organisms utilize as a means of protection against pests and disease-causing organisms. In comparison to synthetic materials, these substances offer superior security and ecological sustainability, which presents an opportunity to address global issues related to agriculture, human health, and environmental sustainability [1]. Pest and disease control have been attempted on a vast array of plants, and currently, a multitude of horticultural mineral oils, plant-based essential oils, and detergents are utilized globally. Neem has emerged as the most effective biopesticide and drought-tolerant plant among a multitude of alternatives. The neem plant (*Azadirachta indica*), a member of the Meliaceae family, is a tropical evergreen with a high insecticidal effect. It is native to regions spanning from East India to Burma and throughout Southeast Asia. Additionally, the plant is native to West Africa, the Caribbean, and Central America, with a particularly high prevalence in Mexico. As an organic compound with a multitude of applications in various sectors, including industry, agriculture, healthcare, and the environment, neem has recently gained significant recognition. Experts from diverse geographical regions are currently investigating the beneficial effects of neem in agriculture. This remarkable tree is utilized in the production of numerous agricultural products, including soil conditioning agents, fertilizers, manures, compost, insecticides (pesticides), and gases [2, 3]. The byproducts of the neem tree have applications in the management of nematodes and fungi, as well as in the regulation of insect growth (IGRs). *Azadirachtin*, the active component of the neem tree, has been observed to repel and eliminate a variety of caterpillars, including those of the thrips and whitefly species. Organic pesticides are defined as those derived from natural substances, including a variety of minerals, plants, microorganisms, and animals. When neem interacts with organisms that are not its intended targets, particularly mammals, natural enemies, aquatic organisms, and the surrounding environment, the effect is found to be minimal. Conversely, neem has been observed to have positive impacts on human health, particularly in the context of blood purification. Neem is emerging as a preferred natural pesticide among growers for integrated pest management and overall pest population regulation. Over time, the Indian subcontinent has recognized neem efficacy in preventing soil-borne and storage-related insect pests. The efficacy of neem oil in managing a diverse array of pests has been demonstrated. These include *Cnaphalocrocis medinalis, Ceraeochrysa claveri, Diaphorina citri* (Hemiptera: Liviidae), *Helicoverpa armigera* (Noctuidae (Lepidoptera): *Mamestra brassicae* (cabbage moth), *Pieris brassicae* (cabbage butterfly), and *Spodoptera frugiperda* (fall armyworm). Neem compositions derived from kernels have demonstrated efficacy against over 105 insect pests belonging to ten orders that infest a range of

DOI: 10.1201/9781003463429-19

fruit crops and vegetables. The bioactive elements present in neem have been demonstrated to be effective in controlling a multitude of crop diseases and pests. It has recently been demonstrated that the bioactive compounds derived from neem are capable of controlling in excess of 400 different kinds of pests of significant importance, in addition to 16 parasitic nematodes which infect crops and cause yield loss. Aerosols produced by neem seed kernel extract and oil have been demonstrated to reduce illnesses caused by *Alternaria* leaf blight, anthracnose, and *Cercospora* leaf spot, which occurs in black gram. Neem oil has been employed to control a variety of plant diseases, including rust (Pucciniales), powdery (ascomycete fungi) and downy mildew (Oomycetes), scab (Streptomyces) and blotch *Taphrina maculans*, twig *Phomopsis vaccinia* and tip blight (fungi and bacteria), and *Botrytis*. The use of neem is beneficial in the management of plant diseases due to the detrimental effects of synthetic insecticides on the environment and human health. A number of scientific investigations have been conducted to ascertain the potential efficacy of *Azadirachtin* as a botanical treatment for diseases and insect pests. Neem represents an environmentally friendly alternative to synthetic pesticides, which have been linked to long-term adverse effects. The vast amount of scientific data on neem and the products derived from it, particularly biopesticides and biofungicides, is currently lacking in organization. Accordingly, the objective of this comprehensive investigation was to collate and highlight the achievements of utilizing neem for the management of insect pests and infectious crop diseases. Accordingly, the present chapter is primarily concerned with evaluating the chemical composition of neem and its applications as a fumigant, pesticide, and disease management measure, as well as the mechanisms of action involved. Furthermore, this assessment will identify potential applications of neem in the agricultural sector and other fields. The findings of this chapter will also serve as a valuable foundation for future neem research, particularly in the development of more cost-effective and environmentally conscious plant biopesticides.

15.2 AZADIRACHTIN

Azadirachtin is one of the secondary metabolic compounds identified in neem seeds. *Azadirachtin* is a limonoid, a class of organic molecule comprising an enol ether, acetal, hemiacetal, tetra-substituted epoxide, and numerous carboxylic esters (Figure 15.1). These compounds are characterized by extensive oxidation of tetranortriterpenoid groups. The molecular configuration of *azadirachtin*

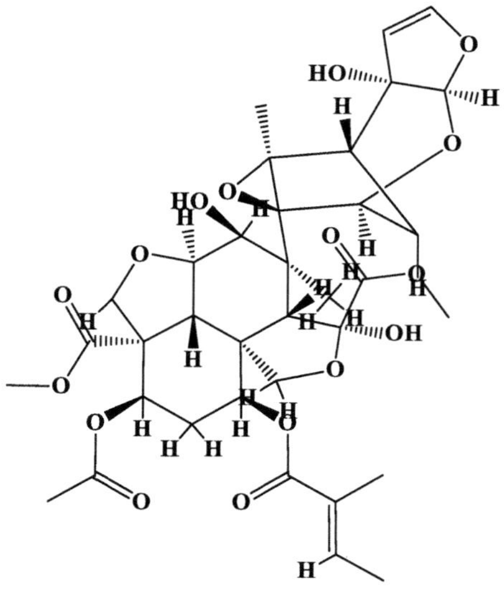

FIGURE 15.1 Structure of *Azadirachtin* molecule [4].

is complex, comprising 16 stereogenic centers, seven of which are tetra-substituted, in addition to secondary and tertiary hydroxyl chains and a tetrahydrofuran ether. These characteristics contribute to the considerable complexity involved in synthesizing this compound from its fundamental elements through synthetic organic chemistry processes. Subsequently, the initial comprehensive production was released in 2007 by Steven Ley's laboratory at the College of Cambridge, over 22 years after its initial synthesis. In the synthesis described, a relaying method was employed to generate the requisite compound, substantially functionalizing the decalin intermediate through the use of a small-scale complete reaction. The requisite gram-scale activities were derived from the plant material itself. The chemical formula of azadirachtin is as follows: (2aR,2a1R,3S,4S,4aR,5S,7aS,8S,10R,10aS)-10-(acetyloxy)-3,5-dihydroxy-4-[(1aR,2S,3aS,6aS,7S,7aS)-6a-hydroxy-7a-methyl-3a, The compound is a 6a,7,7a-tetrahydro-2,7-methanofuro[2,3-b]oxireno[2,3-e]oxepin-1a(2H)-yl]-4-methyl-8-{[(2E)-2-methylbut-2-enoyl]oxy}octahydro-1H,7H-naphtho[1,8-bc: 4,4a-c′]difuran-5,10a(8H)-dicarboxylate.

15.3 BIOSYNTHESIS OF AZADIRACHTIN

Tirucallol is a type of steroid that is produced from two molecules of farnesyl diphosphate (FPP) to generate a C_{30} triterpene. Upon loss of all three methyl groups, the molecule undergoes transformation into a C_{27} steroid. It is hypothesized that the tirucallol molecule serves as a precursor for neem triterpenoid metabolites, which are secondary metabolites. Azadirachtin is then synthesized *via* a complex biosynthetic route. The generation of butyro-spermol is contingent upon the allylic isomerization of tirucallol, which is accompanied by an oxidation process. Moreover, apotirucallol is the result of the oxidation of butyro-spermol, which undergoes a Wagner-Meerwein 1, 2-methyl shift during reorganization. Subsequently, the four terminal carbon atoms from the side chains undergo a process of splitting, resulting in the formation of a tetranortriterpenoid. This is followed by an additional oxidation process, which generates azadirone and *azadiradione*. Concurrently, the remaining side chains undergo a cyclization reaction, forming a furan ring. The C-seco-limonoids, which include *salannin, nimbin*, and nimbidininin, are generated by the release and oxidation of the final ring through esterification with tiglic acid monomer, which is created by L-isoleucine. The current hypothesis posits that the desired molecule is formed through a biosynthetic transformation of azadirone to salanin, which is then subjected to intense oxidation and carbon cycling to generate azadirachtin.

15.4 MECHANISM OF ACTION OF NEEM

15.4.1 Insecticidal Properties of *Azadirachtin*

Derived through the mevalonic acid metabolic route (pathway for producing isoprenoid building blocks needed for synthesis of important cellular constituents) in the neem tree, azadirachtin is a very complex tetranortriterpenoid composed of 16 asymmetric carbon atoms [5]. With regard to limonoids, these are naturally produced, heavily oxidized tetranorterpenoid chemicals that are linked to limonin, the acidic chemical found in the fruit of the citrus tree (Citrus). Commercial preparations of *azadirachtin* for the sustainable control of insects in agriculture are sold worldwide and include a specific quantity of azadirachtin-A (Table 15.1) [6]. The chemical composition of the compound is highly complex, and nearly 20 years after its initial identification, the first comprehensive preparation of the compound was published in 1985. Azadirachtin (Figure 15.1) [4], a broad-spectrum pesticide derived from marijuana, is effective against a range of crop insects, including Coleoptera, Hemiptera, Diptera, Orthoptera, and Isoptera. It functions as a feeding repellent, a developmental disruptor (IGD), and a sterilant [7]. The toxicity of *azadirachtin* varies across different insect categories, with diverse detoxification enzyme levels and rates of penetration influencing this differential toxicity (Table 15.2). The chemical composition of azadirachtin reduces the

TABLE 15.1
***Azadirachtin*-based Biopesticide Products in the Global Market**

Product	Manufacturer	% of a.i. (*Azadirachtin*)
Agroneem plus	Agro logistic Systems Inc.	0.15
Azagro	Indian Mart	1.0
Azaguard	BioSafe Systems	3.0
Azamax	EID Parry Ltd.	1.2
AzaPRO	CannCare Company	1.2
Azasol	Arborjet Inc.	6.0
Debug TRES	ARGOLogistic Systems	3.0
EcoZin Plus	AMVAC Chemicals corp.	1.2
Fortune Aza	Fortune Biotech Ltd.	3.0
MOLT-X	BioWorks Inc.	3.0
Neem azal TS	Trifolio-M GmbH	1.0
Neemarin	Aza-Direct Gowan Company LLC	0.15
Neemfol	Gasain Pierrie	5.0
Neemix	Certis	0.25
Omazin	SERPO Corporation	3.0

TABLE 15.2
Lethal Concentration (LC 50%) of Azadirachtin against Various Insect Pests under Field Conditions

Species	Mode of Application	Stage of Development	LC50 (ppm)
Diamond back moth	Orally	3rd instar	0.6310
Tomato leaf miner	Orally	2nd instar	5.620
Coconut spike moth	Orally	Larval	28.790
Pea pod borer	Orally	3rd instar	12.950
Dinidorid bugs	Orally	Adults	0.0031
Coffin fly	Dipping	1st instar	0.2200
Pistachio psylla	Dipping	5th instar	13.790
Grasshoppers	Topically	4th instar	101.200
European grapevine moth	Orally	1st instar	2.100
Bee Wax moth	Topically	Immature stage	16.560

probability of insect resistance development. The peach aphid (*Myzus persicae*) exhibited resistance to active ingredients of *azadirachtin* throughout its 40 cyclic generations. However, the resistance to neem extract obtained from neem seeds was minimal. However, rather than avoiding neem without an equivalent quantity of *azadirachtin*, it was demonstrated that the acclimation of pure azadirachtin from tobacco cutworms (*Spodoptera litura*) resulted in decreased susceptibility to the antifeedant properties of the azadirachtin compound. This may elucidate the mechanism by which commercially available neem-based insecticides lacking additional non-AZA chemicals have no effect on desensitization [5]. It has been demonstrated that azadirachtin-A is extensively absorbed from the underground root system of the plant. Following this, it is transported systematically to the vegetative parts of plant cells via the xylem and preserved in an undamaged form in the leaves of the plant. Moreover, it has been demonstrated that plants with minimal levels of azadirachtin-A are particularly vulnerable to vegetative damage from phytophagous pest larvae. Moreover, as documented in, *azadirachtin* exhibited remarkable sensitivity with minimal human toxicity. It was indicated in reference that *azadirachtin* demonstrated an LD_{50} value exceeding 5,000 mg/kg and was classified in the class U (least likely to cause acute danger) of the WHO toxicity assessment report [8]. Moreover, the *azadirachtin* has been classified by the Environmental Protection Agency (EPA) as a category IV insecticide, indicating its relatively low toxicity. It has been approved for use as a general-purpose insecticide in the United States. The perceived threat posed by *azadirachtin* has recently been subject to scrutiny, particularly in relation to natural predators and insect pollinators [9]. However, *azadirachtin* has been demonstrated to be a selective, non-mutagenic, and readily biodegradable compound. Additionally, it has been identified as a safer alternative for organisms that are not the intended target, as well as for beneficial organisms, including mammals. Nevertheless, the combined approach of semi-field and field investigations may prove effective in anticipating any potential adverse effects of *azadirachtin* on non-target insects with a high degree of certainty. Moreover, azadirachtin is widely recognized as a highly attractive and effective chemical for managing insect pests in the environment. It is also regarded as one of the most effective alternatives to traditional pesticides in integrated pest management (IPM) strategies. Despite the advancement in our understanding of azadirachtin's physiological and biological properties, as well as its agricultural applications, the precise mechanism of action at the fundamental level remains elusive.

15.5 EFFECTS ON NEURO-ENDOCRINE ACTIVITY

It has been proposed that both juvenile hormone (JH) (acyclic sesquiterpenoids) and 20-hydroxyecdysone (also referred to as ecdysterone or 20E) play a crucial role in regulating insect development and growth. The equilibrium of these two hormones is believed to determine the precise progression of each developmental phase [10]. Consequently, disruption of the hormonal equilibrium leads to a cessation of development and is regarded as an appropriate approach for targeted pest management. As a receptor antagonist of all three primary endocrine compounds, *azadirachtin* has been demonstrated to prevent the biosynthesis of morphogenetic peptides (PTTH) from prothoracic glands and allatotropins across the corpus cardiacum complex. This provides further insight into the widely reported influence on IGD, which is predominantly characterized as a decreased pupation malformation process or an adult unsuccessful development [11]. In accordance with the findings of several researchers, the chemical compound in question also causes irreversible structural alterations in the nuclei of all hormone-producing glands, including the thoracic gland, corpus allatum, and corpus cardiacum, which regulate the occurrence of insect eruption and molting. Subsequently, malfunctioning of the neuroendocrine system may propagate extensively. When administered at a concentration of 74 parts per million to larvae of the *Ostrinia furnacalis* Guenée species, *azadirachtin* has been observed to decelerate the pace of growth and development, inhibit the synthesis of ecdysteroids, and reduce ecdysis [12].

A single milligram of *azadirachtin* has been demonstrated to be effective in decreasing immune-reactive ecdysteroids in recently ecdysed *Tenebrio molitor* pupae. This, in turn, had a significant impact on the levels of 20-hydroxyecdysone, which suppressed the ecdysteroid peak that typically occurs in the middle of the instar. Additionally, a significant reduction in hemolymph ecdysteroid titers was observed in *Rhodnius prolixus* following the administration of an individualized *azadirachtin* dose. In addition to its impact on morphogenetic PTTH, a compound affects the functioning of ecdysone 20-monooxygenase (EC 1.14.99.22), as well as cytochrome P450-dependent hydroxylase, which converts the growth hormone and thereby generates steroid ecdysone, and its metabolic product 20E [13]. Indeed, when these pests were incubated with a radiolabeled hormone called *azadirachtin*, homogenates of stumbling mitochondria from the second instar of *Drosophila melanogaster* larvae, or midguts from *Manduca sexta* larvae, were observed. The larvae in their conclusive instar, as well as the abdomens from adult female larvae of the *Aedes aegypti*, have been demonstrated to be dependent on dosage limitations of the ecdysone 20-monooxygenase. However, *azadirachtin* dosage or consumption had a minimal impact on the enzyme (20-monooxygenase) activities of *Spodoptera frugiperda*. In addition to its deleterious impact on the molting process, *azadirachtin* has also been demonstrated to influence the levels of juvenile hormone titers, either by reducing them or by promoting their accumulation. The inability of azadirachtin to degrade allatotropins, which impeded the subsequent generations and elimination processes of JH, was the primary reason for this phenomenon [14]. Furthermore, it has been demonstrated that azadirachtin suppresses the growth of insects and the process of molting, as well as significantly retards development at every stage of growth and development, from the adolescent stage to the pupal stage and finally to adulthood. This has the effect of undermining the insect's capacity for continued existence. Furthermore, the growth, development, and muscle mass of an organism are directly affected by the digestive process, which is primarily regulated by the insulin/insulin-like signaling (IIS) for development variables. Physiological processes associated with development include those of nutrition absorption and growth. In accordance with the findings of a few researchers [15], azadirachtin treatment has been observed to inhibit the growth and development of *D. melanogaster* in a manner that is similar to the suppression of the IIS pathway. Furthermore, *azadirachtin* has been shown to interfere with a number of endocrinological and physiological processes in insects due to its ability to diminish cholinergic communication among neurons and to partially inhibit calcium signaling pathways and circuits [16].

15.6 EFFECTS ON REPRODUCTION

A substantial body of research from a range of insect orders has demonstrated that azadirachtin has a detrimental impact on reproductive processes. The effects of azadirachtin on insect reproduction have been demonstrated in a number of species, including *S. littoralis Tobacco caterpillar*, *D. melanogaster* (fruit fly), *Galleria mellonella* (wax moth), *Dysdercus cingulatus* (Pyrrhocoridae bug), *Tuta absoluta* (tomato leaf miner), and *Helicoverpa armigera* (pea pod borer). These studies have revealed a reduction in fecundity and fertility in these insects following exposure to azadirachtin. This may have been caused by azadirachtin interfering with the synthesis of egg yolk proteins or acting through the absorption into oocytes of multiple insects [17].

Azadirachtin ($C_{35}H_{44}O_{16}$) has been demonstrated to have a detrimental impact on oviposition, to diminish and inhibit vitellogenin, and to significantly hinder egg formation in leaf-cutting queen ants *Atta sexdens*. Furthermore, it has been demonstrated that any disruption to the production and absorption of vitellogenin (the precursor to egg yolk) results in sterility in females. An individual injection of 10 micrograms of azadirachtin resulted in the sterilization of *Locusta migratoria* migratorioides, which led to the interruption of the developmental process and oviposition of intermediate oocytes. In accordance with some authors [18], the reproductive organs of female *Heteracris littoralis* individuals treated with azadirachtin exhibited substantial shrinkage, which triggered oocyte development arrest and follicular cell and mitochondrial disintegration and elimination. *Azadirachtin* has been demonstrated to significantly reduce the number of ovarian cysts and their apical nuclei in *D. melanogaster*. Furthermore, studies have demonstrated that the *Mylabris indica* beetle (Meloidae) and the *Heteracris littoralis* grasshopper (Orthoptera: Acrididae) exhibit repression of sperm production. The adverse effects on reproductive parameters are the result of azadirachtin's antagonistic action on JH and 20E, both of which are essential for the appropriate development of oogenesis and spermatogenesis. Indeed, the exogenous production of 20E has been demonstrated to counteract the depressive effects of azadirachtin on *D. melanogaster*, thereby restoring protein synthesis from egg yolk in ovaries and from fat body following neem treatments [19]. Moreover, it has been demonstrated that *azadirachtin* alters the reproductive cycle of *D. melanogaster* by reducing the probability of mating. In accordance with the scientific literature, azadirachtin has been demonstrated to modulate reproductive activity and sex habits in the large milkweed bug, *Oncopeltus fasciatus*, and the potential predator, *Neoseiulus baraki*, a predatory mite (Acari: Phytoseiidae) [20].

Once the bio-insecticide has been identified on the surface to which it has been applied, oviposition areas that have been subjected to treatments involving *azadirachtin* and other neem-based molecules develop an oviposition repellence, a deterrent, or suppression of populations in a variety of insect species under field conditions. This is evidenced by a number of sources, including references [21]. Furthermore, it has been demonstrated that purified azadirachtin prevents the eggs of *Nezara viridula* from being laid. It has been demonstrated that exposure of immature insects (larvae) to azadirachtin, a compound found in the commercial formulation Neem Azal, has resulted in a reduction in reproduction in female fruit flies (*D. melanogaster*) and an increased avoidance of treated surfaces. These outcomes, which are often employed as disincentives in pest management initiatives, have been observed in subsequent generations that did not experience exposure [22].

15.7 ANTIFEEDANCY AND ANTI-FEEDING IMPACTS

Neem has been observed to disrupt the sense of smell to such an extent that many insects would prefer to starve than consume food laced with *azadirachtin*. This effect acts as a potent antifeedant and deterrent to feeding. The chemical inhibits the activity of digestive enzymes following ingestion, and may induce an aversive taste memory by stimulating neurons that synthesize dopamine. Moreover, *azadirachtin* has been demonstrated to exert a pronounced antifeedant effect and engender psychological avoidance of substrates in a multitude of insect species, particularly those belonging to the orders *Hemiptera, Lepidoptera, Orthoptera, Coleoptera*, and *Diptera* [3, 5]. In order to identify potential food sources, insects rely on their olfactory abilities. However, they are then subjected to chemoreception, which acts as a fundamental antifeedant, verifying the nutritional content of the source and thereby influencing their subsequent food choices.

The communication to the central nervous system (CNS) causes the avoidance of approaching or consuming more neem biopesticides. The findings of numerous scientific studies indicate that neem biochemical exert a primary anti-feeding effect, which is mediated by the disruption of taste chemosensilla and is associated with a limitation on the frequency of stimulation of sugar-sensitive cells from gustatory chemical receptors. This excessive triggering of bitter-sensitive gustatory cells is caused by the major component of neem, azadirachtin. Indeed, numerous species have been demonstrated to be highly susceptible to azadirachtin's fundamental antifeedant effects. Rather than consuming the neem-based chemicals, these arthropods die of starvation [23]. Further observations have revealed the existence of a system of internal feedback, designated as secondary antifeedancy, which involves a sustained and reduced dietary intake, thereby exerting an adverse impact on multiple insect cells. The most significant tissues affected include the muscles, fat body, and gut epithelial cells. The administration of sub-lethal amounts of azadirachtin orally to the third larval stage of the tobacco caterpillar resulted in a reduction in food consumption, diminished capacity for conversion, and altered consumption patterns. In the case of *Spodoptera eridania* larvae, the subsequent instar demonstrated a reduction in relative consumption following immediate nourishment treatment with Azatrol, a commercial azadirachtin formulation. This was observed alongside a reduction in the rate of development, an improvement in digestive efficiency and an increase in the rate of food assimilation over the duration of the larval growth period. The administration of azadirachtin to larvae throughout the pro-3rd instar stage of *D. melanogaster* resulted in a substantial decrease in food consumption and interference with their natural ability to digest food. This was achieved by preventing the actions of enzymes that break down food. In the study [24], the response was additionally observed in adults who had undergone the pre-imaginal treatment (newly emerged adults with olfactory tissues that had been affected). This indicates that the antifeedancy is prolonged and that the consequences are delayed throughout the phase of development, with an eventual enhancement of azadirachtin's insecticidal action. Similarly, azadirachtin demonstrated an agonistic effect on neurons linked to dopamine, prompting *D. melanogaster* to develop an aversive taste memory. This specific form of memory can be influenced by dopamine-producing signals within the neural network, which result in a reduction of the proboscis extended reflex (PER) [24, 25].

15.8 CELLULAR AND MOLECULAR EFFECTS

In addition to the aforementioned behaviors, the literature describes a wide range of functions that *azadirachtin* exhibits, particularly in the form of p53 overexpression in the *Spodoptera litura* S1-1 cell line. This process mediates the cell death and increase of cell suppression. It was demonstrated that in other species of the Lepidoptera order, *azadirachtin* induced alterations in the structure of the larval midgut through the activation of apoptosis, which included the discharge of cytochrome-C (commonly known as an electron-carrying mitochondrial protein) from mitochondria to the plasma membrane and a greater expression of caspase-related enzymes (responsible for proinflammatory cytokines) and RNA-based apoptosis-binding motif-1 [25]. It is plausible that this modification may have had an impact on the digestive process and the absorption of vitamins and minerals. Furthermore, it has been demonstrated that *azadirachtin* promotes death *via* caspase-dependent mechanisms within *S. frugiperda* cell line Sf9. Recent proteomic investigations have revealed that numerous amino acids in the focal adhesive route can be monitored to influence the separation of cell attachment, disappearance of cell-cell signaling interactions, and the emergence of apoptosis throughout the pupal stage (pre-emergence of adults). This may initially correspond to the cellular response and the mechanism of infertility among male *S. litura* resulting from *azadirachtin* treatment [21].

Moreover, following the administration of *azadirachtin* during the larval stage, a number of proteins involved in the adenosine-monophosphate (MoA, RoA, molecule type)-activated protein kinase (AMPK) cascade demonstrated alterations at early developmental stage. *Azadirachtin* treatment has been demonstrated to induce caspase-independent necrosis and actin polymerization in *D. melanogaster*, resulting in mitochondrial arrest. *Azadirachtin* has been demonstrated to interfere with the production and release of enzymes at the cellular level. The ingestion of 3 μg neem (*azadirachtin*) per gram of body weight in *Schistocerca gregaria* has been observed to impede the incorporation of radio-labeled glycine with locust proteins. This finding has been documented in sources [20]. The literature indicates that *azadirachtin A* binds to heat-shock amino acids, such as hsp-60, which assist protein folding and function as molecular chaperones. This phenomenon has been well-documented in Drosophila Kc-167 cells, specifically in embryonic hemocytes. This may contribute to an interruption in protein production and secretion. *Azadirachtin* has been demonstrated to modulate or limit gene transcription and the activity of various proteins at the molecular structural level. The administration of 10 ppm azadirachtin to third-instar *Ostrinia furnacalis* larvae resulted in a significant disruption to the synthesis of enzymes associated with hemolymph lipid, leading to the accumulation of fatty tissue in the body (Table 15.2).

As reported, *azadirachtin* has been observed to enhance the level of expression of the genes odorant-binding protein 99b (Obp99b) and to decrease the transcription of genes encoding cuticular protein and amylase in *D. melanogaster*. This discovery may have implications for the development of bio pesticides, as well as for understanding molting issues and antifeedancy characteristics [20, 26]. It has been demonstrated that azadirachtin treatment enhances the production of malondialdehyde (MDA), as well as superoxide dismutase-related activity (SOD) in *D. melanogaster*. Furthermore, it has been demonstrated that *azadirachtin* induces the amplification of mutations that result in the production of antioxidant enzymes, including SOD, catalase (CAT), and GST. These enzymes play a crucial role in protecting against damage caused by oxidation, which is induced by increased and accumulated reactive oxygen species (ROS) triggered by *azadirachtin*. In the *Ipomoea batatas* plant afflicted with whitefly infestations, *azadirachtin* also inhibits the development of genes associated with oxidative stress protection, including ferritin and thioredoxin dehydrogenase. *Azadirachtin* has recently been demonstrated to exert a negative regulatory effect on the developmental process of *S. frugiperda*. This may be attributed to the influence it exerts on the insect's chitin production pathway, which encompasses the diminished expression of 31 cuticular proteins and numerous additional genes that encode pivotal enzymes involved in the biosynthesis of insect chitin and hormones, including trehalase, chitin synthesis, chitin deacetylase, and chitinase [27].

The biochemical basis for observed developmental and molting slowdown may be attributed to azadirachtin's inhibitory effect on chitosan production and cuticle gene regulation. Additionally, *azadirachtin* influenced alleles encoding enzymes essential for the synthesis of hormones [28]. Moreover, *azadirachtin* modified alleles that express essential enzymes involved in chemical synthesis, including those coding for farnesol dehydrating enzymes, which degrade farnesol, a precursor to JH (farnesal). The JH epoxide hydrolase enzyme gene initially breaks down JH epoxide, causing JH degeneration. The enzyme known as aldehyde dehydrogenase gene, which stimulates the breakdown of farnesal to farnesoic acids and CYP15A1-C1 to JH-III acids, is also affected. The gene family that encodes the two cytochrome oxidase-related protein molecules, CYP307A1 and CYP314A1, is similarly impacted, as they catalyze the 20-hydroxyecdysone. The restriction of growth has been partially triggered by variations in the expression levels of essential genes, which additionally impair the production of JH and ecdysone, thereby tipping the hormonal equilibrium [29].

15.9 TAXONOMICAL CLASSIFICATION OF NEEM

15.9.1 Morphological Description of Neem

Azadirachta indica, commonly referred to as neem, margosa, nimtree, or Indian lilac, is a perennial evergreen plant belonging to the Meliaceae family (Table 15.3). It is classified within the genus *Azadirachta* and is indigenous to various regions of the Southeast Asian area and the Indian subcontinent. The neem tree is a rapidly expanding species with comparatively thin and long, extending limbs; a straight trunk; and thick, rough, and longitudinal fissured skin. A fully developed tree may attain a height of up to 7.15 meters and may survive for more than two centuries. It typically begins producing yellowish elliptical fruits, known as drupes, approximately four years after germination. In approximately ten years, the tree attains its maximum level of productivity, exhibiting approximately 15 leaflets distributed in alternated pairs containing a terminal leaflet. The leaves

TABLE 15.3
Taxonomical Classification of Neem and Its Chemical Structure and Weight (g/mol)

Rank	Neem
Kingdom	Plantae
Division	Magnoliophyte
Class	Dipsacales
Order	Rutales
Suborder	Rutinae
Family	Meliacease
Superfamily	Melioideae
Tribe	Meliae
Genus	Azadirachta
Common name	Indica
Species	*Azadirachta indica*
Chemical formula	$C_{35}H_{44}O_{16}$
Molecular mass	720.721 $g{\cdot}mol^{-1}$
Origin	Indian subcontinent Southeast Asia
Authority *Azadirachta indica*	Adrien-Henri de Jussieu in 1830

are multifaceted and imparipinnate in arrangement. The blades are minute and lanceolate, reaching a length of up to 6 cm. The white, fragrant clusters, which are readily visible in the foliage axils, are the flowers. The tree typically reaches maturity during the month of May. The seeds are oval in shape, measuring 2 cm in width, and are yellow in color once they have reached maturity. They are a rich source of neem oil, which is extracted from the fruits and seeds of the tree. This oil is commonly referred to as *nimba*, derived from the Hindi term *nim*, which is the root of the word used in Hindi to describe the tree [30].

15.10 CHEMICAL CONSTITUENTS/BIOACTIVE COMPOUNDS OF NEEM

It is possible that a number of bioactive substances may be identified in neem. *Azadirachtin* is the primary component of neem, representing the highest percentage (>80%) of the total composition. As the azadirachtin ingredient in neem is responsible for over 90% of the pest-control characteristics, it has been determined that this component has the greatest biological interest [31]. The biochemical agent inhibits the growth, mating, oviposition, and overall reproduction of insects. Additionally, neem biopesticides act as a deterrent, discouraging insects from feeding, which in turn facilitates rapid mortality. A variety of biopesticides derived from neem include nimbin, gedunin, salannin, meliantriol, nimbidin, mahmoodin, nimbolinin, and sodium nimbinate. Additionally, other neem-based biopesticides include 22, 23-dihydronimocinol, gallic acid, and quercetin, which comprise further chemical compounds that are crucial for insect control, in conjunction with the azadirachtin component. A significant number of these essential bioactive plant substances are responsible for the remarkable biological characteristics of neem, including its efficacy as a pesticide and biofungicide.

15.11 ACTIVE INGREDIENTS IN NEEM AND THE MECHANISM OF ACTION

The treatment of plant diseases and insect pests has been achieved through the use of a variety of neem biopesticide formulations. The biological processes of neem leaves and their components have been observed to significantly impact the growth and development of microorganisms. This is achieved by hindering the division of cells and tissues, which subsequently impairs their subsequent development. The principal chemical constituent of neem is a member of the limonoid family [6]. *Azadirachtin* has been identified as the primary active ingredient responsible for the antifeedant and toxic effects observed in pests [32]. In soil immersion procedures, plants absorb the toxic elements of neem through their root system on a systemic level. The physiological function of the active ingredients is not altered; rather, it is distributed consistently throughout the plant, affecting areas of active proliferation (meristematic tissues). Crop plants treated systemically exhibit a range of physiological effects, including growth inhibition, antifeedant properties, and oviposition deterrence, sporulation, hypha growth and fungi cell division which can have a detrimental impact on pests and diseases. The degree of action and subsequent impact on insect pests varies depending on the test organism under consideration [2].

15.12 NEEM IN INSECT PEST CONTROL

The diverse range of compounds present in neem-based products have been demonstrated to possess insecticidal properties. Neem elements impede the development and proliferation of insect resistance in subsequent generations by acting on the insect's hormonal systems, as opposed to its central nervous system or gastrointestinal system, which are the primary targets of synthetic pesticides [5].

Neem has the potential to be utilized in the production of a biopesticide, which has been identified as beneficial for plants and soil, as well as for its numerous environmentally friendly benefits. *Azadirachtin* is the primary active ingredient of neem that has been employed in the production of these biological pesticides. During the larval stage of insect development, the *azadirachtin*-containing element in neem demonstrated characteristic insect growth regulator (IGR) effects.

The limonoid allomones present in neem have been observed to exert a profound influence on the physiological processes of insects, particularly those belonging to the Homoptera order and the Aphididae genus. The insecticide effects of neem (*azadirachtin*) against two important and economically significant cowpea insect pests, namely *Marcuca testulalis* Geyer (maruca pod borer, Lepidoptera: Pyraloidae) and *Clavigralla tomentosicollis* Stal (brown pod sucking bug, Hemiptera: Coreidae), have been demonstrated. The neem plant has been demonstrated to have efficacy against the brown pod-sucking bug (*Igralla tomentosicollis* Stal, Hemiptera: Coreidae) when applied in the form of pulverized neem seeds and kernels, as well as water-soluble extracts, at various dosages. Neem has been demonstrated to effectively retard the development of insect eggs at concentrations as low as 9.0% (w/v). It has been observed that the application of neem to vegetables in the growing environment or throughout storage affects pests like insects by way of certain biochemical reactions. It has been established that the application of neem to products in the field or throughout storage inhibits the feeding and oviposition of insect pests and reduces their metamorphosis, among additional biological implications [33].

15.13 BIOLOGICAL IMPACTS OF NEEM ON INSECT PESTS

In the next paragraphs the major insecticidal and other properties of neem based insecticides are discussed.

15.14 ANTIFEEDANT

Plants synthesize chemical compounds, designated as antifeedants, which serve to deter scavenging and insect attacks. As these substances enable an extension of life, rather than being required for the plant's metabolic processes, they are frequently referred to as secondary metabolites. *Azadirachtin* is a chemical compound extracted from neem that has been demonstrated to possess significant antifeedant qualities towards a diverse array of insect species. The two systems that coordinate antifeedant effects of neem include internal response mechanisms that function as a secondary antifeedant and contact sensory reception [34], which functions as an essential antifeedant. It has been demonstrated that neem compounds are effective as antifeedant insecticides against a range of insects, including bunch-footed and psychid species, citrus blackflies, houseflies, rice, pulse beetles, leaf folders, leaf borers, and gall midges [35]. The interaction of insects of various genera and orders with neem occurs in a multitude of ways at the psychological level. Moreover, an investigation was conducted to ascertain the sensitivity of different insect orders to neem biopesticides. The results demonstrated that the Lepidoptera order exhibited a particularly high sensitivity, with potent antifeedant impacts ranging from 1 to 50 ppm, as compared to the major insect orders [20]. A trial conducted in an enclosed house revealed that cowpea seedlings injected with neem exhibited an antifeeding action when offered with diet to *Zonocerus variegatus* L. (Order: Orthoptera: Family: Pyrgomorphidae). In a study examining the antifeedant properties of neem, it was also found that neem-based products significantly inhibited the development of the locust, *Schistocerca gregaria*, as well as the *Spodoptera litoralis* moth larvae and nymphs.

15.15 OVIPOSITION DETERRENCE

The term "oviposition" is used to describe the process by which insects select a specific location to deposit their eggs. Due to their remarkable reproductive abilities, female insect pests are capable of depositing a vast number of eggs on available substrates. Insects have the potential to cause significant damage to plants due to their high reproductive rate. It has recently been demonstrated that all of the bioactive compounds present in neem possess significant oviposition deterrent properties, preventing insects from depositing their eggs on surfaces that have been in contact with the plant extract. The veracity of this assertion has been corroborated by a multitude of investigations conducted with a diverse array of insect species. The impact of neem biopesticides at varying concentrations, from 18.7 to 600 ppm, on both sterilizing (radiation sterility) and oviposition and reproductive functions of *Bactrocera zonata* has been examined [36] through the use of the filtration paper immersion method. These findings indicated that the neem oil formulation significantly reduced the number of *B. zonata* ova at all tested concentrations when compared to the control group. The reproductive potential of *Myzus persicae* that fed on azadirachtin-treated food items was found to be just over 50% that of *M. persicae* that fed on an untreated diet during the first 36 hours of the study. Furthermore, nymph output ceased entirely after 50 hours of treatment [37]. Also, an investigation was conducted to ascertain whether the development of oviposition on previously stored items could be prevented. The results demonstrated that, in contrast with products infused with neem, those with no neem treatments exhibited a greater average number of eggs. The aforementioned findings substantiate the capacity of neem to impede oviposition and discourage insect establishment on a specific surface [38].

15.16 METAMORPHOSIS INHIBITION

The metamorphosis of an insect or amphibian is the process through which an organism undergoes a complete morphological transformation, progressing from a juvenile stage to an advanced adult stage. This transition occurs over the course of multiple distinct stages. These phases typically encompass all stages of development, commencing from three (egg, nymph, and adult) to four (egg, larva, pupa, and adult) stages, depending on the species of insect in question. It has recently been demonstrated that neem inhibits these transformational modifications that occur in insects. It has been demonstrated that *azadirachtin* in neem exerts a suppressive effect on the neuroendocrine system, regulating the activity of prothoracicotropic hormones (a neurosecretory polypeptide) and allotropic hormones (a hormone responsible for growth stimulation and cuticle formation). This, in turn, controls the synthesis of juvenile hormones in insects [39]. The neuroendocrine substances, known as "ecdysones", which originate from the central nervous system in insects, regulate the process of metamorphosis as the insect progresses from the larval stage to the pupal stage and ultimately attains a mature form [40]. It is important to note that neem does not destroy insect biological cycles; rather, it merely disrupts them. The application of neem oil and supplemental neem kernels of seed to two different species of rice planthoppers (*Nilaparvata lugens* and *Sogatella furcifera*) resulted in the interruption of the molting procedure, an extension of the larval duration, and a dose-dependent premature death of the insect pests [41]. Furthermore, it was demonstrated that the administration of 0.1 g of *azadirachtin* to the pupae of the yellow mealworm (*Tenebrio molitor*) influenced the life history of this species. The *T. molitor* larvae were unable to undergo the molting process, although they retained their immature characteristics. The application of neem leaf extract has been observed to induce certain abnormalities in the metamorphosis process of lepidopteran insect larvae [38].

15.17 INSECT REPELLENCE

The life history of *Tenebrio molitor* was identified shortly after the administration of yellow mealworm larvae and 0.1 g of *azadirachtin* [6, 18]. Despite retaining some immature characteristics, *T. molitor* larvae are unable to undergo the molting process. The administration of neem leaves has been demonstrated to induce specific alterations in the metamorphosis process in lepidopteran insects. The insect-repelling effects of neem have been demonstrated in research against an extensive array of plant bug types. The administration of 3% neem seed oil via ultralow-volume spray (ULVS) treatment resulted in a reduction in the number of adult grasshoppers (*Nilaparvata lugens*) [42]. The application of 1% neem oil to red gram seeds has been demonstrated to serve as an effective surface protector against insect infestation, resulting in a notable reduction in fruit production

[40–42]. These published results collectively demonstrate the potential efficacy and efficiency of neem biopesticides in repelling insect pests of various agricultural plants.

15.18 NEEM AS INSECT PEST FUMIGANT

Fumigation agents are organic compounds that are employed for the purpose of disinfection. These agents include volatile, poisonous substances that are applied as sprays to vegetation, food storage facilities, residences, and other goods with the objective of eradicating insects and other living creatures. The efficacy of neem as a fumigant has been demonstrated in the elimination of various pests, including potato moths, flour insects, bean-seed beetles, and weevils. The toxic substances present in neem function as antifeedants, oviposition deterrents, mating competitors, and developmental inhibitors, thereby eliminating insects [43]. A substantial body of scientific literature has demonstrated the efficacy of neem in generating fumes that are effective against a wide range of insect pests. The investigation demonstrated the efficacy of phosphine and neem oil as fumigants against the pulse beetle. The investigation findings indicated that substantial quantities of neem oil, a naturally occurring fumigant, were as effective as low levels of synthetic phosphine in eliminating mature pulse beetles. The evaluation of the potential and effectiveness of NSKE and neem oil from seeds as a fumigant on insects that destroy wood is presented in [44]. In the experiment, six doses of neem oil extracted from seeds were utilized as a fumigant for wooden structures prior to exposure to *Lyctus africanus* larvae.

The results demonstrated that neem oil extracted from seeds was highly effective against *L. africanus* larvae at all concentrations tested. The efficacy of neem was evaluated and found to have potential as a fumigant for *Sitophilus oryzae* and *Rhyzopertha dominica*, two significant pests responsible for damage to maize storage under laboratory settings [29]. The research investigation revealed that both insect pests were entirely susceptible to the effects of the neem product, with *S. oryzae* demonstrating a greater susceptibility to fumigation than *R. dominica*. Additionally, this biological fumigant leaves no residual effect on vegetation, in contrast to developed synthetic fumigants.

15.19 NEEM IN DISEASE CONTROL

A number of defensive mechanisms present in plants have the potential to be triggered in response to infectious agents, particularly bacteria, viruses, and fungi. The use of agricultural chemicals for the treatment of plant diseases is neither economically nor environmentally sustainable [39]. Consequently, for the realization of profitable agricultural production, comprehensive disease management approaches that integrate environmentally friendly and culturally compatible bioagents are of paramount importance. Neem and its constituent elements have been demonstrated to be highly effective in limiting the growth of a wide range of microbial organisms, including bacteria, fungi, and viruses. A brief overview of the documented antimicrobial effectiveness of neem for plant disease management is provided in the following. Additionally, various products and formulations of neem employed for the management of phytopathogenic diseases and abnormal conditions are addressed in the following sections [45].

15.20 ANTIFUNGAL ACTIVITY

The rising utilization of chemical-based fungicides for combating fungal diseases has resulted in detrimental effects on ecosystems and animal health over time. This has led to an increase in resistance to fungicides, underscoring the necessity for alternative, sustainable strategies for managing fungal diseases. Consequently, a considerable number of researchers are currently concentrating their efforts on the development of innovative approaches to the management of fungal diseases in an environmentally sustainable manner. Among these, the use of neem as an organic biofungicide has been the subject of considerable attention [23–25]. The antimicrobial properties of neem seeds and leaves were examined toward two fungal species, *Fusarium solani* and *Rhizoctonia solani*. It was found that both species exhibited significantly inhibited growth, with a reduction of 52.4 and 37.5% crop loss, respectively. The researchers typically utilize water and ethanol extracts of neem seeds and leaves in a 30:70 (v/v) ratio. The fungicidal qualities of neem against *Alternaria alternata*, the causal agent of potato early blight, were examined, and it was discovered that neem extracts exhibited the second-highest inhibition of *A. alternata*, accounting for more than 54% of spore inhibition. The antifungal activities of neem were also evaluated in relation to *Alternaria solani* Sorauer. The investigation additionally indicated that the neem component containing ethyl acetate was the most efficacious in preventing fungal growth, at a minimum inhibitory dose of 0.19 mg/liter. Furthermore, the neem component in the formulation was found to be more effective than the synthetic fungicide (metalaxyl + mancozeb) even at the minimum inhibitory concentration of 0.78 mg/liter. The antifungal activities of neem seed powder on a tomato infected with *Fusarium oxysporum* and *Meloidogyne incognita* were also examined [46], where the host plants were gathered and studied for almost 60 days after the inoculation. The findings showed that *Solanum lycopersicum* treated with neem had significantly less infection from the fungi *Fusarium* and *Meloidogyne* than did the control group. The efficacy of neem as a biofungicide in the management of plant diseases caused by fungi is supported by the numerous antifungal properties of plant extracts, which have been the subject of investigation by researchers worldwide [23, 24].

15.21 ANTIBACTERIAL ACTIVITY

The treatment of bacterial infections in crops is frequently a challenging endeavor. A novel strategy that employs natural substances, such as botanical extracts, is essential for effectively safeguarding plants against diseases. Neem has been extensively employed for the purpose of combating a multitude of bacterial pathogens that are detrimental to plants. The antibacterial effect of neem extracts was observed against 21 strains of foodborne bacteria [47]. The findings indicated that an extract of neem contains antimicrobial components are highly beneficial in reducing foodborne pathogens and organisms responsible for spoilage. The topical antibacterial properties of neem extracts, including those derived from the leaves, seeds, fruits, and bark, have been demonstrated in the context of their reducing damage in agriculture production. The findings of this study demonstrated that the neem bark and foliage extracts exhibited a pronounced antibacterial activity against the diverse bacterial strains. The antimicrobial effect of the fruit and seed juice extracts of neem was only discernible at elevated concentrations, whereas the leaf extract was demonstrated to have significant antimicrobial activity toward *Xanthomonas axonopodis* [32–37]. It has been demonstrated that neem seed extract exhibits potent inhibitory effects on the replication of *Pseudomonas* pv. *syringae*, *Xanthomonas arboricola* pv. corylina, and *Agrobacterium tumefaciens*. The ability of neem to prevent the growth and proliferation of pathogenic bacteria, such as phytobacteria, is currently supported by evidence derived from the use of neem extracts derived from specific plant parts [27].

15.22 ANTIVIRAL ACTIVITY

In addition to causing a notable reduction in agricultural yield and quality, viruses are responsible for a multitude of severe plant diseases. Viruses from plants represent a significant threat to a wide range of crops, with the financial impact of virus-related damage ranking third behind that of other pathogen-related losses [35]. Leaf deformations, leaf discoloration, and other growth disorders are common indicators of infection. However, the specific manifestations can vary depending on the plant in question [28]. Given the difficulty of eradicating infectious diseases in agricultural crops once they have taken hold, unlike fungal and bacterial illnesses, there has been limited investigation into the efficacy of neem in addressing infectious viruses in plants. In light of these considerations, the most effective approach is to eliminate the disease-spreading insect vector through the application of an efficacious biological control agent or technology. A number of different living organisms, including aphids, whiteflies, mites (gall mites), and leafhoppers, have been identified as potential vectors for the transmission of viruses. The present research has demonstrated the interesting potential of neem extract to neutralize these transmission vectors. The evaluation's numerous recommendations provide compelling evidence that neem and its preparations are effective treatments for exhausted plants [35]. The present research has demonstrated the intriguing potential of neem extract to neutralize these transmission vectors. The evaluation's numerous recommendations provide compelling evidence that the use of neem, as well as its preparations, can be effective treatments for exhausted plants [38].

15.23 THE POTENTIAL AND PROSPECTS OF NEEM TO CONTROL DISEASES

The use of a neem-based insecticide as an acceptable replacement in pest management strategies is a viable option, given the potential dangers associated with synthesized pesticides and their link to adverse effects on human health and the environment. As an increasing number of individuals opt for environmentally friendly alternatives, and given that neem-based insect repellents have been shown to have minimal to no impact on agricultural produce, neem may represent the most viable option for managing insects [39]. The present research has confirmed that neem biological pesticides are systemic and provide plants with long-term insect protection. As documented in [29], 413 insect species are susceptible to neem treatment. Derivatives of neem have been shown to exhibit remarkable efficacy in combating a range of harmful microbes that cause plant diseases. This encompasses bacteria and fungi, in particular. A total of 14 common fungal species have been demonstrated to be susceptible to neem-based formulations. Additionally, neem leaf extracts have been demonstrated to be highly effective in controlling a range of other plant diseases, including anthracnose, downy mildew, rust, and black spot [9–15]. A comprehensive investigation of neem is warranted to assess its value in combating phytopathogenic microbes and pests. There have been numerous reports on the physiological properties of its leaves and its compositions in controlling plant diseases and pests [46]. The potential of neem is considerable, spanning numerous sectors, including industry, medicine, and the environment. Its biocidal characteristics are particularly noteworthy [43].

15.24 CONCLUSION

The neem tree (*Azadirachta indica*) has been demonstrated to exert a range of effects on plant pests and diseases due to its high concentration of bioactive compounds. These compounds contribute to its ability to reduce insect growth, deter oviposition, act as an anti-feedant, and repel insects. Additionally, neem has been demonstrated to be an effective fumigant and to reduce the activities of plant pathogens, including bacteria, viruses, and fungi. The research demonstrates the efficacy, reliability, cost-effectiveness, and environmental compatibility of neem in the control of plant pests and diseases. The review indicates a need for further research into the biological properties of neem and its potential for synergistic interactions with other biological products, with the aim of enhancing its efficacy against pests and diseases. Neem and its derivatives present a secure and affordable alternative to synthetic biocides, particularly for low-income farmers. Recently, azadirachtin-based insecticides have emerged as a subject of interest as safer pest control methods. Although azadirachtin has been demonstrated to be effective against a wide range of pests, further research is required to ascertain its long-term and transgenerational effects on insects. This will enable the optimization of its use in integrated pest management practices, which may result in a reduction of the environmental impact.

REFERENCES

[1] Oguh, C. E.; Okpaka, C. O.; Ubani, C. S.; Okekeaji, U.; Joseph, P. S.; Amadi, E. U. Natural pesticides (bio-pesticides) and uses in pest management-a critical review. *Asian J Biochem Genet Eng.* **2019**, *2*(3), 1–8.

[2] Kumari, P.; Geat, N.; Maurya, S.; Meena, S. Neem: Role in leaf spot disease management. *J Pharmacognosy Phytochem.* **2020**, *9*(1), 1995–2000.

[3] Egwu, O. C.; Dickson, M. A.; Gabriel, O. T.; Okai, I. R.; Amanabo, M. Risk assessment of heavy metals level in soil and jute leaves (*Corchorus olitorius*) treated with azadirachtin neem seed solution and organochlorine pesticides. *Int J Env Agr Biotechnol (IJEAB).* **2019**, *4*(3), 756–776.

[4] de Castro, E.A.S.; de Oliveira, D.A.B.; Farias, S.A.S. Structure and electronic properties of azadirachtin. *J Mol Model.* **2014**, *20*, 2084. https://doi.org/10.1007/s00894-014-2084-0

[5] Pereira, V.V.; Kumar, D.; Agiwal, M.; Prasad, T.G. Stability of azadirachtin: A tetranortriterpenoid from Neem tree. *Int. J. Chem. Studies.* **2019**, *7*(6), 412–419.

[6] Muhammad, A.; Kashere, M. A. NEEM, *Azadirachta indica* L. (A. Juss): An eco-friendly botanical insecticide for managing farmers' insects pest problems-a review. *FUDMA J. Sci.* **2021**, *4*(4), 484–491.

[7] Gupta, A.; Ansari, S.; Gupta, S.; Narwani, M.; Gupta, M.; Singh, M. Therapeutics role of neem and its bioactive constituents in disease prevention and treatment. *J. Pharmacognosy Phytochem.* **2019**, *8*(3), 680–691.

[8] Herrera-Calderon, O.; Ejaz, K.; Wajid, M.; et al. *Azadirachta indica*: Antibacterial activity of neem against different strains of bacteria and their active constituents as preventive in various diseases. *Pharmacognosy J.* **2019**, *11*(6), 1597–1604.

[9] Ferdenache, M.; Bezzar-Bendjazia, R.; Marion-Poll, F.; Kilani-Morakchi, S. Transgenerational effects from single larval exposure to azadirachtin on life history and behavior traits of *Drosophila melanogaster. Sci. Rep.* **2019**, *9*(1), Article ID 17015.

[10] Gandhi, B. K.; Srivastava, C. Efficacy of phosphine and neem oil as fumigant against pulse beetle. *Indian J. Entomol.* **2019**, *81*(3), 585–588.

[11] Muhammad, A.; Malgwi, A. M.; Nahunnaro, H. Effect of sowing dates, intra-row spacings and pesticides on *Maruca vitrata* (fab.) (*lepidoptera: pyralidae*) damage on cowpea in Samaru, northern Guinea Savanna. *Nigerian J. Entomol.* **2018**, *34*(1), 87–98.

[12] Jain, A.; Sarsaiya, S.; Wu, Q.; Lu, Y.; Shi, J. A review of plant leaf fungal diseases and t environment speciation. *Bio-engineered*, **2019**, *10*(1), 409–424.

[13] Tlak Gajger, I.; Svečnjak, L.; Bubalo, D.; Žorat, T. Control of *Varroa destructor* Mite Infestations at Experimental Apiaries Situated in Croatia. *Diversity* **2020**, *12*, 12. https://doi.org/10.3390/d12010012

[14] Tlak Gajger, I.; Ribarić, J.; Smodiš Škerl, M.; Vlainić, J.; Sikirić, P. Stable gastric pentadecapeptide BPC 157 in honeybee (Apis mellifera) therapy, to control Nosema ceranae invasions in apiary conditions. *J Vet Pharmacol Ther.* **2018** Aug, *41*(4), 614-621. doi: 10.1111/jvp.12509. Epub 2018 Apr 23. PMID: 29682749.

[15] Tlak Gajger, I.; Vlainić, J.; Šoštarić, P.; Prešern, J.; Bubnič, J.; Smodiš Škerl, M. I. Effects onsome therapeutical, biochemical, and immunologicalparameters of honey bee (Apis mellifera) exposed toprobiotic treatments, in field and laboratory conditions. *Insects*, **2020**, *11*(9), 638. doi:10.3390/insects11090638.36.

[16] Tlak Gajger, I.; Sušec, P. Efficacy of varroacidalfood additive appliance during summer treatment ofhoneybee colonies (Apis mellifera). *Vet. Arhiv.* **2019**, *89*(1), 87–96. DOI: 10.24099/vet.arhiv.0441

[17] Ullah, A.; Tlak Gajger, I.; Majoros, A.; Dar, S.A.; Khan, S.; Kalimullah, Haleem Shah A.; Nasir Khabir, M.; Hussain, R.; Khan, H.U.; Hameed, M.; Anjum, S.I. Viral impacts on honey bee populations: A review. *Saudi J Biol Sci.* **2021** Jan, *28*(1), 523–530. doi: 10.1016/j.sjbs.2020.10.037. Epub 2020 Oct 28. PMID: 33424335; PMCID: PMC7783639.

[18] Tlak Gajger, I.; Sakač, M.; Gregorc, A. Impact of thiamethoxam on honey bee queen (*Apis mellifera carnica*) reproductive morphology and physiology. *Bull Environ Contam Toxicol.* **2017**, *99*, 297–302. https://doi.org/10.1007/s00128-017-2144-0

[19] Xu, Q.S.; He, Y.X.; Yan, X.M.; Zhao, S.Q.; Zhu, J.Y.; Wei, C.L. Unraveling a crosstalk regulatory network of temporal aroma accumulation in tea plant (*Camellia sinensis*) leaves by integration of metabolomics and transcriptomics. *Environ. Exp. Bot.* **2018**, *149*, 81–94. [Google Scholar] [CrossRef]

[20] Kitaoka, N.; Zhang, J.; Oyagbenro, R.K.; Brown, B.; Wu, Y.S.; Yang, B.; Li, Z.H.; Peters, R.J. Interdependent evolution of biosynthetic gene clusters for momilactone production in rice. *Plant Cell* **2021**, *33*, 290–305.

[21] Richter, A.; Powell, A.F.; Mirzaei, M.; Wang, L.J.; Movahed, N.; Miller, J.K.; Piñeros, M.A.; Jander, G. Indole-3-glycerolphosphate synthase, a branchpoint for the biosynthesis of tryptophan, indole, and benzoxazinoids in maize. *Plant J.* **2021**, *106*, 245–257.
[22] Song, J.; Liu, H.; Zhuang, H.F.; Zhao, C.X.; Xu, Y.X.; Wu, S.B.; Qi, J.F.; Li, J.; Hettenhausen, C.; Wu, J.Q. Transcriptomics and alternative splicing analyses reveal large differences between maize lines b73 and mo17 in response to aphid *Rhopalosiphum padi* infestation. *Front. Plant Sci.* **2017**, *8*, 1738.
[23] Zhou, P.; Li, Z.; Magnusson, E.; Cano, F.G.; Crisp, P.A.; Noshay, J.M.; Grotewold, E.; Hirsch, C.N.; Briggs, S.P.; Springer, N.M. Meta gene regulatory networks in maize highlight functionally relevant regulatory interactions. *Plant Cell* **2020**, *32*, 1377–1396.
[24] Batyrshina, Z.S.; Shavit, R.; Yaakov, B.; Bocobza, S.; Tzin, V. The transcription factor TaMYB31 regulates the benzoxazinoid biosynthetic pathway in wheat. *J. Exp. Bot.* **2022**, *73*, 5634–5649.
[25] Sun, W.J.; Zhan, J.Y.; Zheng, T.R.; Sun, R.; Wang, T.; Tang, Z.Z.; Bu, T.L.; Li, C.L.; Wu, Q.; Chen, H. The jasmonate-responsive transcription factor CbWRKY24 regulates terpenoid biosynthetic genes to promote saponin biosynthesis in *Conyza blinii* H. Lév. *J. Genet.* **2018**, *97*, 1379–1388.
[26] Hernández-Aparicio, F.; Lisón, P.; Rodrigo, I.; Bellés, J.M.; López-Gresa, M.P. Signaling in the tomato immunity against *Fusarium oxysporum*. *Molecules* **2021**, *26*, 1818.
[27] Zhang, H.; Chen, J.L.; Lin, J.H.; Lin, J.T.; Wu, Z.Z. Odorant-binding proteins and chemosensory proteins potentially involved in host plant recognition in the Asian citrus psyllid, *Diaphorina citri*. *Pest Manag. Sci.* **2020**, *76*, 2609–2618.
[28] Clark, J.; Bennett, T. Cracking the enigma: Understanding strigolactone signalling in the rhizosphere. *J. Exp. Bot.* **2024**, *75*, 1159–1173.
[29] Jia, K.P.; Baz, L.; Al-Babili, S. From carotenoids to strigolactones. *J. Exp. Bot.* **2018**, *69*, 2189–2204.
[30] Lahari, Z.; van Boerdonk, S.; Omoboye, O.O.; Reichelt, M.; Höfte, M.; Gershenzon, J.; Gheysen, G.; Ullah, C. Strigolactone deficiency induces jasmonate, sugar and flavonoid phytoalexin accumulation enhancing rice defense against the blast fungus *Pyricularia oryzae*. *New Phytol.* **2024**, *241*, 827–844.
[31] Han, Z.; Zhang, C.; Zhang, H.; Duan, Y.; Zou, Z.; Zhou, L.; Zhu, X.; Fang, W.; Ma, Y. CsMYB transcription factors participate in jasmonic acid signal transduction in response to cold stress in tea plant (*Camellia sinensis*). *Plants* **2022**, *11*, 2869.
[32] Qi, P.F.; Zhang, Y.Z.; Liu, C.H.; Chen, Q.; Guo, Z.R.; Wang, Y.; Xu, B.J.; Jiang, Y.F.; Zheng, T.; Gong, X.; et al. Functional analysis of FgNahG clarifies the contribution of salicylic acid to wheat (*Triticum aestivum*) resistance against fusarium head blight. *Toxins* **2019**, *11*, 59.
[33] Sarkar, S.; Das, A.; Khandagale, P.; Maiti, I.B.; Chattopadhyay, S.; Dey, N. Interaction of Arabidopsis TGA3 and WRKY53 transcription factors on Cestrum yellow leaf curling virus (CmYLCV) promoter mediates salicylic acid-dependent gene expression in plants. *Planta* **2018**, *247*, 181–199.
[34] Song, G.C.; Ryu, C.M. Evidence for volatile memory in plants: Boosting defence priming through the recurrent application of plant volatiles. *Mol. Cells* **2018**, *41*, 724–732.
[35] Ye, M.; Liu, M.M.; Erb, M.; Glauser, G.; Zhang, J.; Li, X.W.; Sun, X.L. Indole primes defence signalling and increases herbivore resistance in tea plants. *Plant Cell Environ.* **2021**, *44*, 1165–1177.
[36] Ye, M.; Glauser, G.; Lou, Y.G.; Erb, M.; Hu, L.F. Molecular dissection of early defense signaling underlying volatile-mediated defense regulation and herbivore resistance in rice. *Plant Cell* **2019**, *31*, 687–698.
[37] Liao, P.; Ray, S.; Boachon, B.; Lynch, J.H.; Deshpande, A.; McAdam, S.; Morgan, J.A.; Dudareva, N. Cuticle thickness affects dynamics of volatile emission from petunia flowers. *Nat. Chem. Biol.* **2021**, *17*, 138–145.
[38] Wang, L.; Erb, M. Volatile uptake, transport, perception, and signaling shape a plant's nose? *Essays Biochem.* **2022**, *66*, 695–702.
[39] Nagashima, A.; Higaki, T.; Koeduka, T.; Ishigami, K.; Hosokawa, S.; Watanabe, H.; Matsui, K.; Hasezawa, S.; Touhara, K. Transcriptional regulators involved in responses to volatile organic compounds in plants. *J. Biol. Chem.* **2019**, *294*, 2256–2266.
[40] Adebesin, F.; Widhalm, J.R.; Boachon, B.; Lefèvre, F.; Pierman, B.; Lynch, J.H.; Alam, I.; Junqueira, B.; Benke, R.; Ray, S.; et al. Emission of volatile organic compounds from petunia flowers is facilitated by an ABC transporter. *Science* **2017**, *356*, 1386–1388.
[41] Meents, A.K.; Mithöfer, A. Plant-plant communication: Is there a role for volatile damage-associated molecular patterns? *Front. Plant Sci.* **2020**, *11*, 583275.
[42] Yin, W.C.; Wang, X.H.; Liu, H.; Wang, Y.; van Nocker, S.; Tu, M.X.; Fang, J.H.; Guo, J.Q.; Li, Z.; Wang, X.P. Overexpression of VqWRKY31 enhances powdery mildew resistance in grapevine by promoting salicylic acid signaling and specific metabolite synthesis. *Hortic. Res.* **2022**, *9*, uhab064.

[43] Manna, M.; Rengasamy, B.; Sinha, A.K. Revisiting the role of MAPK signalling pathway in plants and its manipulation for crop improvement. *Plant Cell Environ.* **2023**, *46*, 2277–2295.
[44] Dombrowski, J.E.; Kronmiller, B.A.; Hollenbeck, V.G.; Rhodes, A.C.; Henning, J.A.; Martin, R.C. Transcriptome analysis of the model grass Lolium temulentum exposed to green leaf volatiles. *BMC Plant Biol.* **2019**, *19*, 222.
[45] Zhang, M.M.; Su, J.B.; Zhang, Y.; Xu, J.; Zhang, S.Q. Conveying endogenous and exogenous signals: MAPK cascades in plant growth and defense. *Curr. Opin. Plant Biol.* **2018**, *45*, 1–10.
[46] Song, N.; Wu, J.S. Synergistic induction of phytoalexins in *Nicotiana attenuata* by jasmonate and ethylene signaling mediated by NaWRKY70. *J. Exp. Bot.* **2024**, *75*, 1063–1080.
[47] Ding, M.L.; Xie, Y.F.; Zhang, Y.H.; Cai, X.N.; Zhang, B.; Ma, P.D.; Dong, J.E. Salicylic acid regulates phenolic acid biosynthesis via SmNPR1-SmTGA2/SmNPR4 modules in *Salvia miltiorrhiza*. *J. Exp. Bot.* **2023**, *74*, 5736–5751.

Section 4

Natural Herbicides

16 Terpenes and Terpenoids as a Sustainable Alternative for Weed Control

Mozaniel Santana de Oliveira, Karyme do Socorro de Souza Vilhena, Marcilene Paiva da Silva, Ravendra Kumar, Eliza de Jesus Barros dos Santos, Oberdan Oliveira Ferreira, and Eloisa Helena de Aguiar Andrade

16.1 INTRODUCTION

In tropical regions, where acidic and low-fertility soils predominate, and environmental conditions are highly favorable to the development of biotic agents that are harmful to crops, the success of agricultural activities has always been linked to the use of growth stimulants and agricultural pesticides [1]. Although these techniques have been successful in terms of productivity and meeting market needs, the agricultural landscape has undergone significant changes in recent decades. This has led to the need for new paradigms that consider the values of modern society and the requirements of responsible agriculture in relation to the preservation of natural resources, wildlife, and human health. There is an increasing demand for food that is free of chemical residues [2].

In this context, one of the most pervasive issues that impede agricultural productivity and, subsequently, the returns on invested capital, are weeds. [3–5]. The most prevalent species infesting agricultural areas are those with broad leaves, particularly those belonging to the families Leguminosae, Malvaceae, Lamiaceae, Convolvulaceae, and Asteraceae [6], Additionally, species with narrow leaves, particularly those belonging to the Cyperaceae and Poaceae families, are of interest. [7–9]. The species belonging to these families are distinguished by their proclivity for aggression and their remarkable competitive ability, particularly when it comes to plants of economic interest. This competitive nature contributes significantly to the overall maintenance costs associated with crops [10, 11]. The management of these species is a crucial aspect of crop productivity. However, the control methods employed by producers have been shown to generate dissatisfaction and insecurity within the sector, particularly with regard to chemical products [12, 13].

The advent of herbicide-resistant weed species has prompted the development of new herbicides to combat these evolving challenges. In recent decades, there has been a notable increase in the number of resistant plant breeds and species observed in various geographical regions worldwide [14–16]. Similarly, in Brazil, the prevalence of herbicide-resistant plants has increased as a consequence of the routine utilization of herbicides with analogous modes of action. The use of allelochemicals in the formulation of innovative products may present a challenge in controlling plants that have developed resistance to current products. However, the development of new products could improve the agricultural system and mitigate social dissatisfaction arising from the use of herbicides [17–20].

Allelochemicals can also offer new and innovative molecules with the potential for direct use in the management of weeds or even make it possible to obtain products as efficient as commercial herbicides [21, 22] without posing any risk to the environment or even to humans, since they have

DOI: 10.1201/9781003463429-21

a low permanence rate in the environment and are quickly degraded by soil microorganisms [23]. Among the various possibilities for this purpose, the terpenoid class deserves to be highlighted due to the wide chemical diversity of its components, which can be classified as monoterpenes (C_{10}), sesquiterpenes (C_{15}), diterpenes (C_{20}), sesterterpenes (C_{25}), triterpenes (C_{30}), tetraterpenes (C_{40}), and polyterpenes ($>C_{40}$) [24]. These compounds have showed phytotoxic activity on invasive plants [25–28], which can constitute an advantageous tool to be considered in the strategies of the current agriculture model. Compounds with phytotoxic activity are referred to in the literature as allelochemicals [29–34], and in Figure 16.1, it is possible to observe a form of interaction between plants called allelopathy, in which one of the species produces allelochemicals capable of inhibiting the development of the other one.

Allelochemicals are naturally occurring compounds produced by plants that exert an influence on the growth and germination of neighboring plants. These substances can either inhibit or promote the development of other species, thereby playing a crucial role in plant competition and ecosystem dynamics. Figure 16.1 depicts a number of terpenes and terpenoids with the potential to exert phytotoxic effects, which can in turn affect germination rates. An understanding of these interactions allows for the potential of allelochemicals to be harnessed for the sustainable management of weeds and the improvement of crops. Figure 16.1 illustrates the manner in which these chemical interactions influence the structure and function of plant communities, as well as the maintenance of ecological balance.

Accordingly, this study aims to collate the latest data that elucidates the full potential of terpenoids in diverse weed management strategies.

FIGURE 16.1 Illustrative interactions between plants.

16.1.1 Terpenes and Terpenoids

Terpenoids represent a significant class of chemically diverse compounds produced by plants. These compounds play a pivotal role in plant defense mechanisms and offer potential for the development of chemical molecules with applications in pest and disease management, as well as in other human needs, including hygiene and health [35–38]. A review of the literature reveals a plethora of reports on terpenoid-producing plants. Table 16.1 provides a representative sample of notable examples. The terpenoid class encompasses essential oils (monoterpenes, diterpenes, and sesquiterpenes) and, in higher molecular weight compounds, triterpenes and tetraterpenes [39–41].

Essential oils are constituted by intricate mixtures of volatile compounds, predominantly comprising terpenoids [42], which are responsible for the flavor of plants, and can be found in different families and parts, such as leaves, barks, roots, flowers, fruits, and seeds [43]. Additionally, they are among the primary components that play a crucial role in plant defense and serve as sources of molecules with significant potential for the development of novel products for direct use in agriculture or in the formulation of new defensive agents [44].

The composition of the oils is highly variable, as is the concentration of each component. Furthermore, these characteristics may fluctuate depending on the species, plant fraction, and time of year. It is notable that plant species belonging to the same family and genus exhibit differences in chemical composition and in the concentration of each component [45–47]. Table 16.1 presents the chemical composition of essential oils rich in terpenoids derived from various species. These bioactive products offer significant potential for regulating seed germination and weed growth, representing a valuable opportunity for integration into the current agricultural model.

16.2 MONOTERPENES

Monoterpenes are formed through the condensation of two C5 isoprene units, resulting in a C_{10} structure. The two-dimensional chemical structures are illustrated in Figure 16.2. The phytotoxic activity of essential oils on the germination and development of weeds is dependent on three factors: (i) the chemical composition of the oils, (ii) the concentration of each component, and (iii) the bioactivity of the major compounds. Oils with pronounced phytotoxic activity are predominantly composed of monoterpenes and diterpenes [62–64].

Among monoterpenes, oxygenated and non-oxygenated ones have shown phytotoxic activity [62, 65]; however, in some cases, oxygenated monoterpenes have shown greater phytotoxic activity in relation to hydrocarbon monoterpenes [66]. Thus, the identification of chemical components, the concentration of each compound, and which are the major substances is a good indicator to predict the phytotoxic potential of certain oils.

In a recent study with three species of Piperaceae, Jaramillo-Colorado et al. [67] analyzed the chemical composition and phytotoxic activity of four species: *P. dilatatum*, *P. divaricatum*, *P. hispidum*, and *P. sanctifelici*. The compounds identified in the highest concentration were the monoterpenes eugenol, methyl eugenol, γ-elemene, apiol, (E)-caryophyllene, δ-3-carene, limonene, p-cymene, β-pinene, nerolidol, limonene, δ-3-carene, and p-cymene, and the species that showed the highest phytotoxic activity were *P. dilatatum* and *P. divaricatum*. In addition, monoterpenes, both oxygenated and hydrocarbons, may have different specificities when tested on the same recipient plant species, suppressing specific parts, such as germination, and radicle and hypocotyl elongation [62]. In Table 16.2, different monoterpenes with potential phytotoxic activity can be seen.

It is possible to observe that monoterpenes can be considered phytotoxins alone or in complex mixtures (Table 16.2). Major compounds present in essential oils have been evaluated individually, as part of the understanding of the overall phytotoxicity effects of these oils on weeds [44]. In this sense, other components have already been isolated and tested for phytotoxic activity, as are the cases of camphor, 1,8-cineol, nerol, and neryl isovalerate [68], among others [69–71].

TABLE 16.1
Main Volatile Compounds Present in Essential Oils of Different Species

Species	Family	Plant Fraction	Compound	Ref.
Origanum vulgare L.	Lamiaceae	Leaves and branches	Sabinene, germacrene D, carvacrol, caryophyllene oxide, γ-terpinene, linalool, β-caryophyllene, spathulenol, thymol, myrcene, sabinene hydrate, linalyl acetate, carvacrol methyl ether, germacrene D-4-ol, p-cymene, α-terpineol, 1,8-cineole, e β-ocimene	[48]
Thymus daenensis and *Thymus vulgaris*	Lamiaceae	Aerial parts	Thymol, carvacrol, p-cymene, and terpinene	[49]
Salvia sclarea	Lamiaceae	Aerial parts	Linalyl acetate, linalool, and germacrene D	[50]
Cinnamomum verum	Lauraceae	Bark	Cinnamaldehyde	[51]
Laurus nobilis	Lauraceae	Leaves	Sabinene, 1,8-cineole, and linalool	[52]
Rosemary officinalis	Lamiaceae	Not informed	1,8-cineole	[53]
Pogostemon cablin	Lamiaceae	Leaves	β-patchoulene, cariofilene, γ-patchoulene, α-patchoulene, β-guaiene, and α-selinene	[54]
Coriandrum sativum L	Apiaceae	Seeds, flowers, and leaves	Camphor, γ-terpinene, decanal, linalool, benzofuran, 2,3-dihydro, 2-methoxy-4-vinylphenol, geranyl acetate, 2,4a-epioxy-3,4,5,6,7,8-hexahydro-2,5,5,8a-tetramethyl-2H-1-benzofuran, dodecanoic acid, α-pinene, trans-2-decenal, hexadecanoic acid methyl ester, 2,6-dimethyl-3-aminobenzoquinone, p-cymene, dodecanal, cyclodecane, 2-decen-1-ol, dodecan-1-ol, cis-2-dodecene, and 2,3,5,6-tetrafluoroanisole.	[55]
Origanum vulgare L.	Lamiaceae	Leaves	Thymol, β-pinene, cineole, camphene, linalool, ethyl caprate, α-pinene, carvacrol, γ-terpinene, ledol, 1,8-cineole, myrcene, (E)-β-farnesene, α-terpinene, ρ-cymene, and β-bisabolene.	[56]
Piper aduncum L.	Piperaceae	Leaves	1,8-cineole	[57]
Lavanda	Lamiaceae	Aerial parts	D-limolene, Eucalyptol, linalyl acetate, camphora, linalol, and caryophyllene	[58]
Zingiber officinale Rosc	Zingiberaceae	Aerial parts	6-Gingerol	[59]
Piper corcovadensis (Miq.) C. DC	Piperaceae	Fresh leaves	1-butyl-3,4-methylenedioxybenzene, terpinolene, *trans*-caryophyllene, α-pinene, δ-cadinene, and Limonene	[60]
Piper cernuum	Piperaceae	Aerial parts	Camphen.	[61]

Monoterpenes with potential phytotoxic activity represent a diverse group of compounds with promising applications for managing invasive plants and improving agricultural outcomes. Notable examples include 1,8-cineole, which may influence plant growth through its unique epoxide structure, and camphor, characterized by its ketone group that could disrupt various plant processes. Pulegone and borneol also hold potential due to their distinct chemical arrangements; pulegone's structure might contribute to its phytotoxic effects, while borneol, as a monoterpene alcohol, could impact plant development through its functional groups. Additionally, limonene is recognized for its citrus aroma and potential effects on plant processes, offering another avenue for exploration in phytotoxic applications.

Further, α-pinene and β-pinene are known for their diverse effects on plant growth due to their complex ring structures, while linalool and carvone present opportunities for targeted applications

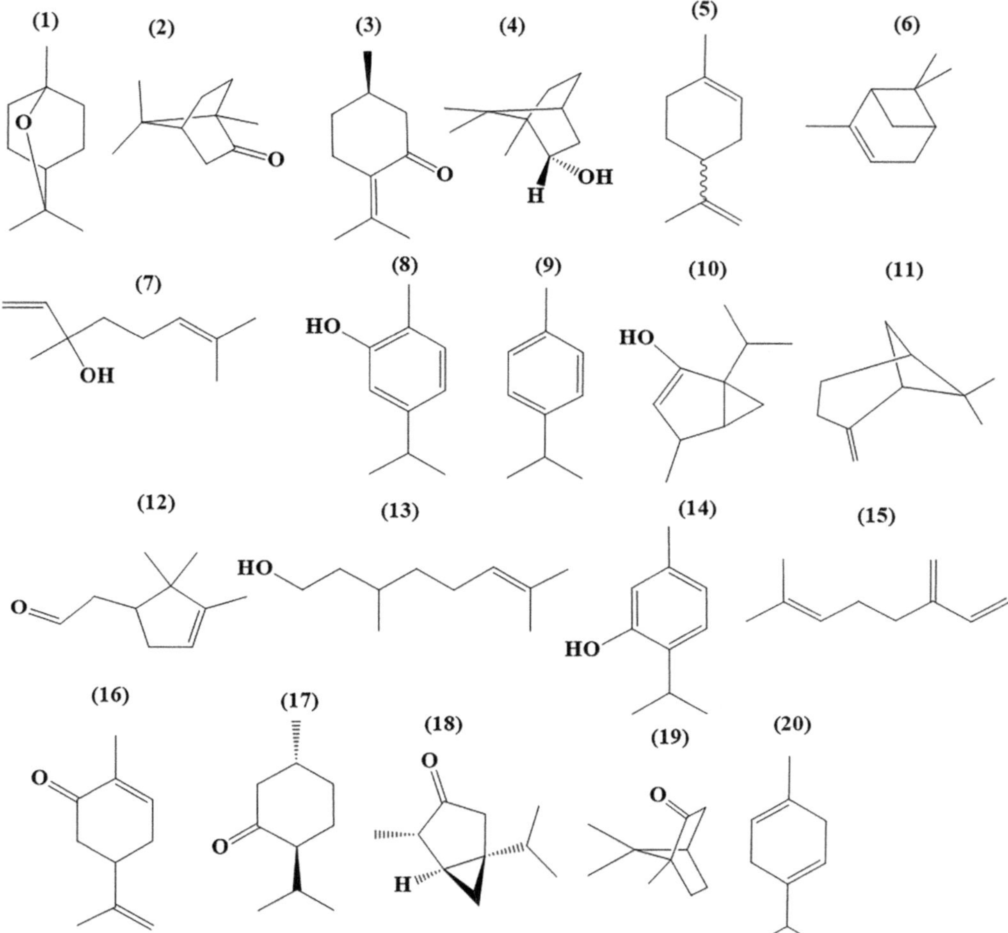

FIGURE 16.2 Monoterpenes with potential phytotoxic activity. (1) = 1,8 cineole, (2) = camphor, (3) = pulegone, (4) = borneol, (5) = limonene, (6) = α-pinene, (7) = linalool, (8) = carvocrol, (9) = p-cymene, (10) = thujenol, (11) = β-pinene, (12) = α-campholenal, (13) = citronellol, (14) = thymol, (15) = β-myrcene, (16) = carvone, (17) = menthone, (18) = α-thujene, (19) = camphor, (20) γ-terpinene.

due to their distinctive functional groups. Thymol and carvacrol are notable for their phenolic structures, which may contribute to their phytotoxic potential. Other monoterpenes such as p-cymene, α-campholenal, citronellol, menthone, α-thujene, and γ-terpinene add to the expanding pool of monoterpenes with potential phytotoxic effects. Together, these compounds offer a rich resource for developing innovative and ecologically sustainable approaches to weed management and crop protection.

16.3 SESQUITERPENES

Sesquiterpenes are composed of three isoprene structures, forming a 15-carbon molecule [81]. These compounds have several industrial applications [82], and in agriculture, are widely known for being potential bioherbicides. For instance, species like *Copaifera duckei* (Dwyer), *Copaifera martii* (Hayne), and *Copaifera reticulata* (Ducke) have essential oils rich in sesquiterpenes like germacrene D, β-caryophyllene, α-humulene, δ-elemene, and δ-cadinene, and these essential oils have been shown to have phytotoxic potential against invasive Amazonian species present in management

TABLE 16.2
Oxygenated and Non-Oxygenated Monoterpenes with Potential Phytotoxic Activity

Species with Phytotoxic Potential	Compounds	Recipient Species	Ref.
Comercial monoterpene	Citronellal, linalool, citronellol, and 1,8-cineole	*Cassia occidentalis*	[72]
Comercial monoterpene	Citronellol	*Chenopodium album* L., *Cassia occidentalis* L., *Malvastrum coromandelianum* (L.) *Garcke, Phalaris* minor Retz., *Parthenium hysterophorus* L., and *Ageratum conyzoides* L.	[73]
Eucalyptus citriodora	Citronelal and citronelol	*Triticum aestivum* L., *Oryza sativa* L.) *Amaranthus viridis* L. *Echinochloa crus-galli* (L.) Beauv.	[74]
Hyptis suaveolens	α-phellandrene, α-pinene, allo-ocimene, limonene, β-thujene, γ-terpinene, and o-cymene	*Oryza sativ*. and *Echinochloa crus-galli*	[75]
A. rosaeodora, C. odorata, C. sativa inflorescences, C. citratus, C. limon, C. nobilis, C. sempervirens, C. longa rhizomes, H. perforatum, I. verum, M. alternifolia, Mentha × piperita, M. spicata, M. fistulosa, O. micranthum, O. basilicum, O. quixos, P. capitatum, P. nigra, P. cablin, S. aromaticum, T.s vulgari, V. zizanoidess, Z. officinale	α-pinene, camphene, sabinene, myrcene, α-phellandrene, limonene, and γ-terpinene	*Solanum lycopersicum* L.	[76]
Citrus aurantiifolia	Limonene and citral	*Avena fatua, Echinochloa crus-galli,* and *Phalaris minor*	[66]
Artemisia scoparia Waldst. & Kit	β-myrcene, p-cymene, and dl-limonene	*Avena sativa,* and *Triticum aestivum*	[77]
Origanum acutidens	carvacrol, thymol, and *p*-cymene	*Amaranthus retroflexus, Chenopodium album,* and *Rumex crispus*	[78]
Achillea millefolium, Acorus calamus, Carum carvi, Chamomilla recutita, Foeniculum vulgare, Lavandula angustifolia, Melissa officinalis, Mentha × piperita, Salvia officinalis, Solidago canadensis, Tanacetum vulgare, and *Thymus vulgaris*	carvone, limonene, thymol, menthone, menthol, α-thujone, and camphor	*Avena sativa, Brassica napus,* and *Zea mays*	[79]
Comercial monoterpene	(±)-β-citronellol, (±)-citronellal, (-)-α-pinene, (-)-β-pinene, α-terpinene, γ-terpinene, α-terpineol, 1,8-cineole, citral, thymol, carvacrol, α+β-thujone, camphene, (±)-camphor, (-)-borneol, p-cymene, myrcene, menthone, (±)-menthol, geraniol, geranyl acetate, linalool, linalyl acetate, (R)-(-)-α-phellandrene, estragole, (R)-(-)-carvon, limonene	*Raphanus sativus* L. (radish) and *Lepidium sativum* L.	[71]

(*Continued*)

TABLE 16.2 ***(Continued)***
Oxygenated and Non-Oxygenated Monoterpenes with Potential Phytotoxic Activity

Species with Phytotoxic Potential	Compounds	Recipient Species	Ref.
Hyssopus officinalis, Lavandula angustifolia, Majorana hortensis, Melissa officinalis, Ocimum basilicum, Origanum vulgare, Salvia officinalis and Thymus vulgaris (Lamiaceae), Verbena officinalis (Verbenaceae), Pimpinella anisum, Foeniculum vulgare, and Carum carvi (Apiaceae)	β-pinene, carvone, linalyl acetate, *iso*-pinocamphone, borneol, *o*-cymene, geraniol, (-)-citronellal, linalol, 1,8-cineole, and cis-anethole	Raphanus sativus, Lactuca sativa, and Lepidium sativum	[80]
S. aromaticum	Eugenol, eugenol acetate	*M. pudica* and *S. obtusifolia*	[44]

areas, such as *Mimosa pudica* L. and *Senna obtusifolia* (L.), at different intensity values depending on the recipient plant [83].

The phytotoxic potential of various allelochemicals, including sesquiterpenes, holds significant promise for developing effective strategies to manage invasive plant species and improve agricultural productivity. Compounds such as italicene epoxide, guaiol, and 1,10-di-epi-cubenol exhibit unique chemical structures that may influence their ability to disrupt plant growth and development. Sesquiterpenes like 8-cedren-13-ol, (z)-α-trans-bergamotol, and α-copaene, with their diverse functional groups and ring structures, offer potential as phytotoxic agents due to their distinctive modes of action. Additionally, β-costol, spathulenol, and β-bourbonene, along with δ-cadinene and β-caryophyllene, are notable for their varied chemical frameworks, which could enhance their efficacy in controlling unwanted plant species. B-farnesene and isoperezone, with their unique molecular configurations, also contribute to the expanding pool of potential phytotoxic compounds. Furthermore, sesquiterpene lactones, characterized by their lactone rings, are known for their robust phytotoxic properties. Collectively, these allelochemicals represent a valuable resource for developing innovative and ecologically sustainable approaches to weed management, offering new opportunities for enhancing crop yields and protecting natural ecosystems [86–95].

In essential oils of *Vitex negundo* L, the sesquiterpene hydrocarbon β-caryophyllene is the major component. Such oils demonstrated cytotoxic activity against seed germination of two plant species: *Avena fatua* L. and *Echinochloa crus-galli* (L.) [84]. The compounds isolated from *Inula viscosa* were as follows: (7R,8R)-1,4-dimethyl-4-hydroxy-secoeudesm-5(10),11(13)-dien-8β-12-olide, (5R,7R,8R,10R)-1,15-methylene-5β-hydroxy-eudesm-1(15). The compounds isolated from *Inula viscosa* included (5R,7R,8R,10R)-1,15-methylene-5β-hydroxy-eudesm-1(15), (7R,8R)-1,4-dimethyl-4-hydroxy-secoeudesm-5(10),11(13)-dien-8β-12-olide, (4E,7R,8R,10S)-3-oxo-germacra-4,11(13)-dien-8β-12-olide, and its analogue 11,13-dihydro. The authors observed that the isolated molecules exhibited inhibition of germination in *C. campestris* and *O. crenata*, with inhibition rates reaching 100% in some cases [85].

To summarize the data, in Table 16.3, different species of plants and their major volatile and non-volatile components (sesquiterpenoids) can be observed as promoters of phytotoxic activity. Their 2D chemical structures are shown in Figure 16.3.

16.4 DITERPENES

Diterpenes are highly diverse molecules, formed by four isoprene structures, resulting in a 20-carbon structure, and they exhibit diverse functions in mediating antagonistic and/or beneficial interactions

TABLE 16.3
Volatile and Non-Volatile Sesquiterpenes with Potential Phytotoxic Activity

Species with Phytotoxic Potential	Compounds	Recipient Species	Ref.
Cynara cardunculus L.	Aguerin B, grosheimin, 11,13-dihydroxy-8-deoxygrosheimin, deacylcynaropicrin, and 11,13-dihydro-deacylcynaropicrin	*Amaranthus retroflexus* L. and *Portulaca oleracea* L	[86]
Coprophilous Fungus Penicillium	3R,6R-dihydroxy-9,7(11)-dien-8-oxoeremophilane, isopetasol, sporogen AO-1, and dihydrosporogen AO-1	*Amaranthus hypochondriacus*	[87]
Saussurea lappa	Lappalone	*Lepidium sativum* L., *Alium cepa* L., *Lactuca sativa* L., and *Solanum lycopersicum* L.	[88]
Bidens sulphurea L.	10-hydroxy-6,10-epoxy-4(14)-isodaucane, 1β,4β-dihydroxyarbusculin, reynosin (2), 10-oxoisodauc-3-en-15-al, 11-epiartesin, 1β,6α-dihydroxyeudesm-4(5)-ene, caryolane-1,9β-diol, clovane-2β-9α-diol, (9Z,12S,13E,15Z)-12-hydroxyoctadeca-9,13,15-trienoic acid, alloromadendrane-4β,10α-diol, (9Z,12S,13E,15Z)-12-hydroxyoctadeca-9,13,15-trienoic acid methyl ester, alloromadendrane-4β,10β-diol, stigmasterol, α-dimorphecolic acid, and 10-hydroxy-15-oxo-α-cadinol	*Amaranthus viridis*, and *Panicum maximum*	[89]
Ferula pseudalliacea	Methyl galbanate, ethyl galbanate, fekrynol acetate, farnesiferol B, and kamonolol acetate	*Nicotiana tabacum* L. cv. Burley 21	[90]
Stereum complicatum	Sterostrein X, sterostrein Y, hirsutenol G, sterpurol C, sterostrein H, sterostrein P., and sterostrein Q	*Lactuca sativa, Agrostis stolonifera,* and *Lemna paucicostata*	[91]
Ligularia cymbulifera	Ligulacymirin A and ligulacymirin B	*A. thaliana*	[92]
Eupatorium adenophorum Spreng	Γ-cadinene, γ-muurolene, 3-acetoxyamorpha-4,7(11)-diene-8-one, and bornyl acetate	*Phalaris minor*	[93]
Plant Essential Oils	Γ-cadinene, γ-muurolene, and 3-acetoxyamorpha-4,7(11)-diene-8-one, β-caryophyllene, (Z)-caryophyllene, germacrene D, hexahydrofarnesyl acetone, β-caryophyllene, caryophyllene oxide, and β-bisabolene	Systematic review	[94]
Drimys brasiliensis Miers	Polygodial, polygodial acetal, dendocarbin L, and (+)-fuegin	*Barbarea verna* (Mill.), *Echinochloa crus-galli* (L.) and *Ipomoea grandifolia (Dammer)*	[95]

in, and among, organisms [96]. In the study performed by Berrueab and Russell [97], the authors described 602 new compounds. In addition, several potential biological activities are reported in the literature [98, 99], such as the phytotoxic potential of harzianelactone A, harzianelactone B, harzianone A, harzianone B, harzianone C, harzianone D, and harziane [100].

In the study conducted by Cimmino et al. [101], the researchers investigated the phytotoxicity of the diterpene (1S,2S,3S,4S,5S,9R,10S,12S,13S)-1,12-acetoxy-2,3-hydroxy-6-oxopimara-7(8),15-dien-18-oic acid-2,18-lactone on various weed species. At a concentration of 2 mg/mL, the compound was observed to induce necrotic lesions in *Mercurialis annua*, *Cirsium arvense*, and *Setaria viride*. Furthermore, other diterpenes, including 2-oxokovalenic acid and 12,19-hydroxyferruginol, which were isolated from *Tectona grandis*, exhibited notable phytotoxic effects against *Allium cepa* (onion), *Lycopersicon esculentum* (tomato), *Lepidium sativum* (cress), and *Lactuca sativa* (lettuce).

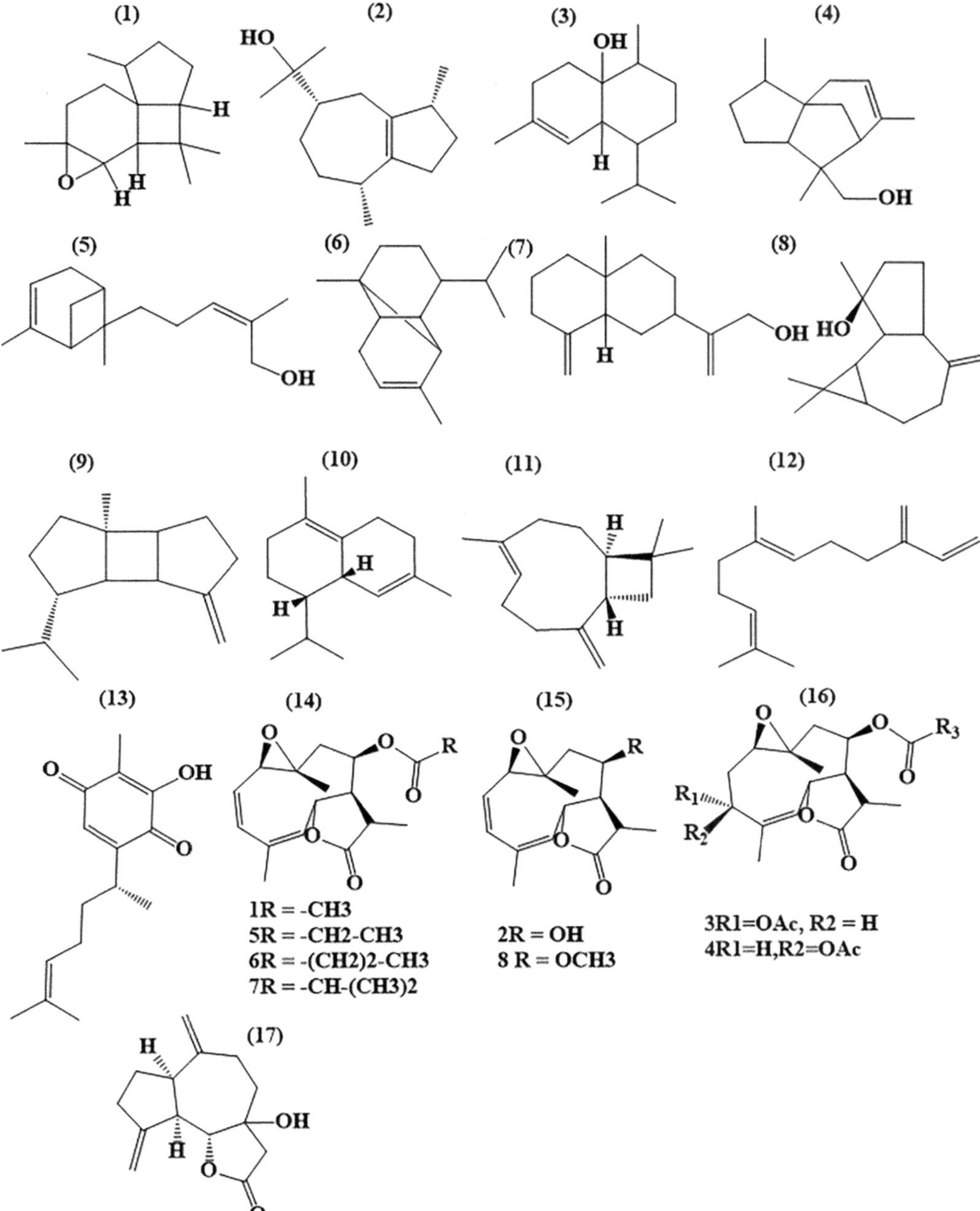

FIGURE 16.3 Chemical structures of sesquiterpenes with potential phytotoxic activity, (1) = Italicene epoxide, (2) = Guaiol, (3) 1,10-di-epi-Cubenol, (4) = 8-Cedren-13-ol, (5) = (Z)-α-trans-Bergamotol, (6) = α-Copaene, (7) = β-Costol, (8) = Spathulenol, (9) = ß-Bourbonene, (10) = δ-Cadinene, (11) = ß-Caryophyllene, (12) = ß-Farnesene, (13) = Isoperezone, (14-17) = sesquiterpene lactones.

Several other articles report the possibility of using terpenoids isolated from the most different species as molecules with potential use in agriculture for the control of invasive plant species. They can be seen in Table 16.4 and are summarized in different studies that demonstrated the effectiveness of this important class of phytochemicals. In addition, the chemical structures of some diterpenoids can be seen in Figure 16.4.

TABLE 16.4
Diterpenoids with Potential Phytotoxic Activity

Species with Phytotoxic Potential	Compounds	Recipient Species	Ref.
Sphaeropsis sapinea f. sp. Cupressi and *Diplodia mutila*	Sphaeropsidins B and C	*Cupressus sempervirens* L	[102]
Salvia miniata Fernald	Ten clerodane diterpenoids	*Papaver rhoeas* L. and *Avena sativa* L.	[103]
Blakiella bartsiifolia (S.F. Blake)	Z-15,16-dihydroxy-3,13-clerodien-20-oic acid (barthydrolic acid), 1,20-epoxy-1,3(20),10(E),14-phytapentaen-18-methyl-19-oic acid (blakifolic acid), 15,16-epoxy-2-hydroxy-3,13(16),14-clerodatrien-20-oic acid (bartsiifolic acid), junceic acid, and 1,20-epoxy-1,3(20),6(E),10(E),14-phytapentaen-18-methyl-19-oic acid (blakielic acid). Another compound is 1,20-epoxy-1,3(20),6,14-phytatetraen-19-methyl-18-oic acid (dihydrocentipedic acid)	*Allium cepa* and *Lactuca sativa*	[103]
Eragrostis plana	Neocassa-12(17),15-dien-3-oneand, and neocassa-12(13),15-diene-3,14-dione	*Ipomoea grandifolia* and *Euphorbia heterophylla*	[104]
Eragrostis plana	Neocassa-1,12(13),15-triene-3,14-dione, 19-norneocassa-1,12(13),15-triene-3,14-dione, and 14-hydroxyneocassa-1,12(17),15-triene-3-one.	*Lemna paucicostata (L.)* Hegelm	[105]
Leucas áspera	(rel 5S,6R,8R,9R,10S,13S,15S,16R)-6-acetoxy-9,13;15,16-diepoxy-15-hydroxy-16-methoxylabdane, and (rel 5S,6R,8R,9R,10S,13S,15R,16R)-6-acetoxy-9,13;15,16-diepoxy-15-hydroxy-16-methoxylabdane	*Lepidum ativum* and *Echinochloa crus-galli*	[106]
Euphorbia esula L	Ellagic acid, 3,3′-di-O-methylellagic acid and 3,3′,4′-tri-O-methylellagic acid	*Arabidopsis thaliana*	[107]
Cupressus sempervirens L	Sphaeropsidins D and E	*Cupressus macrocarpa, C. sempervirens* and *C. arizonica*	[108]

The investigation of diterpenoids is of great significance in the identification of novel allelochemicals that can be employed for the management of invasive plant species and the enhancement of agricultural practices. The chemical structures of compounds such as 19-norneocassa-1,12(13),15-triene-3,14-dione, and 14-hydroxyneocassa-1,12(17),15-triene-3-one are of particular significance due to their influence on the potential phytotoxic effects of these compounds. These diterpenoids exhibit intricate molecular frameworks that may contribute to their efficacy in suppressing the growth of unwanted plant species. Another noteworthy compound (rel 5S,6R,8R,9R,10S,13S,15R,16R)-6-acetoxy-9,13;15,16-diepoxy-15-hydroxy-16-methoxylabdane) possesses a diverse array of functional groups, including epoxy and hydroxy groups, which could enhance its biological activity and effectiveness as an allelochemical. These compounds present valuable opportunities for the development of targeted weed management strategies and underscore the potential of diterpenoids in advancing more sustainable agricultural practices [102–108]. Furthermore, diterpenoids such as perezone, isoperezone, dihydroperezone, and dihydroisoperezone are of significant interest due to their diverse hydroxyl and epoxy functionalities. These structural features may impact their

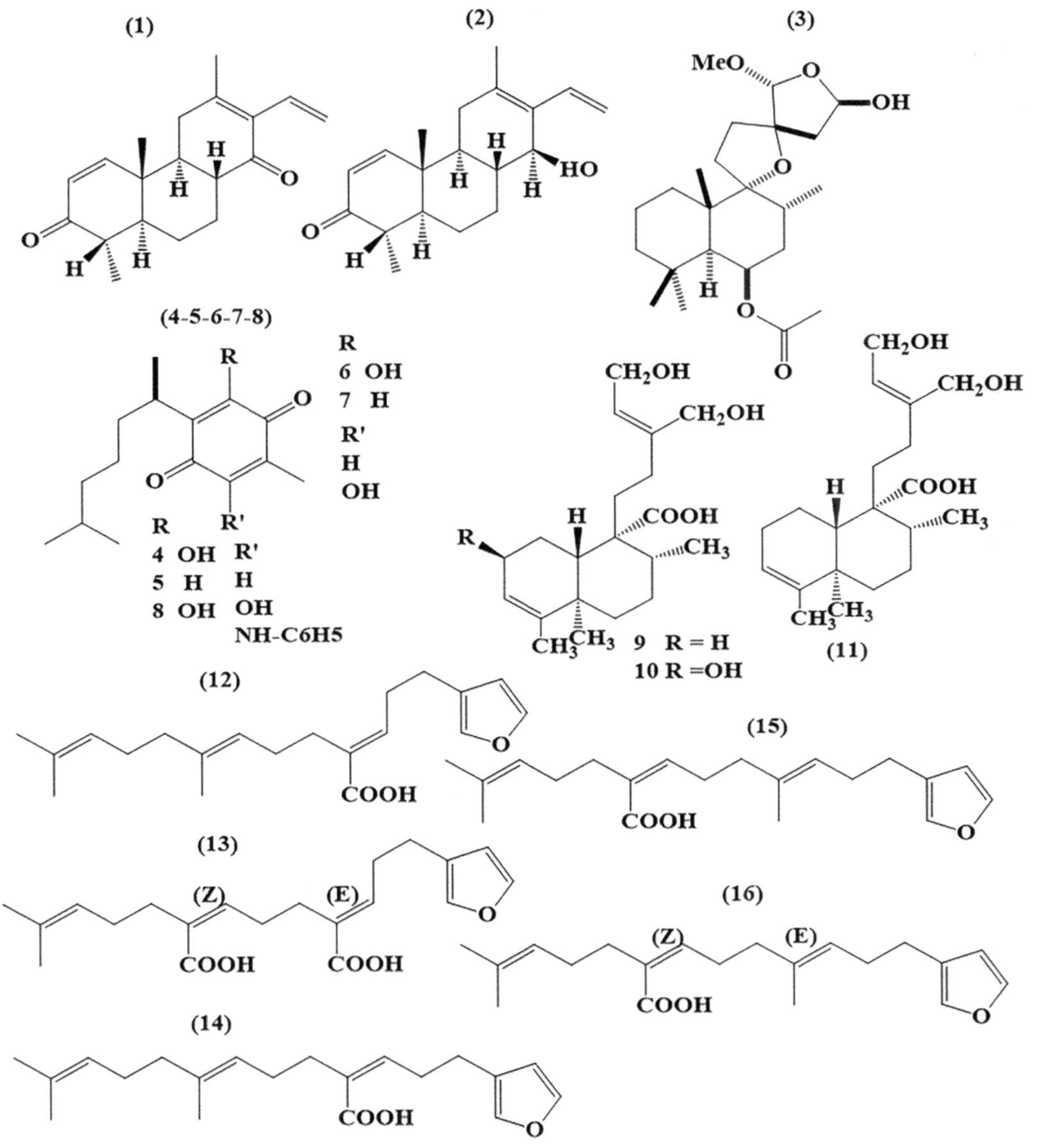

FIGURE 16.4 Chemical structures of some diterpenoids. (1) = 19-norneocassa1,12(13),15-triene-3,14-dione,(2)=14-hydroxyneocassa 1,12(17),15-triene-3-one, (3) = rel 5S,6R,8R,9R,10S,13S,15R,16R)-6-acetoxy-9,13;15,16-diepoxy-15-hydroxy-16-methoxylabdane, (4) = Perezone; (5)= Isoperezone, (6)= Dihydroperezone; (7) = Dihydroisoperezone; (8) = Anilidoperezone, (9) = junceic acid, (10) = 5,16-epoxy-2-hydroxy-3,13(16),14-clerodatrien-20-oic acid, (11) = Z-15,16-dihydroxy-3,13-clerodien-20-oic acid, (11) = barthydrolic acid, (12) = 1,20-epoxy-1,3(20),6(E),10(E),14-phytapentaen-18-methyl-19-oic acid, (13) = blakielic acid, (14) = 1,20-epoxy-1,3(20),10(E),14-phytapentaen-18-methyl-19-oic, and (15) = 1,20-epoxy-1,3(20),6,14-phytatetraen-19-methyl-18-oic acid.

phytotoxicity, rendering them suitable candidates for further research and potential application in invasive plant control. Moreover, compounds such as anilidoperezone and junceic acid introduce further chemical diversity to the range of diterpenoids with potential allelopathic properties. The distinctive characteristics of these compounds offer promising avenues for the development of more effective and ecologically sustainable solutions for the management of invasive plants. By investigating the diverse chemical properties and potential applications of these diterpenoids, researchers can advance the field of allelochemicals and develop innovative strategies to enhance agricultural productivity and environmental conservation.

16.5 CONCLUSION

The utilization of allelochemicals derived from the secondary metabolism of plants represents an ecologically sustainable approach for the management of invasive plant species. Essential oils, which are known for their complex and diverse chemical profiles, have been demonstrated to be effective in this context. The efficacy of these compounds is largely attributed to their diverse chemical composition, which includes monoterpenes, sesquiterpenes, and diterpenes, all of which exhibit significant phytotoxic potential. Of particular note are monoterpenes, including linalool and 1,8-cineole, which have been demonstrated to disrupt plant growth and development. Their potent allelopathic effects render them invaluable for the control of invasive plant species. Similarly, sesquiterpenes such as β-caryophyllene have demonstrated efficacy in phytotoxicity assays, indicating their potential utility in weed management strategies. Diterpenes, including sphaeropsidins B and C, also demonstrate considerable phytotoxic activity, making them essential for further research and potential applications. These compounds have the potential to target specific weed species effectively due to their distinctive chemical structures and modes of action.

It is important to recognize that the efficacy of these allelochemicals may be influenced by a number of factors beyond their intrinsic chemical properties. The concentration of the compound employed in experimental contexts is of paramount importance in determining its efficacy. Furthermore, the osmotic activity of water can influence the bioavailability and efficacy of these compounds in practical applications. It can be reasonably concluded that monoterpenes, sesquiterpenes, and diterpenes represent promising categories of allelochemicals. However, further research is needed to optimize their application. This includes the determination of the most efficacious concentrations and an understanding of the interplay between these compounds and environmental factors. By addressing these aspects, we can enhance the practical utility of allelochemicals in the control of invasive plant species and the promotion of ecological balance.

REFERENCES

[1] M.M.X. de Carvalho, E.S. Nodari, R.O. Nodari, "Defensivos" ou "agrotóxicos"? História do uso e da percepção dos agrotóxicos no estado de Santa Catarina, Brasil, 1950-2002, *História, Ciências, Saúde-Manguinhos*. 24 (2017) 75–91. https://doi.org/10.1590/s0104-59702017000100002.

[2] J.J. Beck, H.T. Alborn, A.K. Block, S.A. Christensen, C.T. Hunter, C.C. Rering, I. Seidl-Adams, C.J. Stuhl, B. Torto, J.H. Tumlinson, Interactions among plants, insects, and microbes: Elucidation of inter-organismal chemical communications in agricultural ecology, *J. Agric. Food Chem.* 66 (2018) 6663–6674. https://doi.org/10.1021/acs.jafc.8b01763.

[3] R. Radhakrishnan, A.A. Alqarawi, E.F. Abd_Allah, Bioherbicides: Current knowledge on weed control mechanism, *Ecotoxicol. Environ. Saf.* 158 (2018) 131–138. https://doi.org/10.1016/j.ecoenv.2018.04.018.

[4] F.K. van Evert, S. Fountas, D. Jakovetic, V. Crnojevic, I. Travlos, C. Kempenaar, Big Data for weed control and crop protection, *Weed Res.* 57 (2017) 218–233. https://doi.org/10.1111/wre.12255.

[5] F.A. Macías, F.J. Mejías, J.M. Molinillo, Recent advances in allelopathy for weed control: From knowledge to applications, *Pest Manag. Sci.* 75 (2019) 2413–2436. https://doi.org/10.1002/ps.5355.

[6] S. Mallick, R. Plant, R. Centre, Weed flora of Rourkela and adjoining areas of Sundargarh district, Odisha, Weed Flora of Rourkela and adjoining areas of, *J. Econ. Taxon. Bot.* 39 (2016) 7.

[7] F.T. de Vries, Chufa (cyperus esculentus, cyperaceae): A weedy cultivar or a cultivated weed? *Econ. Bot.* 45 (1991) 27–37. https://doi.org/10.1007/BF02860047.

[8] P.E. Arriola, N.C. Ellstrand, Crop-to-weed gene flow in the genus Sorghum (Poaceae): Spontaneous interspecific hybridization between johnsongrass, Sorghum halepense, and crop sorghum, *S. Bicolor, Am. J. Bot.* 83 (1996) 1153–1159. https://doi.org/10.1002/j.1537-2197.1996.tb13895.x.

[9] M. Hjorth, L. Mondolot, B. Buatois, C. Andary, S. Rapior, P. Kudsk, S.K. Mathiassen, H.W. Ravn, An easy and rapid method using microscopy to determine herbicide effects in Poaceae weed species, *Pest Manag. Sci.* 62 (2006) 515–521. https://doi.org/10.1002/ps.1194.
[10] T. Jayabarathi, A. Yazdani, V. Ramesh, T. Raghunathan, Combined heat and power economic dispatch problem using the invasive weed optimization algorithm, *Front. Energy.* 8 (2014) 25–30. https://doi.org/10.1007/s11708-013-0276-4.
[11] A.K. Barisal, R.C. Prusty, Large scale economic dispatch of power systems using oppositional invasive weed optimization, *Appl. Soft Comput.* 29 (2015) 122–137. https://doi.org/10.1016/j.asoc.2014.12.014.
[12] H.F. Abouziena, W.M. Haggag, Weed control in clean agriculture: A review, *Planta Daninha.* 34 (2016) 377–392. https://doi.org/10.1590/S0100-83582016340200019.
[13] K. Jabran, G. Mahajan, V. Sardana, B.S. Chauhan, Allelopathy for weed control in agricultural systems, *Crop Prot.* 72 (2015) 57–65. https://doi.org/10.1016/j.cropro.2015.03.004.
[14] A. Forouzesh, E. Zand, S. Soufizadeh, S. Samadi Foroushani, Classification of herbicides according to chemical family for weed resistance management strategies—an update, *Weed Res.* 55 (2015) 334–358. https://doi.org/10.1111/wre.12153.
[15] V.K. Nandula, D.E. Riechers, Y. Ferhatoglu, M. Barrett, S.O. Duke, F.E. Dayan, A. Goldberg-Cavalleri, C. Tétard-Jones, D.J. Wortley, N. Onkokesung, M. Brazier-Hicks, R. Edwards, T. Gaines, S. Iwakami, M. Jugulam, R. Ma, Herbicide metabolism: Crop selectivity, bioactivation, weed resistance, and regulation, *Weed Sci.* 67 (2019) 149–175. https://doi.org/10.1017/wsc.2018.88.
[16] C. Wu, V. Varanasi, A. Perez-Jones, A nondestructive leaf-disk assay for rapid diagnosis of weed resistance to multiple herbicides, *Weed Sci.* 69 (2021) 274–283. https://doi.org/10.1017/wsc.2021.15.
[17] R.F. Lopez Ovejero, H.K. Takano, M. Nicolai, A. Ferreira, M.S.C. Melo, A.L. Cavenaghi, P.J. Christoffoleti, R.S. Oliveira, Frequency and dispersal of glyphosate-resistant sourgrass (digitaria insularis) populations across Brazilian agricultural production areas, *Weed Sci.* 65 (2017) 285–294. https://doi.org/10.1017/wsc.2016.31.
[18] R. Alcántara-de la Cruz, G. Moraes de Oliveira, L. Bianco de Carvalho, M. Fátima das Graças Fernandes da Silva, Herbicide Resistance in Brazil: Status, impacts, and future challenges, in: D. Kontogiannatos, A. Kourti, K.F. Mendes (Eds.), *Pests, Weeds and Diseases in Agricultural Crop and Animal Husbandry Production*, 1st ed., IntechOpen, Londo, 2020: p. 30. https://doi.org/10.5772/intechopen.91236.
[19] C.-H. Kong, T.D. Xuan, T.D. Khanh, H.-D. Tran, N.T. Trung, Allelochemicals and signaling chemicals in plants, *Molecules.* 24 (2019) 2737. https://doi.org/10.3390/molecules24152737.
[20] A. Scavo, C. Abbate, G. Mauromicale, Plant allelochemicals: Agronomic, nutritional and ecological relevance in the soil system, *Plant Soil.* 442 (2019) 23–48. https://doi.org/10.1007/s11104-019-04190-y.
[21] M.J. Reigosa, L. Gonzalez, A. Sanches-Moreiras, B. Duran, D. Puime, D.A. Fernadez, J.C. Bolano, Comparison of physiological effects of allelochemicals and commercial herbicides, *Allelopath. J.* 8 (2001) 211–220.
[22] B. Durán-Serantes, L. González, M.J. Reigosa, Comparative physiological effects of three allelochemicals and two herbicides on Dactylis glomerata, *Acta Physiol. Plant.* 24 (2002) 385–392. https://doi.org/10.1007/s11738-002-0034-4.
[23] C. Das, A. Dey, A. Bandyopadhyay, Allelochemicals: An emerging tool for weed management, in: S.C.M. lRaja Chakraborty, S. Sen (Eds.), *Evidence Based Validation of Indian Traditional Medicine*, 1st ed., Springer Singapore, Singapore, 2021: pp. 249–259. https://doi.org/10.1007/978-981-15-8127-4_12.
[24] E.M. Davis, R. Croteau, Cyclization enzymes in the biosynthesis of monoterpenes, sesquiterpenes, and diterpenes, in: *Biosynthesis Topics in Current Chemistry*, Amisterdan, 2000: pp. 53–95. https://doi.org/10.1007/3-540-48146-X_2.
[25] H.B. He, H.B. Wang, C.X. Fang, Y.Y. Lin, C.M. Zeng, L.Z. Wu, W.C. Guo, W.X. Lin, Herbicidal effect of a combination of oxygenic terpenoids on Echinochloa crus-galli, *Weed Res.* 49 (2009) 183–192. https://doi.org/10.1111/j.1365-3180.2008.00675.x.
[26] F. Araniti, A.M. Sánchez-Moreiras, E. Graña, M.J. Reigosa, M.R. Abenavoli, Terpenoid trans-caryophyllene inhibits weed germination and induces plant water status alteration and oxidative damage in adult Arabidopsis, *Plant Biol.* 19 (2017) 79–89. https://doi.org/10.1111/plb.12471.
[27] J.Y. Li, S.X. Lin, Q. Zhang, L. Li, W.W. Hu, H. Bin He, Phenolic acids and terpenoids in the soils of different weed-suppressive circles of allelopathic rice, *Arch. Agron. Soil Sci.* 66 (2020) 266–278. https://doi.org/10.1080/03650340.2019.1610560.
[28] M. Umehara, A. Hanada, S. Yoshida, K. Akiyama, T. Arite, N. Takeda-Kamiya, H. Magome, Y. Kamiya, K. Shirasu, K. Yoneyama, J. Kyozuka, S. Yamaguchi, Inhibition of shoot branching by new terpenoid plant hormones, *Nature.* 455 (2008) 195–200. https://doi.org/10.1038/nature07272.

[29] N. Chotsaeng, C. Laosinwattana, P. Charoenying, Herbicidal activities of some allelochemicals and their synergistic behaviors toward amaranthus tricolor L., *Molecules.* 22 (2017) 1841. https://doi.org/10.3390/molecules22111841.

[30] X.-F. Yang, K. Lei, C.-H. Kong, X.-H. Xu, Effect of allelochemical tricin and its related benzothiazine derivative on photosynthetic performance of herbicide-resistant barnyardgrass, *Pestic. Biochem. Physiol.* 143 (2017) 224–230. https://doi.org/10.1016/j.pestbp.2017.08.010.

[31] B.K. Ghimire, M.H. Hwang, E.J. Sacks, C.Y. Yu, S.H. Kim, I.M. Chung, Screening of allelochemicals in miscanthus sacchariflorus extracts and assessment of their effects on germination and seedling growth of common weeds, *Plants.* 9 (2020) 1313. https://doi.org/10.3390/plants9101313.

[32] S. Latif, G. Chiapusio, L.A. Weston, Allelopathy and the role of allelochemicals in plant defence, in: G. Becard (Ed.), *Advances in Botanical Research*, 82nd ed., Academic Press, Amisterdan, 2017: pp. 19–54. https://doi.org/10.1016/bs.abr.2016.12.001.

[33] S. Chaïb, J.C.A. Pistevos, C. Bertrand, I. Bonnard, Allelopathy and allelochemicals from microalgae: An innovative source for bio-herbicidal compounds and biocontrol research, *Algal Res.* 54 (2021) 102213. https://doi.org/10.1016/j.algal.2021.102213.

[34] W. Mushtaq, M.B. Siddiqui, K.R. Hakeem, Role of allelochemicals in agroecosystems, in: W. Mushtaq, M.B. Siddiqui, K.R. Hakeem (Eds.), *Allelopathy: Potential for Green Agriculture*, 1st ed., Springer, Cham, 2020: pp. 45–52. https://doi.org/10.1007/978-3-030-40807-7_5.

[35] R.J. Thoppil, Terpenoids as potential chemopreventive and therapeutic agents in liver cancer, *World J. Hepatol.* 3 (2011) 228. https://doi.org/10.4254/wjh.v3.i9.228.

[36] M. Huang, J.-J. Lu, M.-Q. Huang, J.-L. Bao, X.-P. Chen, Y.-T. Wang, Terpenoids: Natural products for cancer therapy, *Expert Opin. Investig. Drugs.* 21 (2012) 1801–1818. https://doi.org/10.1517/13543784.2012.727395.

[37] L. Caputi, E. Aprea, Use of terpenoids as natural flavouring compounds in food industry, *Recent Patents Food, Nutr. Agric.* 3 (2011) 9–16. https://doi.org/10.2174/2212798411103010009.

[38] M. Santana de Oliveira, V.M. Pereira da Silva, L. Cantão Freitas, S. Gomes Silva, J. Nevez Cruz, E.H. Aguiar Andrade, Extraction Yield, Chemical composition, preliminary toxicity of bignonia nocturna (bignoniaceae) essential oil and in silico evaluation of the interaction, *Chem. Biodivers.* 18 (2021) cbdv.202000982. https://doi.org/10.1002/cbdv.202000982.

[39] A. Ludwiczuk, K. Skalicka-Woźniak, M.I. Georgiev, Terpenoids, in: S. Badal, R. Delgoda (Eds.), *Pharmacognosy*, Elsevier, Amisterdan, 2017: pp. 233–266. https://doi.org/10.1016/B978-0-12-802104-0.00011-1.

[40] D. Tholl, Biosynthesis and Biological Functions of Terpenoids in Plants, in: J. Bohlmann (Ed.), *Advances in Biochemical Engineering/Biotechnology*, Cham, 2015: pp. 63–106. https://doi.org/10.1007/10_2014_295.

[41] J.-J. Lu, Y.-Y. Dang, M. Huang, W.-S. Xu, X.-P. Chen, Y.-T. Wang, Anti-cancer properties of terpenoids isolated from Rhizoma Curcumae—A review, *J. Ethnopharmacol.* 143 (2012) 406–411. https://doi.org/10.1016/j.jep.2012.07.009.

[42] M.S. de Oliveira, S.G. Silva, J.N. da Cruz, E. Ortiz, W.A. da Costa, F.W.F. Bezerra, V.M.B. Cunha, R.M. Cordeiro, A.M. de J.C. Neto, E.H. de A. Andrade, R.N. de C. Junior, Supercritical CO2 application in essential oil extraction, in: R.M. Inamuddin, A.M. Asiri (Eds.), *Industrial Applications of Green Solvents—Vol. II*, 2nd ed., Materials Research Foundations, Millersville PA, USA, 2019: pp. 1–28. https://doi.org/10.21741/9781644900314-1.

[43] M. Santana de Oliveira, W. Almeida da Costa, S. Gomes Silva, *Essential Oils—Bioactive Compounds, New Perspectives and Applications*, 1st ed., IntechOpen, London, 2020. https://doi.org/10.5772/intechopen.87266.

[44] M.S. de Oliveira, W.A. da Costa, D.S. Pereira, J.R.S. Botelho, T.O. de Alencar Menezes, E.H. de Aguiar Andrade, S.H.M. da Silva, A.P. da Silva Sousa Filho, R.N. de Carvalho, Chemical composition and phytotoxic activity of clove (Syzygium aromaticum) essential oil obtained with supercritical CO 2, *J. Supercrit. Fluids.* 118 (2016) 185–193. https://doi.org/10.1016/j.supflu.2016.08.010.

[45] S.G. Silva, P.L.B. Figueiredo, L.D. Nascimento, W.A. da Costa, J.G.S. Maia, E.H.A. Andrade, Planting and seasonal and circadian evaluation of a thymol-type oil from Lippia thymoides Mart. & Schauer, *Chem. Cent. J.* 12 (2018) 113. https://doi.org/10.1186/s13065-018-0484-4.

[46] P.L.B. Figueiredo, S.G. Silva, L.D. Nascimento, A.R. Ramos, W.N. Setzer, J.K.R. da Silva, E.H.A. Andrade, Seasonal study of methyleugenol chemotype of ocimum campechianum essential oil and its fungicidal and antioxidant activities, *NPC Nat. Prod. Commun.* 13 (2018) 1055–1058.

[47] T.L.M. Rodrigues, G.L.S. Castro, R.G. Viana, E.S.C. Gurgel, S.G. Silva, M.S. de Oliveira, E.H. de A. Andrade, Physiological performance and chemical compositions of the Eryngium foetidum L.

(Apiaceae) essential oil cultivated with different fertilizer sources, *Nat. Prod. Res.* 34 (2020) 1–5. https://doi.org/10.1080/14786419.2020.1795653.

[48] B. Lukas, C. Schmiderer, J. Novak, Essential oil diversity of European Origanum vulgare L. (Lamiaceae), *Phytochemistry.* 119 (2015) 32–40. https://doi.org/10.1016/j.phytochem.2015.09.008.

[49] M. Askary, M.A. Behdani, S. Parsa, S. Mahmoodi, M. Jamialahmadi, Water stress and manure application affect the quantity and quality of essential oil of Thymus daenensis and Thymus vulgaris, *Ind. Crops Prod.* 111 (2018) 336–344. https://doi.org/10.1016/j.indcrop.2017.09.056.

[50] H. Cui, X. Zhang, H. Zhou, C. Zhao, L. Lin, Antimicrobial activity and mechanisms of Salvia sclarea essential oil, *Bot. Stud.* 56 (2015) 16. https://doi.org/10.1186/s40529-015-0096-4.

[51] Y. Zhang, X. Liu, Y. Wang, P. Jiang, S. Quek, Antibacterial activity and mechanism of cinnamon essential oil against Escherichia coli and Staphylococcus aureus, *Food Control.* 59 (2016) 282–289. https://doi.org/10.1016/j.foodcont.2015.05.032.

[52] L. Caputo, F. Nazzaro, L.F. Souza, L. Aliberti, L. De Martino, F. Fratianni, R. Coppola, V. De Feo, Laurus nobilis: Composition of essential oil and its biological activities, *Molecules.* 22 (2017) 1–11. https://doi.org/10.3390/molecules22060930.

[53] H. Turasan, S. Sahin, G. Sumnu, Encapsulation of rosemary essential oil, *LWT—Food Sci. Technol.* 64 (2015) 112–119. https://doi.org/10.1016/j.lwt.2015.05.036.

[54] G.A. Costa, J.L.S. Carvalho Filho, C. Deschamps, Yield and composition of the patchouli (Pogostemon cablin) essential oil according to the extraction time, *Rev. Bras. Plantas Med.* 15 (2013) 319–324. https://doi.org/10.1590/S1516-05722013000300002.

[55] S. Mandal, M. Mandal, Coriander (Coriandrum sativum L.) essential oil: Chemistry and biological activity, *Asian Pac. J. Trop. Biomed.* 5 (2015) 421–428. https://doi.org/10.1016/j.apjtb.2015.04.001.

[56] H. Cui, C. Zhang, C. Li, L. Lin, Antibacterial mechanism of oregano essential oil, *Ind. Crops Prod.* 139 (2019) 111498. https://doi.org/10.1016/j.indcrop.2019.111498.

[57] G.L. Oliveira, S.K. Cardoso, C.R. Lara Junior, T.M. Vieira, E.F. Guimaraes, L.S. Figueiredo, E.R. Martins, D.L. Moreira, M.A.C. Kaplan, Chemical study and larvicidal activity against Aedes aegypti of essential oil of Piper aduncum L. (Piperaceae), *An. Acad. Bras. Cienc.* 85 (2013) 1227–1234. https://doi.org/10.1590/0001-3765201391011.

[58] G.L. Da Silva, C. Luft, A. Lunardelli, R.H. Amaral, D.A.D.S. Melo, M.V.F. Donadio, F.B. Nunes, M.S. De Azambuja, J.C. Santana, C.M.B. Moraes, R.O. Mello, E. Cassel, M.A.D.A. Pereira, J.R. De Oliveira, Antioxidant, analgesic and anti-inflammatory effects of lavender essential oil, *An. Acad. Bras. Cienc.* 87 (2015) 1397–1408. https://doi.org/10.1590/0001-3765201520150056.

[59] M. Mahboubi, Zingiber officinale Rosc. essential oil, a review on its composition and bioactivity, *Clin. Phytoscience.* 5 (2019) 6. https://doi.org/10.1186/s40816-018-0097-4.

[60] M.F.R. da Silva, P.C. Bezerra-Silva, C.S. de Lira, B.N. de Lima Albuquerque, A.C. Agra Neto, E.V. Pontual, J.R. Maciel, P.M.G. Paiva, D.M. do A.F. Navarro, Composition and biological activities of the essential oil of Piper corcovadensis (Miq.) C. DC (Piperaceae), *Exp. Parasitol.* 165 (2016) 64–70. https://doi.org/10.1016/j.exppara.2016.03.017.

[61] N. Girola, C.R. Figueiredo, C.F. Farias, R.A. Azevedo, A.K. Ferreira, S.F. Teixeira, T.M. Capello, E.G.A. Martins, A.L. Matsuo, L.R. Travassos, J.H.G. Lago, Camphene isolated from essential oil of Piper cernuum (Piperaceae) induces intrinsic apoptosis in melanoma cells and displays antitumor activity in vivo, *Biochem. Biophys. Res. Commun.* 467 (2015) 928–934. https://doi.org/10.1016/j.bbrc.2015.10.041.

[62] M.F. Andrés, G.E. Rossa, E. Cassel, R.M.F. Vargas, O. Santana, C.E. Díaz, A. González-Coloma, Biocidal effects of Piper hispidinervum (Piperaceae) essential oil and synergism among its main components, *Food Chem. Toxicol.* 109 (2017) 1086–1092. https://doi.org/10.1016/j.fct.2017.04.017.

[63] J.E. Parra Amin, L.E. Cuca, A. González-Coloma, Antifungal and phytotoxic activity of benzoic acid derivatives from inflorescences of Piper cumanense, *Nat. Prod. Res.* (2019) 1–9. https://doi.org/10.1080/14786419.2019.1662010.

[64] E. Valarezo, P. Flores-Maza, L. Cartuche, S. Ojeda-Riascos, J. Ramírez, Phytochemical profile, antimicrobial and antioxidant activities of essential oil extracted from Ecuadorian species Piper ecuadorense sodiro, *Nat. Prod. Res.* (2020) 1–6. https://doi.org/10.1080/14786419.2020.1813138.

[65] M. Ibáñez, M. Blázquez, Phytotoxicity of essential oils on selected weeds: Potential hazard on food crops, *Plants.* 7 (2018) 79. https://doi.org/10.3390/plants7040079.

[66] S.K. Fagodia, H.P. Singh, D.R. Batish, R.K. Kohli, Phytotoxicity and cytotoxicity of Citrus aurantiifolia essential oil and its major constituents: Limonene and citral, *Ind. Crops Prod.* 108 (2017) 708–715. https://doi.org/10.1016/j.indcrop.2017.07.005.

[67] B.E. Jaramillo-Colorado, N. Pino-Benitez, A. González-Coloma, Volatile composition and biocidal (antifeedant and phytotoxic) activity of the essential oils of four Piperaceae species from Choco-Colombia, *Ind. Crops Prod.* 138 (2019) 111463. https://doi.org/10.1016/j.indcrop.2019.06.026.
[68] A.R. Jassbi, S. Zamanizadehnajari, I.T. Baldwin, Phytotoxic volatiles in the roots and shoots of artemisia tridentata as detected by headspace solid-phase microextraction and gas chromatographic-mass spectrometry analysis, *J. Chem. Ecol.* 36 (2010) 1398–1407. https://doi.org/10.1007/s10886-010-9885-0.
[69] S. Kaur, H.P. Singh, S. Mittal, D.R. Batish, R.K. Kohli, Phytotoxic effects of volatile oil from Artemisia scoparia against weeds and its possible use as a bioherbicide, *Ind. Crops Prod.* 32 (2010) 54–61. https://doi.org/10.1016/j.indcrop.2010.03.007.
[70] N. Chowhan, H.P. Singh, D.R. Batish, R.K. Kohli, Phytotoxic effects of β-pinene on early growth and associated biochemical changes in rice, *Acta Physiol. Plant.* 33 (2011) 2369–2376. https://doi.org/10.1007/s11738-011-0777-x.
[71] L. De Martino, E. Mancini, L.F.R. de Almeida, V. De Feo, The antigerminative activity of twenty-seven monoterpenes, *Molecules.* 15 (2010) 6630–6637. https://doi.org/10.3390/molecules15096630.
[72] H.P. SINGH, D.R. BATISH, S. KAUR, H. RAMEZANI, R.K. KOHLI, Comparative phytotoxicity of four monoterpenes against Cassia occidentalis, *Ann. Appl. Biol.* 141 (2002) 111–116. https://doi.org/10.1111/j.1744-7348.2002.tb00202.x.
[73] H.P. Singh, D.R. Batish, S. Kaur, R.K. Kohli, K. Arora, Phytotoxicity of the volatile monoterpene citronellal against some weeds, *Zeitschrift Für Naturforsch. C.* 61 (2006) 334–340. https://doi.org/10.1515/znc-2006-5-606.
[74] D.R. Batish, H.P. Singh, N. Setia, S. Kaur, R.K. Kohli, Chemical composition and phytotoxicity of volatile essential oil from intact and fallen leaves of Eucalyptus citriodora, *Zeitschrift Fur Naturforsch.—Sect. C J. Biosci.* 61 (2006) 465–471. https://doi.org/10.1515/znc-2006-7-801.
[75] A. Sharma, H.P. Singh, D.R. Batish, R.K. Kohli, Chemical profiling, cytotoxicity and phytotoxicity of foliar volatiles of Hyptis suaveolens, *Ecotoxicol. Environ. Saf.* 171 (2019) 863–870. https://doi.org/10.1016/j.ecoenv.2018.12.091.
[76] E. Rolli, M. Marieschi, S. Maietti, G. Sacchetti, R. Bruni, Comparative phytotoxicity of 25 essential oils on pre- and post-emergence development of Solanum lycopersicum L.: A multivariate approach, *Ind. Crops Prod.* 60 (2014) 280–290. https://doi.org/10.1016/j.indcrop.2014.06.021.
[77] H.P. Singh, R.K. Kohli, S. Mittala, S. Kaur, D.R. Batish, Phytotoxicity of major constituents of the volatile oil from leaves of artemisia scoparia waldst. & kit, *Zeitschrift Fur Naturforsch.—Sect. C J. Biosci.* 63 (2008) 663–666. https://doi.org/10.1515/znc-2008-9-1009.
[78] S. Kordali, A. Cakir, H. Ozer, R. Cakmakci, M. Kesdek, E. Mete, Antifungal, phytotoxic and insecticidal properties of essential oil isolated from Turkish Origanum acutidens and its three components, carvacrol, thymol and p-cymene, *Bioresour. Technol.* 99 (2008) 8788–8795. https://doi.org/10.1016/j.biortech.2008.04.048.
[79] A. Synowiec, D. Kalemba, E. Drozdek, J. Bocianowski, Phytotoxic potential of essential oils from temperate climate plants against the germination of selected weeds and crops, *J. Pest Sci.* (2004). 90 (2017) 407–419. https://doi.org/10.1007/s10340-016-0759-2.
[80] L.F.R. De Almeida, F. Frei, E. Mancini, L. De Martino, V. De Feo, Phytotoxic activities of mediterranean essential oils, *Molecules.* 15 (2010) 4309–4323. https://doi.org/10.3390/molecules15064309.
[81] F. Le Bideau, M. Kousara, L. Chen, L. Wei, F. Dumas, Tricyclic sesquiterpenes from marine origin, *Chem. Rev.* 117 (2017) 6110–6159. https://doi.org/10.1021/acs.chemrev.6b00502.
[82] R. Kramer, W.-R. Abraham, Volatile sesquiterpenes from fungi: What are they good for?, *Phytochem. Rev.* 11 (2012) 15–37. https://doi.org/10.1007/s11101-011-9216-2.
[83] E.S.C. Gurgel, M.S. de Oliveira, M.C. Souza, S.G. da Silva, M.S. de Mendonça, A.P. da S. Souza Filho, Chemical compositions and herbicidal (phytotoxic) activity of essential oils of three Copaifera species (Leguminosae-Caesalpinoideae) from Amazon-Brazil, *Ind. Crops Prod.* 142 (2019). https://doi.org/10.1016/j.indcrop.2019.111850.
[84] M. Issa, S. Chandel, H. Pal Singh, D. Rani Batish, R. Kumar Kohli, S. Singh Yadav, A. Kumari, Appraisal of phytotoxic, cytotoxic and genotoxic potential of essential oil of a medicinal plant Vitex negundo, *Ind. Crops Prod.* 145 (2020) 112083. https://doi.org/10.1016/j.indcrop.2019.112083.
[85] A. Andolfi, N. Zermane, A. Cimmino, F. Avolio, A. Boari, M. Vurro, A. Evidente, Inuloxins A–D, phytotoxic bi-and tri-cyclic sesquiterpene lactones produced by Inula viscosa: Potential for broomrapes and field dodder management, *Phytochemistry.* 86 (2013) 112–120. https://doi.org/10.1016/j.phytochem.2012.10.003.

[86] A. Scavo, C. Rial, J.M.G. Molinillo, R.M. Varela, G. Mauromicale, F.A. Macías, Effect of shading on the sesquiterpene lactone content and phytotoxicity of cultivated cardoon leaf extracts, *J. Agric. Food Chem.* 68 (2020) 11946–11953. https://doi.org/10.1021/acs.jafc.0c03527.
[87] P. Del Valle, M. Figueroa, R. Mata, Phytotoxic eremophilane sesquiterpenes from the coprophilous fungus penicillium sp. G1-a14, *J. Nat. Prod.* 78 (2015) 339–342. https://doi.org/10.1021/np5009224.
[88] D.M. Cárdenas, A. Cala, J.M.G. Molinillo, F.A. Macías, Preparation and phytotoxicity study of lappalone from dehydrocostuslactone, *Phytochem. Lett.* 20 (2017) 66–72. https://doi.org/10.1016/j.phytol.2017.04.017.
[89] B.P. da Silva, M.P. Nepomuceno, R.M. Varela, A. Torres, J.M.G. Molinillo, P.L.C.A. Alves, F.A. Macías, Phytotoxicity study on bidens sulphurea Sch. Bip. as a preliminary approach for weed control, *J. Agric. Food Chem.* 65 (2017) 5161–5172. https://doi.org/10.1021/acs.jafc.7b01922.
[90] D. Dastan, P. Salehi, F. Ghanati, A.R. Gohari, H. Maroofi, N. Alnajar, Phytotoxicity and cytotoxicity of disesquiterpene and sesquiterpene coumarins from Ferula pseudalliacea, *Ind. Crops Prod.* 55 (2014) 43–48. https://doi.org/10.1016/j.indcrop.2014.01.051.
[91] W.H. Perera, K.M. Meepagala, D.E. Wedge, S.O. Duke, Sesquiterpenoids from culture of the fungus Stereum complicatum (Steraceae): Structural diversity, antifungal and phytotoxic activities, *Phytochem. Lett.* 37 (2020) 51–58. https://doi.org/10.1016/j.phytol.2020.03.012.
[92] J. Chen, G. Zheng, Y. Zhang, H.A. Aisa, X.-J. Hao, Phytotoxic terpenoids from ligularia cymbulifera roots, *Front. Plant Sci.* 7 (2017). https://doi.org/10.3389/fpls.2016.02033.
[93] V. Ahluwalia, R. Sisodia, S. Walia, O.P. Sati, J. Kumar, A. Kundu, Chemical analysis of essential oils of Eupatorium adenophorum and their antimicrobial, antioxidant and phytotoxic properties, *J. Pest Sci.* (2004). 87 (2014) 341–349. https://doi.org/10.1007/s10340-013-0542-6.
[94] A.M. Abd-ElGawad, A.E.-N.G. El Gendy, A.M. Assaeed, S.L. Al-Rowaily, A.S. Alharthi, T.A. Mohamed, M.I. Nassar, Y.H. Dewir, A.I. Elshamy, Phytotoxic effects of plant essential oils: A systematic review and structure-activity relationship based on chemometric analyses, *Plants.* 10 (2020) 36. https://doi.org/10.3390/plants10010036.
[95] S. Anese, L.J. Jatobá, P.U. Grisi, S.C.J. Gualtieri, M.F.C. Santos, R.G.S. Berlinck, Bioherbicidal activity of drimane sesquiterpenes from Drimys brasiliensis Miers roots, *Ind. Crops Prod.* 74 (2015) 28–35. https://doi.org/10.1016/j.indcrop.2015.04.042.
[96] R.K. Devappa, H.P.S. Makkar, K. Becker, Jatropha diterpenes: A review, *J. Am. Oil Chem. Soc.* 88 (2011) 301–322. https://doi.org/10.1007/s11746-010-1720-9.
[97] F. Berrue, R.G. Kerr, Diterpenes from gorgonian corals, *Nat. Prod. Rep.* 26 (2009) 681. https://doi.org/10.1039/b821918b.
[98] A. Vasas, J. Hohmann, Euphorbia diterpenes: Isolation, structure, biological activity, and synthesis (2008–2012), *Chem. Rev.* 114 (2014) 8579–8612. https://doi.org/10.1021/cr400541j.
[99] R. Li, S.L. Morris-Natschke, K.-H. Lee, Clerodane diterpenes: Sources, structures, and biological activities, *Nat. Prod. Rep.* 33 (2016) 1166–1226. https://doi.org/10.1039/C5NP00137D.
[100] D.L. Zhao, L.J. Yang, T. Shi, C.Y. Wang, C.L. Shao, C.Y. Wang, Potent phytotoxic harziane diterpenes from a soft coral-derived strain of the fungus trichoderma harzianum XS-20090075, *Sci. Rep.* 9 (2019) 1–9. https://doi.org/10.1038/s41598-019-49778-7.
[101] A. Cimmino, A. Andolfi, M.C. Zonno, F. Avolio, A. Santini, A. Tuzi, A. Berestetskyi, M. Vurro, A. Evidente, Chenopodolin: A phytotoxic unrearranged ent-pimaradiene diterpene produced by phoma chenopodicola, a fungal pathogen for chenopodium album biocontrol, *J. Nat. Prod.* 76 (2013) 1291–1297. https://doi.org/10.1021/np400218z.
[102] A. Evidente, L. Sparapano, O. Fierro, G. Bruno, F. Giordano, A. Motta, Sphaeropsidins B and C, phytotoxic pimarane diterpenes from Sphaeropsis sapinea f. sp. Cupressi and Diplodia mutila, *Phytochemistry.* 45 (1997) 705–713. https://doi.org/10.1016/S0031-9422(97)00006-X.
[103] A. Bisio, G. Damonte, D. Fraternale, E. Giacomelli, A. Salis, G. Romussi, S. Cafaggi, D. Ricci, N. De Tommasi, Phytotoxic clerodane diterpenes from Salvia miniata Fernald (Lamiaceae), *Phytochemistry.* 72 (2011) 265–275. https://doi.org/10.1016/j.phytochem.2010.11.011.
[104] A.P.P. Klein Hendges, E.F. dos Santos, S.D. Teixeira, F.S. Santana, M.M. Trezzi, A.N.L. Batista, J.M. Batista, V.A. de Lima, F. de Assis Marques, B.H.L.N.S. Maia, Phytotoxic neocassane diterpenes from eragrostis plana, *J. Nat. Prod.* 83 (2020) 3511–3518. https://doi.org/10.1021/acs.jnatprod.0c00324.
[105] A. Favaretto, C.L. Cantrell, F.R. Fronczek, S.O. Duke, D.E. Wedge, A. Ali, S.M. Scheffer-basso, new phytotoxic cassane-like diterpenoids from eragrostis plana, *J. Agric. Food Chem.* 67 (2019) 1973–1981. https://doi.org/10.1021/acs.jafc.8b06832.

[106] A.K.M.M. Islam, O. Ohno, K. Suenaga, H. Kato-Noguchi, Two novel phytotoxic substances from Leucas aspera, *J. Plant Physiol.* 171 (2014) 877–883. https://doi.org/10.1016/j.jplph.2014.03.003.

[107] B. Qin, L.G. Perry, C.D. Broeckling, J. Du, F.R. Stermitz, M.W. Paschke, J.M. Vivanco, Phytotoxic allelochemicals from roots and root exudates of leafy spurge (Euphorbia esula L.), *Plant Signal. Behav.* 1 (2006) 323–327. https://doi.org/10.4161/psb.1.6.3563.

[108] A. Evidente, L. Sparapano, G. Bruno, A. Motta, Sphaeropsidins D and E, two other pimarane diterpenes, produced in vitro by the plant pathogenic fungus Sphaeropsis sapinea f. sp. cupressi, *Phytochemistry.* 59 (2002) 817–823. https://doi.org/10.1016/S0031-9422(02)00015-8.

17 Microbial Herbicides

Alan K. Watson

17.1 INTRODUCTION

Weeds are the major constraint to crop production, but weeds are also prone to disease and there are several ways these natural enemies may be used to suppress weeds. To begin this chapter, the following terms are defined: biological weed control, biopesticides, bioherbicide, microbial herbicide and biochemical herbicide. Biological control of weeds is defined as "the use of an agent, a complex of agents, or biological processes to bring about weed suppression. All forms of macrobial and microbial organisms are considered as biological control agents. Examples of biological control agents include, but are not limited to, arthropods (insects and mites), plant pathogens (fungi, bacteria, viruses, and nematodes), fish, birds, and other animals" (http://wssa. net/wssa/weed/biological-control/). The United States Environmental Protection Agency (EPA) defines biopesticides as "naturally occurring substances that control pests (biochemical pesticides), microorganisms that control pests (microbial pesticides), and pesticidal substances produced by plants containing added genetic material (plant-incorporated protectants) or PIPs" (www.epa.gov/pesticides/biopesticides). Bioherbicides are living phytopathogenic microorganisms (microbial herbicides) or microbial phytotoxins (biochemical herbicides) that are field applied in ways like conventional chemical herbicides.

17.2 BIOLOGICAL WEED CONTROL

Biological weed control can be realized by two main strategies: classical biocontrol and bioherbicide strategies (Templeton et al. 1979; Yandoc 2001; Evans 2013). Classical biological weed control targets exotic, non-native weed species that have arrived from another part of the world without their natural enemies and have become dominant in their new habitats (Wapshere 1974; Watson 1991b). These invasive weeds often infest large areas of marginal lands, such as pastures and rangelands. Host specific natural enemies (mostly insects, but occasionally plant pathogens) are obtained from a target weed's native range, and host specificity and impact are carefully evaluated. Founding populations of the host specific biocontrol agent are released (inoculated) into the weed infested areas of the invaded country. After release, the biocontrol agent populations are monitored to evaluate establishment, spread and level of weed control. Classical biocontrol (CBC) is an ecological approach that can provide sustainable, long-term control.

The bioherbicide strategy or inundative (IBC) approach involves the use of local endemic phytopathogenic microorganisms (fungi, bacteria, viruses and nematodes) to control a native or naturalized weed. These weeds are already in dynamic equilibrium with their natural enemies and can be controlled by manipulation of existing weed-natural enemy relationships. Large inundative populations of an existing natural enemy are mass produced, formulated and applied on the crop fields like a chemical herbicide, following extensive testing to ensure that non-target species are not negatively affected. Weed control is rendered, but the biocontrol agent normally does not survive in high numbers requiring retreatment, and the control is not sustained (like a chemical herbicide). Bioherbicides are a technological approach, like chemical herbicides, and deliver transient, non-sustained weed control.

Biological weed control can also follow a third strategy, herbivory. Fish, tadpole shrimps, ducks and goats provide weed control in Asian rice fields (de Datta and Baltazar 1996; Shibayama 2001).

DOI: 10.1201/9781003463429-22

Rice-fish-duck systems often have significantly higher weed control than other farming systems tested (Men et al. 1999; Zhang et al. 2010). In North America, leafy spurge (*Euphorbia esula* L.) is effectively controlled with grazing of sheep and goats (Landgraf et al. 1984; Sedivec et al. 1995).

17.3 CLASSICAL BIOLOGICAL WEED CONTROL WITH PHYTOPATHOGENS

Much of the effort and success of classical biological weed control has been dominated by insect biocontrol agents (Wapshere 1982; Waterhouse 1994; Julien et al. 2012; Winston et al. 2014; Day and Winston 2016) but many biopesticide papers reviewed for this chapter encompassed reference to obligate fungal pathogens that have been introduced as classical biocontrol (CBC) agents into various countries of the world (Barreto 2007; Barreto et al. 2012; Barton 2004, 2005; Bruckart 2005; Burdon et al. 2000; Cullen 1985; Cullen et al. 1973; Evans 1995, 2013; Evans et al. 2001; Hershenhorn et al. 2016; Watson 1991b; Winston et al. 2014). Weed species targeted with classical biocontrol are primarily weeds of aquatic systems and rangelands; nonetheless success with plant pathogenic fungi has been recorded in several grassland systems (Table 17.1).

TABLE 17.1
Successful Classical Biological Weed Control Projects for Cropland Weeds (adapted from Barreto et al. 2012; Winston et al. 2014)

Weed Target	Pathogen	Country of Introduction	Control Status	References
Acacia saligna (Labill.) H. L. Wendl. (Port Jackson willow)	*Uromycladium tepperianum* (Sacc.) McAlpine	South Africa	Significant	Morris 1987, 1997
Ageratina riparia (Regel) R.M. King & H. Rob (mistflower)	*Entyloma ageratinae* R.W. Barreto & H.C. Evans	USA, New Zealand, South Africa	Significant	Morin et al. 1997; Frohlich et al. 1999
Carduus nutans L. (nodding thistle)	*Puccinia carduorum* Jacky	USA (continental)	Significant	Politis et al. 1984; Bruckart 2005
C. pycnocephalus L. (Italian thistle) and *C. tenuiflorus* Curtis (slender-flower thistle)	*Puccinia cardui-pycnocephali* P. Syd. & Syd.	Australia	Significant	Burdon et al. 2000
Chondrilla juncea L. (skeleton weed)	*Puccinia chondrillina* (Bubak & Syd.)	Australia, USA (continental), Canada	Significant/Partial	Cullen et al. 1973; Cullen 1985
Clidemia hirta (L.) D. Don (Koster's curse)	*Colletotrichum gloeosporioides* f. sp. *clidemiae* E.E. Trujillo, Latterell & A.E. Rossi	USA (Hawaii)	Partial	Trujillo 2005
Cryptostegia grandiflora (Roxb. ex R. Br.) R. Br. (rubber vine)	*Maravalia cryptostegiae* (Cummins) Y. Ono	Australia	Significant	Tomley and Evans 2004
Passiflora tarminiana Coppens & V.E. Barney (banana poka)	*Septoria passiflorae* Louw	USA (Hawaii)	Significant	Trujillo 2005
Rubus constrictus P.J. Müll. & Lefèvre, *R. ulmifolius* Schott (wild blackberry)	*Phragmidium violaceum* (Schultz) G. Winter	Chile	Significant	Oehrens and Gonzales 1977
R. fruticosus aggregate (shrubby blackberry)	*Phragmidium violaceum*	Australia	Partial	Evans et al. 2004

Host specificity is the most important factor when evaluating microorganism for CBC (Wapshere 1982; Watson 1985; Evans 2013). Extensive life-history studies are conducted in the microbe's native range and before release of the microbe, a full life history risk assessment is presented to authorities in a receiving country (Barreto et al. 2012; Evans 2013). This forms the basis of the pest risk assessment, of which the principal objective is to demonstrate specificity to the target weed. CBC safety record and success rate have been very good (Evans 2013).

17.4 MICROBIAL HERBICIDES

Microbial herbicides comprise living plant pathogenic organisms including fungi (mycoherbicides), bacteria (bacto-herbicides), viruses (viral herbicides) and nematodes (nematoda herbicides). Most bioherbicide research activity has been with fungal plant pathogens and much less effort with plant pathogenic bacteria, viruses or nematodes.

17.5 MYCOHERBICIDES

Mycoherbicides "are simply plant-pathogenic fungi developed and used in the inundative strategy to control weeds the way chemical herbicides are used" (TeBeest and Templeton 1985). Interest in mycoherbicide research began with knowledge of the Lubao mycoherbicide in China. Lubao No. 1, a formulated suspension of *Colletotrichum gloeosporioides* (Penz.) Penz. & Sacc. f. sp. *cuscutae* for the control of dodder (*Cuscutae australis* R. Br.), a weed in soybean fields (Gao and Yu 1992). Lubao was discovered in 1963 and by the late 1970s was applied to 670,000 ha of soybean (Zhang et al. 2011).

In 1973, reports of the biological control of milkweed vine (*Morrenia odorata* Lindl.) with a race of *Phytophthora citrophthora* (Butler) Butler (Burnett et al. 1973, 1974) and the biological control of northern jointvetch [*Aeschynomene virginica* (L.) B.S.P.] in rice with *Colletotrichum gloeosporioides* Penz. Sacc. f. sp. *aeschynomene* (Daniel et al. 1973) were published in the scientific literature. Subsequently, *Phytophthora citrophthora* was registered as a microbial pest control product by the United States Environmental Protection Agency (EPA) as DeVine for control of stranglervine (*Morrenia odorata*) in Florida citrus groves in 1981 (Ridings 1986) and the next year, *Colletotrichum gloeosporioides* f. sp. *aeschynomene* was registered as Collego for the control of northern jointvetch (*Aeschynomene virginica*) in fields of rice and soybeans in Arkansas, Louisiana and Mississippi (TeBeest and Templeton 1985; Bowers 1986; Smith 1986).

Templeton (1982a) outlined the discovery, development and deployment phases involved in forming a biological herbicide. The discover phase involves the collection, isolation, identification and culture maintenance of a weed pathogen. The development phase includes inoculum production, culture conditions, disease etiology, field trials and host range determination. Product formulation, mass production scale-up, intellectual property protection, patents, government registration approval and commercialization complete the deployment phase. Later, Bailey et al. (2009) and Bailey and Falk (2011) suggested different approaches were needed to evaluate scientific and commercial potential of a bioherbicide organism because "commercialization is the ultimate goal, then the science must consider factors deemed important to the industry" (Bailey et al. 2009).

Early success of Luboa, Collego and DeVine in the late 1970s and early 1980s was followed by relatively well-funded research in many countries (Templeton 1982b; Charudattan 1991). Hundreds of weeds were targeted with fungi, bacteria, viruses and nematodes resulting in numerous manuscripts and patents being fashioned. Considerable basic knowledge was acquired but success was limited, if measured in number of registrations and commercially viable products. Loss of virulence, limited market size, small specialist market and control persistence were factors responsible for the demise of Luboa, Collego and DeVine mycoherbicides, all single weed target-restricted products.

There have been numerous comprehensive regional and world reviews on the progress and listings of microbial herbicide research and development projects (Templeton et al. 1979; Templeton

1982b; TeBeest and Templeton 1985; Charudattan 1991, 2005; Watson 1991a, 1994, 1999, 2017; Yoo 1991; TeBeest et al. 1992; Evans 1995, 2013; Cother 1996; Fujimori 1999; Morris et al. 1999; Rosskopf et al. 1999; Auld 2000; Charudattan and Dinoor 2000; Müller-Schärer et al. 2000; Evans et al. 2001; Li et al. 2003; Boyetchko and Peng 2004; Barton 2005; Trujillo 2005; Chutia et al. 2007; Vurro and Evans 2007; Ash 2010, 2011; Bailey et al. 2010; Barreto et al. 2012; Stubbs and Kennedy 2012; Aneja et al. 2013; Bailey 2014; Winston et al. 2014; Patel and Patel 2015; Harding and Raizada 2015; Pacanoski 2015; Cordeau et al. 2016; Hershenhorn et al. 2016; Cai and Gu 2016; Gaddeyya et al. 2017). After DeVine and Collego were marketed, very few commercial microbial herbicide products were registered during the past 35 years (Table 17.2). A commercial product must perform under field conditions and be economically produced and formulated to retain sufficient shelf-life during commercial distribution (Zorner et al. 1993; Charudattan 1991; Bailey and Falk 2011). Mycoherbicides have not met earlier expectations and have contributed little to weed management in cropping systems (Zhang et al. 2011; Barreto et al. 2012; Evans 2013; Hershenhorn et al. 2016).

Questions arise after examining the number of potential microbial herbicides cited in the literature, and realizing that very few microbial herbicides have made it to the marketplace. Why the limited success? Was the wrong target selected? Would a different pathogen be preferred? What factors instigated inconsistent field results? Disease development involves interplay between the host, the pathogen and the environment, a susceptible host, a virulent pathogen and favorable moisture and temperature conditions are essential for disease to occur. Constraints to bioherbicide development including biological (low virulence), environmental (temperature and dew period requirements), technological (mass production, formulation issues), and commercial (patent, registration, market analysis) factors have been considered over the past 25 years to explain the lack of commercial success (Watson and Wymore 1990; Auld and Morin 1995; Makowski and Mortensen 1998; Pacanoski 2015).

Considerable efforts to overcome these constraints were instigated in many research groups.

Several reviews covering improvements in adjuvants, formulation and application techniques were published (Boyette et al. 1991, 1996; Green et al. 1998). Innovative approaches including vegetable oil suspensions (Auld 1993), sodium alginate granules (Walker and Connick 1983), pesta-like formulation (Connick et al. 1991; Elzein et al. 2008) and invert emulsions (Womack et al. 1996) were used to overcome dew period requirements.

Some mycoherbicide failures have been attributed to poor virulence of the pathogen. Weed pathogens have coevolved with their plant host; they are in biological balance; and if the necrotrophic fungal pathogen were hypervirulent, it would lead to self-extinction (Gressel 2001). Efforts to increase virulence of bioherbicide candidate pathogens have included interfering with host plant defense mechanisms (Sharon et al. 1992; Ahn et al. 2005), using synergies to enhance virulence (Hodgson et al. 1988; Wymore et al. 1987; Gressel 2010), genetically enhancing virulence (Tiourebaev et al. 2001; Thompson et al. 2007; Nzioki et al. 2016) and engineering hypervirulence (Sharon et al. 2001; Amsellem et al. 2002; Cohen et al. 2002a; Gressel et al. 2007; Meir et al. 2009). These, and other advances in molecular biology, may help resolve bioherbicide deficiencies and the realization of functional, commercial microbial herbicides (Gressel et al. 2007; Ash 2011).

Many well-funded bioherbicide search and development research projects were carried out in many countries in the world from the 1970s and 1980s to early 2000s, but few have matured into commercial products. George Templeton (1992a, 1992b) coined the term "orphaned mycoherbicides" for ones that provided effective weed control of their target weed but did not become commercialized due to low market potential, mass production difficulties or other concerns (Table 17.3).

Microbial herbicide success has occurred with virulent, broad host range pathogens *Chondrosterum purpureum* and *Sclerotinia* spp. Chontrol (*Chondrosterum purpureum*) is one of the few microbial herbicides available today and is used to control re-sprouting of hardwood species (Hintz 2007). *Sclerotinia minor* has been developed as the Sarritor microbial herbicide for control of dandelion, broadleaved plantain (*Plantago major* L.) and other broadleaved weeds (Abu-Dieyeh and Watson 2007; Health Canada 2010; Watson and Bailey 2013).

Government pesticide restriction and bans in Canada expedited the research and commercialization of Chontrol and Sarritor.

Sclerotinia sclerotiorum is a voracious, virulent pathogen, an ideal microbial herbicide for broadleaf weed control, but its broad host range is expanded due to the sporogenic (ascospores) phase endangering broadleaf crops (Watson 2007). In New Zealand, *S. sclerotiorum* is being used

TABLE 17.2
World List of Registered Microbial Herbicides for Cropland Weeds

Product	Microorganism	Weed Target	Reference	Present Status
Luboa	*Colletotrichum gloeosporioides* f. sp. *cuscutae* T.Y. Zhang,	*Cuscuta* spp. (dodder)	Gao and Yu 1992	Unknown, maybe local cottage industry
DeVine	*Phytophthora palmivora* (E.J. Butler) E.J. Butler	*Morrenia odorata* (Hook. & Arn.) Lindl. (strangler vine)	Ridings 1986	No longer available
Collego (reregistered as Lockdown)	*Colletotrichum gloeosporioides* f. sp. *aeschynomene* = *C. aeschynomenes* B. Weir & P.R. Johnst (ATCC 20358)	*Aeschynomene virginica* (L.) B.S.P (Northern jointvetch)	Bowers 1986; TeBeest et al. 1992 Cartwright et al. 2010	May still be available, but small market
Casst	*Alternaria cassiae* Jurair & A. Khan (NRRL #12553)	*Cassia obtusifolia* L. (sicklepod), *C. occidentalis* L. (coffee senna), *Crotalaria spectabilis* Roth. (showy crotalaria)	Bannon 1988; Walker and Riley 1982	Never commercialized
BioMal	*Colletotrichum gloeosporioides* (Penzig) Penzig & Saccardo f. sp. *malvae* Mortensen	*Malva pusilla* Sm. (roundleaf mallow)	Boyetchko et al. 2007; Mortensen 1988	Not marketed. Unable to be economically mass produced
Dr. BioSedge	*Puccinia canaliculata* (Schwein.) Lagerh.	*Cyperus esculentus* L. (yellow nutsedge)	Phatak et al. 1983	Never marketed, no production
Biochon	*Chondrostereum purpureum* (Pers.) Pouzar	*Prunus serotina* Ehrh. (black cherry)	De Jong et al. 1990	Removed from the market
StumpOut (registered)	*Cylindrobasidium laeve* (Pers.) Chamuris	*Acacia mearnsii* De Wild. (black wattle)	Morris et al. 1999	Seldom produced
Hakatak (not registered)	*Colletotrichum acutatum* J. H. Simmonds	*Hakea sericea* Schrad. & J.C. Wendl. (silky hakea)	Morris 1989	Occasionally produced on request
Camperico (JT-P482)	*Xanthomonas campestris* pv. *poae* (Pammel 1895) Dowson 1939 emend. Vauterin et al. 1995	*Poa annua* L. (annual bluegrass)	Fujimori 1999; Nishino and Tateno 2000	Not available, difficult to produce
Chontrol	*Chondrostereum purpureum* (PFC2139)	*Alnus rubra* Bong. and *A. sinuata* (Regel) Rydb. (red and sitka alders)	Becker et al. 2005; Hintz 2007	Mycologic Inc. Product available (?)
Myco-Tech Paste	*Chondrostereum purpureum* (HQ1).	Brush weeds in rights-of-ways and forest plantations	Bailey 2014	Company closed

(Continued)

TABLE 17.2 *(Continued)*
World List of Registered Microbial Herbicides for Cropland Weeds

Product	Microorganism	Weed target	Reference	Present status
Woad Warrior	*Puccinia thlaspeos* Ficinus & C. Schub. 1823 (strain woad)	*Isatis tinctoria* L. (Dyer's woad)	Kropp et al. 1996, 2002	Not marketed
Smolder	*Alternaria destruens* E.G. Simmons	*Cuscuta* spp. (dodder)	Bewick et al. 2000	Project terminated
Sarritor	*Sclerotinia minor* Jagger	*Taraxacum officinale* (L.) Weber ex F.H. Wigg (dandelion) and other broadleaf weeds	Abu-Dieyeh and Watson 2007; Health Canada 2010	Company restructuring
Wilson Lawn Bio-Phoma, Premier Tech	*Phoma macrostoma* Montagne	Broadleaf turf weeds	Bailey and Falk 2011; Bailey 2014	Market pending, manufacturing difficulties
SolviNix Strain	*Tobamovirus*, Group IV ((+) ss	*Solanum viarum* Dunal (tropical soda	Charudattan and Hiebert 2007	Commercial product
U2	RNA), Virgaviridae, Tobacco Mild Green Mosaic Tobamovirus	apple)	Charudattan 2016	available

to control Canada thistle (*Cirsium arvense* L.) in pastures, and risk analysis simulates the dispersal of ascospores (de Jong et al. 2002). Safety zones for susceptible horticultural crops away from a *S. sclerotiorum*-based mycoherbicide treatment have been determined regarding variations in regional and yearly climate (Bourdôt et al. 2006).

Charles Wilson's (1970) "commencement" paper, "Plant Pathogens in Weed Control", mentioned *Sclerotium rolfsii* Sacc., another virulent, aggressive, broad host range crop pathogen, could be considered a biocontrol agent. Several groups have reported bioherbicide research interest with *Sclerotium rolfsii*: Mishra et al. (1995) for the control of *Parthenium* in India; Tang et al. (2011) for broadleaf weed control in dry, direct-seeded rice fields; and Gibson et al. (2014) for control of swallowworts (*Vincetoxicum* spp.) in eastern North America. In Australia, three virulent, broad host range, destructive fungi, *Lasiodiplodia pseudotheobromae*, *Neoscytalidium novaehollandiae* and *Macrophomina phaseolina*, are combined in a capsule that is injected into *Parkinsonia* shrub trunks (Cripps 2017). Weed control should be very good with these pathogens, but regulatory acceptance may be challenging. This Australian effort is reminiscent of reports of a local research foundation providing farmers with spores of the fungus *Acremonium* (*Cephalosporium*) *diospyri* (Crand.) W. Gams for application to cut stumps for control of common persimmon (*Diospyros virginiana* L.), an invasive weed in Oklahoma grasslands (Wilson 1965; Griffith 1970).

Early on, the bioherbicide industry indicated major efforts were needed in mass production, formulation and delivery technologies before commercialization of additional bioherbicides could occur (Zorner et al. 1993). A "bioherbicide innovation chain" was proposed by Bailey et al. (2009) to assist researchers and industry to work together to increase microbial herbicide product commercialization. The nine-step process, from discovery to technology adoption, requires involvement of scientists, market experts and a solid industrial partner to link research activities with business models. Future microbial herbicide projects are encouraged to follow the bioherbicide innovation chain proposed and tested by Bailey et al. (2009).

17.6 BACTO-HERBICIDES

Interest in using soil-borne phytopathogenic bacteria for biological weed control has been strong (Johnson et al. 1996; Kremer and Kennedy 1996), but commercial success of a bacterial herbicide

TABLE 17.3
Orphaned Microbial Herbicide Candidates Targeting Crop Weeds

Code	Pathogen Agents	Weed Targets	References
VELGO	*Colletotrichum coccodes* (Wallr.) S. Hughes	*Abutilon theophrasti* Medik (velvetleaf)	Wymore et al. 1988; DiTommaso and Watson 1995
IMI 48942	*Colletotrichum orbiculare* (Berk. and Mont.) v. Arx	*Xanthium spinosum* L. (Bathurst burr).	Auld and Say 1999; Chittick and Auld 2001
NRRL 13737	*Colletotrichum truncatum* (Schw.) Andrus et Moore	*Sesbania exaltata* (Raf.) Rydb. ex A.W. Hill. (hemp sesbania)	Jackson and Bothast 1990; Jackson and Schisler 1995; Boyette et al. 2007
Myco-herb	*Lewia chlamidosporiformans* B.S. Vieira & R.W. Barreto	*Euphorbia heterophylla* L. (wild poinsettia)	Vieira et al. 2008; Vieira and Barreto 2010
MTB-951	*Drechslera monoceras* (Drechsler) Subram. et Jain (=*Exserohilum monoceras* [Drechsler] Leonard et Suggs)	*Echinochloa crus-galli* (L.) P. Beauv. (barnyardgrass)	Fujimori 1999; Hirase et al. 2004, 2006
JTB-808	*Exserohilum monoceras* (Drechsler) K. J. Leonard & Suggs	*Echinochloa crus-galli* (barnyardgrass)	Tsukamoto et al. 1997, 1998, 2001
QZ-2000	*Curvularia eragrostidis* (Henn.) J.A. Mey.	*Digitaria sanguinalis* (L.) Scop. (large crabgrass)	Zhu and Qiang 2004; Wang et al. 2013
FOXY 2, M12-4A, PSM197	*Fusarium oxysporum* Schlecht. emend. Snyder & Hansen f. sp. *strigae* Elzein & Thines	*Striga hermonthica* (Del.) Benth. (witchweed)	Ciotola et al. 1995; Marley et al. 1999; Elzein and Kroschel. 2004; Venne et al. 2009; Watson 2013
FOG	*Fusarium oxysporum* Schltdt.	*Phelipanche ramosa* (L.) Pomel (branched broomrape).	Müller-Stöver et al. 2009a; Kohlschmid et al. 2009
FT2	*Fusarium oxysporum*	*P. ramosa*	Boari and Vurro 2004; Cipriani et al. 2009
FOXY	*Fusarium oxysporum*	*Phelipanche aegyptiaca* (Pers.) Pomel (Egyptian broomrape), *P. ramosa*, *Orobanche cernua* Loefl. (nodding broomrape)	Amsellem et al. 2001a, 2001b; Cohen et al. 2002b
FARTH	*Fusarium arthrosporioides* Sherb.	*P. aegyptiaca*, *P. ramosa*, *O. cernua*	Amsellem et al. 2001a, 2001b; Cohen et al. 2002b
FOO	*Fusarium oxysporum* Schlecht. f. sp. *orthoceras* (Appel & Wollenw) Bilay	*Orobanche cumana* Wallr (sunflower broomrape), *O. cernua*, *P. aegyptiaca*	Thomas et al. 1998; Müller-Stöver et al. 2004, 2009b

has not been realized (Barreto et al. 2012). Several deleterious rhizobacteria (DRB), *Pseudomonas fluorescens* strain D7, *Pseudomonas fluorescens* strain BRG100, *Pseudomonas fluorescens* strain G2-11 and *Pseudomonas trivialis* X33d are being evaluated for weed control. DRB suppress weed seed germination and early growth of the weed and function as plant growth promoting rhizobacteria (PGPR) favoring crop plant development to the detriment of weed growth. *Pseudomonas fluorescens* strain D7 was shown to suppress downy brome (*Bromus tectorum* L.) infesting winter wheat crops in the Pacific Northwest U.S. (Kennedy et al. 1991, 2001). *Pseudomonas fluorescens* strain BRG100 adversely affects germination and root growth of green foxtail (*Setaria viridis* (L.) P. Beauv. and wild oat (*Avena fatua* L.) (Caldwell et al. 2012). Economic and technical analyses of large scale production of pre-emergent *Pseudomonas fluorescens* microbial bioherbicide for green foxtail and wild oat control were conducted in Canada to support bioherbicide research and development investment and commercialization strategies (Mupondwa et al. 2015). *Pseudomonas fluorescens* strain G2-11 was isolated from roots of giant foxtail (*Setaria faberi* Herrm.), and herbicidal performance was affected by formulation and soil properties (Zdor et al. 2007). A semolina-kaolin granular formulation (Pesta) improved *Pseudomonas trivialis* X33 biocontrol of ripgut brome (*Bromus diandrus* Roth) in durum wheat (Mejri et al. 2012).

One plant pathogenic bacteria, *Xanthomonas campestris* pv. *poae* P-482, Camperico, was commercialized by Japan Tobacco Ltd as a microbial bioherbicide for the control of the turfgrass weed, annual bluegrass (*Poa annua* L.) in Japan (Fujimori 1999; Imaizumi et al. 1997; Nishino and Tateno 2000). Research with *Xanthomonas campestris* pv. *poannua* was also conducted in the U.S. (Zhou and Neal 1995), but Camperico is not available in Japan nor elsewhere. Even though, another isolate of *Xanthomonas campestris* has recently been studied in the United States for the control of horseweed (*Conyza canadensis* (L.) Cronquist) as a major glyphosate herbicide-resistant weed in limited and no tillage cropping systems (Boyette and Hoagland 2015).

17.7 VIRAL HERBICIDES

Plant viruses have seldom been evaluated as potential microbial herbicides as they are not ideal candidates for biocontrol agents, due to their general broad host ranges and their need for vectors like surface abrasion or injection/transmission of viral particles by insects, fungi or nematodes into host plant cells. Nevertheless, one plant virus, tobacco mild green mosaic tobamovirus (TMGMV), *Tobamovirus,* Group IV ((+) ss RNA), Virgaviridae, SolviNix, was recently registered as a microbial herbicide for the control of tropical soda apple, *Solanum viarum* Dunal (Solanaceae) (EPA 2015; Charudattan 2016).

Solanum viarum is native to southeastern Brazil, northeastern Argentina, Paraguay and Uruguay but recently arrived in the USA, becoming a serious, invasive weed of rangeland in Florida (Medal et al. 2012). Once introduced, *S. viarum* rapidly invades cropland, establishing large impenetrable, monotypic stands (Charudattan and Hiebert 2007). SolviNix is usually spot-spray applied with high-pressure sprayers, providing excellent control of tropical soda apple (Ferrell et al. 2008). SolviNix can be mixed with herbicides to control other weeds as well. The commercialization of SolviNix by BioProdex Inc. is another example of a university–private company collaboration for successful microbial herbicide development. BioProdex is partnering with a Brazilian company interested in applying SolviNix for native Solanaceae weeds, and expansion into other countries is planned.

17.8 NEMATODA HERBICIDES

17.8.1 Silverleaf Nightshade

Silverleaf nightshade (*Solanum elaeagnifolium* Cav.) is an important native perennial weed in western U.S., and this weed has also invaded Australia, India and South Africa (Parker 1991a). Silverleaf nightshade plants are commonly parasitized by a leaf and stem galling nematode (*Orrina phyllobia* (Thorne) Brezeski, Anguinidae: Nematoda). Adult nematodes and larvae infect and initiate gall formation in fresh juvenile leaves and stems (Notham and Orr 1982). The galled leaves and stems soon become dry and are abscised. Adults die, but second generation infective larvae enter a state of anhydrobiosis and can remain in that state for several years. Dried galled plant debris can simply be collected, and large numbers of infective larvae can be readily distributed to other silver nightshade populations. With the arrival of moisture, the anhydrobiosis state is overcome, and infective larvae search for nightshade shoots to invade. The nematode can reduce the biomass and density of silverleaf nightshade. A government-funded pilot project developed and implemented the mass rearing of *O. phyllobia*, but when compared to other weed control options, mass rearing was not competitive (Parker 1991a, 1991b). Perhaps with improved, cost-effective mass rearing technology, this augmentation tactic could contribute to an integrated weed management approach for silverleaf nightshade suppression.

17.8.2 Russian Knapweed

Russian knapweed, *Acroptilon repens* (L) DC [*Rhaponticum repens* (L.) Hidalgo], Asteraceae, is a deep-rooted, aggressive perennial, native to Eurasia, that rapidly spreads, forming persistent dense monotypic stands degrading native range and crop land. Russian knapweed arrived in Canada as a contaminant of Turkestan alfalfa seed in the early 1900s and was recorded in the U.S. in California in 1910. Russian knapweed infestations are common in western Canada, widespread in the western and central regions of the U.S., and problematic in Afghanistan, Argentina, Australia, India, Iran, Turkey and South Africa (http://www.cabi. org/isc/datasheet/2946). Biological control research activities on Russian knapweed involved collaboration amongst Canadian and Soviet scientists evaluating several insects and one nematode, as potential biocontrol agents (Watson and Harris 1984). The stem-gall nematode, *Subanguina picridis* (Kirj.) Brezeski (Anguinidae: Nematoda), was shown to be host limited and damaging to Russian knapweed and was released on Russian knapweed infestations in Canada and in the United States (Watson and Harris 1984). The life cycle is almost indistinguishable from the silverleaf nightshade nematode, including the anhydrobiosis capabilities of the infective larvae, facilitating the collection of galls from infected plants and re-distribution to additional sites. Soviet scientists developed a process to extract larvae from field collected galls and prepared water suspensions for sprayer applications (Kovalev et al. 1973). In efforts to augment this biocontrol organism in the United States, nematodes extracted from galls were encapsulated in calcium alginate granules, oil coated, dried and frozen (–20°C) to provide nine months' shelf life (Caesar-Ton That et al. 1995). *Subanguina* fecundity and gall numbers were low, limiting distribution. To resolve this problem, an *in vitro* mass culture system on callus, excised roots and shoot tissues of *Acroptilon repens* was developed (Ou and Watson 1992). Mass cultured *S. picridis* were virulent and rapidly increased in population size. In three months, the initial 50 larvae increased to 7,000–10,000 per petri dish, a 140- to 200-fold increase (Ou and Watson 1993). Mass rearing of *S. picridis* is achievable, but there has been limited interest in commercialization.

17.9 SUSTAINABILITY, SAFETY, HAZARDS AND RISKS OF MICROBIAL HERBICIDES

Webster's New World dictionary defines sustainability as the ability to "keep in existence, keep up, maintain or prolong" (Neufeldt 1988). The goal of microbial herbicide research is development of commercially acceptable weed control products that effectively suppress weed growth and promote crop growth. Microbial herbicides must be economically produced, formulated with lengthy shelf life and perform consistently under field conditions. Microbial herbicides would seemingly be sustainable in terms of human health, environment pollution and social aspects, but not in economic terms presently, as there has been no successful, widely marketed, microbial herbicide to date.

Plant pathogens used as microbials may cause risks to non-target organisms, including plants, animals, microbes and humans, that must be rigorously scrutinized (Hoagland et al. 2007; Saharan and Mehta 2008; Bailey et al. 2013). All microorganisms proposed as microbial herbicides must be tested and registered for use following various national or regional guidelines for registration as microbial pest control products. Prior to marketing and using a microbial herbicide, Australian Pesticides and Veterinary Medicines Authority (APVMA), Health Canada's Pest Management Regulatory Agency (PMRA), United States Environmental Protection Agency (EPA) and Organization for Economic Co-operation and Development (OECD 2003) agencies rigorously evaluate the proposed microbial herbicide to ensure that its use will not pose unreasonable risks or harm to human health and the environment. Also see Kabaluk et al. (2010) to view the regulation of microbial pesticides in representative jurisdictions worldwide. Once released, microbial herbicides should be monitored for stability, impact and persistence. Genetic markers have been developed for risk assessment and monitoring persistence of both registered broad host range microbial herbicides, Chontrol (Hintz et al. 2001) and Sarritor (Pan et al. 2010).

17.10 CONCLUDING REMARKS

Classical biological weed control with fungal plant pathogens has provided good control of several dominant invasive weeds in grassland ecosystems. Success with microbial herbicides has generally been limited to non-agricultural systems, and no bioherbicide product has been developed for a major crop weed. Many microorganisms have been studied, mass-produced, formulated and field tested and controlled target weeds but were deemed commercially unacceptable. Herbicide resistance, absence of new chemistries, government restrictions, public pressure and expansion of organic agriculture support the need for non-chemical weed control. Future business models, innovative ideas and collaborative efforts will expand the prospects for microbial herbicides, such as witchweed biocontrol on a toothpick (Nzioki et al. 2016).

REFERENCES

Abu-Dieyeh, M. and A.K. Watson. 2007. Efficacy of *Sclerotinia minor* for dandelion control: Effect of dandelion accession, age and grass competition. *Weed Res.* 47: 63–72.

Ahn, B., T. Paulitz, S. Jabaji-Hare and A.K. Watson. 2005. Enhancement of *Colletotrichum coccodes* virulence by inhibitors of plant defence mechanisms. *Biocontrol. Sci. Technol.* 15(3): 299–308.

Amsellem, Z., S. Barghouthi, B. Cohen, Y. Goldwasser, J. Gressel, L. Hornok, Z. Kerenyi, Y. Kleifeld, O. Klein, J. Kroschel, J. Sauerborn, D. Müller-Stöver, H. Thomas, M. Vurro and M.C. Zonno. 2001a. Recent advances in the biocontrol of *Orobanche* (broomrape) species. *Biol. Control.* 46: 211–228.

Amsellem, Z., B.A. Cohen and J. Gressel. 2002. Engineering hypervirulence in a mycoherbicidal fungus for efficient weed control. *Nat. Biotechnol.* 20: 1035–1039.

Amsellem, Z., Y. Kleifeld, Z. Kerenyi, L. Hornok, Y. Goldwasser and J. Gressel. 2001b. Isolation, identification and activity of mycoherbicidal pathogens from juvenile broomrape plants. *Biol. Control.* 21: 274–284.

Aneja, K.R., V. Kumar, P. Jiloha, M. Kaur, C. Sharma, P. Surain, R. Dhiman and A. Aneja. 2013. Potential bioherbicides: Indian perspectives. pp. 197–215. *In:* R.K. Salar, S.K. Gahlawat, P. Siwach and J.S. Duhan (Eds.) *Biotechnology: Prospects and Applications.* © Springer, India, New Delhi.

Ash, G.J. 2010. The science, art and business of successful bioherbicides. *Biol. Control.* 52: 230–240.

Ash, G.J. 2011. Biological control of weeds with mycoherbicides in the age of genomics. *Pest Technol.* 5(Special Issue 1): 41–47. © Global Science Books.

Auld, B.A. 1993. Vegetable oil suspension emulsion reduces dew dependence of a mycoherbicide. *Crop. Prot.* 12: 477–479.

Auld, B.A. 2000. Success in biological control of weeds by pathogens, including bioherbicides. pp. 323–340. *In:* G. Gurr and S. Wratten (Eds.) *Biological Control: Measures of Success.* Kluwer Academic Publishers, Dordrecht, Netherlands.

Auld, B.A. and L. Morin. 1995. Constraints in the development of bioherbicides. *Weed Technol.* 9(3): 638–652.

Auld, B.A. and M.M. Say. 1999. Comparison of isolates of *Colletotrichum orbiculare* from Argentina and Australia as potential bioherbicides for *Xanthium spinosum* in Australia. *Agric. Ecosyst. Environ.* 72: 53–58.

Bailey, K.L. 2010. Canadian innovations in microbial biopesticides. *Can. J. Plant Pathol.* 32: 113–121.

Bailey, K.L. 2014. The bioherbicide approach to weed control using plant pathogens. pp. 245–266. *In:* D.P. Abrol (Ed.) *Integrated Pest Management: Current Concepts and Ecological Perspective.* © 2013 Elsevier Inc., London, UK.

Bailey, K.L., S.M. Boyetchko and T. Langle. 2010. Social and economic drivers shaping the future of biological control: A Canadian perspective on the factors affecting the development and use of microbial biopesticides. *Biol. Control.* 52(3): 221–229.

Bailey, K.L., S.M. Boyetchko, G. Peng, R.K. Hynes, W.G. Taylor and W.M. Pitt. 2009. Developing weed control technologies with fungi. pp. 1–44. *In:* R.K. Mahendrah (Ed.) *Current Advances in Fungal Technology.* I.K. International, New Delhi, India.

Bailey, K.L. and S. Falk. 2011. A Phoma story. *Pest Technol.* 5(Special Issue 1): 73–79. Global Science Books.

Bailey, V.L., S.J. Fansler, J.C. Stegen and L.A. McCue. 2013. Linking microbial community structure to β-glucosidic function in soil aggregates. *International Society for Microbial Ecology.* 7(10): 2044–2053.

Bannon, J.S. 1988. CASST™ herbicide (*Alternaria cassiae*), A case history of a mycoherbicide. *Am. J. Alt. Agr.* 3(2–3): 73–76.

Barreto, R.W. 2007. Latin American weed biological control science at the crossroads. pp. 109–121. *In:* M.H. Julien, R. Sforza, M.C. Bon, H.C. Evans, P.E. Hatcher, H. Hinz and B.G. Rector (Eds.) *Proceedings XII International Symposium on Biological Control of Weeds*. CABI, La Grande Motte, France.

Barreto, R.W., C.A. Ellison, M.K. Seier and H.C. Evans. 2012. Biological control of weeds with plant pathogens: Four decades on. pp. 299–350. *In:* D.P. Abrol and U. Shankar (Eds.) *Integrated Pest Management: Principles and Practice*. CABI, Wallingford, Oxfordshire, UK.

Barton, J. 2004. How good are we at predicting the field host-range of fungal pathogens used for classical biological control of weeds? *Biol. Control*. 31: 99–122.

Barton, J. 2005 November. Bioherbicides: All in a day's work . . . for a superhero. pp. 4–6. *In: What's New in Biological Control of Weeds?* Issue 34. Manaaki Whenua, Landcare Research, Lincoln, New Zealand.

Becker, E., S.F. Shamoun and W.E. Hintz. 2005. Efficacy and environmental fate of *Chondrostereum purpureum* used as a biological control for red alder (*Alnus rubra*). *Biol. Control*. 33: 269–277.

Bewick, T.A., J.C. Porter and R.C. Ostrowski. 2000. Smolder™: A bioherbicide for suppression of dodder (*Cuscuta* spp.). *Proc. Southern Weed Sci. Soc. Abstracts*. 53: 152.

Boari, A. and M. Vurro. 2004. Evaluation of *Fusarium* spp. and other fungi as biological control agents of broomrape (*Orobanche ramosa*). *Biol. Control*. 30: 212–219.

Bourdôt, G.W., D. Baird, G. Hurrell and M.D. de Jong. 2006. Safety zones for a *Sclerotinia sclerotiorum*-based mycoherbicide: Accounting for regional and yearly variation in climate. *Biocontrol. Sci. Techn.* 16: 345–358.

Bowers, R.C. 1986. Commercialization of CollegoTM – an industrialist's view. *Weed Sci.* 34(Suppl.): 24–25.

Boyetchko, S.M., K.L. Bailey, R.K. Hynes and G. Peng. 2007. Development of Bio Mal. pp. 274–283. *In:* C. Vincent, M.S. Goettel and G. Lazarovits (Eds.) *Biological Control: A Global Perspective*. CABI, Wallingford, UK.

Boyetchko, S.M. and G. Peng. 2004. Challenges and strategies for development of mycoherbicides. pp. 111–121. *In:* D.K. Arora (Ed.) *Fungal Biotechnology in Agricultural, Food, and Environmental Applications*. Marcel Dekker, New York, USA.

Boyette, C.D. and R.E. Hoagland. 2015. Bioherbicidal potential of *Xanthomonas campestris* for controlling *Conyza canadensis*. *Biocontrol Sci. Techn.* 25: 229–237.

Boyette, C.D., R.E. Hoagland and M.A. Weaver. 2007. Biocontrol efficacy of *Colletotrichum truncatum* for hemp sesbania (*Sesbania exaltata*) is enhanced with unrefined corn oil and surfactant. *Weed Biol. Mang.* 7(1): 70–76.

Boyette, C.D., P.C. Jr. Quimby, A.J. Caesar, J.L. Birdsall, W.J. Jr. Connick, D.J. Daigle, M.A. Jackson, G.E. Eagley and H.K. Abbas. 1996. Adjuvants, formulations and spraying systems for improvement of mycoherbicides. *Weed Technol.* 10: 637–644.

Boyette, C.D., P.C. Jr. Quimby, W.J. Jr. Connick, D.J. Daigle and F.E. Fulgham. 1991. Progress in the production, formulation and application of mycoherbicides. pp. 209–222. *In:* D.O. TeBeest (Ed.) *Microbial Control of Weeds*. Chapman & Hall, New York, USA.

Bruckart, W.L. 2005. Supplemental risk evaluations and status of *Puccinia carduorum* for biological control of musk thistle. *Biol. Control*. 32: 348–355.

Burdon, J.J., P.H. Thrall, P.H. Groves and P. Chaboudez. 2000. Biological control of *Carduus pycnocephalus* and *C. tenuiflorus* using the rust fungus *Puccinia cardui-pycnocephali*. *Plant Prot. Q.* 15: 14–17.

Burnett, H.C., D.P.H. Tucker, M.E. Patterson and W.H. Ridings. 1973. Biological control of milkweed vine with a race of *Phytophthora citrophthora*. *Proc. FL. State Hortic. Soc.* 85: 111–115.

Burnett, H.C., D.P.H. Tucker and W.H. Ridings. 1974. *Phytophthora* root and stem rot of milkweed vine. *Plant Dis. Rep.* 58: 355–357.

Caesar-Ton That, T.C., W.E. Dyer, P.C. Quimby and S.S. Rosenthal. 1995. Formulation of an endoparasitic nematode *Subanguina picridis* Brezeski, a biocontrol agent for Russian knapweed [*Acroptilon repens* (L.) DC.]. *Biol. Control*. 5(2): 262–266.

Cai, X. and M. Gu. 2016. Bioherbicides in organic horticulture. *Horticulturae*. 2: 1–10.

Caldwell, C.J., R.K. Hynes, S.M. Boyetchko and D.R. Korber. 2012. Colonization and bioherbicidal activity on green foxtail by *Pseudomonas fluorescens* BRG100 in a pesta formulation. *Can. J Microbiol.* 58(1): 1–9.

Cartwright, K., D. Boyette and M. Roberts. 2010. Lockdown: Collego bioherbicide gets a second act. *Phytopathology*. 100: S162.

Charudattan, R. 1991. The mycoherbicide approach with plant pathogens. pp. 24–57. *In:* D.O. TeBeest (Ed.) *Microbial Control of Weeds*. Chapman & Hall, New York, USA.

Charudattan, R. 2005. Ecological, practical, and political inputs into selection of weed targets: What makes a good biological control target? *Biol. Control*. 5(3): 183–196.

Charudattan, R. 2016. SolviNix LC, the first registered bioherbicide containing a plant virus as the active ingredient (Abstract 99). Proceedings 7th International Weed Science Congress, Prague.

Charudattan, R. and A. Dinoor. 2000. Biological control of weeds using plant pathogens: Accomplishments and limitations. *Crop. Prot.* 19(8–10): 691–695.

Charudattan, R. and E. Hiebert. 2007. A plant virus as a bioherbicide for tropical soda apple, *Solanum viarum. Outlooks Pest Manage.* 18: 167–171.

Chittick, A.T. and B.A. Auld. 2001. Polymers in bioherbicide formulation: *Xanthium spinosum* and *Colletotrichum orbiculare* as a model system. *Biocontrol. Sci. Techn.* 11(6): 691–702.

Chutia, M., J.J. Mahanta, N. Bhattacharyya, M. Bhuyan, P. Boruah and T.C. Sarma. 2007. Microbial herbicides for weed management: Prospects, progress and constraints. *Plant Pathol. J.* 6(3): 210– 218.

Ciotola, M., A.K. Watson and S.G. Hallett. 1995. Discovery of an isolate of *Fusarium oxysporum* with potential to control *Striga hermonthica* in Africa. *Weed Res.* 35: 303–309.

Cipriani, M.G., G. Stea, A. Moretti, C. Altomare, G. Mulè and M. Vurro. 2009. Development of a PCR-based assay for the detection of *Fusarium oxysporum* strain FT2, a potential mycoherbicide of *Orobanche ramosa. Biol. Control.* 50: 78–84.

Cohen, B.A., Z. Amsellem, S. Lev-Yadun and J. Gressel. 2002b. Infection of tubercles of the parasitic weed *Orobanche aegyptiaca* by mycoherbicidal *Fusarium* species. *Ann. Bot.* 90: 567–578.

Cohen, B.A., Z. Amsellem, R. Maor, A. Sharon and J. Gressel. 2002a. Transgenically enhanced expression of indole-3-acetic acid confers hypervirulence to plant pathogens. *Phytopathology*. 92(6): 590–596.

Connick, W.J. Jr., C.D. Boyette and J.R. McAlpine. 1991. Formulations of mycoherbicides using a pesta like process. *Biol. Control.* 1: 281–287.

Cordeau, S., M. Triolet, S. Wayman, C. Steinberg and J.P. Guillemin. 2016. Bioherbicides: Dead in the water? A review of the existing products for integrated weed management. *Crop. Prot.* 87: 44–49.

Cother, E.J. 1996. Bioherbicides and weed management in Asian rice fields. pp. 183–200. *In*: R. Naylor (Ed.) *Herbicides in Asian Rice: Transitions in Weed Management*. Institute for International Studies, Stanford University and Manila (Philippines): International Rice Research Institute, Palo Alto, CA, 270 pp.

Cripps, S. 2017. *New Parkinsonia Bioherbicide Control Demonstrated on Aramac Property*. http://www.northqueenslandregister.com.au/story/595596/graziers-gunning-for-parkinsonia

Cullen, J.M. 1985. Bringing the cost benefits of analysis of biological control of *Chondrilla juncea* up to date. pp. 145–152. *In:* E.S. Delfosse (Ed.) *Proceedings IV Symposium on Biological Control of Weeds*. Agriculture Canada, Vancouver.

Cullen, J.M., P.F. Kable and M. Catt. 1973. Epidemic spread of a rust imported for biological control. *Nature*. 244: 462–464.

Daniel, J.T., G.E. Templeton, R.J. Jr. Smith and W.T. Fox. 1973. Biological control of northern jointvetch in rice with an endemic fungal disease. *Weed Sci.* 21: 303–307.

Day, M.D. and R.L. Winston. 2016. Biological control of weeds in the 22 Pacific island countries and territories: Current status and future prospects. *Neo Biota*. 30: 167–192.

De Datta, S.K. and A.M. Baltazar. 1996. Integrated weed management in rice in Asia. pp. 145–165. *In:* R. Naylor (Ed.) *Herbicides in Asian Rice: Transitions in Weed Management*. Institute for International Studies, Stanford University and Manila (Philippines): International Rice Research Institute, 270 pp.

De Jong, M.D., G.W. Bourdôt, G.A. Hurrell and D.J. Saville. 2002. Risk analysis for biological weed control—simulating dispersal of *Sclerotinia sclerotiorum* (Lib.) de Bary ascospores from a pasture after biological control of *Cirsium arvense* (L.) Scop. *Aerobiologia*. 18: 211–222.

De Jong, M.D., P.C. Scheepens and J.C. Zadoks. 1990. Risk analysis for biological control: A Dutch case study in biocontrol of *Prunus serotina* by the fungus *Chondrostereum purpureum*. *Plant Dis.* 74: 189–194.

DiTommaso, A. and A.K. Watson. 1995. Impact of a fungal pathogen, *Colletotrichum coccodes* on growth and competitive ability of *Abutilon theophrasti*. *New Phytol.* 131: 51–60.

Dowson, W.J. 1939. *On the Systematic Position and Generic Names of the Gram-Negative Bacterial Plant Pathogens*. Zentralblatt fur Bakteriologie, Parasitenkunde, Infektionskrankheiten und Hygiene, no. 9–13, pp. 177–193.

Elzein, A. and J. Kroschel. 2004. *Fusarium oxysporum* Foxy 2 shows potential to control both *Striga hermonthica* and *S. asiatica*. *Weed Res.* 44: 433–438.

Elzein, A., J. Kroschel and G. Cadisch. 2008. Efficacy of Pesta granular formulation of Striga-mycoherbicide *Fusarium oxysporumf.* sp. *strigae* Foxy 2 after 5-year of storage. *J. Plant Dis. Prot.* 115(6): 259–262.

EPA (Environmental Protection Agency). 2015. *Biopesticides Registration Action Document: Tobacco Mild Green Mosaic Tobamovirus Strain U2. PC Code: 056705*. United States Environmental Protection Agency.

Evans, H.C. 1995. Fungi as biocontrol agents of weeds: A tropical perspective. *Can. J. Bot.* 73(Suppl. 1): S58–S64.

Evans, H.C. 2013. Biological control of weeds with fungi. pp. 145–172. *In:* F. Kempken (Ed.) *Agricultural Applications, The Mycota XI*, 2nd Edition. Springer Verlag, Berlin Heidelberg.

Evans, H.C., M.P. Greaves and A.K. Watson. 2001. Fungal biocontrol of weeds. pp. 169–192. *In:* T.M. Butt, C.W. Jackson and N. Magan (Eds.) *Fungi as Biocontrol Agents: Progress, Problems and Potential.* CABI Publishing, CABI, Wallingford, Oxfordshire, UK.

Evans, K.J., L. Morin, E. Bruzzese and R.T. Roush. 2004. Overcoming limits on rust epidemics in Australian infestations of European blackberry. p. 514. *In:* J.M. Cullen, D.T. Briese, D.J. Kriticos, W.M. Lonsdale, L. Morin and J.K. Scott (Eds.) *The Proceedings of 9th International Symposium on Biological Control of Weeds.* CSIRO, Canberra, Australia.

Ferrell, J., R. Charudattan, M. Elliott and E. Hiebert. 2008. Effects of selected herbicides on the efficacy of *Tobacco mild green mosaic virus* to control tropical soda apple (*Solanum viarum*). *Weed Sci.* 56: 128–132.

Frohlich, J., S.V. Fowler, A. Gianotti, R.L. Hill, E. Killgore, L. Morin, L. Sugiyama and C. Winks. 1999. Biological control of mist flower (*Ageratina riparia*, Asteraceae) in New Zealand. pp. 6–11. *In:* M. O'Callaghan (Ed.) *The Proceedings of 52nd N.Z. Plant Protection Conference.* The New Zealand Plant Protection Society Inc.

Fujimori, T. 1999. New developments in plant pathology in Japan. *Australas. Plant Pathol.* 28: 292– 297.

Gaddeyya, G., G. Easteru Rani, B. Susmitha and K. Subhashini. 2017. Microbial technology in weed management: A special reference of biological control of horse purslane weed. *Int. J. Current Adv. Res.* 6(7): 4978–4991.

Gao, Z.Y. and J.E. Yu. 1992. Biological control of *Cuscuta* spp. with Lubao #1, a product formulated from *Colletotrichum* sp. *Chin. J. Biocontrol.* 8(4): 173–175.

Gibson, D.M., R.H. Vaughan, J. Biazzo and L.R. Milbrath. 2014. Exploring the feasibility of *Sclerotium rolfsii* VrNY as a potential bioherbicide for control of swallowworts (*Vincetoxicum* spp.). *Invasive Plant Sci. Manage.* 7(2): 320–327.

Green, S., S.M. Stewart-Wade, G.J. Boland, M.P. Teshler and S.H. Liu. 1998. Formulating microorganisms for biological control of weeds. pp. 249–281. *In:* G.J. Boland and L.D. Kuykendall (Eds.) *Plant-Microbe Interactions and Biological Control.* Marcel Dekker, New York.

Gressel, J. 2001. Potential fail safe mechanisms against the spread and introgression of transgenic hypervirulent biocontrol fungi. *Trends Biotechnol.* 19: 149–154.

Gressel, J. 2010. Herbicides as synergists for mycoherbicides, and vice versa. *Weed Sci.* 58: 324–328.

Gressel, J., S. Meir, Y. Herschkovitz, H. Al-Ahmad, I. Greenspoon, O. Babalola and Z. Amsellem. 2007. Approaches to and successes in developing transgenically enhanced mycoherbicides. pp. 297–305. *In:* M. Vurro and J. Gressel (Eds.) *Novel Biotechnologies for Biocontrol Agent Enhancement and Management.* Springer, Dordrecht.

Griffith, C.A. 1970. *Persimmon wilt research. Annual Report 1960–1970.* Noble Foundation Agricultural Division, Ardmore, OK.

Harding, D.P. and M.H. Raizada. 2015. Controlling weeds with fungi, bacteria and viruses: A review. *Front. Plant Sci.* 6: 659. Published online 2015 August 28. doi: 10.3389/fpls.2015.00659.

Health Canada. 2010. *Sclerotinia minor* strain IMI 344141. Registration decision RD 2010-08. 22 September 2010. Pest Management Regulatory Agency. Health Canada HC Pub: 100362. IBBN: 978-1-100-16450-2 (978-1-100-16451-9), 6 p.

Hershenhorn, J., F. Casella and M. Vurro. 2016. Weed biocontrol with fungi: Past, present and future. *Biocontrol. Sci. Techn.* 26(10): 1313–1328.

Hintz, W.E. 2007. Development of *Chondrostereum purpureum* as a mycoherbicide for deciduous brush control. pp. 284–290. *In*: C. Vincent, M.S. Goettel and G. Lazarovits (Eds.) *Biological Control: A Global Perspective.* CAB International, Wallingford, UK, 432 p.

Hintz, W.E., E.M. Becker and S.F. Shamoun. 2001. Development of genetic markers for risk assessment of biological control agents. *Can. J. Plant Pathol.* 23(1): 13–18.

Hirase, K., M. Nishida and T. Shinmi. 2006. Effects of lodging of *Echinochloa crusgalli* L. on the herbicidal efficacy of MTB-951, a mycoherbicide using *Drechslera monoceras* (Drechsler) Subram. et Jain (=*Exserohilum monoceras* [Drechsler] Leonard et Suggs). *Weed Biol. Manage.* 6(1): 30–34.

Hirase, K., S. Yoshigai, M. Nishida, Z. Takanaka and T. Shinmi. 2004. Influence of water management, application timing and temperature on efficacy of MTB-951, a mycoherbicide using *Drechslera monoceras* to control *Echinochloa crus-galli* L. *Weed Biol. Manage.* 4(2): 71–74.

Hoagland, R., C. Boyette, M. Weaver and H. Abbas. 2007. Bioherbicides: Research and risks. *Toxin Rev.* 26: 313–342.

Hodgson, R.H., L.A. Wymore, A.K. Watson, R. Snyder and A. Collette. 1988. Efficacy of *Colletotrichum coccodes* and thidiazuron for velvetleaf (*Abutilon theophrasti*) control in soybean (*Glycine max*). *Weed Technol.* 2: 473–480.

Imaizumi, S., T. Nishino, K. Miyabe, T.M. Fujimori and M. Yamada. 1997. Biological control of annual bluegrass (*Poa annua* L.) with a Japanese isolate of *Xanthomonas campestris* pv. *poae* (JT-P482). *Biol. Control.* 8: 7–14.

Jackson, M.A. and R.J. Bothast. 1990. Carbon concentration and carbon-to-nitrogen ratio influence submerged-culture conidiation by the potential bioherbicide *Colletotrichum truncatum* NRRL 13737. *Appl. Environ. Microbiol.* 56(11): 3435–3438.

Jackson, M.A. and D.A. Schisler. 1995. Liquid culture production of microsclerotia of *Colletotrichum truncatum* for use as bioherbicidal propagules. *Mycol. Res.* 99(7): 879–884.

Johnson, D.R., D.L. Wyse and K.L. Jones. 1996. Controlling weeds with phytopathogenic bacteria. *Weed Technol.* 10: 621–624.

Julien, M., R. McFadyen and J. Cullen (Eds.) 2012. *Biological Control of Weeds in Australia.* CSIRO Publishing, Melbourne, Australia, 620 p.

Kabaluk, J.T., A.M. Svircev, M.S. Goettel and S.G. Woo (Eds.) 2010. *The Use and Regulation of Microbial Pesticides in Representative Jurisdictions Worldwide.* IOBC Global, p. 99.

Kennedy, A.C., L.F. Elliott, F.L. Young and C.L. Douglas. 1991. Rhizobacteria suppressive to the weed downy brome. *Soil Sci. Soc. Am. J.* 55: 722–727.

Kennedy, A.C., B.N. Johnson and T.L. Stubbs. 2001. Host range of a deleterious rhizobacterium for biological control of downy brome. *Weed Sci.* 49: 792–797.

Kohlschmid, E., J. Sauerborn and D. Müller-Stöver. 2009. Impact of Fusarium oxysporum on the holoparasitic weed Phelipanche ramosa: Biocontrol efficacy under field-grown conditions. *Weed Res.* 49: 56–65.

Kovalev, O.V., L.G. Danilov and T.S. Ivanova. 1973. Method of controlling Russian knapweed. *Opisanie Izobreteniia Kavtorskomu Svidetel'stvu Byull.* 38: 2. (Translation—Translation Bureau, Canada Department of Secretary of State, No. 619707, Ottawa).

Kremer, R.J. and A.C. Kennedy. 1996. Rhizobacteria as biocontrol agents of weeds. *Weed Technol.* 10: 601–609.

Kropp, B.R., D.R. Hansen, K.M. Flint and S.V. Thomson. 1996. Artificial inoculation and colonization of Dyer's woad (*Isatis tinctoria*) by the systemic rust fungus *Puccinia thlaspeos*. *Phytopathology.* 86: 891–896.

Kropp, B.R., D.R. Hansen and S.V. Thomson. 2002. Establishment and dispersal of *Puccinia thlaspeos* in field populations of Dyer's woad. *Plant Dis.* 86(3): 241–246.

Landgraf, B.K., P.K. Fay and K.M. Havstad. 1984. Utilization of leafy spurge (*Euphorbia esula* L.) by sheep. *Weed Sci.* 32(3): 348–352.

Li, Y., Z. Sun, X. Zhuang, L. Xu, S. Chen and M. Li. 2003. Research progress on microbial herbicides. *Crop. Prot.* 22(2): 247–252.

Makowski, R.M. and K. Mortensen. 1998. Latent infections and penetration of the bioherbicide agent Colletotrichum gloeosporioides f. sp. malvae in non-target field crops under controlled environmental conditions. *Mycol. Res.* 102(12): 1545–1552.

Marley, P.S., S.M. Ahmed, J.A.Y. Shebayan and S.T.O. Lagoke. 1999. Isolation of *Fusarium oxysporum* with potential for biocontrol of the witch weed (*Striga hermonthica*) in the Nigerian savanna. *Biocontrol. Sci. Techn.* 9: 159–163.

Medal, J., W. Overholt, R. Charudattan, J. Mullahey, R. Gaskalla, R. Díaz and J. Cuda (Eds.) 2012. *Tropical Soda Apple Management Plan, University of Florida-IFAS.* Florida Department of Agriculture and Consumer Services-DPI, Gainesville. https://plants.ifas.ufl.edu/plant-directory/solanum-viarum/

Meir, S., A. Amsellem, H. Al-Ahmad, E. Safran and J. Gressel. 2009. Transforming a NEP1 toxin gene into two *Fusarium* spp. to enhance mycoherbicide activity on *Orobanche* – failure and success. *Pest Manage. Sci.* 65: 588–595.

Mejri, D., E. Gamalero and T. Souissi. 2012. Formulation development of the deleterious rhizobacterium *Pseudomonas trivialis* X33d for biocontrol of brome (*Bromus diandrus*) in durum wheat. *J. Appl. Microbiol.* 114(1): 219–228.

Men, B.X., T.K. Tinh, T.R. Preston, B. Ogle and J.E. Lindberg. 1999. Use of local ducklings to control insect pests and weeds in the growing rice field. *Livestock Res. Rural Dev.* 11: 8.

Mishra, J., A.K. Pandey and S.K. Hasija. 1995. Evaluation of *Sclerotium rolfsii* Sacc. as mycoherbicide for *Parthenium*: Factors affecting viability and virulence. *Indian Phytopath.* 48: 476–479.

Morin, L., R.L. Hill and S. Matayoshi. 1997. Hawaii's successful biological control strategy for mist flower (*Ageratina riparia*)—can it be transferred to New Zealand? *Biocontrol. News Inform.* 18: 77N–88N.

Morris, M.J. 1987. Biology of the Acacia gall rust, *Uromycladium tepperianum. Plant Path.* 36(1): 100– 106.

Morris, M.J. 1989. A method for controlling *Hakea sericea* Shrad. seedlings using the fungus *Colletotrichum gloeosporioides* (Penz.) Sacc. *Weed Res.* 29: 449–454.

Morris, M.J. 1997. Impact of the gall-forming rust fungus *Uromycladium tepperianum* on the invasive tree *Acacia saligna* in South Africa. *Biol. Control.* 10: 75–82.

Morris, M.J., A.R. Wood and A. den Breeÿen. 1999. Plant pathogens and biological control of weeds in South Africa: A review of projects and progress during the last decade. pp. 129–137. *In:* T. Olckers and M.P. Hill (Eds.) *Biological Control of Weeds in South Africa (1990–1998).* Entomological Society of Southern Africa.

Mortensen, K. 1988. The potential of an endemic fungus, *Colletotrichum gloeosporioides*, for biological control of round-leaved mallow (*Malva pusilla*) and velvetleaf (*Abutilon theophrasti*). *Weed Sci.* 36: 473–478.

Müller-Schärer, H., P.C. Scheepens and M.P. Greaves. 2000. Biological control of weeds in European crops: Recent achievements and future work. *Weed Res.* 40: 83–98.

Müller-Stöver, D., R. Batchvarova, E. Kohlschmid and J. Sauerborn. 2009b. Mycoherbicidal management of *Orobanche cumana*: Observations from three years of field experiments. p. 86. *In:* D. Rubiales, J. Westwood and A. Uledag (Eds.) *The Proceedings of the 10th International World Congress of Parasitic Plants.* Kusadasi, Turkey.

Müller-Stöver, D., E. Kohlschmid and J. Sauerborn. 2009a. A novel strain of *Fusarium oxysporum* from Germany and its potential for biocontrol of *Orobanche ramosa. Weed Res.* 49(Suppl. 1): 175–182.

Müller-Stöver, D., H. Thomas, J. Sauerborn and J. Kroschel. 2004. Two granular formulations of *Fusarium oxysporum* f. sp. *orthoceras* to mitigate sunflower broomrape *Orobanche cumana. Bio. Control.* 49(5): 595–602.

Mupondwa, E., X. Li, S. Boyetchko, R. Hynes and J. Geissler. 2015. Technoeconomic analysis of large scale production of pre-emergent *Pseudomonas fluorescens* microbial bioherbicide in Canada. *Bioresour. Technol.* 175: 517–528.

Neufeldt, E. (Ed.) 1988. *Webster's New World Dictionary*, 3rd College Edition. Simon & Schuster, New York, USA.

Nishino, J. and A. Tateno. 2000. Camperico®—a new bioherbicide for annual bluegrass in turf. *Agrochem. Jpn.* 77: 13–16.

Notham, F.E. and E.C. Orr. 1982. Effects of a nematode on biomass and density of silverleaf nightshade. *J. Range Manag.* 35(4): 536–537.

Nzioki, H.S., F. Oyosi, C.E. Morris, E. Kaya, A.L. Pilgeram, C.S. Baker and D.C. Sands. 2016. *Striga* biocontrol on a toothpick: A readily deployable and inexpensive method for smallholder farmers. *Front. Plant Sci.* 08. doi:10.3389/fpls.2016.01121.

OECD. 2003. Guidance for registration requirements for microbial pesticides. *OECD Series on Pesticides.* 18: 51. http://www.oecd.org/officialdocuments/publicdisplaydocumentpdf/?cote= env/jm/mono(2003)5& doclanguage=en

Oehrens, E. and S. Gonzales. 1977. Dispersion, ciclo biologico y danos causados por *Phragmidium violaceum* (Schulz) Winter en zarzamora (*Rubus constrictus* Lef. et M. y *R. ulmifolius* Schott.) en laszonas centro-sur de Chile. *Agro Sur.* 5: 73–85.

Ou, X. and A.K. Watson. 1992. *In vitro* culture of *Subanguina picridis* in *Acroptilon repens* callus, excised roots, and shoot tissues. *J. Nematol.* 24: 199–204.

Ou, X. and A.K. Watson. 1993. Mass culture of *Subanguina picridis* and its bioherbicidal efficacy on *Acroptilon repens. J. Nematol.* 25: 89–94.

Pacanoski, Z. 2015. Bioherbicides. pp. 153–274. *In:* A. Price, J. Kelton and L. Sarunaite (Eds.) *Bioherbicides, Herbicides, Physiology of Action, and Safety.* In Tech. doi: 10.5772/61528.

Pan, L., G.J. Ash, B. Ahn and A.K. Watson. 2010. Development of strain specific molecular markers for the *Sclerotinia minor* bioherbicide strain IMI 344141. *Biocontrol. Sci. Techn.* 20(9): 939–959.

Parker, P.E. 1991a. Nematodes as biological control agents of weeds. pp. 58–68. *In:* D.O. TeBeest (Ed.) *Microbial Control of Weeds.* Chapman & Hall, New York, USA.

Parker, P.E. 1991b. Nematode control of silverleaf nightshade, a biological control pilot project. *Weed Sci.* 34(Suppl. 1): 33–34.

Patel, R. and D.R. Patel. 2015. Biological control of weeds with pathogens: Current status and future trends. *Int. J. Pharm. Life Sci.* 6(6): 4531–4550.

Phatak, S.C., D.R. Summer, H.D. Wells, D.K. Bell and N.C. Glaze. 1983. Biological control of yellow nutsedge with the indigenous rust fungus *Puccinia canaliculata. Science.* 219(4591): 1446–1447.

Politis, D.J., A.K. Watson and W.L. Bruckart. 1984. Susceptibility of musk thistle and related composites to *Puccinia carduorum*. *Phytopathology*. 74: 687–691.

Ridings, W.H. 1986. Biological control of strangler vine in citrus—a researcher's view. *Weed Sci.* 34(Suppl. 1): 31–32.

Rosskopf, E.N., R. Charudattan and J.B. Kadir. 1999. Use of plant pathogens in weed control. pp. 891–918. *In:* T.W. Fisher, T.S. Bellows, L.E. Caltagirone, D.L. Dahlsten, C.B. Huffaker and G. Gordh (Eds.) *Handbook of Biological Control: Principles and Applications of Biological Control*. Academic Press. San Diego, USA.

Saharan, G.S. and N. Mehta. 2008. *Sclerotinia* as Mycoherbicide. pp. 377–381. *In: Sclerotinia Diseases of Crop Plants: Biology, Ecology and Disease Management*. Springer, Dordrecht, 486 p.

Sedivec, K., T. Hanson and C. Heiser. 1995. *Controlling Leafy Spurge Using Goats and Sheep*. North Dakota State University, Fargo, ND 58105 NDSU EXTENSION SERVICE MAY 1995.

Sharon, A., Z. Amsellem and J. Gressel. 1992. Glyphosate suppression of an elicited defense response: Increased susceptibility of *Cassia obtusifolia* to a mycoherbicide. *Plant Physiol.* 98: 654–659.

Sharon, A., S. Barhoom and R. Maor. 2001. Genetic engineering of *Collectotrichum gloeosporioides* f. sp. *aeschynomene*. pp. 240–247. *In:* M. Vurro et al. (Eds.) *Enhancing Biocontrol Agents and Handling Risks*. IOS Press, Washington, DC.

Shibayama, H. 2001. Weeds and weed management in rice production in Japan. *Weed Biol. Manage.* 1(1): 53–60.

Smith, R.J. Jr. 1986. Biological control of northern jointvetch in rice and soybeans—a researcher's view. *Weed Sci.* 34(Suppl. 1): 17–23.

Stubbs, T.L. and A.C. Kennedy. 2012. Microbial weed control and microbial herbicides. pp. 135–166. *In:* R. Alvarez-Fernandez (Ed.) *Herbicides—Environmental Impact Studies and Management*. Intech Open. doi: 10.5772/32705.

Tang, W., Y.-Z. Zhu, H-Q. He, S. Qiang and B.A. Auld. 2011. Field evaluation of *Sclerotium rolfsii*, a biological control agent for broadleaf weeds in dry, direct-seeded rice. *Crop. Prot.* 30(10): 1315–1320.

TeBeest, D.O. and G.E. Templeton. 1985. Mycoherbicides: Progress in the biological control of weeds. *Plant Dis.* 69: 6–10.

TeBeest, D.O., X.B. Yang and C.R. Cisar. 1992. The status of biological control of weeds with fungal pathogens. *Ann. Rev. Phytopathol.* 30: 637–657.

Templeton, G.E. 1982a. Biological herbicides: Discovery, development, deployment. *Weed Sci.* 30: 430–433.

Templeton, G.E. 1982b. Status of weed control with plant pathogens. pp. 29–44. *In:* R. Charudattan and H.L. Walker (Eds.) *Biological Control of Weeds with Plant Pathogens*. John Wiley and Sons, New York, USA.

Templeton, G.E. 1992a. Potential for developing and marketing mycoherbicides. pp. 264–268. *In:* J.H. Combellack, K.J. Levick, J. Parsons and R.G. Richardson (Eds.) *The Proceedings of the. 1st International Weed Control Congress*. Weed Science Society of Victoria Inc., Melbourne, Australia.

Templeton, G.E. 1992b. Some "orphaned" mycoherbicides and their potential for development. *Plant Prot. Q.* 7: 149–150.

Templeton, G.E., D.O. TeBeest and R.J. Jr. Smith. 1979. Biological control of weeds with mycoherbicides. *Ann. Rev. Phytopathol.* 17: 301–310.

Thomas, H., J. Sauerborn, D. Müller-Stöver, A. Ziegler, J.S. Bedi and J. Kroschel. 1998. The potential of *Fusarium oxysporum* f. sp. *orthoceras* as a biological control agent for *Orobanche cumana* in sunflower. *Biol. Control.* 13: 41–48.

Thompson, B.M., M.M. Kirkpatrick, D.C. Sands and A.L. Pilgeram. 2007. Genetically enhancing the efficacy of plant pathogens for control of weeds. pp. 267–275. *In:* M. Vurro and J. Gressel (Eds.) *Novel Biotechnologies for Biocontrol Agent Enhancement and Management.* Springer, Dordrecht.

Tiourebaev, K., G. Semenchenko, M. Dolgovskaya, M. McCarthy, T. Anderson, L. Carsten, A. Pilgeram and D. Sands. 2001. Biological control of infestations of ditchweed (*Cannabis sativa*) with *Fusarium oxysporum* f. sp. *cannabis* in Kazakhstan. *Biocontrol. Sci. Techn.* 11(4): 535–540.

Tomley, A.J. and H.C. Evans. 2004. Establishment of, and preliminary impact studies on, the rust, *Maravalia cryptostegiae*, of the invasive alien weed, *Cryptostegia grandiflora* in Queensland, Australia. *Plant Path.* 53: 475–494.

Trujillo, E.E. 2005. History and success of plant pathogens for biological control of introduced weeds in Hawaii. *Biol. Control.* 33: 113–122.

Tsukamoto, H., M. Gohbara, M. Tsuda and T. Fujimori. 1997. Evaluation of pathogenic fungi for biological control of *Echinochloa* species. *Ann. Phytopathol. Soc. Jpn.* 63(5): 366–372.

Tsukamoto, H., M. Takabayashi, T. Hieda and M. Gohbara. 2001. Strains belonging to *Exserohilum monoceras*, and uses thereof. US Patent # 6,313,069 B1.

Tsukamoto, H., M. Tsuda, M. Gohbara and T. Fujimori. 1998. Effect of water management on mycoherbicidal activity of against *Echinochloa oryzicola*. *Ann. Phytopathol. Soc. Jpn.* 64: 526–531.

Vauterin, L., B. Hoste, K. Kersters and J. Swings. 1995. Reclassification of xanthomonas. *Int. J. Microbiol.* 45(3): 472–489.

Venne, J., F. Beed, A. Avocanh and A.K. Watson. 2009. Integrating *Fusarium oxysporum* f. sp. *strigae* into cereal cropping systems in Africa. *Pest Manage. Sci.* 65(5): 572–580.

Vieira, B.S. and R.W. Barreto. 2010. Liquid culture production of chlamydospores of *Lewia chlamidosporiformans* (Ascomycota: Pleosporales), a mycoherbicide candidate for wild poinsettia. Australas. *Plant Pathol.* 39(2): 154–160.

Vieira, B.S., K.L. Nechet and R.W. Barreto. 2008. *Lewia chlamidosporiformans*, a mycoherbicide for control of *Euphoria heterophylla*: Isolate selection and mass production. pp. 221–226. *In:* M.H. Julien, R. Sforza, M.C. Bon, H.C. Evans, P.E. Hatcher, H.L. Hinz and B.G. Rector (Eds.) *The Proceedings of the 12th International Symposium on Biological Control of Weeds.* © CABI, La Grande Motte, France.

Vurro, M. and H.C. Evans. 2007. Opportunities and constraints for the biological control of weeds in Europe. pp. 455–462. *In:* M.H. Julien, R. Sforza, M.C. Bon, H.C. Evans, P.E. Hatcher, H. Hinz and B.G. Rector (Eds.) *The Proceedings of the 12th International Symposium on Biological Control of Weeds.* CABI, La Grande Motte, France.

Walker, H.L. and W.J. Connick Jr. 1983. Sodium alginate for production and formulation of mycoherbicides. *Weed Sci.* 31(3): 333–338.

Walker, H.L. and J.A. Riley. 1982. Evaluation of *Alternaria cassiae* for the biocontrol of sicklepod (*Cassia obtusifolia*). *Weed Sci.* 30(6): 651–654.

Wang, J., X. Wang, B. Yuan and S. Qiang. 2013. Differential gene expression for *Curvularia eragrostidis* pathogenic incidence in crabgrass (*Digitaria sanguinalis*) revealed by cDNA-AFLP analysis. *PLoS One.* 8(10): e75430. doi: 10.1371/journal.pone.0075430.

Wapshere, A.J. 1974. A strategy for evaluating the safety of organisms for biological weed control. *Ann. Appl. Biol.* 77(2): 201–211.

Wapshere, A.J. 1982. Biological control of weeds. pp. 47–56. *In:* W. Holzner and N. Numata (Eds.) *Biology and Ecology of Weeds.* Junk Publishers, The Hague.

Waterhouse, D.F. 1994. Biological control of weeds: Southeast Asian prospects. *ACIAR Monograph.* 26: 302.

Watson, A.K. 1985. Host specificity of plant pathogens in biological weed control. pp. 577–586. *In:* E.S. Delfosse (Ed.) *The Proceedings of the 6th International Symposium on Biological Control of Weeds.* Agriculture Canada, Vancouver, Ottawa.

Watson, A.K. 1991a. Prospects for bioherbicide development in Southeast Asia. pp. 65–73. *In:* J.T. Swarbrick, R.K. Nishimoto and M. Soerjani (Eds.) *The Proceedings of the 13th Asian Pacific Weed Science Society Conference.* Asian Pacific Weed Science Society and Weed Science Society of Indonesia, Jakarta, Indonesia.

Watson, A.K. 1991b. The classical approach with plants pathogens. pp. 3–23. *In:* D.O. TeBeest (Ed.) *Microbial Control of Weeds.* Chapman & Hall, New York, USA.

Watson, A.K. 1994. Current status of bioherbicide development and prospects for rice in Asia. pp. 195–201. *In:* H. Shibayama, K. Kiritani and J. Bay-Petersen (Eds.) *Integrated Management of Paddy and Aquatic Weeds in Asia. FFTC Book Series No. 45.* Food and Fertiliser Technology Center for the Asian and Pacific Region, Taipei, Taiwan.

Watson, A.K. 1999. Can viable weed control be attainable with microorganisms. pp. 59–63. *In:* L.W. Hong, S.S. Sastroutomo, I.G. Caunter, J. Ali, L.K. Yeang, S. Vijaysegaran and Y.H. Sen (Eds.) *Biological Control in the Tropics: Towards Efficient Biodiversity and Bioresource Management for Effective Biological Control.* CABI. Oxford University Press, Serdang, Malaysia.

Watson, A.K. 2007. *Sclerotinia minor* – Biocontrol target or agent? pp. 205–211. *In:* M. Vurro and J. Gressel (Eds.) *Novel Biotechnologies for Biocontrol Agent Enhancement and Management.* Springer, Dordrecht.

Watson, A.K. 2013. Biocontrol. pp. 469–497. *In:* D.M. Joel, J. Gressel and L.J. Musselman (Eds.) *Parasitic Orobanchaceae, Parasitic Mechanisms and Control Strategies.* Springer, Heidelberg.

Watson, A.K. 2017. Biocontrol and weed management in rice of Asian Pacific region. pp. 113–134. *In:* A. Rao and H. Matsumoto (Eds.) *Weed Management in Rice of Asian-Pacific Region.* Pacific Weed Science Society, @ APWSS.

Watson, A.K. and K. Bailey. 2013. *Taraxacum officinale* Weber, Dandelion (Asteraceae). pp. 383–391. *In:* P. Mason and D. Gillespie (Eds.) *Biological Control Programmes in Canada 2001–2012.* CABI Publishing, Wallingford, UK.

Watson, A.K. and P. Harris. 1984. *Acroptilon repens* (L.) DC, Russian knapweed (Compositae). pp. 105–110. *In:* J.S. Kelleher and M.A. Hulme (Eds.) *Biological Control Programmes against Insects and Weeds in Canada 1969–1980*. Farnham Royal: Commonwealth Agricultural Bureaux, Farnham Royal, Slough, England.

Watson, A.K. and L.A. Wymore. 1990. Identifying limiting factors in the biocontrol of weeds. pp. 305–316. *In:* R. Baker and P. Dunn (Eds.) *New Directions in Biological Control, UCLA Symposia on Molecular and Cellular Biology, New Series 112*. Alan R. Liss, New York.

Wilson, C.L. 1965. Consideration of the use of persimmon wilt as a silvicide for weed persimmons. *Plant Dis. Rep*. 49(9): 789–791.

Wilson, C.L. 1970. Use of plant pathogens in weed control. *PANS*. 16(3): 482–487.

Winston, R.L., M. Schwarzländer, H.L. Hinz, M.D. Day, M.J.W. Cock and M.H. Julien (Eds.) 2014. *Biological Control of Weeds: A World Catalogue of Agents and Their Target Weeds*, 5th Edition. USDA Forest Service, Forest Health Technology Enterprise Team, Morgantown, West Virginia, FHTET-2014-04, 838 p.

Womack, J.G., G.M. Eccleston and M.N. Burge. 1996. A vegetable oil based invert emulsion for mycoherbicides delivery. *Biol. Control*. 6: 23–28.

Wymore, L.A., C. Poirier, A.K. Watson and A.R. Gotlieb. 1988. *Colletotrichum coccodes*, a potential bioherbicide for control of velvetleaf (*Abutilon theophrasti*). *Plant Dis*. 72: 534–538.

Wymore, L.A., A.K. Watson and A.R. Gotlieb. 1987. Interaction between *Colletotrichum coccodes* and thidiazuron for control of velvetleaf (*Abutilon theophrasti*). *Weed Sci*. 35: 377–383.

Yandoc, C.B. 2001. *Biological Control of Cogongrass, Imperata Cylindrica (L.) Beauv*. University of Florida.

Yoo, I.D. 1991. Status and perspectives of microbial herbicide development. *Korean J. Weed Sci*. 11: 47–55.

Zdor, R., C. Alexander and R. Kremer. 2007. Weed suppression by deleterious rhizobacteria is affected by formulation and soil properties. *Comm. Soil Sci. Plant Anal*. 36(9–10): 1289–1299.

Zhang, D., Q.W. Min, S.K. Cheng, H.L. Yang, L. He, W.J. Jiao et al. 2010. Effects of different rice farming systems on paddy field weed community. [Article in Chinese]. *Chin. J. Appl. Ecol*. 21(6): 1603–1608.

Zhang, J., S. Yang, Z. Zhou and L. Yu. 2011. Development of bioherbicides for control of barnyard grass in China. *Pest Technol*. 5(Special Issue 1): 56–60.

Zhou, T. and J.C. Neal. 1995. Annual bluegrass (*Poa annua*) control with *Xanthomonas campestris* pv. *poannua* in New York State. *Weed Technol*. 9: 173–177.

Zhu, Y. and S. Qiang. 2004. Isolation, pathogenicity and safety of *Curvularia eragrostidis* isolate QZ-2000 as a bioherbicide agent for large crabgrass (*Digitaria sanguinalis*). *Biocontrol. Sci. Techn*. 14(8): 769–782.

Zorner, P.S., S.L. Evans and S.D. Savage. 1993. Perspectives on providing a realistic technical foundation for the commercialization of bioherbicides. pp. 79–86. *In:* S.O. Duke, J.J. Menn and J.R. Plimmer (Eds.) *Pest Control with Enhanced Environmental Safety, ACS Symposium Series*, Vol 52.

18 Fungal Biopesticides and Their Uses for Control of Insect Pests and Diseases*

Deepak Kumar, M. K. Singh,
Hemant Kumar Singh, and K. N. Singh

18.1 INTRODUCTION

Agriculture has been facing the vicious activities of many pests like fungi, weeds, and insects from time immemorial, leading to drastic decline in yields. Pests are regularly being introduced to fresh areas either naturally or accidentally, or, in a few cases, organisms that are deliberately introduced become pests. Global employment has resulted in increased numbers of insidious nonnative pest species being introduced to fresh areas. Controlling these invasive species presents an unparalleled challenge worldwide. After many years of successful control by conventional agrochemical insecticides, a number of factors are challenging the effectiveness and continued use of these agents. These include the development of insecticide resistance and use-cancellation or deregistration of some insecticides due to human health and environmental issues. Therefore, an ecofriendly substitute is the need in the present scenario. Advances in pest control approaches are one method of generating elevated quality and a greater quantity of agricultural products. Therefore, there is a requirement to develop biopesticides that are efficient, biodegradable, and not harmful to the environment. Their shield against pests is a main concern, and due to the poor impact of chemical insecticides, use of biopesticides is escalating.

18.2 CONCEPT OF BIOPESTICIDES

According to the US Environmental Protection Agency (EPA), biopesticides are pesticides prepared from natural sources such as animals, plants, bacteria, and minerals. Biopesticides also include live organisms that obliterate agricultural pests. The EPA divides biopesticides into three main classes on the basis of the type of active ingredient used: namely, biochemical, plant-incorporated protectants, and microbial pesticides (USEPA, 2008). Biochemical pesticides are chemicals either isolated from natural source or manufactured to have a similar structure and function to naturally occurring chemicals. Biochemical pesticides differ from conventional pesticides in their construction (source) and mode of action (method by which they kill or control pests) (O'Brien et al., 2009). At the global level, there is a discrepancy in understanding the term *biopesticide* as the aforementioned definition of the term used by the EPA is not followed around the world. That is why the International Biocontrol Manufacturer's Association (IBMA) and the International Organization for Biological Control (IOBC, 2008) endorse the term *biocontrol agents* (BCAs) instead of *biopesticides* (Guillon, 2003). IBMA categorize BCAs into four groups: (1) macrobials, (2) microbials, (3) natural products, and (4) semiochemicals (insect behavior-modifying agents). Among the BCAs, the most important products are

DOI: 10.1201/9781003463429-23

microbials (41%), followed by macrobials (33%), and, finally, other natural products (26%) (Guillon, 2003). This chapter spotlights microbe-based biopesticides, specifically fungal biopesticides.

Biopesticides or biological pesticides rely on pathogenic microorganisms specific to a target pest and present an ecologically effective solution to pest problems. They cause less threat to the environment and to human health. The most frequently used biopesticides are living organisms that are pathogenic for the pest of concern. These include biofungicides (*Trichoderma*), bioherbicides (*Phytopthora*), and bioinsecticides (*Beauvarria bassiana, Metarhizium anisopliae*). The benefits to agriculture and public health programs of the use of biopesticides are substantial. The interest in biopesticides is based on the rewards associated with such products, which include:

- Inherently fewer harmful effects and a lower environmental load
- Designed to affect only one specific pest or, in a few cases, some target organisms
- Often effective in very minute quantities and often decaying quickly, thereby resulting in less exposure and largely avoiding pollution problems
- Contribute much when used components of integrated pest management (IPM) programs

18.2.1 Fungal Biopesticides

Fungal pathogens have a major role in the development of diseases on various important field and horticultural crops, resulting in huge plant-yield losses (Khandelwal et al., 2012). Intensified use of fungicides has amassed toxic compounds potentially hazardous to humans and the environment, as well as the buildup of resistance to the pathogens. Fungal biopesticides, including other fungi, bacteria, nematodes, and weeds, can be used to manage insects and plant diseases. Their means of action are varied and depend on both the pesticidal fungus and the target pest. One benefit of fungal biopesticides, in contrast with many bacterial and all viral biopesticides, is that they do not need to be eaten to be effective. However, they are living organisms that often need a narrow range of situations, including moist soil and cool temperatures, to flourish. BCAs like *Trichoderma* are acclaimed as efficient, ecofriendly, and cheap, nullifying the ill effects of chemicals (Table 18.1). Therefore, of late, these BCAs are recognized as acting against an array of important soil-borne plant pathogens causing severe diseases of crops (Bailey and Gilligan, 2004). Fungal biopesticides used against plant pathogens include *T. harzianum*, which is an antagonist of *Rhizoctonia*, *Pythium*, *Fusarium*, and other soilborne pathogens (Harman, 2005). *Trichoderma* is a fungal antagonist that enters the main tissue of a disease-causing fungus and exudes enzymes that degrade the cell walls of the other fungus and then eat the contents of the cells of the target fungus and multiply its own spores. *Trichoderma* is one of the general fungal BCAs being used worldwide for management of a variety of foliar and soil-borne plant pathogens like *Ceratobasidium*, *Fusarium*, *Rhizoctonia*, *Macrophomina*, *Sclerotium*, *Pythium*, and *Phytophthora* spp. (Dominguesa et al., 2000; Anand and Reddy, 2009). *Trichoderma viride* has shown to be very promising against soil-borne plant parasitic fungi (Khandelwal et al., 2012). A precise strain, *Muscodor albus* QST 20799, is a naturally occurring fungus initially isolated from the bark of a cinnamon tree in Honduras. The *M. albus* strain is reported to generate a number of volatiles, mainly alcohols, acids, and esters, that inhibit and kill specific bacteria and other organisms that generate soil-borne and postharvest diseases in plants. Products containing QST 20799 can be also used in fields, greenhouses, and warehouses (USEPA, 2008) for disease control. Other two fungi—namely *B. bassiana* (Balsamo) Vuillemin and *Metarhizium anisopliae* (Metchnikoff) Sorokin—are naturally occurring entomopathogenic fungi that infect sucking pests, including *Nezara viridula* (L) (green vegetable bug) and *Creontiades* sp. (green and brown mirids) (Sosa-Gómez and Moscardi, 1998). Fungi have the unique ability to attack insects by penetrating the cuticle, making them ideal for the control of sucking pests. *B. bassiana* is currently registered in the US as Mycotrol ES (Mycotech, Butte) and Naturalis L (Troy Biosciences). These biopesticide products are registered against sucking pests, such as whitefly, aphids, thrips, mealybugs, leafhoppers, and weevils. Studies also show that *B. bassiana* is potent against *Lygus hesperus* Knight (Hemiptera: Miridae), a major pest of alfalfa and cotton in the United States (Noma and Strickler, 2000).

TABLE 18.1
Fungal Biopesticides Developed or Being Developed for the Biological Control of Pests

Product	Fungus	Target Pests	Producer Companies and Countries
BIO 1020	*Metarhizium anisopliae*	Vine weevils	Licensed to Taensa, US
Bio-Blast	*Metarhizium anisopliae*	Termites	EcoScience, US
Biogreen	*Metarhizium anisopliae*	Scarab larvae in pasture	Bio-care Technology, Australia
Bio-Path	*Metarhizium anisopliae*	Cockroaches	EcoScience, US
Boverin	*Beauveria bassiana*	Colorado beetles	Former USSR
Conidia	*Beauveria bassiana*	Coffee berry borers	Live Systems Technology, Colombia
Corn Guard	*Beauveria bassiana*	European corn borers	Mycotech, US
Engerlingspilz	*Beauveria brongniartii*	Cockchafers	Andermatt, Switzerland
Jas Bassi	*Beauveria bassiana*	Colorado beetles	Shri Ram Solvent Ext. Pvt., India
Jas Meta	*Metarhizium anisopliae*	Sugarcane spittle bugs, termites	Shri Ram Solvent Ext. Pvt., India
Jas Verti	*Verticillium lecanii*	Whiteflies and thrips	Shri Ram Solvent Ext. Pvt., India
Laginex	*Lagenidium giganteum*	Mosquito larvae	AgraQuest, US
Metaquino	*Metarhizium anisopliae*	Spittle bugs	Brazil
Mycotal	*Verticillium lecanii*	Whiteflies and thrips	Koppert, The Netherlands
Mycotrol WP	*Beauveria bassiana*	Whiteflies, aphids, thrips	Mycotech, US
Naturalis-L	*Beauveria bassiana*	Cotton pests including bollworms	Troy Biosciences, US
Ostrinil	*Beauveria bassiana*	Corn borers	Natural Plant Protection (NPP), France
Pae-Sin	*Paecilomyces fumosoroseus*	Whiteflies	Agrobionsa, Mexico
PFR-97	*Paecilomyces fumosoroseus*	Whiteflies	ECO-tek, US
Proecol	*Beauveria bassiana*	Army worms	Probioagro, Venezuela
Schweizer Beauveria	*Beauveria brongniartii*	Cockchafers	Eric Schweizer, Switzerland
Vertalec	*Verticillium lecanii*	Aphids	Koppert, The Netherlands

Source: Adapted and modified from Butt et al. (2001) and Rai et al. (2014)

18.2.2 *Beauvaria* spp.

The genus *Beauveria* includes at least 49 species of which approximately 22 are regarded as pathogenic (Kirk, 2003). *Beauveria bassiana*, a white muscardine fungus, is one of the most historically central fungus commonly used in this genus. It was originally recognized as *Tritirachium shiotae* and renamed after the Italian lawyer and scientist Agostino Bassi, who first implicated it as the causative agent of a white (later yellowish or occasionally reddish) muscardine disease in domestic silkworms (Furlong and Pell, 2005; Zimmermann, 2007). *Beauveria bassiana* is a fungus that grows naturally in soils all through the world and acts as a pathogen on various insect species belonging to the entomopathogenic fungi (Sandhu et al., 2004; Jain et al., 2008). When the microscopic spores of the fungus get in touch with the body of an insect host, they germinate, penetrate the cuticle, and grow inside, killing the insect within a matter of days. Later, a white mold appears on the cadaver and produces fresh spores. A typical isolate of *B. bassiana* can attack a broad range of insects; various isolates differ in their host range. An attractive attribute of *Beauveria* spp. is the high host specificity of several isolates. Hosts of agricultural and forest implication include the Colorado potato beetle, the codling moth, several genera of termites, and the American bollworm *Helicoverpa armigera* (Thakur et al., 2010). *Beauveria bassiana* can easily be isolated from insect cadavers or from soil in forested regions using media (Beilharz et al., 1982), as well as by baiting soil with insects.

18.2.3 *Metarhizium* spp.

The fungus *Metarhizium anisopliae*, first known by the name *Entomphthora anisopliae*, was first described near Odessa in Ukraine from infected larvae of the wheat cockchafer *Anisopliae austriaca* in 1879 and, later on, *Cleonus punctiventis* by Metschnikoff. It was later renamed *M. anisopliae* by Sorokin in 1883 (Tulloch, 1976). *Metarhizium* causes a disease in insect hosts known as "green muscardine" for the green color of its conidial cells. When these mitotic (asexual) spores (called conidia) of the fungus come in touch with the body of an insect host, they germinate, and the hyphae that emerge penetrate the cuticle. The fungus then expands within the body, ultimately killing the insect after a few days; this lethal outcome is very likely aided by the creation of insecticidal cyclic peptides (destruxins). If the ambient moisture is high enough, a white mold then develops on the cadaver that soon turns green as spores are produced. Several insects living near the soil have advanced natural defenses against entomopathogenic fungi like *M. anisopliae*. The genus *Metarhizium* is pathogenic to a large number of insect species, many of which are agricultural and forest insects (Ferron, 1978). The taxonomy of *Metarhizium* is complex. Using the morphological characteristics of conidia and conidiogenous cells, Tulloch (1976) documented *M. flavoviride* and *M. anisopliae*, of which the latter was further subdivided into two variants: *majus* (short conidia up to 9 μm) and *anisopliae* (long conidia up to 18 μm). Currently, *M. anisopliae* consists of four varieties or genetic groups (Driver et al., 2000).

18.2.4 *Verticillium lecanii*

Verticillium lecanii is a broadly distributed fungus which can cause huge epizootics in tropical and subtropical regions, as well as in warm and humid environments (Nunez et al., 2008). It was reported by Kim et al. (2002) that *V. lecanii* is an effective BCA against *Trialeurodes vaporariorum* in South Korean greenhouses. The conidia (spores) of *V. lecanii* are slimy and connect to the cuticle of insects. The fungus infects the insects by producing hyphae from germinating spores that enter the insect's integument; the fungus then obliterates the internal contents, and the insect dies. The fungus eventually rises out through the cuticle and sporulates on the outside of the insect body. Infected insects emerge as white to yellowish cottony particles. Diseased insects typically appear in seven days. However, due to environmental conditions, there may be considerable lag time from infection to the death of insect. The fungus *V. lecanii* works best at temperatures of 15–25 °C and at a relative humidity of 85–90%. This fungus attacks nymphs and adults and sticks to the leaf underside by means of a filamentous mycelium (Nunez et al., 2008). In the 1970s, *V. lecanii* was developed to manage whitefly and several aphid species, including green peach aphids (*Myzus persicae*), in greenhouse chrysanthemums (Hamlen et al., 1979).

18.2.5 *Paecilomyces* spp.

The fungus *Paecilomyces fumosoroseus* is one of the most important natural enemies of whiteflies worldwide and causes a sickness called yellow muscardine (Nunez et al., 2008). A strong epizootic prospective against *Bemisia* and *Trialeurodes* spp. in both greenhouse and open field environments was reported with *Paecilomyces fumosoroseus*. The ability of this fungus to grow broadly over the leaf surface under humid conditions is a characteristic that certainly enhances its ability to extend rapidly through whitefly populations (Wraight et al., 2000). Kim et al. (2002) accounted that *P. fumosoroseus* is best for managing the nymphs of whitefly. These fungi cover the whitefly's body with mycelial threads and stick them to the bottom of the leaves. The nymphs show a "feathery" aspect and are surrounded by mycelia and conidia (Nunez et al., 2008). *Paecilomyces lilacinus* has been characterized as aggressive, and Dunn et al. (1982) stated, "The fungal egg parasites as group appear to be more promising to investigate as potential biological control agents of nematodes". Research results utilizing this fungus have been contradictory and erratic. In one experiment, the fungus caused a 71% decline in root galls and 90% decline in egg masses on root-knot nematode–infected corn (Ibrahim et al., 1987).

18.2.6 *Nomuraea* spp.

Nomuraea rileyi, another potential entomopathogenic fungi, is a dimorphic hyphomycete that can cause epizootic death in various plant insects. It has been demonstrated that several insect species belonging to Lepidoptera, including *Spodoptera litura*, and few belonging to Coleoptera are susceptible to *N. rileyi* (Ignoffo, 1981). The host specificity of *N. rileyi* and its ecofriendly nature encourage its use in insect pest management. Its mode of infection and development have been reported for several insect hosts, such as *Trichoplusiani*, *Heliothis zea*, *Plathypena scabra*, *Bombyx mori*, *Pseudoplusia includes*, and *Anticarsia gemmatalis*.

18.3 PATHOGENICITY AND MODE OF ACTION OF ENTOMOPATHOGENIC FUNGAL BIOPESTICIDES

Insect pathogenic fungi are different in pathogenicity from bacteria and viruses in that they infect insects by breaching the host cuticle. The cuticle is composed of chitin fibrils embedded in a matrix of proteins, lipids, pigments, and N-acylcatecholamines (Richard et al., 2010). Fungi exude extracellular enzymes proteases, chitinases, and lipases to degrade the major constituents of the cuticle (i.e., protein, chitin, and lipids) and allow hyphal infiltration (Wang et al., 2005; Cho et al., 2006). The success of the infection was reliably proportional to the emission of exo-enzymes (Khachatourians, 1996). It is considered that both mechanical force and enzymatic action are involved in the penetration of fungus to the hemocoel of the insect. There are a large number of toxic compounds in the filtrate of entomopathogenic fungi, such as small secondary metabolites, cyclic peptides, and macromolecular proteins. *B. bassiana* is reported to secrete low molecular weight cyclic peptides and cyclosporins A and C with insecticidal properties such as beauvericin, enniatins, bassianolide (Roberts, 1981; Vey et al., 2001). Unlike other biopesticides such as bacteria and viruses, entomopathogenic fungi do not have to be ingested to cause infection, making them valuable as BCAs. Although little information suggests a mode of infection through the siphon tips or gut of insect larvae (Lacey et al., 1988), entomopathogenic fungi generally infect or penetrate their targets percutaneously (Charnley, 1989). This can occur by adhesion of spores to the insect integument, particularly the intersegmental folds, or by simple tarsal contact (Clarkson and Charnley, 1996).

Normally, germinated conidia manufacture an appressorium, which then forms a contagion peg (St Leger et al., 1991). The penetration of conidia throughout the host's cuticle involves both mechanical pressure and enzymatic degradation (Bidochka et al., 1995; Clarkson and Charnley, 1996). Enzymatic degradation involves the production of large amounts of numerous cuticle-degrading enzymes, which differ according to the species and strains of the fungi. These enzymes show varying levels of pathogenicity toward their hosts. Following successful penetration of the cuticle, the fungus then produces blastospores or hyphae bodies, which are passively distributed in the hemolymph and the fat body (Hajek and Leger, 1994). So as to kill their host, fungal pathogens discharge a wide variety of secondary metabolic compounds, commonly called toxins, inside the insect host, mainly in the hemocoel. Two fungi, *M. anisopliae* and *B. bassiana*, secrete large amounts of a single extracellular protease called chymoelastase protease or Prl to degrade the host cuticle (St Leger et al., 1992, 1996). The endomoprotease Prl, which is the major enzyme secreted by *Metarhizium* throughout the degradation process of the cuticle, also differs among strains in terms of biochemistry. However, the cuticle degrading-enzymes are manufactured sequentially, with the proteolytic enzymes and esterases first, followed by the chitinases; in other words, the proteins surrounding the cuticle must be degraded before the actions of the chitinases begin (Leger et al., 1996).

Time to fatality of an infected insect varies from 2 to 15 days post-infection, depending on the fungal strain and species and the characteristics of the host (Boucias and Pendland, 1998). When the infection procedure followed by the death of the host is complete, the fungus changes back to its hyphal mode. Under relatively humid conditions, the fungus then grows out of the cadaver surface to produce new, external, infective conidial saprophytic growth (Jianzhong et al., 2003; Mitsuaka, 2004). Under very dry conditions, the fungus may remain in the hyphal stage inside the cadaver, where the conidia are produced (Hong et al., 1997). Under positive conditions, sporulated cadavers can infect other individuals from the same target species through horizontal transmission (Meadow et al., 2000; Quesada-Moraga et al., 2004).

18.4 BIOFUNGICIDES OR MYCOFUNGICIDES

A range of fungal species can be used as biological control means and may offer effective activity against a variety of pathogenic microorganisms; these are known as biofungicides or mycofungicides (Table 18.2). Examples of these biofungicides or mycofungicides are *Trichoderma harzianum*, a species with biocontrol potential against *Botrytis cineria*, *Fusarium*, *Pythium*, and *Rhizoctonia* (Khetan, 2001); *Ampelomyces quisqualis*, a hyperparasite of powdery mildew (Liang et al., 2007; Viterbo et al., 2007); *Chaetomium globosum* and *C. cupreum*, which have biocontrol activity against root rot caused by *Fusarium*, *Phytophthora*, and *Pythium* (Soytong et al., 2001); and *Gliocladium virens*, an effective biocontrol of soil-borne pathogens (Viterbo et al., 2007). An efficient BCA should be genetically constant, effective in low concentrations, easy to mass produce in culture on inexpensive media, and be very effective against a wide range of pathogens (Wraight et al., 2001; Irtwange, 2006).

18.4.1 *TRICHODERMA*

Trichoderma species are common in the soil and root ecosystems of several plants, and they are easily isolated from soil, decaying wood, and other organic material (Howell, 2003; Zeilinger and Omann, 2007). *Trichoderma* species have been used as biological control means against a wide range of pathogenic fungi: for example, *Rhizoctonia* spp., *Pythium* spp., *Botrytis cinerea*, and *Fusarium* spp. *Phytophthora palmivora*, *P. parasitica*, and different species can be used: for example, *T. harzianum*, *T. viride*, *T. virens* (Sunantapongsuk et al., 2006; Zeilinger and Omann, 2007). Among them, *Trichoderma harzianum* is reported to be the most widely used as an efficient BCA (Szekeres et al., 2004; Abdel-Fattah et al., 2007). *Trichoderma* has a combination of biocontrol mechanisms (Benítez et al., 2004; Zeilinger and Omann, 2007). The main mechanisms are mycoparasitism and antibiosis (Vinale et al., 2008). Mycoparasitism relies on the recognition, binding,

TABLE 18.2
Fungal Biofungicides/Mycofungicides Developed or Being Developed for the Biological Control of Diseases

Producer Companies					
Product	**Fungus**		**Target Pests and Countries**		
AQ10	*Ampelomyces quisqualis*	Powdery mildews	Ecogen Inc., US		
Biofungicide Aspire		*Candida oleophila*	*Botrytis* spp., *Penicillium* spp. Ecogen Inc., US		
Binab T	*Trichoderma harzianum*	Fungi causing wilt	Bio-Innovation, Sweden		
Biofox C	*Fusarium oxysporium*	*Fusarium oxysporium, F. moniliforme*	SIAPA, Italy		
Bio-Trek,	*Trichoderma harzianum*	*Rhizoctonia solani,*	BioWorks (TGT Inc.)		
RootShield	*Sclerotium rolfsii, Pythium*	Geneva, US			
Cotans WG	*Coniothyrium minitans*	*Sclerotinia* species	Prophyta, Germany KONI, Germany		
Fusaclean	*Fusarium oxysporium*	*Fusarium oxysporium*	Natural Plant Protection, France Thailand		
Ketomium	*Chaetomium* sp.	*Botrytis cinerea, Didymella*			
applanata, Fusarium oxysporum, and *Rhizoctonia solani*					
Neemoderma	*Trichoderma harzianum,Trichoderma viride*		Wide range of fungal diseases Shri Ram Solvent Ext.Pvt., India		
Polygandron,	*PPythium oligandrum*	*PPythium ultimum*	PPlant Protection Institute, Polyversum	Slovak Republic	
Primastop	*Gliocladium catenulatum* Several plant diseases		Kemira, Agro Oy, Finland		
SoilGard	ThermoTrilogy, USA	*Gliocladium virens*	Several plant diseases, (GlioGard)	damping off, and root	
pathogens					
T-22 and	*Trichoderma harzianum*	*Rhizoctonia solani,*	BioWorks (TGT Inc) T-22HB	*Sclerotium rolfsii, Pythium*	Geneva, US
Trichoderma	*Trichoderma harzianum*	*Rhizoctonia solani,*	Mycontrol (EfA1) Ltd, 2000	*Sclerotium rolfsii, Pythium*	Israel
Trichodex	*Trichoderma harzianum cinerea*		Fungal diseases e.g., *Botrytis*	Makhteshim-Agan, several	
European companies e.g., DeCeuster, Belgium.					
Trichodowels, *Trichoderma viride*		*Chondrostereum purpureum* Agrimms Biologicals,			
Trichoject,	and other soil and foliar	New Zealand pathogens			
Trichopel	*Trichoderma harzianum*	Wide range of fungal diseases Agrimm Technologies Ltd, New Zealand YIELDPLUS			
Cryptococcus albidus		*Botrytis* spp., *Penicillium* spp. Anchor Yeast, S. Africa			

Source: Adapted and modified from Butt et al. (2001) and Soytong et al. (2001)

and enzymatic disturbance of the host fungus cell wall (Woo and Lorito, 2007). *Trichoderma* species have been very successfully used as mycofungicides as they are fast growing and have a high reproductive capability. *Trichoderma* also inhibit a broad spectrum of fungal diseases and have a variety of control mechanisms. *Trichoderma* have excellent competitors in the rhizosphere and the capacity to modify it and are tolerant of or resistant to soil fungicides. This fungus also has the ability to survive under unfavorable conditions, is efficient in utilizing soil nutrients, is strong against phytopathogenic fungi, and also promotes plant growth (Benítez et al., 2004; Vinale et al., 2006). Its ability to colonize and produce in association with plant roots is known as rhizosphere competence.

18.4.2 *Chaetomium*

Chaetomium species are usually found in soil and organic compost (Soytong et al., 2001). The application of *Chaetomim* as a biological control for controlling of plant pathogens began in about 1954 when Martin Tviet and M. B. Moor found *C. globosum* and *C. cochliodes* on oat seeds and that these taxa provided some control of *Helminthosporium victoriae* (Tviet and Moor, 1954). *Chaetomium* species have been reported to be potential antagonists of various plant pathogens, especially soil-borne and seed-borne pathogens (Aggarwal et al., 2004; Park et al., 2005). Numerous species of *Chaetomium* with probable BCA capacity suppress the growth of bacteria and fungi through competition (for substrate and nutrients), mycoparasitism, antibiosis, or various combinations of these (Marwah et al., 2007; Zhang and Yang, 2007). *Chaetomium globosum* and *C. cupreum* have been extensively considered and successfully applied to control root rot in citrus, black pepper, and strawberries and have been reported to minimize damping off in sugar beets (Soytong et al., 2001; Tomilova and Shternshis, 2006). These taxa have been prepared in the form of powder and pellets as Ketomium, a broad spectrum mycofungicide. Ketomium has been also registered as a biofertilizer for decaying organic matter and to induce plant immunity and stimulate plant growth (Soytong et al., 2001).

18.4.3 *Gliocladium*

Gliocladium species are general soil saprobes, and numerous species have been reported to be parasites of several plant pathogens (Viterbo et al., 2007): for example, *Gliocladium catenulatum* parasites, *Sporidesmium sclerotiorum*, and *Fusarium* spp. These species demolish the fungal host by direct hyphal contact and form pseudoappressoria (Punja and Utkhede, 2004; Viterbo et al., 2007). *Gliocladium catenulatum* (Strain JI446) has also been employed as a wettable powder called Primastop by Kemira Agro Oy in Finland. This product can be used on soils, roots, and foliage to minimize the frequency of damping-off disease caused by *Pythium ultimum* and *Rhizoctonia solani* in the greenhouse (Paulitz and Belanger, 2001; Punja and Utkhede, 2004). *Gliocladium virens* produces antibiotic metabolites such as gliotoxin, which have important functions, including antibacterial, antifungal, antiviral, and antitumor. Recently, molecular confirmation indicated that *G. virens* is more strongly related to *Trichoderma* than those of *Gliocladium*. This chains suggest that this taxon should be referred to as *Trichoderma virens* (Hebbar and Lumsden, 1999; Punja and Utkhede, 2004).

18.4.4 *Ampelomyces*

Ampelomyces quisqualis is the mycoparasitic anamorphic ascomycete that decreases the growth of and kills powdery mildews. It can affect the pathogen through antibiosis and parasitism (Kiss, 2003; Viterbo et al., 2007). The fungus *A. quisqualis* was the initial organism reported to be a hyperparasite of powdery mildew, and it can be established simply in powdery mildew colonies (Paulitz and Belanger, 2001). Hyphae of *Ampelomyces* pierce the hyphae of powdery mildews and grow inside, then kill all the parasitized cells (Kiss, 2003). *Ampelomyces quisqualis* isolate M-10 has been prepared as AQ10 Biofungicide, developed by Ecogen, Inc. in the United States. This mycofungicide includes conidia of *A. quisqualis* and is formulated as water-soluble granules for the control of powdery mildew on carrots, cucumbers, and mangos (Kiss, 2003; Viterbo et al., 2007).

18.4.5 Other Fungi as Biofungicides

Coniothyrium minitans is an anamorphic coelomycete which has been reported to be a mycoparasite of *Sclerotinia* species: for example, *Sclerotinia minor*, *S. sclerotiorum*, *S. trifoliorum*, and *S. cepivorum* (Viterbo et al., 2007; Whipps et al., 2008). It has been used effectively to control disease in numerous crops including lettuce, (Jones et al., 2004), rapeseed (Li et al., 2006), peanuts (Partridge et al., 2006), and alfalfa (Li et al., 2005). The utilization of nonpathogenic strains of *Fusarium oxysporum* to control Fusarium wilt has been reported for several crops, but there has been little commercial production due of a lack of understanding of their genetics, biology, and ecology (Fravel et al., 2003; Kvas et al., 2009). Nonpathogenic *F. oxysporum* strain Fo47 has been promoted as a liquid formulation called Fusaclean by Natural Plant Products, Nogueres, France, for soilless culture (Khetan, 2001; Paulitz and Belanger, 2001). *Pythium oligandrum* has proven its ability to control soil-borne pathogens both in the laboratory and in the field. *Pythium oligandrum* oospores have been used as seed treatments which minimize damping-off disease caused by *P. ultimum* in sugar beets (Khetan, 2001). *Pythium oligandrum* has been formulated as a granular or powder product called Polygangron by Vyskumny Ustav in the Slovak Republic (Khetan, 2001). This fungus has indirect effects by controlling pathogens in the rhizosphere and/or direct effects by inducing plant resistance after application. It also provokes plants to respond more quickly and efficiently to pathogen infections and enhances phosphorus uptake (Le Floch et al., 2003). Several other fungi that can be employed as mycofungicides are *Aspergillus* and *Penicillium* species. *Aspergillus* species are useful against white-rot basidiomycetes (Bruce and Highley, 1991). The fungal antagonists *Aureobasidium pullulans* and *Ulocladium atrum* have also been analyzed for control of *Botrytis aclada*, which causes onion neck rot in the field (Köhl et al., 1997).

18.5 MECHANISMS OF BIOLOGICAL CONTROL

Biological control may result from direct or indirect communication between the beneficial microorganisms and the pathogen. Direct communication may entail physical contact and synthesis of hydrolytic enzymes, toxic compounds, or antibiotics as well as competition. An indirect communication may result from induced resistance in the host plant or the use of organic soil modifications to advance the activity of antagonists against the pathogens (Benítez et al., 2004; Viterbo et al., 2007). The mechanisms of BCAs and reaction with the pathogen are numerous and complex interactions among the host and microorganisms. These mechanisms are influenced by soil type, temperature, pH, and the moisture of the plant and soil environment, as well as by the existence of other microorganisms (Howell, 2003). There are four principle mechanisms of biological control in plants: namely, antibiosis, competition, mycoparasitism or lysis, and induced resistance (Fravel et al., 2003; Viterbo et al., 2007).

18.5.1 Antibiosis

Antibiosis is described as the inhibition or destruction of the microorganism by substances such as specific or nonspecific metabolites or by the production of antibiotics that inhibit the growth of one or more microorganisms (Benítez et al., 2004; Haggag and Mohamed, 2007). Many BCA agents produce several types of antibiotics (Lewis et al., 1989; Handelsman and Stabb, 1996). Some antibiotics have been reported to play function in disease control (Lewis et al., 1989), either by impeding spore germination (fungistasis) or killing the cells (antibiosis) (Benítez et al., 2004; Haggag and Mohamed, 2007). *Gliocladium* and *Trichoderma* species are well known BCAs which create a wide range of antibiotics and control disease by various mechanisms (Whipps, 2001; Harman et al.,

2004). Gliovirin metabolites produced by *Gliocladium virens* can kill *Pythium ultimum* by causing coagulation of the protoplasm of the cell (Whipps, 2001; Viterbo et al., 2007).

18.5.2 Competition

Competition occurs between microorganisms when space and nutrients are limited (Lewis et al., 1989; Viterbo et al., 2007). The rhizosphere of a plant and the soil zone are where competition for space and nutrients mainly occurs (Viterbo et al., 2007). Competition between the BCA and the pathogen can result in displacement of the pathogen. BCAs can compete with other fungi for food and essential elements in the soil and around the rhizosphere (Chet et al., 1990; Irtwange, 2006) and can narrow the space or change the rhizosphere by acidifying the soil so that pathogens cannot breed (Benítez et al., 2004). For example, *Trichoderma harzianum* T-35's control of the *Fusarium* species on numerous crops arises via competition for nutrients and rhizosphere colonization (Viterbo et al., 2007).

18.5.3 Mycoparasitism

Mycoparasitism involves a complex process that includes the following steps: (1) the chemothophic growth of the antagonist to the host, (2) recognition of the host by the mycoparasite, (3) attachment, (4) excretion of extracellular enzymes, and (5) lysis and exploitation of the host (Whipps, 2001; Benítez et al., 2004; Viterbo et al., 2007). BCAs are able to lyse hyphae of pathogens by releasing lytic enzymes, and this is an important and powerful tool for control of plant disease (Chet et al., 1990, Viterbo et al., 2007) such as chitinases, proteases, and β-1, 3 glucanases (Whipps, 2001). For example, β-1, 3 glucanases produced by *Chaetomium* sp. can degrade the cell walls of plant pathogens including *Rhizoctonia solani*, *Gibberella zeae*, *Fusarium* sp., *Colletotrichum gloeosporioides*, and *Phoma* sp. (Sun et al., 2006). Proteases produced by *Trichoderma harzianum* T-39 are involved in the degradation of pathogen hyphal membranes and cell walls.

18.5.4 Induced Resistance

Induced resistance is reported in large number of plants in response to infestation by phytopathogens (Harman et al., 2004). Induced resistance of host plants can be restricted and/or systematic, depending on the nature, source, and amount of stimuli of pathogens (Pal and Gardener, 2006). Induced resistance by BCAs entails a similar suite of genes and gene products engaged in plant response known as systematic acquired resistance (SAR) (Handelsman and Stabb, 1996; Whipps, 2001). *Trichoderma* strains are proficient at creating interaction-induced metabolic changes in plants that enhance resistance to a broad range of plant pathogenic fungi (Harman et al., 2004).

18.6 BIOHERBICIDES

Phytopathogenic microorganisms or microbial phytotoxins are valuable source of biological weed management and apply in related ways to conventional herbicides (Goeden, 1999; Boyetchko et al., 2002; Boyetchko and Peng, 2004). Bioherbicides serve an additional significant role as a complementary component in flourishing integrated management strategies (Hoagland et al., 2007) and not as a replacement for chemical herbicides and other weed management tactics (Singh et al., 2006). Several microbial components have been under evaluation for their prospects as bioherbicides for horticultural crops, turf, and forest trees, including obligate fungal parasites, soil-borne fungal pathogens, nonphytopathogenic fungi, pathogenic and nonpathogenic bacteria, and nematodes (Kremer, 2005). DeVine (Encore Technologies, Plymouth, MN, US) was one of the first herbicides registered with the active ingredient *Phytophthora palmivora*, which was developed to manage the strangler vine (*Morrenia odorata*) on citrus in Florida (Charudattan, 2005) (Table 18.3).

TABLE 18.3
Fungal Bioherbicides/Mycoherbicides Developed or Being Developed for the Biological Control of Weeds

Supplier and Country Sl. no.	Fungus	Product Name	Target Weed Where Registered
1. *Cercospora rodmanii* (*Eichhornia crassipes*)	ABG 5003	Water hyacinth	Abbott Labs, US
2. *Chondrosterium purpureum* BioChon (*Prunus serotina*)	Black cherry Netherlands in forestry	Koppert, The	
3. *Colletotrichum gloeosporioides* f. sp. *Malvae*	Biomal *pusilla*) in wheat and lentils	Mallow (*Malva*	Canada
4. *Alternaria cassiae obtusifolia*) and coffee senna (*C. occidentalis*) in soybeans and peanuts	Casst	Sicklepod (*Cassia*	US
5. *Colletotrichum gloeosporioides* f. sp *aeschynomene*	Collego (*Aeschynomene virginica*) in rice	Northern jointvetch Encore Technologies, US	
6. *Phytophthora palmivora* (*Morrenia odorata*) in Florida citrus	Devine	Milkweed vine	Sumitomo, Valent, US
7. *Colletotrichum gloeosporioides* f. sp *cuscutae*	Luboa 2 *C. australis* in soybeans	*Cuscuta chinensis,*	PR China
8. *Colletotrichum coccodes theophrasti*) in corn and soybeans	Velgo	Velvetleaf (*Abutilon* US, Canada	

Plant pathogens are used as BCAs and can cause severe damage to target weed species. In order to become appropriate pathogens, they have to be mass produced and their pathogenicity tested on weeds in a variety of environmental conditions, followed by field efficacy and host range tests (Ayres and Paul, 1990). A range of phytotoxins produced by plant pathogens can interfere with plant metabolism, ranging from delicate effects on gene expression to plant mortality (Walton, 1996). A few fungal pathogens are toxic to a broad range of weed species. The early mycoherbicides (DeVine; Collego, with the active ingredient *Colletotrichum gloeosporioides* f. sp. *aeschynomene*; Biomal, with the active ingredient *Colletotrichum gloeosporioides*) had extremely virulent fungal plant pathogens that could be mass cultured to produce large quantities of inoculum for inundative application to the weed host. These fungi infect the aerial portion of weed hosts, resulting in noticeable disease symptoms (Charudattan, 2005).

The rust fungus *Puccinia canaliculata* is a foliar pathogen of yellow nutsedge (*Cyperus esculentus*), and it can be mass cultured on the weed host in little field plots or in greenhouses (Phatak et al., 1983). Application of the fungal pathogen *Chonrotereum purpureum* to damaged branches or stumps of weedy tree species inhibited resprouting and corroded the woody tissues of the plant (Prasad, 1996). Weidemann et al. (1992) reported that the fungal pathogen *Microsphaeropsis amaranthi* controlled a few pigweed (*Amaranthus*) species, while *Phoma proboscis* controlled field bindweed (*Convolvulus arvensis*), and *Colletotrichum capsici* controlled morning glory (*Ipomoea* spp.). The naturally occurring fungus *Phoma macrostoma* has been tested as a control agent for dandelion (*Taraxacum officinale*), Canada thistle (*Cirsium arvense*), chickweed (*Stellaria media*), and scentless chamomile (*Matricaria perforata*), and its result is comparable to the industry standard synthetic herbicide pendimethalin (Bailey and Derby, 2001). A study by Héraux et al. (2005) on *Trichoderma virens* (*Gliocladium virens*) reported that colonization of composted chicken manure significantly reduced the emergence and growth of redroot pigweed (*Amaranthus retroflexus*) and broadleaf weeds in fields of horticulture crops.

18.7 LIMITATIONS IN SUCCESSFUL UTILIZATION OF FUNGAL BIOPESTICIDES

Billions of dollars have been invested by the biopesticide companies for the development of a variety of microbial products so as to eliminate crop diseases. It is impossible to accurately predict the market trends for biopesticides, and there is a substantial difference between predicting global sales and selecting categories of biopesticides. The agriculture market is observing an increase in demand for environmentally friendly, chemical residue–free organic products. However, growth in several regions is hindered by vigorous established chemical pesticide markets. However, there is lack of awareness of benefits of biopesticides and their uneven efficiency. The lack of awareness, knowledge, and confidence by farmers is one of the major reasons for the lagging of these ecofriendly pest control options. Commercial biological control of fungal biopesticides in the present global scenario is a high-tech venture, in terms of both safety and sustenance. The viability and virulence of fungal inoculum (conidia) after field application is the prerequisite threshold for their efficacy (Doust and Roberts, 1983). Diverse isolates of *B. bassiana* and *M. anisopliae* have been the most relied-on entomopathogens, have been extensively researched, and have considerable effect in field tests and their commercial usage for insect pest management (Easwaramoorthy, 2003). Upon field application, the entomopathogens are exposed to a variety of abiotic stresses like temperature (Rangel et al., 2005a), UV radiation (Rangel et al., 2006a), humidity-osmolarity (Lazzarini et al., 2006), edaphic factors, and nutrient sources (Shah et al., 2005) that negatively affect the field use of fungi as BCAs. Solar radiation, which includes visible light, UV radiation, infrared rays, and radio waves, has been the dominant source in which all organisms evolve and adapt.

In a biological context, UV radiation merits a special mention in terms of its impact on life (Bjorn, 2006). Soil temperature is a major factor which affects the success or failure in the establishment and production of fungal inoculum (Thomas and Jenkins, 1997). Entomopathogenic fungi have tolerance to soil temperature and also have the capacity to survive through the thermoregulatory defense response of the host insect (Ouedraogo et al., 2003). It has been confirmed that high/low temperatures change the vegetative growth among isolates of entomopathogenic fungi (Ouedraogo et al., 2004). Dry heat exposure causes DNA breakage through base loss, leading to depurination, and this may cause mutation in several cases (Nicholson et al., 2000). Wet heat (that is, heat in conjunction with high humidity) results in protein denaturation and membrane disorganization of the cell. It has been reported that *M. anisopliae* has a temperature tolerance upper limit of 37–40 °C (Thomas and Jenkins, 1997).

B. bassiana, on the other hand can survive up to a maximum temperature of 37 °C (Fargues et al., 1997). In fungi, the temperature range for germination and mycelial growth has been reported to have a similar pattern. Environmental factors also affect pathogenicity as well as mode of virulence of entomopathogenic fungi (Hasan, 2014).

The efficacy of fungal bioherbicides, often due to environmental factors, is the major limiting feature for their use. There are humidity needs for the establishment and spread of many foliar and stem fungal pathogens for weed control. There are also a few elements necessary for the development of special formulations to ensure the effectiveness of agents applied in the field. An extended dew period is needed by a few pathogens for infection on the aerial surfaces of target weeds (Auld et al., 2003). The bioherbicide application process should be judged by how it enhances the efficacy of the BCA. This includes attention to spray droplet size, droplet retention and distribution, spray application volume, and the equipment used (Charudattan, 2001). Other factors, such as the spectrum of the bioherbicide (whether broad or targeted to specific species), the type of formulation, and whether it involves amino acid–excreting strains, can significantly affect efficacy.

18.8 CONCLUSION

The relevance of fungal biopesticides in biological control is increasing largely because of superior environmental awareness, food safety concerns, and the breakdown of conventional chemicals due to an increasing number of insecticide-resistant species. Whether or not the use of fungal biopesticides has been thriving in pest and disease management, it is necessary to regard each case individually, and direct comparisons with chemical insecticides are usually unsuitable. Fungal biopesticides as components of an integrated system can offer significant and selective insect and disease control. In the near future, we should see synergistic combinations of microbial control agents with other technologies (in combination with semiochemicals, soft chemical pesticides, fungicides, other natural enemies, resistant plants, chemigation, remote sensing, etc.) that will augment the effectiveness and sustainability of integrated control strategies for plant disease.

REFERENCES

Abdel-Fattah, M. G.; Shabana, M. Y.; Ismail, E. A.; Rashad, M. Y. *Trichoderma harzianum*: A Biocontrol Agent against *Bipolaris oryzae*. *Mycopathologia* **2007,** *164*, 81–89.

Aggarwal, R.; Tewari, A. K.; Srivastava, K. D.; Singh, D. V. Role of Antibiosis in the Biological Control of Spot Blotch (*Cochliobolus sativus*) of Wheat by Chaetomium Globosum. *Mycopathologia* **2004,** *157*, 369–377.

Anand, S.; Reddy, J. Biocontrol Potential of *Trichoderma* Sp against Plant Pathogens. *Inter. J. Agri. Sci.* **2009,** *2*, 30–39.

Auld, B. A.; Hethering, S. D.; Smith, H. E. Advances in Bioherbicide Formulation. *Weed Biol. Man.* **2003,** *3*, 61–67.

Ayres, P.; Paul, N. Weeding with Fungi. *New Sci.* **1990,** *732*, 36–39.

Bailey, D. J.; Gilligan, C. A. Modeling and Analysis of Disease Induced Host Growth in the Epidemiology of Take All. *Phytopathology* **2004,** *94*, 535–540.

Bailey, K. L.; Derby, J. Fungal Isolates and Biological Control Compositions for the Control of Weeds. U.S. Patent 60/294,475, May 20, 2001.

Beilharz, V. C.; Parbery, D. G.; Swart, H. J. Dodine: A Selective Agent for Certain Soil Fungi. *Trans. British Mycol. Soc.* **1982**, *79*(3), 507–511.

Benítez, T.; Rincón, M. A.; Limón, M. C.; Codón, C. A. Biocontrol Mechanisms of *Trichoderma* Strains. *Int. Microbiol.* **2004,** *7*, 249–260.

Bidochka, M. J.; St Leger, R. J.; Joshi, L.; Roberts, D. W. An Inner Cell Wall Protein (Cwp1) from Conidia of the Entomopathogenic Fungus *Beauveria bassiana. Microbiology* **1995,** *141*, 1075–1080.

Bjorn, L. O. Stratospheric Ozone, Ultraviolet Radiation, and Cryptogams. *Biol. Conservation* **2006,** *135*, 326–333.

Boucias, D. R.; Pendland, J. C. Entomopathogenic Fungi: Fungi Imperfecti. In *Principles of Insect Pathology*; Boucias, D. R.; Pendland, J. C., Eds.; Kluwer Academic Publishers: Dordrecht, 1998, pp 321–359.

Boyetchko, S. M.; Peng, G. Challenges and Strategies for Development of Mycoherbicides. In *Fungal Biotechnology in Agricultural, Food, and Environmental Applications*; Arora, D. K., Ed.; Marcel Dekker: New York, 2004, pp 11–121.

Boyetchko, S. M.; Rosskopf, E. N.; Caesar, A. J. Charudattan, R. Biological Weed Control with Pathogens: Search for Candidates to Applications. In *Applied Mycology and Biotechnology*, Vol. 2; Khachatourians, G. G.; Arora, D. K. Eds.; Elsevier: Amsterdam, 2002, pp 239–274.

Bruce, A.; Highley, L. T. Control of Growth of Wood Decay Basidiomycetes by *Trichoderma* Spp. and Other Potentially Antagonistic Fungi. *Forest Products J.* **1991,** *41*, 63–67.

Butt, T. M.; Jackson, C. W.; Magan, N., Eds. *Fungal Biological Control Agents: Progress, Problems and Potential*. CABI International: Wallingford, Oxon, 2001.

Charnley, A. K. Mechanisms of Fungal Pathogenesis in Insects. In *The Biotechnology of Fungi for Improving Plant Growth*; Whipps, J. M.; Lumsden, R. D., Eds.; Cambridge University Press: London, 1989, pp 85–125.

Charudattan, R. Biological Control of Weeds by Means of Plants Pathogens: Significance for Integrated Weed Management in Modern Agroecology. *Biocontrol* **2001,** *46*, 229–260.

Charudattan, R. Use of Plant Pathogens as Bioherbicides to Manage Weeds in Horticultural Crops. *Proc. Fla. State Hort. Soc.* **2005,** *118*, 208–214.

Chet, I.; Ordentlich, A.; Shapira, R.; Oppenheim, A. Mechanisms of Biocontrol of Soil-borne Plant Pathogens by *Rhizobacteria. Plant Soil* **1990,** *129*, 85–92.

Cho, E. M.; Boucias, D.; Keyhani, N. O. EST Analysis of CDNA Libraries from the Entomopathogenic Fungus *Beauveria* (Cordyceps) *Bassiana*. II. Fungal Cells Sporulating on Chitin and Producing Oosporein. *Microbiology* **2006,** *152*, 2855–2864.

Clarkson, J. M.; Charnley, A. K. New Insights into the Mechanisms of Fungal Pathogenesis in Insects. *Trends Microbiol.* **1996,** *4*, 197–203.

Dominguesa, F. C.; Queiroza, J. A.; Cabralb, J. M. S.; Fonsecab, L. P. The Influence of Culture Conditions on Mycelial Structure and Cellulose Production by *Trichoderma reesei* rut C-30. *Enz. Microb. Technol.* **2000,** *26*, 394–401.

Doust, R. A.; Roberts, D. W. Studies on the Prolonged Storage of *Metarhizium Anisopliae* Conidia: Effect of Temperature and Relative Humidity on Conidial Viability and Virulence against Mosquitoes. *J. Invertebrate Pathol.* **1983,** *41*, 143–150.

Driver, F.; Milner, R. J.; Trueman, J. W. H. A Taxonomic Revision of *Metarhizium* Based on a Phylogenetic Analysis of RDNA Sequence Data. *Mycol. Res.* **2000,** *104*, 134–150.

Dunn, M. T.; Sayee, R. M.; Carrell, A.; Wergin, W. R. Colonization of Nematodes Eggs by *P. lilacinus* Samson of Observed with SEM. *SEM* **1982,** *3*, 1351–1357.

Easwaramoorthy, S. Entomopathogenic Fungi. In *Biopesticides and Bioagents in Integrated Pest Management of Agricultural Crops*; Srivastava, R. P. Ed.; International Book Distributing Co.: Lucknow, 2003, pp 341–379.

Fargues, J.; Goettel, M. S.; Smits, N.; Ouedraogo, A.; Rougier, M. Effect of Temperature on Vegetative Growth of *Beauveria Bassiana* Isolates from Different Origins. *Mycologia* **1997,** *89*, 383–382.

Ferron, P. Biological Control of Insect Pests by Entomologenous Fungi. *Ann. Rev. Entomol.* **1978,** *23*, 409–442.

Fravel, D.; Olivain, C.; Alabouvette, C. *Fusarium oxysporum* and Its Biocontrol. *New Phytologist* **2003,** *157*, 493–502.

Furlong, M. J.; Pell, K. J. Interactions between Entomopathogenic Fungi and Arthropod Natural Enemies. In *Insect—fungal Associations: Ecology and Evolution*; Vega, F. E.; Blackwell, M., Eds.; Oxford University Press: Oxford, 2005, pp 51–73.

Goeden, R. D. Projects on Biological Control of Russian Thistle and Milk Thistle in California: Failures That Contributed to the Science of Biological Weed Control. In *Abstracts of the 10th International Symposium on Biological Control of Weeds*; Spencer, N.; Noweierski, R., Eds.; Montana State University: Bozeman, MT, 1999, p 27.

Guillon, M. L. Regulation of Biological Control Agents in Europe. In *International Symposium on Biopesticides for Developing Countries*; Roettger, U.; Reinhold, M., Eds.; CATIE: Turrialba, 2003, pp 143–147.

Haggag, W. M.; Mohamed, H. A. A. Biotechnological Aspects of Microorganisms Used in Plant Biological Control. *Am. Eur. J. Sustainable Agric.* **2007,** *1*, 7–12.

Hajek, A. E.; St. Leger, R. J. Interactions between Fungal Pathogens and Insects Hosts. *Ann. Rev. Entomol.* **1994,** *39*, 293–322.

Hamlen, R. A. Biological Control of Insects and Mites on European Greenhouse Crops: Research and Commercial Implementation. *Proc. Florida State Horticultural Soc.* **1979**, *92*, 367–368.

Handelsman, J.; Stabb, V. E. Biocontrol of Soilborne Plant Pathogens. *Plant Cell* **1996,** *8*, 1855–1869.

Harman, G. E. Overview of Mechanisms and Uses of *Trichoderma* spp. 648. *Phytopathology* **2005,** *96*, 190–194.

Harman, G. E.; Howell, C. R.; Viterbo, A.; Chet, I.; Lorito, M. *Trichoderma* Species Opportunistic, Avirulent Plant Symbionts. *Nat. Rev. Microbiol.* **2004,** *2*, 43–56.

Hasan, S. Entomopathogenic Fungi as Potent Agents of Biological Control. *Int. J. Eng. Technol. Res.* **2014,** *2*, 234–237.

Hebbar, P. K.; Lumsden, R. D. Biological Control of Seedling Diseases. In *Methods in Biotechnology Vol. 5: Biopesticides: Use and Delivery*; Frinklin, R. H.; Julius, J. M., Eds.; Humana Press: New York, 1999, pp 103–116.

Héraux, F. M. G.; Hallett, S. G.; Ragothama, K. G.; Weller, S. C. Composted Chicken Manure as a Medium for the Production and Delivery of *Trichoderma virens* for Weed Control. *Hort. Sci.* **2005,** *40*, 1394–1397.

Hoagland, R. E.; Weaver, M. A.; Boyette, C. D. Myrothecium Verrucaria Fungus: A Bioherbicide and Strategies to Reduce Its Non-target Risks. *Allelopathy J.* **2007,** *19*(1), 179–192.

Hong, T. D.; Ellis, R. H.; Moore, D. Development of a Model to Predict the Effect of Temperature and Moisture on Fungal Spore Longevity. *Ann. Botany* **1997,** *79*, 121–128.

Howell, R. C. Mechanisms Employed by *Trichoderma* Species in the Biological Control of Plant Diseases: The History and Evolution of Current Concepts. *Plant Dis.* **2003,** *87*, 4–10.

Ibrahim, I. K. A.; Raza, M. A.; El-Saedy, M. A.; Ibrahim, A. A. M. Control of *Meloidogyne incognita* on Corn, Tomato and Okra with *P. lilacinus* and the Nematicide Aldecarb. *Nematologia Mediterria* **1987,** *15*, 265–268.

Ignoffo, C. M. The Fungus *Nomuraea rileyi* as a Microbial Insecticide. In *Microbial Control of Pests and Plant Diseases*; Burges, H. D., Ed., Academic Press: London, 1981, 513–538.

IOBC. International Organization for Biological Control. *IOBC Newslet.* **2008,** *84*, 5–7.

Irtwange, V. S. Application of Biological Control Agents in Pre- and Postharvest Operations. Agricultural Engineering International: The CIGR Ejournal. *Invited Overview.* **2006,** *3*, 1–12.

Jain, N.; Rana, I. S.; Kanojiya, A.; Sandhu, S. S. Characterization of *Beaveria bassiana* Strains Based on Protease and Lipase Activity and Their Role in Pathogenicity. *J. Basic Appl. Mycol.* **2008,** *I–II*, 18–22.

Jianzhong, S.; Fuxa, J. R.; Henderson, G. Effects of Virulence, Sporulation, and Temperature on *Metarhizium anisopliae* and *Beauveria bassiana* Laboratory Transmission in *Coptotermes formosanus. J. Invertebrate Pathol.* **2003,** *84*, 38–46.

Jones, E. E.; Mead, A.; Whipps, J. M. Effect of Inoculum Type and Timing of Application of *Coniothyrium Minitans* on *Sclerotinia Sclerotio*46 *Rum*: Control of Sclerotinia Disease in Glasshouse Lettuce. *Plant Pathol.* **2004,** *53*, 611–620.

Khachatourians, G. G. Biochemistry and Molecular Biology of Entomopathogenic Fungi. In *Human and Animal Relationships*; Howard, D. H.; Miller, J. D., Eds.; *Mycota* VI, Springer: Heidelberg, 1996, pp 331–363.

Khandelwal, M.; Datta, S.; Mehta, J.; Naruka, R.; Makhijani, K.; Sharma, G.; Kumar, R.; Chandra, S. Isolation, Characterization and Biomass Production of *Trichoderma Viride* Using Various Agro Products—A Biocontrol Agent. *Adv. Appl. Sci. Res.* **2012,** *3*, 3950–3955.

Khetan, S. K. *Microbial Pest Control.* Marcel Dekker Inc.: New York, Basel, 2001, p 300.

Kim, J. J.; Lee, M. H.; Yoon, C. S.; Kim, H. S.; Yoo, J. K.; Kim, K. C. Control of Cotton Aphid and Greenhouse Whitefly with a Fungal Pathogen. *J. Nat. Inst. Agric. Sci. Technol.* **2002,** 7–14.

Kirk, P. M. *Indexfungorum.* 2003. www.indexfungorum.org (accessed Oct 28, 2009).

Kiss, L. A Review of Fungal Antagonists of Powdery Mildews and Their Potential as Biocontrol Agents. *Pest Management Sci.* **2003,** *59*, 475–483.

Köhl, J.; Bélanger, R. R.; Fokkema, N. J. Interaction of Four Antagonistic Fungi with *Botrytis Aclada* in Dead Onion Leaves: A Comparative Microscopic and Ultrastructural Study. *Phytopathology* **1997,** *87*, 634–642.

Kremer, R. J. The Role of Bioherbicides in Weed Management. *Biopestic. Int.* **2005,** *1*, 127–141.

Kvas, M.; Marasas, W. F. O.; Wingfield, B. D.; Wingfield, M. J.; Steenkamp, E. T. Diversity and Evolution of *Fusarium* Species in the Gibberella Fujikuroi Complex. *Fungal Diversity* **2009,** *34*, 1–21.

Lacey, C. M.; Lacey, L. M.; Roberts, D. R. Route of Invasion and Histopathology of *Metrahizium anisopliae* in *Culex quinquefasciatus. J. Invertebrate Pathol.* **1988,** *52*, 108–118.

Lazzarini, G. M. J.; Rocha, L. F. N.; Luz, C. Impact of Moisture on In Vitro Germination of *Metarhizium anisopliae* and *Beauveria bassiana* and Their Activity on *Triatoma infestans*. *Mycol. Res.* **2006,** *100*, 485–492.

Le Floch, G.; Rey, P.; Benizri, E.; Benhamou, N.; Tirilly, Y. Impact of Auxin-compounds Produced by the Antagonistic Fungus *Pythium Oligandrum* or the Minor Pathogen *Pythium* Group F on Plant Growth. *Plant Soil* **2003,** *257*, 459–470.

Leger, R. J.; Joshi, L.; Bidochka, M. J.; Roberts, D. W. Construction of an Improved Mycoinsecticide Overexpressing a Toxic Protease. *Proc. Natl. Acad. Sci. U. S. Am.* **1996,** *93*, 6349–6354.

Lewis, K.; Whipps, J. M.; Cooke, R. C. Mechanisms of Biological Disease Control with Special Reference to the Case Study of Pytium Oligandrum as an Antagonist. In *Biotechnology of Fungi for Improving Plant Growth*; Whipps, J. M.; Lumsden, R. D., Eds.; Cambridge University Press: Cambridge, UK, 1989, pp 191–217.

Li, G. Q.; Huang, H. C.; Acharya, S. N.; Erickson, R. S. Effectiveness of *Coniothyrium minitans* and *Trichoderma atroviride* in Suppression of Sclerotinia Blossom Blight of Alfalfa. *Plant Pathol.* **2005,** *54*, 204–211.

Li, G. Q.; Huang, H. C.; Miao, H. J.; Erickson, R. S.; Jiang, D. H.; Xiao, Y. N. Biological Control of Sclerotinia Diseases of Rapeseed by Aerial Applications of the Mycoparasite *Coniothyrium minitans*. *Eur. J. Plant Pathol.* **2006,** *114*, 345–355.

Liang, C.; Yang, J.; Kovács, G. M.; Szentiványi, O.; Li, B.; Xu, X. M.; Kiss, L. Genetic Diversity of *Ampelomyces Mycoparasites* Isolated from Different Powdery Mildew Species in China Inferred from Analyses of RDNA ITS Sequences. *Fungal Diversity* **2007,** *24*, 225–240.

Marwah, R. G.; Fatope, M. O.; Deadman, M. L.; Al-Maqbali, Y. M.; Husband, J. Musanahol: A New Aureonitol-related Metabolite from a *Chaetomium* sp. *Tetrahedron* **2007**, *63*, 8174–8180.

Meadow, R.; Vandberg, J. D.; Shelton, A. M. Exchange of Inoculum of *Beauveria bassiana* (Bals) Vuil. (Hyphomycetes) between Adult Flies of the Cabbage Maggot *Delia radicum* L. (Diptera: Anthomyiidae). *Biocontrol Sci. Technol.* **2000**, *10*, 479–485.

Mitsuaka, S. Effect of Temperature on Growth of *Beauveria Bassiana* F-263, A Strain Highly Virulent to the Japanese Pine Sawyer, *Monochamus alternatus*, Especially Tolerance to High Temperatures. *Appl. Entomol. Zool.* **2004,** *39*, 469–475.

Nicholson, W. L.; Munakata, N.; Horneck, G.; Melosh, H. J.; Setlow, P. Resistance of *Bacillus* Endospores to Extreme Terrestrial and Extraterrestrial Environments. *Microbiol. Mol. Biol. Rev.* **2000,** *64*, 548–572.

Noma, T.; Strickler, K. Effects of *Beauveria bassiana* on Lygus *Hesperus* (Hemiptera: Miridae) Feeding and Oviposition. *Environ, Entmol.* **2000,** *29*, 394–402.

Nunez, E. J.; Iannacone; Omez, H. G. Effect of Two Entomopathogenic Fungi in Controlling *Aleurodicus cocois* (Curtis, 1846) (Hemiptera: Aleyrodidae). *Chilean J. Agric. Res.* **2008,** *68*, 21–30.

O'Brien, K. P.; Franjevic, S.; Jones, J. Green Chemistry and Sustainable Agriculture: The Role of Biopesticides. *Advancing Green Chem.* **2009**. http://advancinggreenchemistry.org/wp-content/uploads/Green-Chemand-Sus-Ag-the-Role-of-Biopesticides.pdf (accessed May 7, 2016).

Ouedraogo, A.; Fargues, J.; Goettel, M. S.; Lomer, C. J. Effect of Temperature on Vegetative Growth Among Isolates of *Metarhizium anisopliae* and *Metarhizium flavoviride*. *Mycopathologia* **2004,** *137*, 37–43.

Ouedraogo, R. M.; Cusson, M.; Goettel, M. S.; Brodeur, J. Inhibition of Fungal Growth in Thermoregulating Locusts, *Locusta Migratoria*, Infected by the Fungus *Metarhizium anisopliae* var. *acridum*. *J. Invertebr. Pathol.* **2003,** *82*, 103–109.

Pal, K.; Gardener, B. M. Biological Control of Plant Pathogens. *Plant Health Instructor* **2006,** 1–25. doi: 10.1094/PHI-A-2006-1117-02. APSnet.

Park, J. H.; Choi, G. J.; Jang, S. K.; Lim, K. H.; Kim, T. H.; Cho, Y. K.; Kim, J. C. Antifungal Activity against Plant Pathogenic Fungi of Chaetoviridins Isolated from *Chaetomium globosum*. *FEMS Microbiol. Lett.* **2005,** *252*, 309–313.

Partridge, D. E.; Sutton, T. B.; Jordan, D. L.; Curtis, V. L.; Bailey, J. E. Management of Sclerotinia Blight of Peanut with the Biological Control Agent *Coniothyrium minitans*. *Plant Dis.* **2006,** *90*, 957–963.

Paulitz, T. C.; Belanger, R. R. Biological Control in Greenhouse System. *Ann. Rev. Phytopathol.* **2001,** *39*, 103–133.

Phatak, S. C.; Summer, D. R.; Wells, H. D.; Bell, D. K.; Glaze, N. C. Biological Control of Yellow Nutsedge with the Indigenous Rust Fungus *Puccinia canaliculata*. *Science* **1983,** *219*, 1446–1447.

Prasad, R. Development of Bioherbicides for Integrated Weed Management in Forestry. In *Proceedings of the 2nd International Weed Control Congress*; Brown, H., Ed.; Department of Weed Control and Pesticide Ecology: Slagelse, 25–28 June 1996, pp 1197–1203.

Punja, Z. K.; Utkhede, R. S. Biological Control of Fungal Diseases on Vegetable Crops with Fungi and Yeasts. In *Fungal Biotechnology in Agricultural, Food, and Environmental Applications*; Arora, D. K., Ed.; Basel: New York, 2004, pp 157–171.

Quesada-Moraga, E.; Santos-Quirós, R.; Valverde-García, P.; Santiago, Á. C. Virulence, Horizontal Transmission, and Sublethal Reproductive Effects of *Metarhizium Anisopliae* (Anamorphic Fungi) on the German Cockroach (Blattodea: Blattellidae). *J. Invertebr. Pathol.* **2004,** *87*, 51–58.

Rai, D.; Updhyay, V.; Mehra, P.; Rana, M.; Pandey, A. K. Potential of Entomopathogenic Fungi as Biopesticides. *Ind. J. Sci. Res. Tech.* **2014**, *2*(5), 7–13.

Rangel, D. E. N.; Braga, G. U. L.; Anderson, A. J.; Roberts, D. W. Variability in Conidial Thermo Tolerance of *Metarhizium Anisopliae* Isolates from Different Geographic Origins. *J. Invertebr. Pathol.* **2005a**, *88*, 116–125.

Rangel, D. E. N.; Butler, M. J.; Torabinejad, J.; Anderson, A. J.; Braga, G. U. L.; Day, A. W.; Roberts, D. W. Mutants and Isolates of *Metarhizium Anisopliae* are Diverse in Their Relationships between Conidial Pigmentation and Stress Tolerance. *J. Invertebr. Pathol* **2006a**, *93*, 170–182.

Richard, J. S.; Neal, T. D.; Karl, J. K.; Michael, R. K. Model Reactions For Insect Cuticle Sclerotization: Participation of Amino Groups in the Cross-Linking of *Manduca Sexta* Cuticle Protein MsCP36. *Insect Biochem. Mol. Biol.* **2010**, *40*, 252–258.

Roberts, D. W. Toxins of Entomopathogenic Fungi. In *Microbial Control of Pests and Plant Diseases*; Burges, H. D., Ed.; Academic Press: New York, 1981, pp 441–464.

Sandhu, S. S.; Vikrant, P. Myco-insecticides: Control of Insect Pests. In *Microbial Diversity: Opportunities & Challenges*; Gautam, S. P.; Sandhu, S. S.; Sharma, A.; Pandey, A. K., Eds.; Indica Publishers: New Delhi, 2004.

Shah, F. A.; Wang, C. S.; Butt, T. M. Nutrition Influences Growth and Virulence of the Insect Pathogenic Fungus *Metarhizium anisopliae*. *Microbiol. Lett.* **2005**, *251*, 259–266.

Singh, H. P.; Batish, D. R.; Kohli, R. K. *Handbook of Sustainable Weed Management;* Food Products Press: Binghamton, NY, 2006.

Sosa-Gómez, D. R.; Moscardi, F. Laboratory and Field Studies on the Infection of Stink Bugs, Nezara viridula, Piezodorus guildinii, and Euschistus heros (Hemiptera: Pentatomidae) with Metarhizium anisopliae and *Beauveria bassianain*. *Brazil J. Invertebr. Pathol.* **1998,** *2*, 115–120.

Soytong, K.; Kanokmadhakul, S.; Kukongviriyapa, V.; Isobe, M. Application of *Chaetomium* Species (Ketomium®) As a New Broad Spectrum Biological Fungicide for Plant Disease Control: A Review Article. *Fungal Diversity* **2001,** *7*, 1–15.

St Leger, R. J.; Bidochka, M. J. Staples, R. C. Preparation Events During Infection of Host Cuticle by *Metarhizium anisopliae. J. Invertebrate Pathol.* **1991,** *58*, 168–179.

St Leger, R. J.; Joshi, L.; Bidochka, M. J.; Roberts, D. W. Construction of an Improved Mycoinsecticide Overexpressing a Toxic Protease. *Proc. Natl. Acad. Sci. U. S. A.* **1996**, *93*, 6349–6354.

St Leger, R. J.; May, B.; Allee, L. L.; Frank, D. C.; Staples, R. C.; Roberts, D. W. Genetic Differences in Allozymes and in Formation of Infection Structures among Isolates of the Entomopathogenic Fungus *Metarhizium anisopliae. J. Invertebrate Pathol.* **1992,** *60*, 89–101.

Sun, H.; Yang, J.; Lin, C.; Huang, X.; Xing, R.; Zhang, K. Q. Purification and Properties of a B-1,3-Glucanase from *Chaetomium* Sp. That Is Involved in Mycoparasitism. *Biotechnol. Lett.* **2006,** *28*, 131–135.

Sunantapongsuk, V.; Nakapraves, P.; Piriyaprin, S.; Manoch, L. In *Protease Production and Phosphate Solubilization from Potential Biological Control Agents Trichoderma Viride and Azomonas Agilis from Vetiver Rhizosphere*, International Workshop on Sustained Management of Soil-Rhizosphere System for Efficient Crop Production and Fertilizer Use, Land Development Department, Bangkok, Thailand, 2006, pp 1–4.

Szekeres, A.; Kredics, L.; Antal, Z.; Kevei, F.; Manczinger, L. Isolation and Characterization of Protease Overproducing Mutants of *Trichoderma harzianum*. *Microbiol. Lett.* **2004**, *233*, 215–222.

Thakur, R.; Sandhu, S. S. Distribution, Occurrence and Natural Invertebrate Hosts of Indigenous Entomopathogenic Fungi of Central India. *Ind. J. Microbiol.* **2010,** *50*, 89–96.

Thomas, M. B.; Jenkins, N. E. Effect of Temperature on Growth of *Metarhizium Flavoviride* and Virulence to the Variegated Grasshopper, *Zonocerus variegatus*. *Mycol. Res.* **1997,** *101*, 1469–1474.

Tomilova, O. G.; Shternshis, M. V. The Effect of a Preparation from *Chaetomium* Fungi on the Growth of Phytopathogenic Fungi. *Appl. Biochem. Microbiol.* **2006,** *42*, 76–80.

Tulloch, M. The Genus *Metarhizium*. *Trans. Britannic Mycol. Soc.* **1976,** *66*, 407–411.

Tviet, M.; Moor, M. B. Isolates of Chaetomium That Protect Oats from *Helminthosporium victoriae*. *Phytopathology* **1954,** *44*, 686–689.

USEPA. *What Are Biopesticides?* 2008. www.epa.gov/pesticides/biopesticides/whatarebiopesticides.htm (accessed May 7, 2016).

Vey, A.; Hoagland, R.; Butt, T. M. Toxic Metabolites of Fungal Biocontrol Agents. In *Fungi as Biocontrol Agents*; Butt, T. M.; Jackson, C. W.; Magan, N., Eds.; CAB International: Wallingford, 2001, pp 311–345.

Vinale, F.; Marra, R.; Scala, F.; Ghisalbert, E. L.; Lorito, M.; Sivasithamparam, K. Major Secondary Metabolotes Produced by Two Commercial *Trichoderma* Strains Active against Different Phytopathogens. *Lett. Appl. Microbiol.* **2006,** *43*, 143–148.

Vinale, F.; Sivasithamparam, K.; Ghisalberti, E. L.; Marra, R.; Woo, S. L.; Lorito, M. Trichoderma Plant Pathogen Interactions. *Soil Biol. Biochem.* **2008,** *40*, 1–10.

Viterbo, A.; Inbar, J.; Hadar, Y.; Chet, I. Plant Disease Biocontrol and Induced Resistance via Fungal Mycoparasites. In *Environmental and Microbial Relationships*, 2nd edn, *The Mycota IV*; Kubicek, C. P.; Druzhinina, I. S., Eds.; Springer-Verlag: Berlin, Heidelberg, 2007, pp 127–146.

Walton, J. D. Host-selective Toxins: Agents of Compatibility. *Plant Cell* **1996,** *8*, 1723–1733.

Wang, C.; Hu, G.; St Leger, R. J. Differential Gene Expression by *Metarhizium Anisopliae* Growing in Root Exudate and Host (*Manduca Sexta*) Cuticle or Hemolymph Reveals Mechanisms of Physiological Adaptation. *Fungal Genetic Biol.* **2005,** *42*, 704–718.

Weidemann, G. J.; TeBeest, D. O.; Templeton, G. E. Fungal Plant Pathogens Used for Biological Weed Control. *Ark. Farming Res.* **1992,** *41*, 6–7.

Whipps, J. M. Microbial Interactions and Biocontrol in the Rhizosphere. *J. Exp. Botany* **2001,** *52*, 487–511.

Whipps, J. M.; Sreenivasaprasad, S.; Muthumeenakshi, S.; Rogers, C. W.; Challen, M. P. Use of *Coniothyrium minitans* as a Biocontrol Agent and Some Molecular Aspects of Sclerotial mycoparasitism. *Eur. J. Plant Pathol.* **2008,** *121*, 323–330.

Woo, L. S.; Lorito, M. Exploiting the Interactions between Fungal Antagonists, Pathogens and the Plant for Biocontrol. In *Novel Biotechnologies for Biocontrol Agent Enhancement and Management*; Vurro, M.; Gressel, J., Eds.; Springer: Berline, 2007, pp 107–130.

Wraight, S. P.; Carruthers, R. I.; Jaronski, S. T.; Bradley, C. A.; Garza, C. J.; Wraight, S. G. Evaluation of the Entomopathogenic Fungi *Beauveria Bassiana* and *Paecilomyces Fumosoroseus* for Microbial Control of the Silver Leaf Whitefly, *Bemisia argentifolii. Biol. Control* **2000,** *17*, 203–217.

Wraight, S. P.; Jackson, M. A.; de Kock, S. L. Production, Stabilization and Formulation of Fungi Biocontrol Agents. In *Fungi as Biocontrol Agents Progress, Problem and Potential;* Butt, T. M.; Jackson, C. W.; Magan, N., Eds.; CABi Publishing: Wallingford, 2001, pp 253–288.

Zeilinger, S.; Omann, M. Trichoderma Biocontrol: Signal Transduction Pathways Involved In Host Sensing and Mycoparasitism. *Gene Regulation Syst. Biol.* **2007,** *1*, 227–234.

Zhang, H. Y.; Yang, Q. Expressed Sequence Tags-based Identification of Genes in the Biocontrol Agent *Chaetomium cupreum*. *Appl. Microbiol. Biotechnol.* **2007,** *74*, 650–658.

Zimmermann, G. Review on Safety of the Entomopathogenic Fungus *Beauveria bassiana* and *Beauveria brongniartii*. *Biocontrol Sci. Technol.* **2007,** *17*, 553–596.

19 Allelopathy
A Green Sustainable Weed Management Approach

Puneet Kaur, SS Rana, Avnee, Randeep Kumar, Niraj Guleria, Chandini and Baljinder Singh

19.1 INTRODUCTION

Weeds are among the most challenging problems facing agricultural production across the world. They are recognized as the most significant biotic constraints on crop production, with crop yield losses ranging from 45 to 95%, depending on agronomic practices used and the environment [1]. As weeds compete for nutrients, water, light, and space, they reduce crop growth and yield. Additionally, weeds harbor insect pests and fungal, bacterial, and viral pathogens, further reducing agricultural quality and market value. By 2050, the world's population is estimated to reach more than 9 billion, so weed competition is likely to significantly affect crop yields [2]. The majority of weed control in developed countries is done with synthetic herbicides. No herbicides with new mechanisms of action have been introduced since the 1980s. Using herbicides with similar modes of action has resulted in 520 unique cases of herbicide-resistant weeds globally across 268 species [3]. In addition, there are a number of other drawbacks associated with using herbicides frequently, such as high chemical expenses, possible groundwater contamination from runoff and leaching, or issues with irrigation water recycling [4]. Therefore, it has become essential to look for sustainable alternative weed management methods. The manipulation of allelopathic mechanisms is one of the most effective low-input and eco-friendly weed management strategies, as evidenced by the growing scientific interest in this area [5]. The utilization of plant-based natural compounds known as allelochemicals to control weeds is a great area for finding such tools. These allelopathic substances have phytotoxic properties and may be used as natural pesticides to control some weeds. Several plant species have been identified as potential sources of allelochemicals for weed control. For example, allelochemicals released by crops such as wheat, sorghum, sunflower, and rice have been shown to suppress the growth of common weed species, including barnyard grass and johnson grass. One advantage of using allelopathy for weed control is that it can target specific weed species without affecting non-target plant species. This can help to promote greater biodiversity and reduce the potential for unintended environmental impacts. Overall, allelopathy has the potential to be an effective and sustainable approach to weed control in agriculture, but it should be carefully integrated into a broader weed management strategy that considers the specific conditions and goals of each agricultural system.

19.2 CONCEPT OF ALLELOPATHY

The word "allelopathy" is derived from two separate Greek words, *allelon*, which means "of each other", and *pathos*, which means "to suffer". Thus, allelopathy refers to the chemical inhibition of one species by another. The "inhibitory" chemical, when released into the environment, interferes with the development and growth of neighboring plants. In other words, allelopathy has been recognized as one of the many stresses which impact plants in their environment. Allelopathy is an ecological phenomenon of plant–plant interference associated with the release of organic chemicals

DOI: 10.1201/9781003463429-24

(allelochemicals) in the environment. Allelopathy is defined as the direct or indirect effect of plants on their neighboring plants or their associated microflora or microfauna by the production of allelochemicals that interfere with the growth of the plant [6]. The effect on the growth of the surrounding plants can be either positive or negative.

19.2.1 History

The concept of allelopathy, or the ability of one plant to affect the growth or survival of another through the release of chemicals, has been recognized for thousands of years. The term "allelopathy" was first used in 1937 by Austrian scientist Hans Molisch in the book *Der Einfluss einer Pflanze auf die andere—Allelopathie* (The Effect of Plants on Each Other) [7], which described the phenomenon as "the chemical inhibition of one plant by another through the release of toxic substances". Theophrastus, a Greek botanist, noted for the first time the detrimental effect of chickpeas on other plants in 300 B.C. Later, Roman scholar Pliny (1 A.D.) noted the inhibitory effect of the walnut tree (*Juglans* spp.) on adjacent crops.

19.2.2 Types of Allelopathy

There are four main types of allelopathy. See Figure 19.1.

19.2.3 Forms of Allelopathic Interactions

There are several forms of allelopathic interactions, including:

Direct allelopathy: This occurs when allelochemicals are released directly from one plant to another, through volatilization, root exudation, or other means.

Indirect allelopathy: This occurs when allelochemicals are released by one plant but interact with other organisms or environmental factors before affecting another plant. For example, allelochemicals released by a plant may be absorbed by soil particles, where they can interact with soil microbes or other chemical compounds before affecting another plant.

Positive allelopathy: This occurs when allelopathic interactions have a positive effect on plant growth or survival. For example, some allelochemicals released by certain plant species can stimulate the growth of other plant species.

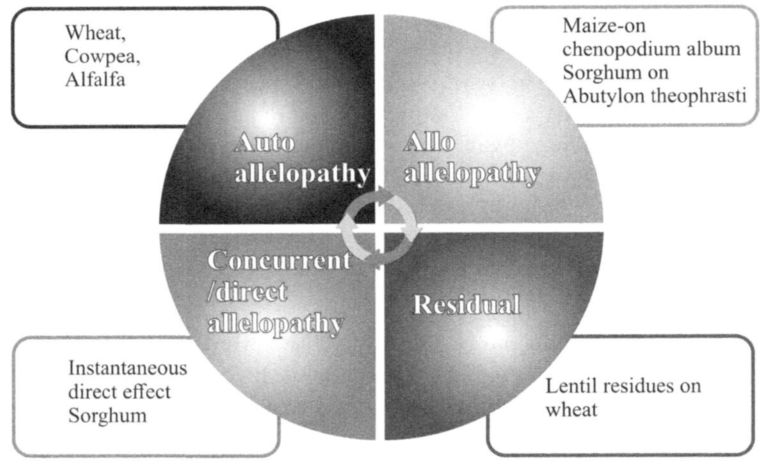

FIGURE 19.1 Allelopathy types.

Negative allelopathy: This occurs when allelopathic interactions have an adverse or negative effect on plant growth or survival. For example, some allelochemicals released by certain plant species can inhibit the growth of other plant species.

Density-dependent allelopathy: This occurs when allelopathic interactions are influenced by the density of plants in a given area. For example, a plant species may release allelochemicals that inhibit the growth of other plants but only at high densities, when competition for resources is most intense.

Overall, the forms of allelopathic interactions can be complex and depend upon a variety of factors, including the specific allelochemicals involved, the organisms and environmental factors present, and the density and diversity of plant communities.

19.2.4 Factors Affecting Allelopathic Effects

The allelopathic effect of one plant on another can be influenced by a variety of factors:

Allelochemical concentration: The concentration of allelochemicals released by a plant can have a significant effect on their allelopathic activity. Higher concentrations of allelochemicals may have stronger inhibitory effects on other plants.

Timing of allelochemical release: The timing of allelochemical release can also affect their allelopathic activity. For example, allelochemicals released during the initial stages of growing season may have a higher impact on competing plant species than those released later in the season.

Environmental conditions: Environmental factors, such as temperature, moisture, and soil fertility, can also influence the allelopathic activity of plants. For example, high temperatures and low soil moisture can enhance the release and activity of some allelochemicals.

Genetic variability: The allelopathic activity of plants can also vary due to genetic differences between individuals or populations. Some plant species may produce more or different allelochemicals than others, which can affect their interactions with other plant species.

Soil microorganisms: Soil microorganisms such as fungi and bacteria can interact with allelochemicals released by plants, affecting their activity and persistence in the soil. Some microorganisms can break down or degrade allelochemicals, reducing their allelopathic activity.

Plant community composition: The composition and diversity of plant communities can also influence the allelopathic activity of plants. For example, the presence of certain plant species may enhance or reduce the allelopathic activity of other species.

Overall, the allelopathic effect of one plant on another can be influenced by a range of factors, making it a complex and dynamic process.

19.3 ALLELOCHEMICALS

Allelochemicals produced by plants have the ability to affect the growth and development of other plants, as well as other organisms. Allelochemicals are released from plant parts through a variety of processes, including root exudation, volatilization, residue decomposition, leaching from leaves or ground litter, and others [8,9]. Following release, the allelochemicals may inhibit the germination, growth, and establishment of nearby plants or alter the rhizosphere's soil characteristics by modifying the microbial population [10,11].

Some common allelochemicals include alkaloids, terpenoids, phenolics, and fatty acids. The alkaloid nicotine produced by tobacco plants has been shown to have allelopathic effects on other plants, while the phenolic compound tannin found in various plant species can inhibit the growth of microorganisms and herbivores.

Allelochemicals are divided into 14 groups based on their chemical similarities [8]. These are water soluble organic acids, straight-chain alcohols, aliphatic aldehydes, and ketones; simple unsaturated lactones; long chain fatty acids and polyacetylenes; anthraquinone, benzoquinone, and complex quinones; simple phenols, benzoic acid and its derivatives; cinnamic acid and its derivatives; coumarin; flavonoids; tannins; terpenoids and steroids; amino acids and peptides; alkaloids and cyanohydrins; sulfide and glucosinolates; and purines and nucleosides [12]. Allelochemicals include plant growth regulators such as ethylene, salicylic acid, and gibberellic acid. The mode of action, absorption, and effectiveness of allelochemicals differ [13,14].

Allelochemicals have been studied as potential tools for weed management in agriculture and forestry, as well as for understanding ecological interactions between plant species. However, it is important to note that allelopathy is just one of many factors that influence plant growth and interactions in natural ecosystems and should be considered in conjunction with other environmental factors when analyzing plant communities.

19.3.1 Ways of Releasing Allelochemicals

Allelochemicals are chemicals produced by plants, fungi, and microorganisms that can have effects on the growth, survival, and reproduction of other organisms. Here are some ways that allelochemicals can be released:

Volatilization: Some allelochemicals can be released into the air as gases. For example, eucalyptus leaves can volatilize certain compounds that inhibit the growth of nearby plants.

Leaching: Allelochemicals can be released from plant tissues into the surrounding soil or water through leaching. This can happen when plant tissues break down or when plant roots exude chemicals into the soil.

Root exudation: Plants can release allelochemicals through their roots, which can inhibit the growth of other plants in the vicinity. For example, some grasses produce chemicals that inhibit the growth of neighboring grasses.

Decomposition: When plant tissues break down, they can release allelochemicals into the surrounding soil or water. This can happen when plant leaves, stems, or roots decay.

Seed exudation: Some plants can release allelochemicals from their seeds. For example, black walnut trees release a chemical called juglone from their seeds, which can inhibit the growth of nearby plants.

Insect herbivory: When insects feed on plant tissues, they can release allelochemicals from their saliva or feces. These allelochemicals can affect the growth and survival of other organisms in the area.

Overall, the release of allelochemicals can be a complex process that involves a variety of mechanisms, and the effects of these chemicals can vary depending on the specific compounds involved and the organisms that are exposed to them.

19.4 PLANTS SHOWING ALLELOPATHIC POTENTIAL

19.4.1 Black Walnut (*Juglans nigra* L.)

The walnut species is one of the most ancient and well-studied allelopathic species. The allelopathic substance found in black walnut is juglone (5-hydroxy-1,4-napthoquinone), a phenolic compound; which is produced by the roots and released into the soil. Juglone is toxic to many plant species and can cause wilting, yellowing, and stunted growth in sensitive plants. Juglone, earlier known as nucin, was isolated and discovered in 1856 [15], but Davis demonstrated the toxicity of synthetic juglone on plants of tomato and alfalfa in 1928. The allelopathic compounds in black walnut trees are primarily found in their leaves, bark, and roots, and they can persist in the soil for several years. Numerous studies have now reported juglone's allelopathic effects on a variety of vegetables, agricultural crops, medicinal plants, fruit trees, and ornamental species [16]. In addition to juglone, black walnut trees also produce other allelopathic compounds such as hydrojuglone and juglanin. These compounds can restrict seed germination and root growth in sensitive plants. Juglone's herbicidal effectiveness against four weed species was tested: wild mustard (*Sinapis arvensis* L.), creeping thistle (*Cirsium arvense* L.), field poppy (*Papaver rhoeas* L.), and henbit (*Lamium amplexicaule* L.). Juglone at high concentrations of 1.15–5.74 mM fully restricted field poppy development and significantly lowered elongation and fresh weight of all remaining weed species [17]. Black walnut extract has the potential to be used as pre- and post-emergent bioherbicide against hairy fleabane (*Conyza bonariensis* L.), horseweed (*Conyza canadensis* L.), tall annual morning glory (*Ipomoea purpurea* L.), and purslane (*Portulaca oleracea* L.) [18]. Juglone has a significant potential for usage as a natural herbicide because research has shown that it effectively controls a variety of weed species.

19.4.2 Tree of Heaven (*Ailanthus altissima*)

The tree of heaven is known for its allelopathic potential; Voigt and Mergen in 1962 identified the tree of heaven's toxic activity by observing that the extract from the leaf and stem was toxic to nearby plants [19]. It releases chemicals into the soil that can inhibit the growth of other plants, similar to black walnut trees. Alkaloids, steroids, terpenoids, flavonoids, phenolic derivatives, and quassinoids are found in the plant extracts and essential oils from various portions of the plant [20–22]. The allelopathic compounds in the tree of heaven are primarily found in its leaves, bark, and roots. These compounds are known to inhibit the germination and growth of other plants by interfering with various physiological processes such as photosynthesis, respiration, and protein synthesis.

Research has shown that the allelopathic potential of the tree of heaven can vary depending on the age and health of the tree, as well as the species of plant being affected. For example, certain plants, such as tomato and bean plants, are particularly susceptible to the allelopathic effects of the tree of heaven, while others may be less affected. The allelopathic potential of the tree of heaven can also vary depending on the environment in which it is growing. For example, the tree may release more allelopathic compounds in soils with low nutrient levels, while in soils with high nutrient levels, the effects may be less pronounced.

Ailanthone, a quassinoid compound, is a significant allelochemical found in the tree of heaven [23,24]. Ailanthone exhibits herbicidal activity both pre- and post-emergence, limiting the germination and growth of monocots and dicots. At application rates as low as 0.5 kg/ha, a greenhouse trial with pure ailanthone demonstrated strong pre- and post-emergence control. Even at a low rate of 0.5 $kgha^{-1}$, post-emergence activity was strong, inhibiting growth or killing young seedlings of garden cress, redroot pigweed (*Amaranthus retroflexus* L.), barnyard grass (*Echinochloa crusgalli* L.), foxtail (*Setaria glauca* L.), and corn (*Zea mays* L.) [23]. It was noteworthy that the tree of heaven seedlings showed no damage from ailanthone post-emergence application, indicating the availability of a protective mechanism to prevent autotoxicity [23]. Ailanthone is not commercially employed as a herbicide [25] because its separation and purification costs are considerable, it degrades quickly in soil, and it is non-selective [23]. However, recent study indicates that it has a high potential for usage as a natural herbicide, particularly for home gardeners and organic farmers.

19.4.3 Eucalyptus (*Eucalyptus* spp.)

Eucalyptus trees are known to possess allelopathic potential, which means that they can release certain chemicals or compounds that inhibit the growth of nearby plants or even kill them. The allelopathic potential of eucalyptus is attributed to the presence of various secondary metabolites, including terpenes, phenolics, and flavonoids, which are synthesized and stored in different parts of the plant. Some studies have shown that eucalyptus extracts can inhibit the germination and growth of certain crop plants, such as maize, wheat, and soybean, by interfering with their physiological processes, such as seed germination, root growth, and nutrient uptake. Similarly, eucalyptus leaf extracts have been shown to restrict the germination and growth of several weed species, including barnyard grass, johnsongrass, and velvetleaf.

The allelopathic effects of many eucalyptus species have been thoroughly studied [26,27]. Allelochemicals found in *Eucalyptus* spp. leaves, bark, and roots include phenolic acids and volatile terpenes [28]. The eucalyptus species' foliage also includes a range of oils and resins that may have an effect on nearby plants, seeds, or microorganisms [29]. The eucalyptus species contains volatile terpenes such as 1,8-cineol, limonene, and a- and b- pinene that function as allelochemicals [30]. 1,8-cineol is one of the main allelopathic substances found in eucalyptus leaves. 1,8-cineol has been shown to suppress mitosis, slow down root growth, and decrease germination [31]. In a study, bluegum eucalyptus essential oil at a concentration of 25% (v/v) inhibited bermudagrass growth by 66% [32]. Volatile essential oils of eucalyptus have a strong inhibitory capability against *Cyndon dactylon* and *Amaranthus blitoides* that could be used for weed management. These oils can also be suggested as a biological herbicide for weed control [33]. *Cyperus rotundus* and *Convolvulus arvensis* had their roots and shoots, as well as their fresh and dry weight, drastically reduced when treated with eucalyptus powder (30 g/kg soil) and eucalyptus water extract (30 ml/kg soil).

19.4.4 Mango (*Mangifera indica*)

The allelopathic potential of mango fruit has been attributed to the presence of phenolic compounds, particularly mangiferin, which has been found to exhibit phytotoxic effects on other plant species. These compounds have been found to inhibit the germination and growth of several weeds, such as barnyardgrass, pigweed, and goosegrass. Purple nutsedge tubers' ability to sprout was entirely inhibited by dried mango leaf powder [34]. In an experiment, results showed that the number of mother shoots per tuber, number of leaves of mother shoots per tuber, number of leaves of daughter shoots per tuber, number of basal bulbs and tubers, number of rhizomes per tuber, length of rhizomes, dry weight of foliage, and total dry weight were all reduced after applying mango leaf extract at 25% or leaf powder at 100 g kg^{-1} of soil and that it increased the total phenol contents in the foliage and in underground parts of purple nutsedge compared to control [35].

19.5 CROPS WITH ALLELOPATHIC POTENTIAL

19.5.1 Sorghum (*Sorghum bicolor*)

Sorghum is a good source of allelochemicals due to root exudation and root and stem residual biomass [36]. To make allelopathic weed management more feasible and successful, sorghum's allelopathic activity must be optimized. Sorghum's weed suppressive action is enhanced by the presence of hydrophilic compounds such as phenolic acids and their aldehyde derivatives, as well as hydrophobic chemicals such as sorgoleone [37]. Sorghum can be used to manage weeds in a variety of ways, including surface mulch, soil mixing, extract spray, and intercropping. Sorghum roots, stems, and leaves have been proven to decrease weed biomass by 25–50% in the soil [38]. Netzly and Butler discovered sorgoleone from sorghum hydrophobic root exudates in 1986 [39]. Sorgoleone accounts for approximately 90% of the chemicals found in sorghum root exudates [37]. Sorgoleone is produced in sorghum root hair cells. Sorgoleone has been described as a powerful bioherbicide due to its ability to control a wide range of weed species. It has been proven to be more active than other allelochemicals like juglone, as well as other phenolics and terpenoids [40]. Investigations on sorghum underground parts show that sorgoleone is primarily responsible for weed control activity, while phenolic substances play an important role in its herbicidal activity on weed aerial parts [41]. Sorgaab, a foliar application of sorghum water extract, decreased purple nutsedge density and dry weight by 44 and 67%, respectively, while boosting maize grain production by 44% [42]. Overall, these studies suggest that sorghum has the potential to be an effective tool in weed management, especially when used in combination with other weed control methods. A multidisciplinary approach incorporating sorghum crops, residues, or allelochemicals for strategic weed management has significant potential.

19.5.2 Rice (*Oryza sativa* L.)

Allelopathy in rice could become an essential tool in integrated weed management strategies in rice cultivation, and rice cultivars can reduce both mono and dicot weed species by maximizing competitive and chemical interference. Many laboratory and field investigations have been undertaken to investigate rice's allelopathic potential. In 2004, the allelopathic activities of 749 rice cultivars were investigated, and results showed that rice cultivar japonica had the strongest allelopathic capacity against barnyard grass root growth [43]. Rice extracts contain a variety of phytotoxic chemicals from many chemical classes, including fatty acids, benzoxazinoids, indoles, phenolic acids, phenylalkanoic acids, and terpenoids [44]. Tricin and momilactone B are the main allelochemicals discovered in allelopathic rice cultivars [45]. Allelopathic chemicals produced by rice could aid in weed suppression and reduce the requirement for synthetic herbicides in rice cropping systems. Rice straw, rice flour, and rice hull can be used directly as components of an integrated weed management program. More research on allelochemical extraction from allelopathic rice cultivars and determining their mode of action will increase the potential of using rice allelopathy for weed control.

19.5.3 Wheat (*Triticum aestivum* L.)

Wheat allelopathy has the potential to be used in sustainable weed management. There is growing consensus that allelpathic wheat cultivars could provide a competitive advantage against weeds. Wheat allelopathy has piqued the interest of researchers because it has the potential to be used in plant protection and resistance breeding. According to Steinsiek et al. (1982), wheat grass water extract was found to phytotoxic to ivyleaf morning glory (*Ipomoea hederacea* L.), velvetleaf (*Abutilon theophrasti*), and pitted morning glory (*Ipomoea lacunose*) [46]. Polyphenols and hydroxamic acids are two major allelochemical groups found in wheat [47]. p-hydroxybenzoic, vanillic, pcoumaric, syringic, and ferulic acids are the most common phenolic acids found in wheat mulch and its surrounding soil [48]. Additionally, hydroxamic acids (benzoxazinoids) and lactams such as 2,4-dihydroxy-1,4-benzoxazin-3-one (DIBOA), 2,4-dihydroxy7-methoxy-1,4-benzoxazin-3-one (DIMBOA), 2-hydroxy-1,4-benzoxazin-3-one (HBOA), and 2-hydroxy-7-methoxy-1,4-benzoxazin-3-one (HMBOA) have been identified as the main allelochemicals in wheat.

Weeds, including resistant biotypes, can be managed using both wheat residue allelopathy and wheat seedling allelopathy. The equal-compartment-agar-method (ECAM) was used to assess seedling allelopathy against *Lolium rigidum* in a worldwide collection of 453 wheat accessions from 50 countries. Wheat accessions vary considerably in seedling allelopathy, reducing ryegrass root development by 10% to 91% [49]. Thus, wheat varieties differ in their allelopathic potential against weeds, implying that selecting allelopathic varieties could be a valuable method in integrated weed management. Wheat has allelopathic potential due to the presence of numerous allelochemicals, but additional research on the genetic control of wheat allelopathy is needed. Furthermore, more attention should be paid to selection and breeding of wheat cultivars with a higher allelopathic effect against weed species.

19.5.4 Rapeseed (*Brassica napus* L.)

Rapeseed (*Brassica napus*, var. *oleifera*) contains allelochemicals that prevent the germination and growth of *Solanum nigrum* (black nightshade), *Amaranthus retroflexus* L. (redroot pigweed), *Portulaca oleracea* (common purslane), *Physalis angulata* (cut leafground cherry), and *Echinochloa colona*. Between rapeseed cultivars, inhibition rates greatly differed [50]. On the weed species under examination, cultivars Tobin, Jumbuck, Lisoune, and Galant were less allelopathic than cultivar Westar, which was determined to be severely allelopathic [51]. The amounts of isothiocyanate benzyl and isothiocyanate allyl varied between rapeseed varieties. Stronger allelopathic potential was found in cultivars with greater levels of isothiocyanate benzyl and isothiocyanate allyl [52]. The examined weed species responded differently to extracts from different parts of rapeseed, and they did so in the following order: shoot extracts > root extracts > root exudates [53].

19.5.5 Sunflower (*Helianthus annus* L.)

Sunflower, an annual oleaginous plant native to North America, exhibits allelopathic activity against weeds. Some broadleaf weeds may respond well to its use as a natural herbicide [54]. A number of substances with allelopathic characteristics, including phenolic compounds, diterpenes, and triterpenes, have been isolated and chemically described [55]. The characteristics of sunflowers are well known, and numerous weeds and crops have been shown to be negatively impacted by them. The allelopathic potential of eight sunflower cultivars against problematic weed species in wheat was assessed by Alsaadawi and his co-workers in 2012. The allelopathic potential of the sunflower cultivars used in the study varied, and they reduced overall weed biomass and total weed density by 34 to 81% and 10 to 87%, respectively [56]. The most important allelopathic compounds identified from sunflower are heliannuols, terpenoids, and flavonoids [57]. Under field and laboratory conditions, the application of sunflower as green manure reduced the population of *Phlaris minor* by 42 and 100 %, respectively. Annuionone obtained from Suncross-42 leaves aqueous extract inhibited the growth of *Phlaris minor*, *Chenopodiun album*, *Coronopsis didymus*, *Medicago polymorpha*, and *Rumex dentatus* [58].

Table 19.1 enumerates allelopathic plant/crop/weed examples with their major allelochemicals against targeted plants.

19.6 WEED SPECIES WITH ALLELOPATHIC POTENTIAL

There are numerous weed species that contain unique allelochemicals that can be employed to prevent the germination and proliferation of other weed species. Parthenin, a sesquiterpene lactone of pseudoguanolide origin, is present in several portions of the ragweed parthenium (*Parthenium hysterophorus* L.), with the highest concentration found in the leaves [68]. Green amaranth (*Amaranthus viridis*), canary grass (*Phalaris minor*), coffee senna (*Cassia occidentalis*), and barnyard grass all had their seedling development and dry weight reduced by parthenin when applied as pre- and post-emergence [66]. Several *Artemisia* spp. plants have shown allelopathic activity in various agricultural environments. The main constituents of common mugwort (*Artemisia vulgaris*) include alpha thujone, a monoterpene, and 20 minor components, including many sesquiterpenes [69]. Morning glory (*Ipomoea tricolor*) is utilized as green manure for controlling weeds in tropical areas of Mexico [70]. The main allelopathic chemical found in morning glory is tricolorin A. The carotenoid sesquiterpene 1aangeloyloxycarotol found in giant ragweed (*Ambrosia trifida*) functions as a potent allelochemical [71]. The common lantana (*Lantana camara*) is an obnoxious weed that contains allelochemicals in its stems, leaves, flowers, fruits, and roots. Flavonoids, mono- and sesquiterpenes, iridoid glycoside, furanonaphoquinones, steroids, triterpenes, and diterpenes are among the allelochemicals found in common lantana [72]. Lantadene A and lantadene B are the more potent

TABLE 19.1
Allelopathic Plant/Crop/Weed Examples with Their Major Allelochemicals against Targeted Plants

Allelopathic Plant/Crop/Weed	Major Allelochemical	Target Weeds	Reference
Tree of heaven (*Ailanthus altissima*)	Ailanthone	*Amaranthus retroflexus, Lepidium sativum, Setaria glauca, Echinochloa crus-galli, and Zea mays*	[23]
Fine fescue grasses (*Festuca* spp.)	M-tyrosine	*Brassica nigra, Lotus corniculatus, Digitaria sanguinalis, Trifolium repens,* and *Taraxacum officinale*	[59,60]
Eucalyptus (*Eucalyptus* spp.)	1,8-cineol	*Amaranthus retroflexus* and *Portulaca oleracea*	[61]
Black Walnut (*Juglans nigra*)	Juglone	*Conyza Canadensis, Conyza bonariensis, Portulaca oleracea,* and *Ipomoea purpurea*	[18]
Sorghum (*Sorghum bicolor*)	Sorgoleone	*Galium spurium, Rumex japonicas, Aeschynomene indica, and Amaranthus retroflexus*	[40]
Rice (*Oryza sativa*)	Tricin and momilactone B	*Echinochloa colona, Ammannia baccifera, and Phyllanthus fraternus, Cyperus iria*	[62,63]
Sunflower (*Helianthus annus*)	Heliannuols	*Digitaria sanguinalis, Sida spinosa,* and *Amaranthus album*	[64]
Common lantana (*Lantana camara*)	Lantadene A and lantadene B	*Pontederia crassipes, Microcystis aeruginosa, Phalaris minor, Avena fatua, Rumex dentatus, Chenopodium album*	[65]
Ragweed parthenium (*Parthenium hysterophorus*)	Parthenin	*Amaranthus viridis, Phalaris minor, Echinochloa crus-galli, Cassia occidentalis*	[66]
Morning glory (*Ipomoea tricolor*)	Tricolorin A	*Lolium mutliflorum, Triticum vulgare, Physalis ixocarpa* and *Trifolium alexandrinum*	[67]

allelochemicals found in lantana. Many other weed species possess unique allelochemicals. These allelochemicals may have novel mechanisms of action and can be used to develop future herbicides as herbicide templates.

19.7 UTILIZATION OF ALLELOPATHY IN WEED MANAGEMENT

Allelopathy can be used to control weeds in agroecosystems by

(i) s+using allelopathic crop cultivars
(ii) incorporating allelopathic crops into rotational sequences
(iii) intercropping allelopathic crops near a cash crop
(iv) cover cropping as living or dead mulches
(v) allelopathic crop residue incorporation into the soil
(vi) using allelochemicals as natural herbicides or bioherbicides

Allelopathy can be used in a variety of cropping systems, but it is most effective in organic agriculture and in conservation, minimum, and no-tillage agricultural systems, where weed management is frequently a challenge.

19.7.1 Allelopathic Crop Cultivars

Crop cultivars that exhibit higher productivity in farmers' fields are commercially acceptable. At the same time, crop cultivars' ability to suppress weeds is becoming a favored criterion for cultivar selection in many regions of the world. Crop plants' allelopathic potential contributes to cultivars' weed suppression abilities. Preferring weed-suppressive allelopathic cultivars over non-allelopathic cultivars can reduce weed infestation without incurring additional costs and can help to increase the efficacy of inputs and weed management methods.

19.7.2 Allelopathic Crops in Rotational Sequence

Planting multiple crops in succession in the same field is a traditional agricultural practice that provides many benefits in the cropping system, such as weed and insect pest control, reduced auto allelopathy or soil sickness due to monoculture, improved soil organic matter and soil microorganisms, reduced nutrient leaching, and increased soil fertility and crop yields. Including an allelopathic crop within rotational sequences can help reduce weeds in both current and following crops by releasing allelochemicals into the soil via root exudation, breakdown of plant wastes, and leaching from plant foliage. By the release of allelochemicals in the soil via decomposition of plant wastes, root exudation and leaching from the plant foliage through the inclusion of an allelopathic crop within rotational sequences can help in reducing weeds in both current and subsequent crops. Allelochemicals, once released into the rhizosphere, can hinder seed germination and reduce weed density and biomass directly or indirectly by microbial modification into more active, less active, or completely inactive substances. Integration of allelopathic crops as smothering crops that can grow quick with thick canopy can provide additional weed control. Crops such as sudan grass (*Sorghum sudanense*), common buckwheat (*Fagopyrum esculentim*), barley, rye, sunflower (*Helianthus annuus*), cowpea, and sweet clover can effectively smother various weed species [73]. In conservative agricultural systems, the beneficial impacts of allelopathic crops within crop rotations are frequently accentuated. For example, Shahzad et al. [74] discovered that the sorghum-wheat rotation had the strongest weed-suppressive effect in terms of density and dry biomass reduction in all tillage systems, especially during the second year, due to the accumulation of sorghum allelochemicals (sorgoleone) in the soil.

19.7.3 Intercropping with Weed-Suppressing Plants

Compatible crops are planted together to maximize net yield and economic benefits. Growing crops in combinations also enhances resource use efficiency (light, water, land, and nutrients). Aside from these advantages, intercropping can be employed to reduce weeds for environmentally friendly and cost-effective weed control. Crops with allelopathic potential, in particular, help to lower weed intensity and thus boost productivity from crops when intercropped with other crop plants. Kandhro et al. [75] studied the intercropping of two allelopathic crops (sorghum and sunflower) for cotton weed management. Both intercrops reduced weeds by 60 to 62% in cotton, resulting in a 17 to 22% increase in seed cotton yield. Sunflower and sorghum were additionally harvested for grain, resulting in increased crop yield, better land utilization, and economic benefits. *Orobanche* spp. are well-known weed parasites that cause significant crop loss. Berseem's allelopathic activity can be used to inhibit *Orobanche* spp. through intercropping. Intercropping maize and cowpea on alternate ridges reduced weed intensity (*Echinochloa colona, Portulaca oleracea, Dactyloctenium aegyptium, Chorchorus olitorius*) by 50% while improving land use efficiency [76]. Intercropping allelopathic crops with the main crop can thus help to lower weed intensity while increasing yield grains.

19.7.4 Cover Cropping

Cover crops have been grown to ensure the sustainability of an agro-ecosystem. Growing cover crops has several objectives, including enhancing soil fertility and quality, as well as reducing weeds and plant pathogens. Weeds can be suppressed by cover crops with allelopathic potential. Wheat, canola, rapeseed, cereal rye, brown mustard, crimson clover, red clover, oats, cowpea, fodder radish, annual ryegrass, buckwheat, hairy vetch, and black mustard are among the key cover crops.

Some cropping systems (for example, organic cropping) rely extensively on cover cropping for weed management. The findings of studies and observations from farmers' fields indicated that the release of allelochemicals from allelopathic cover crops, as well as their physical impacts, was responsible for weed suppression in conservation organic farm fields [77]. Bernstein et al. [78] studied the performance of a rye cover crop in suppressing weeds before planting soybean in a no-till system and concluded that soybean could be successfully seeded in no-till soil with a standing rye cover crop. Planting soybean in a standing rye field provided long-term and excellent weed control while causing no damage to the soybean crop.

In Table 19.2, examples of allelopathic cover crops, main crops, and targeted weeds by cover crops are given.

19.7.5 Allelochemicals or Modified Allelochemicals as Herbicides

Bioherbicides are natural biological products derived from living organisms or their secondary metabolites that decrease target weed populations without damaging the environment. Plant-based allelochemical bioherbicides are gaining popularity as a means of reducing the usage of synthetic herbicides, overcoming weed resistance, and minimizing their environmental impact. These benefits include water solubility, an environmentally friendly chemical structure, high degradability in soil and water, the possibility of new molecular targets in weeds, and public acceptance. However, the chemical characteristics of allelochemicals raise the cost of their production, especially as their identification and setup as bioherbicides is more complicated than that of synthetic herbicides. Recent improvements in metabolomic techniques and chemical analytic instruments (for example, liquid or gas chromatography paired with mass spectrometry) are making it easier to rediscover allelochemicals and identify and quantify them directly in crude plant extracts. Anjum and Bajwa [84] tested the bioherbicidal activity of sunflower leaf extracts (at 100 mL m^{-2}) applied three times in post-emergence at 7-day intervals, reporting that extracts at the highest concentration reduced

TABLE 19.2
Examples of Allelopathic Cover Crops, Main Crops and Targeted Weeds by Cover Crops

Allelopathic Cover Crops	Main Crop	Weeds Suppressed	Reference
Rye	Cotton	*E. indica, A. palmeri, I. lacunosa*	[79]
Sunflower, maize, rice, sorghum	Wheat	*Phalaris minor* Retz.	[80]
Wheat	Cotton	*E. indica, Amaranthus palmeri, Ipomoea lacunosa* L.	[79]
Rye	Soybean	*C. album, Abutilon theophrasti Medik*	[78]
Hairy vetch, oat	Maize	*Digitaria sanguinalis, E. indica, A. retroflexus, Datura stramonium*	[81]
Sorghum sudangrass [*Sorghum bicolour* x *Sorghum sudanense*]	Broccoli	Broad leaved weeds	[82]
Bristle oat, Hairy vetch	Cotton	*A. palmeri, P. oleracea, Helianthus annuus*	[83]

lambsquarters (*Chenopodium album*) by 70% and increased wheat biomass and harvest index compared to a weedy control due to the elimination of weed-crop competition. Radhakrishnan et al. [85] examined the impact of plant-based bioherbicides on weed physiology, highlighting major metabolic pathways that result in the inhibition of seed germination and seedling growth.

19.7.6 Application of Allelopathic Crop Residues as Mulch

Allelopathic plant residues, whether deliberately or unintentionally left in the field, express their ability to inhibit weeds. Plant residues of rye, barley, and triticale were tested for their allelopathic activity against *Echinochloa crus-galli* and *Setaria verticillata* in a maize field [86]. When compared to the no-mulch condition, allelopathic mulches reduced the establishment of *Setaria verticillata* (0–67%) and *Echinochloa crus-galli* (27–80%). The applied mulches had no negative impact on the maize plants. Khaliq et al. [87] used a combination of allelopathic plant residues including sunflower, sorghum, and canola at 7.5 t ha^{-1} for weed control in a maize crop. The applied mulch reduced *Cyperus rotundus* and *Trianthema portulacastrum* densities and biomass by 90% while increasing maize grain yield, 1000-grain weight, and harvest index by 54%, 13%, and 29%, respectively. There was no negative effect of the applied mulch material on maize growth or development, implying that allelopathic materials might be utilized for weed management without harming maize plants.

Similarly, in an organically grown maize-broccoli rotation, maize residues were found to lower weed biomass in the next crop (broccoli) by 22 to 47% [88]. The presence of various allelochemicals in allelopathic materials can boost allelopathic effectiveness against weeds. Allelochemicals' synergistic action can also boost their efficacy against target weeds. In conclusion, allelopathic plant remains can be used as mulch to control weeds and to increase yields of grains. Other advantages of employing mulch for weed control include improved moisture conservation and nutrient availability.

19.7.7 Green Manuring

Several studies have been reported in recent years on the incorporation of allelopathic plant wastes into soil (i.e., green manuring) for the control of weeds under field conditions. Puig et al. [89] studied the allelopathic potential of *Eucalyptus globules* leaf green manure in corn fields over two seasons and at two different locations. Eucalyptus green manure significantly decreased weed biomass, particularly that of *Digitaria sanguinalis* and *Chenopodium album*, and particularly during the early phases of corn establishment, while corn was not negatively affected. Masilionyte et al. [90] studied the weed-suppressive capacity of different cover crops cultivated for green manure in a 6-year field experiment and found that white mustard (*S. alba*), especially when combined with buckwheat (*F. esculentum*), reduced the number of weeds and weed biomass more than narrow-leafed lupine (*Lupinus angustifolius*) in a mixture with oil radish (*Raphanus sativus*).

19.8 MODIFICATION OF CROPS FOR ENHANCING THE PRODUCTION OF ALLELOCHEMICALS

Allelochemicals are natural compounds produced by plants that can have positive or negative effects on the growth and development of other plants. Some allelochemicals can enhance the growth of nearby crops by promoting root growth, increasing nutrient uptake, and suppressing weed growth. Thus, modifying crops for enhancing the production of allelochemicals can potentially lead to increased productivity in agriculture. Although many allelochemicals have low efficacy and specificity, they are a good alternative to synthetic herbicides because they do not have harmful or lasting effects [91]. Due to allelochemicals' relatively high degradability, the technology that alters them to produce ecologically acceptable insecticides and plant growth regulators permits successful agricultural production management while causing few environmental concerns in the soil [92].

19.8.1 Modification through Breeding

One way to enhance the production of allelochemicals in crops is through genetic modification. By inserting genes that code for the production of specific allelochemicals, crops can be engineered to produce higher levels of these compounds. The primary metabolites thought to be involved in this process include indoles (benzoxazinoids and its derivatives generated by maize, rice, wheat, and so on), phenylpropanoids (for example, cinnamic acid derivatives), and terpenoids (momilactones a and b produced by rice) [93]. One of the primary issues with allelopathic control of weeds is the low concentration of allelochemicals in the donor plant and the related synthesis of these chemicals for commercial usage [94]. The genes responsible for the production of allele chemicals in certain crops can be overexpressed to increase their production. For example, if a crop has a known allelopathic chemical that inhibits the growth of nearby weeds, its gene responsible for producing that chemical can be overexpressed to boost its production, resulting in increased allelopathic effects. Biosynthetic pathways in crops can be modified to enhance the production of allelechemicals. By manipulating the enzymes and genes involved in the biosynthesis of allelochemicals, their production can be increased, resulting in higher levels of these chemicals in the crop. Multiple allelopathic traits from different crops or species can be stacked into a single crop through genetic modification. This can result in a crop that produces a combination of allelochemicals, providing a broader spectrum of allelopathic effects against various weeds or pests.

It's important to note that genetic modification of crops is a complex and controversial topic, and it raises ethical, environmental, and regulatory concerns. Additionally, careful consideration must be given to potential impacts on non-target organisms, biodiversity, and the environment. Proper risk assessment and thorough evaluation of the potential benefits and risks of genetically modified crops for enhancing the production of allelochemicals should be conducted before implementing any such practices in the field.

Another approach is to use traditional breeding methods to select and crossbreed crops that naturally produce high levels of desirable allelochemicals. This can involve screening large numbers of plants to identify those with high allelochemical production and then crossing them to create new cultivars with even higher levels of the desired allelochemicals. Holt and Lovett (1990) discovered allelopathic features in wild accessions of a variety of crop plants, which conferred resistance to pests and weeds [95]. However, due to breeding for high yields and rapid growth, genes encoding for allelochemical production were mistakenly degraded throughout the culture process [96]. However, these genes can be restored through a variety of genetic improvement initiatives, and these crops may serve as instruments for biological weed control [97].

19.9 FUTURE PROSPECTS

Despite the numerous hurdles associated with using the concept of allelopathy for management of weeds, there is enormous potential for investigating allelopathy as a novel tool for weed management. There is a lack of research on the safety of employing known extracted allelochemicals on various crops. Future research should concentrate on the mechanisms of allelochemical selectivity, their modes of action, interactions with diverse species, and methods of implementation. We have the opportunity to learn a lot about allelochemicals thanks to advances in genetics, molecular biology, and biochemistry. Furthermore, concentrating on development of transgenic allelopathy in crops by genetic engineering can provide a novel way to implement the concept of allelopathy. More research on allelopathy is required to help farmers integrate sustainable weed management strategies into ongoing crop production practices. Enhanced collaborations and interactions will be required among weed scientists, ecologists, molecular biologists, and plant breeders to promote the utilization of allelopathy in agriculture and natural settings for weed suppression.

REFERENCES

1. Oerke EC. Crop losses to pests. *J. Agric. Sci.* 2006, 144, 31–43.
2. Chauhan BS. Grand challenges in weed management. *Front. Agron.* 2020, 1.
3. International survey of herbicide resistant weeds. Available at: www.weedscience.org (Accessed April 26, 2023).
4. Poudyal S, Cregg BM. Irrigation nursery crops with recycled runoff: A review of potential impact of pesticides on plant growth and physiology. *HortTechnology* 2019, 29, 716–729.
5. Scavo A, Restuccia A, Mauromicale G. Allelopathy: General principles and basic aspects for agroecosystem control. In *Sustainable agriculture reviews*, Gaba S, Smith B, Lichtfouse E, Eds. (Cham, Switzerland: Springer, 2018), 28, 47–101.
6. International allelopathy society. 2018. Available at: http://allelopathysociety.osupytheas.fr/about/.
7. Willis RJ. *The history of allelopathy* (Dordrecht, The Netherlands: Springer Science & Business Media), 2007.
8. Rice EL. *Allelopathy* (Orlando, FL: Academic Press), 2nd edn, 1984.
9. Anaya AL. Allelopathy as a tool in the management of biotic resources in agroecosystems. *Crit. Rev. Plant Sci.* 1999, 18(6), 697–739.
10. Weir TL, Park SW, Vivanco JM. Biochemical and physiological mechanisms mediated by allelochemicals. *Curr. Opin. Plant Biol.* 2004, 7(4), 472–479.
11. Zhou B, Kong CH, Li YH, Wang P, Xu XH. Crabgrass (Digitaria sanguinalis) allelochemicals that interfere with crop growth and the soil microbial community. *J. Agric. Food Chem.* 2013, 61(22), 5310–5317.
12. Cheng F, Cheng Z. Research progress on the use of plant allelopathy in agriculture and the physiological and ecological mechanisms of allelopathy. *Front. Plant Sci.* 2015, 6.
13. Weston LA, Duke SO. Weed and crop allelopathy. *Crit. Rev. Plant Sci.* 2003, 22(3–4), 367–389.
14. Rice EL. *Allelopathy* (New York, NY: Academic Press), 2012, 104–125.
15. Vogel A, Reischauer C. Buchner Neues. *Rep. Fur Pharm.* 1856, 5, 106.
16. Strugstad M, Despotovski S. A summary of extraction, synthesis, properties, and potential uses of juglone: A literature review. *J. Ecosyst. Manage.* 2012, 13(3), 1–16.
17. Topal S, Kocacaliskan I, Arslan O, Tel AZ. Herbicidal effects of juglone as an allelochemical. *Phyton.* 2007, 46, 259–269.
18. Shrestha A. Potential of a black walnut (Juglans nigra) extract product (Naturecur®) as a pre- and post-emergence bioherbicide. *J. Sustain. Agric.* 2009, 33(8), 810–822.
19. Voigt G, Mergen F. Seasonal variation in toxicity of ailanthus leaves to pine seedlings. *Bot. Gaz.* 1962, 123(4), 262–265.
20. Albouchi F, Hassen I, Casabianca H, Hosni K. Phytochemicals, antioxidant, antimicrobial and phytotoxic activities of Ailanthus altissima (Mill.) swingle leaves. *South Afr. J. Bot.* 2013, 87, 164–174.
21. Kim HM, Kim SJ, Kim HY, Ryu B, Kwak H, Hur J, et al. Constituents of the stem barks of Ailanthus altissima and their potential to inhibit LPS-induced nitric oxide production. *Bioorg. Medicinal Chem. Lett.* 2015, 25(5), 1017–1020.
22. Ni J, Shi J, Tan Q, Chen Q. Two new compounds from the fruit of Ailanthus altissima. *Nat. Prod. Res.* 2018, 33(1), 101–107.
23. Heisey R. Identification of an allelopathic compound from Ailanthus altissima (Simaroubaceae) and characterization of its herbicidal activity. *Am. J. Bot.* 1996, 83(2), 192–200.
24. Sladonja B, Sušek M, Guillermic J. Review on invasive tree of heaven (Ailanthus altissima (Mill.) swingle) conflicting values: Assessment of its ecosystem services and potential biological threat. *Environ. Manage.* 2015, 56(4), 1009–1034.
25. Bhowmik P, Inderjit S. Challenges and opportunities in implementing allelopathy for natural weed management. *Crop Prot.* 2003, 22(4), 661–671.
26. Sasikumar K, Vijayalakshmi C, Parthiban KT. Allelopathic effects of eucalyptus on blackgram (Phaseolus mungo l.). *Allelopathy J.* 2002, 9(2), 205–214.
27. Bajwa R, Nazi I. Allelopathic effects of Eucalyptus citriodora on growth, nodulation and AM colonization of Vigna radiata (L) wilczek. *Allelopathy J.* 2005, 15(2), 237–246.
28. May FE, Ash JE. An assessment of the allelopathic potential of eucalyptus. *Aust. J. Bot.* 1990, 38(3), 245–254.
29. Ruwanza S, Gaertner M, Esler K, Richardson D. Allelopathic effects of invasive Eucalyptus camaldulensison germination and early growth of four native species in the Western Cape, South Africa. *South. For. A J. For. Sci.* 2014, 77(2), 91–105.

30. Muller C, Muller W, Haines B. Volatile growth inhibitors produced by aromatic shrubs. *Science* 1964, 143(3605), 471–473.
31. Romagni JG, Allen SN, Dayan FE. Allelopathic effects of volatile cineoles on two weedy plant species. *J. Chem. Ecol.* 2000, 26(1), 303–313.
32. Daneshmandi MS, Azizi MAJID. Allelopathic effect of Eucalyptus globulus labill. on bermuda grass (Cynodon dactylon (L.) pers.) germination and rhizome growth. *Iranian J. Med. Aromat. Plants Res.* 2009, 25(3), 333–346.
33. Rassaeifar M, Hosseini N, Haji Hasani Asl N, Zandi P, Moradi Aghdam A. Allelopathic effect of *Eucalyptus globulus* essential oil on seed germination and seedling establishment of *Amaranthus blitoides* and *Cyndon dactylon*. *Trakia J. Sci.* 2013, 11(1), 73–81.
34. James JF, Bala R. *Allelopathy: How plants suppress other plants* (The Horticultural Sciences Department, University of Florida Institute of Food and Agricultural Science), 2003. Available at: http://edis.ifas.ufl.edu.
35. El-Rokiek, Kowthar G, Rafat R, El-Masry, Nadia K, Messiha, Ahmed SA. The allelopathic effect of mango leaves on the growth and propagative capacity of purple nutsedge (*Cyperus rotundus* L.). *J. Am. Sci.* 2010, 6(9), 151–159.
36. Alsaadawi IS, Khaliq A, Lahmod NR, Matloob A. Weed management in broad bean (*Vicia faba* L.) through allelopathic *Sorghum bicolor* (L.) Moench residues and reduced rate of a pre-plant herbicide. *Allelopathy J.* 2013, 32, 203–212.
37. Czarnota M, Paul R, Weston L, Duke S. Anatomy of sorgoleone-secreting root hairs of sorghum species. *Int. J. Plant Sci.* 2003, 164(6), 861–866.
38. Cheema ZA. *Weed control in wheat through sorghum allelochemicals*. Ph.D. Thesis (Faisalabad, Pakistan: Agronomy Department, University of Agriculture), 1988.
39. Netzly D, Butler L. Roots of sorghum exude hydrophobic droplets containing biologically active components. *Crop Sci.* 1986, 26(4), 775–778.
40. Uddin M, Park S, Dayan F, Pyon J. Herbicidal activity of formulated sorgoleone, a natural product of sorghum root exudate. *Pest Manage. Sci.* 2013, 70(2), 252–257.
41. Sene M, Dore T, Pellisier F. Effects of phenolic acids in soil under and between rows of a prior sorghum (*Sorghum bicolor*) crop on germination, emergence and seedling growth of peanut (*Arachis hypogaea*). *J. Chem. Ecol.* 2000, 26, 625–637.
42. Cheema Z, Khaliq A, Saeed S. Weed control in maize (Zea mays l.) through sorghum allelopathy. *J. Sustain. Agric.* 2004, 23(4), 73–86.
43. Lee SB, Ku YC, Kim KH, Hahn SJ, Chung IM. Allelopathic potential of rice germplasm against barnyardgrass. *Allelopathy J.* 2004, 13(1), 17–28.
44. Belz R. Allelopathy in crop/weed interactions—an update. *Pest Manage. Sci.* 2007, 63(4), 308–326.
45. Kato-Noguchi H, Ino T. Rice seedlings release momilactone b into the environment. *Phytochem.* 2003, 63(5), 551–554.
46. Steinsiek JW, Oliver LR, Collins FC. Allelopathic potential of wheat (Triticum aestivum) straw on selected weed species. *Weed Sci.* 1982, 30(5), 495–497.
47. Krogh SS, Mensz SJ, Nielsen ST, Mortensen AG, Christophersen C, Fomsgaard IS. Fate of benzoxazinone allelochemicals in soil after incorporation of wheat and rye sprouts. *J. Agric. Food Chem.* 2006, 54(4), 1064–1074.
48. Lodhi MAK, Bilal R, Malik KA. Allelopathy in agroecosystems: Wheat phytotoxicity and its possible roles in crop rotation. *J. Chem. Ecol.* 1987, 13(8), 1881–1891.
49. Dudai N, Putievshy E, Lerner HR, Ravid U, Lewinsohn E, Mayer AR. Biotransformation of constituents of essential oils by germinating wheat seed. *Phytochem.* 2000, 55, 375–382.
50. Uremis I, Arslan M, Sangun MK, Uygur V, Isler N. Allelopathic potential of rapeseed cultivars on germination and seedling growth of weeds. *Asian J. Chem.* 2009, 21, 2170–2184.
51. Younesabadi M. *The fourth world congress on allelopathy* (Wagga Wagga, Australia: Charles Sturt University), 2005, 86–93.
52. Gardiner JB, Morra MJ, Eberlein CV, Brown PD, Borek V. Allelochemicals released in soil following incorporation of rapeseed (*Brassica napus*) green manures. *J. Agric. Food Chem.* 1999, 47, 3837–3842.
53. Bones AM, Rossiter JT. The myrosinase-glucosinolate system, its organisation and biochemistry. *Physiol. Plant.* 1996, 97, 194–208.
54. Anjum T, Bajwa R. Field appraisal of herbicide potential of sunflower leaf extract against *Rumex dentatus*. *Field Crops Res.* 2007, 100, 139–142.
55. Macias FA, Molinillo JMG, Chinchilla D, Galindo JCG. Heliannanes—a structure-activity relationship (SAR) study. In *Allelopathy: Chemistry and mode of action of allelochemicals*, Macias FA, Galindo, JCG, Molinillo JMG, Cuttler HG, Eds. (CRC Press Publisher), 2004, Chapter 5, 103–124.

56. Alsaadawi IS, Sarbout AK, Al-Shamma LM. Differential allelopathic potential of sunflower (*Helianthus annuus* L.) genotypes on weeds and wheat (*Triticum aestivum* L.) crop. *Arch. Agron. Soil Sci.* 2012, 58, 1139–1148.
57. Vyvyan JR. Allelochemicals as leads for new herbicides and agrochemicals. *Tetrahedron* 2002, 58(9), 1631–1646.
58. Anjum T, Bajwa R. Screening of sunflower varieties for their herbicidal potential against common weeds of wheat. *J. Sustain. Agric.* 2008, 32, 213–229.
59. Stephenson R, Posler G. The influence of tall fescue on the germination, seedling growth and yield of birdsfoot trefoil. *Grass Forage Sci.* 1988, 43(3), 273–278.
60. Bertin C, Weston L, Huang T, Jander G, Owens T, Meinwald J, Schroeder FC. Grass roots chemistry: Meta-tyrosine, an herbicidal nonprotein amino acid. *Proc. Natl. Acad. Sci.* 2007, 104(43), 16964–16969.
61. Azizi M, Fuji Y. Allelopathic effect of some medicinal plant substances on seed germination of *Amaranthus retroflexus* and *Portulaca oleraceae*. *Acta Hortic.* 2006, 699, 61–68.
62. Khan AH, Vaishya RD. Allelopathic effects of different crop residues on germination and growth of weeds. In *Proceedings of national symposium allelopathy in agroecosystems*, Tauro P, Narwal SS, Eds. (Hisar, India: Indian Society of Allelopathy), 1992, 59–60.
63. Lin J, Smith RJ, Dilday RH. Allelopathic activity of rice germplasm on weed. *Proc. South. Weed Sci. Soc.* 1992, 45, 99.
64. Frans RE, Semidey N. The role of allelopathic sunflower in cotton production [weed suppression]. In *1st international weed control congress* (Melbourne, VIC: Weeds Science Society of Victoria Inc.), 17 Feb 1992, 2, 171–174.
65. Kong CH, Wang P, Zhang CX, Zhang MX, Hu F. Herbicidal potential of allelochemicals from *Lantana camara* against *Eichhornia crassipes* and the alga *microcystis aeruginosa*. *Weed Res.* 2006, 46(4), 290–295.
66. Batish D, Pal Singh H, Kohli R, Kaur S, Saxena D, Yadav S. Assessment of phytotoxicity of parthenin. *Z. Für Naturforschung C.* 2007, 62(5–6), 367–372.
67. Lotina-Hennsen B, King-Díaz B, Pereda-Miranda R. Tricolorin as a natural herbicide. *Molecules* 2013, 18(1), 778–788.
68. Kanchan SD. Growth inhibitors from *Parthenium hysterophorus* Linn. *Curr. Sci.* 1975, 44, 358–359.
69. Dung NX, Nam VV, Huóng HT, Leclercq PA. Chemical composition of the essential oil of *Artemisia vulgaris* L. var. indica Maxim. from Vietnam. *J. Essent. Oil Res.* 1992, 4(4), 433–434.
70. Anaya A, Calera M, Mata R, Pereda-Miranda R. Allelopathic potential of compounds isolated from *Ipomoea tricolor* cav. (Convolvulaceae). *J. Chem. Ecol.* 1990, 16(7), 2145–2152.
71. Kong C, Wang P, Xu X. Allelopathic interference of *Ambrosia trifida* with wheat (*Triticum aestivum*). *Agri. Ecosyst. Environ.* 2007, 119(3–4), 416–420.
72. Sharma OP, Sharma S, Pattabhi V, Mahato SB, Sharma PD. A review of the hepatotoxic plant *Lantana camara*. *Crit. Rev. Toxicol.* 2007, 37(4), 313–352.
73. Cheema ZA, Farooq M, Wahid A. *Allelopathy: Current trends and future applications* (Berlin, Heidelberg: Springer Science & Business Media), 2012.
74. Shahzad M, Farooq M, Jabran K, Hussain M. Impact of different crop rotations and tillage systems on weed infestation and productivity of bread wheat. *Crop. Prot.* 2016, 89, 161–169.
75. Kandhro MN, Tunio S, Rajpar I, Chachar Q. Allelopathic impact of sorghum and sunflower intercropping on weed management and yield enhancement in cotton. *Sarhad J. Agric. Sci.* 2014, 30, 311–318.
76. Saudy HS. Maize-cowpea intercropping as an ecological approach for nitrogen use rationalization and weed suppression. *Arch. Agron. Soil Sci.* 2015, 61, 1–14.
77. Altieri MA, Lana MA, Bittencourt HV, Kieling AS, Comin JJ, Lovato PE. Enhancing crop productivity via weed suppression in organic no-till cropping systems in Santa Catarina, Brazil. *J. Sustain. Agric.* 2011, 35, 855–869.
78. Bernstein ER, Stoltenberg DE, Posner JL, Hedtcke JL. Weed community dynamics and suppression in tilled and no-tillage transitional organic winter rye-soybean systems. *Weed Sci.* 2014, 62, 125–137.
79. Norsworthy JK, McClelland M, Griffith G, Bangarwa SK, Still J. Evaluation of cereal and Brassicaceae cover crops in conservation-tillage, enhanced, glyphosate-resistant cotton. *Weed Technol.* 2011, 25, 6–13.
80. Abbas T, Nadeem MA, Tanveer A, Ali HH, Farooq N. Role of allelopathic crop mulches and reduced doses of tank-mixed herbicides in managing herbicide-resistant *Phalaris minor* in wheat. *Crop. Prot.* 2018, 110, 245–250.
81. Dube E, Chiduza C, Muchaonyerwa P, Fanadzo M, Mthoko T. 2012. Winter cover crops and fertiliser effects on the weed seed bank in a low-input maize-based conservation agriculture system. *South Afr. J. Plant Soil* 2012, 29, 195–197.

82. Finney DM, Creamer NG, Schultheis JR, Wagger MG, Brownie C. Sorghum sudangrass as a summer cover and hay crop for organic fall cabbage production. *Renew. Agric. Food Syst.* 2009, 24, 225–233.
83. Moran P, Greenberg S. Winter cover crops and vinegar for early-season weed control in sustainable cotton. *J. Sustain. Agric.* 2008, 32, 483–506.
84. Anjum T, Bajwa R. The effect of sunflower leaf extracts on *Chenopodium album* in wheat fields in Pakistan. *Crop. Prot.* 2007, 26, 1390–1394.
85. Radhakrishnan R, Alqarawi AA, Abd Allah EF. Bioherbicides: Current knowledge on weed control mechanism. *Ecotoxicol. Environ. Saf.* 2018, 158, 131–138.
86. Dhima K, Vasilakoglou I, Eleftherohorinos I, Lithourgidis A. Allelopathic potential of winter cereals and their cover crop mulch effect on grass weed suppression and corn development. *Crop Sci.* 2006, 46, 345–352.
87. Khaliq A, Matloob A, Irshad MS, Tanveer A, Zamir MSI. Organic weed management in maize (Zea mays L.) through integration of allelopathic crop residues. *Pak. J. Weed Sci. Res.* 2010, 16, 409–420.
88. Bajgai Y, Kristiansen P, Hulugalle N, McHenry M. 2013. Comparison of organic and conventional managements on yields, nutrients and weeds in a corn-cabbage rotation. *Renew. Agric. Food Syst.* 2015, 30(2), 132–142.
89. Puig CG, Revilla P, Barreal ME, Reigosa MJ, Pedrol N. On the suitability of *Eucalyptus globulus* green manure for field weed control. *Crop. Prot.* 2019, 121, 57–65.
90. Masilionyte L, Maiksteniene S, Kriauciuniene Z, Jablonskyte-Rasce D, Zou L, Sarauskis E. Effect of cover crops in smothering weeds and volunteer plants in alternative farming systems. *Crop. Prot.* 2017, 91, 74–81.
91. Bhadoria P. Allelopathy: A natural way towards weed management. *Am. J. Exp. Agric* 2011, 1, 7–20.
92. Ihsan MZ, Khaliq A, Mahmood A, Naeem M, El-Nakhlawy F, Alghabari F. Field evaluation of allelopathic plant extracts alongside herbicides on weed management indices and weed-crop regression analysis in maize. *Weed Biol. Manage.* 2015, 15, 78–86.
93. Aci MM, Sidari R, Araniti F and Lupini A. Emerging trends in allelopathy: A genetic perspective for sustainable agriculture. *Agron.* 2022, 12(9), 2043.
94. Scavo A, Mauromicale G. Crop allelopathy for sustainable weed management in agroecosystems: Knowing the present with a view to the future. *Agron.* 2021, 11(11), 2104.
95. Holt JS, Lovett HN. Significance and distribution of herbicide resistance. *Weed Technol.* 1990, 4, 141–149.
96. Singh HP, Daizy RB, Kohli RK. Allelopathic interactions and allelochemicals: New possibilities for sustainable weed management. *Crit. Rev. Plant Sci.* 2003, 22(3–4), 239–311.
97. Ebana K, Yan W, Dilday RH, Namai H, Okuno K. Analysis of QTL associated with the allelo-pathic effect of rice using water-soluble extracts. *Breed. Sci.* 2001, 51, 47–51.

Section 5

Seaweeds and Vermicompost

20 Marine Products
Key Source for Novel Pesticidal Agents

Himani Karakoti, Tanuja Kabdal, Pooja Bargali, Ravendra Kumar, and Om Prakash

20.1 INTRODUCTION

The vast and largely unexplored marine environment is a treasure trove of biodiversity, hosting a myriad of organisms that have adapted to some of the planet's most extreme and varied conditions. This rich biodiversity has led to the evolution of unique biochemical pathways and the production of a diverse array of natural products with novel structures and potent biological activities. As the search for sustainable and environmentally friendly pest control solutions intensifies, marine products are emerging as a promising source of novel pesticidal agents. Marine ecosystems cover more than 70% of the Earth's surface and are home to an estimated 50-80% of all life forms. The organisms that inhabit these ecosystems, from the shallow coastal waters to the deep-sea trenches, have developed a wide range of chemical defenses to survive and thrive in their specific environments [1,2]. This chemical diversity offers an untapped reservoir of compounds that can be harnessed for various applications, including pest control.

The widespread use of synthetic pesticides in agriculture and public health has led to significant problems, including the development of resistance in pest populations, negative impacts on non-target organisms, and environmental contamination. These issues highlight the urgent need for new, effective, and sustainable pesticidal agents that can address these challenges [3].

Marine-derived natural products offer several advantages over conventional synthetic pesticides. Their novel modes of action can provide effective control of resistant pest populations, and their biodegradability reduces the risk of environmental persistence and bioaccumulation. Additionally, many marine natural products are less toxic to non-target species, making them a safer alternative for integrated pest management (IPM) strategies. The search for marine-derived natural products with pesticidal properties involves the exploration of various marine habitats, including coral reefs, mangroves, seagrass beds, and the deep sea. Each of these habitats hosts a unique community of organisms with distinct chemical profiles [4].

Sponges, for instance, are a rich source of bioactive compounds known for their potent anti-predatory and antimicrobial properties [5]. Algae, particularly macroalgae, produce a range of secondary metabolites that have shown promise as herbicides and insecticides [6]. Mollusks, such as marine snails and bivalves, produce neurotoxins and other defensive chemicals that can be harnessed for pest control. Marine bacteria and fungi, often found in symbiotic relationships with other marine organisms, produce secondary metabolites with diverse biological activities, including pesticidal properties [7,8].

The pesticidal activity of marine natural products can be attributed to various mechanisms, including neurotoxicity, enzyme inhibition, and disruption of cellular processes. For example, certain marine-derived compounds target the nervous system of pests, leading to paralysis and death. Others inhibit key enzymes involved in the metabolic processes of pests, effectively starving them or disrupting their development [5]. A comparative analysis of marine-derived and terrestrial

DOI: 10.1201/9781003463429-26

pesticidal agents reveals that many marine compounds offer unique modes of action that are not present in conventional pesticides. This uniqueness not only enhances their effectiveness but also reduces the likelihood of cross-resistance with existing pesticides.

The exploration of marine biodiversity for novel pesticidal agents holds great promise for the development of sustainable and effective pest control solutions. The unique chemical diversity of marine organisms, coupled with the urgent need for new pest management strategies, makes the marine environment a key source of innovative and environmentally friendly pesticides. As research in this field progresses, it is expected that more marine-derived compounds will be identified, characterized, and developed into commercially viable products, contributing to the global effort to achieve sustainable agriculture and pest management practices.

This chapter will delve into the various marine organisms that produce pesticidal compounds, the mechanisms by which these compounds exert their effects, and the potential applications of these compounds in pest management. By understanding the vast potential of marine-derived pesticidal agents, we can unlock new pathways to sustainable pest control and environmental conservation.

20.2 KEY MARINE ORGANISMS AND THEIR PESTICIDAL COMPOUNDS

20.2.1 Marine Sponges: Biotoxins and Their Applications

Marine sponges are a vast reservoir of bioactive compounds, many of which exhibit potent pesticidal properties. These compounds have evolved primarily as chemical defenses against natural predators, pathogens, and competitors in their environment. Given the growing need for alternatives to synthetic pesticides, marine sponge-derived biotoxins present a promising avenue for developing eco-friendly pest control agents. Their specific modes of action, along with their environmental adaptability, make these biotoxins suitable candidates for integrated pest management systems. Marine sponges synthesize a variety of biotoxins that exhibit insecticidal, nematicidal, and antifungal activities. Following are key examples of biotoxins isolated from marine sponges and their applications in pest control (Table 20.1).

20.2.2 Marine Algae: Bioactive Metabolites with Pesticidal Properties

Marine algae are increasingly recognized for their bioactive metabolites, many of which exhibit pesticidal properties. These secondary metabolites play a crucial role in the defense mechanisms of algae and are being investigated as potential biopesticides. Due to their unique environment, marine algae produce compounds that show strong pesticidal, antifeedant, larvicidal, and insect growth regulatory activities. Following is a comprehensive overview of these compounds, focusing on their mode of action, target pests, and potential applications in sustainable pest management (Table 20.2).

20.2.3 Marine Bacteria and Fungi: Secondary Metabolites with Potential Pesticidal Activity

Marine fungi and bacteria have emerged as an important source of novel bioactive natural products with significant potential for pharmaceutical development. These organisms, thriving in unique and often extreme marine environments, are subject to varying conditions such as high salinity, fluctuating temperatures, elevated pressures, and intense competition from bacteria, viruses, and other microorganisms. Such environmental pressures may have driven the evolution of specialized secondary metabolic pathways, distinguishing them from their terrestrial counterparts. Over the years, numerous antibiotics have been isolated from the culture broths of filamentous marine fungi, underscoring their role in the discovery of new therapeutic agents. Recent research continues to reveal the immense potential of these marine fungi, particularly in the search for biologically active secondary metabolites that could serve as the basis for innovative medicines [21].

TABLE 20.1
Biotoxins from Marine Sponges and Their Pesticidal Properties

Biotoxin	Sponge Source	Target Pest	Mode of Action	Pesticidal Application	References
Manzamine alkaloids	*Haliclona* spp.	Insects, nematodes	Inhibits cyclin-dependent kinases, disrupting cell cycle progression in pests.	Effective against mosquito larvae and soil nematodes	[9,10]
Aeroplysinin-1	*Aplysina aerophoba*	Insects, mollusks, fungi	Disrupts cellular signaling pathways, leading to apoptosis in pests.	Antifungal and molluscicidal, effective in crop protection	[11]
Halichondrins	*Lissodendoryx* spp.	Insects, nematodes	Inhibits microtubule formation, preventing cell division in pests.	Potential insecticide and nematicide	[12]
Theonellamides	*Theonella* spp.	Fungi, insects	Binds to sterols in cell membranes, causing membrane disruption and cell death.	Antifungal agent with applications in agricultural pest control	[9]
Swinholide A	*Theonella swinhoei*	Insects, mollusks	Destabilizes actin filaments, impairing cytoskeletal integrity and causing cell death.	Insecticide and molluscicide	[13]
Palytoxin	*Palythoa toxica* (found in association with sponges)	Insects, mollusks	Binds to the sodium-potassium ATPase pump, transforming it into a non-selective ion channel.	Neurotoxic, effective in mollusk and insect pest control	[12]
Avarol	*Dysidea avara*	Insects, nematodes, fungi	Inhibits detoxification enzymes and disrupts fungal cell walls.	Broad-spectrum pesticide and fungicide	[14]
Okadaic acid	*Halichondria* spp.	Insects, mollusks	Inhibits protein phosphatases, disrupting essential metabolic processes in pests.	Potential for molluscicidal and insecticidal use	[12]

TABLE 20.2
List of Some Marine Algae Compounds Active against Pests

Compound Name	Marine Algae Source	Target Pest	Mode of Action	Reference
Caulerpin	*Caulerpa racemosa*	*Culex pipiens* (mosquito)	LC_{50} = 0.86 ppm, disrupts nervous system	[15]
Caulerpinic acid	*Caulerpa racemosa*	*Culex pipiens* (mosquito)	LC_{50} = 1.69 ppm, disrupts nervous system	[15]
Octadecadienoic acid	*Laurencia brandenii*	*Sitophilus oryzae* (rice weevil)	Inhibits development and growth	[16]
Palytoxin-like CA II	*Chondria armata*	*Periplaneta americana* (American cockroach)	MLD = 1.8×10^{-12} mol, neurotoxin	[17]
Methyl 10,13-dimethyltetradecanoate	*Caulerpa veravalensis*	*Dysdercus cingulatus* (cotton stainer)	Affects growth and development	[18]
Luteolin 4'-glucuronide	*Enhalus acoroides*	*Spodoptera litura* (armyworm)	Antifeedant activity, AR = 69.85%	[19]
Dimethylsulfoxonium formylmethylide	*Sargassum wightii*	*Culex tritaeniorhynchus* (mosquito)	Larvicidal activity	[20]

TABLE 20.3
Marine Products Obtained from Bacteria and Fungi against Pests

Name	Producing Organism	Anti-Pests	Efficacy	References
Penicixanthenes A	*Penicillium* sp. JY246	*Culex quinquefasciatus*	LC_{50} = 38.5 μg/mL	[23]
Penicixanthenes B	*Penicillium* sp. JY246	*Helicoverpa armigera*	IC_{50} = 100 μg/mL	[23]
Penicixanthenes C	*Penicillium* sp. JY246	*Culex quinquefasciatus*	LC_{50} = 11.6 μg/mL	[23]
Penicixanthenes D	*Penicillium* sp. JY246	*Culex quinquefasciatus*	LC_{50} = 23.5 μg/mL	[23]
Sporyzin A	*Aspergillus oryzae*	*Artemia salina*	61.9% mortality	[24]
Sporyzin B	*Aspergillus oryzae*	*Artemia salina*	42.3% mortality	[24]
Sporyzin C	*Aspergillus oryzae*	*Artemia salina*	32.8% mortality	[24]
JBIR-03	*Aspergillus oryzae*	*Artemia salina*	74.2 % mortality	[24]
Emindole SB	*Aspergillus oryzae*	*Artemia salina*	60.3% mortality	[24]
Emeniveol	*Aspergillus oryzae*	*Artemia salina*	31.4% mortality	[24]
Anhydride aspergide	*Aspergillus fumigatus*	*Spodoptera litura*	76.7% mortality (20 ppm)	[25]
Altemicidin	*Streptomyces sioyaensis*	*Artemia salina*	LC_{50} = 3 ppm	[26]
Cyclopentanepropanoic acid,3,5 bis(acetyloxy)-2-[3 (methoxyimino) octyl], methyl ester	*Streptomyces*VITSTK7 sp.	*Culex quinquefasciatus*	LC_{50} = 430.06 ppm	[27]
5-azidomethyl-3-(2-ethoxy carbonyl-ethyl)-4-ethoxycarbo nylmethyl-1hpyrrole-2-carboxylic acid, ethyl ester	*Streptomyces*VITSTK8 sp.	*Culex quinquefasciatus*	LC_{50} = 881.59 ppm	[27]
Akuammilan-16-carboxylic acid, 17-(acetyloxy)-10 methoxy, methyl ester	*Streptomyces*VITSTK7 sp.	*Culex quinquefasciatus*	LC_{50} = 195.70 ppm	[27]
Chloramphenicol D1	*Acremonium vitellinum*	*Helicoverpa armigera*	LC_{50} = 930 ppm	[28]
Chloramphenicol D2	*Acremonium vitellinum*	*Helicoverpa armigera*	LC_{50} = 560ppm	[28]
Chloramphenicol D3	*Acremonium vitellinum*	*Helicoverpa armigera*	LC_{50} = 910 ppm	[28]
Okalaminei B	*Aspergillus* sp.	*Spodoptera exigua*	LD_{50} = 0.2 μg/g	[28]
Communesin B	*Penicillium* sp.	*Bombyx mori*	LD_{50} = 5 μg/g	[29]
Communesin E	*Penicillium* sp.	*Bombyx mori*	LD_{50} = 80 μg/g	[29]
Cristatumin B	*Eurotium cristatum* EN220	*Artemia salina*	LD_{50} = 74.4 μg/g	[30]
Isoechinulin A	*Eurotium cristatum* EN220	*Artemia salina*	LD_{50} = 16.9 μg/g	[30]
Variecolorin G	*Eurotium cristatum* EN220	*Artemia salina*	LD_{50} = 42.6 μg/g	[30]

Recent research has identified several novel nematicides from *Aspergillus fumigatus*, including fumiquinones A and B, spinulosin, LL-S490β, and pseurotin A. These compounds demonstrated effective nematicidal activity against *Bursaphelenchus xylophilus* and *Pratylenchus penetrans*, with minimal plant growth inhibition [22]. The various reports showing the marine products obtained from bacteria and fungi exhibiting potent pesticidal activity are presented in Table 20.3.

20.3 MECHANISMS OF ACTION

Marine ecosystems provide an extensive array of bioactive compounds produced by various organisms, including algae, sponges, mollusks, bacteria, and fungi. These compounds have demonstrated significant potential as biopesticides, with multiple modes of action, such as neurotoxicity, enzyme inhibition, phytotoxicity, and fungicidal activity. Understanding these mechanisms is crucial for harnessing marine biodiversity in sustainable pest management strategies. Neurotoxicity is one of the primary modes of action exhibited by marine-derived pesticides. Many marine organisms produce neurotoxic compounds that interfere with neuronal function in target pests. For instance, metabolites from brown algae, such as *Sargassum* and *Ascophyllum nodosum*, can block sodium channels or inhibit acetylcholinesterase (AChE), leading to a dangerous accumulation of neurotransmitters [31]. This disruption results in paralysis and death of pests, illustrating a potent mechanism reminiscent of conventional neurotoxic pesticides. Molluscan extracts have also shown neurotoxic effects against agricultural pests, further highlighting the efficacy of marine-derived bioactive compounds. In addition to neurotoxicity, many marine compounds exhibit enzyme inhibition, disrupting critical metabolic processes in pests. For example, compounds isolated from marine fungi and algae often inhibit chitinase, an enzyme essential for the development and integrity of exoskeletons in insects and nematodes [32]. The brominated diterpenes derived from red algae, such as *Jania rubens*, effectively disrupt nematode development through enzyme inhibition, showcasing the potential for targeted pest control [33, 5]. Moreover, marine compounds are increasingly recognized for their herbicidal properties. Algal metabolites can suppress the growth of unwanted plants and weeds, with mechanisms that often involve allelopathy—where certain chemicals inhibit the growth of competing species. For instance, cyanotoxins from species like *Synechocystis aquatilis* exhibit herbicidal effects by inhibiting photosynthesis in competing plant species, thus reducing their biomass and growth [34]. This action not only controls weeds but also enhances crop yields by reducing competition for resources. Fungicidal activity is another significant mode of action for marine-derived compounds. Various marine organisms, including fungi and macroalgae, produce metabolites with potent antifungal properties. Extracts from brown algae, such as *Laminaria* and *Fucus*, have shown effectiveness in controlling phytopathogenic fungi like *Botrytis* and *Fusarium*. These fungicidal compounds often act by inducing plant defense mechanisms or directly damaging fungal cell walls, preventing infection and promoting plant health [35]. When compared to terrestrial pesticides, marine compounds offer several advantages, including a broader spectrum of action and a reduced risk of resistance development. Traditional synthetic pesticides often face challenges with resistance due to their repetitive use, leading to diminished efficacy. In contrast, the unique chemical diversity of marine metabolites, derived from organisms adapted to competitive marine environments, provides a wide range of mechanisms that pests have yet to develop resistance against. Additionally, marine biopesticides tend to be environmentally safer, with lower toxicity to non-target organisms and a faster biodegradation rate.

20.4 EXAMPLES OF MARINE-DERIVED PESTICIDES IN USE

Marine-derived pesticides are bioactive compounds sourced from marine organisms, such as algae, sponges, and marine bacteria, which have shown potential in sustainable agriculture (pest control). These natural products often exhibit unique modes of action and can offer environmentally friendly alternatives to synthetic pesticides. Several specific marine-derived compounds are emerging in the agricultural sector for their pest control properties. For example, compounds extracted from marine plants, animals, and microbes have shown insecticidal and antifungal activities essential for enhancing crop protection. These natural bioactive agents are gaining attention as viable alternatives to traditional synthetic pesticides due to their lower environmental impact [**36, 7**].

20.4.1 Pesticides from Marine Actinomycetes

Marine actinomycetes, particularly from the genus *Streptomyces*, produce a range of bioactive compounds with pesticidal properties. These compounds exhibit antibacterial, antifungal, and antiviral activities, making them effective in managing agricultural pests and diseases. Recent studies have highlighted the potential of these microorganisms to produce antibiotics that could be harnessed as biopesticides. Salinosporamides is one example of a pesticide derived from the marine actinomycete *Salinispora tropica*. These compounds inhibit proteasomes, leading to cell cycle arrest and apoptosis in pests [6].

20.4.2 Compounds Derived from Sea-Snail

The cone snail, particularly of the genus *Conus*, is known for producing venom that contains potent neurotoxins. These neurotoxins have been studied for their insecticidal properties, capable of paralyzing prey. The compounds extracted from these snails can be utilized in developing novel insect control agents that target the nervous systems of pests, providing an alternative to synthetic insecticides [7].

20.4.3 Algal Antifungal Agents

Marine algae, especially seaweeds, are prolific producers of bioactive metabolites. Compounds such as fucoidans from brown algae and carrageenans from red algae have shown potential as biopesticides due to their antifungal and antibacterial properties. These polysaccharides not only protect plants from pathogens but also stimulate plant defense mechanisms. Certain species of marine macroalgae are recognized for their antifungal properties. For example, extracts from red algae such as *Gracilaria* and *Kappaphycus* have shown effectiveness in inhibiting fungal pathogens that threaten crops. The bioactive compounds isolate in these algae can serve as natural antifungal agents, thus promoting healthier crop growth and yield while reducing reliance on chemical fungicides [36,37].

20.4.4 Coral-Derived Natural Products

Coral reefs and their associated organisms have been found to produce a range of bioactive compounds with pesticidal properties. For example, the soft coral *Sarcophyton* can yield extracts that exhibit antibacterial and antifungal activity. These extracts can contribute to pest management strategies, particularly in protecting crops from bacterial and fungal infections [7].

20.4.5 Bioactive Compounds from Marine Bacteria

Marine bacteria are another promising source of biopesticidal compounds. Extracts from bacteria such as *Pseudomonas* and *Bacillus* species, found in marine environments, are known to produce secondary metabolites that exhibit insecticidal and antifungal activities. These compounds can disrupt the life cycles of various agricultural pests, making marine bacteria valuable for developing sustainable pest control solutions [36–38].

Spinosad is derived from the soil-dwelling bacterium *Saccharopolyspora spinosa,* employed in controlling a wide range of insects, including caterpillars, thrips, and leafminers. Spinosad affects the nervous system of insects, causing rapid excitation of the insect's nervous system, leading to paralysis and death. Therefore, it is widely used in organic farming and as a topical treatment for pets against fleas.

Avermectins produced by the bacterium *Streptomyces avermitilis* are effective against a variety of mites, nematodes, and insects. They interfere with neural transmission in invertebrates by binding to glutamate-gated chloride channels, causing paralysis and death. This is utilized in agriculture, veterinary medicine, and human pharmaceuticals for treating parasitic infections.

Milbemycins are obtained from the fermentation products of *Streptomyces hygroscopicus*. They are primarily used against mites and insect pests. Similar to avermectins, milbemycins target the glutamate-gated chloride channels, disrupting the neural and muscular system of pests.

Polyhydroxyalkanoates (PHAs) are biopolymers produced by various marine bacteria such as *Pseudomonas* species. They serve as biopesticides and bioplastics used in developing environmentally friendly pesticide formulations.

Marinopyrroles: The source of marinopyrroles is *Streptomyces* sp., which exhibits strong antibacterial and antifungal properties. It is responsible for the inhibition of bacterial cell wall synthesis and disruption of fungal cell membranes.

Cyclodipeptides: The source of this pesticide is marine *Pseudoalteromonas* spp., which strongly exhibit antifungal and antibacterial activity. The function of cyclodipeptides is inhibition of fungal spore germination and bacterial cell division.

20.4.6 Bryozoan-Derived Compounds

Compounds derived from bryozoans, small aquatic invertebrates, have shown potential for use as bioactive agents in agricultural pest management. Some bryozoan-derived extracts demonstrate significant efficacy against herbivorous insects and fungal pathogens. Their utilization in integrated pest management programs could promote a more sustainable approach to agriculture. Bryostatins is derived from the marine bryozoan *Bugula neritina*. Bryostatins are known to modulate protein kinase C, which can affect cellular processes in pests [7].

20.4.7 Nereistoxin and Its Analogues

Nereistoxin and its analogues, including thiocyclam, bensultap, and cartap, are some of the few significant marine-derived agrochemical agents currently utilized in agriculture. These compounds originate from marine organisms and have established efficacy as pesticides, especially in controlling pest populations [39].

20.4.8 Fungal Strains as Pesticides

Marine-derived fungi are another promising source of biopesticides. They produce a variety of secondary metabolites that can be used to control plant diseases. Research has identified a substantial number of fungal strains isolated from coastal marine ecosystems that demonstrate potential as pesticides. Specifically, 133 fungal strains were isolated, among which 37 distinct isolates from 20 different genera showed promise in agricultural pest control applications. These findings highlight the rich biodiversity of marine-derived microbial species that can contribute to developing effective bio-pesticides [40,41].

20.5 ENVIRONMENTAL AND ECONOMIC BENEFITS

The marine environment, rich in chemical and bioactive diversity, represents a largely untapped reservoir for the discovery of new agrochemical agents. Marine plants, animals, and microorganisms possess unique metabolic pathways and secondary metabolites that can serve as leads for the development of novel pesticides, herbicides, and fungicides. These resources are particularly valuable in the search for insecticidal agents, as insects, being primarily terrestrial or freshwater organisms, have had little exposure to compounds produced in marine ecosystems. As a result, there has been minimal selection pressure for resistance development, making marine-derived compounds especially promising for sustainable pest management strategies [39].

The environmental benefits of utilizing marine resources for agrochemicals are considerable. Marine-derived compounds are often biodegradable, minimizing the risk of long-term environmental contamination, and they tend to have targeted bioactivity, reducing harm to non-target species and beneficial organisms in agricultural ecosystems. This can lead to healthier ecosystems and biodiversity preservation, which are critical to long-term agricultural sustainability.

Economically, the potential of marine bioprospecting could have wide-reaching implications. The development of marine-derived agrochemical agents offers an opportunity for innovation, particularly in coastal regions where marine resources are abundant. Local economies could benefit from the sustainable harvesting of marine organisms, coupled with investment in research and development. Moreover, marine agrochemicals may provide more cost-effective solutions by reducing the reliance on synthetic pesticides, which are often costly to produce and apply. Despite their vast potential, many marine-derived structural classes remain largely unexamined for their agrochemical activities. Further research could not only lead to more effective pest control solutions but also reduce the risks associated with traditional pesticides, such as environmental pollution and pesticide resistance. In this way, marine natural products offer both environmental and economic benefits, presenting a promising path for future agrochemical innovation.

20.6 CHALLENGES AND FUTURE PROSPECTS

Despite their potential, the development and commercialization of marine-derived pesticides face several challenges, such as the following:

- Efficient methods for the extraction and large-scale production of marine-derived bioactive compounds need to be developed.
- Marine-derived pesticides must undergo rigorous testing and regulatory approval before they can be used in agriculture.
- The long-term ecological impacts of these compounds need to be studied to ensure they do not harm marine ecosystems.

Therefore, future research in the field of related to marine-derived novel pesticides should focus on exploring the genomes of marine bacteria to identify new biosynthetic pathways and novel pesticidal compounds, utilizing synthetic biology to enhance the production of marine-derived compounds and optimize their pesticidal properties, and conducting extensive field trials to assess the effectiveness and safety of marine-derived pesticides in real-world agricultural settings.

20.6.1 Challenges in Large-Scale Production and Consistency

The large-scale production of marine-derived pesticidal agents faces several significant challenges. One major issue is the limited supply of raw materials, as marine organisms often produce these bioactive compounds in small quantities, making it difficult to generate sufficient amounts for agricultural use. Additionally, achieving consistency in efficacy is critical for their acceptance in the market, but variations in environmental conditions and the inherent biological variability of marine sources complicate this endeavor. The sustainability of harvesting these marine organisms is also a concern, as over-exploitation can lead to depletion of marine resources, further impacting the availability of these natural products for pest management strategies. The utilization of agro-industrial waste and solid-state fermentation (SSF) presents an alternative method for bio-pesticide production, offering several benefits over traditional submerged fermentation. These advantages include lower costs, reduced energy usage, minimal wastewater generation, and the production of more stable products [42]. The production of carbon-neutral biodiesel from microalgae is seen as the most promising alternative to dwindling petro-diesel resources. This is due to its high lipid content, resource efficiency, economic sustainability, and overall benefits compared to other biofuel sources [43].

20.6.2 Regulatory and Safety Considerations

Agencies such as the U.S. Environmental Protection Agency (EPA) heavily regulate the approval and utilization of marine-derived pesticides. These regulatory bodies ensure that pesticides do not cause unreasonable adverse environmental effects and are safe for non-target species, including endangered and threatened species. This comprehensive evaluation process involves assessing the direct and indirect impacts of the pesticides and their interactions with existing species protection laws. The stringent requirements can delay the introduction of potential marine-derived pesticides to the market and may require extensive data on the environmental fate of these substances.

20.7 FUTURE RESEARCH DIRECTIONS AND POTENTIAL BREAKTHROUGHS

Future research in marine-derived pesticides holds exciting potential for breakthroughs that can significantly enhance pest management practices. One promising direction is the exploration of bioactive metabolites produced by marine sediment-derived microorganisms, which may lead to the discovery of novel compounds with unique modes of action. Additionally, advancements in bioactivity screening techniques can facilitate the identification of effective marine-derived agents more efficiently. Sustainable harvesting methods and aquaculture practices are also areas ripe for innovation, ensuring that marine resources can be utilized responsibly while meeting agricultural demands. Finally, increasing collaboration between researchers, regulatory agencies, and the agricultural sector can improve the pathways for integrating marine-derived agents into conventional pest management strategies [44].

20.8 CONCLUSION

The introduction of marine-derived pesticides into fields such as organic farming and sustainable agriculture offers significant advantages. Many marine-derived pesticides are biodegradable and less toxic to non-target organisms compared to synthetic pesticides. The novel modes of action of marine-derived compounds reduce the likelihood of pests developing resistance. Marine bacteria can be cultivated sustainably, providing a renewable source of pesticidal compounds. These products not only provide pest control but also contribute to maintaining higher levels of environmental health and soil quality. As awareness of the ecological impacts of synthetic pesticides grows, the demand for natural alternatives is increasing. This deepening understanding of marine-derived compounds underscores their potential in both pest management and broader agricultural applications, marking a promising avenue for future research and development. Marine-derived pesticides represent a promising frontier in pest management, offering environmentally friendly alternatives with unique modes of action. As research advances and challenges are addressed, these natural products have the potential to play a significant role in sustainable agriculture and pest control strategies. The continued exploration and utilization of marine biodiversity are likely to yield even more innovative solutions in the field of agricultural biotechnology. However, further research and development are necessary to overcome production challenges and ensure the safe and sustainable use of these compounds in agriculture.

REFERENCES

1. Mora, C., Tittensor, D. P., Adl, S., Simpson, A. G., & Worm, B. (2011). How many species are there on Earth and in the ocean? *PLoS Biology*, *9*(8), e1001127.
2. Dolbeth, M., & Arenas, F. (2022). Marine ecosystems: Types, their importance, and main impacts. In *Life below water* (pp. 591–607). Cham: Springer International Publishing.
3. Himani, P. U., Mahawer, S. K., Kumar, R., & Prakash, O. (2022). Plant protection through agrochemicals and its consequences. *Plant Protection: From Chemicals to Biologicals*, *25*.
4. Carroll, A. R., Copp, B. R., Davis, R. A., Keyzers, R. A., & Prinsep, M. R. (2023). Marine natural products. *Natural Product Reports*, *40*(2), 275–325.
5. Asimakis, E., Shehata, A. A., Eisenreich, W., Acheuk, F., Lasram, S., Basiouni, S., & Tsiamis, G. (2022). Algae and their metabolites as potential bio-pesticides. *Microorganisms*, *10*(2), 307.
6. Jagannathan, S. V., Manemann, E. M., Rowe, S. E., Callender, M. C., & Soto, W. (2021). Marine actinomycetes, new sources of biotechnological products. *Marine Drugs*, *19*(7), 365.
7. Song, C., Yang, J., Zhang, M., Ding, G., Jia, C., Qin, J., & Guo, L. (2021). Marine natural products: The important resource of biological insecticide. *Chemistry & Biodiversity*, *18*(5), e2001020.

8. Hong, D. D., Thom, L. T., Ha, N. C., Thu, N. T. H., Hien, H. T. M., Tam, L. T., & Ambati, R. R. (2023). Isolation of fucoxanthin from *Sargassum oligocystum* Montagne, 1845 seaweed in Vietnam and its neuroprotective activity. *Biomedicines*, *11*(8), 2310.
9. Edrada, R. A., Proksch, P., Wray, V., Witte, L., Müller, W. E. G., & Van Soest, R. W. (1996). Four new bioactive manzamine-type alkaloids from the Philippine marine sponge *Xestospongia ashmorica*. *Journal of Natural Products*, *59*(11), 1056–1060.
10. Lee, S., & Sperry, J. (2022). Isolation and biological activity of azocine and azocane alkaloids. *Bioorganic & Medicinal Chemistry*, *54*, 116560.
11. Teeyapant, R., Woerdenbag, H. J., Gross, H. J., & Proksch, P. (1993). Biotransformation of the brominated compounds in the marine sponge Verongia aerophoba: Evidence for an induced chemical defense? *Planta Medica*, *59*(S1), A641–A642.
12. Martínez, A., Garrido-Maestu, A., Ben-Gigirey, B., Chapela, M.-J., González, V., Vieites, J. M., & Cabado, A. G. (2015). Marine biotoxins. In S. K. Kim (Ed.) *Springer handbook of marine biotechnology*. Springer Handbooks. Berlin, Heidelberg: Springer.
13. El Sayed, K. A., Dunbar, D. C., Perry, T. L., Wilkins, S. P., Hamann, M. T., Greenplate, J. T., & Wideman, M. A. (1997). Marine natural products as prototype insecticidal agents. *Journal of Agricultural and Food Chemistry*, *45*(7), 2735–2739.
14. Singh, A., & Thakur, N. L. (2016). Significance of investigating allelopathic interactions of marine organisms in the discovery and development of cytotoxic compounds. *Chemico-Biological Interactions*, *243*, 135–147.
15. Alarif, W. M., Abou-Elnaga, Z. S., Ayyad, S. E. N., & Al-Lihaibi, S. S. (2010). Insecticidal metabolites from the green alga Caulerpa racemosa. *CLEAN-Soil, Air, Water*, *38*(5–6), 548–557.
16. Manilal, A., Sujith, S., Sabarathnam, B., Kiran, G. S., Selvin, J., Shakir, C., & Lipton, A. P. (2011). Biological activity of red alga *Laurencia brandenii*. *Acta Botanica Croatica*, *70*(1), 81–90.
17. Mori, S., Sugahara, K., Maeda, M., Shimamoto, K., Iwashita, T., & Yamagaki, T. (2018). A truncated palytoxin analogue, palytoxin carboxylic acid, isolated as an insecticidal compound from the red alga, *Chondria armata*. *Tetrahedron Letters*, *59*(50), 4420–4425.
18. Sahayaraj, K., Asharaja, A., Ponsankar, A., Rathi, J. M., & Senthil-Nathan, S. (2019). Behavioral response and relative toxicity for the active compounds of *Caulerpavera veravalensis* (Thivy and Chauhan) against nymph of *Dysdercus cingulatus* (Fab.)(Hemiptera: Pyrrhocoridae). *Journal of Asia-Pacific Entomology*, *22*(2), 417–426.
19. Qi, S. H., Zhang, S., Qian, P. Y., & Wang, B. G. (2008). Antifeedant, antibacterial, and antilarval compounds from the South China Sea seagrass *Enhalus acoroides*. *Botanica Marina*, *51*(5).
20. Suganya, S., Ishwarya, R., Jayakumar, R., Govindarajan, M., Alharbi, N. S., Kadaikunnan, S., & Vaseeharan, B. (2019). New insecticides and antimicrobials derived from *Sargassum wightii* and *Halimeda gracillis* seaweeds: Toxicity against mosquito vectors and antibiofilm activity against microbial pathogens. *South African Journal of Botany*, *125*, 466–480.
21. Zain, M. E., Awaad, A. S., Al-Othman, M. R., Alafeefy, A. M., and El-Meligy, R. M. (2014). Biological activity of fungal secondary metabolites. *International Journal of Chemistry and Applied Biological Science*, *1*(1), 14–22.
22. Hayashi, A., Fujioka, S., Nukina, M., Kawano, T., Shimada, A., & Kimura, Y. (2007). Fumiquinones A and B, nematicidal quinones produced by *Aspergillus* fumigatus. *Bioscience, Biotechnology, and Biochemistry*, *71*(7), 1697–1702.
23. Bai, M., Zheng, C. J., Nong, X. H., Zhou, X. M., Luo, Y. P., & Chen, G. Y. (2019). Four new insecticidal xanthene derivatives from the mangrove-derived fungus Penicillium sp. JY246. *Marine Drugs*, *17*(12), 649.
24. Qiao, M. F., Ji, N. Y., Liu, X. H., Li, K., Zhu, Q. M., & Xue, Q. Z. (2010). Indoloditerpenes from an algicolous isolate of *Aspergillus oryzae*. *Bioorganic & Medicinal Chemistry Letters*, *20*(19), 5677–5680.
25. Guo, Z., Gai, C., Cai, C., Chen, L., Liu, S., Zeng, Y., & Dai, H. (2017). Metabolites with insecticidal activity from *Aspergillus fumigatus* JRJ111048 isolated from mangrove plant *Acrostichum specioum* endemic to Hainan Island. *Marine Drugs*, *15*(12), 381.
26. Takahashi, A., Kurasawa, S., Ikeda, D., Okami, Y., & Takeuchi, T. (1989). Altemicidin, a new acaricidal and antitumor substance I. Taxonomy, fermentation, isolation and physico-chemical and biological properties. *The Journal of Antibiotics*, *42*(11), 1556–1561.
27. Thenmozhi, M., Gopal, J. V., Kannabiran, K., Rajakumar, G., Velayutham, K., & Rahuman, A. A. (2013). Eco-friendly approach using marine actinobacteria and its compounds to control ticks and mosquitoes. *Parasitology Research*, *112*, 719–729.

28. Chen, D., Zhang, P., Liu, T., Wang, X. F., Li, Z. X., Li, W., & Wang, F. L. (2018). Insecticidal activities of chloramphenicol derivatives isolated from a marine alga-derived endophytic fungus, *Acremonium vitellinum*, against the cotton bollworm, *Helicoverpa armigera* (Hübner)(Lepidoptera: Noctuidae). *Molecules*, *23*(11), 2995.
29. Zuo, Z., & Ma, D. (2011). Synthetic studies toward communesins. *Israel Journal of Chemistry*, *51*(3–4), 434–441.
30. Du, F. Y., Li, X. M., Li, C. S., Shang, Z., & Wang, B. G. (2012). Cristatumins A-D, new indole alkaloids from the marine-derived endophytic fungus *Eurotium cristatum* EN-220. *Bioorganic & Medicinal Chemistry Letters*, *22*(14), 4650–4653.
31. Hong, L. L., Ding, Y. F., Zhang, W., & Lin, H. W. (2022). Chemical and biological diversity of new natural products from marine sponges: A review (2009–2018). *Marine Life Science & Technology*, *4*(3), 356–372.
32. Guo, X. C., Zhang, Y. H., Gao, W. B., Pan, L., Zhu, H. J., & Cao, F. (2020). Absolute configurations and chitinase inhibitions of quinazoline-containing diketopiperazines from the marine-derived fungus Penicillium polonicum. *Marine Drugs*, *18*(9), 479.
33. Lever, J., Brkljača, R., Kraft, G., & Urban, S. (2020). Natural products of marine macroalgae from South Eastern Australia, with emphasis on the Port Phillip Bay and heads regions of Victoria. *Marine Drugs*, *18*(3), 142.
34. Akmukhanova, N. R., Leong, Y. K., Seiilbek, S. N., Konysbay, A., Zayadan, B. K., Sadvakasova, A. K., & Allakhverdiev, S. I. (2023). Eco-friendly biopesticides derived from CO2-Fixing cyanobacteria. *Environmental Research*, 117419.
35. Shukla, P. S., Borza, T., Critchley, A. T., & Prithiviraj, B. (2021). Seaweed-based compounds and products for sustainable protection against plant pathogens. *Marine Drugs*, *19*(2), 59.
36. Avhad, A. B., & Bhangale, C. J. (2023). Marine natural products and derivatives. *RPS Pharmacy and Pharmacology Reports*, *2*(2), rqad008.
37. Eckstien, D., Maximov, N., Margolis, N., & Raanan, H. (2024). Towards sustainable biocontrol: Inhibition of soil borne fungi by microalgae from harsh environments. *Frontiers in Microbiology*, *15*, 1433765.
38. Ghosh, S., Sarkar, T., Pati, S., Kari, Z. A., Edinur, H. A., & Chakraborty, R. (2022). Novel bioactive compounds from marine sources as a tool for functional food development. *Frontiers in Marine Science*, *9*, 832957.
39. Peng, J., Shen, X., El Sayed, K. A., Dunbar, D. C., Perry, T. L., Wilkins, S. P., & Wideman, M. A. (2003). Marine natural products as prototype agrochemical agents. *Journal of Agricultural and Food Chemistry*, *51*(8), 2246–2252.
40. de Sá, J. D., Kumla, D., Dethoup, T., & Kijjoa, A. (2022). Bioactive compounds from terrestrial and marine-derived fungi of the genus *Neosartorya*. *Molecules*, *27*(7), 2351.
41. Huang, R. H., Gou, J. Y., Zhao, D. L., Wang, D., Liu, J., Ma, G. Y., Li, Y. Q., & Zhang, C. S. (2018). Phytotoxicity and anti-phytopathogenic activities of marine-derived fungi and their secondary metabolites. *RSC Advances*, *8*(66), 37573–37580.
42. De la Cruz Quiroz, R., Roussos, S., Hernández, D., Rodríguez, R., Castillo, F., & Aguilar, C. N. (2015). Challenges and opportunities of the bio-pesticides production by solid-state fermentation: Filamentous fungi as a model. *Critical Reviews in Biotechnology*, *35*(3), 326–333.
43. Rawat, I., Kumar, R. R., Mutanda, T., & Bux, F. (2013). Biodiesel from microalgae: A critical evaluation from laboratory to large scale production. *Applied Energy*, *103*, 444–467.
44. Putra, N. R., Rizkiyah, D. N., Che Yunus, M. A., & Qomariyah, L. (2024). Towards a greener future: Bioactive compounds extraction from shrimp shells using eco-friendly techniques. *eFood*, *5*(3), e146.

21 Sustainable Agriculture Practices

Impact of Vermicompost on Soil Health and Ornamental Plant Production

Saiqa Andleeb, Irsa Shafique, and Iram Liaqat

Abbreviations: Nitrogen, phosphorous, and potassium (**NPK**); Total organic carbon (**TOC**); Organic matter (**OM**); Cation exchange capacity (**CEC**); Indole acetic acid (**IAA**); Hydrogen cyanide (**HCN**); Plant growth–promoting bacteria (**PGPR**); Chromium (**Cr**); Manganese (**Mn**); Copper (**Co**); Nickel (**Ni**); Selenium (**Se**); Cobalt (**Cu**); Molybdenum (**Mo**); Plant growth–promoting vermibacteria (**PGPVB**)

21.1 INTRODUCTION

The use of agrochemicals to boost food productivity due to rapid growth in the human population is a major global problem (Muhibbullah and Sarwar, 2017). To overcome the gap between food consumption and production, different agrochemicals are being used in agriculture and because human health all over the world has been adversely affected by this chemically grown food (Sinha et al., 2011). "Agrochemical" is a broad term used for different chemical products (hormones, fertilizers, diverse pesticides, or soil treatments) that have a role in improving crop production (Muhibbullah and Sarwar, 2017). Previous literature indicated that agrochemicals have been used to attain the maximum crop yield in both developed and developing countries (Carvalho, 2017).

Currently, various agrochemicals such as fungicides, herbicides, insecticides, molluscicides, nematicides, and rodenticides are being used that have adverse effects not only on humans but also on the environment and beneficial soil biota (Onder et al., 2011; Chandra and Mani, 2011; Meena et al., 2016). Soil and water pollution are the major risks linked with the regular usage of agrochemicals, along with pesticide-resistant strain development of weeds, herbs, and pests and toxicity to humans and other organisms (Majeed et al., 2017). Natural nutrients on the soil surface are also being reduced due to overuse of agro-chemicals and thus contaminating the biotic and aquatic environment, associated with some human health diseases (Jayasumana et al., 2015; Parks et al., 2016).

Agrochemicals (pesticides) involved in the reduction of soil enzyme activities include nitrate reductase, hydrolyzes, oxidoreductases, urease, nitrogenase, and dehydrogenase, which affect the soil nutrient profile. All biotransformation methods, that is, biological nitrogen fixation, ammonification, denitrification, nitrification, S-oxidation, and phosphorus solubilization, are also influenced by the usage of agrochemicals. Previous literature illustrated the consequences of agrochemicals on soil micro-organisms involved in soil fertility (Bano and Iqbal, 2016). These organisms are the decomposers of organic matter and provide phosphorous, nitrogen, and potassium through

DOI: 10.1201/9781003463429-27

mineralization and fixation processes which help in plant growth and development. These microbes have a role in the suppression of disease-causing organisms and detoxification of harmful chemicals and produce plant hormones that encourage the growth of plants and productivity (Reitz et al., 2015; Bano and Iqbal, 2016).

Many farmers do not have sufficient knowledge and information regarding health issues associated with the use of pesticides and handling (Boateng and Amuzu, 2013). Inappropriate use and disposal of pesticide containers are mainly caused by insufficient knowledge, improper equipment, and storage; application of non-approved pesticides; and the use of an overdosage (Okoffo et al. 2016). There are three ways by which agrochemicals can enter the human body, either by direct flow by mouth (ingestion), by infiltration via the skin (contact), or by breathing (inhalation) (Bashed et al., 2012). These pesticides may be excreted, metabolized, and stored or bio-transformed in body fat within a human or animal body (Pirsaheb et al., 2015). Various negative health issues linked with chemical pesticides include gastrointestinal, reproductive, dermatological, respiratory, neurological, and endocrine effects (Seralini et al., 2014; Thakur et al., 2014).

Furthermore, accidental, high occupational, or intentional exposure to agrochemicals may cause affected individuals to have asthma or lung cancer that may sometimes result in hospitalization and even death (Reile et al., 2015). Moreover, agrochemicals can cause many health hazards to people who are directly exposed to synthetic chemical fertilizers. At present, chemical fertilizers have become much more serious for short-term and long-term effects on the human body as well as in foods such as fruits, meat, vegetables, and other products through pesticide residues (Bashed et al., 2012). Runoff from these chemicals continues to deteriorate food crops and even spread to affect water bodies; thus, pesticide residues are found in the air, soil, and surface and groundwater across the globe. Okoffo et al. (2016) illustrated that chemical pesticides have a role in the destabilization of agro-ecological systems and biodiversity by deteriorating the environment.

21.2 VERMICOMPOSTING

Vermicomposting is a biochemical, non-thermophilic, and eco-biotechnological way to transform organic waste materials, that is, food scraps, agro-industrial waste, household waste, and animal waste into valuable fertilizer called vermicompost via combined action of earthworms and mesophilic microbes after 90 days (Shafique et al., 2023; Ravindran et al., 2016; Adhikary, 2012) (Figure 21.1). It is a sustainable, cheap, effective, and eco-friendly approach for agricultural production and waste management. During vermicomposting different types of organic waste such as grass clippings, paper, manure waste (cattle dung, sheep dung, goat dung), crop residues, vegetable waste, eggshells, and coco fiber and earthworm species including *Eisenia fetida, Perionyx excavates*, and *Eudrilus eugeniae* are used (Shafique et al., 2023; Shafique et al., 2021; Chitrapriya et al., 2013; Fu et al., 2014). Vermicompost is an organic manure composed of worm castings, live earthworms and their cocoons, organic materials, and microorganisms (Prajapati et al., 2023; Shafique et al., 2023).

21.3 SUITABLE VERMICOMPOSTING SPECIES

The important earthworm species used for the composting process are *Eisenia andrei*, *Metaphire Californica*, *Perionyx excavates, Eudrilus eugeniae*, and *Eisenia fetida* (Shafique et al., 2023; Ferber et al., 2019; Singh et al., 2014). Various studies have illustrated that *Eisenia fetida* is a hermaphrodite and needs aerobic conditions and moisture for reproduction. Their mating occurs during the night and produces a capsule. From this capsule, 4–20 worms are born after 14–21 days (Singh et al., 2014). Earthworms are usually capable of the decomposition of various wastes like garden waste, municipal solid waste, animal waste, household waste, and urban and industrial waste (Asgharzadeh et al., 2014).

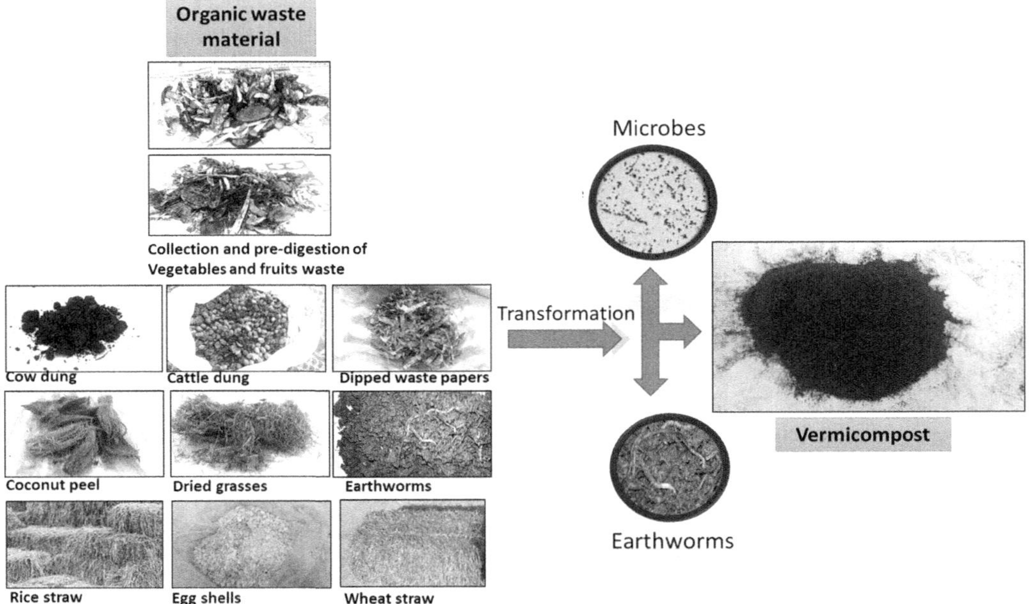

FIGURE 21.1 Combined action of earthworms and microbes in conversion of solid waste to valuable fertilizer (vermicompost).

21.4 PHYSICOCHEMICAL PROPERTIES OF VERMICOMPOST

Vermicompost is blackish brown in color and has a porous and granular appearance, earthy smell, 25°C temperature, 7.8 pH, 2.82 electrical conductivity (mS/cm), and 25% moisture content. Vermicompost is rich in nitrogen (2.28%), phosphorus (0.030%), potassium (0.033%), manganese (0.2038%), iron (1.3313%), copper (0.0048%), zinc (0.110%), calcium (1.35%), sulphates (0.55%), and magnesium (0.78%). The total organic carbon (OC) and organic matter (OM) was 25.2% and 43%, and the cation exchange capacity (CEC) of vermicompost was 72.48 me/100g. Furthermore, vermicompost also comprises helpful or beneficial soil bacteria, humic substances, vermiwash (leachate), enzymes, and plant growth hormones (Shafique et al., 2023; Ayyobi et al., 2014).

21.5 CHARACTERIZATION OF EARTHWORM-ASSOCIATED VERMIBACTERIA

The gastrointestinal track of earthworm species is a suitable habitat for microbes such as bacteria, actinomycetes, and fungi (Munnoli, 2007). Earthworms help to increase microbial activities in the gut by releasing mucus with active and easily digestible compounds (Martin et al., 1987) and a favorable physico-chemical environment like high moisture, neutral pH, and ideal temperature (Barois and Lavelle, 1986). The gut environment has a pH of 6.9, is anoxic, has 50% water contents. This water contents is enriched in nitrogen, total carbon, and organic carbon (Horn et al., 2003; Edwards et al., 2004). Isolation and characterization of vermibacteria are carried out using different cultural media, microscopic techniques, biochemical tests, and molecular methods (Andleeb et al., 2022; Naseer et al., 2022), as shown in Figure 21.2. *Bacillus thuringiensis, Bacillus aryabhattai, Staphylococcus hominis, Bacillus toyonensis, Bacillus cabrialesii, Bacillus tequilensis, Bacillus mojavensis, Bacillus amyloliquefaciens, Bacillus anthracis, Bacillus paranthracis, Bacillus mycoides, Bacillus megaterium, Bacillus subtilis, Bacillus spizizenii, Bacillus licheniformis,* and *Bacillus cereus* were found in the gut of *E. fetida* and identified through microscopic studies, biochemical

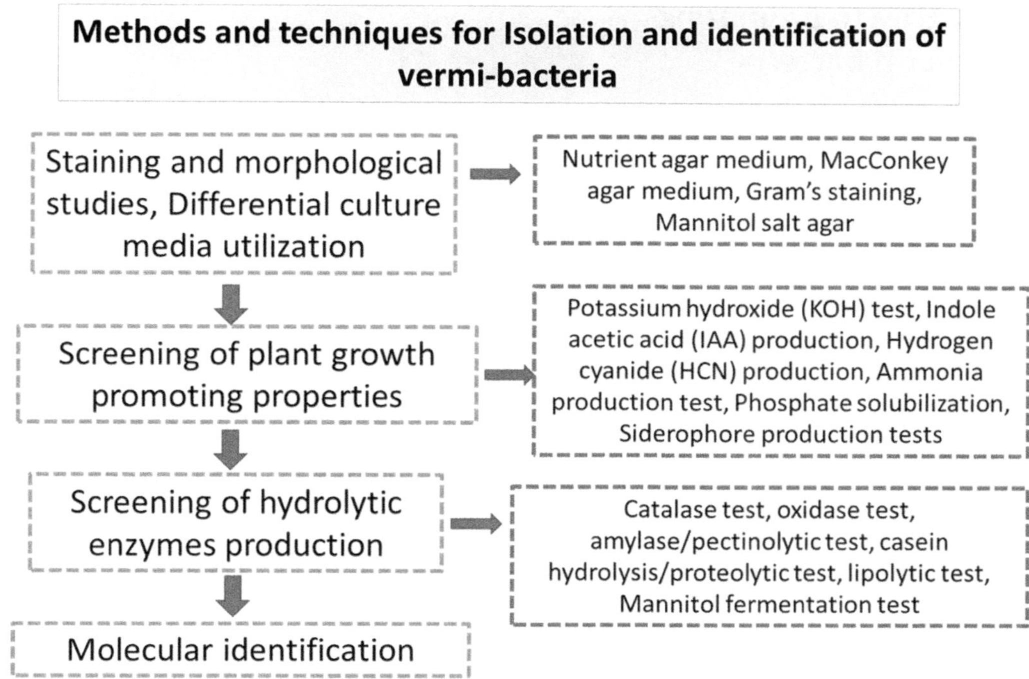

FIGURE 21.2 Screening, isolation, and characterization of vermibacteria.

tests, and molecular analysis. All vermibacteria possessed agricultural traits; that is, they produce different compounds such as indole acetic acid, siderophore, oxidase, catalase, proteases, lipases, and amylases and also act as phosphate solubilizers (Andleeb et al., 2022; Naseer et al., 2022).

21.6 VERMIBACTERIA AS PLANT GROWTH BACTERIA

The phytomicrobiome is the major part of the soil ecosystem and is associated with plants (Lebeis, 2014; Chaparro et al., 2014; Bulgarelli et al., 2015). The phytomicrobiome, along with the plant, is called the holobiont (Berg et al., 2016; Smith et al., 2017). This makes the soil sustainable for crop production and is involved in various biotic activities (Ahemad et al., 2009). It improves the soil structure and is crucial for soil fertility (Glick, 2012). The phytomicrobiome is found around the root surface (called rhizosphere) and inside the roots. Those microbes living on the root surface are called the rhizoplane, while those living inside the root are known as endophytes (Zhang et al., 2017), as shown in Figure 21.3. A wide range of rhizospheric bacteria are explored for agricultural traits such as pesticide degradation (Ahemad and Khan, 2012a, 2012b), heavy metal detoxifying potential (Ma et al., 2011), salinity tolerance (Mayak et al., 2004), and biocontrol of insects and phytopathogens (Russo et al., 2008; Hynes et al., 2008), as well as plant growth–promoting properties like phytohormones (Tank and Saraf, 2010), siderophore production (Jahanian et al., 2012), phosphate solubilization (Ahemad and Khan, 2012a), IAA, HCN, ammonia production, and nitrogenase activity (Glick, 2012; Esitken et al., 2010; Kumar et al., 2015a, 2015b) (Figure 21.3).

Bacillus thuringiensis, B. aryabhattai, Staphylococcus hominis, B. toyonensis, B. cabrialesii, B. tequilensis, B. mojavensis, B. amyloliquefaciens, B. anthracis, B. paranthracis, B. mycoides, B. licheniformis, B. subtilis, B. megaterium, B. spizizenii, and *B. cereus* present in the *E. fetida* gut produce all plant growth traits such as siderophores, IAA, and phyto-hormones and also act as

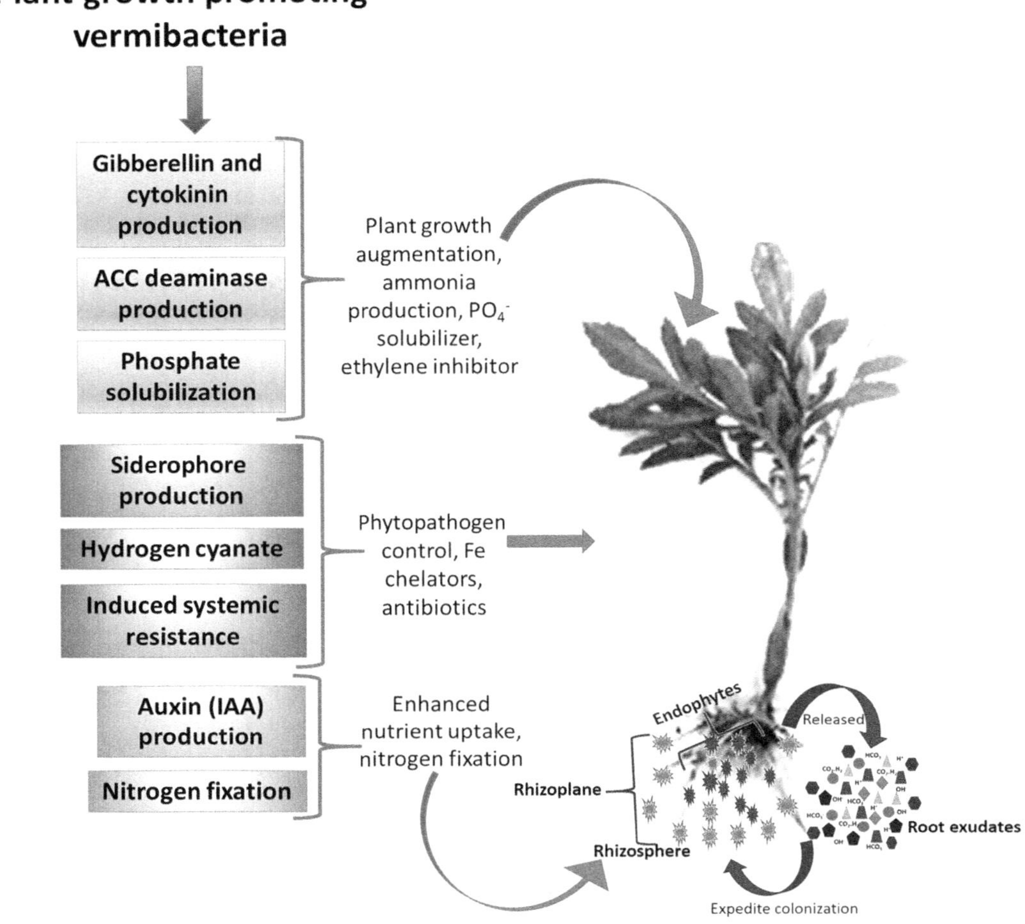

FIGURE 21.3 Vermibacteria promote plant growth and development.

phosphate solubilizers, as well as enzymes (catalase, oxidase, proteases, and lipases) (Andleeb et al., 2022). These vermibacteria play an important role as plant growth–promoting rhizobacteria in the phytomicrobiome. PGPR inoculation significantly enhances germination characteristics, and PGPR effects could be attributed to their ability to produce growth regulators (Mangmang et al. 2014). Several phytomicrobiomes, such as *Rhizobium, Mesorhizobium,* and *Bradyrhizobium* (symbiotic) and *Bacillus, Pseudomonas, Klebsiella, Azospirillum, Azotobacter,* and *Azomonas* (non-symbiotic) rhizobacteria are now being used as microbial bio-fertilizers worldwide to endorse plant growth and development under various conditions like agrochemicals and abiotic factors.

21.7 VERMIBACTERIA AS BACTOREMEDIATORS

Heavy metals like chromium (Cr), manganese (Mn), copper (Co), nickel (Ni), selenium (Se), cobalt (Cu), and molybdenum (Mo) affect crop productivity (Rascio and Navari-Izzo, 2011). These metals can reduce crop production when accumulated in higher ratios (Foucault et al., 2013; Xiong et al., 2014). Several studies reported that heavy metal accumulation in food crops happened through the roots while grown in polluted soil (Foucault et al., 2013; Schreck et al., 2013; Xiong et al., 2014), and major accumulation through plants can lead to reducing crop production via interfering with nutrient uptake, uptake of water, photosynthesis, and nitrogen metabolism (Austruy et al., 2014); reducing growth and biomass; reducing root and shoot development; reducing the biosynthesis of chlorophyll' causing disorganization of the grana structure; interfering with the respiration process (Austruy et al., 2014); and causing programmed cell death/necrosis (Sabir et al., 2015) (Figure 21.4). Physical methods are done by replacing the polluted soil with uncontaminated soil, but this method is time consuming, laborious, and not cost effective (Dhaliwal et al., 2020). Chemical remediation involves verification technology (Meuser, 2012), chemical fixation by adding reagents to the contaminated soil (Jinadasa et al., 2016; Huang et al., 2017, 2020), and electrokinetic remediation (Figueroa et al., 2016).

Bioremediation technology is used to transform hazardous metals into a less hazardous form using microorganisms (algae, fungi, bacteria), worms, and plant species or their enzymes to clean the environmental hosts (Abbas et al., 2014; Akcil et al., 2015; Ndeddy Aka and Babalola, 2016; Kaur et al., 2018; Kumar & Dwivedi, 2021). These techniques are cost effective, environmentally friendly, and easy to approach (Okoduwa et al., 2017). Vermiremediation approaches could include the introduction of earthworms into polluted soils directly, application of earthworms with other biological media like compost to the soil, application of earthworm-associated bacteria (vermibacteria), application of contaminants to earthworms for feeding, and application of earthworms indirectly by the application of vermidigested materials (Ahmad et al., 2021).

Bacteria are important and powerful metal biosorbents (live or dead biomass) because of their distinctive size, as well as their capacity to develop under measured and ecological situations (Srivastava and Kumar, 2015). *Enterobacter, Flavobacterium, Pseudomonas, Bacillus,* and *Micrococcus* spp. have been utilized to remediate toxic metals and biosorption capacity due to their active cell wall chemosorption sites (teichoic acid), along with a great surface-to-volume ratio (Kapahi and Sachdeva, 2019). The cell wall is the outer protective covering connecting bacteriological biomass with external metallic pollution. The negative charge of anionic functional groups (sulfate, amine, hydroxyl, carboxyl, phosphate) present in Gram-negative bacteria (in peptidoglycan, lipopolysaccharides, and phospholipids) and in Gram-positive bacteria (in peptidoglycan, teichoic acids) imparts the metal-binding capacity of the cell wall (Joo et al., 2010) (Figure 21.4).

Bacillus thuringiensis, B. aryabhattai, Staphylococcus hominis, B. toyonensis, B. cabrialesii, B. tequilensis, B. mojavensis, B. amyloliquefaciens, B. anthracis, B. paranthracis, B. mycoides, B. megaterium, B. subtilis, B. spizizenii, B. licheniformis, and *B. cereus* helped in the bioregulation of heavy metals, reducing and tolerating their toxicity (Figure 21.4). Previous literature found that microbes not only play a crucial role in soil fertility maintenance but also improve crop productivity (Kosev and Vasileva, 2014) by developing many strategies (methylation and demethylation, bioaccumulation/biosorption, biotransformation, oxidation and reduction, and expulsion of heavy metals) to alleviate heavy metal toxicity (Akcil et al., 2015) (Figure 21.4).

A microbial-centered strategy for the recovery and elimination of heavy metals stands out as a more effective, cost-efficient, and environmentally friendly alternative when compared to physicochemical methods (Medfu Tarekegn et al., 2020). In this investigation, all identified vermibacterial species exhibited notable efficacy in removing cadmium, lead, and chromium. Consequently, they hold promise as potential sources for bactoremediation. Previous studies by Medfu Tarekegn et al. (2020) demonstrated the effectiveness of bacterial species such as *B. megaterium, B. subtilis, Penicillium,* and *A. niger* in remediating cadmium, lead, and chromium. Additionally, Njoku et al. (2020) highlighted the efficacy of *Rhizopus stolonifer* and *Bacillus megaterium* in the remediation of heavy metals, including Pd, Cd, and Ni.

Microbes employ a variety of mechanisms for remediating heavy metals, including adsorption, biosorption, detoxification, chelation, precipitation, enzyme regulation, methylation, and complexation, as illustrated in Figure 21.4 (Ma et al., 2016). The interplay and removal of heavy metals are significantly influenced by the bacterial cell physiology, cell wall composition, and genome composition (Li et al., 2019) (Figure 21.4).

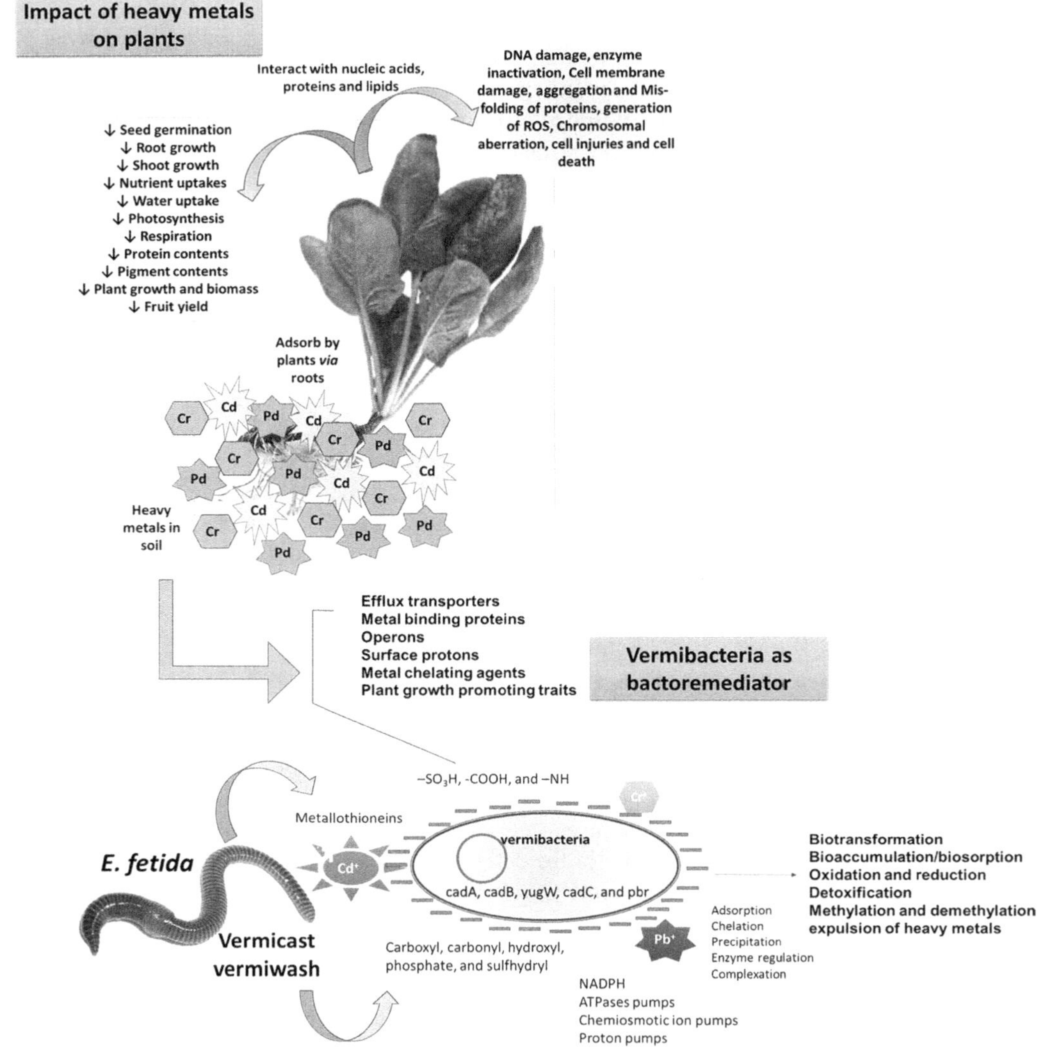

FIGURE 21.4 Toxic effects of heavy metals on plants and their remediation through vermibacteria. ↓ reduced/declined.

The bacterial cell wall plays a crucial role in preventing the entry of heavy metals into bacterial cells through extracellular substances and anionic functional groups, such as polysaccharides, proteins carbonyl, humic substances, carboxyl, phosphate, sulfonate, hydroxyl, amide groups, amines, and sulfhydryl, which impart a negative charge to the cell wall (Ahemad and Kibret, 2014). All vermibacterial isolates exhibit a Gram-positive nature, and owing to the presence of these substances, they demonstrate heightened resistance to heavy metals and efficiently detoxify them compared to Gram-negative bacteria.

Bacillus species, endowed by nature with specific operons and enzymes, have evolved to counteract the toxicity of heavy metals. Operons such as cadA, cadB, cadC, yugW, and pbr play a role in accumulating cadmium and lead from contaminated environments (Li et al., 2019; Naseer et al., 2022) (Figure 21.4). Cadmium regulation occurs through the uptaking operons, efflux mechanisms, and chelation of cadmium with cell surface protons, represented by –SO3H, –COOH, and –NH groups (Li et al., 2019). Similarly, *Bacillus* species employ the pbr operon to counteract lead toxicity. Additionally, ATPases and proton/chemiosmotic pumps facilitate the expulsion of toxic metals outside the bacterial cell (Ahmed et al., 2013) (Figure 21.4).

21.8 VERMICOMPOST IMPROVES SOIL HEALTH

Vermicompost enhances soil structure by increasing aggregate stability and reducing compaction. This allows for better aeration, root penetration, and water infiltration. Vermicompost is rich in essential nutrients, including nitrogen, phosphorus, potassium, and micronutrients. When incorporated into the soil, it provides a slow-release source of these nutrients, promoting plant growth and health (Figure 21.5). Vermicompost encourages beneficial microbial activity in the soil. Earthworms' digestive processes help break down organic matter, creating a nutrient-rich environment for microorganisms. This microbial activity can suppress harmful pathogens and boost nutrient cycling (Rehman et al., 2023). Vermicompost has a natural pH buffering capacity, helping to maintain soil pH within an optimal range for plant growth. Vermicompost enhances the soil's ability to retain and exchange essential nutrients with plant roots. This is particularly important for ion retention, which benefits plant nutrient uptake. Improved soil structure and root growth encouraged by vermicompost can reduce soil erosion, preserving valuable topsoil. The continuous use of vermicompost can promote sustainable soil fertility without relying on synthetic fertilizers, reducing the risk of soil degradation (Shafique et al., 2023). Earthworms have the ability to detoxify certain substances in the soil. This can help reduce the concentration of certain harmful chemicals and heavy metals in the soil (Naseer et al., 2022). The use of vermicompost often reduces the need for chemical fertilizers and pesticides, which can have positive environmental effects by reducing the risk of groundwater contamination and pollution (Moustafa et al., 2023) (Figure 21.5).

21.9 VERMICOMPOST AND INTEGRATED MANAGEMENT SYSTEMS STIMULATE GROWTH OF ORNAMENTAL PLANTS

Vermicompost positively influences plant growth and development of ornamental plants by providing essential nutrients, improving root development, enhancing water retention, boosting disease resistance, increasing flowering and fruiting, promoting stress tolerance, regulating soil pH, and supporting sustainable growth practices (Figure 21.5) due to increased levels of NPK and other micronutrients (Serri et al., 2021; Ebrahimi et al., 2021; Shafique et al., 2021; Van Nguyen et al., 2022; Shafique et al., 2023). Vermi-compost could be used for a sustainable horticulture system as a natural bio-fertilizer to increase ornamental plant production instead of inorganic fertilizer. Shafique et al. (2021) and Shafique et al. (2023) showed the significant impact of vermicompost on the seed sprouting and vegetative growth of *Viola wittrockiana* (pansy) and *Tagetes erectus* (Marigold) (Figure 21.6).

An integrated management system (PGPVB+vermicompost) indicated a prominent effect on seed germination and seedling parameters of both marigold and pansy plants. PGPVB inoculated seeds promoted a significant increase in growth due to plant growth–promoting traits (Figure 21.7). Synergistic relationships that improved plant growth, reduced plant mortality, and increased microbial biomass have been reported (Sahni et al., 2008; Song et al., 2015). This could be because vermicompost contains humus which allows PGPVB to thrive well and multiply in the population. Shahzad et al. (2014) reported that PGPB with P-enriched compost increased N, P, and K contents of soil and showed improved soil fertility and plant productivity. Synergism between vermicompost and PGPR improved the soil quality as well as the crop yield of spinach and tomato (Song et al., 2015). In addition to the soil properties, compost also expands PGPR activity by providing resources and habitats for microbial growth (Song et al., 2015). This response was more effective in terms of increased plant growth compared to uninoculated treatments.

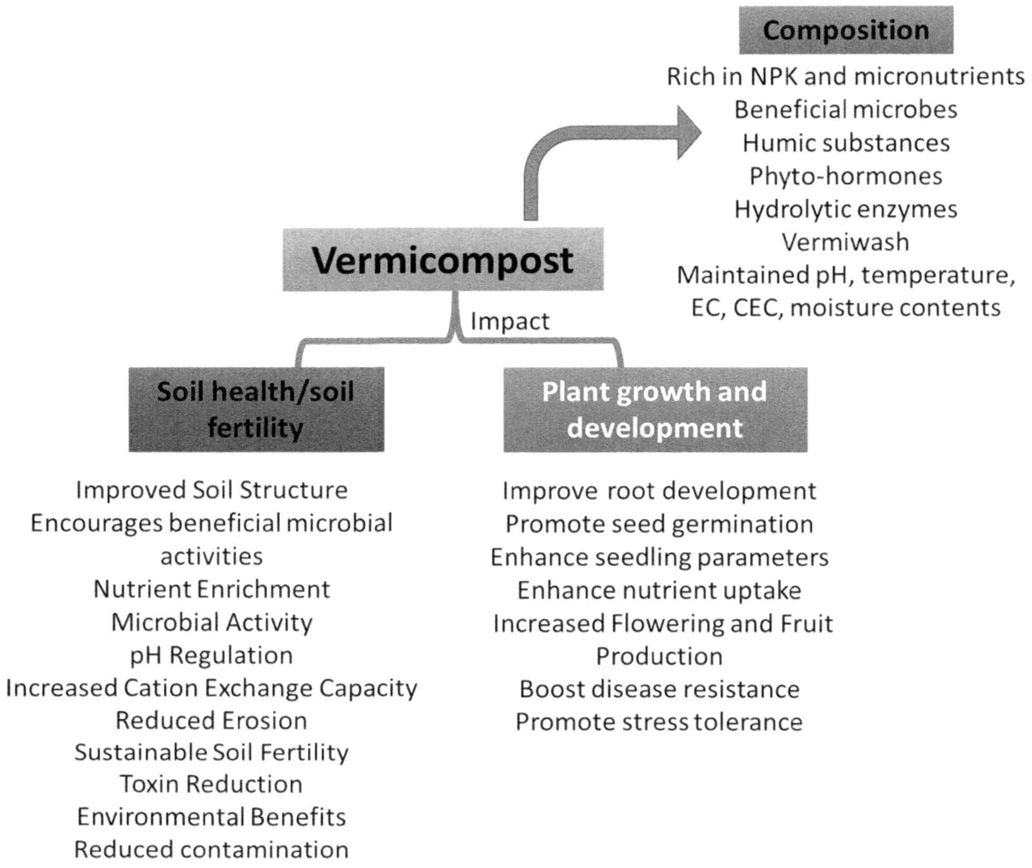

Composition

Rich in NPK and micronutrients
Beneficial microbes
Humic substances
Phyto-hormones
Hydrolytic enzymes
Vermiwash
Maintained pH, temperature, EC, CEC, moisture contents

Vermicompost

Impact

Soil health/soil fertility

Improved Soil Structure
Encourages beneficial microbial activities
Nutrient Enrichment
Microbial Activity
pH Regulation
Increased Cation Exchange Capacity
Reduced Erosion
Sustainable Soil Fertility
Toxin Reduction
Environmental Benefits
Reduced contamination

Plant growth and development

Improve root development
Promote seed germination
Enhance seedling parameters
Enhance nutrient uptake
Increased Flowering and Fruit Production
Boost disease resistance
Promote stress tolerance

FIGURE 21.5 Impact of vermicompost on soil health and growth of ornamental plants.

FIGURE 21.6 Impact of vermicompost on sprouting and vegetative growth of ornamental plants.

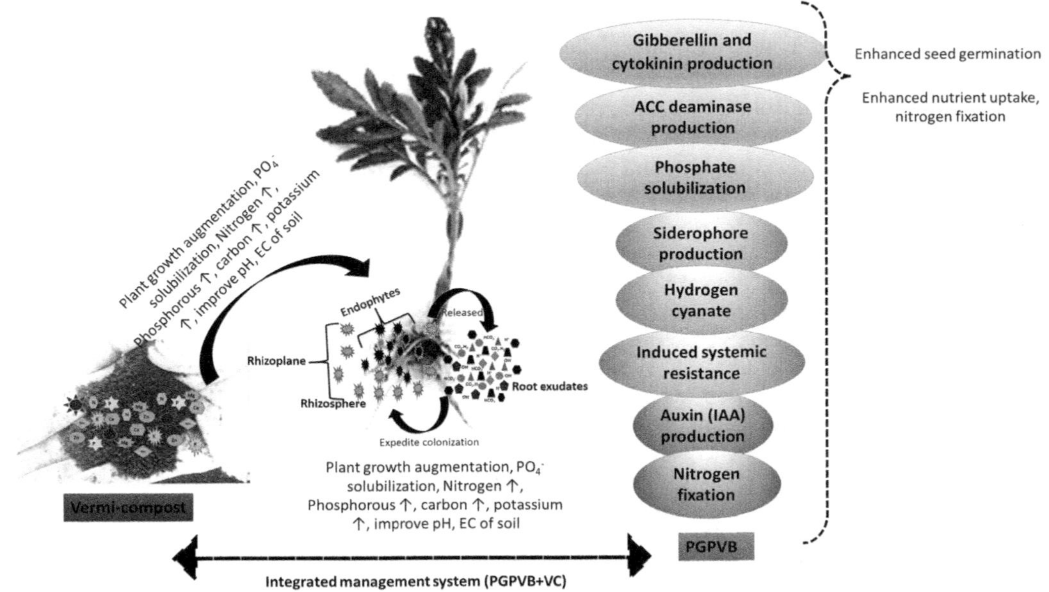

FIGURE 21.7 Impact of integrated management system on ornamental plant production.

The positive effect of *Bacillus paranthracis*, *Staphylococcus hominis*, *Bacillus mycoides*, and *Bacillus licheniformis* on the seed germination and growth of marigold and pansy was illuminated by endorsing the capabilities of PGPVB for N_2-fixation, phytohormone production, and phosphate solubilization (Sahin et al., 2004; Cakmakci et al., 2001; Esitken et al., 2006; Aslantas et al., 2007). Significant increases in germination parameters in the current research may be attributed to the production of plant growth regulators by PGPVB such as indole acetic acid (auxin), gibberellins, and cytokinins (Omer et al., 2004; Gupta et al., 2015; Kumar et al., 2015a). Auxin plays a vital role in plant–rhizobacteria interactions (Spaepen and Vanderleyden, 2011; Ahemad and Kibret, 2014; Afzal et al., 2015). IAA stimulates tuber and seed germination, enhances xylem and root development and biosynthesis of various metabolites, initiates fluorescence, enhances root surface area and root length, and gives resistance to biotic and abiotic stresses (Spaepen and Vanderleyden, 2011; Glick, 2012; Jha and Saraf, 2015; Ruzzi and Aroca, 2015; Llorente et al., 2016). IAA is involved in cell division, cell differentiation, and vascular bundle formation, which are necessary for nodule formation (Spaepen et al., 2007; Glick, 2012). Most *Rhizobium* species, *Agrobacterium tumefaciens*, *Pseudomonas syringae*, *Pseudomonas putida,* and *Pseudomonas fluorescens* produce IAA (Omer et al., 2004; Leveau and Lindow, 2005; Gupta et al., 2015). Costa et al. (2014) showed that *Grimontella, Enterobacter, Pantoea, Escherichia, Klebsiella,* and *Rahnella* are IAA producers.

Nitrogen, phosphorous, and iron are the most vital nutrients for plant growth and production. Iron is involved in photosynthesis, chlorophyll synthesis, respiration, and biological nitrogen fixation (Kobayashi and Nishizawa, 2012). Rhizobacteria have a chelating agent (siderophore) for the uptake of Fe. Souza et al. (2013) and Souza et al. (2014) illustrated that *Enterobacter* and *Burkholderia* species produced high levels of siderophore. In current research, all PGPVBs produced siderophore. We can say that vermibacteria might be involved in the stimulation of plant growth and improved nutrient availability. Phosphorous is another essential nutrient that is important, particularly for nitrogen fixation and photosynthesis, as well as participating as a structural component of ATP, phospholipid, and nucleic acids (Khan et al., 2009; Richardson and Simpson, 2011). Phosphorous is found in soil in an insoluble form which is not sufficient for plant nutrition. Hence, some phosphate solubilizing bacteria (PSB) solubilize inorganic phosphates through the action of siderophore, organic acid, and hydroxyl ion production (Rodriguez et al., 2006; Sharma et al., 2013) that promotes plant growth. Several PSBs have been isolated from the rhizospheric region and the roots (Ambrosini et al., 2012; Farina et al., 2012; Costa et al., 2013; Granada et al., 2013; Souza et al., 2014). The current research indicated that isolated vermibacteria are also phosphate-solubilizing bacteria.

REFERENCES

Abbas, S. H., Ismail, I. M., Mostafa, T. M., & Sulaymon, A. H. (2014). Biosorption of heavy metals: A review. *Journal of Chemical Science and Technology*, *3*(4), 74–102.

Adhikary, S. (2012). Vermicompost, the story of organic gold: A review. *Agricultural Sciences*, *3*, 905–917.

Afzal, I., Shinwari, Z. K., & Iqrar, I. (2015). Selective isolation and characterization of agriculturally beneficial endophytic bacteria from wild hemp using canola. *Pakistan Journal of Botany*, *47*, 1999–2008.

Ahemad, M., & Khan, M. S. (2012a). Evaluation of plant growth promoting activities of *rhizobacterium* and *Pseudomonas putida* under herbicide-stress. *Annals of Microbiology*, *62*, 1531–1540.

Ahemad, M., & Khan, M. S. (2012b). Alleviation of fungicide-induced phytotoxicity in greengram (*Vigna radiata* (L.) Wilczek) using fungicide-tolerant and plant growth promoting *Pseudomonas* strain. *Saudi Journal of Biology Science, 19*, 451–459.

Ahemad, M., & Kibret, M. (2014). Mechanisms and applications of plant growth promoting rhizobacteria: Current perspective. *Journal of King Saud University Science, 26*, 1–20.

Ahemad, M., Khan, M. S., Zaidi, A., & Wani, P. A. (2009). Remediation of herbicides contaminated soil using microbes. In M. S. Khan, A. Zaidi, & J. Musarrat (Eds.), *Microbes in sustainable agriculture*. New York: Nova Science Publishers.

Ahmad, A., Aslam, Z., Bellitürk, K., Iqbal, N., Idrees, M., Nawaz, M., Nawaz, M.Y., Munir, M.K., Kamal, A., Ullah, E., Jamil, M.A., Akram, Y., Abbas, T., & Aziz, M. M. (2021). Earth worms and vermicomposting: A review on the story of black gold. *Journal of Innovative Sciences*, *7*(1), 167–173.

Ahmed, M. Z., Shimazaki, T., Gulzar, S., Kikuchi, A., Gul, B., Khan, M. A., . . . Watanabe, K. N. (2013). The influence of genes regulating transmembrane transport of Na+ on the salt resistance of Aeluropus lagopoides. *Functional Plant Biology*, *40*(9), 860–871.

Akcil, A., Erust, C., Ozdemiroglu, S., Fonti, V., & Beolchini, F. (2015). A review of approaches and techniques used in aquatic contaminated sediments: Metal removal and stabilization by chemical and biotechnological processes. *Journal of Cleaner Production*, *86*, 24–36.

Ambrosini, A., Beneduzi, A., Stefanski, T., Pinheiro, F. G., Vargas, L. K., & Passaglia, L. M. P. (2012). Screening of plant growth promoting rhizobacteria isolated from sunflower (*Helianthus annuus* L.). *Plant Soil, 356*, 245–264.

Andleeb, S., Shafique, I., Naseer, A., Abbasi, W. A., Ejaz, S., Liaqat, I., et al. (2022). Molecular characterization of plant growth-promoting vermibacteria associated with Eisenia fetida gastrointestinal tract. *PLoS ONE*, *17*(6), e0269946.

Asgharzadeh, F., Ghaneian, M. T., Amouei, A. I., & Barari, R. (2014). Evaluation of cadmium, lead and zinc contents of compost producted in babol composting plant. *Iranian Journal of Health Sciences*, *2*, 62–7.

Aslantas, R., Cakmakci, R., & Sahin F. (2007). Effect of plant growth promoting rhizobacteria on young apples trees growth and fruit yield under orchard conditions. *Scientia Horticulturae*, *111*(4), 371–377.

Austruy, A., Shahid, M., Xiong, T., Castrec, M., Payre, V., Niazi, N. K., et al. (2014). Mechanisms of metal-phosphates formation in the rhizosphere soils of pea and tomato: Environmental and sanitary consequences. *Journal of Soils and Sediments*, *14*(4), 666–678.

Ayyobi, H., Olfati, J. A., & Peyvast, G. A. (2014). The effects of cow manure vermicompost and municipal solid waste compost on peppermint (Mentha piperita L.) in Torbat-e-Jam and Rasht regions of Iran. *International Journal of Recycling of Organic Waste in Agriculture*, *3*, 147–153.

Bano, S. A., & Iqbal, S. M. (2016). Biological nitrogen fixation to improve plant growth and productivity. *International Journal of Agriculture Innovations and Research*, *4*, 2319–1473.

Barois, I., & Lavelle, P. (1986). Changes in respiration rate and some physico-chemical properties of a tropical soil during transit through *Pontoscolex corethrurus* (Glossoscolecidae, Oligochaeta). *Soil Biology and Biochemistry*, *18*, 539–541.

Bashed, M. A., Alam, G. M., Kabir, M. A., & Amin, A. Q. A. (2012). Male infertility in Bangladesh: What serve better pharmacological help or awareness programme? *International of Pharmacology*, *8*(8), 687–694.

Berg, G., Rybakova, D., Grube, M., & Koberl, M. (2016). The plant microbiome explored: Implications for experimental botany. *Journal of Experimental Botany, 67*, 995–1002.

Boateng, G. O., & Amuzu, K. K. (2013). A survey of some critical issues in vegetable crops farming along river oyansia in opeibea and dzorwulu, Accra-Ghana. *Global Advanced Research Journal of Physical and Applied Sciences*, *2*, 24–31.

Bulgarelli, D., Garrido-Oter, R., Munch, P. C., Weiman, A., Droge, J., Pan, Y., et al. (2015). Structure and function of the bacterial root microbiota in wild and domesticated barley. *Cell Host & Microbe*, *17*, 392–403.

Cakmakci, R., Kantar, F., & Sahin, F. (2001). Effect of N2-fixing bacterial inoculations on yield of sugar beet and barley. *Journal of Plant Nutrition and Soil Science, 164*, 527–531.

Carvalho, F. P. (2017). Pesticides, environment and food safety. *Food Energy Security, 6*, 48–60.

Chandra, T., & Mani, P. (2011). A study of 2 rapid tests to differentiate Gram positive and Gram negative aerobic bacteria. *Journal of Medical Allied Sciences*, 2, 84–85.

Chaparro, J. M., Badri, D. V., & Vivanco, J. M. (2014). Rhizosphere microbiome assemblage is affected by plant development. *ISME Journal, 8*, 790.

Chitrapriya, K., Asokan, S., & Nagarajan, R. (2013). Estimating the level of Phosphate solubilizing bacteria and Azobacter in vermicompost of Eudrilus eugeniae and Perionyx excavates with various combinations of Cow Dung and Saw Dust. *International Journal of Scientific and Research Publication*, *3*, 2250–3153.

Costa, P., Beneduzi, A., Souza, R., Schoenfeld, R., Vargas, L. K., & Passaglia, L. M. P. (2013). The effects of different fertilization conditions on bacterial plant growth promoting traits: Guidelines for directed bacterial prospection and testing. *Plant Soil, 368*, 267–280.

Costa, P. B., Granada, C. E., Ambrosini, A., Moreira, F., Souza, R., Passos, J. F. M., Arruda, L., & Passaglia, L. M. P. (2014). A model to explain plant growth promotion traits: A multivariate analysis of 2,211 bacterial isolates. *PLoS One, 9*, 1160–1180.

Dhaliwal, S. S., Singh, J., Taneja, P. K., & Mandal, A. (2020). Remediation techniques for removal of heavy metals from the soil contaminated through different sources: A review. *Environmental Science and Pollution Research*, *27*, 1319–1333.

Ebrahimi, M., Mousavi, A., Souri, M. K., & Sahebani, N. (2021). Can vermicompost and biochar control Meloidogyne javanica on eggplant? *Nematology*, *23*, 1053–1064.

Edwards, C. A., Dominguez, J., & Arancon, N. Q. (2004). The influence of vermicompost on plant growth and pest incidence. In S. H. Shakir & W. Z. A. Mikhail (Eds.), *Soil zoology for sustainable development in the 21st century.* Cairo: Self-Publisher, 397–420.

Esitken, A., Pirlak, P., Turan, M., & Sahin F. (2006). Effects of floral and foliar application of plant growth promoting rhizobacteria (PGPR) on yield, growth of nutrition of sweet cherry. *Scientia Horticulturae, 1*(10), 324–327.

Esitken, A., Yildiz, H. E., Ercisli, S., Donmez, M. F., Turan, M., & Gunes, A. (2010). Effects of plant growth promoting bacteria (PGPB) on yield, growth and nutrient contents of organically grown strawberry. *Scientia Horticulturae*, *124*, 62–66.

Farina, R. A., Beneduzi, A., Ambrosini, A., Campos, S. B., Lisboa, B. B., Wendisch, V., Vargas, L. K., & Passaglia, L. M. P. (2012). Diversity of plant growth-promoting rhizobacteria communities associated with the stages of canola growth. *Applied Soil Ecology*, *55*, 44–52.

Ferber, T., Slaveykova, V. I., Sauzet, O., & Boivin, P. (2019). Upward mercury transfer by anecic earthworms in a contaminated soil. *European Journal of Soil Biology*, *91*, 32–37.

Figueroa, A., Cameselle, C., Gouveia, S., & Hansen, H. K. (2016). Electrokinetic treatment of an agricultural soil contaminated with heavy metals. *Journal of Environmental Science and Health, Part A*, *51*(9), 691–700.

Foucault, Y., Lévèque, T., Xiong, T., Schreck, E., Austruy, A., Shahid, M., & Dumat, C. (2013). Green manure plants for remediation of soils polluted by metals and metalloids: Ecotoxicity and human bioavailability assessment. *Chemosphere*, *93*(7), 1430–1435.

Fu, K., Huang, F., Li., & Chen, X. (2014). The biochemical properties and microbial profiles of vermicomposts affected by the age groups of earthworms. *Pakistan Journal of Zoology*, *46*, 1205–1214.

Glick, B. R. (2012). Plant growth-promoting bacteria: Mechanisms and applications. *Scientifica,* 1–15.

Granada, C., Costa, P. B., Lisboa, B. B., Vargas, L. K., & Passaglia, L. M. P. (2013). Comparison among bacterial communities present in arenized and adjacent areas subjected to different soil management regimes. *Plant Soil*, *373*, 339–358.

Gupta, G., Parihar, S. S., Ahirwar, N. K., Snehi, S. K., & Singh, V. (2015). Plant growth promoting rhizobacteria (PGPR): Current and future prospects for development of sustainable agriculture. *Journal of Microbial Biochemistry and Technology, 7*, 096–102.

Horn, M. A., Schramma, A., & Draka, H. (2003). The earthworm gut, An ideal habitat for ingested N2O producing microorganisms. *Applied & Environmental Microbiology*, *69*, 1662–1669.

Huang, C., Wang, W., Yue, S., Adeel, M., & Qiao, Y. (2020). Role of biochar and Eisenia fetida on metal bioavailability and biochar effects on earthworm fitness. *Environmental Pollution*, *263*, 114586.

Huang, M., Zhu, H., Zhang, J., Tang, D., Han, X., Chen, L., et al. (2017). Toxic effects of cadmium on tall fescue and different responses of the photosynthetic activities in the photosystem electron donor and acceptor sides. *Scientific reports*, *7*(1), 1–10.

Hynes, R. K., Leung, G. C., Hirkala, D. L., & Nelson, L. M. (2008). Isolation, selection, and characterization of beneficial rhizobacteria from pea, lentil and chickpea grown in Western Canada. *Canadian Journal of Microbiology*, *54*, 248–258.

Jahanian, A., Chaichi, M. R., Rezaei, K., Rezayazdi, K., & Khavazi, K. (2012). The effect of plant growth promoting rhizobacteria (pgpr) on germination and primary growth of artichoke (*Cynara scolymus*). *International Journal of Agriculture and Crop Science*, *4*, 923–929.

Jayasumana, C., Fonseka, S., Fernando, A., Jayalath, K., Amarasinghe, M., et al. (2015). Phosphate fertilizer is a main source of arsenic in areas affected with chronic kidney disease of unknown etiology in Sri Lanka. *Springer Plus*, *4*, 90.

Jha, C. K., & Saraf, M. (2015). Plant growth promoting rhizobacteria (PGPR): A review. *E3 Journal of Agricultural Research Development*, *5*, 108–119.

Jinadasa, N., Collins, D., Holford, P., Milham, P. J., & Conroy, J. P. (2016). Reactions to cadmium stress in a cadmium-tolerant variety of cabbage (Brassica oleracea L.): Is cadmium tolerance necessarily desirable in food crops?. *Environmental Science and Pollution Research*, *23*, 5296–5306.

Joo, J. H., Hassan, S. H., & Oh, S. E. (2010). Comparative study of biosorption of Zn2+ by Pseudomonas aeruginosa and Bacillus cereus. *International Biodeterioration & Biodegradation*, *64*(8), 734–741.

Kapahi, M., & Sachdeva, S. (2019). Bioremediation options for heavy metal pollution. *Journal of Health and Pollution*, *9*(24), 191203.

Kaur, P., Singh, N., Pal, P., & Kaur, A. (2018). Variation in composition, protein and pasting characteristics of different pigmented and non pigmented rice (Oryza sativa L.) grown in Indian Himalayan region. *Journal of Food Science and Technology*, *55*(9), 3809–3820.

Khan, M. S., Zadi, A., & Wani, P. A. (2009). Role of phosphate solubilizing microorganisms insustainable agriculture—A review. *Agronomy and Sustainable Development*, *27*, 29–43.

Kobayashi, T., & Nishizawa, N. K. (2012). Iron uptake, translocation, and regulation in higher plants. *Annual Review of Plant Biology*, *63*, 131–152.

Kosev, V., & Vasileva, V. (2014). Some studies on the selection of forage pea (Pisum sativum L.) to increase the symbiotic nitrogen fixing potential. *International Journal of Pharmacy and Life Sciences*, *5*, 3570–3579.

Kumar, J., Bahadur, A., Maurya, I., Raghuwanshi, B., Meena, R., et al. (2015a). Does a plant growth promoting rhizobacteria enhance agricultural Sustainability. *Journal of Pure Applied Microbiology*, *9*, 715–724.

Kumar, P. G., Suseelendra, D., Amalraj, E. L. D., & Reddy, G. (2015b). Isolation of Fluorescent *Pseudomonas* spp. From diverse agro-ecosystems of India and characterization of their PGPR traits. *Journal of Bacteriology*, *5*, 13–24.

Kumar, V., & Dwivedi, S. K. (2021). Bioremediation mechanism and potential of copper by actively growing fungus Trichoderma lixii CR700 isolated from electroplating wastewater. *Journal of Environmental Management*, *277*, 111370.

Lebeis, S. L. (2014). The potential for give and take in plant-microbiome relationships. *Frontier Plant Science*, *5*, 287.

Leveau, J. H. J., & Lindow, S. E. (2005). Utilization of the plant hormone indole-3-acetic acid for growth by *Pseudomonas putida* strain 1290. *Applied Environmental Microbiology*, *71*, 2365–2371.

Li, H., Xu, W., Dai, M., Wang, Z., Dong, X., & Fang, T. (2019). Assessing heavy metal pollution in paddy soil from coal mining area, Anhui, *China. Environ. Monit. Assess.*, *191*, 1–11.

Llorente, B. E., Alasia, M. A., & Larraburu, E. E. (2016). Biofertilization with Azospirillum brasilense improves in vitro culture of Handroanthus ochraceus, a forestry, ornamental and medicinal plant. *New Biotechnology Journal*, *33*, 32–40.

Ma, Y., Rajkumar, M., Luo, Y., & Freitas, H. (2011). Inoculation of endophytic bacteria on host and non-host plants-effects on plant growth and Ni uptake. *Journal of Hazardous Material*, *195*, 230–237.

Ma, Y., Rajkumar, M., Zhang, C., & Freitas, H. (2016). Beneficial role of bacterial endophytes in heavy metal phytoremediation. *Journal of Environmental Management*, *174*, 14–25.

Majeed, A., Muhammad, Z., Ullah, Z., Ullah, R., & Ahmad, H. (2017). Late blight of potato (Phytophthora infestans) I: Fungicides application and associated challenges, Turkish. *Journal of Agriculture—Food Science and Technology*, *5*, 261–266.

Mangmang, J., Deaker, R., & Rogers, G. (2014). Effects of plant growth promoting rhizobacteria on seed germination in characteristics of tomato and lettuce. *Journal of Tropical Crop Science*, *1*, 35–40.

Martin, A., Cortez, J., Barosis, I., & Lavelle, P. (1987). Les mucus intestinaux de Var de Terre, moteur de leurs interactions avec la microflore. *Revue D'ecologie et de biologie du sol, 24*, 549–558.

Mayak, S., Tirosh, T., & Glick, B. R. (2004). Plant growth-promoting bacteria confer resistance in tomato plants to salt stress. *Plant Physiology and Biochemistry, 42*, 565–572.

Medfu Tarekegn, M., Zewdu Salilih, F., & Ishetu, A. I. (2020). Microbes used as a tool for bioremediation of heavy metal from the environment. *Cogent Food & Agriculture, 6*(1), 1783174.

Meena, H. Meena, R. S., Rajput, B. S., & Kumar, S. (2016). Response of bio-regulators to morphology and yield of clusterbean (*Cyamopsis tetragonoloba* (L.) Taub.) under different sowing environments. *Journal of Applied Natural Science, 8*, 715–718.

Meuser, H. (2012). *Soil remediation and rehabilitation: Treatment of contaminated and disturbed land* (Vol. 23). Springer Science & Business Media.

Moustafa, Y., Mustafa, N., El-Dahshou, M., EL-Sawy, S., Haggag, L., Zhang, L., & Zuhair, R. (2023). Role of vermicompost types (fish sludge and cow dung) in improving agronomic behavior and soil health of tomato crop. *Asian Journal of Plant Sciences, 22*(1), 1–12. https://doi.org/10.3923/ajps.2023.1.12.

Muhibbullah, M., & Sarwar, M. I. (2017). Using behavior of agrochemicals and pesticides and their impacts on human health: A perception based rural study in Bangladesh. *Asian Journal of Agricultural Extension, Economics & Sociology, 21*(4), 1–15.

Munnoli, P. M. (2007). *Management of industrial organic solid wastes through vermiculture biotechnology with special reference to microorganisms*. PhD Thesis, Goa University, India, 1–334.

Naseer, A., Andleeb, S., Basit, A., Abbasi, W. A., Ejaz, S., Ali, S., & Ali, N. M. (2022, August 4). Phylogenetic illustration of Eisenia fetida associated vermi-bacteria involved in heavy metals remediation and retaining plant growth promoting traits. *Journal of Oleo Science, 71*(8), 1241–1252.

Ndeddy Aka, R. J., & Babalola, O. O. (2016). Effect of bacterial inoculation of strains of Pseudomonas aeruginosa, Alcaligenes feacalis and Bacillus subtilis on germination, growth and heavy metal (Cd, Cr, and Ni) uptake of Brassica juncea. *International journal of Phytoremediation, 18*(2), 200–209.

Njoku, K. L., Asunmo, M. O., Ude, E. O., Adesuyi, A. A., & Oyelami, A. O. (2020). The molecular study of microbial and functional diversity of resistant microbes in heavy metal contaminated soil. *Environmental Technology & Innovation, 17*, 100606.

Okoduwa, S. I. R., Igiri, B., Udeh, C. B., Edenta, C., & Gauje, B. (2017). Tannery effluent treatment by yeast species isolates from watermelon. *Toxics, 5*(1), 6.

Okoffo, E. D., Mensah, M., & Fosu-Mensah, B. Y. (2016). Pesticides exposure and the use of personal protective equipment by cocoa farmers in Ghana. *Environmental Systems Research, 5*(17), 1–9.

Omer, Z. S., Tombolini, R., Broberg, A., & Gerhardson, B. (2004). Indole-3-acetic acid production by pink-pigmented facultative methylotrophic bacteria. *Plant Growth Regulation, 43*, 93–96.

Onder, M., Ceyhan, E., & Kahraman, A. (2011). Effects of agricultural practices on environment. *Journal of Biology and Environmental Chemistry, 24*, 28–32.

Parks, C. G., Hoppin, J. A., De Roos, A. J., Costenbader, K. H., Alavanja, M. C., & Sandler, D. P. (2016). Rheumatoid arthritis in agricultural health study spouses: Associations with pesticides and other farm exposures. *Environmental Health Perspectives, 124*, 1728.

Pirsaheb, M., Limoee, M., Namdari, F., & Khamutian, R. (2015). Organochlorine pesticides residue in breast milk: A systematic review, *Medical Journal of Islamic Repubic of Iran, 29*, 228.

Prajapati, S. K., Soni, R. L., Patel, K., & Prajapati, B. K. (2023). Vermicomposting method and its importance in sustainable crop production. *Food and Scientific Reports, 4*(5), 51–60.

Rascio, N., & Navari-Izzo, F. (2011). Heavy metal hyperaccumulating plants: How and why do they do it? And what makes them so interesting? *Plant Science, 180*(2), 169–181.

Ravindran, B., Wong, J. W., Selvam, A., & Sekaran, G. (2016). Influence of microbial diversity and plant growth hormones in compost and vermicompost from fermented tannery waste. *Bioresource Technology, 217*, 200–204.

Rehman, S. U., De Castro, F., Aprile, A., Benedetti, M., & Fanizzi, F. P. (2023). Vermicompost: Enhancing plant growth and combating abiotic and biotic stress. *Agronomy, 13*, 1134.

Reile, E., Jors, E., Bælum, J., Huici, O., Alvarez, Caero, M. M., & Cedergreen, N. (2015). The influence of tomato processing on residues of organochlorine and organophosphate insecticides and their associated dietary risk. *Science Total Environment, 527–528*, 262–269.

Reitz, M. U., Gifford, M. L., & Schäfer, P. (2015). Hormone activities and the cell cycle machinery in immunity-triggered growth inhibition. *Journal of Experimental Botany, 66*, 2187–2197.

Richardson, A. E., & Simpson, R. J. (2011). Soil microorganisms mediating phosphorus availability. *Plant Physiology, 156*, 989–996.

Rodriguez, H., Fraga, R., Gonzalez, T., & Bashan, Y. (2006). Genetics of phosphate solubilization and its potential applications for improving plant growth-promoting bacteria. *Plant Soil, 287*, 15–21.

Russo, A., Vettori, L., Felici, C., Fiaschi, G., Morini, S., & Toffanin, A. (2008). Enhanced micropropagation response and biocontrol effect of *Azospirillum brasilense* Sp245 on *Prunus cerasifera* L. clone Mr. S 2/5 plants. *Journal of Biotechnology, 134*, 312–319.

Ruzzi, M., & Aroca, R. (2015). Plant growth-promoting rhizobacteria act as biostimulants in horticulture. *Science and Horticulture, 196*, 124–134.

Sabir, M., Ali, A., Zia-Ur-rehman, M., & Hakeem, K. R. (2015). Contrasting effects of farmyard manure (FYM) and compost for remediation of metal contaminated soil. *International Journal of Phytoremediation, 17*(7), 613–621.

Sahin, F., Cakmakci, R., & Kantar, F. (2004). Sugar beet and barley yields in relation to inoculation with N2-fixing and phosphate solubilizing bacteria. *Plant Soil, 265*, 123–129.

Sahni, S., Sarma, B., Singh, D. P., & Singh, H. B. (2008). Vermicompost enhances performance of plant growth-promoting rhizobacteria in Cicer arietinum rhizosphere against Sclerotium rolfsii. *Crop Protection, 27*(3), 369–337.

Schreck, E., Laplanche, C., Le Guédard, M., Bessoule, J. J., Austruy, A., Xiong, T., et al. (2013). Influence of fine process particles enriched with metals and metalloids on Lactuca sativa L. leaf fatty acid composition following air and/or soil-plant field exposure. *Environmental Pollution, 179*, 242–249.

Seralini, G. E., Clair, E., Mesnage, R., Gress, S., Defarge, N., & Malatesta, M. (2014). Republished study: Long-term toxicity of a Roundup herbicide and a Roundup-tolerant genetically modified maize. *Environmental Sciences Europe, 26*, 14.

Serri, F., Souri, M. K., & Rezapanah, M. (2021). Growth, biochemical quality and antioxidant capacity of coriander leaves under organic and inorganic fertilization programs. *Chemical and Biological Technologies in Agriculture, 8*, 33–50.

Shafique, I., Andleeb, S., Aftab, M. S., Naeem, F., Ali, S., Yahya, S., et al. (2021, January 6). Efficiency of cow dung based vermi-compost on seed germination and plant growth parameters of Tagetes erectus (Marigold). *Heliyon, 7*(1), e05895.

Shafique, I., Andleeb, S., Naeem, F., Ali, S., Tabassam, T., Sultan, T., et al. (2023). Cow dung putrefaction via vermicomposting using Eisenia fetida and its influence on seed sprouting and vegetative growth of Viola wittrockiana (pansy). *PLoS ONE, 18*(2), e027982.

Shahzad, S. M., Khalid, A., Arif, M. S., Riaz, M., Ashraf, M., Iqbal, Z., & Yasmeen, T. (2014). Co-inoculation integrated with P-enriched compost improved nodulation and growth of Chickpea (Cicer arietinum L.) under irrigated and rainfed farming systems. *Biology and Fertility of Soils, 50*, 1–12. ISO 690.

Sharma, S. K., Ramesh, A., & Johri, B. N. (2013). Isolation and characterization of plant growth promoting Bacillus amyloliquefaciens strain sks_bnj_1 and its influence on rhizosphere soil properties and nutrition of soybean (Glycine max L. Merrill). *Journal of Virology and Microbiology, 2013*, 1–19.

Shrivastava, P., & Kumar, R. (2015). Soil salinity: A serious environmental issue and plant growth promoting bacteria as one of the tools for its alleviation. *Saudi Journal of Biological Sciences, 22*(2), 123–131.

Singh, A., Singh, R. V., Saxcena, A. K., Shivay, Y. S, & Nain, L. (2014). Comparative studies on composting efficiency of *Eisenia foetida* and *Perionyx excavates*. *Journal of Experimental Biology and Agricultural Sciences, 2*, 508–17.

Sinha, K., Valani, D., Soni, B., & Chandran, V. (2011). Earthworm vermicompost: A sustainable alternative to chemical fertilizers for organic farming. In *Agriculture issues and policies*. New York: Nova Science Publishers Inc, 71.

Smith, D. L., Gravel, V., & Yergeau, E. (2017). Editorial: Signaling in the Phytomicrobiome. *Frontier Plant Science, 8*, 611.

Song, X., Liu, M., Wu, D., Griffiths, B. S., Jiao, J., Li, H., et al. (2015). Interaction matters: Synergy between vermicompost and PGPR agents improves soil quality, crop quality and crop yield in the field. *Applied Soil Ecology, 89*, 25–34.

Souza, R., Beneduzi, A., Ambrosini, A., Costa, P. B., Meyer, J., Vargas, L. K., et al. (2013). The effect of plant growth-promoting rhizobacteria on the growth of rice (*Oryza sativa* L.) cropped in southern Brazilian fields. *Plant Soil, 366*, 585–603.

Souza, R., Meyer, J., Schoenfeld, R., Costa, P. B., & Passaglia, L. M. P. (2014). Characterization of plant growth-promoting bacteria associated with rice cropped in iron-stressed soils. *Annals of Microbiology, 65*, 951–964.

Spaepen, S., & Vanderleyden, J. (2011). Auxin and plant-microbe interactions. *Cold Spring Harbor Perspectives in Biology*. http://dx.doi.org/10.1101/cshperspect.a001438.

Spaepen, S., Vanderleyden, J., & Remans, R. (2007). Indole-3-acetic acid in microbial and microorganism-plant signaling. *FEMS Microbiology Reviews*, *31*, 425–448.

Tank, N., & Saraf, M. (2010). Salinity-resistant plant growth promoting rhizobacteria ameliorates sodium chloride stress on tomato plants. *Plant Interactions*, *5*, 51–58.

Thakur, D. S., Khot, R., Joshi, P. P., Pandharipande, M., & Nagpure, K. (2014). Glyphosate poisoning with acute pulmonary edema. *Toxicology International*, *21*, 328–30.

Van Nguyen, S., Chikamatsu, S., Kato, R., Chau, K. M., Nguyen, P. K. T., Ritz, K., & Toyota, K. (2022). A biochar improves the efficacy of green manure-based strategies to suppress soybean cyst nematode (Heterodera glycines) and promotes free-living nematode populations. *Journal of Soil Science and Plant Nutrition*, *22*(3), 3414–3427.

Xiong, X., Grunwald, S., Myers, D. B., Ross, C. W., Harris, W. G., & Comerford, N. B. (2014). Interaction effects of climate and land use/land cover change on soil organic carbon sequestration. *Science of the Total Environment*, *493*, 974–982.

Zhang, R., Vivanco, J. M., & Shen, Q. (2017). The unseen rhizosphere rootsoil-microbe interactions for crop production. *Current Opinion in Microbiology*, *37*, 8–14.

Index

For Product Safety Concerns and Information please contact our EU representative GPSR@taylorandfrancis.com Taylor & Francis Verlag GmbH, Kaufingerstraße 24, 80331 München, Germany

Batch number: 10392095

Printed by Printforce, the Netherlands